AF767774

The 3rd IEEE International Conference on Electronic Communications, Internet of Things and Big Data 2023

The 3rd IEEE International Conference on Electronic Communications, Internet of Things and Big Data 2023

Editors

Teen-Hang Meen
Hsin-Hung Lin
Cheng-Fu Yang

MDPI

Basel • Beijing • Wuhan • Barcelona • Belgrade • Novi Sad • Cluj • Manchester

Editors

Teen-Hang Meen
National Formosa University
Yunlin
Taiwan

Hsin-Hung Lin
Asia University
Taichung
Taiwan

Cheng-Fu Yang
National University of Kaohs
Kaohsiung
Taiwan

Editorial Office
MDPI AG
Grosspeteranlage 5
4052 Basel, Switzerland

This is a reprint of articles from the Proceedings published online in the open access journal *Engineering Proceedings* (ISSN 2673-4591) (available at: https://www.mdpi.com/2673-4591/38/1).

For citation purposes, cite each article independently as indicated on the article page online and as indicated below:

Lastname, A.A.; Lastname, B.B. Article Title. *Journal Name* **Year**, *Volume Number*, Page Range.

ISBN 978-3-7258-1447-3 (Hbk)
ISBN 978-3-7258-1448-0 (PDF)
doi.org/10.3390/books978-3-7258-1448-0

© 2024 by the authors. Articles in this book are Open Access and distributed under the Creative Commons Attribution (CC BY) license. The book as a whole is distributed by MDPI under the terms and conditions of the Creative Commons Attribution-NonCommercial-NoDerivs (CC BY-NC-ND) license.

Contents

engineering proceedings

MDPI

Editorial

Preface of 3rd IEEE International Conference on Electronic Communications, Internet of Things and Big Data 2023 (IEEE ICEIB 2023) †

Teen-Hang Meen [1,*], Hsin-Hung Lin [2] and Cheng-Fu Yang [3,4,*]

1 Department of Electronic Engineering, National Formosa University, Yunlin 632, Taiwan
2 Department of Creative Product Design, Asia University, Taichung 413, Taiwan; hhlin@asia.edu.tw
3 Department of Chemical and Materials Engineering, National University of Kaohsiung, Kaohsiung 811, Taiwan
4 Department of Aeronautical Engineering, Chaoyang University of Technology, Taichung 413, Taiwan
* Correspondence: thmeen@gs.nfu.edu.tw (T.-H.M.); cfyang@nuk.edu.tw (C.-F.Y.)
† All proceeding papers published in the volumes are presented at the 3rd IEEE International Conference on Electronic Communications, Internet of Things and Big Data Conference 2023, Taichung, Taiwan, 14–16 April 2023.

Citation: Meen, T.-H.; Lin, H.-H.; Yang, C.-F. Preface of 3rd IEEE International Conference on Electronic Communications, Internet of Things and Big Data 2023 (IEEE ICEIB 2023). *Eng. Proc.* **2023**, *38*, 88. https://doi.org/10.3390/engproc2023038088

Published: 27 July 2023

Copyright: © 2023 by the authors. Licensee MDPI, Basel, Switzerland. This article is an open access article distributed under the terms and conditions of the Creative Commons Attribution (CC BY) license (https://creativecommons.org/licenses/by/4.0/).

This volume represents the proceedings of the 3rd IEEE International Conference on Electronic Communications, Internet of Things, and Big Data 2023 (IEEE ICEIB 2023). This conference was organized by Asia University, the Institute of Electrical and Electronics Engineers (IEEE), and the International Institute of Knowledge Innovation and Invention (IIKII), and was held at Asia University, Taichung, Taiwan, on 14–16 April 2023. The conference provided a unified communication platform for researchers on a wide range of topics such as big data and cloud computing, technologies and applications of artificial intelligence, robotics science and engineering, the Internet of Things and sensor technology, intelligent big data analysis and applications, and other related fields. In recent years, the rapid development of electronic technology and microelectronic technology has fundamentally and universally brought about the new technology revolution. The continuous revolution of electronic technology has not only appeared in very large-scale integrated circuits (VLSIs) and computers, but has also contributed to the development of modern communication technologies. Electronics and communication engineering involve information and communication systems, electronic science and technology, communication and information systems, signal and information processing, circuits and systems, electromagnetic fields and microwave technology, physical electronics and optoelectronics, microelectronics, and solid-state electronics, to name a few. Related research content includes information transmission, information exchange, information processing, signal detection, integrated circuit design and manufacture, electronic components, microwave and antenna, instrumentation technology, computer engineering, application, and others. The analysis and application of big data improve the efficiency of existing products and allow new products to be developed. Recent technological breakthroughs have greatly reduced the cost of data storage and computing, making the storage of huge amounts of data considerably less complicated and expensive than in the past. With the volume of data increasing day by day and being coupled with lower costs and easier access, using big data helps in making more precise business decisions that previously have been difficult.

IEEE ICEIB 2023 was held in a hybrid form, featuring on-site and online presentations. Figure 1 shows the group photo from the conference's opening. The first keynote speech was "Towards Net Zero: Opportunities and Challenges from a Materials Scientist's Perspective" presented by Professor Li-Chyong Chen, an academician of Academia Sinica, Taiwan. He emphasized that the discovery and development of materials for clean energy were essential to accelerate the transition toward a carbon-neutral economy. For instance, photocatalytic CO_2 conversion to hydrocarbon fuels, or so-called artificial photosynthesis, enables simultaneous solar energy harvesting and CO_2 reduction reactions (CO2RRs), and so it is considered a killing-two-birds-with-one-stone approach to solving the energy and environmental problems. Advanced future points were made by showing selective cases of scanning probe-based microscopies and in situ/operando synchrotron radiation-based spectroscopies along with vibrational spectroscopies. These enable scientists to probe the geometric, bonding, and electronic information of the catalyst and obtain atomic insights into the catalytic surfaces and reaction mechanisms. The second keynote speech was about "Asia University's Smart Campus Development", and was presented by Chair Professor Shian-Shyong Tseng, the Vice President of Asia University, Taiwan. Firstly, he briefly introduced the smart campus of Asia University and explained how the smart campus was built using the artificial Internet of Things (AIoT) as well as data and knowledge engineering skills. With such a well-established strategy, Asia University started to build the infrastructure and then developed the application and service processes. In the meantime, Asia University enhanced the quality, performance, and interactivity of campus services. The speech presented several examples of the various applications, especially those focusing on smart learning.

Figure 1. Group photo at the opening ceremony of IEEE ICEIB 2023.

IEEE ICEIB 2023 provided five regular sessions and seven invited sessions, covering various cutting-edge IoT technology fields, including big data, artificial intelligence (AI), robotics, IoT, cloud computing, global concerns about sustainable wisdom, and other issues. Figures 2 and 3 show examples of on-site and online oral presentation sessions.

Many substantial results were shared at IEEE ICEIB 2023 by enthusiastic participants, and 89 excellent papers on engineering fields related to the conference were selected through peer review for the publication of the proceedings of IEEE ICEIB 2023. The proceedings are expected to accelerate the interdisciplinary collaboration of science and engineering technologists in the academic and industrial fields and encourage international networking.

Figure 2. Presentation at an on-site session of IEEE ICEIB 2023.

Figure 3. Online presentation of IEEE ICEIB 2023.

Author Contributions: All authors contributed equally to this paper. All authors have read and agreed to the published version of the manuscript.

Funding: This editorial received no external funding.

Institutional Review Board Statement: Not applicable.

Informed Consent Statement: Not applicable.

Data Availability Statement: Data is available at https://www.mdpi.com/2673-4591/38/1.

Conflicts of Interest: The authors declare no conflict of interest.

Disclaimer/Publisher's Note: The statements, opinions and data contained in all publications are solely those of the individual author(s) and contributor(s) and not of MDPI and/or the editor(s). MDPI and/or the editor(s) disclaim responsibility for any injury to people or property resulting from any ideas, methods, instructions or products referred to in the content.

engineering proceedings

Editorial

Statement of Peer Review †

Teen-Hang Meen [1,*], Hsin-Hung Lin [2] and Cheng-Fu Yang [3,4,*]

1 Department of Electronic Engineering, National Formosa University, Yunlin 632, Taiwan
2 Department of Creative Product Design, Asia University, Taichung 413, Taiwan; hhlin@asia.edu.tw
3 Department of Chemical and Materials Engineering, National University of Kaohsiung, Kaohsiung 811, Taiwan
4 Department of Aeronautical Engineering, Chaoyang University of Technology, Taichung 413, Taiwan
* Correspondence: thmeen@gs.nfu.edu.tw (T.-H.M.); cfyang@nuk.edu.tw (C.-F.Y.)

† All proceeding papers published in the volumes are presented at the 3rd IEEE International Conference on Electronic Communications, Internet of Things and Big Data Conference 2023, Taichung, Taiwan, 14–16 April 2023.

In submitting conference proceedings of the 3rd IEEE International Conference on Electronic Communications, Internet of Things and Big Data 2023 (IEEE ICEIB 2023) to *Engineering Proceedings*, the volume editors of the proceedings certify to the publisher that all papers published in this volume were subjected to peer review arranged by the volume editors. The reviews were conducted by expert referees according to the professional and scientific standards expected of a proceedings journal:

- Type of peer review: two to three reviews by single-blinded reviewers.
- Conference submission management system: http://www.iceib.asia/.
- Number of submissions sent for review: 196.
- Number of submissions accepted: 91.
- Acceptance rate: 0.464.
- Average number of reviews per paper: two.
- Total number of reviewers involved: 50.
- Any additional information on the review process: Please refer to Figure 1.

Citation: Meen, T.-H.; Lin, H.-H.; Yang, C.-F. Statement of Peer Review. *Eng. Proc.* **2023**, *38*, 1. https://doi.org/10.3390/engproc2023038001

Published: 16 June 2023

Copyright: © 2023 by the authors. Licensee MDPI, Basel, Switzerland. This article is an open access article distributed under the terms and conditions of the Creative Commons Attribution (CC BY) license (https://creativecommons.org/licenses/by/4.0/).

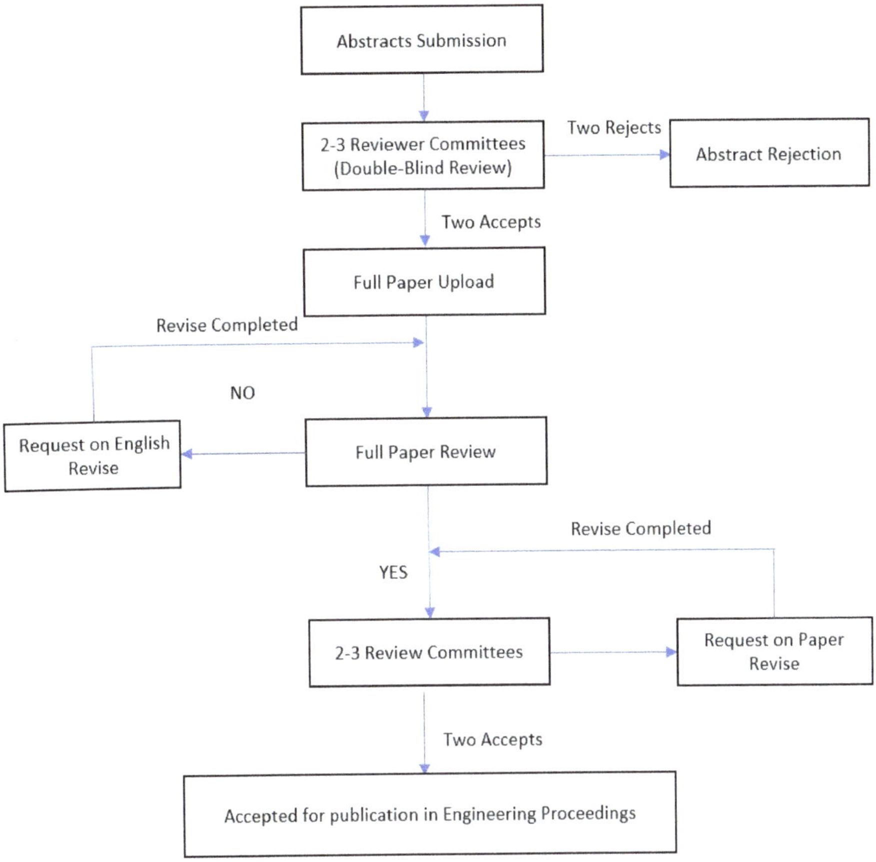

Figure 1. Review process for conference proceedings of the 3rd IEEE International Conference on Electronic Communications, Internet of Things and Big Data 2023 (IEEE ICEIB 2023).

Conflicts of Interest: The authors declare no conflict of interest.

Disclaimer/Publisher's Note: The statements, opinions and data contained in all publications are solely those of the individual author(s) and contributor(s) and not of MDPI and/or the editor(s). MDPI and/or the editor(s) disclaim responsibility for any injury to people or property resulting from any ideas, methods, instructions or products referred to in the content.

engineering proceedings

MDPI

Proceeding Paper

How to Prepare High-Level Massive Online Open Courses for the Metaverse: Tools and Needs †

Saverio Iacono and Gianni Vercelli *

Department of Computer Science and Technology, Bioengineering, Robotics, and Systems Engineering, University of Genova, 16145 Genova, Italy; saverio.iacono@unige.it
* Correspondence: gianni.vercelli@unige.it

† Presented at the 3rd IEEE International Conference on Electronic Communications, Internet of Things and Big Data Conference 2023, Taichung, Taiwan, 14–16 April 2023.

Abstract: The adoption of methodologies and practices from distance learning (MOOC) and gamification is a promising basis to facilitate the design of a new generation of MOOCs (massive online open courses) and well suited for XR-based immersive shared social spaces, i.e., educational metaverses. Therefore, we discuss the tools and requirements to assist teachers to acquire two levels of competence, "beginner-level creator" and "advanced-level creator", within the current ecosystem of metaverses. The adoption of such tools and XR platforms within STEM university educational settings is discussed to introduce the novel concept of the edu-metaverse.

Keywords: interoperability; metaverse; gamification; e-learning; stem education; MOOC; XR platforms; edu-metaverse

Citation: Iacono, S.; Vercelli, G. How to Prepare High-Level Massive Online Open Courses for the Metaverse: Tools and Needs. *Eng. Proc.* **2023**, *38*, 2. https://doi.org/10.3390/engproc2023038002

Academic Editors: Teen-Hang Meen, Hsin-Hung Lin and Cheng-Fu Yang

Published: 19 June 2023

Copyright: © 2023 by the authors. Licensee MDPI, Basel, Switzerland. This article is an open access article distributed under the terms and conditions of the Creative Commons Attribution (CC BY) license (https://creativecommons.org/licenses/by/4.0/).

1. Introduction

It is now required to discuss how to prepare for the new needs created by the announcement of the ethereal and self-defined "metaverse" of Facebook/Meta [1]. The metaverse first took its shape as a concept of Science Fiction in 1992 by Neal Stephenson in "Snow Crash" [2]. Subsequently, the Acceleration Studies Foundation announced in 2006 a roadmap identifying it as a result of the encounter of Web 2.0 with the world of video games [3] as observed by Kye et al. [4]. It deals with our digital life by adding, ironically, rules of real-life physics (and beyond) [5] and the possibilities of ubiquity. The concept is still "in its infancy" despite its mature roots, in the same way as Qualcomm theorized about Extended Reality (XR) technology in 2017 [6]. Kye et al. analyzed the possibilities of the metaverse in the world of education by extrapolating an interesting definition:

"Metaverse means a world in which virtual and reality interact and co-evolve, and in it social, economic and cultural activities take place to create value."

This study focuses on the importance of interoperability and gamification concerning the design of a new generation of MOOCs (massive online open courses) that are well suited for XR-based immersive shared social spaces, i.e., educational metaverses. According to Hwang and Lee [7], the main features of the emerging ecosystem of metaverses can be described using the recently proposed SPICE model. This model uses seamlessness, presence, interoperability, concurrence, and economic flow, as depicted in Figure 1, to capture the factors, customer satisfaction, and purchase intention of users within metaverses.

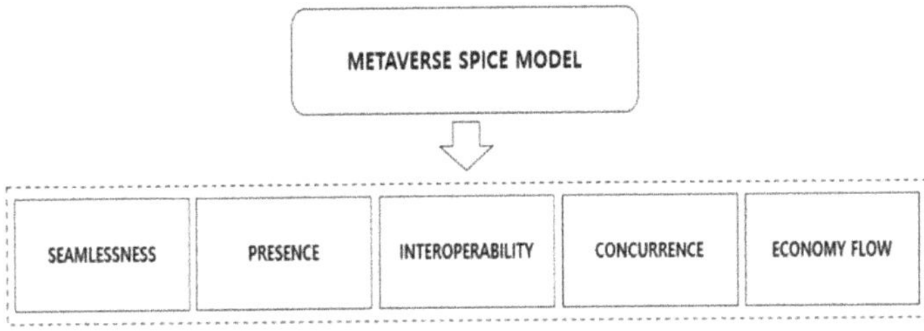

Figure 1. SPICE model of a metaverse [7].

Smart et al. in 2007 defined four different types of metaverse: Augmented Reality, Life Logging, Mirror World, and Virtual Reality. Each of these metaverses has dimensions that can be identified with a reality already in place, but which exist precisely and separately. Augmented Reality and Virtual Reality are well-established research fields for the scientific community, and Life Logging means recording one's life using smart or wearable devices and software platforms such as current users do on Instagram and Facebook. Mirror World is referred to as a more or less faithful reflection of the real world in the metaverse, defined as an "efficient expansion", i.e., a virtual place where certain tasks are performed efficiently to improve the players' quality of life. It is worth noting how interconnected these four types of metaverse can be and how much thery are so, thanks to technological innovation. According to this classification, there are platforms developed in the last decade to which it is legitimate to apply the "metaverse" label. For example, gamified social learning platforms such as Mondly [8] can be listed as augmented reality, virtual reality, and lifelogging metaverses.

Moreover, educational settings in schools and campuses have been successfully mirrored into metaverses [4], allowing to recreate educational contexts and situations. The idea of having multiple interconnected real/augmented/mixed/virtual worlds would lead us to think about the necessity of a holistic continuum of educational experiences; hence, interoperability is a key aspect. If we look at the recent definition given by the Meta group through a training course on Coursera [9], specific aspects are highlighted to understand the metaverse. "A series of immersive and interconnected digital spaces and an embodied Internet" is, therefore, presented as an Internet 3.0 that is mostly used with virtual reality or mixed reality viewers, the use of which gradually decreases as they approach widely used tools such as computers and smartphones.

Hence, again, the key aspect to interconnecting real to virtual and vice versa through the user's embodiment is interoperability, which requires heterogeneous platforms to refer to a unique ID from several linked IDs (from legal ID cards to biometric fingerprints for a real person, as well as digital identities and avatar IDs). This unique ID, according to Meta's plans, will be used to access any metaverse system.

It is beyond the scope of this study to present a deeper discussion about all the possible philosophical and ethical issues or legal consequences of having one or multiple avatars within a metaverse(s). The general idea of having a unique avatar ID is reasonable to guarantee the same physical body (the real user) has suitable access to every linked platform. For instance, within formal educational institutions, this digital embodiment is necessary to guarantee a proper learning path. On the contrary, within social networks, the possession of multiple identities is widely accepted (even debated) either for practical or more subtle reasons. Again, interoperability among platforms allows real users to access their different avatars to manage different social contexts (such as, in real life, wearing different suits in different life situations).

It is therefore evident that interoperability is the key element to implementing the convergence of different metaverses into one. The potential of this new metaverse accessed through a single "gate", as highlighted by Zhong and Zheng [10], is "a sort of educational environment beyond reality, which has the immersive characteristics of the real world and the open and free characteristics of the virtual world. It supports students to carry out interdisciplinary, interregional, and shared learning activities instantly using the avatar", that is, an edu-metaverse.

In this perspective, a research question is raised about the specific issues of creating an innovative educational MOOC experience, methodologically and technically valuable and embodied into an engaging XR experience. Gaming and gamification seem to be a natural support for metaverse experiences because most software implements them in the gaming world. Gamification theory [11] can be used to define different types of "learners as gamers". Even if this proposal is criticized in a traditional teaching framework, there is long-lasting evidence from situated learning [12] and GBL methodologies that the adoption of different gaming patterns is effective in an edu-metaverse context, where each proposed learning activity can be easily perceived as an engaging gaming situation. The best way is to convey this feeling of positive engagement by identifying the type of learners through the Bartle test [13], which gives teachers an idea of how to build privilege gamification-based approaches connected to four categories of gamers: achievers, explorers, killers, and socializers, as depicted in Figure 2. We investigated the application of gamification techniques in an e-learning setting [14–16], and it is reasonable to expect experiments on this topic within current metaverses.

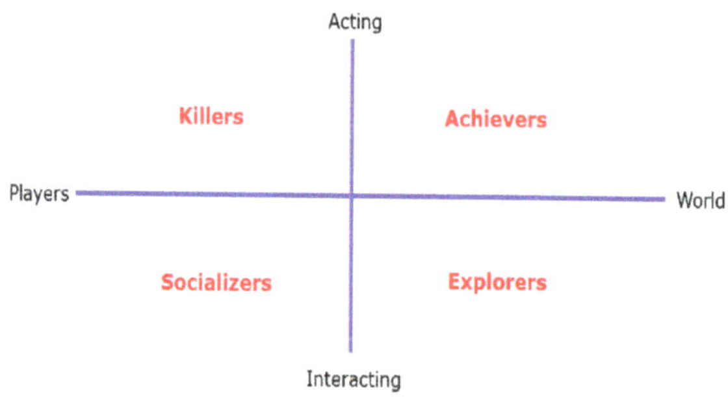

Figure 2. Bartle model of player types.

2. MOOCs and the Edu-Metaverse

A key pillar of early MOOCs (also known as cMOOCs) when they appeared in 2008 [17] was connectivism pedagogy. Taking its roots from the open educational resources (OER) movement with the MIT Open Courseware project and the Open University, the methodological approach was primarily based on the aggregation of reusable multimedia content and learning resources. The paradigmatic course on "Connectivism and Connective Knowledge" (CCK08) at Athabasca University, led by George and Stephen in 2008, enrolled 2200 online students based on active learning methodologies applied on the open-source e-learning platform Moodle [18] and the virtual world Second Life [19]. So far, the importance of a shared virtual immersive environment has been highlighted from the very beginning to promote engagement, communication, and the social sense of presence. On the other hand, starting in 2011, the emergent ecosystem of e-learning platforms offering MOOCs emphasized the importance of self-paced personalized learning, imposing a "de facto" model for less interactive and more narration-centered courses. This shift from connectivism to digital storytelling led to the so-called xMOOCs (extended MOOCs), where the most used pedagogical approach is a syllabus of self-paced or session-based well-defined lists of recorded video lectures intertwined with self-assessment interactive activities (quizzes/tests/choices).

In the last three years, during the COVID-19 pandemic period, another huge shift happened on a global level: the global adoption of distance learning platforms at every educational level. During the tragic lockdown period in March 2020, more than 1.3 billion students moved in 1 month from traditional classrooms to online platforms [20]. Most lectures became conference calls managed by millions of teachers, suddenly pushed from face-to-face to online teaching. Even if e-learning platforms and online/blended courses were already available and used worldwide in almost all universities before the pandemic, the number of people enrolled on MOOCs exploded as a sustainable educational scaffold. The long-term impact of this shift needs to be deeply analyzed at pedagogical and organizational levels. As a matter of fact, another emerging consequence of the pandemic put forward in 2021 is the idea of an educational metaverse, or edu-metaverse [4,21].

According to Kye et al., the educational possibilities of the mirror world, i.e., "a type of simulation of the external world that refers to an informationally enhanced virtual model of reflection of the real world" [3], arise from realistic digital reproductions of educational settings such as laboratories and classrooms. Video-conferencing systems are playing the role of classrooms for synchronous communication, but they are not yet well suited for recreating socially shared spaces. One of the most interesting examples is Gathertown, based on a 2D pixel art RPG style. The simplified interface and the automatic activation of audio–video connections (in case of virtual proximal distances in-between players) represent effective well-known methods to promote social interactions.

Another interesting aspect of metaverses from an educational perspective is that "the design of metaverse in education has evolved over generations, where generation Z is more targeted with AI (and XR) technologies compared to generation X or Y" [22]. In this bibliometric analysis, the findings provide a roadmap of future research directions to foster an effective edu-metaverse. Most studies on edu-metaverses still rely on virtual-world-based educational settings, where the reduced availability of low-cost 3D HMDs is a key limiting factor to widespread adoption. Hence, a new "digital divide" is emerging related to having an adequate set of digital devices for XR. Many university institutions, which are also strongly committed to the expansion of STEM disciplines, must take into adequate consideration the investment necessary to allow their students (regardless of gender, ethnicity, social conditions, and disabilities) access to these new types of virtual educational spaces.

3. Tools and Needs

3.1. Softwares

We define two levels of competence as "beginner-level creator" and "advanced-level creator". Beginner-level creators are, for instance, teachers who are trying to adopt new teaching methods but lack the skills needed to create something with highly personalized content from scratch. Tools such as Wikitude [23] and Zappar [24] represent acceptable solutions for them. These software have online studio creators that allow a beginner to create simple marker-based and object-based augmented reality experiences, adding informative layers, videos, or sounds to the target object. Zappar starts with two types of starting tools called Studio and Designer, and Wikitude adds the possibility of World Track and Face tracking. To create augmentation based on 3D objects, the simplest choice is to use 3D libraries such as Sketchfab or CGtrader that provide a large number of meshes, with a free section which is ready to use.

Currently, the simplest platform to create metaverse-like experiences without requiring great effort is Eon XR [25], which promises the possibility to create multi-object environments explorable in "walking and talking" modes, starting with the creation of avatars and creating the metaverse environment. The metaverse builder video demo shows how it is possible to create a simple environment and add objects, enabling voice research between internal and external libraries such as Sketchfab. Furthermore, taking advantage of the AI integrated into the builder, all notes inserted and translated based on the context give suggestions and recommendations to propose elements or information to insert in the scene.

"Advanced-level creators" are teachers with specific competence in digital tools, knowledge, or programming. They handle more complex tools. Trying to define a scale of complexity, they create with all previous tools in a deeper way and look forward to more complex tools. A preliminary example of a more complex software is Cospaces Edu [26], which allows the creation of an experience with visual code or script code and sharing XR lessons with students in metaverse-like environments. In some ways, we can assume that the structure is an "easy mode" of game engines such as Unreal [27] and Unity [28]. Furthermore, similar to Eon XR, it provides class management and assignment management.

In general, game engines have the highest possibilities of personalization for metaverse experiences, but at the cost of greater effort. Quixel MetaHumans and the Megascan free library for Epic customers are a big leap forward, and game engines offer the chance to make use of digital twins [29], which are an important block for the creation of deeper experiences in extended reality, giving strength and consistency to the metaverse concept. For instance, the new Unreal Engine 5 is making a big leap forward in reproducing the reality of virtual environments. Unreal with the brand new nanite system allows hand meshing with millions of polygons, considerably optimizing the real-time rendering system. In Unreal 5, the documentation explains that Nanite mesh is still essentially a triangle mesh at its core with a high level of detail and compression applied to its data. This optimization allows complex meshes with less optimization work to be created and directly imported into the engine. For instance, a photogrammetry mesh is recorded and recreated from the original object or original environment such as an antique sculpture or an entire heritage site. The Open XR standard is another tool for the interoperability of XR experiences, and it fits into the mission of an accessible and inclusive metaverse.

3.2. *Gamification as Tool*

As previously anticipated, gamification can be expected in the metaverse experience, and MOOCs are not excluded. Gamification can be considered in a middle space between tools and methodology, but in any case, we prefer to classify it as a tool, because it is affordable in the right context and not for every situation.

Why gamification? It is a strong tool to increase learners' engagement. There is a possibility at a certain point that the initial push of the "metaverse wow-effect" will run out, and thus maintaining engagement can be an issue, so gamification is a valid and complementary option. Motivation can be empowered. Deci and Ryan [30] developed a comprehensive approach to human motivation, called self-determination theory (SDT), which describes a motivational spectrum. Amotivation or total indifference to an activity is the first element of the motivational spectrum of SDT. Amotivation is followed by four levels of extrinsic motivation. In the first level, the behavior is regulated externally, which means that it is motivated by a request from external sources or by a reward imposed by external sources. Introjection is the next level of extrinsic motivation. People act to maintain or enhance their self-esteem or to reduce social pressure. A more self-determined form of extrinsic motivation enables the identification of activities that are important to the individual. Integrating a person's goals with the activity is the most self-determined form of extrinsic motivation. The person wants to perform the activity not because of a fondness for the activity. Intrinsic motivation is the final step, enhanced by conditions that lead to psychological need satisfaction, whereas it is undermined by conditions tending to thwart need satisfaction.

Gamification techniques specifically aim to create a path through the spectrum and change the learning motivation from extrinsic to the nearest step of intrinsic motivation. Werbach and Hunter, Chou, and Kapp well described how to achieve these goals with different types of gamification approaches [11,31,32].

4. Criticism and Conclusions

A recent study [33] has shown how seemingly simple it is to exploit software platforms to commit sexual misconduct, inducing mental distress in harassed users and causing them to feel less at ease in the metaverse. MIT has already confirmed that these incidents are not just isolated cases. In the future, this will be an issue to constantly deal with and solutions will need to be found to prevent these episodes from being crucial [34]. Ensuring the preservation of the digital uniqueness of learner and teacher avatars is still an open issue, and the ethical and legal aspects must be deeply investigated. This is not only to guarantee tools to monitor and ensure the safety of users, but also to protect them from bullying and impractical behavior. AI, Machine Learning, and Deep Learning techniques can be useful to detect and classify possible threats through sentiment analysis, for example. In every educational and social setting, real or virtual, these misbehaviors and cybercrime incidents can be reduced, but not a priori excluded. Following the approach of Wang et al., within the edu-metaverse, each participant must have the right to revoke their image and digital body even if it is granted, which is a problem that is already having implications in the use/abuse of deep fakes.

Author Contributions: Both authors equally contributed to all parts of the published version of the manuscript. All authors read and agreed to the published version of the manuscript.

Funding: This research received no external funding.

Institutional Review Board Statement: Not applicable.

Informed Consent Statement: Not applicable.

Data Availability Statement: No new data were created or analyzed in this study. Data sharing is not applicable to this article.

Acknowledgments: The authors would like to thank Edoardo Bellanti, Alice Corsi, Mario Vallarino, and Daniele Zolezzi, members of the 3DLabFactory research group, for their suggestions and deeper discussions on the topics.

Conflicts of Interest: The authors declare no conflict of interest.

References

1. Milmo, D. D.M.G. (Technology Ed.). Enter the Metaverse: The Digital Future Mark Zuckerberg is Steering US toward. *The Guardian*. 28 October 2021. Available online: https://www.theguardian.com/technology/2021/oct/28/facebook-mark-zuckerberg-meta-metaverse (accessed on 25 March 2022).
2. Stephenson, N. *Snow Crash*; Penguin: London, UK, 1994.
3. Smart, J.; Cascio, J.; Paffendorf, J. Metaverse Roadmap: Pathway to the 3D web, Ann Arbor (MI): Acceleration Studies Foundation. 2007. Available online: https://metaverseroadmap.org/MetaverseRoadmapOverview.pdf (accessed on 14 November 2022).
4. Kye, B.; Han, N.; Kim, E.; Park, Y.; Jo, S. Educational applications of metaverse: Possibilities and limitations. *J. Educ. Eval. Health Prof.* **2021**, *18*, 32. [CrossRef] [PubMed]
5. Kakalios, J. *The Physics of Superheroes*; Gotham Books: Sheridan, WY, USA, 2005.
6. Extended Reality XR | Immersive VR | Qualcomm. Available online: https://www.qualcomm.com/research/extended-reality (accessed on 5 November 2022).
7. Hwang, R.; Lee, M. The Influence of Music Content Marketing on User Satisfaction and Intention to Use in the Metaverse: A Focus on the SPICE Model. *Businesses* **2022**, *2*, 141–155. [CrossRef]
8. Mondly—Learn Languages Online for Free. Available online: https://www.mondly.com/ (accessed on 15 November 2022).
9. What Is the Metaverse?—Who Is Creating the Metaverse? *Coursera*. Available online: https://www.coursera.org/learn/what-is-the-metaverse/home/welcome (accessed on 11 November 2022).
10. Zhong, J.; Zheng, Y. Empowering Future Education: Learning in the Edu-Metaverse. In Proceedings of the 2022 Intl. Symposium on Educational Technology (ISET), Hong Kong, China, 19–22 July 2022; pp. 292–295. [CrossRef]
11. Werbach, K.; Hunter, D. *For the Win, Revised and Updated Edition: The Power of Gamification and Game Thinking in Business, Education, Government, and Social Impact*; Wharton School Press: Upper Saddle River, NJ, USA, 2020.
12. Lave, J.; Wenger, E. *Situated Learning: Legitimate Peripheral Participation*; Cambridge University Press: Cambridge, UK, 1991.
13. Bartle, R. Hearts, clubs, diamonds, spades: Players who suit MUDs. *J. MUD Res.* **1996**, *1*, 19.
14. Coccoli, M.; Iacono, S.; Vercelli, G. Applying gamification techniques to enhance effectiveness of video-lessons. *J. e-Learn. Knowl. Soc.* **2015**, *11*, 73–84. [CrossRef]
15. Iacono, S.; Vercelli, G.; Bellanti, E.; Corsi, A. A Lesson Learned About Gamification and Engagement In A Master Degree Course. In Proceedings of the 11th International Conference of Education, Research, and Innovation ICERI2018, Seville, Spain, 12–14 November 2018; pp. 2898–2902.
16. Iacono, S.; Vallarino, M.; Vercelli, G.V. Gamification in Corporate Training to Enhance Engagement: An Approach. *Int. J. Emerg. Technol. Learn. (IJET)* **2020**, *15*, 69–84. [CrossRef]
17. Baturay, M.H. An Overview of the World of MOOCs. *Procedia—Soc. Behav. Sci.* **2015**, *174*, 427–433. [CrossRef]

18. Dougiamas, M.; Taylor, P. Moodle: Using Learning Communities to Create an Open Source Course Management System. In *EdMedia+ Innovate Learning*; AACE: Honolulu, HI, USA, 2003; pp. 171–178.
19. Linden Lab. Available online: https://www.lindenlab.com/releases/start-your-second-life-in-june-2003 (accessed on 10 November 2022).
20. Infographic: COVID-19's Staggering Impact On Global Education», Statista Infographics. Available online: https://www.statista.com/chart/21224/learners-impacted-by-national-school-closures/ (accessed on 15 October 2022).
21. Wang, M.; Yu, H.; Bell, Z.; Chu, X. Constructing an Edu-Metaverse Ecosystem: A New and Innovative Framework. *IEEE Trans. Learn. Technol. (TLT)* **2022**, *15*, 685–696. [CrossRef]
22. Tlili, A.; Huang, R.; Shehata, B.; Liu, D.; Zhao, J.; Metwally, A.H.S.; Wang, H.; Denden, M.; Bozkurt, A.; Lee, L.; et al. Is Metaverse in education a blessing or a curse: A combined content and bibliometric analysis. *Smart Learn. Environ.* **2022**, *9*, 24. [CrossRef]
23. Stenina, M. Augmenting the Future: Interview with Martin Herdina. *Wikitude*. 10 November 2021. Available online: https://www.wikitude.com/blog-headworn-ar-interview-martin-herdina/ (accessed on 25 October 2022).
24. Zappar: Quick, Get Me in the Metaverse! The Reality of the Latest Virtual Trend. *Zappar*. Available online: https://www.zappar.com/ar-pioneers/2022/quick-get-me-in-the-metaverse-the-reality-of-the-l/ (accessed on 15 November 2022).
25. Chesler, J. Announcing EON Reality's Knowledge Metaverse Project, EON Reality, 30 July 2021. Available online: https://eonreality.com/knowledge-metaverse-project-xr/ (accessed on 6 November 2022).
26. Introducing Your Classroom to the Metaverse», The Mind Lab, 21 Marzo 2022. Available online: https://themindlab.com/2022/03/21/introducing-your-classroom-to-the-metaverse/ (accessed on 6 November 2022).
27. Real-Time Round-up 2022: The Metaverse and Other Emerging Trends, Unreal Engine. Available online: https://www.unrealengine.com/en-US/blog/real-time-round-up-the-metaverse-and-emerging-2022-trends (accessed on 7 November 2022).
28. Introducing the Metaverse Minute: How AR, VR, and XR are Changing Every Industry, Unity Blog. Available online: https://blog.unity.com/manufacturing/introducing-the-metaverse-minute (accessed on 8 November 2022).
29. Jeong, H.; Yi, Y.; Kim, D. An Innovative E-Commerce Platform Incorporating Metaverse To Live Commerce. *Int. J. Innov. Comput. Inf. Control.* **2022**, *18*, 221–229. [CrossRef]
30. Ryan, R.M.; Deci, E.L. Self-determination theory and the facilitation of intrinsic motivation, social development, and well-being. *Am. Psychol.* **2000**, *55*, 68–78. [CrossRef] [PubMed]
31. Chou, Y. *Actionable Gamification: Beyond Points, Badges, and Leaderboards*; Packt Publishing Ltd.: Birmingham, UK, 2019.
32. Kapp, K.M. *The Gamification of Learning and Instruction: Game-Based Methods and Strategies for Training and Education*; John Wiley & Sons: Hoboken, NJ, USA, 2012.
33. Sexual Harassment in the Metaverse? *Woman Alleges Rape in Virtual World», USA TODAY*. Available online: https://www.usatoday.com/story/tech/2022/01/31/woman-allegedly-groped-metaverse/9278578002/ (accessed on 10 November 2022).
34. The Metaverse Has a Groping Problem Already», MIT Technology Review. Available online: https://www.technologyreview.com/2021/12/16/1042516/the-metaverse-has-a-groping-problem/ (accessed on 10 November 2022).

Disclaimer/Publisher's Note: The statements, opinions and data contained in all publications are solely those of the individual author(s) and contributor(s) and not of MDPI and/or the editor(s). MDPI and/or the editor(s) disclaim responsibility for any injury to people or property resulting from any ideas, methods, instructions or products referred to in the content.

engineering proceedings

MDPI

Proceeding Paper

Evaluation of User Satisfaction in Using the Emergency Department Information System (EDIS) from User's Perspective †

Shiow-Luan Wang [1,*] and Jo-Yun Tsui [2]

1 Department of Information Management, National Formosa University, 64 Wunhua Rd., Huwei, Yunlin 632, Taiwan
2 National Taiwan University Hospital Yunlin Branch, Douliu, Yunlin 640, Taiwan; chilan0507@gmail.com
* Correspondence: slwang@nfu.edu.tw
† Presented at the 3rd IEEE International Conference on Electronic Communications, Internet of Things and Big Data Conference 2023, Taichung, Taiwan, 14–16 April 2023.

Abstract: The service industry depends on Internet development that always changes. The changing service environment provides a range of opportunities to develop new information technology to communicate with customers and add new channels to reach all service providers. Hospital administration pays attention to improving medical service quality and using information systems for clinical care. A nursing information system is the most important part of the whole medical care information system. The aim of this study is to explore the key factors of user interface (UI) and the function of the emergency department information system (EDIS) from the viewpoint of the user's operation. The questionnaire design is based on literature reviews and invited academic and industry scholars to conduct a pre-test for the validity of the content. The staff of a regional teaching hospital in central Taiwan were taken as the research subjects The study was based on the information system success model proposed by DeLong and McLean in 2003. In order to understand the use of information systems in hospitals, the TAM (technology acceptance model) is used to investigate the factors that influence the acceptance of new technology. A survey was conducted on the employees who are using the EDIS at present. There are four categories of participants: attending physicians, nurses, specialist nurses, and secretaries/registers. The results indicate that "user satisfaction" affects the success of the EDIS. For the question, "emergency medical information system has more benefits than traditional paper", respondents gave good marks. For the question, "the stability of the emergency medical information system is very satisfactory" was given the lowest. The aspects of the information system success model for the terms "system quality", "information quality", "service quality", etc. were analyzed. User satisfaction is positively related, although user education is different. The reference data allow hospital administrators to consider the needs of users while developing or conversing about new systems for the future.

Keywords: medical service; emergency information system; information systems success model

Citation: Wang, S.-L.; Tsui, J.-Y. Evaluation of User Satisfaction in Using the Emergency Department Information System (EDIS) from User's Perspective. *Eng. Proc.* **2023**, *38*, 5. https://doi.org/10.3390/engproc2023038005

Academic Editors: Teen-Hang Meen, Hsin-Hung Lin and Cheng-Fu Yang

Published: 19 June 2023

Copyright: © 2023 by the authors. Licensee MDPI, Basel, Switzerland. This article is an open access article distributed under the terms and conditions of the Creative Commons Attribution (CC BY) license (https://creativecommons.org/licenses/by/4.0/).

1. Introduction

The measurement of success in an introduced system relies on many indicators. User satisfaction is an important indicator of information system success. The introduction of the emergency information system in clinical practice not only simplifies the work and the procedure of clinical care for medical personnel, but it also improves work productivity, reduces paperwork and workload, provides real-time information, and even increases the accuracy of medical orders; thus, it effectively boosts the level of satisfaction in patients and medical staff.

The four aspects of this study are as follows:

- To investigate the impact of the emergency information system on users.
- To investigate the overall impact of user satisfaction on the company.
- To review the relationship between the impact of emergency information systems and the impact on user satisfaction including their influences.
- To find out the factors of users' relationships with the emergency information system.

2. Related Studies

2.1. Influential Factors of Satisfaction to the Emergency Information System

Based on the individual case study by Zhang, she found cognitive awareness of system characteristics, quality of informative product, cognition of user environment, user knowledge, and professionalism of the information personnel were the relevant factors for affecting the user satisfaction of an information system [1]. In 2005, Smith, Smith, Krugman, and Oman suggested a focus group of nursing staff for improvement, so that users' attitudes could be turned in a more positive direction [2].

2.2. Information System Success Model (ISSM)

To assess the effectiveness of an information system, the most frequently discussed model of system quality and user satisfaction is the information system success model (ISSM). DeLone and McLean proposed an information system success model (ISSM) in 1992 based on an empirical study which includes system quality (for its usability, usefulness, and portability), information quality (for accuracy, timeliness, and integrity), service quality (for personnel properly understanding, helping, and solving users' problems), user willingness and satisfaction (for the level of assistance by the system in completing an individual task), and net benefit (to see if the system has improved work performance and saved decision time) [3].

DeLone and McLean further modified their proposed information model in accordance with the advancement of information technology and electronic commerce in 2003 by adding a new construct of service quality and suggesting "user intention" as the representative aspect. Moreover, "net benefits" would replace and combine "individual impact" and "organizational impact" into one single construct by representing the impact of the information system in terms of its overall influence or benefit [4].

2.3. Technology Acceptance Model (TAM)

The Technology Acceptance Model (TAM) is a currently popular approach to studying the acceptance of certain technologies by users, such as evaluating the acceptance of a newly introduced medical technology by medical caregivers. It is by far the most common research model in foreign studies of technology acceptance [5]. Individuals will affect one's attitude (A) toward an event and such attitudes determine the cognitive behavioral control. TAM can be used in the domestic medical industry. Similarly, perceived usefulness (PU) and ease of use (PEOU) continue to serve as a directing force on the user's will (AU) and intention (BI). The order of influence is, first, the perceived usefulness, then followed by perceived ease of use and satisfaction.

3. Methods

3.1. Research Framework

In this study, 130 medical staff in a teaching hospital in the central region of Taiwan were the subjects. The subjects worked in four different departments, used the emergency information system, and received the questionnaire. They were, respectively, attending physicians, nurses, nurse practitioners, registrars, and cashier who actually use the system and these questionnaires would hopefully gather information for the purpose of this study. The initial draft of the questionnaire was discussed, designed, and revised by various experts and professors of different fields advised on the expert validity. The pretest was first completed by nurses in the emergency department who would answer the questionnaire as a pilot test before the actual official survey began. In order to examine the effect of the success of the information system in the emergency room, a research model is developed in Figure 1.

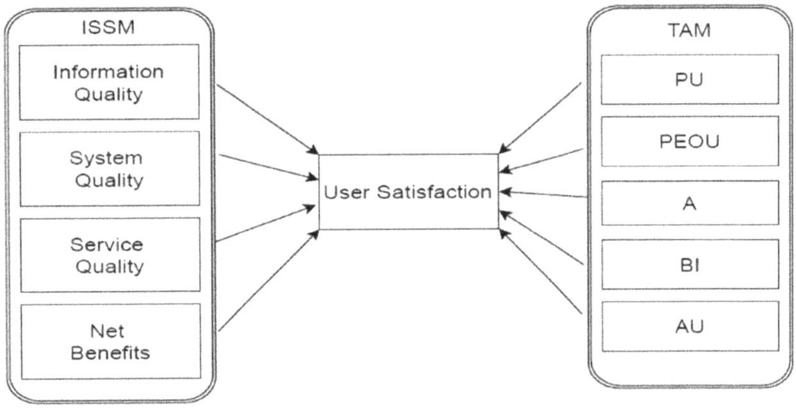

Figure 1. Research Model.

3.2. Research Hypothesis

Based on the purpose of study and statistics in the literature, we propose:

Hypothesis 1. System quality positively affects the information system success model and user satisfaction.

Hypothesis 2. *Information quality positively affects the information system success model and user satisfaction.*

Hypothesis 3. *Service quality positively affects the information system success model and user satisfaction.*

Hypothesis 4. *Net benefits positively affect the information system success model and user satisfaction.*

Hypothesis 5. *Subjective usability positively affects the technology acceptance model and user satisfaction.*

Hypothesis 6. *Subjective ease of use positively affects the technology acceptance model and user satisfaction.*

Hypothesis 7. *User attitude positively affects the technology acceptance model and user satisfaction.*

Hypothesis 8. *User intention positively affects the technology acceptance model and user satisfaction.*

Hypothesis 9. *Actual usage positively affects the technology acceptance model and user satisfaction.*

A total of valid surveys were collected after removing the survey lack of information, not match or empty. The SPSS software version 20 (IBM Corp. Released 2011. IBM SPSS Statistics for Windows, Version 20.0. Armonk, NY, USA: IBM Corp.) is used to perform the analysis.

4. Data Analysis

In this survey, we only considered participants who had experience in using the emergency department information system (EDIS). For the verification of reliability analysis and substantive relationship, the method of correlation variance analysis and multiple regression analysis were adopted, with the results explained as follows.

4.1. Reliability Analysis

In the pilot study and according to George and Mallery (2003), the Cronbach's alpha (α) coefficient is considered as the most frequently used estimate of internal consistency reliability [6]. An α value above 0.7 ($\alpha \geq 0.7$) is acceptable. Accordingly, the inter-item reliability for each of the multi-item variables were assessed. The results are given in Table 1. They indicate that the Cronbach's alpha coefficient is high in all scales, ranging from the lowest of 0.701 (Information Quality) to the highest of 0.966 (Perceived ease of use). This confirms that the measurement scales of the constructs are stable and consistent in Table 1.

Table 1. Reliability testing results.

Construct	Factors	(α)	Qualified
ISSM	System Quality	0.812	Yes
	Information Quality	0.701	Yes
	Service Quality	0.892	Yes
	User Satisfaction	0.811	Yes
	Net Benefits	0.901	Yes
TAM	PU	0.901	Yes
	PEOU	0.966	Yes
	A	0.811	Yes
	BI	0.798	Yes
	AU	0.829	Yes

4.2. The Correlation between the Various Aspects of the Information System Success Model and User Satisfaction

For the verification of substantive relationship, the method of correlation variance analysis and multiple regression analysis were adopted, with the results explained as follows. The research model and hypotheses were examined as follows "Tables 2–5".

Table 2. Correlation variance analysis of various aspects of the information system success model and user satisfaction (H1~H4).

Aspect	Average	Standard Deviation	Variance	F-Value	Significance	Hypothetical Result
User satisfaction	3.73	0.692	0.479			
System quality	3.62	0.628	0.394	13.292	0.000 ***	Valid
Information quality	3.76	0.709	0.502	8.510	0.000 ***	Valid
Service quality	3.52	0.809	0.654	8.516	0.000 ***	Valid
Net benefits	3.98	0.669	0.448	13.295	0.000 ***	Valid

*** $p < 0.01$

Table 3. Regression analysis of various aspects of the information system success model and user satisfaction (H1~H4).

	User Satisfaction					
	Unstandardized Coefficient		Standardized Coefficient	T-Value	Significance	Hypothetical Result
Aspect	Regression Coefficient	Standard Deviation				
System quality	0.348	0.192	0.315	4.327	0.000 ***	Valid
Information quality	0.118	0.066	0.121	1.798	0.004 ***	Valid
Service quality	0.189	0.056	0.221	3.386	0.001 ***	Valid
Net benefits	0.368	0.069	0.356	5.318	0.000 ***	Valid
			R2 = 0.799, Adjust R2 = 0.792			

*** $p < 0.01$.

Table 4. Correlation of the variance analysis of various aspects of the information system success model and user satisfaction (H5~H9).

Aspect	Mean	Standard Deviation	Variance	F	Significance	Hypothetical Results
User satisfaction	3.73	0.692	0.479			
PU	4.09	0.672	0.452	7.897	0.000 ***	Valid
PEOU	3.88	0.815	0.664	8.420	0.000 ***	Valid
A	4.01	0.666	0.443	8.155	0.000 ***	Valid
BI	4.29	0.685	0.470	4.765	0.000 ***	Valid
AU	4.00	0.698	0.487	7.760	0.000 ***	Valid

*** $p < 0.01$.

Table 5. Regression analysis of each aspect of the technology acceptance model and user satisfaction (H5~H9).

	User Satisfaction					
	Unstandardized Coefficient		Standardized Coefficient	T-Value	Significance	Hypothetical Result
Aspect	Regression Coefficient	Standard Deviation				
Subjective usability	0.167	0.119	0.162	1.401	0.164 *	Invalid
Subjective ease of use	0.203	0.077	0.239	2.650	0.009 **	Invalid
User attitude	0.197	0.122	0.190	1.621	0.108 *	Invalid
User intention	−0.014	0.091	−0.013	−0.149	0.882 *	Invalid
Actual usage	0.325	0.083	0.328	3.899	0.000 ***	Valid
			R2 = 0.648, Adjust R2 = 0.632			

*** $p < 0.01$ ** $p < 0.05$ * $p < 0.1$

5. Conclusions and Suggestion

5.1. Conclusions

This study analyzed the different functions of the emergency information system under different categorizations used by the users of different occupations to investigate more in depth the actual needs of various users. It was found that users of different job positions showed significant differences in "system quality", "information quality", "service quality", "user satisfaction", and "actual usage", where doctors were less contemptous in every aspect than other medical administrative staff. Most medical staff require higher education, with at least a college degree or above, but in terms of the education level, users showed no significant difference in factorial aspects of the information system success model and the technology acceptance model.

5.2. Suggestions

- Improve the management's knowledge and support of information innovation technology;
- Provide user-friendly system interfaces;
- Information establishment must conform with users' working habits and external cooperation of information technology.

5.3. Limitations and the Future

Although the authors expected a more rigorous and comprehensive research framework and procedure because of the time limit, the questionnaire design, and sampling constraint, this study still had many limitations and issues. The limitations and possible directions for the futures will be discussed as follows.

5.3.1. Improve the Research Tool

This research adopted the questionnaire as the tool which has its primary structure based on the information system success model proposed by DeLone and McLean (2003). Information quality, system quality and service quality were the three constructs used to understand the correlation with "user satisfaction", while "technology acceptance" was the intervening variable. The goal was to evaluate the impact of system quality, information quality and service quality on user satisfaction when the user's technology acceptance was considered, even though it has been pretested and verified for reliability. However, the question of whether the content of the questionnaire encompassed all aspects of the emergency information system would require further investigation.

5.3.2. Limited Research Subjects

The study was based on subjects from a district city teaching hospital in the central region of Taiwan. In the future, subjects should be included from other medical institutions with emergency information systems throughout the country in order to provide a more complete insight into the effective improvement of nursing staff's acceptance and satisfaction of the emergency information system.

5.3.3. Addition of Qualitative and Quantitative Research Methods

The study primarily used a survey questionnaire as the main approach. Thus, when inferring the research results, it was limited due to incomplete data and a lack of inmdepth quantitative data. Therefore, the combination of both quantitative and qualitative approaches can be adopted to increase to a more in depth qualitative data collection process and to increase understanding, such as when explaining users' attitudes, strengths vs. weaknesses, and operational difficulties of the emergency information system, which would provide a more complete picture of system users. By incorporating the actual feeling and thoughts of users in the questionnaire, as well as a way to convert information into a quantitative model, it would allow a reflection of the actual situation.

Author Contributions: Conceptualization, all authors; methodology, S.-L.W.; software, S.-L.W.; formal analysis and data extraction, S.-L.W. and J.-Y.T.; writing—original draft preparation, J.-Y.T.; writing—review and editing, S.-L.W.; visualization, S.-L.W.; supervision, S.-L.W. All authors have read and agreed to the published version of the manuscript.

Funding: This research received no external funding.

Institutional Review Board Statement: Not applicable.

Informed Consent Statement: Informed consent was obtained from all subjects involved in the study.

Data Availability Statement: https://www.ylh.gov.tw/english/?aid=624 (accessed on 16 June 2023).

Conflicts of Interest: The authors declare no conflict of interest.

References

1. Zhang, L. Thinking Styles and Cognitive Development. *J. Genet. Psychol.* **2002**, *163*, 179–195. [CrossRef] [PubMed]
2. Smith, K.; Smith, V.; Krugman, M.; Oman, K. Evaluating the impact of computerized clinical documentation. *Comput. Inf. Nurs.* **2005**, *23*, 132–138. [CrossRef] [PubMed]
3. DeLone, W.H.; McLean, E.R. Information systems success: The quest for the dependent variable. *Inf. Syst. Res.* **1992**, *3*, 60–95. [CrossRef]
4. DeLone, W.H.; McLean, E.R. The DeLone and McLean model of information Systems success: A ten-year update. *J. Manag. Inf. Syst.* **2003**, *19*, 9–30.
5. Aggelidis, V.P.; Chatzoglou, P.D. Using a modified technology acceoptance model in hospitals. *Int. J. Med. Inform.* **2009**, *78*, 115–126. [CrossRef] [PubMed]
6. George, D.; Mallery, P. *SPSS for Windows Step by Step: A Simple Guide and Reference*, 4th ed.; Allyn & Bacon: Boston, MA, USA, 2003.

Disclaimer/Publisher's Note: The statements, opinions and data contained in all publications are solely those of the individual author(s) and contributor(s) and not of MDPI and/or the editor(s). MDPI and/or the editor(s) disclaim responsibility for any injury to people or property resulting from any ideas, methods, instructions or products referred to in the content.

engineering proceedings

MDPI

Proceeding Paper

Research on the Learning Performance and Communication Networking of Online Analytical Processing Courses †

Yung-Cheng Liao

Department of Insurance and Finance Management, Chaoyang University of Technology, Taichung 413310, Taiwan; peterliao@gm.cyut.edu.tw

† Presented at the 3rd IEEE International Conference on Electronic Communications, Internet of Things and Big Data Conference 2023, Taichung, Taiwan, 14–16 April 2023.

Abstract: Most schools do not have a large insurance database or OLAP platform for students' learning, operating, and communication. The university courses investigated in this research were insurance information system courses. Those courses were conducted in a project-based learning approach and mutual communication networking. To facilitate the student's practice, these courses were divided into several groups to create an Excel micro-database and OLAP simulation analysis with pivot analysis charts was executed. In these courses, students needed to overcome the challenge of software operation. After being taught by teachers or classmates, all students were able to complete the designated project reports in those courses. A questionnaire survey result showed that students' course satisfaction was higher than average. The flexibility for students to choose the topic of projects is the key success indicator. Additionally, integrating group and individual communication networking was also important to enhance students' performance.

Keywords: online analytical processing (OLAP); pivot analysis charts; project-based learning; communication networking

Citation: Liao, Y.-C. Research on the Learning Performance and Communication Networking of Online Analytical Processing Courses. *Eng. Proc.* **2023**, *38*, 6. https://doi.org/10.3390/engproc2023038006

Academic Editors: Teen-Hang Meen, Hsin-Hung Lin and Cheng-Fu Yang

Published: 19 June 2023

Copyright: © 2023 by the author. Licensee MDPI, Basel, Switzerland. This article is an open access article distributed under the terms and conditions of the Creative Commons Attribution (CC BY) license (https://creativecommons.org/licenses/by/4.0/).

1. Introduction

Most insurance companies have built big data to conduct online multi-dimensional analysis and real-time processing of reports through online analytical processing (OLAP) platforms in Taiwan. OLAP platforms perform customer-related statistical analysis through a simple click-and-drag operation without writing programs, which is convenient.

Most universities do not have insurance databases, policyholder consumption records, and OLAP platforms. The related insurance information system and customer relationship management (CRM) courses need to be learned through a simulation approach. Furthermore, the communication networking of online analytical processing (OLAP) courses is the key success factor for learning performance. Efficient communication networking may be helpful to enhance student's learning performance.

To facilitate students' learning, students create an Excel customer micro-database and operate the pivot analysis table or chart to conduct OLAP multi-dimensional statistical analysis in the course. The meaningful customer analysis results are obtained for marketing campaign planning and are presented to every member of the classes. During the teaching and learning process of database application and pivot analysis courses, students need to overcome learning frustrations and difficulties.

According to the teaching and learning practice in the insurance information system course, this research is carried out to explore the key success indicator to improve learning interests and satisfaction. Furthermore, the shortcomings are reviewed based on the learning process and satisfaction scores for reference.

The research scope and methods for this study are as follows. The research object is the course of insurance information systems. The course adopts project-based learning and a group cooperation approach. Through the research and analysis during the learning process, the study tries to find the key improvements during the learning process. To understand the learning satisfaction of students, the courses accompany a satisfaction survey to monitor the learning results. The courses are communicated with students through Line software, an e-learning system, and the team's oral communication. It is reconfirmed whether the communication networking between students and teachers are a key success indicator for learning performance.

2. Literature Review

2.1. Project-Based Learning

Hsu and Hsu (2020) investigated student satisfaction and acceptance of the project-based learning course [1]. The results of their study showed that students' satisfaction with the courses was significantly positively correlated. Gary (2015) stated that project-based learning was an approach particularly well-suited to achieve better learning results for computing students [2]. Krajcik and Blumenfeld (2006) stated that project-based learning had the potential to help most students learn science [3]. Solomon (2003) stated that project-based learning comprises students working in groups to solve challenging problems [4].

2.2. OLAP-Related Literature

Kumar, Verma, and Saxena (2012) stated that the OLAP platform had a strong function in the interaction and cross-analysis between data and summary figures [5]. Tardío, Maté, and Trujillo (2020) determined that modern technologies enabled OLAP processing platforms using data pre-aggregation techniques to show multidimensional reports [6]. Chia and Liao (2021) stated that the OLAP platform conducted relevant statistical analyses through the online real-time report and multidimensional customer analyses [7].

3. OLAP Course Design and Conduct

Students' learning ability for customer data analysis and software operation is significantly different depending on the learning process. The process of completing the project requires an understanding of the product, price, and customer profile to establish a customer database. In the research course, students need to operate an Excel customer micro-database and then analyze customer figures through pivot analysis tables and charts. After students completed their customer analysis charts, they need to summarize key company profiles and business findings to develop marketing campaigns using PowerPoint and present them to all members of their respective classes. The details of the research are shown in Figure 1.

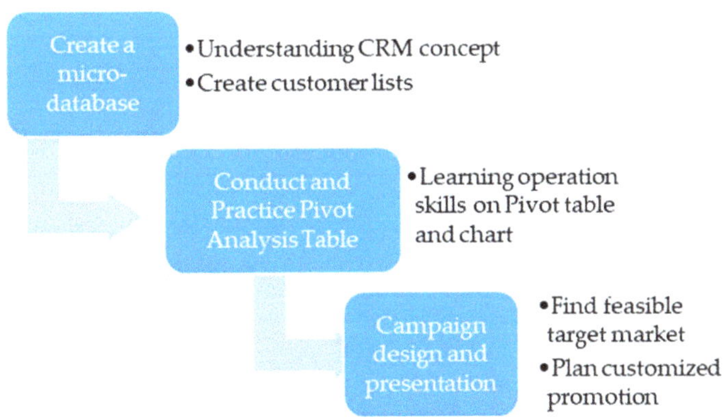

Figure 1. Learning procedure of researched course.

During the learning process, students usually have learning difficulties and frustrations with the establishment of the Excel micro-database. While operating pivot analysis tables and charts, students usually express countless common and individual learning difficulties. Therefore, the project-based learning course needs to integrate group and individual teaching and demonstration to accurately assist students to complete their projects. In this study, the communication networking between students and teachers is the key success factor for learning performance. The common learning difficulties and individual learning difficulties can be summarized as follows.

3.1. Common Learning Difficulties: Adopting Group Teaching and Demonstration

While students have common learning difficulties, teachers adopt group teaching and demonstration approaches. In general, teachers need to explain common knowledge, demonstrate, and communicate the operation process in advance. Items are described as follows and details are shown in Table 1:

1. To improve students' learning interests, students are allowed to choose companies for projects in which they are interested. Flexibility is needed for students to choose the topic of the project as one of the key success indicators to enhance students' learning interests.
2. Students need to position their company and discuss with team members of their projects. If their company positioning changes, the contents of columns in the customer database are different.
3. Mutual learning among members produces assistance during the learning procedure. Communication networking for classmates and teams is helpful to enhance learning performance.
4. Students can add new customer fields in the customer database for meaningful findings.
5. After the teacher's demonstration on the pivot analysis of table and chart, students need to operate by themselves.

Table 1. Common learning difficulties.

Items	Main Problem Items
Create a micro-database	1. What field name needs to be included?
	2. How many customer lists are needed to be included?
	3. How to quickly create customer lists?
	4. How to classify selected fields by level of range?
	5. How to fill in the value or text in Excel quickly?
How to conduct and practice pivot analysis table or chart	6. How to insert a pivot analysis chart or table?
	7. How to choose, drag, or filter fields?
	8. How to convert accumulated figure in pivot analysis table?
	9. How to execute the range analysis?
	10. Which graph should I choose? Pie chart, bar chart or line chart?

3.2. Individual Learning Difficulties: Adopting Individual Teaching and Demonstration

While students have individual learning difficulties, teachers adopt individual teaching and demonstration approaches. In general, teachers need to explain and communicate specific instructions and demonstrate the operation process for the project. The related items are summarized as follows, as shown in Table 2:

1. Teachers need to guide students to complete the creation of the database by copying, pasting, dragging, and inserting formulas in the Excel software.
2. Teachers explain individual knowledge and demonstration for individual and team members.
3. During the procedure of courses, team cooperation is helpful for students' learning among classmates or teams. Especially while the teacher is busy instructing or

demonstrating to other students, mutual learning and interaction among members and students play an important role in learning improvement. Communication networking among students, teams, and teachers is the key success factor for enhancing learning performance.

Table 2. Individual learning difficulties.

Items	Main Problem Items
How to decide and choose the topic	1. Can teachers suggest topics for other kinds of companies?
Create a micro-database	2. How to operate specific formulas?
	3. Can I fill in an accumulative value in customer name lists?
How to conduct and practice pivot analysis table or chart	4. How to set the age range or product range in customer database, pivot analysis table, and chart?
	5. Why cannot specific fields be shown in the pivot table?
	6. Why did the pivot analysis table disappear?
	7. How to complete meaningful findings?
	8. How to draw and edit the presented product, channel name, and value in pivot analysis chart?
How to plan promotion campaign	9. Are there any promotion campaign suggestions for my project?

Under project-based learning, students need to cooperate and interact with classmates to learn how to draw and operate the pivot analysis tables and charts, as shown in Figure 2. Furthermore, students need to filter important analysis findings and plan promotion campaigns. In the final step, every member needs to present to other students in the class. After being guided by teachers or learning among classmates, all students complete the designated project reports in those courses. The details of the research are shown in Figure 2.

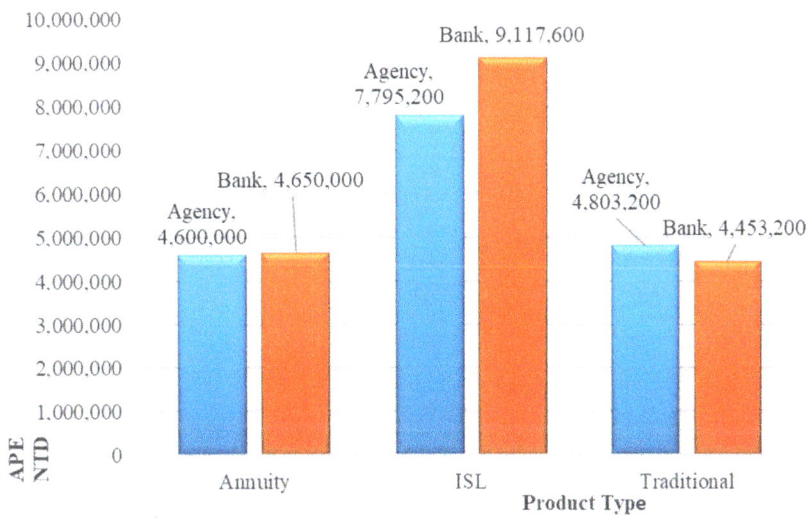

Figure 2. Example of pivot analysis charts for customers.

4. Student Satisfaction Analysis

4.1. Learning Satisfaction

The course investigated in this study offered two classes of insurance information systems, and the number of students was 20 and 14 persons. The students of the courses were university students with jobs. A student learning satisfaction survey was conducted at the end of the semester. The highest score of learning satisfaction was 5 points. The results of student satisfaction scores were higher than college average scores. For example, the average college satisfaction score was 4.45. The scores for A and B courses were 5 and 4.9, which were higher than the college average score. It was found that the course designed and operated by project-based learning and group cooperation approach enhanced students' satisfaction with the learning procedure.

Project-based learning approach allowed students to focus on the operation of the project. A variety of methods to evaluate students' scores is important. In the courses, students were able to communicate through Line software, an e-learning system, and oral communication. The combination of oral and the e-platform were important for the courses. Furthermore, teachers established a scoring system for project tasks and announced it to students in the beginning. Those methods and communications were beneficial to improve the interests and satisfaction of students, as shown in Figure 3.

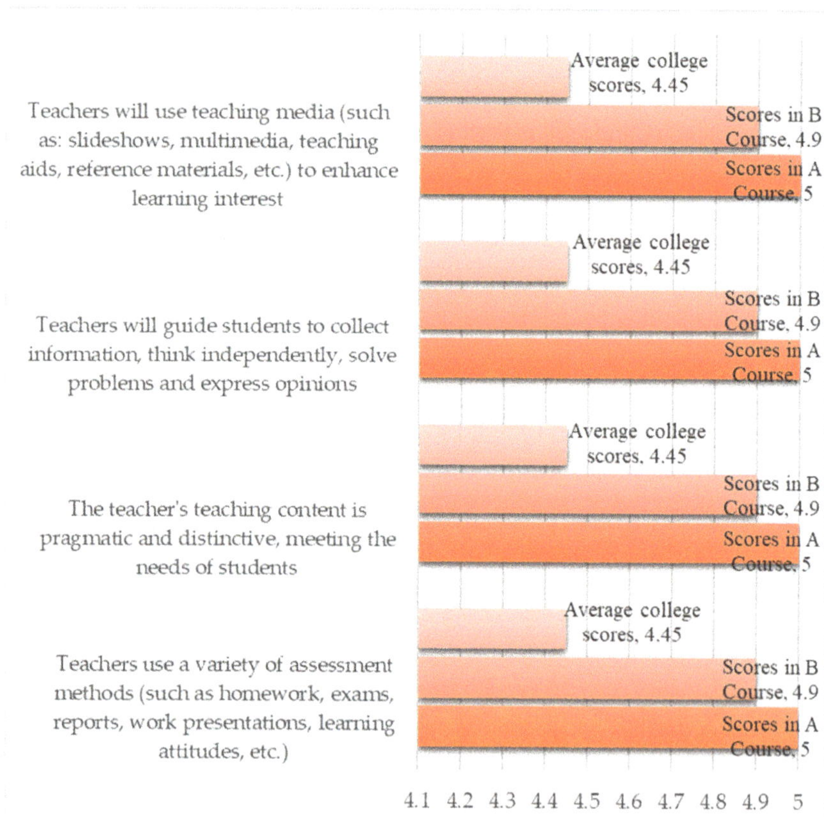

Figure 3. Students' learning satisfaction scores by selected items.

4.2. Shortcomings

After reviewing the learning process, the following shortcomings were identified:

1. When the course enhanced the interaction between teachers and students during the process, the learning performance improved. Especially during the COVID-19 period, many courses were taught by remote video teaching, and so enhancing interaction is critical.
2. Teachers were usually busy helping students during the learning process. The number and composition of persons in every project team needed to be considered because of teachers' time allocation.

5. Conclusions and Recommendations

5.1. Conclusions

The investigated information course adopted the application of big data and the online analytical processing (OLAP) platform based on project-based learning and a group cooperation approach. The course had students in two classes of insurance information system courses. In the insurance information system courses, the flexibility for students to choose the topic of projects is one of the key success indicators to enhance students' interest and satisfaction. For students' different learning abilities, it is also important to integrate group and individual teaching and demonstration.

Communication networking among students, teams, and teachers is also critical for enhancing learning performance. Additionally, it is required that the courses be communicated with students through a variety of methods, such as Line software, an e-learning system, and the team's oral communication.

Students completed their projects and then obtained higher scores in the courses. Project-based learning approach allowed students to focus on the operation of projects. Based on students' satisfaction and learning processes, insurance information system courses designed and operated by project-based learning, group cooperation, and communication approaches were successfully in enhancing students' interest and satisfaction.

5.2. Recommendations

In this study, there was a limitation in the small number of students, which was 20 or 14 in every class. Subject to the number of students, it was not feasible to execute the comparison analysis between project-based learning and the general learning approach. Based on the results, the following is recommended:

1. The number of students in every course needs to be considered because they affect the teacher's time allocation and procedure.
2. The number and composition of students in every team need to be considered. For example, one team is composed of higher-aged and lower-aged students for their cooperation. It is better for communication networking and students' learning.
3. Enhancing the interaction between teacher and students during the process may result in a better learning performance.

Funding: This research received no external funding.

Institutional Review Board Statement: Not applicable.

Informed Consent Statement: Not applicable.

Data Availability Statement: Not available.

Conflicts of Interest: The author declares no conflict of interest.

References

1. Hsu, L.I.; Hsu, Y.H. Integration Scaffolding Theory and Flipped Teaching Approaches into Project-Based Course. *J. Teach. Pract. Pedagog. Innov.* **2020**, 3, 129–163.
2. Gary, K. Project-Based Learning. *Computer* **2015**, *48*, 98–100. [CrossRef]
3. Krajcik, J.S.; Blumenfeld, P.C. *Project-Based Learning*; Cambridge University Press: Cambridge, UK, 2006; pp. 317–333.
4. Solomon, G. Project-Based Learning: A Primer. *Technol. Learn. Mag.* **2003**, 23, 20.

5. Kumar, N.; Verma, V.; Saxena, V. Data Cube Representation for Vehicle Insurance Policy System. *Int. J. Comput. Appl.* **2012**, *58*, 1–4. [CrossRef]
6. Tardío, R.; Maté, A.; Trujillo, J. A New Big Data Benchmark for OLAP Cube Design, Using Data Pre-Aggregation Techniques. *J. Appl. Sci.* **2020**, *10*, 8674. [CrossRef]
7. Chia, T.H.; Liao, Y.C. *Life Insurance Marketing and Management*; Xinfule Culture and Education Published: Taichung, Taiwan, 2021; pp. 263–265.

Disclaimer/Publisher's Note: The statements, opinions and data contained in all publications are solely those of the individual author(s) and contributor(s) and not of MDPI and/or the editor(s). MDPI and/or the editor(s) disclaim responsibility for any injury to people or property resulting from any ideas, methods, instructions or products referred to in the content.

engineering proceedings

Proceeding Paper

Research on Big Data Ad Hoc Query Technology Based on an Accident Insurance Campaign †

Yung-Cheng Liao and Mei-Su Chen *

Department of Insurance and Finance Management, Chaoyang University of Technology, Taichung 413310, Taiwan; peterliao@gm.cyut.edu.tw
* Correspondence: chmeisu@cyut.edu.tw

† Presented at the 3rd IEEE International Conference on Electronic Communications, Internet of Things and Big Data Conference 2023, Taichung, Taiwan, 14–16 April 2023.

Abstract: Lots of Insurance companies have constructed databases for ad hoc query software in Taiwan that combines customer relationship management and marketing campaign management. An ad hoc query is a non-routine and specific query performed in real time to filter specific customer information from big data. Ad hoc query has the strength to retrieve customer information more quickly and conveniently than by filtering target customer lists using a mainframe or OLAP. In this study, the strengths and weaknesses of ad hoc query, online analytical processing (OLAP), and general query using a mainframe are analyzed. The results indicate that ad hoc query has the advantage of flexibility for users' specific needs. Ad hoc query has obstacles and challenges for users regarding how to learn its system fields and writing programs. It is concluded that the design between individual assured suggestions and a convenient operation process is critical for raising the response rate. Additionally, precisely filtering technology for target customers is the key success factor for an accident insurance campaign.

Keywords: ad hoc query; big data; accident insurance campaign

Citation: Liao, Y.-C.; Chen, M.-S. Research on Big Data Ad Hoc Query Technology Based on an Accident Insurance Campaign. *Eng. Proc.* **2023**, *38*, 8. https://doi.org/10.3390/engproc2023038008

Academic Editors: Teen-Hang Meen, Hsin-Hung Lin and Cheng-Fu Yang

Published: 19 June 2023

Copyright: © 2023 by the authors. Licensee MDPI, Basel, Switzerland. This article is an open access article distributed under the terms and conditions of the Creative Commons Attribution (CC BY) license (https://creativecommons.org/licenses/by/4.0/).

1. Introduction

1.1. Research Motivation and Objective

Originally, most insurance companies query and filter customer data using an existing mainframe. However, mainframe queries are not convenient for users and marketers. Today, lots of insurance companies have constructed customer relationship management (CRM) using big data. There are many analyses and query tools in CRM such as ad hoc query and online analytical processing (OLAP) tools [1–3].

The staff of all non-IT departments execute customer management and marketing campaign management using ad hoc query, an OLAP platform, and CRM. There are different strengths and weaknesses of ad hoc query, OLAP query, and general query when using a mainframe system. We study what query needs are suitable for ad hoc query campaigns and their key success factors via accident insurance campaigns. It is helpful to realize feasible and specific campaign needs to summarize the critical success points of a direct mail campaign.

1.2. Research Scope and Methods

This research compares the strengths and weaknesses of ad hoc query, online analytical processing (OLAP), and general query using a mainframe system. It is helpful to realize which of these methods is feasible for specific campaign needs.

This research endeavors to summarize the critical success points of a direct mail campaign by reviewing campaign experience. The details of the research are shown in Figure 1.

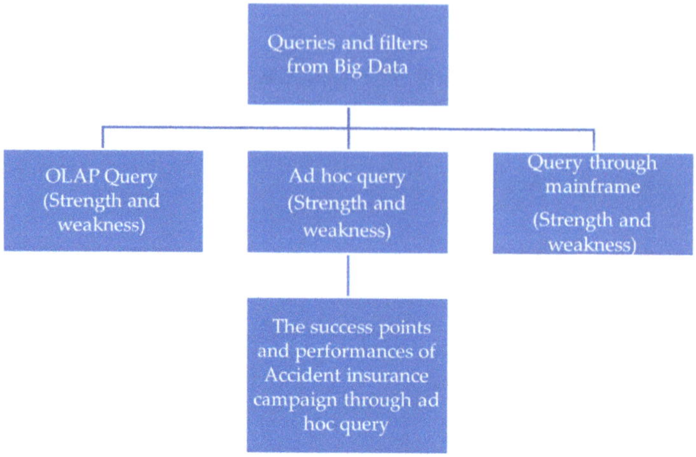

Figure 1. Structure of research.

1.3. Research Restrictions

1. There are few studies on the campaign performance of direct mail in the insurance industry. The research can only compare the performance of a general product campaign rather than an accident insurance campaign.
2. The research compares campaign performance by evaluating response rate rather than other indicators.
3. Data mining or other query tools are not within the scope of the research.

2. Literature Review

There are many approaches to consider for customer relationship management databases, including ad hoc queries, OLAP queries, and general queries using mainframe systems. A general query is the original approach used to retrieve customer and policy information before the construction of CRM. Employees of insurance companies need to prepare the system requirement documents and send them out to the information and technology department for system program development and permission. An ad hoc query is a one-time or specific query performed in real time to retrieve specific information from big data. Users need to write SQL or related programs to filter data from the database [1–3].

The OLAP platform is related to relevant statistical analysis by the performance of online real-time and multidimensional customer analysis via the user's simple click-and-drag operation. The OLAP platform is an interactive business intelligence and data pre-aggregation application for data querying [1,4,5]. However, the system fields and functions of OLAP are restricted to defined functions.

The response rate of direct mail marketing campaigns has been on a downtrend in the past few decades in several areas [6]. According to the related literature, traditional product-directed mail campaigns typically receive a response rate of around 0.5—2% [6–10].

3. Comparison among Ad Hoc Query, OLAP Query, and Query Using a Mainframe

3.1. The Comparison among Ad Hoc Query, OLAP Query, and Query Using a Mainframe

We compare the strengths and weaknesses of ad hoc query, online analytical processing (OLAP), and general query using a mainframe system. Although mainframe query is not convenient for the operations of users and marketers, it has the strength of the widest range of data. However, it needs complicated programs and system knowledge.

Ad hoc query filters customer name lists more quickly and conveniently than a general data query using a mainframe. Ad hoc query has flexibility in writing SQL programs. On the other hand, ad hoc query has obstacles and challenges for writing programs and system fields. An example SQL program is shown in Table 1.

Table 1. An example of a SQL program in ad hoc query.

Items	Content
Tables	• Database Table 1: applicant_table • Database Table 2: acct_life_table • Product: personal accident insurance • Channel: agency channel
SQL ProgramExample	sel applicant_ID, Name, address, total_assured from applicant_table inner join acct_life_table on insured_id = insured_id where total_assured < '1,000,000' and channel = '1' and product_code like any ('IPA%', 'PAR%')

Source: this research.

OLAP query has the strength of being easier to operate. However, the OLAP platform is for regular and defined requirements, so the scope and function are highly restricted. Ad hoc queries and OLAP queries are based on real time, but the database must be updated regularly.

3.2. The Comparison Summary for Three Approaches

1. The most flexible and feasible campaign management approach is the ad hoc query approach.
2. The approach that allows the easiest operation of the drag and drill process is the OLAP query approach.
3. The approach with the widest range of data, which are updated immediately, is the mainframe query approach. The details of these approaches are shown in Table 2.

Table 2. Comparison among ad hoc query, OLAP query, and general query using a mainframe.

Approach	Strengths	Weaknesses
Ad hoc query	• Flexible and non-regular needs. • It is feasible and convenient to plan marketing campaigns. • Wider data range. • The recency rule is easily contained. • Authorized users can query by themselves.	• The obstacles of learning and writing SQL programs. • The obstacles of learning system fields.
OLAP query	• Easier operation of the drag and drill process. • Regular and defined query needs. • Authorized users can query by themselves.	• The system fields and functions are highly restricted. • The recency rule cannot be easily contained for a specific campaign.
General query using a mainframe	• The widest range of data, which are updated immediately.	• Complicated professions of the program are necessary. • Only specific staff of the department can query by themselves.

Source: this research.

4. Design and Performance of Insurance Campaign

4.1. Planning and Design of Campaign

After a CRM system is constructed, the customer management and the marketing campaign management become convenient for the operations and the launch. Big data marketing needs more precise marketing to enhance the campaign performance. If the campaign is launched while target customers need a timely and automatically combined management process, the campaign performance may be better. The planning and design processes of the direct mail campaign are as follows [1]:

1. Querying and filtering target customers: after comparing the mainframe query, OLAP query, and ad hoc query, the ad hoc query approach is selected to filter target customers.
2. The recency rule of target customers is defined.
3. The appropriate insurance and product recommendations are provided for each target customer.
4. Personalized customer letters are prepared with touching, warmer, and life-oriented care.
5. A simple insurance application and premium withholding process is planned for target customers.
6. Follow-up reminders and statistics tracking management are prepared.

4.2. Filtering Criteria

The filtering criteria of the accident insurance campaign are summarized as follows:

1. Target customers need to be existing and effective individual customers.
2. The insured must be the same person as the applicant of the main contract for every responding target customer.
3. The range of insured ages is restricted between 20 and 60 years old.
4. The occupation level of target customers is restricted to level 1.
5. Target customers must have no claim records.

After filtering from the CRM database, the campaign retrieves around 41,700 customer name lists. Precisely filtering technology for target customers is the key success factor for the accident insurance campaign.

4.3. Direct Mail Planning of Accident Insurance Campaign

The accident insurance campaign is focused on direct mail marketing based on target customers via ad hoc queries. The accident insurance marketing campaigns are combined with individual assured suggestions and a convenient operation process. The key process of the direct mail accident insurance campaign is listed as follows:

1. Adopting two-stage direct mail marketing.
2. Offering appropriate accident insurance rider suggestions: suggestions for accidental death, dismemberment, major burn, and disability coverage are included for every customer.
3. Health statements: customers do not need any health statements.
4. Premium payment method: the same premium payment method as the main insurance contract.
5. Simplified insurance application process: Every policyholder directly signs the pre-authorized insurance application form and sends it back to the company. The company automatically underwrites the insurance application for every customer.

4.4. Campaign Performances and Recommendations

The campaign sent out around 41,700 direct letters, and around 1300 physical letters received back. The total response rate of the campaign was around 3.1%. The average annualized paid premium per policy was around NTD 3100, and the total annualized premium was around NTD 4 million.

Traditional product-directed mail campaigns typically receive a response rate of around 0.5–2% according to related studies [6–10]. The response rate of the researched insurance campaign, thus, has a better response rate than a traditional direct mail campaign. There are some recommendations for the insurance campaign, which are listed as follows:

1. If the budget is sufficient, the control group and the test group may be included for comparison purposes.
2. Several customers complained about higher premium expenses compared to other accident insurance companies in the insurance market. It is feasible to launch a campaign with less coverage and a lower premium expense basis while the number of target customers is sufficient.

5. Conclusions and Recommendations

5.1. Conclusions

Although mainframe query is not convenient for the user's customer management and campaign management, it has the strength of a complete range of data. Additionally, ad hoc query filters customer name lists more quickly and conveniently than a general data query. Ad hoc query has the strengths of flexibility and a wider range. On the other hand, it has obstacles and challenges for users regarding how to learn its writing programs and system fields. OLAP query has the strength of an easier operation process for defined query needs. However, the system fields and functions are restricted in the OLAP query platform.

The researched campaign sent out around 41,700 direct letters with a total response rate of around 3.1%. The average annualized premium per policy was around NTD 3100, and the total annualized premium was around NTD 4 million. The response rate of the direct mail campaign using ad hoc query was higher than the response rate of a traditional direct mail campaign.

The research finds that the the design of personalized assured suggestions and a convenient application operation process are critical for the performance of direct mail campaigns. Additionally, precisely filtering technology for target customers is also important for the researched campaign.

5.2. Recommendations

1. It is recommended to include other kinds of insurance products in the direct mail campaign.
2. There are many measure indicators for marketing campaigns. It is recommended to include comprehensive campaign performance indicators.
3. It is recommended to include comparisons with other data mining or other query tools.
4. If the budget is sufficient, the control group and test group may be included for comparison purposes.
5. It might be feasible to launch a campaign with less coverage and a lower premium basis while the number of customers is sufficient.

Author Contributions: Writing—original draft preparation: Y.-C.L.; writing—review and editing: M.-S.C. All authors have read and agreed to the published version of the manuscript.

Funding: This research received no external funding.

Institutional Review Board Statement: Not applicable.

Informed Consent Statement: Not applicable.

Data Availability Statement: Not available.

Conflicts of Interest: The authors declare no conflict of interest.

References

1. Chia, T.H.; Liao, Y.C. *Life Insurance Marketing and Management*; Xinfule Culture and Education: Taichung, Taiwan, 2021; pp. 263–265.
2. Hurd, M.; Nyberg, L. *The Value Factor*; McGraw-Hill Education (Asia): Berkshire, UK, 2004; pp. 121–123.
3. Oracle Website, Running Ad Hoc. Queries. Available online: https://docs.oracle.com/cd/E13167_01/aldsp/docs25/samples_tutorial/adhoc_queries.html#:~:text=Run%20AD%20Hoc%20Query%201%20Create%20a%20StringBuffer,Mediator%20API%20or%20ALDSP%20Control.%20.%20See%20More (accessed on 13 February 2023).

4. Liao, Y.C. *Research on Insurance Information Courses Using Pivot Analysis Charts to Simulate the Application of Online Analytical Processing Platform*; IEEE: Taichung, Taiwan, 2022.
5. Tardío, R.; Maté, A.; Trujillo, J. A New Big Data Benchmark for OLAP Cube Design, Using Data Pre-Aggregation Techniques. *J. Appl. Sci.* **2020**, *10*, 8674. [CrossRef]
6. Lau, K.N.; Chow, H.; Liu, C. A database approach to Cross Selling in the Banking Industry: Practices, Strategies and Challenges. In *Database Marketing & Customer Strategy Management*; Henry Stewart Publications: London, UK, 2003; pp. 216–234.
7. Feld, S.; Frenzen, H.; Krafft, M.; Peters, K.; Verhoef, P.C. *The Effects of Mailing Design Characteristics on Direct Mail Campaign Performance*; Elsevier: Amsterdam, The Netherlands, 2013.
8. Richards, J. What Is the Average Rate of Return on a Direct Mail Campaign? Available online: https://smallbusiness.chron.com/average-rate-return-direct-mail-campaign-23974.html (accessed on 13 February 2023).
9. Data Marketing Association (DMA). What Is the Response Rate from Direct Mail Campaigns? May 2021. Available online: https://dma.org.uk/article/what-is-the-response-rate-from-direct-mail-campaigns (accessed on 13 February 2023).
10. Direct Mail Advertising Global Market Report 2022. Available online: https://www.reportlinker.com/p04442212/Direct-Mail-Advertising-Global-Market-Briefing.html?utm_source=GNW (accessed on 13 February 2023).

Disclaimer/Publisher's Note: The statements, opinions and data contained in all publications are solely those of the individual author(s) and contributor(s) and not of MDPI and/or the editor(s). MDPI and/or the editor(s) disclaim responsibility for any injury to people or property resulting from any ideas, methods, instructions or products referred to in the content.

engineering proceedings

MDPI

Proceeding Paper

Classification Crisis Communication: Semiotic Approach with Latent Semantic Analysis †

Richard G. Mayopu, Long-Sheng Chen and Venkateswarlu Nalluri *

Department of Information Management, Chaoyang University of Technology, Taichung 413310, Taiwan; s11014905@gm.cyut.edu.tw (R.G.M.); lschen@cyut.edu.tw (L.-S.C.)
* Correspondence: nallurivenkey7@gmail.com

† Presented at the 3rd IEEE International Conference on Electronic Communications, Internet of Things and Big Data Conference 2023, Taichung, Taiwan, 14–16 April 2023.

Abstract: Previous crisis communication research has been based on qualitative methods such as interviews or questionnaires, which require considerable manpower, material resources, and time to focus on specific topics. The current situation needs to be reflected timelier. With the rise of social communities, community users' comments have gradually become an important reference for other community members. Twitter is one of the most popular social media in the world. During the COVID-19 pandemic, people were restricted by rules and government policies, such as wearing masks, maintaining social distancing, and avoiding crowding. This led people to spend time on devices. By using devices, most people are involved in social media activities. This study aims to discover the awareness Indonesians display in the text they upload to Twitter. Using the Twitter crawling technique, we collected data. We also analyzed the text with text mining techniques and latent semantic analysis (LSA) with semiotic methods. The crisis communication was classified, and the definition of crisis terminology was improved in social media.

Keywords: text mining; latent semantic analysis; crisis communication; social media; semiotic

Citation: Mayopu, R.G.; Chen, L.-S.; Nalluri, V. Classification Crisis Communication: Semiotic Approach with Latent Semantic Analysis. *Eng. Proc.* **2023**, *38*, 9. https://doi.org/10.3390/engproc2023038009

Academic Editors: Teen-Hang Meen, Hsin-Hung Lin and Cheng-Fu Yang

Published: 19 June 2023

Copyright: © 2023 by the authors. Licensee MDPI, Basel, Switzerland. This article is an open access article distributed under the terms and conditions of the Creative Commons Attribution (CC BY) license (https://creativecommons.org/licenses/by/4.0/).

1. Introduction

Crisis communication has become very important in the last three years, during which the world struggled with the COVID-19 pandemic since the first time this virus appeared and affected society, the economy, culture, politics, and technology. During the pandemic, people often used technology to communicate with families and friends and work from home.

Social media is one of the most used technologies in daily life in modern society. Workers use it for their jobs, students for studying, and parents for parenting. Technologies support many activities and change people's behavior. At the same time, people offer advice to the developer to improve the technology. Social media has rapidly increased the interaction of humans and the awareness of understanding how to use technology wisely. People can use social media for e-commerce to promote their products and share content. Social media helps a community build close relationships. The interaction among the members of the community is an interesting phenomenon to observe. This interaction occasionally becomes a new approach or a new habit in society, which is named the "new normal".

The new normal is the concept provided by scientists to describe social activity during the pandemic. People had to do social distancing and wear masks often. In this situation, billions of people went online where they could interact without facing their friends or family directly, to feel safer and healthier. The new normal has been created to describe the world on the "crisis road". Crisis during the pandemic became a global issue. This is the challenge in the digital and technology era.

A crisis must be communicated to the public with a strategy. This can be conducted by using technology as the channel. In conventional communication theory, the channel can be the newspaper, television, or radio. However, in the modern era, the channel can be the new social media such as Twitter, Facebook, YouTube, and many others. Twitter is one of the most popular social media that was used during the pandemic. Crisis communication is the term for messages delivered from the communicator to the communicant (recipient), considering crisis requirements such as fast response communication, clear communication, and informative fact. The crisis messages contain meaning based on the understanding of the text and context. In this view, the semiotic approach is the best approach to finding the meaning behind the text related to context. Previous research in crisis communication has always used qualitative approaches [1–3] with semiotics. Thus, we aim to provide a combination of semiotics and latent semantic analysis as a new approach in this field of research.

Our research question is "how to classify crisis communication using a semiotic approach with semantic analysis". This research was focused on crisis communication on a large scale not limited to corporations, institutions, or organizations. We attempted to find out the awareness that Indonesians display in the text they upload to Twitter. Using the Twitter crawling technique, we collected data and analyzed the text with latent semantic analysis (LSA) combined with semiotic methods. The results provide classified crisis communication to redefine crisis terminology in social media.

2. Literature Review

2.1. Crisis Communication

Scholars have defined crisis communication for decades and discussed it through an interdisciplinary approach. Several have described crisis communications from a management perspective, while others have explained it through a cultural approach. Crisis communication is also dealt with in communication science. Based on this background, we reviewed crisis communication from the communication and technology perspective. This approach is expected to enrich the understanding of crisis communication for the development of knowledge. We classified crises into small and large scopes depending on the crisis. Most crises occur in the institution or corporation [4] and have local impacts. Nevertheless, a crisis with a serious impact, such as the global crisis of COVID-19, is a large-scope crisis, and the strategy to deal with it is different.

2.2. Social Media

Social media is a digital platform to enable users to communicate or share content such as texts, photos, videos, and files [5]. Earlier, social media was applied in web-based technology; however, nowadays, most users use social media on mobile apps [6] such as smartphones or tablets.

2.3. Text Mining

In the past few decades, text mining has become popular in computer science as a part of deep learning focusing on how to detect or identify text. Text mining is used to find the meaning of the text by extracting meaningful information from it. According to experts, text mining is used to find new information about human character [7]. Moreover, text mining provides flexible analysis depending on the research objects. In this research, we investigated crisis communication.

2.4. Latent Semantic Analysis

Many published works utilized LSA to identify relevant semantic spaces. For example, Das and Sultana [8] performed a semantic analysis of the Bengali language to propose a predictive model [8]. Hsiao and Hsiao [9] employed LSA for online hotel reviews to understand how satisfied or dissatisfied guests felt. We employed LSA to extract important concepts from a huge number of documents.

3. Methodology

We used a semiotic approach combined with latent semantic analysis (LSA). Previous research usually used qualitative research to find models or strategies for crisis communication. We mixed semiotics in social research with latent semantic analysis for information technology research.

3.1. Pre-Processing

We pre-processed [10] collected data using Python for tokenization and lemmatization and obtained the TDM using MATLAB.

3.2. Natural Language Processing

Natural language processing (NLP) usually considers the interesting applications of text mining and is able to synthesize information from text. Messages on Twitter have an intention for objective communication. However, if a complicated sentence appears, information can be extracted through NLP [11]. The implementation procedure is described as follows.

3.2.1. Step 1: Tokenization

Textual tokenization uses the Natural Language Toolkit (NLTK) of the Python language.

3.2.2. Step 2. Data Cleaning

We deleted stop words such as "the", "and", and other low-importance words in this step.

3.2.3. Step 3. Lemmatization

This step is to reduce complex forms of a single word to its most basic form, e.g., "ate" to "eat".

3.2.4. Step 4. Counting Word Frequency

We counted word frequency and deleted words with a frequency of less than 5.

3.2.5. Step 5. Constructing a Term–Document Matrix (TDM)

In this step, we used TF–IDF (term frequency–inverse document frequency) in Equation (1) to create a term–document matrix (TDM) for further analysis.

$$\text{TF–IDF} = TF(t_i, d_i) \times \log\left(\frac{N}{N(t_i)}\right) \quad (1)$$

3.3. Latent Semantic Analysis

We performed singular value decomposition (SVD) on TDM to investigate the relationship between words and their concepts.

SVD

The process of SVD is shown in Figure 1. The SVD function results in three matrices "terms matrix $U_{t\times r}$", "orthogonal matrix $S_{r\times r}$", and "document matrix $V^T_{r\times n}$", where t refers to the number of words, n is the document term, and r is the number of concepts in the semantic space.

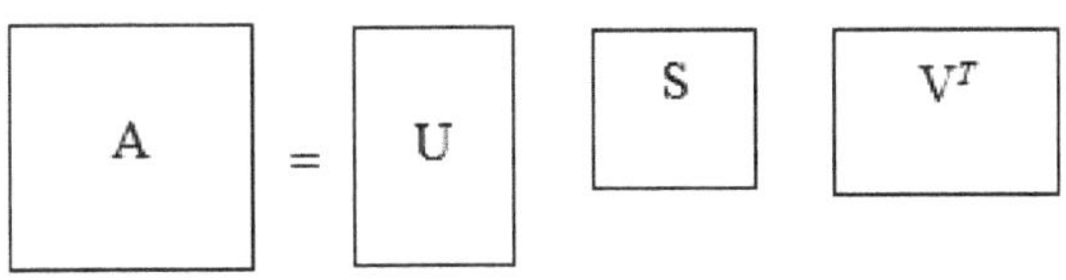

Figure 1. Singular Value Decomposition.

Even after running SVD, it still contains a lot of unimportant information, so it is necessary to reduce the dimensional space. To affect the original characteristics, the feature value k must be chosen. In this study, the scree test was used to determine the k value. In Figure 2, we selected k concepts, so we took the S_k matrix to reduce the dimensionality.

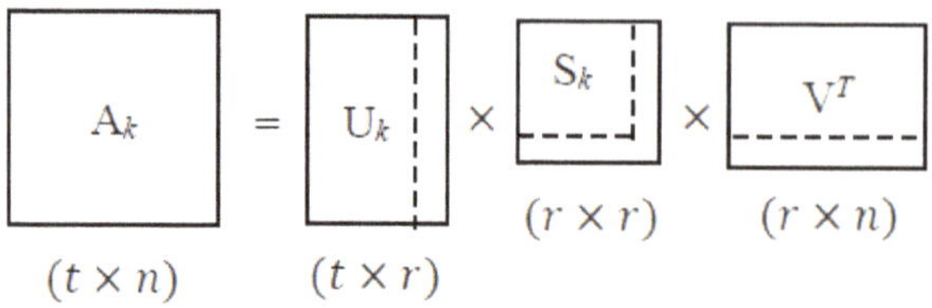

Figure 2. SVD after dimension reduction.

3.4. Ortoghonal Rotation of Axes

The concept load L_T was then calculated by multiplying the dimensionally delimited concept U_k with the concept S_k (Equation (2)) to obtain the concept load L_T. Each feature word was ranked according to the load, and the word concept was named.

$$TL_T = U_k \times S_k \quad (2)$$

4. Results and Discussion

4.1. Data Processing

Crisis communication has increased for decades, especially in the last three years during the COVID-19 pandemic, going beyond conventional crisis communication. Traditional crisis communication usually uses conventional media such as television, radio, newspaper, and word of mouth. Nowadays, crisis communication uses the internet. There are many platforms to deliver and receive messages that contain crisis communication. Moreover, it can be an interaction medium between the communicator and the communicant.

The data has been investigated with SVD and charted to show the information. We selected the k value from the top five (k = 5) as follows. It is possible to ignore the variation gap after the decision point since it decreases after the plot slope, as shown in Figure 3.

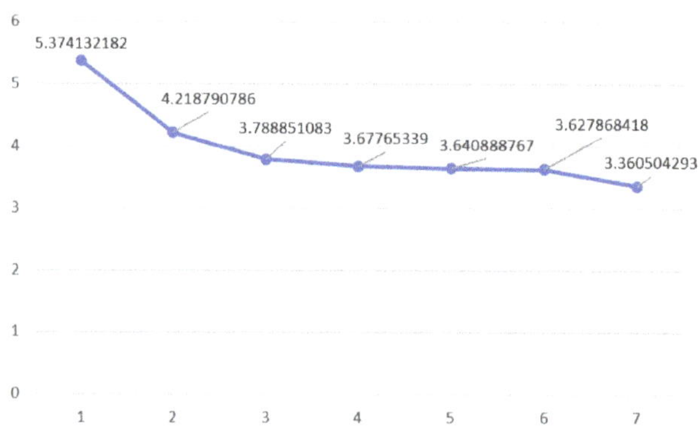

Figure 3. Scree plot.

4.2. LSA and Concepts

In the LSA process, we obtained the five concepts using the semiotic approach. The data are shown in Table 1.

Table 1. Keywords and Loadings.

1st Concept		2nd Concept		3rd Concept	
Keyword	**Loading**	**Keyword**	**Loading**	**Keyword**	**Loading**
Abramovic	1.7554	Crisis	0.2504	Injury	0.0863
Army	0.3468	Destroyed	0.1601	Damage	0.0824
Battle	0.2258	Pandemic	0.0773	Fighting	0.0750
Lockdown	0.0645	Death	0.0637	Military	0.0346
4th Concept		**5th Concept**			
Keyword	**Loading**	**Keyword**	**Loading**		
Civilian	0.0264	Rescuer	0.0223		
Killing	0.0248	Volunteer	0.0212		
Suffered	0.0242	Community	0.0199		
Violence	0.0222	Protester	0.0023		

Based on Table 1, we formulated the concepts of the LSA experiments with five concepts. The top 50 words from each loading were chosen to name the concept after ranking the defining words based on their loadings. Furthermore, we found the fixed concepts to be analyzed further. Table 2 shows the concepts for crisis communication.

Table 2. Concepts of crisis communication using latent semantic analysis.

No.	Concept	Representative Terms
1st	The Antagonist	Abramovic, army, battle, lockdown
2nd	The Phenomenon	Crisis, destroyed, pandemic, death
3rd	The Destructions	Injury, damage, fighting, military
4th	The Victims	Civilian, killing, suffered, violence
5th	The Protagonists	Rescuer, volunteer, community, protester

4.3. Semiotics in Crisis Communication

Roland Barthes' model is used in this research to find the signifier in crisis communication. Figure 4 shows the model to explain the concept.

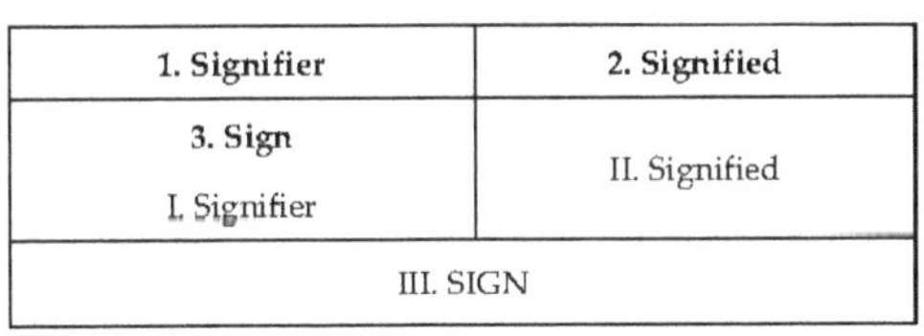

Figure 4. Barthes' model of semiotics.

Figure 4 shows the base model of the semiotics. We put the concept of LSA into the diagram for analysis. The dataset is about the crisis caused by the invasion of Russia into Ukraine. Therefore, the context of the data is about invasion. The context must be clear to analyze the text, and we found several concepts in this research.

In this research, the signifier was the people who provided the text to the mass media, the victims, the soldiers, and the politicians regarding the invasion. The signified is the text that the signifier has produced. Based on the text we found, the sign was used for analyzing the text. In the model, denotation and connotation were defined. Denotation is the system of level signification in Barthes' semiology at the first level, while connotation is at the second level. It means that once readers see the sign (text or image), they consider using denotation to analyze the sign. Then, they find the denotation from the sign.

In this research, denotation has a more closed meaning. With the theory of signs, we understand that literal meaning is natural. According to this theory, the meaning of signs is expanded with meaning that takes place in two stages as shown in Figure 5.

<table>
<tr><td>1. Signifier</td><td>2. Signified</td><td rowspan="2">Denotation: Invasion, civil war, economic restriction</td></tr>
<tr><td>3. Sign
I. Signifier: Mass media, Politicians, etc.</td><td>II. Signified: text</td></tr>
<tr><td colspan="2">III. SIGN</td><td>Connotation: Conspiration, Intelligent movement, etc.</td></tr>
</table>

Figure 5. Implementation of Barthes' models of semiotics.

5. Conclusions and Future Directions

We found a classification of crisis communication in a large scope. Previous research found a small scope of crisis communication, while a large scope was found on Twitter. We found several concepts from LSA that divided texts into small and large scopes. Semiotic analysis was conducted to provide a new approach as a research method. Barthes' model was implemented to analyze the concepts, and semiotics was used in this research. For future research, researchers can consider the semiotic approach in text mining to develop the utilization of semiotic models and provide new models of the denotation and connotation concepts.

Author Contributions: Conceptualization, R.G.M. and V.N.; methodology, R.G.M. and V.N.; software, R.G.M. and V.N.; validation, R.G.M.; formal analysis, V.N. and L.-S.C.; writing—original draft preparation, R.G.M.; writing—review and editing, V.N. and L.-S.C.; visualization, R.G.M.; supervision, L.-S.C.; project administration, L.-S.C.; funding acquisition, L.-S.C. All authors have read and agreed to the published version of the manuscript.

Funding: This work was supported in part by the National Science and Technology Council, Taiwan (Grant No. MOST 111-2410-H-324-006).

Institutional Review Board Statement: Not applicable.

Informed Consent Statement: Not applicable.

Data Availability Statement: The secondary data that were used for analysis in this study are available from the corresponding author upon request.

Conflicts of Interest: The authors declare no conflict of interest.

References

1. Arendt, C.; LaFleche, M.; Limperopulos, M.A. A qualitative meta-analysis of apologia, image repair, and crisis communication: Implications for theory and practice. *Public Relat. Rev.* **2017**, *43*, 517–526. [CrossRef]
2. Coady, A.; Lainchbury, K.; Godard, R.; Holtzman, S. What twitter can tell us about user experiences of crisis text lines: A qualitative study. *Internet Interv.* **2022**, *28*, 100526. [CrossRef] [PubMed]
3. Lu, X.; Jin, Y. Information vetting as a key component in social-mediated crisis communication: An exploratory study to examine the initial conceptualization. *Public Relat. Rev.* **2020**, *46*, 101891. [CrossRef]
4. Huynh-Cam, T.T.; Nalluri, V.; Chen, L.S.; Yang, Y.Y. IS-DT: A New Feature Selection Method for Determining the Important Features in Programmatic Buying. *Big Data Cogn. Comput.* **2022**, *6*, 118. [CrossRef]
5. Wang, H.; Xiong, L. Understanding the antecedents of mobile social media acceptance during the public health crisis maintenance stage. *Int. J. Disaster Risk Reduct.* **2022**, *74*, 102925. [CrossRef]
6. Morgan, A.; Wilk, V. Social media users' crisis response: A lexical exploration of social media content in an international sport crisis. *Public Relat. Rev.* **2021**, *47*, 102057. [CrossRef]

7. Jung, H.; Lee, B.G. Research trends in text mining: Semantic network and main path analysis of selected journals. *Expert Syst. Appl.* **2020**, *162*, 113851. [CrossRef]
8. Das, P.; Sultana, N. Sentiment Analysis on Comments in Bengali Language Using Text Mining & Machine Learning Approach. In Proceedings of the IEEE 7th International conference for Convergence in Technology, Mumbai, India, 7–9 April 2022; pp. 1–4.
9. Hsiao, Y.-H.; Hsiao, Y.-T. Online review analytics for hotel quality at macro and micro levels. *Ind. Manag. Data Syst.* **2020**, *121*, 268–289. [CrossRef]
10. Öztürk, N.; Ayvaz, S. Sentiment analysis on Twitter: A text mining approach to the Syrian refugee crisis. *Telemat. Inform.* **2018**, *35*, 136–147. [CrossRef]
11. Martinelli, D.D. Evolution of Alzheimer's disease research from a health-tech perspective: Insights from text mining. *Int. J. Inf. Manag. Data Insights* **2022**, *2*, 100089. [CrossRef]

Disclaimer/Publisher's Note: The statements, opinions and data contained in all publications are solely those of the individual author(s) and contributor(s) and not of MDPI and/or the editor(s). MDPI and/or the editor(s) disclaim responsibility for any injury to people or property resulting from any ideas, methods, instructions or products referred to in the content.

engineering proceedings

Proceeding Paper

Importance-Performance Analysis of Online Insurance: Communication and Networking †

Ming-Hsi Tang * and Yu-Lin Niu

Department of Insurance, Chaoyang University of Technology, Taichung 413310, Taiwan; s10821041@cyut.edu.tw
* Correspondence: mhtang@cyut.edu.tw

† Presented at the 3rd IEEE International Conference on Electronic Communications, Internet of Things and Big Data Conference 2023, Taichung, Taiwan, 14–16 April 2023.

Abstract: Over the past few decades, drastic advances in technology have forced insurers to seek transformation in order to survive in a changing environment. Along with the outbreak of COVID-19, online insurance has become a much-needed distribution channel for insurers, not only as a cost-effective option, but also in line with pandemic prevention measures. However, in contrast to insurers' vast investment in AI, chatbots, etc., online insurance only accounted for less than 2% of the total premium in Taiwan in 2022. This article uses importance-performance analysis (IPA) to identify disparities between client perceptions of crucial elements of online insurance services and their actual performance. The results show that among constructs such as perceived usefulness, ease of use, security, and service quality, customer experience can be greatly enhanced if insurers prioritize improving website security and service quality.

Keywords: InsurTech; importance-performance analysis; communication and networking

Citation: Tang, M.-H.; Niu, Y.-L. Importance-Performance Analysis of Online Insurance: Communication and Networking. *Eng. Proc.* **2023**, *38*, 7. https://doi.org/10.3390/engproc2023038007

Academic Editors: Teen-Hang Meen, Hsin-Hung Lin and Cheng-Fu Yang

Published: 19 June 2023

Copyright: © 2023 by the authors. Licensee MDPI, Basel, Switzerland. This article is an open access article distributed under the terms and conditions of the Creative Commons Attribution (CC BY) license (https://creativecommons.org/licenses/by/4.0/).

1. Introduction

With the swift advancement in technology, the Internet has changed people's lifestyles with its convenience and immediacy. More than five billion people had used the Internet as of January 2023, making up 64.4% of the world's population. Asia continued to be the region with the largest number of online users in 2022 [1]. Advances in electronic service technology have presented both enormous opportunities and challenges in a variety of business and service sectors at the same time, especially with the advancements in mobile devices and wireless infrastructure. Organizations have implemented innovative customer communication strategies in an effort to cut costs and retain profitability [2]. The digital transformation of financial services and the related growth of Fintech are made possible by information and communication technology (ICT).

It is particularly challenging for sector incumbents to enhance current value propositions due to the rapid pace of technology development and the restrictions imposed by legislation. Examples include the use of chatbots in banking services, quick response (QR) codes on credit cards, and insurances that hope to gain from blockchain [3]. However, companies that can effectively navigate these challenges may be able to gain a competitive advantage by offering innovative products and services that meet customer needs and expectations.

InsurTech is the design and delivery of insurance products and services through technology. The insurance sector is striving to use technology in response to the threats posed by disruptive new financial services. Among them, communication and networking technologies that allow customers and insurers to interact in the digital world are crucial for the development of online insurance. Although specific communication and networking technologies for insurers are outside the purview of this study, it is important to explore whether insurers can meet consumers' changing needs with them.

Taiwan's insurance industry comprises a huge portion of financial institutions. The asset ratio of the insurance industry to financial institutions was 36.07% in 2021 [4]. However, compared to the thriving of mobile banking [3,5,6], Taiwan's insurers still struggle to move forward with their online distribution channels. Moreover, 84.3% of Taiwan's population had home Internet access in 2022 [7]. Despite the prevalence of the Internet, people are hesitant to provide important personal information on websites due to the lack of confidence in e-commerce security [8]. This is a big challenge for insurers, since extensive private financial and health information must be submitted when purchasing an insurance policy. As a result, less than 2% (1.96%) of the premium insurance received from newly issued insurance policies in 2021 came from online channels [4].

Based on the above, this study aims to identify the causes of the sluggishness of online insurance in Taiwan. The following are this study's major contributions. First, it investigates the relative importance of attributes to assist insurers in reinforcing and refining the evaluation of value-added web services and functions. Second, as a reference for customer relationship management (CRM) and marketing management strategies in the insurance sector, this study explores the perceived performance of consumers' online insurance experiences. Third, this study fills a gap in the insurance literature by applying importance-performance analysis (IPA), which is rarely addressed in the insurance sector. Based on research results, the authors provide suggestions concerning communication and networking technologies that can be applied to improve consumers' online insurance experiences.

2. Literature Review

2.1. Perceived Usefulness and Perceived Ease of Use

In order to provide a vigorous explanation for behavior prediction in the area of information technology adaptation, Davis (1989) proposed the technology acceptance model (TAM) [9]. The TAM's main contention is that a person's behavioral intention to use a system or piece of technology is influenced by its perceived usefulness and perceived ease of use [9,10]. The degree to which a person thinks using an information system would increase their productivity is known as perceived usefulness. The extent to which a person believes that using a certain system would be devoid of effort is the definition of perceived ease of use [9].

Applying TAM to insurance mobile application usage, Tang and Yeh (2018) identified perceived usefulness (PU) and perceived ease of use (PE) as significant factors [11]. Due to the complexity of human behavior and the limitations of models, there is no one framework that dominates most of the components concerning users' behavior toward technology usage. In an effort to reduce theories' level of constraint, Venkatesh et al. (2003) combined eight prior established theories of technology acceptance and developed a unified theory of acceptance and usage of technology (UTAUT) [12]. Customer satisfaction was positively impacted by consumers' perceptions of the usefulness and ease of use, according to an empirical study from the life insurance sector [10].

2.2. Website Security and Service Quality

Measures of safeguarding and discouraging hackers from violating customer's private information can be defined as website security. Studies have suggested that Internet users have serious concerns regarding their private information, therefore website security becomes a deciding factor in the development of e-commerce. Online customers will increase their tendency to participate in online transactions if they believe that websites offer solid security measures in protecting transactions and data safety [13].

The quality of services provided and perceived can be affected by customer expectations and service comprehension. Customer satisfaction is elevated if the service quality perceived exceeds the expected level. There is a lot of evidence showing a direct connection between customer satisfaction and service quality [14,15]. For e-commerce to be successful and to attract clients, the service quality of the system will be crucial. Website security and service quality, which have been discovered in other studies, are characteristics we have considered when analyzing the factors influencing consumers' use of online insurance [16,17].

2.3. *Importance-Performance Analysis*

Martilla and James proposed importance-performance analysis (IPA) in 1977 [18]. In a conventional IPA, customers evaluate and determine the average value of importance and performance of various service qualities using the defined system, where the horizontal axis denotes performance and the vertical axis denotes importance. The average values of the importance and performance of various services and product elements are graphically displayed in a two-dimensional coordinate system, primarily in the region divided into four quadrants (Figure 1).

Figure 1. Importance-Performance Grid, Source: Martilla and James, 1977 [18].

In today's environment of rapid technological innovation and worldwide competition, managers must regularly evaluate how satisfied customers are with the services and goods they are supplied. IPA is a helpful and practical tool that can assist decision-makers in identifying service/product components, and hence, allocating resources to where they may boost customer satisfaction. For example, attributes in quadrant II indicate high importance to customers, yet the perceived performance is low. Therefore, these attributes should receive immediate attention and effort, so that customers are not drawn away.

For IPA, choosing the appropriate collection of attributes is essential because the subsequent managerial decisions are based on the scores assigned to these attributes. It is advised that performance and importance be evaluated using the same set of criteria [19]. Based on the previous literature, this article uses perceived usefulness, ease of use, security, and service quality to identify disparities between client perceptions of crucial elements of online insurance services and their actual performance.

3. Methodology

The statistical population in the present research consisted of online insurance customers in Taiwan. This research utilized random sampling to collect data using a questionnaire survey. Please see Appendix A for sample questionnaire items. A five-point Likert scale questionnaire assessed each attribute's importance and the performance of each attribute was distributed to gather data to determine the relationship between the study variables. In all, 300 people were randomly selected. After removing the unusable and incomplete ones, 263 questionnaires were ready for analysis. The effective rate of the questionnaire was 87.67%.

In Table 1, a sample demographic summary is presented. The sample was 62.74% female and 37.26% male, respectively. Since 92.4% of the respondents were under the age of 50, the sample was mostly made up of young and middle-aged people. Given that the sample was only comprised of people with experience in online insurance, the age distribution was fair. The majority of responders had a college or university degree or higher (95.06%).

Table 1. Description of the sample.

Item	Category	Number of Sample	(%)
Gender	Male	98	37.26
	Female	165	62.74
Age	20–29	105	39.92
	30–39	82	31.18
	40–49	56	21.29
	Over50	20	7.60
Education	Senior high and under	13	4.94
	College/university	207	78.71
	Graduate school	43	16.35

Scale reliability in this study was measured using Cronbach's alpha method. Cronbach's alpha is a standard measure of reliability, with values above 0.7 indicating acceptable reliability [20]. There were two sections to the survey questionnaire. The respondents' demographic data, such as gender, age, and educational attainment, were tallied in the first section. Each respondent was asked to rate how much they agreed or disagreed with each statement in the second section. A five-point Likert scale was used to gather the data to facilitate the measurement of the constructs. Twenty attributes measuring the four constructs were included in the questionnaire. The sources and definitions of the constructs are compiled in Table 2.

Table 2. Definition of key variables.

Construct	Definition	Sources
Perceived usefulness (PU)	The degree to which a person believes that using an online insurance channel would enhance or improve his or her condition.	Davis [21] Elseidi [22]
Perceived ease of use (PE)	The degree to which a person believes that using an online insurance channel would be free of effort.	Davis [21] Elseidi [22]
Website security (WS)	The degree to which respondents perceived regarding the security of the insurers' online insurance channel.	Mecovac [13] Li et al. [16]
Service quality (SQ)	A person's subjective evaluation of the service quality of the insurers' online insurance channel.	Haussen et al. [14] Li et al. [16]

4. Results

4.1. Reliability Analyses

Statistical software SPSS 18 was used for the questionnaire analysis. The calculation of the questionnaire's reliability was based on Cronbach's alpha coefficient. The questionnaire proved satisfying in terms of content validity, criterion-related validity, and construct validity (Table 3).

Table 3. Reliability Statistics.

Construct	Cronbach's Alpha
Perceived usefulness	0.884
Perceived ease of use	0.865
Website security	0.905
Service quality	0.924

4.2. Importance-Performance Scores

To establish criteria for the importance-performance analysis, this study combined the diagonal approach with the data-centered quadrants approach (IPA). We used the differences between the performance score and the importance score to plot and arrange the attributes within each quadrant after applying the grand mean of importance and performance scores to divide the attributes into four quadrants.

The average relevance and performance ratings given by survey respondents are shown in Table 4 for each of the 20 online insurance channel attributes. Of the 20 attributes, the top level of performance, on average, was easy access, time saving, and a secure infrastructure platform. However, the scale averages of performance were lower than the scale averages of importance for all attributes, indicating that customers' evaluation of importance was higher than performance.

Table 4. Rank, Means of Importance, and Performance.

	Importance			Performance			Gap	
Items	Mean	Rank	Std. Dev.	Mean	Rank	Std. Dev.	P-I	Rank
PU1	4.457	11	0.621	4.295	6	0.678	−0.162	10
PU2	4.324	17	0.815	4.171	12	0.727	−0.152	9
PU3	4.438	13	0.692	4.343	4	0.691	−0.095	4
PU4	4.514	8	0.667	4.390	2	0.714	−0.124	7
PU5	4.505	9	0.695	4.162	13	0.845	−0.343	15
PE1	4.543	5	0.621	4.467	1	0.651	−0.076	3
PE2	3.733	20	0.902	3.876	20	0.958	0.143	1
PE3	4.133	19	0.844	4.029	19	0.882	−0.105	6
PE4	4.343	16	0.691	4.238	9	0.766	−0.105	5
PE5	4.381	15	0.726	4.333	5	0.716	−0.048	2
WS1	4.629	3	0.593	4.352	3	0.720	−0.276	11
WS2	4.648	2	0.554	4.248	8	0.757	−0.400	19
WS3	4.667	1	0.566	4.238	9	0.803	−0.429	20
WS4	4.629	3	0.542	4.286	7	0.743	−0.343	16
WS5	4.533	7	0.637	4.133	16	0.760	−0.400	17
SQ1	4.286	18	0.781	4.152	14	0.794	−0.133	8
SQ2	4.419	14	0.732	4.114	18	0.812	−0.305	13
SQ3	4.448	12	0.707	4.133	16	0.785	−0.314	14
SQ4	4.486	10	0.681	4.200	11	0.813	−0.286	12
SQ5	4.543	5	0.636	4.143	15	0.786	−0.400	17

Note: Performance and importance scores are based on a 1–5 Likert scale.

The IPA results for online insurance attributes are graphically shown in Figure 2. The average values of the importance and performance of various attributes are graphically displayed in a two-dimensional coordinate system, primarily in the region divided into four quadrants. As mentioned above, attributes fall within the second quadrant, indicating a high expectation yet low perception.

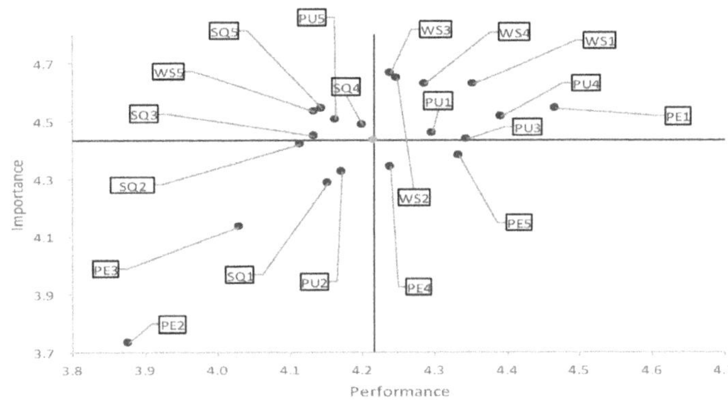

Figure 2. IPA Chart for Online Insurance Attributes.

Based on the results of the above IPA chart, we organized the attributes in quadrant II, as shown in Table 5. The attributes in quadrant II indicate that consumers value these services, but are not satisfied with the performance of the services provided and they need to be improved as a matter of priority. Three out of five are attributes concerning service quality, and the remaining two attributes pertain to perceived usefulness and website security, respectively.

Table 5. Attributes in Quadrant II.

Attributes	Measures
PU5	I can save money by using online insurance websites.
WS5	Online insurance website security is increased by the use of digital signatures.
SQ3	Online insurance channel technology boosts competitiveness while enhancing customer service.
SQ4	Online insurance channels are accessible, user-friendly, and simple to use.
SQ5	Technical support is excellent, and the quality of the services is adequate.

5. Discussion and Conclusions

The purpose of this study was to understand consumers' online insurance behavior, and analyze customers' perceptions of the importance and performance of online insurance services, in order to find ways to improve customer satisfaction and customer loyalty, and to provide suggestions for business improvement in the insurance industry. We used importance-performance analysis (IPA) to identify disparities between client perceptions of crucial elements of online insurance services and their actual performance

It demonstrated that customers consider service quality to be crucial while using online insurance websites. Customers' evaluations of performance, however, fell short of what they had hoped for. As a result, this study makes the case for insurance companies to use communication and networking technologies, such as artificial intelligence, to support customer care so that customers can receive prompt assistance when they run into issues with online insurance. Additionally, to address security issues, insurance companies can implement blockchain technology and conduct a routine security scan at least once every three months or once a year so that customers will feel more at ease providing information online.

Nevertheless, for our study and future research areas, some limitations may be worth attention. First, the interpretation of attributes near the thresholds of discrimination is a problem in IPA. It is challenging to interpret these characteristics with an acceptable confidence level [23,24]. Because the mean, which is often applied to indicate the significance and effectiveness of a characteristic, does not reveal information regarding a distribution's variability, the results must be evaluated with great care. Second, while the ease of use and understanding of IPA may explain its widespread acceptance and application, this does not necessarily ensure its validity [19].

Furthermore, given the emerging and changing nature of the insurance technology field, it may not be appropriate to directly apply the constructs and attributes validated in the past literature. Researchers should exercise discretion in the process of attribute selection. We suggest that the major components of online insurance attributes could be identified using Exploratory Factor Analysis (EFA), and the findings of the two analyses should be compared in future studies to determine the convergent validity of IPA.

Author Contributions: Both authors contributed equally to this work. All authors have read and agreed to the published version of the manuscript.

Funding: This research received no external funding.

Institutional Review Board Statement: Not applicable.

Informed Consent Statement: Not applicable.

Data Availability Statement: Data will be made available upon request.

Conflicts of Interest: Authors declare no conflict of interest.

Appendix A

Table A1. Sample questionnaire items.

Attributes	Measures
	Perceived Usefulness (PU)
PU1	I can find products and make purchases more quickly by using online insurance websites.
PU2	Online insurance websites make it easier to look for and buy insurance.
PU3	Online insurance websites improve my performance when looking for and buying products.
PU4	I can save time by using online insurance websites.
PU5	I can save money by using online insurance websites.
	Perceived Ease of Use (PE)
PE1	It is ease to access online insurance websites.
PE2	My interaction with the online insurance websites is simple and straightforward.
PE3	I find the online insurance websites are easy to use.
PE4	The layouts of online websites are ease to use.
PE5	The online insurance websites are flexible to interact with.
	Website Security (WS)
WS1	There are secure infrastructure platforms in online insurance websites.
WS2	Stability in the way services are provided in online insurance websites creates security.
WS3	The use of electronic and online insurance services is safe.
WS4	Confidentiality of information is protected in online insurance websites.
WS5	Online insurance website security is increased by the use of digital signatures.

Table A1. *Cont.*

Attributes	Measures
	Service Quality (SQ)
SQ1	The speed of service delivery in online insurance channel is satisfactory.
SQ2	Online insurance channel technology reduces the cost of purchasing insurance.
SQ3	Online insurance channel technology boosts competitiveness while enhancing customer service.
SQ4	Online insurance channels are accessible, user-friendly, and simple to use.
SQ5	Technical support is excellent, and the quality of the services is adequate.

References

1. Worldwide Digital Population 2023. Available online: https://www.statista.com/statistics/617136/digital-population-worldwide/ (accessed on 12 April 2023).
2. Baker, E.W.; Hubona, G.S.; Srite, M. Does "being there" matter? The impact of web-based and virtual world's shopping experiences on consumer purchase attitudes. *Inf. Manag.* **2019**, *56*, 103153. [CrossRef]
3. Breidbach, C.; Keating, B.; Lim, C. Fintech: Research directions to explore the digital transformation of financial service systems. *J. Serv. Theory Pract.* **2020**, *30*, 79–102. [CrossRef]
4. Taiwan Insurance Institute (TII) Statistics. Available online: https://www.tii.org.tw/tii/information/information1/000001.html (accessed on 20 May 2022).
5. Cheng, E.; Lam, D.; Yeung, A. Adoption of internet banking: An empirical study in Hong Kong. *Decis. Support Syst.* **2006**, *42*, 1558–1572. [CrossRef]
6. Koenig-Lewis, N.; Palmer, A.; Moll, A. Predicting young consumers' take up of mobile banking services. *Int. J. Bank Mark.* **2010**, *28*, 410–432. [CrossRef]
7. Taiwan Network Information Center Statistics. Available online: https://report.twnic.tw/2022/TrendAnalysis_internetUsage.html (accessed on 25 September 2022).
8. Kim, D.J.; Ferrin, D.L.; Rao, H.R. A trust-based consumer decision-making model in electronic commerce: The role of trust, perceived risk, and their antecedents. *Decis. Support Syst.* **2008**, *44*, 544–564. [CrossRef]
9. Davis, F.D. Perceived usefulness, perceived ease of use, and user acceptance of information technology. *MIS Q.* **1989**, *13*, 318–341. [CrossRef]
10. Shen, J.; Eder, L.B. Exploring intentions to use virtual worlds for business. *J. Electron. Commer. Res.* **2009**, *10*, 94–103.
11. Tang, M.H.; Yeh, C.C. The acceptance of mobile applications in health insurance claim procedure. In Proceedings of the 2nd International Conference on E-Society, E-Education and E-Technology, Taipei, Taiwan, 13–15 August 2018; pp. 73–77. [CrossRef]
12. Venkatesh, V.; Morris, M.G.; Davis, G.B.; Davis, F.D. User acceptance of information technology: Toward a unified view. *MIS Q.* **2003**, *27*, 425–478. [CrossRef]
13. Mekovec, R.; Hutinski, Ž. The role of perceived privacy and perceived security in online market. In Proceedings of the 35th International Convention MIPRO, Opatija, Croatia, 21–25 May 2012; pp. 1883–1888.
14. Hussain, R.; Al Nasser, A.; Hussain, Y.K. Service quality and customer satisfaction of a UAE-based airline: An empirical investigation. *J. Air Transport Manag.* **2015**, *42*, 167–175. [CrossRef]
15. Parasuraman, A.; Grewal, D. The impact of technology on the quality-value-loyalty chain: A research agenda. *J. Acad. Mark. Sci.* **2000**, *28*, 168–174. [CrossRef]
16. Li, F.; Lu, H.; Hou, M.; Cui, K.; Darbandi, M. Customer satisfaction with bank services: The role of cloud services, security, e-learning and service quality. *Technol. Soc.* **2021**, *64*, 101487. [CrossRef]
17. Siddiqui, M.; Sharma, T. Analyzing customer satisfaction with service quality in life insurance services. *J. Target. Meas. Anal. Mark.* **2010**, *18*, 221–238. [CrossRef]
18. Martilla, J.A.; James, J.C. Importance-performance analysis. *J. Mark.* **1977**, *41*, 77–79. [CrossRef]
19. Oh, H. Revisiting importance–performance analysis. *Tour. Manag.* **2001**, *22*, 617–627. [CrossRef]
20. Cronbach, L.J. Coefficient alpha and the internal structure of tests. *Psychometrika* **1951**, *16*, 297–334. [CrossRef]
21. Davis, F. User acceptance of information technology: System characteristics, user perceptions, and behavioral impacts. *Int. J. Man-Mach. Stud.* **1993**, *38*, 475–487. [CrossRef]
22. Elseidi, R.I. Understanding egyptian consumers' intentions in online shopping. *Int. J. Online Mark.* **2018**, *8*, 1–18. [CrossRef]

23. Bacon, D.R. A comparison of approaches to importance-performance analysis. *Int. J. Mark. Res.* **2003**, *45*, 1–15. [CrossRef]
24. Tarrant, M.A.; Smith, E.K. The use of a modified importance-performance framework to examine visitor satisfaction with attributes of outdoor recreation settings. *Manag. Leis.* **2002**, *7*, 69–82. [CrossRef]

Disclaimer/Publisher's Note: The statements, opinions and data contained in all publications are solely those of the individual author(s) and contributor(s) and not of MDPI and/or the editor(s). MDPI and/or the editor(s) disclaim responsibility for any injury to people or property resulting from any ideas, methods, instructions or products referred to in the content.

engineering proceedings

Proceeding Paper

Effects of Integrating Brain Training Digital Game for Improving Learning Gains †

Pornpimol Sampanyen and Niwat Srisawasdi *

Faculty of Education, Khon Kaen University, Khon Kaen 40002, Thailand; pornpimol_s@kkumail.com
* Correspondence: niwsri@kku.ac.th

† Presented at the 3rd IEEE International Conference on Electronic Communications, Internet of Things and Big Data Conference 2023, Taichung, Taiwan, 14–16 April 2023.

Abstract: Mathematics is a fundamental subject for learning at all levels of education. Thus, developing effective methods for increasing children's learning performance in mathematics is important. In this study, first graders who were identified to have difficulties with mathematics experienced a digital game-based intervention as a parallel part of normally scheduled class activities for one semester. The results revealed that the students showed positive results in their mathematical learning and numerical fluency after receiving the game intervention.

Keywords: digital game; brain training; mathematical disabilities; mathematics learning; numeracy training

Citation: Sampanyen, P.; Srisawasdi, N. Effects of Integrating Brain Training Digital Game for Improving Learning Gains. *Eng. Proc.* **2023**, *38*, 10. https://doi.org/10.3390/engproc2023038010

Academic Editors: Teen-Hang Meen, Hsin-Hung Lin and Cheng-Fu Yang

Published: 19 June 2023

Copyright: © 2023 by the authors. Licensee MDPI, Basel, Switzerland. This article is an open access article distributed under the terms and conditions of the Creative Commons Attribution (CC BY) license (https://creativecommons.org/licenses/by/4.0/).

1. Introduction

One of the core subjects in schools is mathematics, as it offers helpful knowledge for everyday life and is crucial for personal growth [1,2]. However, many students in basic and secondary education consider mathematics to be depressing. For example, it was demonstrated that students frequently categorized mathematics as a disliked course because they thought it was boring, challenging, and pointless [3]. Those who struggle to master mathematics may have lower aspirations for their professional futures [4]. Therefore, it is crucial to develop effective teaching strategies to increase students' interest in math, enhance their conceptual comprehension, and develop their arithmetic skills [4,5]. To address the problem, Muñez et al. proposed a digital game-based intervention to facilitate students' mathematics learning performance [6]. The present study was conducted based on the guidelines of Ref. [6] to see if students could successfully improve their mathematical learning using the proposed digital game-based intervention. The result was discussed to answer the following research questions.

- Do students with mathematical disabilities demonstrate better mathematical fluency with digital game-based intervention?
- Do students with mathematical disabilities demonstrate better mathematics learning achievement with digital game-based intervention?

2. Literature Review

2.1. Digital Game-Based Learning (DGBL)

Recently, much research has been done in the context of DGBL as an enjoyable environment that helps students in education. Considering this, Prensky underlined that merging serious learning with interactive entertainment is the primary characteristic of DGBL environments [7]. Digital games make learning enjoyable, facilitate task-focused attention [8], encourage learning, boost success [9], and influence behavior and attitude [10]. These findings suggest that instructional video games ought to have a set of features that must not be created at random. However, different scholars have different ideas about what characteristics of a digital game are needed to be instructive.

2.2. DGBL in Mathematics Learning

The educational potential of digital games has been generally acknowledged for a while [11–13]. Digital games provide delight and pleasure for children to progress in their learning and performance in mathematics by reducing their stress and frustration [4,5,11,14,15]. Digital games offer students numerous options for learning, such as repeating activities to not make the same mistakes [16]. They are not just for enjoying [12,15,17], but for allowing students to participate and explore mathematical knowledge [18,19]. Digital game-based learning creates such an environment that fosters students' learning interest [12,20], learning confidence [15], and eagerness to learn [5,12,19].

3. Brain Training with Digital Game Intervention

Muñez et al. developed a computerized game-based brain training program in which the game intervention was translated into the Thai language, as shown in Figure 1 [6]. This tablet game uses narratives and themes to simulate a video game environment. There are twelve minigames in each game, and each one has a different game concept, theme, action required, and type of stimuli. The difficulty of each minigame increases as a function of the number of stimuli, working memory capacity, and numerical aspects. Each completed task gives rewards, and progressing through the missions earns the player badges to upgrade the player's status.

Figure 1. A screenshot of the game intervention.

3.1. Participants

A total of 10 first-grade students were recruited from 67 students as participants in this study. They were screened for mathematics fluency and prior mathematics knowledge from the students of two classes in a public primary school in Thailand's northeast. The selected students had poorer working memory and performed at or below the 20th percentile on the aforementioned exams in the same age group of the two classes. The participants had satisfactory skills in using a tablet and had yet to experience using it in mathematics classes. The group of participants was heterogeneous in terms of their backgrounds in mathematics before interacting with the game intervention in the experimental study.

3.2. Training Intervention

The game intervention was given to the participants for 16 weeks, which was a whole regular semester. Children participated in the training games on similar 7-inch touchscreen tablets and recorded their responses. During 16 weeks of the intervention, the participants spent 800 training minutes in total, experiencing the intervention for 20 min twice or three times a week.

3.3. Questionnaire Survey

Each participant's mathematics learning achievement was measured through a questionnaire survey with 25 items using a multiple-choice method. Surveys were conducted before and after the game intervention. In addition, a math fluency test consisting of 52 arithmetic addition problems was used to measure their mathematical fluency. The participants finished as many calculations in each set as they could in one minute. The total number of correct calculations within the allotted time was their raw scores.

3.4. Data Collection and Analysis

Before having the game intervention, the participants were pre-tested for mathematics learning achievement and math fluency for 31 min. For the post-test, the same test was conducted. The results from pre- and post-tests were analyzed to obtain descriptive statistics including arithmetic mean, standard deviation, frequency, and ratios. Individual actual gains were calculated as the percent absolute gain as follows.

$$\text{Gain} = [(\text{post-test score\%} - \text{pre-test score\%})/100\% - \text{pre-test score\%}] \quad (1)$$

This is the actual gain divided by the maximum gain achievable by an individual participant. The participant's gain is used to understand course effectiveness.

4. Results

4.1. Mathematical Fluency

The descriptive statistics of the pre-test and post-test mathematical fluency scores of the participants are presented in Table 1. The mathematical fluency scores after experiencing the game intervention (mean = 13.11, standard deviation (SD) = 7.08) were considerably higher than those before the intervention (mean = 1.44, SD = 1.33). (The total score of the mathematical fluency test was 52). All participants' scores for pre- and post-tests are shown in Figure 2. The participants' scores were significantly increased.

Table 1. Descriptive statics on students' mathematical fluency.

Test	Mathematical Fluency Score			
	Mean	**SD**	**Highest**	**Lowest**
Pre-test	1.44	1.33	3	0
Post-test	13.11	7.08	22	4

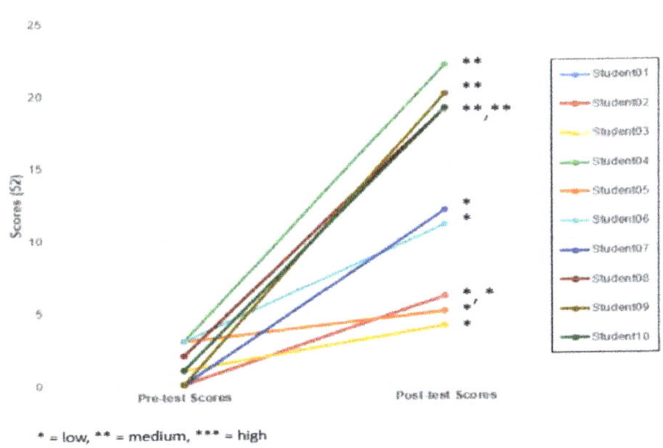

Figure 2. Improvement in the mathematical fluency test score for participants.

4.2. Mathematics Learning Achievement

The descriptive statistics of the scores of the mathematics learning achievement in the pre- and post-tests are presented in Table 2. The mathematical fluency scores of the participants after the game intervention (mean = 20.00, SD = 2.62) were significantly higher than before the intervention (mean = 9.80, SD = 5.39). (The total score is 25). The pre- and post-test scores of all participants are shown in Figure 3, showing considerable improvement in test scores.

Table 2. Descriptive statics on students' mathematics learning achievement.

Test	Mathematics Learning Achievement			
	Mean	**SD**	**Highest**	**Lowest**
Pre-test	9.80	5.39	17	2
Post-test	20.00	2.62	23	15

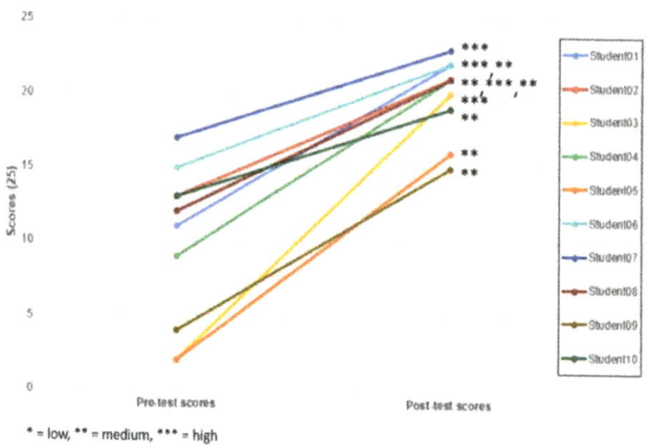

Figure 3. Improvement in the mathematics learning achievement test score for participants.

5. Discussions and Conclusions

Although studies on working memory and numeracy training have been carried out over the past ten years, it has been unclear whether the training helps children with learning impairments in mathematics [6]. The purpose of the present study was to confirm the effect of the digital game intervention on the increase in mathematics learning in terms of learning achievement and mathematical fluency for primary school students with mathematical disabilities.

Our first research question was whether digital game-based intervention enhances the mathematical fluency of students with mathematical disabilities. The findings of the present study showed the positive effects of the game intervention on enhancing mathematical fluency. For the second research question, whether mathematics learning gains are modified by the digital game-based intervention, the findings proved that students with mathematical disabilities improved their mathematics learning achievements. These findings coincided with the results of Refs. [6,21] which stated that such instruction had a positive impact on children with math learning problems. The training effects on mathematics performance have been confirmed as children who underwent the numerical training demonstrated improvements in their aptitude for learning mathematics [22,23].

Considering the limitation of the small number of participants and the short intervention period, further study is required with a large number of participants and a longer intervention period than 16 weeks. The students' experiences with the games and the scope of the impacts in this study may be also limited by their limited gaming times. However, the findings of the present study are the basis for research on how to design interventions and training programs to enhance instructional efforts and overcome the inherent constraints of real classroom environments.

Author Contributions: Conceptualization, N.S.; methodology, N.S.; software, N.S.; validation, N.S.; formal analysis, N.S. and P.S.; investigation, N.S. and P.S.; resources, N.S.; writing—original draft preparation, N.S. and P.S.; writing—review and editing, N.S.; visualization, N.S. and P.S.; supervision, N.S.; project administration, N.S. All authors have read and agreed to the published version of the manuscript.

Funding: This research received no external funding.

Institutional Review Board Statement: Not applicable.

Informed Consent Statement: Not applicable.

Data Availability Statement: The data used in this research are available upon request.

Acknowledgments: The authors would like to express their gratitude to Kerry Lee, Department of Early Childhood Education, Faculty of Education and Human Development, Education University of Hong Kong, for providing the digital game.

Conflicts of Interest: The authors declare no conflict of interest.

References

1. Kiili, K.; Devlin, K.; Multisilta, J. Is game-based math learning finally coming of age? *Int. J. Serious Games* **2015**, *2*, 1–4. [CrossRef]
2. Huang, Y.M.; Huang, S.H.; Wu, T.T. Embedding diagnostic mechanisms in a digital game for learning mathematics. *Educ. Technol. Res.* **2014**, *62*, 187–207. [CrossRef]
3. Luhan, J.; Novotna, V.; Kriz, J. ICT support for creative teaching of mathematics disciplines. *Interdiscip. Stud. J.* **2013**, *2*, 89–100.
4. Kiili, K.; Devlin, K.; Perttula, A.; Tuomi, P.; Lindstedt, A. Using video games to combine learning and assessment in mathematics education. *Int. J. Serious Games* **2015**, *2*, 37–55. [CrossRef]
5. Pope, H.; Mangram, C. The influence of a digital math game on student number sense. *Int. J. Serious Games* **2015**, *2*, 5–17. [CrossRef]
6. Muñez, D.; Lee, K.; Bull, R.; Khng, K.H.; Cheam, F.; Rahim, R.A. Working memory and numeracy training for children with math learning difficulties: Evidence from a large scale implementation in the classroom. *J. Educ. Psychol.* **2022**, *114*, 1866–1880. [CrossRef]
7. Prensky, M. The Games Generations: How Learners Have Changed. In *Digital Game–Based Learning*; McGraw–Hill: New York, NY, USA, 2001; pp. 02-1–02-26.
8. Chen, Z.H.; Liao, C.C.; Cheng, H.N.; Yeh, C.Y.; Chan, T.W. Influence of game quests on pupils' enjoyment and goal-pursuing in math learning. *J. Educ. Technol. Soc.* **2012**, *15*, 317–327.
9. Hung, C.M.; Huang, I.; Hwang, G.J. Effects of digital game-based learning on students' self-efficacy, motivation, anxiety, and achievements in learning mathematics. *Comput. Educ.* **2014**, *1*, 151–166. [CrossRef]
10. Ke, F.; Grabowski, B. Game playing for mathematics learning: Cooperative or not? *J. Educ. Technol.* **2007**, *38*, 249–259.
11. Chang, M.; Evans, M.A.; Kim, S.; Norton, A.; Deater-Deckard, K.; Samur, Y. The effects of an educational video game on mathematical engagement. *Educ. Inform. Technol.* **2016**, *21*, 1283–1297. [CrossRef]
12. Iten, N.; Petko, D. Learning with serious games: Is fun playing the game a predictor of learning success? *Br. J. Educ. Technol.* **2014**, *47*, 151–163. [CrossRef]
13. Nousiainen, T.; Kangas, M.; Rikala, J.; Vesisenaho, M. Teacher competencies in game-based pedagogy. *Teach. Teach. Educ.* **2018**, *74*, 85–97. [CrossRef]
14. Barkatsas, A.T.; Kasimatis, K.; Gialamas, V. Learning secondary mathematics with technology: Exploring the complex interrelationship between students' attitudes, engagement, gender and achievement. *Comp. Educ.* **2009**, *52*, 562–570. [CrossRef]
15. Huang, X.; Ye, Y.; Guo, H.; Zhang, H.; Li, Y. DSKmeans: A new kmeans-type approach to discriminative subspace clustering. *Knowl. Based Syst.* **2014**, *70*, 293–300. [CrossRef]
16. Bakker, M.; van den Heuvel-Panhuizen, M.; Robitzsch, A. Effects of mathematics computer games on special education students' multiplicative reasoning ability. *J. Educ. Technol.* **2016**, *47*, 633–648. [CrossRef]
17. Drijvers, P.; Doorman, M.; Kirschner, P.; Hoogveld, B.; Boon, P. The effect of online tasks for algebra on student achievement in grade 8. *Technol. Knowl. Learn.* **2014**, *19*, 1–18. [CrossRef]
18. Chen, C.-H.; Law, V. Scaffolding individual and collaborative game-based learning in learning performance and intrinsic motivation. *Comput. Human Behav.* **2016**, *55*, 1201–1212. [CrossRef]
19. Chen, M.-P.; Wong, Y.-T.; Wang, L.-C. Effects of type of exploratory strategy and prior knowledge on middle school students' learning of chemical formulas from a 3D roleplaying game. *Educ. Technol. Res.* **2014**, *62*, 163–185. [CrossRef]
20. Siklander, P.; Kangas, M.; Ruhalahti, S.; Korva, S. Exploring triggers for arousing interest in the online learning. In Proceedings of the 11th International Technology, Education and Development Conference, Valencia, Spain, 6–8 March 2017.
21. Nemmi, F.; Nymberg, C.; Helander, E.; Klingberg, T. Grit is associated with structure of nucleus accumbens and gains in cognitive training. *J. Cogn. Neurosci.* **2016**, *28*, 1688–1699. [CrossRef]

22. Sella, F.; Tressoldi, P.; Lucangeli, D.; Zorzi, M. Training numerical skills with the adaptive videogame "The Number Race": A randomized controlled trial on preschoolers. *Trends Neurosci. Educ.* **2016**, *5*, 20–29. [CrossRef]
23. Toll, S.W.; Van Luit, J.E. The development of early numeracy ability in kindergartners with limited working memory skills. *Learn. Individ. Differ.* **2013**, *25*, 45–54. [CrossRef]

Disclaimer/Publisher's Note: The statements, opinions and data contained in all publications are solely those of the individual author(s) and contributor(s) and not of MDPI and/or the editor(s). MDPI and/or the editor(s) disclaim responsibility for any injury to people or property resulting from any ideas, methods, instructions or products referred to in the content.

engineering proceedings

MDPI

Proceeding Paper

Design of Interactive Learning Materials with Concept of Sustainability Integrated into Macau Drunken Dragon Dance †

Yan-Jing Luo and Peng-Wei Hsiao *

Faculty of Arts and Design, Macau Polytechnic University, Macao 999078, China; p2109185@mpu.edu.mo
* Correspondence: pwhsiao@mpu.edu.mo; Tel.: +853-6698-7175

† Presented at the 3rd IEEE International Conference on Electronic Communications, Internet of Things and Big Data Conference 2023, Taichung, Taiwan, 14–16 April 2023.

Abstract: The Feast of Drunken Dragon is a special intangible cultural heritage of Macau, which is also known as the Drunken Dragon and Lion Dance Gala. Every year, on the eighth day of the fourth month of the lunar calendar, members of the Macau Fish Traders Association hold a drunken dragon and lion dance activity at the Macau Sanjie Guild Hall. Festivals play a part in cultural designing. They not only attract a large number of tourists but also promote local culture. The Drunken Dragon Dance is currently encountering many problems such as the inability to introduce new performance methods as technicians are getting older, and young people are unwilling to inherit it. This cultural heritage tends to gradually decline. The Cultural Affairs Bureau of the Macao Government has paid attention to the problem. However, the publicity methods are still mainly based on videos, simple online graphics, and cultural and creative products, and the impact of the COVID-19 epidemic harmed the promotion. With the rapid development of technology, mobile devices provide digital reading and augmented reality (AR) to assist learning. Thus, learning is not limited in space and time, and individuals continue to use it without hindrance. Using the concept of experiential learning, we design interactive teaching materials with dynamic images to advertise the historical stories of the Feast of Drunken Dragon in Macau, folk rituals, and traditional dragon performances through the AR interactive experience of the drunken dragon festival. The concept of learning by doing is an experiential way to understand relevant intangible cultural heritage knowledge. Using digital content, teaching material design is provided based on user experience to achieve sustainable education.

Keywords: Macau Drunken Dragon; experiential learning; sustainable education; augmented reality; electronic picture book

Citation: Luo, Y.-J.; Hsiao, P.-W. Design of Interactive Learning Materials with Concept of Sustainability Integrated into Macau Drunken Dragon Dance. *Eng. Proc.* **2023**, *38*, 12. https://doi.org/10.3390/engproc2023038012

Academic Editors: Teen-Hang Meen, Hsin-Hung Lin and Cheng-Fu Yang

Published: 20 June 2023

Copyright: © 2023 by the authors. Licensee MDPI, Basel, Switzerland. This article is an open access article distributed under the terms and conditions of the Creative Commons Attribution (CC BY) license (https://creativecommons.org/licenses/by/4.0/).

1. Introduction

The Drunken Dragon Festival in Macau has a history of more than 400 years. This festival originated from the traditional folk sacrificial activities in the Xiangshan area of Guangdong. After the Cultural Revolution, the Drunken Dragon Dance on the Mainland was lost. As marine resources are being exhausted by overfishing and climate change, Macau's fishery industry is shrinking, and the number of fishers is decreasing significantly. The younger generation lacks a sense of commitment to traditional culture. To solve this problem, the Macau Fisherman's Association holds the Drunken Dragon Festival every year to create a unique festival brand, allowing visitors from Macau and other places to experience the traditional folk festival. Recently, the government has also begun to pay attention to the event and the lack of professionals for it. The promotion for the event can be launched in Macao's primary and secondary schools to encourage more Macao youth to learn about Macao's traditional culture and customs [1].

With the rapid development of information technology, many traditional cultures are preserved and disseminated in digital ways. In 1997, Thalmann and others transformed the terracotta warriors into 3D, including weapons, costumes, and scene environments, and integrated and produced a 90 min animation film for display and testing [2]. Korean scholars also used the interactive operation of sound visualization and gesture recognition to invite visitors to experience Korean traditional royal court music and manuscript illustrations. This program enhanced the curiosity of the participants, as well as their understanding of traditional royal court music. Overall, this study is carried out to provide evidence that multisensory digital media facilitate learning and cultural appreciation of historical artifacts [3]. Based on the result, traditional culture can be combined with digital interaction and various educational puzzle-type APPs and virtual reality (VR) experiences.

2. Literature Review

2.1. Macau Drunken Dragon Dance

Every year on the eighth day of the fourth month of the lunar calendar, the "drunken dragon dance" team performs in groups of two with one holding the dragon's head and the other holding the dragon's tail. They tour seven major vegetable markets in Macau. The dancing posture of the dancers expresses drunkenness. During the dance, the dancers enter the side and spray wine into the air, implying that the dragon sprays water. In 2009, it was officially included in the Macao Intangible Cultural Heritage Preparatory List, and in 2011, it was officially included in the third batch of the national intangible cultural heritage list. The Drunken Dragon Festival, as a traditional cultural festival in Macao, contains rich cultural connotations and represents the intangible cultural heritage of humans. It is a characteristic traditional culture of Macao and is of great significance to the popularization and promotion of traditional culture [4]. Therefore, the Macau government has focused on promoting and organizing related groups and projects, making the Drunken Dragon Festival a precious traditional folk dance art event and providing participants with generous bonuses.

2.2. Experiential Learning with AR Interaction

Organizational psychologist Kolb put forward the experiential learning cycle model in 1983 by sorting out the experiential learning theories of Dewey, Levin, and Piaget. The model is composed of four stages: concrete experience, observation and reflection, conceptual abstraction, and active practice [5]. Related research used the experience mode and AR for design. Oleksy and Wnuk used augmented reality (AR) technology to allow users to view past historical photos in a real space. The results of the study found that AR enhanced the local significance of multiculturalism and reduced racial prejudice [6]. Unger and Kvetina used AR in Prague archeology and constructed a virtual museum for users to experience the archaeological site. From the research and application of the above literature, it was found that the technology of augmented reality combined with experiential visual interaction achieved better learning effects and a sense of experience for learners [7].

2.3. Digital Design of Intangible Cultural Heritage, Sustainable Education, and Management

With the continuous advancement of digital technology, 50% of young people around the world use 3C products on average every week, and the frequency of use is as high as 25% every day. Therefore, sustainable digitalization is important for investors to understand the company's current situation and grasp the latest data in real-time. Sustainable digitalization is more convenient for third-party verification units because it greatly improves efficiency while ensuring data quality [8]. The concept of digital sustainability is combined with cultural heritage, and related projects were launched recently such as the "Dong Wooden Architecture Construction Technique" and "Guangxi Cultural Field" in Guangxi in 2015. Due to the dwindling number of inheritors of traditional skills, lack of successors, fragmented graphic materials, and lack of systematic organization, there are worries of traditions being forgotten, destroyed, or even gradually disappearing. Therefore, the pilot project of the digital management system which is professionally organized to preserve traditions is required. In 2015, the United Nations Organization for the Promotion of World Cultural Heritage (CIPA) held a seminar at the University of Science and Technology of China to discuss the application of technologies such as data acquisition and recording and technology digital processing and models to sustainable preservation of culture.

3. Research Methods

In addition to using digital technology for the Drunken Dragon Festival culture, we explored the mode of sustainable inheritance through digital art, experiential learning, and AR interaction. To overcome the constraints of time and space, a system was developed and ported on the mobile APP. For the development of the system, we carried out field surveys, literature reviews, and expert interviews. Based on this, the digital design of the historical background, shape, and color matching of the Drunken Dragon Festival was established. The overall process comprises three stages as follows.

3.1. First Stage: Data Collection, Collation, and Analysis

In 2022, Wei-Ming Guan, the vice president of the Macau Fresh Fish Traders Association was interviewed. He said that the Drunken Dagon Dance in Macau is different from the traditional dragon dance. There are only two parts, the dragon head and the dragon tail (Figure 1). The dragon head is generally 3 feet long, and the dragon tail is about 2 feet long. Both parts are made of teak and painted with gold or silver for color matching. Dragon horns are made of antlers from Taiwan or China. The overall shape of the drunken dragon is rounded and streamlined, and the scales are drawn on the surface.

Figure 1. The shape of drunken dragon (photographed in this research).

3.2. Second Stage: KOLB Experiential Learning Framework for Digital Content Design

With the historical background data of the Drunken Dragon Festival organized, its stories are reinterpreted with illustration effects (Figures 2 and 3). Photoshop, Illustrator, and After Effect were used for digital content production, and Unity 3D, C Sharp, and Android platforms were adopted to develop the AR Drunken Dragon (Figures 4 and 5). The digital content was designed according to the KOLB experiential learning framework.

1. Animation display of the historical stories of the Drunken Dragon Festival (specific experience).
2. AR interactive experience, understanding, and observing the shape and color matching of Drunken Dragon (observation and reflection).

3. Concept formation of Drunken Dragon Festival culture, sacrificial activities, Drunken Dragon dance, and Drunken Dragon modeling (an abstract).
4. Practice and evaluate the final interactive experience (active practice).

Figure 2. Sacrificial activities in the Drunken Dragon Festival (drawn in this research).

Figure 3. Drunken Dragon Dance Troupe–The local cultural characteristics are highlighted in Chinese on costumes and musical instruments (drawn in this research).

Figure 4. 3D modeling design of drunken dragon (made in this research).

Figure 5. AR interactive experience and operation (made by researchers).

Based on the experience mentioned above, digital humanities were developed to help users have a better understanding of intangible cultural heritage on smartphones, computers, and other devices. Using this model, the preservation function of skills inheritance learning and activity materials was provided.

3.3. Third Stage: Feasibility Study of Sustainable Education

The "AR Drunken Dragon Interactive Experience Textbook" was provided to users for experiments, and a questionnaire survey was provided to explore their learning motivation, satisfaction, and effectiveness and observe whether users were interested in the interactive mode and had the desire to continue learning and exploring persistent behavior.

4. Conclusions

Dynamic images in the AR interactive design and experienced learning mode were created for intangible cultural heritages. Users observed and understood the traditional shapes, colors, and materials of the drunken dragon with 360 stereoscopic effects. Then, they understood the cultural value, development, and current situation of the Drunken Dragon Festival. The learning motivation of the users with the interactive materials was explored to understand the satisfaction, effectiveness, and sustainability of the design. The interaction was mainly for the single-user experience. In the future, extended production can be added for multi-person modes. Moving pictures may allow users to choose the level of experience interaction at different levels, which needs to be discussed and analyzed for the feasibility of teamwork learning.

Author Contributions: Conceptualization, Y.-J.L. and P.-W.H.; Supervision, P.-W.H.; software, Y.-J.L.; writing—original draft preparation, Y.-J.L., writing—review and editing, P.-W.H. All authors have read and agreed to the published version of the manuscript.

Funding: Not applicable.

Institutional Review Board Statement: Not applicable.

Informed Consent Statement: Not applicable.

Data Availability Statement: Not applicable.

Acknowledgments: The authors thank W. M. Kwan and C. W. Kwan for their contributions to the provision of troupe materials.

Conflicts of Interest: The authors declare no conflict of interest.

References

1. Shi, L.Y. *Protection, Inheritance and Promotion: Stories of Macao's "Intangible Cultural Heritage"*; Xinhua News Agency: Macau, China, 2019.
2. Thalmann, N.M.; Pandzic, I.S.; Moussaly, J.C.; Thalmann, D.; Huang, Z.; Shen, J. The Making of the Xian terra-cotta Soldiers. In *Computer Graphics*; Academic Press: London, UK, 1995; pp. 281–295.
3. Rhee, B.; Pianzola, F.; Oh, N.; Choi, G.; Kim, J. Remediating tradition with technology: A case study of From Tangible to Intangible: A Media Showcase of Kisa chin p'yori chinch'an uigwe. *Digit. Creat.* **2021**, *32*, 56–70. [CrossRef]
4. Tan, S.B. *The History of Taipa and Jiu'ao Inscribed in Gold and Stone: A Study on the Collection of Bell Inscriptions on Temple Inscriptions in Taipa and Jiu'ao in the Qing Dynasty*; Guangdong People's Publishing House: Guangzhou, China, 2011.
5. Kolb, D.A. *The Process of Experiential Learning. Experiential Learning: Experience as the Source of Learning and Development*; Prentice-Hall: Hoboken, NJ, USA, 1984; pp. 20–84.
6. Oleksy, T.; Wnuk, A. Augmented places: An impact of embodied historical experience on attitudes towards places. *Comput. Hum. Behav.* **2016**, *57*, 11–16. [CrossRef]
7. Unger, J.; Kvetina, P. An on-site presentation of invisible prehistoric landscapes. *Internet Archaeol.* **2017**, *43*, 1–12. [CrossRef]
8. Lin, A. *From Paper to Digital Sustainability Report Deepens Instant Communication*; Global Reporting Initiative: Amsterdam, The Netherlands, 2018.

Disclaimer/Publisher's Note: The statements, opinions and data contained in all publications are solely those of the individual author(s) and contributor(s) and not of MDPI and/or the editor(s). MDPI and/or the editor(s) disclaim responsibility for any injury to people or property resulting from any ideas, methods, instructions or products referred to in the content.

engineering proceedings

MDPI

Proceeding Paper

Design and Development of Interactive Moodle-Based E-Learning Platform for Competency Training †

Sakkarach Muangbangyung and Niwat Srisawasdi *

Faculty of Education, Khon Kaen University, Khon Kaen 40002, Thailand; sakkarach_muang@kkumail.com
* Correspondence: niwsri@kku.ac.th

† Presented at the 3rd IEEE International Conference on Electronic Communications, Internet of Things and Big Data Conference 2023, Taichung, Taiwan, 14–16 April 2023.

Abstract: During the COVID-19 pandemic, the global education system changed its context of instruction to be online. Teachers need to interact with students in learning management systems (LMSs) to overcome the limitation of instruction during the pandemic. To promote interactive online learning with LMSs, moodle-based e-learning is proposed in this study. In the e-learning system, the instructional design was proposed using the integration of Moodle and the H5P engine. A six-step inquiry learning approach was included in the design as an effective solution for using online platforms.

Keywords: online platform; interactive learning; competency training

1. Introduction

The COVID-19 pandemic disrupted the global education system and changed the context of students' learning styles. Normally, most students were educated in a learning environment at school, but now the COVID-19 pandemic has changed it to online learning or e-learning environments. Therefore, the learning management system (LMS) has become necessary to support teachers on online learning platforms to change teaching and learning mechanisms. Therefore, designing and developing an interactive Moodle-based e-learning platform to support students' learning with LMS is necessary. Through this study, we provided a solution for various organizational challenges to improve the learning platform effectiveness of regular classes based on human-centered design (HCD) and LMS. The solution offers students self-regulated learning (SRL) via scientific inquiry in the six-step model on the Moodle platform, affecting students' development in reasoning. Integrating the six-step scientific inquiry and interactive Moodle-based e-learning platform improves the student's scientific explanation ability on the engine-driven online learning platform.

2. Literature Review

2.1. LMS

LMS is a technology platform that allows distance and flexible learning through e-education with software and supports virtual academic activities without location and time limitations [1]. Habits and social pressure affect people's use of LMSs [2]. Students and teachers depend on the LMS in online education as all course materials and assignments are sent through its platform in the learning process. E-learning platforms like Moodle are one type of LMS. The main idea is to change the human-made learning mechanism on site into an engine-driven online learning mode with H5P, Phet interactive simulation, and online resources.

2.2. HCD

HCD is a set of techniques to invent innovative solutions for various societal challenges, including goods, services, spaces, institutions, and methods of communication [3]. HCD

Citation: Muangbangyung, S.; Srisawasdi, N. Design and Development of Interactive Moodle-Based E-Learning Platform for Competency Training. *Eng. Proc.* **2023**, *38*, 11. https://doi.org/10.3390/engproc2023038011

Academic Editors: Teen-Hang Meen, Hsin-Hung Lin and Cheng-Fu Yang

Published: 20 June 2023

Copyright: © 2023 by the authors. Licensee MDPI, Basel, Switzerland. This article is an open access article distributed under the terms and conditions of the Creative Commons Attribution (CC BY) license (https://creativecommons.org/licenses/by/4.0/).

is used in engineering design for students to understand and practice the way of the engineering design process and techniques to solve open-ended engineering problems [4]. The model is used to identify the open-ended problem with the scope of ideas and concepts to select the prototype. After building the prototype, it is used in the experiment to check the constraints. In HCD, a suitable method to improve the learning platform can be found to support teaching and learning.

2.3. Self-Regulated Learning (SRL)

SRL effectively promotes students' motivation to learn, reflects on their learning, and improves their understanding of subjects [5,6]. By engaging in self-regulated learning, students are better equipped to comprehend challenging topics in-depth during the learning process [7]. Research about flipped classroom integrated SRL revealed that students who adopted the self-regulated flipped classroom approach demonstrate improved performance in goal setting, strategy planning, time management, seeking assistance, and self-regulation [8]. Thus, we designed an e-learning platform to promote SRL to drive students to engage and be active in learning.

3. Design and Development

3.1. Human-Made Learning to Engine-Driven Online Learning

A prototype of the engine-driven online learning platform was developed in this study to improve the student's competencies by incorporating MS in regular classrooms. Previously, teachers' pedagogies depended on their classroom context. They gave normal instruction with their instructional media in classrooms. Normally, the teacher directs scientific content, scientific activity, experiments, and data summaries for students to learn. As shown in Figure 1, human-made learning starts with content-related videos to motivate the student. After the video, the teacher asks questions to the student. The student takes a few minutes to think about their answers before the activity. To find the questions, the teacher sets equipment or information to guide the student to find the answer. When the student completes the activity or experiment, the teacher starts to explain and conclude the lesson. Even though this technique provides students with understanding, all processes are offered and led only by the teacher. The student receives information rather than discovers it. Therefore, the upgraded method is necessary to change the learning mechanism.

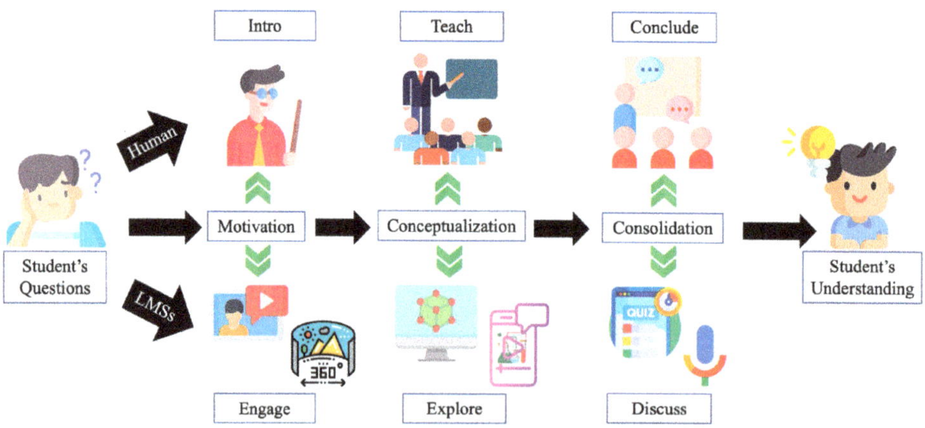

Figure 1. The comparison of the human-made learning mechanism with the LMSs moodle-based e-learning platform with the MC^2 model.

LMS is implemented to assist students' online learning in classrooms with a focus on an interactive Moodle-based e-learning platform combined with the pedagogies known as MC^2. In LMS, students are motivated and are engaging in learning with curiosity to find questions to ask. Conceptualization is the process to explore and discover a shred of evidence, and consolidation is the stage of conclusion and discussion through the student's communication to find reasoning. In the Moodle-based e-learning platform, MC^2 is used to conduct a scientific inquiry process with 6 steps to develop students' competencies. The first step is a scientific investigation to define the circumstance. Step 2 is the inquiry question to find the background knowledge in the third step. Developing conceptualized content and data analysis is carried out in the fourth step. Step 5 is for the analysis of the results, followed by step 6 to conclude. Such a process allows students to learn independently as personalized learning and be motivated. Then, SRL is realized in the student's online learning process.

3.2. Interactive Moodle-Based E-Learning Platform

In the six steps of scientific inquiry, we looked for the proper H5P hub, H5P-specific plugins, and web applications such as the Phet interactive simulation for each sequence on the Moodle platform. In interactive videos, we offer a virtual 360° tour and picture hotspots, course presentation, scientific interactive video or scientific simulation, quiz or set of questions, and audio records. The first phase of scientific inquiry is to build an interactive movie with motivating questions to engage students in the event. An interactive video may lack content knowledge of the solution for the inquiry. As shown in Figure 2A, 3 to 5 sub-questions with multiple interactions are asked to attract the student's attention.

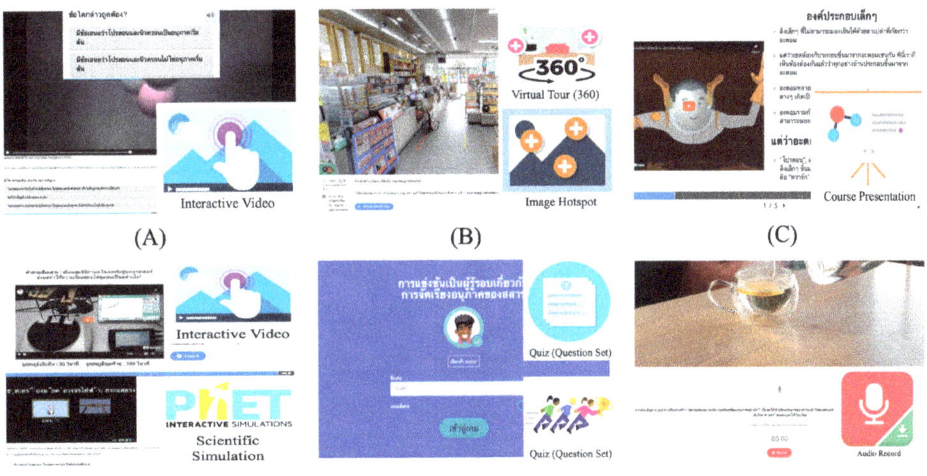

Figure 2. The sequence of six-step scientific inquiry with the platform: (**A**) presenting learning scenario using interactive video, (**B**) discovering essential inquiry question in virtual tour 360°, (**C**) providing related scientific information in a multimedia presentation, (**D**) interacting content-specific learning materials and analyzing obtained primary data, (**E**) sharing and monitoring the results, and (**F**) producing a scientific explanation-based conclusion.

The goal of the inquiry question was asked during an interactive virtual tour to conceal the inquiry question within the image. To create the virtual tour and picture hotspots, we must provide a proper setting for the knowledge point. On student devices, the virtual tour of the environment outside of the classroom was displayed. They explored the area and found the inquiry question to move to the next phase as shown in Figure 2B. To understand background information, the interactive course presentation is provided to recheck the student's understanding of the material. In the presentation and the video, scientific contents are given with caution not to provide a solution to an inquiry question. A scientific foundation is required to comprehend the material and present their investigation. These activities are streamlined as shown in Figure 2C.

Depending on the topic, two types of interactions occur in scientific simulation and interactive science experiments. The student must pay attention to the procedure to find answers to the inquiry question and conduct data analysis. In the simulation, we integrate the interactive simulation from Phet, the University of Colorado Boulder, which contains simulations with scientific and mathematical contents. The simulation's parameter can be changed to represent the scientific phenomena at microscopic and macroscopic scales for observation and data collection. The interactive scientific experimental video gives authentic laboratory experiments. Variables are found in each experiment, and data are collected to explain macroscopic-level phenomena. It prompts a subquestion to encourage them to concentrate on the process. In this step, we interactively assess the student's performance in H5P using fill-in-the-blank, multiple-choice, and drag-and-drop questions. This assessment is to evaluate the collected data and information interactively as shown in Figure 2D. The investigation results of the student are assessed through quizzes or questions to find out the level of the student's understanding. The quiz is given to provide various questions in various formats such as multiple choice, drag and drop, and fill-in-the-blanks. In addition, the questions in a quiz set are made for students to challenge and communicate with their peers. When the student has misconceptions, the teacher interrupts and corrects them as shown in Figure 2E. The teacher's role is emphasized more than in other steps. Finally, the student's audio records are made in the conclusion stage as shown in Figure 2F. The student explains the question, evidence, and reasoning from their learning and concludes their knowledge.

4. Conclusions

The design of online instruction on LMS, Moodle, and the H5P engine was developed to overcome the teaching and learning limitations. We combined the MC^2 and the interactive Moodle-based e-learning platform in online learning. The online learning platform was designed to improve students' competencies in a six-step process. (1) Based on the circumstances, an interactive film with questions is created to motivate the student. (2) The inquiry question is found in an interactive virtual tour or interactive image, and (3) the theoretical background and information are given in an interactive presentation with questions to check the student's comprehension. (4) Content and data analysis are performed in the scientific simulation and interactive experimental video. This stage requires the student to pay attention to the evidence of the answers to the question through data analysis. (5) The results are given in quizzes or questions to test the student's comprehension. (6) The conclusion is made by the student who explains evidence and reasoning. This scientific inquiry approach and interactive Moodle-based e-learning platform enhance the students' learning online.

Author Contributions: Conceptualization, N.S.; methodology, N.S.; software, N.S.; validation, N.S.; formal analysis, N.S. and S.M.; investigation, N.S. and S.M.; resources, N.S.; writing—original draft preparation, N.S. and S.M.; writing—review and editing, N.S.; visualization, N.S. and S.M.; supervision, N.S.; project administration, N.S. All authors have read and agreed to the published version of the manuscript.

Funding: This research received no external funding.

Institutional Review Board Statement: Not applicable.

Informed Consent Statement: Not applicable.

Data Availability Statement: The data used in this research is available upon request.

Acknowledgments: The authors would like to express their gratitude to Khon Kaen University Smart Learning Innovation Research Center (KKU SLIRC) for providing an opportunity to meet and collect data with teachers under the collaboration of the institution and preparing the e-learning platform.

Conflicts of Interest: The authors declare no conflict of interest.

References

1. Chaw, L.Y.; Tang, C.M. What Makes Learning Management Systems Effective for Learning. *J. Educ. Technol.* **2018**, *47*, 152–169. [CrossRef]
2. Meyliana, H.A.E.; Widjaja, S.W.; Fernando, S.E.; Condrobimo, A.R. Improving the Quality of Learning Management System (LMS) based on Student UTAUT2 and Trust Perspectives Using Model. In Proceedings of the 4th International Conference on Informatics and Computational Sciences, Semarang, Indonesia, 10–11 November 2020.
3. Zoltowski, C.B.; Oakes, W.C.; Cardella, M.E. Students' ways of experiencing human-centered design. *J. Eng. Educ.* **2012**, *101*, 28–59. [CrossRef]
4. Dickrell, P.; Virguez, L. Engineering Design & Society: A First-Year Course Teaching Human-Centered Design. In Proceedings of the 2018 World Engineering Education Forum—Global Engineering Deans Council (WEEF-GEDC), Albuquerque, NM, USA, 12–16 November 2018.
5. Michalsky, T.; Schechter, C. Preservice teachers' capacity to teach self-regulated learning: Integrating learning from problems and learning from successes. *Teach. Teach. Educ.* **2013**, *30*, 60–73. [CrossRef]
6. Ingkavara, T.; Panjaburee, P.; Srisawasdi, N.; Sajjapanroj, S. The use of a personalized learning approach to implementing self-regulated online learning. *Comput. Educ. Artif. Intell.* **2022**, *3*, 100086. [CrossRef]
7. Chaipidech, P.; Srisawasdi, N.; Kajornmanee, T.; Chaipah, K. A personalized learning system-supported professional training model for teachers' TPACK development. *Comput. Educ. Artif. Intell.* **2022**, *3*, 100064. [CrossRef]
8. Lai, C.L.; Hwang, G.J. A self-regulated flipped classroom approach to improving students' learning performance in a mathematics course. *Comput. Educ.* **2016**, *100*, 126–140. [CrossRef]

Disclaimer/Publisher's Note: The statements, opinions and data contained in all publications are solely those of the individual author(s) and contributor(s) and not of MDPI and/or the editor(s). MDPI and/or the editor(s) disclaim responsibility for any injury to people or property resulting from any ideas, methods, instructions or products referred to in the content.

engineering proceedings

MDPI

Proceeding Paper

Augmented Learning: Case Study of Gamified and Extended Reality Courses †

Saverio Iacono

Department of Computer Science and Technology, Bioengineering, Robotics, and Systems Engineering, University of Genova, 16145 Genova, Italy; saverio.iacono@unige.it

† Presented at the 3rd IEEE International Conference on Electronic Communications, Internet of Things and Big Data Conference 2023, Taichung, Taiwan, 14–16 April 2023.

Abstract: The maturation of gamification techniques and virtual reality technologies has progressed differently. Today, gamification has been consolidated, and virtual reality is becoming a widespread technology. These two are united in the university course of "virtual reality, augmented reality and gamification" as the Msc curriculum in Digital Humanities. This is a STEM course where an original gamification system was studied to teach complex topics with different bachelor backgrounds as engineering, design, communication, and literature studies. This research was conducted starting in the academic year 2016–2017 using TV series to create a sense of engagement during the course. The results of the 6 year teaching of the gamified course were presented in this article with constant feedback from involved students.

Keywords: gamification; enhanced learning; extended reality; flipped classroom; e-learning; stem education; computer graphics

Citation: Iacono, S. Augmented Learning: Case Study of Gamified and Extended Reality Courses. *Eng. Proc.* **2023**, *38*, 13. https://doi.org/10.3390/engproc2023038013

Academic Editors: Teen-Hang Meen, Hsin-Hung Lin and Cheng-Fu Yang

Published: 20 June 2023

Copyright: © 2023 by the author. Licensee MDPI, Basel, Switzerland. This article is an open access article distributed under the terms and conditions of the Creative Commons Attribution (CC BY) license (https://creativecommons.org/licenses/by/4.0/).

1. Introduction

Playing is an activity that is often perceived as a waste of time and an unseemly activity that hinders the maturity of children. With video games, this perception is further spread as numerous teenagers spend many hours playing games such as Fortnite, Minecraft, and Clash of Clans. However, this negative phenomenon cannot ignore the positive aspects of games [1]. Game-based learning is an important step prior to gamification. Pelling used the term for the first time in 2003 [2], and the term was then defined better by Deterding [3]. Unlike game-based learning, in which a game is used to educate content, gamification is applied to non-game contexts with techniques and tools to improve engagement and experience. In 2012, the MOOC on Coursera by Werbach [4] and the books by Kapp, McGonigal, and Chou [5–7] triggered the spread of the term on a larger scale and brought excitement in related research. Gamification is often assimilated into the triad "PBL" (Points, Badges and Leaderboard) as the necessary elements. They also argued that gamification cannot be integrated into every context and always be useful for the purpose, but rather it must be considered a tool to be associated wisely.

The adoption of competitive approaches makes it difficult for those who do not feel activated by the competition and motivated by ranking. Good gamification promotes positive behavior to transform extrinsic motivation into intrinsic motivation [8]. After the wave of aggressive marketing on the subject ended, it became possible to do sensible research without having to chase a trending topic. The narrative approach to gamification can be more engaging but also needs a lot of effort to produce story-driven gamified experiences. Thus, the approach used for this course is based on a story-driven experience with a light approach to PBL. Gamification in STEM higher education is studied to investigate the effectiveness, engagement level, and major tools [9,10].

In this study, a different gamified approach was created and analyzed in the context of a STEM course in Digital Humanities, where the majority of data was collected. Furthermore, here is presented a preliminary analysis of data collected in the STEM context of an engineering course.

2. Course and Methodology

The course "Virtual Reality, Augmented Reality and Gamification" (hereinafter VRARGAM) is gamified in a deliberately self-referential way, aiming to provide students with the knowledge of the fundamentals of 3D graphics and animation and the practical skills to design applications and systems based on virtual/augmented/extended reality simulation and gamification. Using a methodological approach based on inter/cross-disciplinarity (web programming, computer graphics, biomechanics, sensory perception, robotics, and video games), the expected learning outcomes have always been acquiring design methods and operational tools applicable in creative contexts related to Digital Humanities. In a preliminary article [11], the 2 year experience was described, and the educational experience was built with an innovative method using the theme of a TV series such as Stranger Things (Netflix), Westworld (HBO), Altered Carbon (Netflix), Psycho-Pass (Fuji TV), Upload (Amazon Prime Video), The Good Place (NBC), and Cyberpunk Edgerunner (Netflix). In the principle of inclusive and participatory gamification, the learning of skills and motivation was improved. The gamification slogan was "when students are not just students, but the cast of a TV show." Each participation in a course activity (lectures and seminars) was rewarded based on the schedule of the course and series TV to produce a dynamic cast adaptation every week. This motivated the students to participate in the course. The context of a master's degree program in Digital Humanities provides a strong interdisciplinary component. Students may have different technical-scientific fields but normally have learned Communication Sciences for 3 years. In a nutshell, there might be students with a bachelor's degree in computer engineering with students with degrees in literature and philosophy.

2.1. First Year: Stranger Things and First Rule Setting

The series, in the first year of experimentation, was used to generate interest and a theme important for the course. With its commercial success, the first season of "Stranger Things" was chosen. Taking into consideration the topics of the series compared to those of the course, it was the poorest choice, while the component of creating engagement was much stronger to finalize the general learning. The opening theme of the series was reconstructed starting from the reference font used, namely Benguiat Bold for the title of the series and the macro letters, while Century Gothic Bold was used for the names of the participants. The final rendering of the opening theme was produced with Adobe Premiere and Adobe After Effects. The original musical theme of the series by Kyle Dixon and Michele Stein was also kept. At this stage, we introduced theming, simply trying to stimulate the core drive of the "Epic Meaning" by its framework [7].

The final exam consisted of the creation of a demo of the game and a game design document on the proposed theme with a new story that was parallel or inspired by the original story. At the beginning of each lesson, the opening theme was sent "on air" with personalized credits and titles because each title was inspired by the series and the topics of each lesson. For example, the lesson on 3D modeling was called "The Big Max and the Little Maya," with reference to 3D Studio Max and Maya. The lesson on orientation systems (left hand and right hand) was entitled "The Upside Down." The students were informed that their names included in the opening were not only aesthetic but for changing their position based on their presence during the lessons after the first 2 lessons. In a real show of any genre, the more important actor's name appears before others.

The shifting of names was never really put into practice for a number of reasons. The number of students involved was 33, and the opening theme was simply too short (50 s) and had too little space. Updating the opening with After Effects was difficult to manage because of the inherent complexity of the software. Therefore, scrolling through all the names could not be handled optimally. For the final exam, the students were divided into small groups of 3 or 4 so that they could work together on a single project and then produce their game design document and a demo/alpha version of their idea using modeling ad game engine skills acquired during the course.

2.2. 5 Years of Experimentation

In the second year, after the good results in the exams were obtained, we chose to reuse the same principle. However, selected something less commercial and closer to the themes of the course and following the new technological trend of A.I.

Thanks to the theme of the "amusement park where anything can happen" safely, like in virtual reality (VR), we proposed the Westworld TV series for the high quality of the narrative production. However, given the complexity of the original opening theme, this has not been entirely recreated, but we eliminated the original credits with After Effects. Students entered their names, and the leaderboard has been updated based on active presence. Students were better organized with a longer theme (1 min 43 s), and fewer students (18) helped the monitoring process. The order of the names was changed at each lesson, and those who did not reach half of the registered attendance were "fired" from the cast and took the exam as "not attending." At the production level, the solution adopted to remove "credits" from the original theme and replace them with course participants has become the standard to be followed for future productions. As a new addition to the course, each student had to produce a pitch document on a personal game proposal and present it: the best idea selected by "the board" (the faculty) became the big project for all students to work on. However, students failed to work together as a large team, challenged by personal, social and productive demands. The competition to impose the best idea produced very interesting document pitches, but in the end, the one selected was really developed only by a small group of 6, thus slightly larger than the previous year, while most of the others ended up creating their own personal project by themselves or in smaller groups. Therefore, they independently recreated the approach given in the previous year.

In the 2018–2019 academic year, the Altered Carbon TV Series was chosen. The number of students had almost doubled (33), and the series provided, in addition to topics close to the course, an opening theme of a similar length to the previous one (1 min and 42 s). The production process was standardized with After Effects. There was the only initial difficulty in recreating the effects of the appearance and disappearance of the names that were harmonious with the original theme. The proposed methodology remains the same, and the projects returned to being performed in small groups of a maximum of 3 or 4 people. The themes of Altered Carbon remained rather connected to the course, thanks to the dynamics of the "change of sleeves," which allowed changing one's body in VR. The theme of possible interaction with A.I. and the normalization of the XR tools was used.

In the 2019–2020 academic year, the chosen series was Psycho-pass, a 2012 anime that maintains a connection with the course topics by showing a possible integration of XR tools into daily life. The course was taught during the COVID-19 pandemic, and as a result, it was converted into an online course. This context, in addition to forcibly changing the starting condition of the experiment, provided an opportunity to understand whether the perception of gamification could change.

The 2020–2021 academic year, characterized by the second and third waves of the COVID-19 pandemic, marked a major increase in the number of students and coincided for the first time with the use of the same series in two different degree programs. The TV series chosen was the first season of Upload. Variations beyond the theme of the series, which offered numerous insights and connections to virtual reality, mainly involved the inclusion of e-learning methodologies: the Online Flipped Classroom [12] and Online Peer Instruction [13]. The new approach required making materials available in advance of class and evaluating their effectiveness in the more technical topics. It also presented an opportunity to test the methodology for a diverse group of students while keeping the main themes of the course unchanged. In addition to the Digital Humanities course (56 students), gamification was provided for the "Virtual Reality for Robotics" course (86 students) for the Dibris Robotics Engineering course. The teaching, in addition to having students only with an engineering background, was delivered completely in English since the students could be of any nationality. For the 6th year, data is still collected but is currently positively in line, as expected.

3. Data Collection

3.1. Questionnaire and Data

In the first year, we tried to understand if the idea of using a TV series for the course was appreciated by students through interviews after the exams. The opinion of students was positive about having a specific thematic task to do. From this first opinion, we were encouraged to repeat the experience again and refine the previous version of the course with a stronger gamified principle using a hidden ranking. During the 2nd year, discussions emerged among the students about the correct placement of their names in the opening theme. After the second release of the gamified course, a test was conducted based on the Likert Scale to record their liking and perception of the course through six sentences. The survey was to understand if their opinions were in agreement with our statements. The questionnaire contained the following statements: (1) "You have positively perceived the experience of the gamified course," (2) "You would do a gamified course like this again," (3) "You were positively impressed to see your name in the series opening credits," (4) "Having a theme to finalize the course activities has inspired you to try harder," (5) "The presence of your name in the series opening credits has positively influenced your participation in the course," and (6) "You were interested in the series chosen for the course gamification."

The rating scale consisted of five steps ranging from "strongly disagree" to the intermediate step, "neutral," to the last step, "strongly agree." The questionnaire was provided to students via Google Forms in a strictly anonymous form. The questionnaire was distributed to the students after the exam. In the academic years 2019–2020 and 2020–2021, to limit the dispersion, students were asked to fill in the questionnaire immediately after the exam or, at the latest, within 24 h.

After the fourth year, we revised the questionnaire to have a more complete and clear result: (1) "You have positively perceived the experience of the gamified course," (2) "You were positively impressed to see your name in series openings credits," (3) "The movement of your name in opening credits has stimulated you to intervene more frequently during the classes," (4) "You positively perceived that each class began with the opening credits as an introduction to the themes of the day and as a guiding thread of the course," (5) "Having a theme to finalize the course activities inspired you to try harder," (6) "You were interested in the TV series chosen for the gamification of the course," (7) "You would do again a gamified course with the opening credits and names that move, but with different contents and/or objectives."

3.2. Results

The results in 2018 and 2019—2020 are shown in Table 1 (4 years). For (1), an approval rate of 81% of the interview was obtained. For (2), the total satisfaction rate was 46%. A positive feedback of 61% and a neutral feedback of 23% were observed for (3). A positive feedback rate of 75% was obtained for (4). A positive feedback rate of 46% and a neutral feedback rate of 23% were obtained for (5). The overall satisfaction for (6) was 81%, with a neutral feedback rate of 19% and a disagreement rate of 4%.

Table 1. Percentages calculated based on 48 responses.

	☹️	🙁	😐	🙂	😃
1	2%	2%	15%	46%	35%
2	6%	29%	19%	27%	19%
3	3%	13%	23%	27%	34%
4	0%	10%	15%	46%	29%
5	6%	12%	23%	19%	27%
6	2%	2%	19%	6%	75%

There was a high level of satisfaction with the course gamified experience (1), but students did not want to repeat an experience of this type (2). It was, therefore, decided to change the positioning of the sentence by clearly clarifying that it was a different course from VRARGAM. The teaching group of the course was aware of the complexity and commitment that required the exam related to VRARGAM. It was necessary to understand if this type of gamification could be favored. Two questions were added: "Has it happened that your name has receded in the opening credits?" and "if so, was it a stressful experience for you?".

In Table 2, the results for the academic year 2020/2021 based on the new questionnaire for VRARGAM were presented. For the last two Yes/No questions, 56% of those interviewed answered yes, and 88% said they did not find it stressful to see their names removed from the opening credits for the first question, "Did it happen that your name lost its position in opening credits?" and "If Yes, was it a stressful experience for you?". As expected, for (7), a high agreement rate of 72% was observed.

Table 2. Percentages calculated based on 38 responses.

	☹️	🙁	😐	🙂	😃
1	0%	0%	6%	44%	50%
2	0%	0%	6%	44%	50%
3	10%	9%	31%	22%	28%
4	0	10%	6%	34%	50%
5	3%	3%	12%	38%	44%
6	3%	10%	6%	6%	75%
7	0%	9%	19%	34%	38%

Results of VR for Robotics Course

The last part of this research was for a group of students extremely different from the original sample. As anticipated, the gamified experience was also provided to students of the "VR for Robotics" teaching (the first semester of the academic year 2021/2022) of the master's degree in Robotics Engineering. In this way, it was possible to further investigate on a sample of international students whether the gamification model implemented could have the same impact in this case. The course, in addition to being dedicated to students with a background in engineering studies, was held in English and open to students with heterogeneous cultural origins compared to that experienced previously. We had no idea how this type of gamification would be welcomed or experienced. There were 82 students enrolled in the course, of which 25 took the exam on the final date of February 2022, and an average of 20 students followed the lessons on Microsoft Teams, thus participating in gamification. The data collected are, therefore, few but interesting (Table 3).

Table 3. Percentages calculated based on 21 responses.

	☹️	🙁	😐	🙂	😀
1	5%	14%	0%	57%	24%
2	14%	5%	29%	33%	19%
3	19%	14%	34%	14%	19%
4	19%	19%	9%	24%	29%
5	19%	9%	10%	38%	24%
6	19%	14%	19%	14%	34%
7	5%	19%	14%	33%	29%

The overall positive perception remained high, with 81% for (1). In the VR for Robotics sample, with the revised version of the questionnaire, the results appeared more distributed. Despite the 18.3% increase in negative opinions, the positive responses remained at a higher percentage. The only exception was for (3). Compared to Digital Humanities students (27.7% of disagreements in the VR for Robotics and 9.4% in the VRARGAM), students showed overall positive opinions.

4. Conclusions and Future Works

The result highlighted that the goal of creating gamification based on participation was achieved. Despite the ranking system, it generated engagement and improvement in involvement in class (46% and 48% of the Digital Humanities). There was a need to induce competition stress (88% did not feel stressed by falling back in the rankings), but students had the pleasure of participating in the gamified experience (81% and 94% for the first and second questionnaires). The 6th year data are still collected (connected to The Good Place TV series), but the trend is still positive. Therefore, we are going to proceed for the next year in Digital Humanities with the already announced theme "Cyberpunk Edgerunner," which allows us to mix Sci-Fi literature and video games. The next step is to find external partners to test this experiment and consolidate the approach. Furthermore, it is necessary to measure the high engagement that seems to be confirmed and the learning potential.

Funding: This research received no external funding.

Institutional Review Board Statement: Not applicable.

Informed Consent Statement: Informed consent was obtained from all subjects involved in the study.

Data Availability Statement: The survey data can be provided upon request to the corresponding author of this manuscript.

Acknowledgments: I would like to thank Gianni Vercelli, Edoardo Bellanti, Alice Corsi, Mario Vallarino, and Daniele Zolezzi as the 3D Lab Factory group of teachers who have helped me on this journey and the students who have participated from year to year in this experiment.

Conflicts of Interest: The author declares no conflict of interest.

References

1. Huizinga, J. *Homo Ludens*; Taylor & Francis: London, UK, 1998; ISBN 978-0-415-17594-4.
2. Pelling, N. The (Short) Prehistory of "Gamification" ... Funding Startups (& Other Impossibilities). 2011. Available online: https://nanodome.wordpress.com/2011/08/09/the-short-prehistory-of-gamification/ (accessed on 20 February 2023).
3. Deterding, S.; Dixon, D.; Khaled, R.; Nacke, L. From Game Design Elements to Gamefulness: Defining "Gamification". In Proceedings of the 15th International Academic MindTrek Conference: Envisioning Future Media Environments, New York, NY, USA, 28 September 2011; pp. 9–15.
4. Werbach, K.; Hunter, D. *For the Win, Revised and Updated Edition: The Power of Gamification and Game Thinking in Business, Education, Government, and Social Impact*; Wharton School Press: Philadelphia, PA, USA, 2020; ISBN 978-1-61363-105-8.
5. Kapp, K.M. *The Gamification of Learning and Instruction: Game-Based Methods and Strategies for Training and Education*; John Wiley & Sons: Hoboken, NJ, USA, 2012; ISBN 978-1-118-09634-5.

6. McGonigal, J. *Reality Is Broken: Why Games Make Us Better and How They Can Change the World*; Vintage: New York, NY, USA, 2012; ISBN 978-0-09-954028-1.
7. Chou, Y. *Actionable Gamification: Beyond Points, Badges, and Leaderboards*; Packt Publishing Ltd.: Birmingham, UK, 2019; ISBN 978-1-83921-077-8.
8. Deci, E.L.; Ryan, R.M. The "What" and "Why" of Goal Pursuits: Human Needs and the Self-Determination of Behavior. *Psychol. Inq.* **2000**, *11*, 227–268. [CrossRef]
9. Venter, M. Gamification in STEM Programming Courses: State of the Art. In Proceedings of the 2020 IEEE Global Engineering Education Conference (EDUCON), Porto, Portugal, 27–30 April 2020; pp. 859–866.
10. Pérez-Jorge, D.; Martínez-Murciano, M.C. Gamification with Scratch or App Inventor in Higher Education: A Systematic Review. *Future Internet* **2022**, *14*, 374. [CrossRef]
11. Iacono, S.; Vercelli, G.V.; Bellanti, E.; Corsi, A. A lesson learned about gamification and engagement in a master degree course. In Proceedings of the 11th annual International Conference of Education, Research and Innovation, Seville, Spain, 12–14 November 2018; IATED: Seville, Spain, 2018; pp. 2898–2902.
12. Che, Y.; Pan, Q.; Yang, J.; Ding, D.; Wan, J. Design and Practice of the Online Flipped Classroom under the Guidance of Learning Drive Based on a Network Teaching Platform. In Proceedings of the 2021 2nd International Conference on Information Science and Education (ICISE-IE), Chongqing, China, 26–28 November 2021; pp. 1273–1276.
13. Vallarino, M.; Iacono, S.; Zolezzi, D.; Vercelli, G.V. Online Peer Instruction on Moodle to Foster Students' Engagement at the Time of COVID-19 Pandemic. *IEEE Trans. Educ.* **2022**, *65*, 628–637. [CrossRef]

Disclaimer/Publisher's Note: The statements, opinions and data contained in all publications are solely those of the individual author(s) and contributor(s) and not of MDPI and/or the editor(s). MDPI and/or the editor(s) disclaim responsibility for any injury to people or property resulting from any ideas, methods, instructions or products referred to in the content.

engineering proceedings

MDPI

Proceeding Paper

Using a Text Mining Approach to Identify Important Factors Influencing the Performance of Programmatic Advertising †

Yi-Yun Wang, Venkateswarlu Nalluri and Long-Sheng Chen *

Department of Information Management, Chaoyang University of Technology, Taichung 413310, Taiwan; winona25544@gmail.com (Y.-Y.W.); nallurivenkey7@gmail.com (V.N.)
* Correspondence: lschen@cyut.edu.tw

† Presented at the 3rd IEEE International Conference on Electronic Communications, Internet of Things and Big Data Conference 2023, Taichung, Taiwan, 14–16 April 2023.

Abstract: Programmatic advertising uses big data to spread personalized marketing materials to target audiences, which is a major driving force for the growth of digital advertising. Among them, in-application advertisements (in-app ads) are an important part of programmatic advertising. In in-app advertising, which is highly related to application revenue, ads are delivered to customers through mobile devices at any time and place based on personal needs. Due to the power of electronic Word-of-Mouth (e-WOM), text comments from social media are becoming a new mode of advertising, influencing consumers' purchase behavior. Text reviews on social media are more powerful than traditional ads. However, relatively little research has studied this issue. Therefore, using text mining and latent semantic analysis techniques, we aimed to discover the advertising elements of text reviews in the social community. Based on the results, suggestions were made to advertising companies to improve the performance of text reviews when employing key opinion leaders (KOL) to write commercial comments that promote products or services.

Keywords: text mining; programmatic advertising; in-app advertising; text mining; latent semantic analysis

Citation: Wang, Y.-Y.; Nalluri, V.; Chen, L.-S. Using a Text Mining Approach to Identify Important Factors Influencing the Performance of Programmatic Advertising. *Eng. Proc.* **2023**, *38*, 15. https://doi.org/10.3390/engproc2023038015

Academic Editors: Teen-Hang Meen, Hsin-Hung Lin and Cheng-Fu Yang

Published: 20 June 2023

Copyright: © 2023 by the authors. Licensee MDPI, Basel, Switzerland. This article is an open access article distributed under the terms and conditions of the Creative Commons Attribution (CC BY) license (https://creativecommons.org/licenses/by/4.0/).

1. Introduction

As technology continues to advance, programmatic advertising is an emerging and rapidly evolving information technology with cost-effectiveness that uses big data to deliver personalized marketing topics to target audiences [1]. In-app advertising (also known as mobile advertising) is a type of advertising that is displayed using applications (apps). In recent years, with the popularity of mobile devices, apps have been growing rapidly, causing marketers to pay more attention to the promotion and marketing capabilities of apps. Manufacturers promote their products and expand their user base through various advertising channels, and the medium of advertising has crossed over from traditional website browsing and social networking to applications [2]. According to e-Marketer in 2022, programmatic advertising is becoming an increasingly important part of the advertising landscape [3]. As a result, in-app advertising is gaining more attention.

Although programmatic advertising has become common in advertising technology, the rapid development of programmatic advertising and the immediate exposure of commercial content has fundamentally changed the traditional marketing model [4]. Many advertisers currently use programmatic advertising techniques, but there is a risk that ads may be presented on non-quality sites. Reference [1] suggested that when premium brand advertisements appeared on non-premium websites, users' negative perceptions of the advertisements and brands increased. Therefore, it has become an important issue to understand the key factors of in-app advertising in user experience to enhance the effectiveness of advertising [5].

In addition, the importance of electronic Word-of-Mouth (e-WOM) in e-commerce is gradually increasing. Consumers make purchase decisions by reading reviews. Therefore, marketers can use e-WOM to build product awareness, increase sales, enhance brand value, and build customer loyalty [6]. Many brands use high prices to purchase reviews to increase brand awareness [7,8]. Therefore, text comments from online community users are becoming a new mode of alternative advertising communication. Reference [8] also confirmed that the advertising behavior of purchasing reviews does influence consumers' perceptions.

Previous studies related to mobile advertising used questionnaires to examine the effect of personalization, entertainment, trustworthiness, and information content on users' attitudes toward receiving mobile advertising [2]. However, questionnaire methods are prone to problems such as sampling errors and do not provide a timely understanding of consumer perceptions [9]. Consequently, we used text mining and latent semantic analysis (LSA) techniques to identify the elements of text reviews. The used methodology can improve traditional questionnaire survey methods and also address the limitation of the human inability to analyze a large number of text reviews.

Due to the power of electronic e-WOM, text comments from social media have become a new mode of ads. Text reviews on social media are more powerful than traditional ads. However, relatively little works have studied this issue. Therefore, using text mining and latent semantic analysis techniques, we discovered the advertising elements of text reviews in the social community. Based on the results, we provide advertising companies with suggestions to improve the performance of text reviews when employing key opinion leaders (KOL) to write commercial comments to promote products or services.

2. Literature Review

2.1. *Programmatic Advertising*

Programmatic advertising is a type of marketing behavior that uses big data and artificial intelligence to precisely target advertisements to a highly personalized target audience [10]. A 10% increase in the number of programmatic ads and a 24% increase in the average price per ad were reported in a 2021 annual report [11], which shows the importance of programmatic advertising.

Reference [10] argued that programmatic advertising was a new and lesser-known advertising technology that delivered advertising messages to target audiences in real-time via the internet. Conversely, programmatic advertising may pose risks [4], focusing on programmatic advertising for alcohol. Programmatic advertising changes the original marketing paradigm beyond the regulation of commercial content exposure, undermines public regulation of alcohol marketing, and presents a new challenge to any model of public marketing oversight [4]. With the high use of programmatic advertising behavior, advertisers are also presenting their ads in various forms. Thus, companies are paying more attention to programmatic advertising and analyzing the performance of programmatic advertising is one of the important issues.

2.2. *In-App Advertising*

Mobile devices offer a wide variety of apps, with global app installs in the App Store and Google Play reaching nearly 37 billion in the first quarter of 2022 [12], and the ubiquity of mobile phones allows marketers to advertise regardless of time and geography [2]. Therefore, the revenue source of apps is more from in-app advertising than from sales [13], and the advertising medium ranges from traditional website browsing, social networking, and even cross-platform apps.

2.3. *Text Mining*

Text mining is the process of organizing and analyzing large amounts of textual data to find useful information that can be used by specific users to make decisions. There are many studies on text mining techniques for online reviews. Reference [14] used text mining to analyze feedback from online shopping reviews. Reference [15] proposed an easy-to-implement online review text analysis procedure through text mining for studying brand image and brand positioning. Reference [16] also used text exploration to build four predictors to analyze the interrelationships between reviewer credibility, review age, and review variance. Most of the existing in-app advertising studies adopted questionnaires and interviews. However, this approach requires a lot of time and capacity, and it is not possible to obtain immediate responses from users [9]. Thus, we used text mining instead of traditional questionnaires to understand the important contents of a large number of reviews.

2.4. Latent Semantic Analysis (LSA)

Natural language processing (NLP) is used to automatically represent and analyze human language [17]. Latent semantic analysis is a natural language processing approach according to the meaning of the text context to provide concepts by extracting words from the text through dimensionality reduction techniques to enable analysis, give features, and describe key themes [18]. We referred to ref. [19] for LSA. Reference [20] performed a semantic analysis of the Bengali language and proposed a predictive model for training datasets to test and evaluate the accuracy of this model against other machine learning methods [20]. Reference [21] used LSA for online hotel reviews to understand how satisfied or dissatisfied guests feel.

3. Methodology

We used text comments on social media to determine the elements of ads and collected relevant online comments from game websites. NLP and LSA were used to discover the key elements of product reviews that affected advertising. The implemental procedure contained 5 steps as follows.

Step1: Data Collection and Preprocessing

We collected online reviews through the social center on the Steam website (https://store.steampowered.com/). Comments with less than 300 words were deleted from the collected data and non-English comments were also deleted.

Step 2: Natural Language Processing

Step 2.1: Tokenization: Textual tokenization used the Natural Language Toolkit (NLTK) of Python language.

Step 2.2: Clean data: We deleted stop words such as "the", "and", and other less important words in this step.

Step 2.3: Lemmatization: This step reduced complex forms of a single word to its most basic form, e.g., "ate" to "eat".

Step 2.4: Count Word Frequency: We conducted a word frequency count and deleted words with a frequency of less than 5.

Step 2.5: Build Term Document Matrix (TDM): We used TF-IDF (term frequency-inverse document frequency) in Equation (1) to create a Term Document Matrix (TDM) for further analysis.

$$TF - IDF = TF(t_i, d_i) \times \log\left(\frac{N}{N(t_i)}\right) \quad (1)$$

Step 3: Latent Semantic Analysis (LSA)

In this study, MATLAB was used to perform Singular Value Decomposition (SVD) on word document matrices to investigate the relationship between words and their contexts.

Step 3.1: SVD

The values were brought into the SVD function in MATLAB, where the equation of SVD is shown in Figure 1.

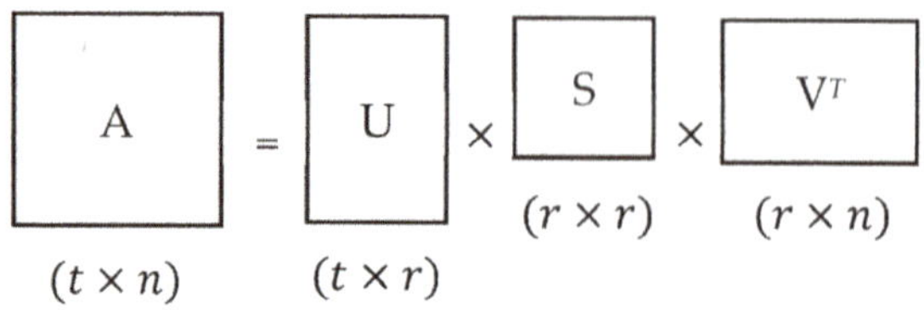

Figure 1. Singular value decomposition.

The SVD function has three matrixes: "matrix $U_{t\times r}$", "orthogonal matrix $S_{r\times r}$", and "document matrix $V^T_{r\times n}$", where t refers to the number of words, n is the document term, and r is the number of concepts in the semantic space [19].

Step 3.2: Dimension Reduction

After running SVD, it is necessary to reduce the dimensional space as it may still contain a lot of unimportant information. So as to not to affect the original characteristics, it is important to choose the feature value k. In this study, the Scree Test was used to determine the k value, as shown in Figure 2. The point before leveling was the decision point ($k = 5$), because the variation gap becomes smaller after the flat slope, and can be ignored after the decision point.

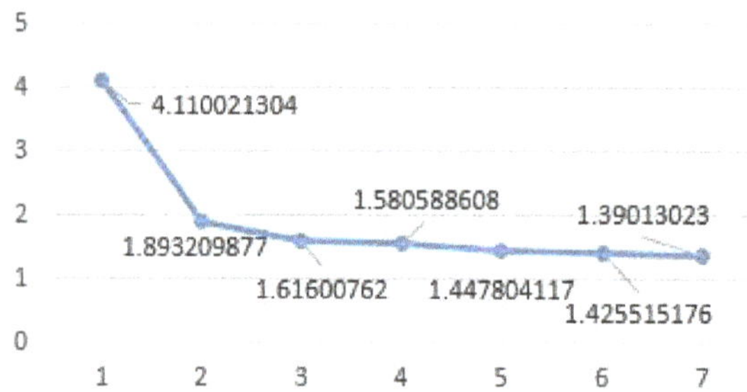

Figure 2. Scree plot.

Figure 3 shows that from the fifth value ($k = 5$), the eigenvalue does not vary much. Thus, we took the first five columns of the S matrix to reduce the dimensionality.

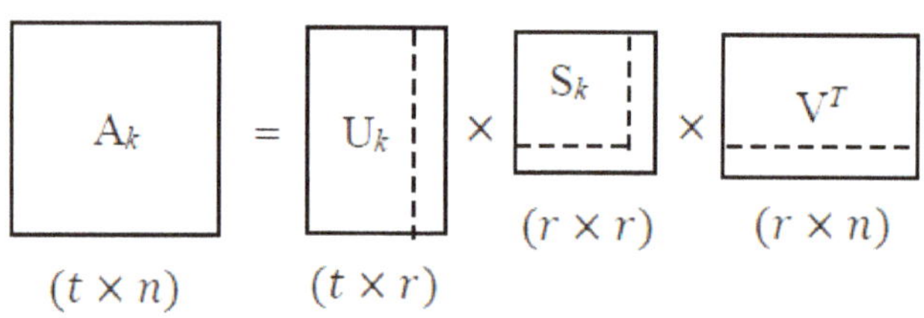

Figure 3. SVD after dimension reduction.

Step 3.3: Orthogonal rotation of axes

The concept load L_T was then calculated by multiplying the dimensionally delimited concept U_k with the concept S_k, which was calculated with Equation (2) to obtain the concept load L_T, then each feature word was ranked according to the load and the word concept was named.

$$L_T = U_k \times S_k \tag{2}$$

Step 4: Concept naming

We selected the top 30 terms in each concept. Then, based on selected important terms, we named the concepts.

Step 5: Conclusion and Discussion

Finally, based on the result of the concept naming, we presented suggestions for the reference of players and advertisers.

4. Experimental Results

4.1. Employed Data and Text Data Processing

In this study, online reviews on the Steam website were used as analysis data. We only considered the "Top Rated" games in the "EARLY ACCESS TITLES", as shown in Figure 4, because vendors were more likely to advertise unreleased games than already released games.

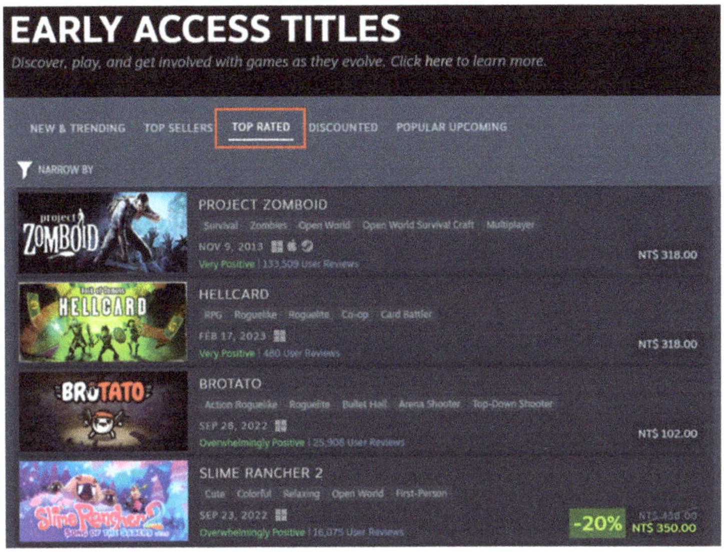

Figure 4. Highest rated game in the sneak preview version.

In addition, we selected games with "overwhelmingly positive reviews" and "recently released" among the preemptive games as shown in Figure 5. Previous studies found that the manufacturers who purchased reviews were usually unknown brands and the products were mostly young products [8].

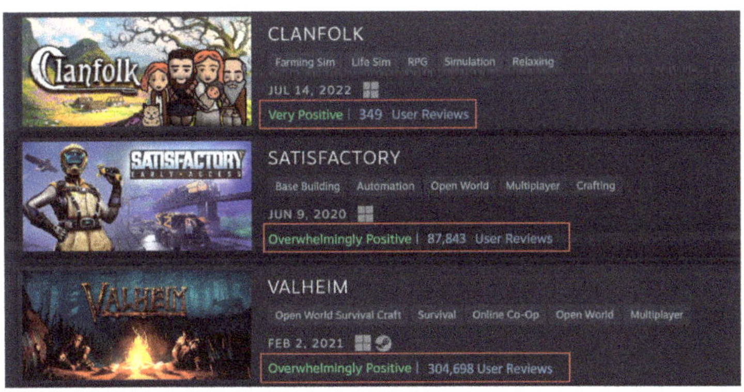

Figure 5. Collection of three games.

We selected three strategy-based games, "clanfolk", "satisfactory", and "valheim" (Figure 5) with 349, 87,843, and 304,689 reviews, respectively. Longer comments (>100 words) were more likely to be correctly classified by machines than shorter comments (<100 words) [22]. We eliminated comments with a total text word count of less than 300 and selected a final dataset of 136 comments. After processing, a 137 × 572 TDM was built.

4.2. LSA and Concept Naming

Based on the results of the LSA experiments, five concepts were determined. Keywords and their loadings in each concept are shown in Table 1 After ranking the characteristic words according to their high loadings, the top 30 words of each loading were selected to name the concept. Table 2 summarizes the selected five concepts in text reviews.

Table 1. Extracted Concepts of Keywords and Their Loadings.

Concept 1		Concept 2		Concept 3	
Keywords	Loadings	Keywords	Loadings	Keywords	Loadings
game	1.7437	rimworld	0.6508	factory	0.5840
like	0.7756	clanfolk	0.6214	factorio	0.3856
get	0.6231	game	0.4205	machine	0.3637
rimworld	0.6086	factory	0.3614	build	0.3045
thing	0.5534	friend	0.3395	velheim	0.2922
⋮	⋮	⋮	⋮	⋮	⋮
Concept 4		**Concept 5**			
Keywords	Loadings	Keywords	Loadings		
rimworld	0.3471	like	0.3795		
survival	0.3348	get	0.3022		
need	0.3191	factory	0.2704		
clanfolk	0.3092	clanfolk	0.2495		
love	0.2965	bad	0.2405		
day	0.2885	feel	0.2249		
⋮	⋮	⋮	⋮		

Table 2. Constructed Concepts.

Concept	Concept Naming	Representative Terms
1	Personal feelings	like, get, really, much, love, need, fun, feel, great, good
2	Game Content Settings	game, friend, family, winter, world, medieval, colony, multiplayer, Scottish, food
3	System Construction	factory, machine, build, server, bad, resource, production
4	Game Props	priority, food, alien, iron, tree, focus, plant
5	Playtime	bad, time, end, spend, started, since, base, playing, access, early, way, day

5. Conclusions

Using online reviews from game social media, the degree of relevance between words and meanings was determined through text mining and LSA to conceptualize the names of words and determine five elements of commercial product reviews. First of all, "personal feelings" were mentioned to make online players mistakenly think that it was a real player's comment and believe them as the commercial commentary of advertisements. In addition, the concepts of "game content settings", "system construction", and "game props" were recommended as game details to give players professional guidance. Game recommendations attracted professional players for reference. The last concept was "playtime" to recommend players to play games freely and without time restrictions. According to the results of this study, the connotations of reviews as a positive concept were mostly related to "recommendation and promotion". If a review showed a negative concept, the review contained more information about the player's "real thoughts".

Author Contributions: Conceptualization, Y.-Y.W. and L.-S.C.; methodology, V.N. and Y.-Y.W.; software, V.N. and Y.-Y.W.; validation, Y.-Y.W.; formal analysis, Y.-Y.W. and L.-S.C.; writing—original draft preparation, Y.-Y.W.; writing—review and editing, V.N. and L.-S.C.; visualization, Y.-Y.W.; supervision, L.-S.C.; project administration, L.-S.C.; funding acquisition, L.-S.C. All authors have read and agreed to the published version of the manuscript.

Funding: This work was supported in part by the National Science and Technology Council, Taiwan (Grant No. MOST 111-2410-H-324-006).

Informed Consent Statement: Not applicable.

Data Availability Statement: The secondary data that were used for analysis in this study are available from the corresponding author upon request.

Conflicts of Interest: The authors declare no conflict of interest.

References

1. Shehu, E.; Nabout, N.; Clement, M. The risk of programmatic advertising: Effects of website quality on advertising effectiveness. *Int. J. Res. Mark.* **2021**, *38*, 663–677. [CrossRef]
2. Maseeh, H.A.; Jebarajakirthy, C.; Pentecost, R.; Ashaduzzaman, M.D. A meta-analytic review of mobile advertising research. *J. Bus. Res.* **2021**, *136*, 33–51. [CrossRef]
3. eMarketer. Digital Advertising in 2022: Market Trends & Predictions. 2022. Available online: https://www.insiderintelligence.com/insights/digital-advertising-market-trends-predictions/ (accessed on 20 April 2022).
4. Goodwin, I. Programmatic alcohol advertising, social media and public health: Algorithms, automated challenges to regulation, and the failure of public oversight. *Int. J. Drug Policy* **2022**, *169*, 103826. [CrossRef] [PubMed]
5. Gao, C.; Zeng, J.; Lo, D.; Xia, X.; King, I.; Lyu, M.R. Understanding in-app advertising issues based on large scale app review analysis. *Inf. Softw. Technol.* **2022**, *142*, 106741. [CrossRef]
6. Mishra, A.; Satish, S.M. eWOM: Extant Research Review and Future Research Avenues. *Vikalpa* **2016**, *41*, 222–233. [CrossRef]
7. Martens, D.; Maalej, W. Towards understanding and detecting fake reviews in app stores. *Empir. Softw. Eng.* **2019**, *24*, 3316–3355. [CrossRef]
8. He, S.; Hollenbeck, B.; Proserpio, D. The Market for Fake Reviews. *Mark. Sci.* **2022**, *41*, 896–921. [CrossRef]
9. Chang, J.; Chen, M.; Chen, L.; Tseng, S. Why Customers Don't Revisit in Tourism and Hospitality Industry? *IEEE Access* **2019**, *7*, 146588–146606. [CrossRef]
10. Samuel, A.; White, G.R.; Thomas, R.; Jones, P. Programmatic advertising: An exegesis of consumer concerns. *Comput. Hum. Behav.* **2021**, *116*, 106657. [CrossRef]
11. Meta. Fourth Quarter and Full Year 2021 Results. 2022. Available online: https://investor.fb.com/investor-news/press-release-details/2022/Meta-Reports-Fourth-Quarter-and-Full-Year-2021-Results/default.aspx (accessed on 2 February 2022).
12. Sensor Tower. Q1 2022 Data Digest: Worldwide App Downloads Reached 37 Billion Last Quarter. 2022. Available online: https://sensortower.com/blog/q1-2022-data-digest (accessed on 1 April 2022).
13. Liu, H.; Liu, S. Considering In-App Advertising Mode, Platform-App Channel Coordination by a Sustainable Cooperative Advertising Mechanism. *Sustainability* **2020**, *12*, 145. [CrossRef]
14. Dinesh, L.; Kaviya, G.; Kavya, N.C.; Kousalya, A.; Lavanya, N. Product Reviews on Opinion Mining using NLP Techniques. In Proceedings of the 2022 7th International Conference on Communication and Electronics Systems, Coimbatore, India, 22–24 June 2022; pp. 952–957.
15. Alzate, M.; Arce-Urriza, M.; Cebollada, J. Mining the text of online consumer reviews to analyze brand image and brand positioning. *J. Retail. Consum. Serv.* **2022**, *67*, 102989. [CrossRef]
16. Biswas, B.; Sengupta, P.; Kumar, A.; Delen, D.; Gupta, S. A critical assessment of consumer reviews: A hybrid NLP-based methodology. *Decis. Support Syst.* **2022**, *159*, 113799. [CrossRef]
17. Liu, Q.; Lu, J.; Zhang, G.; Shen, T.; Zhang, Z.; Huang, H. Domain-specific meta-embedding with latent semantic structures. *Inf. Sci.* **2021**, *555*, 410–423. [CrossRef]
18. Hutchison, P.D.; Daigle, R.J.; George, B. Application of latent semantic analysis in AIS academic research. *Int. J. Account. Inf. Syst.* **2018**, *31*, 83–96. [CrossRef]
19. Chen, W.; Chen, L.; Pa, Y. A text mining-based framework to discover the important factors in text reviews for predicting the views of live streaming. *Appl. Soft Comput.* **2021**, *111*, 107704. [CrossRef]

20. Das, P.; Sultana, N. Sentiment Analysis on Comments in Bengali Language Using Text Mining & Machine Learning Approach. In Proceedings of the IEEE 7th International conference for Convergence in Technology, Pune, India, 7–9 April 2022; pp. 1–4.
21. Hsiao, Y.; Hsiao, Y. Online review analytics for hotel quality at macro and micro levels. *Ind. Manag. Data Syst.* **2020**, *121*, 268–289. [CrossRef]
22. Salminen, J.; Kandpal, C.; Kamel, A.M.; Jung, S.; Jansen, B.J. Creating and detecting fake reviews of online products. *J. Retail. Consum. Serv.* **2022**, *64*, 102771. [CrossRef]

Disclaimer/Publisher's Note: The statements, opinions and data contained in all publications are solely those of the individual author(s) and contributor(s) and not of MDPI and/or the editor(s). MDPI and/or the editor(s) disclaim responsibility for any injury to people or property resulting from any ideas, methods, instructions or products referred to in the content.

engineering proceedings

MDPI

Proceeding Paper

Research and Practice of the Construction of an Online Education Community in Higher Education for Disabled Individuals †

Di Min * and Chundong Wang

Technical College for the Deaf, Tianjin University of Technology, Tianjin 300384, China; michael3769@163.com
* Correspondence: tyrongwen@tjut.edu.cn

† Presented at the 3rd IEEE International Conference on Electronic Communications, Internet of Things and Big Data Conference 2023, Taichung, Taiwan, 14–16 April 2023.

Abstract: Since the establishment of 'Careers for the Disabled', China has been committed to promoting the program. Special higher education for the disabled is regarded as important for the protection of the rights of the disabled and for the achievement of equity and promotion of the development of special higher education. Currently, many colleges and universities in China provide higher education for the disabled. However, special higher education has problems such as a lack of capable teachers, which requires active inter-school cooperation. Thus, we propose the construction of an online education community for the disabled to overcome the existing problems and propose a methody of promoting the development of special higher education to a higher standard.

Keywords: special higher education; construction of the community; online learning platform

Citation: Min, D.; Wang, C. Research and Practice of the Construction of an Online Education Community in Higher Education for Disabled Individuals. *Eng. Proc.* **2023**, *38*, 14. https://doi.org/10.3390/engproc2023038014

Academic Editors: Teen-Hang Meen, Hsin-Hung Lin and Cheng-Fu Yang

Published: 20 June 2023

Copyright: © 2023 by the authors. Licensee MDPI, Basel, Switzerland. This article is an open access article distributed under the terms and conditions of the Creative Commons Attribution (CC BY) license (https://creativecommons.org/licenses/by/4.0/).

1. Introduction

China has promoted the development of careers and emphasized the level of special higher education for the disabled. In 1985, Binzhou Medical University established a medical department for the disabled. This was the beginning of higher education for the disabled in China [1]. In 1987, the Special Education College of Changchun University accepted disabled students for the first time as a higher education institution [2]. In 1991, the Technical College for the Deaf of Tianjin University of Technology enrolled students and became the first engineering college of special higher education for the disabled in China [3]. Higher education for the disabled has developed from scratch and is now comprehensive, requiring refinement. The healthy development of special higher education cannot be achieved without relevant laws and regulations, as special higher education for the disabled can be guaranteed through these policies, such as "Several Opinions on the Development of Special Education" and "Regulations on Education for Individuals with Disabilities" [4]. Currently, special higher education in China is facing problems and challenges in accelerating innovative developments and improving quality, including the poor quality of education for students with hearing impairments, the lack of capable teachers, the lag of teaching methods and models, and the paucity of teaching resources and auxiliary teaching equipment. These problems hinder colleges and universities engaged in special higher education in China.

At the 18th National Congress of the Communist Party of China (CPC), General Secretary Xi Jinping put forward "a community of shared future for mankind", "a community of shared interests", "a community of cooperation", and "a community of shared future in cyberspace". As a concept of holistic development, "community" has become deeply rooted in the hearts of people [5]. The concept of "community" implies the integration of the high-quality resources of all members in that community with mutual support and integration, as well as common development by solving the community's problems and promoting common interests. In special higher education, the common goal is to pursue education equity and improve education levels. Wisdom along with brainstorming among people allows for better results with less effort in solving practical problems. Thus, the construction of an online education community for the disabled is required in special higher education.

2. Goal of Community in Special Higher Education

2.1. Basic Connotation

The construction of an online education community in higher education for the disabled (Figure 1) is not only necessary for the development of an online learning platform for the disabled but also for the innovation and reform of teaching methods in special higher education. The online education community for higher education of the disabled aims to improve the teaching level and quality. In the construction process, modern information technology is used to provide advantages in regard to the existing educational resources, such as notable teaching resources. The members of the community find teaching resources, information, and technology to secure "perfect offline teaching equipment and rich online teaching resources". The integration of such elements enriches special higher education by improving the level of teaching.

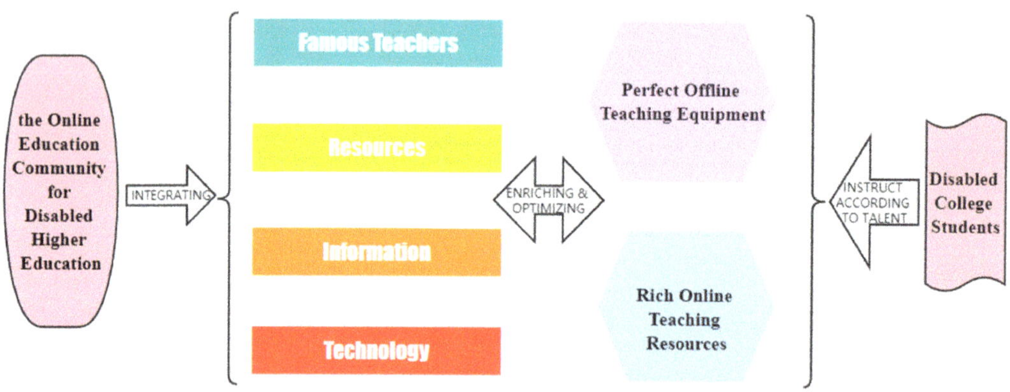

Figure 1. The construction of an online education community for higher education.

2.2. Co-Constructed Goal

Capable teachers help to build a training community for special higher education. With their help, the community innovates teaching methods and researches various activities to promote teachers' abilities, share teaching resources, improve education for the disabled, and help the disabled to enjoy high-quality education. This helps to realize the humanization of education, the ubiquity of educational resources, and the universality of educational models, eliminating regional differences in special higher education. The community can set the following goals for this realization (Figure 2):

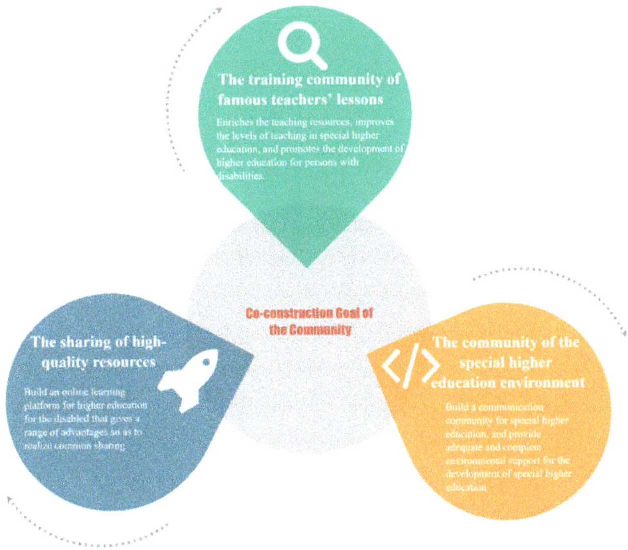

Figure 2. Co-constructed goal of the community in special higher education.

2.2.1. Training Community

Special higher education requires dedicated educational teams which do not rely on special education alone. To avoid the isolation of teaching in special higher education, teachers in general higher education need to engage in special higher education. The community in special higher education can then learn teaching experience and methods used in general higher education, and under the guidance of experienced teachers, appropriate training can be provided for "backbone leadership, discipline linkage, team cooperation, and overall improvement". In addition, the community forms student groups for effective teaching with enriched teaching resources.

2.2.2. Sharing High-Quality Resources

The community integrates various teaching resources from high-quality professional courses and online courses. Advanced teaching devices and technologies and platforms also need to be shared to build an online learning platform for special higher education through sharing, common use, and common maintenance.

2.2.3. Special Higher Education Environment

The community members need to cooperate to build a rich and humanistic environment for the development of special higher education. In addition to building a "friendly and inclusive" humanistic environment, the community needs to have a top-level design for education in order to cultivate professional teams and develop relevant systems in special higher education. Community members need to be active in online exchanges and offline visits to cooperate closely in special higher education and provide an adequate and complete environment for disabled students.

3. Motivation and Experience of Teachers and Members

Most experienced teachers at all education levels have more than 20 years of experience in education. They may have their own habits and traditional concepts and prefer offline classroom teaching. In offline classroom teaching, teachers play a dominant role in teaching, but they understand the teaching process and efficiently conduct their teaching with students face-to-face. In online teaching, the status of teachers and students tends to be equal. Teachers cannot communicate with students effectively in real time. The lack of interaction makes it difficult to receive feedback from students and to satisfy their teaching needs. In addition to preparing teaching content, teachers also need to master relevant technological skills, which takes time and energy [6]. This reduces the enthusiasm and motivation of teachers. Thus, the community needs to encourage the full participation of experienced teachers so that they can use and share their teaching experience for the education of disabled students, as such students may have difficulties in understanding the key points effectively and accurately and achieving their goals in special higher education.

Although the community establishes a cooperating mechanism, it relies on emotional and cognitive identity without a superior–subordinate relationship between members [7]. Therefore, all members focus on achieving their goals with integrity and communication. Feedback from each other results in the common progress of members.

Traditional special education is segregated from special higher education, which also segregates disabled students and general students. The members of the special higher education community actively invite teachers and students of general higher education.

4. Solving Problems

4.1. Online Teaching

The advent of the information age, coupled with the impact of COVID-19, has forced the need for informational reform in education. Teaching information technology has become an important part of educational reform, with an important role. The community can conduct online teaching seminars to motivate teachers and demonstrate the necessity and inevitability of special higher education for the disabled. Teaching models of experienced teachers can also be shared to promote an information-based teaching environment. In addition, the community can organize support groups to strengthen ties within the community and with other communities.

4.2. Sharing Experience

To improve the teaching level of special higher education and enhance the relevance of this education and teaching, it is necessary to investigate the situation of special education student groups and summarize the characteristics of the students to determine if the physical defects or impairments of students make them less able to acquire information and less capable of interpersonal communication and being sensitive to others' thoughts and feelings. They may be prone to unhealthy attitudes such as loneliness, inferiority, suspicion, and dependence. Moreover, they may have poor learning ability, initiative taking, creativity, abstraction, and logical thinking skills. Therefore, the community needs to establish a team of experienced teachers who engage in special higher education and general higher education to gain teaching experience and student communication experience and learn educational methods. Then, the syllabus and program of special education can be prepared effectively for online courses. Figure 3 shows a diagram of the online education system for disabled college students.

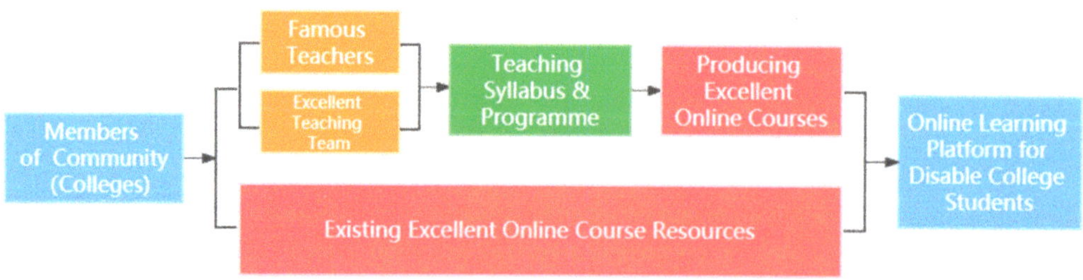

Figure 3. Diagram of an online education system for disabled college students.

The construction of the curriculum aims to improve special higher education for the continuous development of the community. In addition to coordinating and deliberating on the structure of the community, multiple communication channels, including WeChat groups, Tencent video conferences, and online community platforms, can be used to communicate, extend the scope of communications, and discuss core ideas. The community needs to actively build a friendly and inclusive humanistic environment for learning. On the social level, the construction of the community is supported by the Education and Employment Department of the China Disabled Persons' Federation under the framework of the Special Education Research Branch of the China Higher Education Society. On the school level, functional departments are responsible for the coordination and management of special education. On the institutional level, the management of "whole school participation, joint training, and special integration", with the concept of "co-prosperity and sharing", is required. In the training system, evaluation methods and quality management with the concept of inclusive education are in demand. On the teacher level, full-time teachers and counselors work for inclusive education. On the student level, a peer support system needs to be established to encourage students to volunteer to support hearing-impaired students. For example, "Chinese Sign Language" can be offered for general students as an elective, and sign language clubs and lounges are encouraged to form and to enhance communication and acceptance among students. Then, all students can learn together and help each other. This creates a friendly, tolerant, mutually helpful environment on campus.

5. Conclusions

The construction of a community for special higher education has started in China. However, it faces problems and challenges. The community requires the cooperation of colleges and universities. In the future, the development of special higher education will require the participation of the government and enterprises, with universities playing a leading role.

Author Contributions: Conceptualization, methodology, investigation, data curation, writing—original draft preparation, D.M.; writing—review and editing, supervision, C.W. All authors have read and agreed to the published version of the manuscript.

Funding: This research was funded by the Educational Programming Projects Key Program of CAHE (China Association of Higher Education) (No. 22TS0302) and Education Research Fund Key Program of Tianjin University of Technology (No. ZD19-03).

Institutional Review Board Statement: Not applicable.

Informed Consent Statement: Not applicable.

Data Availability Statement: The data presented in this study are available on request from the corresponding author.

Conflicts of Interest: The authors declare no conflict of interest.

References

1. Zhang, X.; Zhang, Y.; Zhu, G. The Exploration Process and Prospects of Higher Education for the Disabled in the 70 Years of New China. *China High. Educ.* **2019**, *19*, 10–12.
2. Xiong, Q. Development, Problems and Suggestions of Higher Education for the Disabled in China under the Concept of Integration. *J. Suihua Univ.* **2016**, *36*, 5–10.
3. Yang, J. Development Characteristics of Special Education for Higher Engineering in China. *Mod. Spec. Educ.* **2001**, *03*, 9–10.
4. Zhang, M.; Wang, N. 70 Years of Development of China's Education for the Disabled. *Chin. Soc. Sci. Today Beijing* **2019**, *11*, 4.
5. Sheng, T. Constructing Practical Education Community in Universities and Colleges: Retrospect and Prospect. *China Educ. Light Ind.* **2019**, *4*, 15–18.

6. Yang, L.; Dai, X. Research on Informatization Teaching and the Construction of a Community of Ideological and Political Teachers in Colleges and Universities. *Tea Fujian* **2020**, *42*, 199–200.
7. Shen, H.; Jiang, H. Research on the Construction and Practice of the Cross-campus Cooperative Community of English Teachers in University. *Coll. Engl. Teach. Res.* **2021**, *1*, 86–90.

Disclaimer/Publisher's Note: The statements, opinions and data contained in all publications are solely those of the individual author(s) and contributor(s) and not of MDPI and/or the editor(s). MDPI and/or the editor(s) disclaim responsibility for any injury to people or property resulting from any ideas, methods, instructions or products referred to in the content.

engineering proceedings

MDPI

Proceeding Paper

Decision-Tree-Computing-Based Usage Intention Prediction of School Social Media †

Huai-Te Huang [1] and Hao-En Chueh [2,*]

1 Department of Industrial Education, National Taiwan Normal University, Taipei 106209, Taiwan; whiter0819@gmail.com
2 Department of Information Management, Chung Yuan Christian University, Taoyuan 320314, Taiwan
* Correspondence: hechueh@cycu.edu.tw
† Presented at the 3rd IEEE International Conference on Electronic Communications, Internet of Things and Big Data Conference 2023, Taichung, Taiwan, 14–16 April 2023.

Abstract: Social media is the main channel that teenagers use to exchange information. The purpose of this study is to construct a prediction model of school social media usage intention. To collect training data for modeling, we conducted a questionnaire survey on students of a senior high school in Taoyuan City, Taiwan. The training data processing method was decision tree computing. In this study, the decision tree computing software, Weka, was used to analyze the training data to extract the key factors affecting high school students' intentions to use school social media. The research results showed that perceived usefulness was the most important factor affecting school social media usage intentions, and trust was the second most important factor affecting school social media usage intention. The prediction model was proposed in this study to predict students' intentions to use school social media. It serves as a guide for schools to use social media as a channel for distributing important information.

Keywords: social media; decision tree computing; prediction model; usage intention

1. Introduction

Following the advancement and innovation of information technology (IT), more social media platforms [1], such as Facebook (FB), Instagram (IG), and Twitter, have been launched. Users not only obtain information, receive comments, and share information via these social media, but they also promote interpersonal relationships through them.

Among all social media, FB is the most used and its users are of all ages. The emergence of FB was due to its use by young people, but gradually more and more middle-aged people are using it, as extensive news and content is available on FB regardless of the correctness of the information. This is convenient for businesspeople and office workers who need to monitor the news every day. Although the use time of young people is decreasing, many of them still use FB to read the news or receive messages. Thus, FB is still the most prominent social media platform [2,3].

IG is a popular platform for the younger generation, and almost everyone in the younger generation has an IG account. In other words, the usage time that is slowly decreasing on FB is being transferred to the IG platform. This generation of young people uses social media to build their styles, share their lives, and allow more people to know and see them. IG is the type of platform that best allows young people to express their characteristics [4,5].

Twitter is also a platform used mostly by young people, especially in the United States and Japan. Unlike IG, which contains mainly pictures, Twitter is a text-based platform, and much of the content is not about sharing knowledge, but simply expressing emotions. Hence, many politicians and entertainers like to use the platform to express their opinions.

Citation: Huang, H.-T.; Chueh, H.-E. Decision-Tree-Computing-Based Usage Intention Prediction of School Social Media. *Eng. Proc.* **2023**, *38*, 16. https://doi.org/10.3390/engproc2023038016

Academic Editors: Teen-Hang Meen, Hsin-Hung Lin and Cheng-Fu Yang

Published: 20 June 2023

Copyright: © 2023 by the authors. Licensee MDPI, Basel, Switzerland. This article is an open access article distributed under the terms and conditions of the Creative Commons Attribution (CC BY) license (https://creativecommons.org/licenses/by/4.0/).

Compared to IG, which is mainly used for personal image management, Twitter is more expressive of the user's real personality [6].

According to analysis of the three most popular social media, it is evident that the main functions of FB are providing information and news. IG is becoming more commercialized every year as the number of young users increases. Twitter has more text-based communication than the above two, but it is also free in terms of usage, leading to difficulties in controlling the dimensions and quality of content [7].

In recent years, the issue of digital transformation has been widely discussed, and every industry, including schools, continues to invest in IT hardware and software to reform education and strengthen students' knowledge in response to the digital trend. In addition, schools are using various IT platforms to collect and analyze student data and actively communicate with parents and students to provide student learning profile data. The latter enables parents to understand important information on students and schools. Such data not only enhance parents' and students' recognition of the school but also serve as an important resource for school development and marketing [8].

Because of the functions and features of FB, many schools are using it as a channel to disseminate important information to parents and students to inform them of the latest policies, events, and school-related educational data. If parents and students can access important school information on FB in real time, the information asymmetry between schools and parents and students reduces.

Information asymmetry means that the parties involved in a transaction do not have the same information that affects the transaction, and the party with information advantage uses inappropriate means to gain more benefits, thereby impairing the interests of the party with an information disadvantage. Even if the party with an information advantage does not intend to hide important information, the party with an information disadvantage may not have enough information to make the best decision for itself. When information asymmetry occurs among schools, parents, and students, it affects the responsibility for the educational performance of schools and the right of choice of parents and students [9].

Although social media is an important channel for adolescents to obtain and transmit information, it is not known whether they are willing to use this channel to obtain important information from schools. Therefore, it is necessary to construct a prediction model of school social media usage intention to extract the key factors that influence high school students' use of school social media as a channel to obtain important information.

To construct the prediction model of school social media usage intention, the decision tree computing software, Weka, was used to extract the key factors affecting the usage intention of school social media. The rest of the paper is organized as follows: In Section 2, the concepts of the usage intention model and decision tree computing are introduced. In Section 3, the research method is presented. In Section 4, the computing results and the discussions are displayed. The conclusions are given in Section 5.

2. Literature Review

2.1. Usage Intention Model

The basic theoretical model of the Technology Acceptance Model (TAM) was developed by Davis et al. [10] based on the Theory of Reasoned Action (TRA), which analyzes users' intention to use new information technology (IT) based on users' perceived usefulness (PU) and perceived ease-of-use (PEOU) of IT. The TAM is used as the theoretical basis for many empirical studies, and a considerable amount of empirical support has accumulated for it. It is also used to evaluate and predict users' acceptance of new IT systems, and therefore the model is widely used in IT-related research.

PU refers to the user's psychological perception of whether a particular system can make their work more efficient. The higher the PU, the higher the intention to use the system. PEOU refers to the user's perception of the ease of use of a particular system, and the higher the PEOU, the higher the intention to use the system. Attitude refers to the positive or negative evaluation of an individual's performance of a specific behavior.

According to TAM, attitudes are influenced by both PU and PEOU. When users perceive the system to have higher PU and PEOU, their attitudes toward the system tend to be positive. Usage Intention is the degree to which users are willing to use a particular system, and according to TAM, usage intention is influenced by both attitude and PU at the same time.

The Theory of Planned Behavior (TPB) [11] was proposed by Ajzen to explain how people change their behavioral patterns. TPB assumes that human behavior is the result of deliberate planning, and the variables in its theoretical model include behavioral attitudes, subjective norm (SN), perceived behavioral control (PBC), and behavior intention (BI). SN refers to the perceived positive or negative perceptions by significant others that an individual performs a particular behavior. PBC refers to the degree of control or mastery that an individual expects to have when adopting a particular behavior.

Trust is a key issue in the relationship between people and technology because it is one of the most important factors influencing users' use of any information technology, especially in the use of social media. When users use social media, they unknowingly reveal personal information, and social media platforms also collect this information from users all the time. Therefore, whether social media can gain the trust of users and effectively protect their information affects their intention to use social media [12].

Based on the above, we suggest that the five factors of PU, PEOU, SN, PBC, and trust have a significant impact on students' intentions to use school social media as an information access channel.

2.2. Decision Tree Computing

Machine learning (ML) is an important branch of artificial intelligence (AI) that focuses on building systems that can learn or improve performance based on the data they use. ML algorithms look for patterns and associations in large amounts of data and make optimal decisions or predictions based on the patterns and associations identified. The algorithms continue to improve as they learn, and the more data they use, the more accurate they become. ML consists of different types of learning models, and the user can decide to use supervised or unsupervised learning models depending on the nature of the data and the desired outcome [13].

The main techniques of ML include classification analysis, prediction analysis, cluster analysis, association rule analysis, and sequential pattern analysis [14,15]. The main purpose of classification analysis is to categorize a new paradigm, which does not have an explicit category, into a predefined category. This technique has been widely used to solve problems in various fields, such as identifying the key attributes of each type of user from their past usage records to find out what attributes make them loyal users and potential churners.

The main purpose of prediction analysis is to use information or conditions recorded in the past or at the present stage to determine possible future outcomes. Most of the methods used for classification analysis can also be used for prediction analysis. While classification analysis is used to make judgments about the present, prediction analysis is used to make judgments about the future. Cluster analysis is the analysis of the similarity of a large number of records to produce a combination of clusters. The combination of clusters produced by cluster analysis is characterized by the high similarity between records within a cluster and the low similarity between records in different clusters. Cluster analysis is mainly used in situations where it is unclear how to classify the data, so it is suitable for analyzing medical images and social networks, or for finding anomalies.

Association rule analysis is mainly used to identify possible relationships between attributes in a database. It is commonly used in the analysis of sales databases. Hence, it is referred to as shopping basket analysis in the field of marketing data science, which focuses on identifying combinations of items in sales databases that are frequently purchased by customers at the same time. Sequential pattern analysis is a technique similar to association rule analysis, which emphasizes the temporal order relationship of the set of items. Therefore, the main purpose of sequential pattern analysis is to identify the sequential relationship between the sets of items that frequently appear in the database during a specific time interval.

This study aimed to predict the intention of high school students to use school social media as an important channel to obtain school information. Therefore, the questionnaire data of high school students' usage intention on school social media was analyzed by using the decision tree technique to build a decision-tree-based usage intention model of school social media.

The decision tree technique is a supervised learning method in ML that is mainly used to deal with classification problems. Decision trees use the hierarchical conception of an inverted tree structure to express the classification process. Starting from the root node at the top level, a research variable is selected at each node, and the data are divided into subsets based on the possible values of the research variable to form the next level of the hierarchy. This process stops when there are no more suitable research variables to be selected or the value of the target variable can be decided. Common decision tree techniques include ID3, C4.5, and CART. ID3 is the first proposed decision tree algorithm, which uses information gain to determine the selected research variable at each node of the tree structure. C4.5 is an improved version of ID3, which does not use the information gain directly but uses the information gain ratio as the basis for research variable selection. CART can be used for both classification and regression problems with the Gini coefficient instead of the information entropy model. In this study, C4.5 was used to construct the prediction model of the school social media usage intention for high school students.

3. Research Method

3.1. Training Data for Modeling

To construct the prediction model of school social media usage intention, we conducted a questionnaire survey of the students in a senior high school in Taoyuan, Taiwan. The questionnaire was divided into two parts. The first part was used to understand the background information, including gender, age, frequency of visiting FB, awareness of the school's official FB fan page, and frequency of visiting the school's official FB fan page. The second part of the questionnaire consisted of six items on the six variables of this study (perceived usefulness, perceived ease-of-use, subjective norm, perceived behavioral control, trust, and usage intention). Each questionnaire item was measured on a five-point Likert scale with the numbers 5 to 1, with 5 indicating strongly agree and 1 indicating strongly disagree.

Before the questionnaire survey, each participant was asked to browse the school's official FB fan page on their smartphones for one minute and then to fill out the questionnaire. The questionnaire was administered from 1 June to 5 July 2022, and completed questionnaires were obtained from 223 participants.

3.2. Decision Tree Computing Software

Weka was used in this study to analyze the collected questionnaire data. Weka is a free software that provides various ML technologies, including data pre-processing, classification analysis, cluster analysis, and association rule analysis, and visualizes the data. Weka was developed by the University of Waikato, and is distributed under the GNU License in addition to open source [16].

There are two ways to use Weka: one is to analyze the data directly with the GUI of the Weka software, and the other is to analyze the data by calling the library provided by Weka in the program code. This study used the GUI of Weka to analyze the data directly (Figure 1). To do this, in the GUI of Weka, the Explorer option is chosen to enter the preprocess tab page (Figure 2), then the Open file option is selected in the Preprocess tab page to enter the questionnaire data to be analyzed. After entering the questionnaire data to be analyzed (Figure 3), the user proceeds to the decision tree analysis.

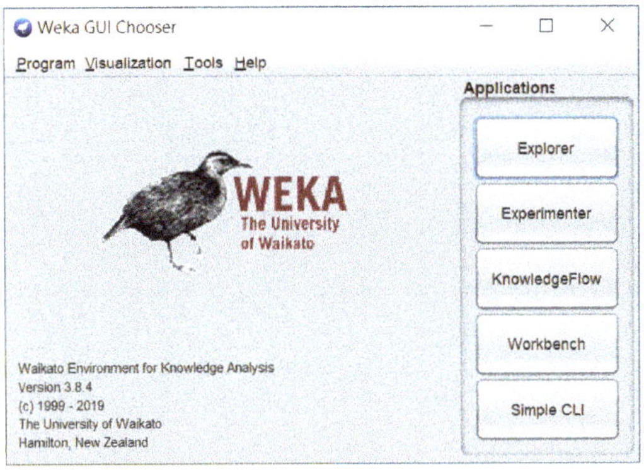

Figure 1. GUI of Weka.

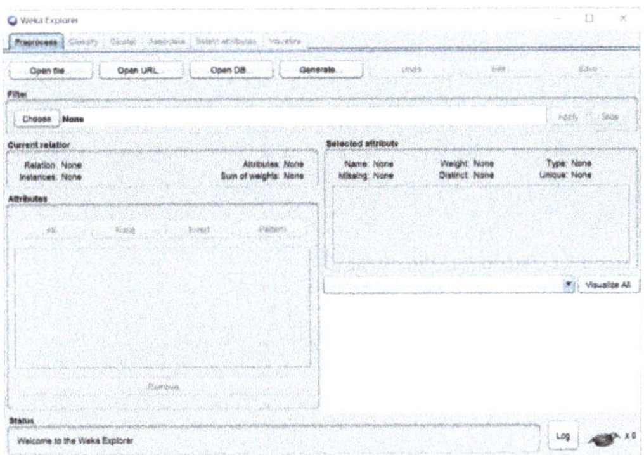

Figure 2. Preprocess tab page.

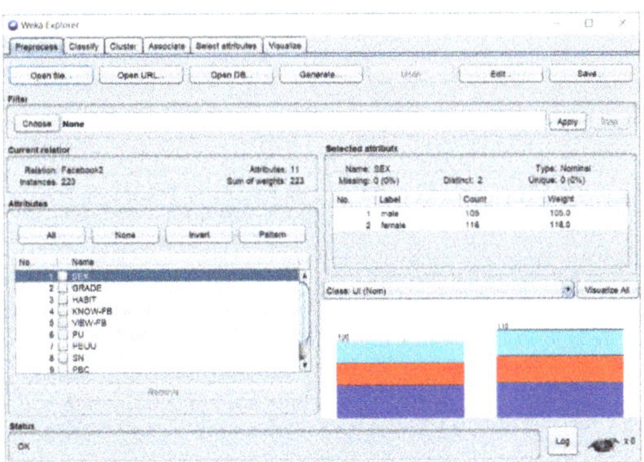

Figure 3. Preprocess tab page after data input.

In Weka, the decision tree analysis is in the Classify tab page (Figure 4), and the decision tree algorithm can be selected by clicking the Choose option on the Classify tab page. In the options of the decision tree algorithm of Weka, the J48 algorithm is the C4.5 decision tree algorithm. After selecting the J48 algorithm, the Test options are set to use the training set, and the target attribute is selected as usage intention (UI), then the Start option is selected to start the decision tree analysis (Figure 5).

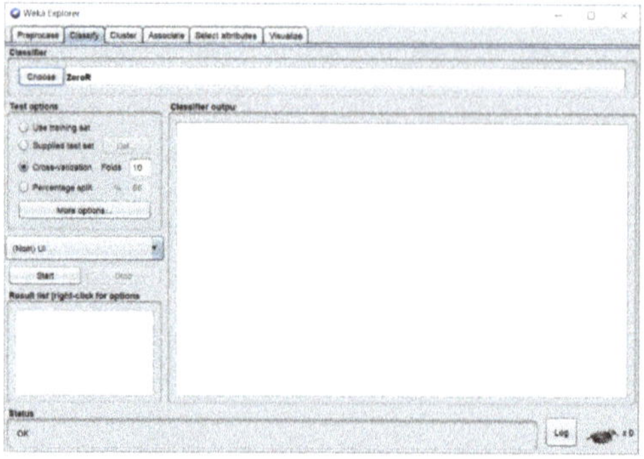

Figure 4. Classify tab page.

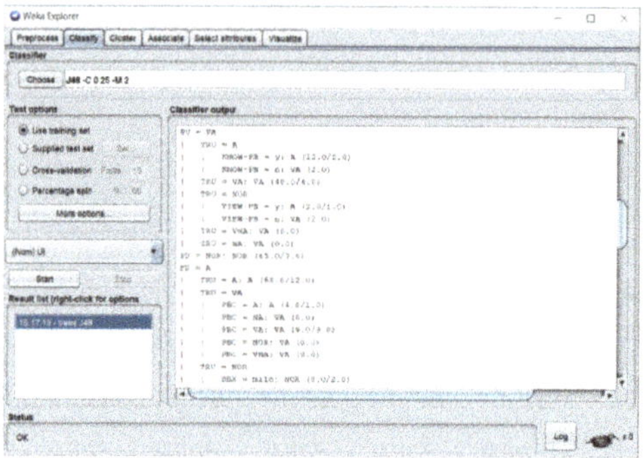

Figure 5. Classify tab page after executing J48.

4. Computing Results and Discussions

4.1. Description of Training Data

The training data for modeling are shown in Tables 1 and 2. According to Table 1, the percentage of students who frequently visited FB did not reach 50%, but most students were aware that the school had an official FB fan page, and most of them had visited the school's official FB fan page.

Table 1. Background information of participants.

Question	Option	Frequency
Gender	Male	105
	Female	118
Age	15 years old	102
	16 years old	71
	17 years old	5
	18 years old	45
Do you visit Facebook frequently?	Strongly disagree	9
	Disagree	16
	Normal	102
	Agree	46
	Strongly agree	50
Do you know that your school has an official Facebook fan page?	Yes	199
	No	24
Have you visited the school's official Facebook fan page?	Yes	170
	No	53

Table 2. Usage intentions of participants.

Question	Option	Frequency
Perceived usefulness (PU): I think it is useful to get important information about the school from school social media.	Strongly disagree	3
	Disagree	4
	Normal	54
	Agree	105
	Strongly agree	66
Perceived ease of use (PEOU): I think it is easy to use school social media to get important information about the school.	Strongly disagree	1
	Disagree	8
	Normal	63
	Agree	84
	Strongly agree	67
Subjective norm (SN): Friends and family around me think I should use school social media to get important school information.	Strongly disagree	3
	Disagree	3
	Normal	66
	Agree	92
	Strongly agree	59
Perceived behavioral control (PBC): I think I can get important school information from school social media.	Strongly disagree	2
	Disagree	4
	Normal	55
	Agree	86
	Strongly agree	76
Trust (TRU): I think it is trustworthy to get important information about the school from school social media.	Strongly disagree	1
	Disagree	4
	Normal	63
	Agree	91
	Strongly agree	64
Usage Intention (UI): I would like to receive important school information from school social media.	Strongly disagree	2
	Disagree	5
	Normal	67
	Agree	89
	Strongly agree	60

According to Table 2, the proportion of students who strongly agreed with PBC was the highest, followed by the proportion of students who strongly agreed with PEOU. This showed that students generally had no difficulty in using school social media to access important information about the school, either in terms of using the system or searching for information. A possible reason is that nowadays, high school students are generally digital natives, who have grown up in the era of the rapid development of IT. Their information skills are better than those of the previous generations of X and Y. Therefore, most high school students think they can quickly learn to use any information platform.

4.2. Decision-Tree-Computing-Based Usage Intention Prediction

The prediction model of school social media usage intention based on decision tree computing is shown in Figure 6. According to Figure 6, five decision rules can be obtained.

Rule 1: if perceived usefulness is "strongly agree" and trust is "strongly agree", then the usage intention is mostly "strongly agree".

Rule 2: if perceived usefulness is "strongly agree" and trust is "agree", then the usage intention is mostly "agree".

Rule 3: if perceived usefulness is "agree", trust was "strongly agree", then the usage intention is mostly "strongly agree" or "agree".

Rule 4: if perceived usefulness is "agree", trust is "agree", then the usage intention is mostly "agree".

Rule 5: if perceived usefulness is "agree", trust is "agree", and gender is female, and PBC is "strongly agree" or "agree", then the usage intention is mostly "agree".

```
PU = VA
|   TRU = VA: VA (48.0/4.0)
|   TRU = A: A (12.0/4.0)
|   TRU = NOR
|   |   VIEW-FB = y: A (2.0/1.0)
|   |   VIEW-FB = n: VA (2.0)
PU = A
|   TRU = VA
|   |   PBC = VA: VA (9.0/3.0)
|   |   PBC = A: A (4.0/1.0)
|   TRU = A: A (68.0/12.0)
|   TRU = NOR
|   |   SEX = male: NOR (8.0/2.0)
|   |   SEX = female
|   |   |   PBC = VA: A (1.0)
|   |   |   PBC = A: A (6.0/1.0)
|   |   |   PBC = NOR: NOR (5.0/1.0)
|   |   |   PBC = NA: NOR (1.0)
|   TRU = NA: NOR (3.0/1.0)
PU = NOR: NOR (45.0/7.0)
PU = NA: NOR (4.0)
PU = VNA: NOR (3.0/2.0)
```

Figure 6. Prediction model of school social media usage intention.

The above results show that perceived usefulness is the key factor that influenced high school students' usage intention of school social media, while trust is the secondary factor. Therefore, if schools want to use social media as the main channel to announce important information, they need to take measures to ensure that students think that it is useful to learn important information from school social media. In addition, schools must protect students' personal information and not allow social media to become a channel for the disclosure of students' personal information.

5. Conclusions

With the rapid advancement of technology and the greatly reduced costs of computing and storage, diversified data, such as images, text, sound, and maps, can be provided to parents in compliance with regulations. This allows schools to market themselves or provide parents and the public with a clear overview of the school to increase their recognition of the school and build satisfaction and loyalty to the school brand. In the past, when there was no low birth rate and information was relatively unavailable, most schools focused only on the quality of teaching and learning, attaching importance to student growth and teacher professionalism. However, with the professionalization of education, many schools are researching school administration, emphasizing data, and using the empirical evidence provided by data and information for decision-making and management. With the rapid development of new media, smartphones have become an ideal carrier for schools to deliver valuable information to parents and students through social media in a fast and low-cost way. We constructed a prediction model of school social media usage intention and produced five decision rules. The findings of this study can be used to predict students' intentions to use school social media, and can also be used as a guide for schools to use social media as an important channel for disseminating information.

Author Contributions: Conceptualization, H.-T.H. and H.-E.C.; methodology, H.-T.H.; software, H.-E.C.; investigation, H.-T.H.; writing—original draft preparation, H.-T.H. and H.-E.C.; writing—

review and editing, H.-T.H. and H.-E.C.; visualization, H.-E.C. All authors have read and agreed to the published version of the manuscript.

Funding: This research received no external funding.

Institutional Review Board Statement: Not applicable.

Informed Consent Statement: Not applicable.

Data Availability Statement: Not applicable.

Conflicts of Interest: The authors declare no conflict of interest.

References

1. Carr, C.T.; Hayes, R.A. Social media: Defining, developing, and divining. *Atl. J. Commun.* **2015**, *23*, 46–65. [CrossRef]
2. Levy, R.E. Social media, news consumption, and polarization: Evidence from a field experiment. *Am. Econ. Rev.* **2021**, *111*, 831–870. [CrossRef]
3. Lamot, K. What the metrics say. The softening of news on the Facebook pages of mainstream media outlets. *Digit. J.* **2022**, *10*, 517–536. [CrossRef]
4. Wahid, R.M.; Gunarto, M. Factors driving social media engagement on Instagram: Evidence from an emerging market. *J. Glob. Mark.* **2022**, *35*, 169–191. [CrossRef]
5. Fagundes, L.S.; Marot, T.A.; Natividade, J.C. Use of Instagram, social comparison, and personality as predictors of self-esteem. *Psico-USF* **2021**, *25*, 711–724. [CrossRef]
6. Foroughi, B.; Griffiths, M.D.; Iranmanesh, M.; Salamzadeh, Y. Associations between Instagram addiction, academic performance, social anxiety, depression, and life satisfaction among university students. *Int. J. Ment. Health Addict.* **2021**, *20*, 2221–2242. [CrossRef]
7. Rhee, L.; Bayer, J.B.; Lee, D.S.; Kuru, O. Social by definition: How users define social platforms and why it matters. *Telemat. Inform.* **2021**, *59*, 101538. [CrossRef]
8. Dhiman, D. The Practice of Media Education and Media Research: A Review on Five Asian Countries. The Practice of Media Education and Media Research: A Review on Five Asian Countries. *Glob. Media J.* **2021**, *19*, 1–7.
9. Goldstein, A.; Eaton, C. Asymmetry by design? Identity obfuscation, reputational pressure, and consumer predation in US for-profit higher education. *Am. Sociol. Rev.* **2021**, *86*, 896–933. [CrossRef]
10. Davis, F.D.; Bagozzi, R.P.; Warshaw, P.R. User acceptance of computer technology: A comparison of two theoretical models. *Manag. Sci.* **1989**, *35*, 982–1003. [CrossRef]
11. Ajzen, I. The theory of planned behavior. *Organ. Behav. Hum. Decis. Process.* **1991**, *50*, 179–211. [CrossRef]
12. De Meo, P.; Messina, F.; Rosaci, D.; Sarné, G.M. Combining trust and skills evaluation to form e-Learning classes in online social networks. *Inf. Sci.* **2017**, *405*, 107–122. [CrossRef]
13. Badillo, S.; Banfai, B.; Birzele, F.; Davydov, I.I.; Hutchinson, L.; Kam-Thong, T.; Siebourg-Polster, J.; Steiert, B.; Zhang, J.D. An introduction to machine learning. *Clin. Pharmacol. Ther.* **2020**, *107*, 871–885. [CrossRef]
14. Mahesh, B. Machine learning algorithms—A review. *Int. J. Sci. Res.* **2020**, *9*, 381–386.
15. Ray, S. A quick review of machine learning algorithms. In Proceedings of the 2019 International Conference on Machine Learning, Big Data, Cloud and Parallel Computing, Faridabad, India, 14–16 February 2019.
16. Attwal, K.P.S.; Dhiman, A.S. Exploring data mining tool-Weka and using Weka to build and evaluate predictive models. *Adv. Appl. Math. Sci.* **2020**, *19*, 451–469.

Disclaimer/Publisher's Note: The statements, opinions and data contained in all publications are solely those of the individual author(s) and contributor(s) and not of MDPI and/or the editor(s). MDPI and/or the editor(s) disclaim responsibility for any injury to people or property resulting from any ideas, methods, instructions or products referred to in the content.

engineering proceedings

Proceeding Paper

Information, Communication, and Technology in the Field of Tourism and Hospitality: A Bibliometric Approach †

Pahrudin Pahrudin [1,2], Tsung-Hua Hsieh [1,3] and Li-Wei Liu [1,*]

1 Department of Leisure and Service Management, Chaoyang University of Technology, Taichung 413310, Taiwan
2 Faculty of Social Science and Economic, Hamzanwadi University, Selong 83611, West Nusa Tenggara, Indonesia
3 Department of Industrial Education and Technology, National Changhua University of Education, Changhua 50007, Taiwan
* Correspondence: llouis@cyut.edu.tw
† Presented at the 3rd IEEE International Conference on Electronic Communications, Internet of Things and Big Data Conference 2023, Taichung, Taiwan, 14–16 April 2023.

Abstract: Recently, information, communication, and technologies (ICTs) and the tourism sector have gained attention from researchers and practitioners. Thus, we examined the trends in using ICTs in the tourism sector using a scientific document approach based on bibliometric analysis. A bibliometric analysis was conducted on 102 selected papers from the Scopus database from 2000 to 2021. The results revealed that the most discussed keywords in the papers were ICT and destination management organization, smart tourism destination, and smart city. Co-occurrence analysis was performed to analyze the trending topic in ICTs, tourism, and hospitality used by academics, while authorship network collaboration examined the partnership between different countries around the world. Statistics such as the number of documents per country, number of citations related to each country, ICT practice studies, and research methods were also discussed in this study. Limitations and implications were also indicated to provide deeper insights to researchers and their future research in terms of ICTs, tourism, and hospitality.

Keywords: scientific approach; bibliometric; ICT; tourism; hospitality

Citation: Pahrudin, P.; Hsieh, T.-H.; Liu, L.-W. Information, Communication, and Technology in the Field of Tourism and Hospitality: A Bibliometric Approach. *Eng. Proc.* **2023**, *38*, 4. https://doi.org/10.3390/engproc2023038004

Academic Editors: Teen-Hang Meen, Hsin-Hung Lin and Cheng-Fu Yang

Published: 19 June 2023

Copyright: © 2023 by the authors. Licensee MDPI, Basel, Switzerland. This article is an open access article distributed under the terms and conditions of the Creative Commons Attribution (CC BY) license (https://creativecommons.org/licenses/by/4.0/).

1. Introduction

Information and communication technologies (ICTs) have benefits in the tourism sector, particularly in tourism destinations, influencing tourist satisfaction. Studies showed that media and technologies have influenced tourist organizations to encourage and assist stakeholders in reshaping business processes and changing how customers and visitors communicate [1,2]. ICTs have overgrown and are now becoming an urgent issue for the tourism industry while they increase operational efficiency [3]. The basic form of tourism has shifted from traditional tourism destinations to smart tourism destinations due to the growth of ICTs [4,5]. Here, ICT has a role in management and marketing to improve the tourism industry [6] and the education industry [7]. Therefore, technology plays an essential role and function in making strategies for enhancing sustainable tourism development [8–10].

Nowadays, ICTs are the most researched issue in tourism [11]. Many researchers from academics and practitioners have been researching tourism from different perspectives of case studies [12], a concept or literature review [13], or methodology/technology [14,15], and behavioral study [16]. Scholars and academics have established and spread the notion of ICT and tourism destinations with the concept of the smart tourism destination (STD) [17–19]. The concept of smart tourism has developed from the concept of the smart ecosystem such as smart technology, smart cities, and smart tourism [20]. As a result, the concept of an intelligent tourism destination has a new definition related to STD [13].

Several years ago, researchers and academics debated the notion of tourist destinations from many perspectives and meanings [17]. Since then, smart tourism destinations' ideas and ground theory have developed [17,21]. However, research on incorporating ICT into a tourist site is insufficient in terms of making it smart [22]. Furthermore, the term STD refers to the integration between technology, tourism, and the environment to satisfy tourists [23,24]. In addition, the concept of STD integrates technology and the business ecosystem in tourism [23]. As a result, tourism destinations must have a new model framework and plan to integrate all components, including economic, technical, and social infrastructures, that make smartness possible and convenient for tourists when visiting a location.

From the theoretical perspective, research on the ICTs, tourism, and hospitality have been discussed by scholars from several perspectives such as emerging tourist behavior on technology in the technology acceptance model and theory-driven technology [25,26]. Several authors or scholars have conducted systematic literature reviews on ICTs, tourism sectors, and other fields [27–29]. However, the topic of ICTs, tourism, and hospitality industries needs more attention and discussion to discover the trends in the literature based on the bibliometric approach. The bibliometrics analysis was proposed to investigate the convergence of ICT and tourism destinations. Because technology and its applications are rapidly evolving, reviewing the current literature is necessary to offer firm ground and direction for future studies.

To fill out the research gap in this study related to ICTs, tourism, and hospitality, the bibliometric analysis is one of the methods to explore the scientific documents based on the authorship, author's collaboration, countries, names of journals, and keyword analysis trend. Studies were conducted [30,31] to apply bibliometric analysis in tourism. This study aims to provide scientific documents such as articles, books, and conference papers from the Scopus database. In addition, it is proposed to explore the scientific documents on the topics of ICTs, tourism, and hospitality through the bibliometric analysis approach with several objectives such as mapping the research trend on ICTs, tourism, and hospitality (O1), the number of publication (O2), authorship (O3), institution (O4), and research methods (O5). The study result provides research trends in ICTs, tourism, and hospitality fields, the conceptual framework of the scientific document in the bibliometric approach, and a direction for future research on ICTs, tourism, and hospitality.

2. Methodology

In this study, bibliometric analysis was used with several processes such as selection, data collection, and analyzing the documents. There are two processes in bibliometric analysis: mapping of the document with several techniques analysis such as Bradford's law, Lotka's Law, and the number of H-index journals and mapping of the scientific reports and social scientific collaboration such as author's collaboration, co-word analysis, and citation network [32].

The purpose of scientific mapping analysis is to investigate the global collaboration among the authors from a statistical perspective and scientific studies from several perspectives such as the trend of publication among the authors, topics discussed by the authors, collaboration among authors in publications, and the document's impact on the scientific community [33]. In this study, several research questions were proposed.

RQ1: what are the main keywords that have been discussed by scholars in the field of ICT, tourism, and hospitality industry?

RQ2: what are the methods used by authors in terms of technology, tourism, and hospitality fields?

RQ3: how do authors collaborate around the world in terms of technology, tourism, and hospitality?

2.1. Data Analysis

One of the methods to manage the publication and number of publications is bibliometric analysis. Nowadays, many researchers use bibliometric analysis in their research [34], with several topics of study, the number of institutions, and countries where authors are based on bibliometric analysis [35]. We used a bibliometric analysis for research related to ICT and tourism destinations. The measurement in this research was conducted for the number of publications, name of the journal, mapping the methods that have been used, the authorship, research trend, literature cited, and mapping of the document around the world. To achieve the aims of this study, we analyzed the literature based on the co-citation analysis and bibliometric analysis of authors, keywords, co-occurrence, and the method in the literature. Figure 1 shows the data analysis process for analyzing the scientific documents based on the bibliometric analysis.

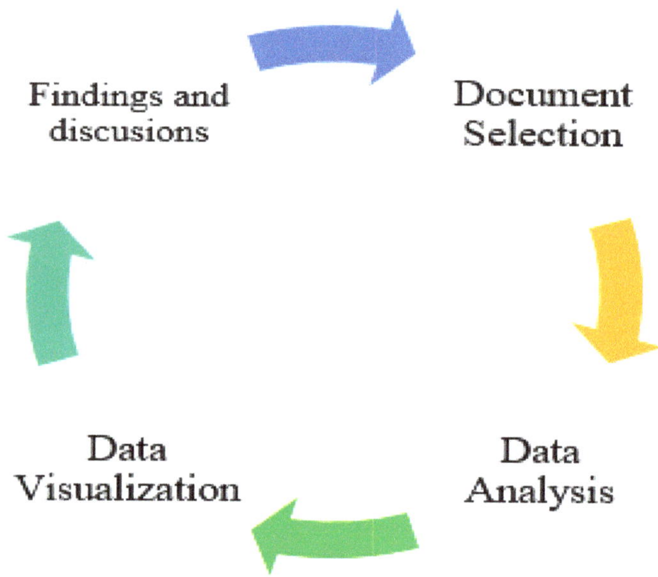

Figure 1. Data analysis process.

2.2. Data Collection

We collected relevant information from the Scopus database to find the focus of the research. In this study, the data were collected in May 2022 based on the Scopus database with several keywords such as "technology", "tourism", and "hospitality". The title, abstract, and keywords were searched for the keywords.

2.3. Data Visualization

We used several analysis methods. First, the qualitative method was applied to analyze the research focus of the selected papers. Second, the number of publications was identified to examine the articles based on the year of publication, authors, types of research used, research region, and theories applied in the papers. Third, we investigated the trend in keywords used over the last six years and the number of collaborations in various geographies, co-occurrence (keyword), and co-authorship (countries) studies. We used VOSviewer software to analyze all the selected articles' keywords, authors, and countries. Several indicators were used to measure statistical descriptive and bibliometric indicators such as the annual publication growth, collaboration among authors and institutions, sources, authors, institutions, keywords, and country production. These indicators were obtained from the analysis [36]. Visualization was used to find the mapping of collaboration such as collaboration network analysis of the authors, institution, countries, citations, and keyword networks. Vos-viewer software was employed for analyzing and visualizing the documents as it has been used for mapping scientific documents (www.vosviewer.com (accessed on 5 May 2022)) [37].

3. Results

This section explains the results and findings of the study based on the topic of research in the fields of ICT, tourism and hospitality.

3.1. General Overview and Number of Publications

Descriptive statistics explained the number of publications and the journal's name based on the research topic information, communication and technologies, and tourism destination. Figure 2 presents the number of publications in the journal distributed yearly. The number of publications related to ICT and tourism destinations commenced in the year 2003. The number of publications during 2000–2011 increased from one article in 2003 to six articles in 2010. The number of publications from 2011 to 2020 increased from 6 to 20 according to the Scopus database.

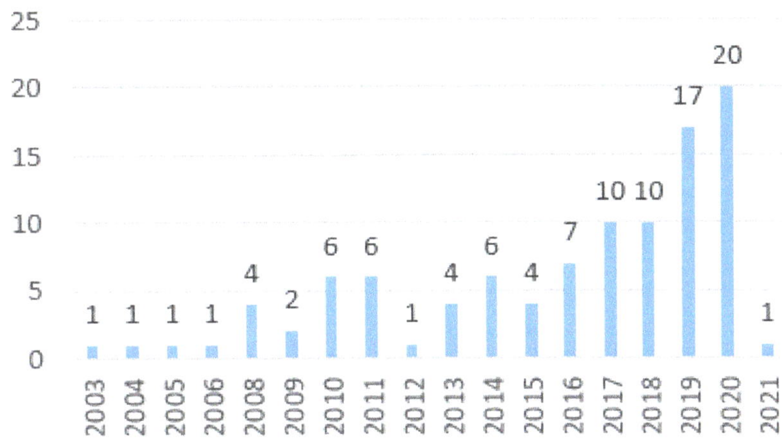

Figure 2. Number of publications by year.

3.2. Research Trends in ICT, Tourism and Hospitality Based on Keywords

Keyword analysis based on co-occurrence was conducted to analyze research trends (2012–2020) related to ICT and tourism destinations using the VOSviewer software. The results in Figure 3 present the occurrence of the keywords that academics or researchers have conducted. The colors in Figure 3 represent the 'year' highlighting the keyword. ICT was the most frequently used keyword within four years compared to other keywords. The other keywords frequently used by researchers included smart tourism, e-tourism, destination management, and tourism. The most commonly used keywords in 2020 are highlighted in yellow such as ecotourism, stakeholder collaboration, and tourist destination. Each line represents the network among the keywords from 2012 to 2020.

3.3. Co-Authorship around the World

Co-authorship analysis was conducted concerning the countries of authors that impacted the research focus. The results of VOS-viewer software explained the countries in which authors collaborated around the world. The text sizes show writers' relationships and collaboration with other nations, where the largest font size indicates that the authors from this particular country collaborated more with other countries. Figure 4 represents the links around countries represent the strength of the network between the counties. Each color represents the strength of the network cluster, and the same colors explain the same research field.

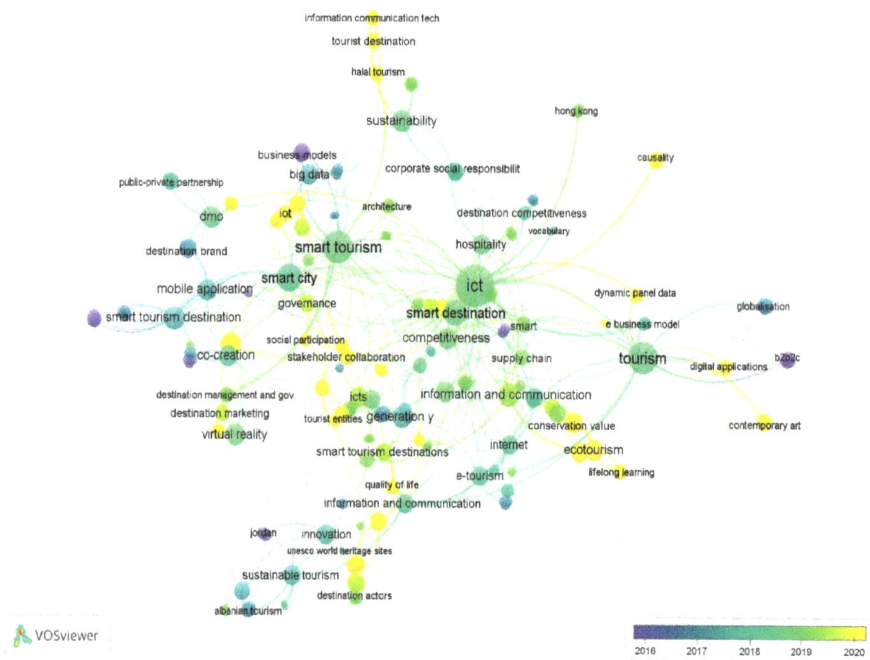

Figure 3. Research trend of keywords.

Figure 4. Collaboration of countries around the world.

3.4. Distribution of Publications

The research on ICTs and tourism destinations was conducted in more than 30 countries. Figure 5 shows the distribution of publications or articles by the countries related to ICTs and tourism destinations. Spain had the maximum number of publications with seventeen articles. The United Kingdom had eleven articles, followed by Italy with six articles. The United States of America published five articles, followed by Portugal, Malaysia, and Austria with four articles.

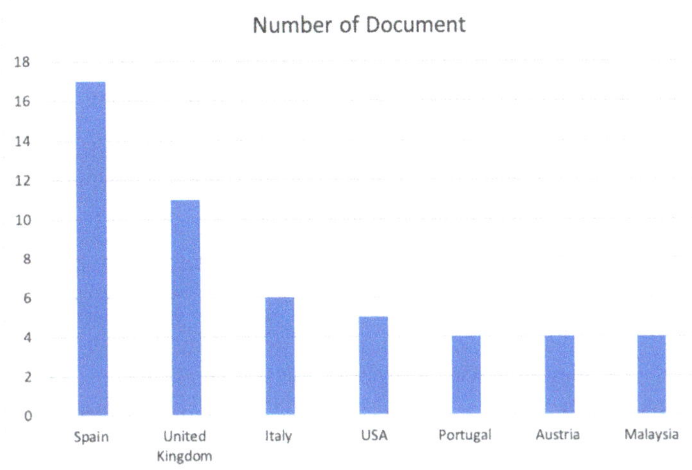

Figure 5. Number of documents per country.

3.5. Research Methods

The authors' use of the methodology in the research topic of ICTs and tourism destinations was investigated. The method was divided into quantitative, qualitative, and mixed methods to research ICT and tourism destinations. As shown in Figure 6, 65 papers (65%) used qualitative methods in the research of ICT and tourism destinations. The quantitative method were adopted by 25% of the research papers. The mixed research method was used in 11% of the research papers.

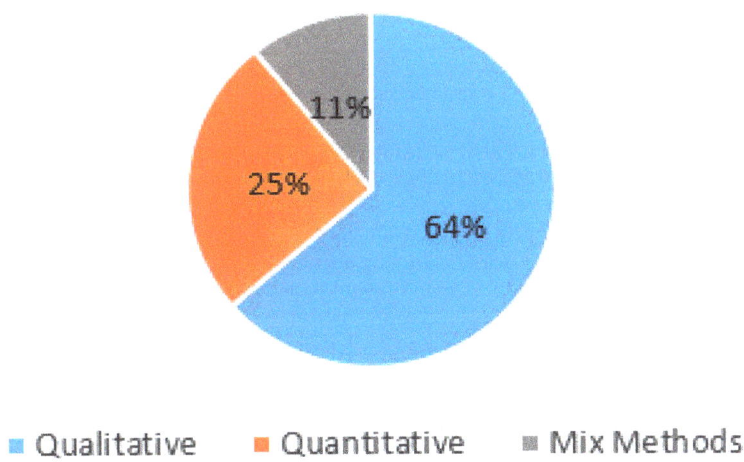

Figure 6. Research methods.

4. Discussion

The result provided an understanding of the trend research on ICTs, tourism, and hospitality. The trend in the publication from 2003 to 2021 showed an increase. In 2019, the publication was 17 articles, and in 2020, it reached 20. The number of publications in ICTs and tourism destinations continued to increase in 2021. The geographical areas were mainly located in European countries such as Spain, Italy, United Kingdom, Switzerland, Portugal, and Austria. The countries that generally discussed ICT and tourism destinations were developed countries because of technological advancement [38]. European countries and the United States of America had made an effort to promote tourism destinations and city development based on technological advancements to help the tourism sector expand [13,24]. Europe and the USA have the technology to support the development of the destination such as networks, infrastructures, and government supports [13,24,39]. Concerning Asian countries, only a few countries discussed ICTs and tourism destinations such as Japan, South Korea, Malaysia, India, Thailand, China, and Hong Kong. The countries in Asia that focused on the research of ICTs and destinations were also advanced in technologies [24]. Mainland China and South Korea have developed infrastructure and technology to support STDs [13,24]. For example, in 2015, the Authority Of China Tourism released the concept of "Guidance to promote Smart Tourism Destination" [11]. Infrastructures and the new system are the keys to building a new technology based on the tourist experience, assuming that all stakeholders and providers have the experience to use the technologies [39]. Therefore, ICT is the key to the tourist experience and enhancing competitiveness in tourism destinations [40].

Researchers and scholars linked ICTs and destination management organizations (DMOs) in 21 papers. The studies mainly emphasized that ICTs impacted consumer or traveler behavior for searching information, services, and products in the tourism industry [41]. DMOs include various functions such as identifying markets, promotions, and serving information about tourism destination and service quality, which become necessities [42]. DMO enhances the function of smart infrastructure and data analysis and helps brainstorm knowledge for the stakeholders [17]. Implementing ICTs in digital technology is the key to implying the advantages and opportunities that digital system and technology offer in destination management [43–45].

Other topics discussed by the researchers related to ICTs and tourism destinations included smart cities and smart tourism destinations. ICTs can develop a smart city to increase citizens' quality of life using ICT infrastructure, supporting smart tourism destinations [46]. STD integrates destination and digitalization in ICT context into physical infrastructure [47]. ICTs impact tourism to shape a new construct for destination management [47]. Researchers were interested in smart ecosystems and smart cities. These studies focused on the potential evolution of destination imagery evaluation criteria. Thus, STD becomes a concept of tourism with ICTs and smart cities in enhancing the tourist experience.

Keyword analysis results based on co-occurrence showed that ICT was the most frequently used keyword within the last four years. Other frequently used keywords by researchers included smart tourism, e-tourism, destination management, and tourism. This finding implied that future potential research is connected to the concept of ICTs and tourism for the destination and building fundamental theories. In earlier studies, ICTs and tourism were focused on several ideas such as tourism destination, sustainability, and innovation, as indicated by the color of each keyword.

In terms of co-authorship (country) network analysis, Spain and Italy were the most powerful, followed by the United States of America, Mexico, and Malaysia. Other countries emphasized local and regional cooperation (e.g., within the same continent). While ICTs and tourist destination practices differ by country, sharing ideas, intelligence, and best practices globally is beneficial. As a result, new worldwide research networks and research centers emerged, resulting in the proliferation of literature.

ICTs are crucial for enhancing sustainability-related tourism destinations for tourist satisfaction. Research into government policies on ICTs may help tourism destinations expand. The link between ICTs and tourism destinations must be promoted to countries worldwide, especially in the countries where tourism is the primary source of income. Future investigation must address how governments can combine the ICT infrastructure and tourism destinations in urban and rural areas to promote tourist destination.

More academics or researchers need to conduct empirical research related to ICTs and tourism destinations in developing countries. Many researchers studied ICTs and tourism destinations or smart tourism destinations in developed countries. The study of ICTs and tourism has been conducted in countries or cities such as Europe, South Korea, Japan, and the United States of America. Recently, several countries have applied ICTs to promote tourism destinations in pandemics with several online or virtual methods. The researchers need to analyze the role that ICTs can play in tourism destinations during pandemic situations.

5. Conclusions

The advancement of technology has impacted tourism. Nowadays, the issue of ICTs and tourism destinations is one issue in developing the tourism sector considering the technology to improve tourist satisfaction. ICTs and tourism are drivers to enhance economic growth in many countries worldwide [1]. The bibliometric analysis on ICTs and tourism destinations shows that this has been discussed from several different perspectives by researchers. Researchers have discussed different dimensions and components related to a research focus on ICTs and tourism destinations including research trends, research methods, research topic, authors, and the number of publications. ICT integrates innovation and technology, productivity, and development, and influences businesses in general and small and medium businesses [48]. Advancement in ICTs give competitive bargain and provides the tourism sector opportunities such as a tourism destination [49]. Advanced technology in tourism is a solution for dealing with tourism development [50] to enhance tourist satisfaction. Hence, the case study must be related to ICTs and tourism destinations.

The bibliometric analysis approach in this study provides insight into the study of ICTs, tourism and hospitality from several perspectives. The primary purpose of conducting research based on bibliometric analysis is to research ICTs and tourism destinations. The result also contributes to extending the literature of tourism and hospitality as a guide for future research.

Author Contributions: Conceptualization, P.P. and L.-W.L.; methodology, P.P.; software, P.P.; validation, T.-H.H. and P.P.; formal analysis, P.P.; investigation, L.-W.L.; resources, P.P.; data curation, T.-H.H.; writing—original draft preparation, P.P.; writing—review and editing, L.-W.L.; visualization, T.-H.H.; supervision, L.-W.L.; project administration, T.-H.H.; funding acquisition, T.-H.H. All authors have read and agreed to the published version of the manuscript.

Funding: This research received no external funding.

Institutional Review Board Statement: Not applicable.

Informed Consent Statement: Not applicable.

Data Availability Statement: Data can be downloaded in this study on the Scopus Database.

Conflicts of Interest: The authors declare no conflict of interest.

References

1. Buhalis, D.; Law, R. Progress in information technology and tourism management: 20 years on and 10 years after the Internet—The state of eTourism research. *Tour. Manag.* **2008**, *29*, 609–623. [CrossRef]
2. Yuan, Y.; Tseng, Y.-H.; Ho, C.-I. Tourism information technology research trends: 1990–2016. *Tour. Rev.* **2019**, *74*, 5–19. [CrossRef]
3. Ip, C.; Leung, R.; Law, R. Progress and development of information and communication technologies in hospitality. *Int. J. Contemp. Hosp. Manag.* **2011**, *23*, 533–551. [CrossRef]
4. de Esteban Curiel, J.; Jalón, M.L.D.; Herráez, B.R.; Antonovica, A. Smart tourism destination in Madrid. In *Sustainable Smart Cities*; Springer: Berlin/Heidelberg, Germany, 2017; pp. 101–114.

5. Jovicic, D.Z. From the traditional understanding of tourism destination to the smart tourism destination. *Curr. Issues Tour.* **2019**, *22*, 276–282. [CrossRef]
6. Mandić, A.; Garbin Praničević, D. Progress on the role of ICTs in establishing destination appeal. *J. Hosp. Tour. Technol.* **2019**, *10*, 791–813. [CrossRef]
7. Pahrudin, P.; Liu, L.-W.; Chang, C.-Y. The influencing factors of ict use in online learning during COVID-19 pandemic in Indonesia. *Eng. Lett.* **2021**, *29*, 382–390.
8. Padash, A.; Ghatari, A.R. Toward an Innovative Green Strategic Formulation Methodology: Empowerment of corporate social, health, safety and environment. *J. Clean. Prod.* **2020**, *261*, 121075. [CrossRef]
9. Bustard, J.R.T.; Bolan, P.; Devine, A.; Hutchinson, K. The emerging smart event experience: An interpretative phenomenological analysis. *Tour. Rev.* **2019**, *74*, 116–128. [CrossRef]
10. Liburd, J.J. Sustainable tourism and innovation in mobile tourism services. *Tour. Rev. Int.* **2005**, *9*, 107–118. [CrossRef]
11. Ye, B.H.; Ye, H. Systematic Review of Smart Tourism Research. *Sustainability* **2020**, *12*, 3401. [CrossRef]
12. Khan, M.S.; Woo, M.; Nam, K.; Chathoth, P.K. Smart city and smart tourism: A case of Dubai. *Sustainability* **2017**, *9*, 2279. [CrossRef]
13. Li, Y.; Hu, C.; Huang, C.; Duan, L. The concept of smart tourism in the context of tourism information services. *Tour. Manag.* **2017**, *58*, 293–300. [CrossRef]
14. Hassannia, R.; Vatankhah Barenji, A.; Li, Z.; Alipour, H. Web-based recommendation system for smart tourism: Multiagent technology. *Sustainability* **2019**, *11*, 323. [CrossRef]
15. Yuan, H.; Xu, H.; Qian, Y.; Li, Y. Make your travel smarter: Summarizing urban tourism information from massive blog data. *Int. J. Inf. Manag.* **2016**, *36*, 1306–1319. [CrossRef]
16. Han, H.; Park, A.; Chung, N.; Lee, K.J. A near field communication adoption and its impact on Expo visitors' behavior. *Int. J. Inf. Manag.* **2016**, *36*, 1328–1339. [CrossRef]
17. Buhalis, D.; Amaranggana, A. Smart tourism destinations. In *Information and Communication Technologies in Tourism 2014*; Springer: Berlin/Heidelberg, Germany, 2013; pp. 553–564.
18. Buhalis, D.; Amaranggana, A. Smart tourism destinations enhancing tourism experience through personalisation of services. In *Information and Communication Technologies in Tourism 2015*; Springer: Berlin/Heidelberg, Germany, 2015; pp. 377–389.
19. Lopez de Avila, A. Smart destinations: XXI century tourism. In Proceedings of the ENTER2015 Conference on Information and Communication Technologies in Tourism, Lugano, Switzerland, 3–6 February 2015; pp. 4–6.
20. Gretzel, U.; Werthner, H.; Koo, C.; Lamsfus, C. Conceptual foundations for understanding smart tourism ecosystems. *Comput. Hum. Behav.* **2015**, *50*, 558–563. [CrossRef]
21. Zhu, W.; Zhang, L.; Li, N. Challenges, function changing of government and enterprises in Chinese smart tourism. *Inf. Commun. Technol. Tour.* **2014**, *10*, 553–564.
22. Leung, R. Smart hospitality: Taiwan hotel stakeholder perspectives. *Tour. Rev.* **2019**, *74*, 50–62. [CrossRef]
23. Gretzel, U.; Sigala, M.; Xiang, Z.; Koo, C. Smart tourism: Foundations and developments. *Electron. Mark.* **2015**, *25*, 179–188. [CrossRef]
24. Gretzel, U.; Zhong, L.; Koo, C. Application of smart tourism to cities. *Int. J. Tour. Cities* **2016**, *2*. [CrossRef]
25. Chung, N.; Han, H.; Joun, Y. Tourists' intention to visit a destination: The role of augmented reality (AR) application for a heritage site. *Comput. Hum. Behav.* **2015**, *50*, 588–599. [CrossRef]
26. Ghaderi, Z.; Hatamifar, P.; Ghahramani, L. How smartphones enhance local tourism experiences? *Asia Pac. J. Tour. Res.* **2019**, *24*, 778–788. [CrossRef]
27. Caldevilla-Domínguez, D.; Martínez-Sala, A.M.; Barrientos-Báez, A. Tourism and ICT. Bibliometric Study on Digital Literacy in Higher Education. *Educ. Sci.* **2021**, *11*, 172. [CrossRef]
28. Pertheban, S.; Samy, G.N. A Systematic Literature Review: Information Accuracy Practices in Tourism A Systematic Literature Review: Information Accuracy Practices in Tourism. *J. Qual. Assur. Hosp. Tour.* **2019**, *21*, 1–30. [CrossRef]
29. Devi, P.; Dupre, K.; Wang, Y. Rural tourism: A systematic literature review on definitions and challenges. *J. Hosp. Tour. Manag.* **2021**, *47*, 134–149.
30. Pahrudin, P.; Liu, L.-W.; Li, S.-Y. What Is the Role of Tourism Management and Marketing toward Sustainable Tourism? A Bibliometric Analysis Approach. *Sustainability* **2022**, *14*, 4226. [CrossRef]
31. Pahrudin, P.; Liu, L.W.; Li, S.Y.; Fahrurrozi, M.; Ali, M. Investigating The COVID-19 Outbreak and Tourism Sector for Future Research Agenda: A Bibliometric Analysis Approach. *Geo J. Tour. Geosites* **2022**, *42*, 824–831. [CrossRef]
32. Cuccurullo, C.; Aria, M.; Sarto, F. Foundations and trends in performance management. A twenty-five years bibliometric analysis in business and public administration domains. *Scientometrics* **2016**, *108*, 595–611. [CrossRef]
33. Aria, M.; Cuccurullo, C. bibliometrix: An R-tool for comprehensive science mapping analysis. *J. Informetr.* **2017**, *11*, 959–975. [CrossRef]
34. Farrukh, M.; Meng, F.; Raza, A.; Tahir, M.S. Twenty-seven years of Sustainable Development Journal: A bibliometric analysis. *Sustain. Dev.* **2020**, *28*, 1725–1737. [CrossRef]
35. Nawaz, K.; Aslam, T.; Saeed, H.A. A Bibliometric Analysis of International Journal of Sports Marketing & Sponsorship. *Int. J. Bus. Psychol.* **2020**, *2*, 45–60.

36. Sweileh, W.M.; Al-Jabi, S.W.; AbuTaha, A.S.; Zyoud, S.H.; Anayah, F.; Sawalha, A.F. Bibliometric analysis of worldwide scientific literature in mobile-health: 2006–2016. *BMC Med. Inform. Decis. Mak.* **2017**, *17*, 72. [CrossRef] [PubMed]
37. Waltman, L.; Van Eck, N.J.; Noyons, E.C.M. A unified approach to mapping and clustering of bibliometric networks. *J. Informetr.* **2010**, *4*, 629–635. [CrossRef]
38. Mohsin, A. Tourist attitudes and destination marketing—The case of Australia's Northern Territory and Malaysia. *Tour. Manag.* **2005**, *26*, 723–732. [CrossRef]
39. Shafiee, S.; Rajabzadeh Ghatari, A.; Hasanzadeh, A.; Jahanyan, S. Smart tourism destinations: A systematic review. *Tour. Rev.* **2021**, *76*, 505–528. [CrossRef]
40. Ballina, F.J.; Valdes, L.; Del Valle, E. The Phygital experience in the smart tourism destination. *Int. J. Tour. Cities* **2019**, *5*, 656–671. [CrossRef]
41. Pawłowska-Legwand, A. One-way Communication or Co-Creating Tourism Experience Already? Polish Perspective on the Use of ICT by DMOs. *e-Rev. Tour. Res.* **2020**, *18*, 342–362.
42. Wang, X.; Kim, M.J.; Love, P.E.D.; Kang, S.-C. Augmented Reality in built environment: Classification and implications for future research. *Autom. Constr.* **2013**, *32*, 1–13. [CrossRef]
43. Gretzel, U.; Yuan, Y.-L.; Fesenmaier, D.R. Preparing for the new economy: Advertising strategies and change in destination marketing organizations. *J. Travel Res.* **2000**, *39*, 146–156. [CrossRef]
44. Yuan, Y.-L.; Gretzel, U.; Fesenmaier, D.R. The role of information technology use in American convention and visitors bureaus. *Tour. Manag.* **2006**, *27*, 326–341. [CrossRef]
45. Petti, C.; De Maggio, M.; Lanna, M.L.; Mele, G.; Solazzo, G. e-business case as a tool for facilitating ICT adoption in local government agencies. In Proceedings of the 4th International Conference on Theory and Practice of Electronic Governance, Beijing, China, 25–28 October 2010; pp. 55–60.
46. Popova, K.K.; Malcheva, M.M. Assessment of the potential of Varna city as a smart tourist destination. *Tour. Int. Interdiscip. J.* **2020**, *68*, 308–321.
47. Ivars-Baidal, J.A.; Celdrán-Bernabeu, M.A.; Mazón, J.-N.; Perles-Ivars, Á.F. Smart destinations and the evolution of ICTs: A new scenario for destination management? *Curr. Issues Tour.* **2019**, *22*, 1581–1600. [CrossRef]
48. Giotopoulos, I.; Kontolaimou, A.; Korra, E.; Tsakanikas, A. What drives ICT adoption by SMEs? Evidence from a large-scale survey in Greece. *J. Bus. Res.* **2017**, *81*, 60–69. [CrossRef]
49. Agyeiwaah, E.; Mensah, E.A. The role of global south volunteer NGOS in home-stay arrangements in Ghana: The parallel of Simmel's mediator hypothesis? *Anatolia* **2017**, *28*, 93–95. [CrossRef]
50. World Tourism Organization (UNWTO). Madrid, Spain, 2018. Available online: https://www.unwto.org/world-tourism-day-2018 (accessed on 20 May 2022).

Disclaimer/Publisher's Note: The statements, opinions and data contained in all publications are solely those of the individual author(s) and contributor(s) and not of MDPI and/or the editor(s). MDPI and/or the editor(s) disclaim responsibility for any injury to people or property resulting from any ideas, methods, instructions or products referred to in the content.

engineering proceedings

MDPI

Proceeding Paper

A Research on Diversified Applications of Technological Education in the Development of Rural District Community Development Associations †

Jiin-Chyuan Mark Lai [1], Chiung-Ling Wang [2] and Ming-Yuan Hsieh [3,*]

[1] Department of Applied Foreign Languages, TransWorld University, Douliu 640302, Taiwan; marklai07@gmail.com
[2] Department of Industrial Education, National Taiwan Normal University, Taipei 5671455, Taiwan; lin8332317@mail.edu.tw
[3] Department of International Business, National Taichung University of Education, Taichung 40306, Taiwan
* Correspondence: cpawisely@mail.ntcu.edu.tw; Tel.: +886-975-118-922
† Presented at the 3rd IEEE International Conference on Electronic Communications, Internet of Things and Big Data Conference 2023, Taichung, Taiwan, 14–16 April 2023.

Abstract: Due to the rapid innovative applications of technological education as a result of the swift development and popularity of telecommunication and wireless technologies, ever-more diversified methods of technological education play a critical role in various education and lecturing activities, without time and distance restrictions. Currently, in order to bridge the urban–rural development gap, the Taiwanese Government has introduced a lot of policies to encourage teenagers to return their hometowns; however, these policies have not achieved their goals, as the most of Taiwanese community development associations faced scarcities of professional knowledge without any educational support during the development of their various schemes, such as economic development, environmental protection, cultural heritage, public services, etc. As a result, local development depends on entire community development associations, especially in the most mountain regions of Taiwan. Moreover, community development associations have a lot of social responsibilities, such as age-related caring activities, driving local economies and industrial development, environmental protection and education, maintaining and developing traditional cultures and arts, etc. Due to the digital characteristics of technological education, most Taiwanese rural community development associations are able to obtain professional information and data about courses without space and time restrictions. Thus, after strengthening and satisfying professionals' demands, local economies will achieve growth and, therefore, younger persons would be encouraged to stay their original area, thus stimulating the development of rural community development associations. Eventually, the positive development cycle is predicted to increased and, as a result, the urban–rural disparities will be directly diminished by the diversified application of technological education in Taiwan.

Keywords: urban–rural disparities; technological education; community development association

Citation: Lai, J.-C.M.; Wang, C.-L.; Hsieh, M.-Y. A Research on Diversified Applications of Technological Education in the Development of Rural District Community Development Associations. *Eng. Proc.* **2023**, *38*, 18. https://doi.org/10.3390/engproc2023038018

Academic Editors: Teen-Hang Meen, Hsin-Hung Lin and Cheng-Fu Yang

Published: 21 June 2023

Copyright: © 2023 by the authors. Licensee MDPI, Basel, Switzerland. This article is an open access article distributed under the terms and conditions of the Creative Commons Attribution (CC BY) license (https://creativecommons.org/licenses/by/4.0/).

1. Introduction

Here is an introduction to the human and physical geography of Taiwan: (1) area: 36,000 square kilometers; (2) population: 23 million; (3) capital: Taipei City; (4) main ethnic group: Chinese; (5) spoken languages: Mandarin, Taiwanese, and Hakka; and (6) religion: Buddhism, Taoism, Christianity, and Islam. All together, the Taiwanese total land area range is approximately 36,000 square kilometers; its shape is similar to a traditional Chinese tobacco leaf that is narrow at both ends. Taiwan is located near to the southeastern coast of mainland Asia, being separated from Mainland China by the Taiwan Strait, while also bordering the western edge of the South China Sea and the Pacific Ocean. Its other island neighbors include Japan to the north and the Philippines to the south. In terms of temperature, Taiwan is suitable for human living and traveling because the island's annual average temperature is 22 degrees Celsius, with lowest temperatures ranging from 12 to 17 degrees Celsius (54–63 Fahrenheit). This hospitable climate is associated with Taiwan's elevated terrain; the highest mountain is the Jade Mountain, which is over 3000 m high, and there are a total of 268 mountains on the island of Taiwan. To reduce the population loss and urban–rural disparities between the majority of large cities (e.g., Taipei Taichung, Tainan, and Taitung) and rural areas, the Executive Yuan, which is the most senior government department, instituted a series of development programs and created national budgets to encourage Taiwanese citizens to "live and work in peace and contentment", as well as to ensure "continuous reproduction breed in an endless succession" and "balanced development in Taiwan". One of these regulations and laws is Article 12 of the Regulations on Community Development Work, which outlined the essential rules governing of Taiwanese community development associations. Based on the basic goals of community development association outlined in Article 12 [1] of the Regulations on Community Development Work, these projects, when using government funds, must respect the government's policies.

By creating a comprehensive survey of technological education research [2–6] in association with a series of diversified applications, technological education has been applied in formal and informal educational systems, such as online education, massive open online courses (MOOCs), digital lectures, and temporary teaching method used during periods of global disease transmission, such as Severe Acute Respiratory Syndrome (SARS) and Coronavirus Disease (COVID-19) outbreaks, because the most critical characteristics of technological education can directly break through space and time restrictions. Continuously, most professionals are able to conveniently upload their information, data, and video courses onto the various technological education platforms at anytime and anywhere. Moreover, the learners are able to easily view and download professional information, data, and video courses to gain diversified professional knowledge [7–11]. Regarding the development and growth of Taiwanese community development associations, there are five critical barriers to overcome, which are:

(1) Population movement: due to economic underdevelopment of Taiwanese rural community development associations, most younger people leave their hometowns in order to make more money and pursue a higher life quality. Unfortunately, this population movement has, eventually, resulted in the ageing of rural populations.

(2) Long-term infrastructure underdevelopment: as rural populations migrated to cities, the Taiwanese Government prioritized using resources to improve infrastructure in urban areas without considering rural areas.

(3) Resource backward: due to the long-term urban–rural economic and infrastructure disparities, rural areas became the overlooked regions in terms of the Taiwanese Government's allocation of budgets and resources.

(4) Age-related budget requirements: due to their ageing populations, most rural areas require larger budgets and resources to support social schemes that support the elder residents, such as medical support, etc.

(5) Lack of environmental protection and education: due to long-term resource misallocation, most Taiwanese rural governments must use their entire budgets to maintain the local population's basic needs, which resulted in environmental protections and education being ignored in budget spending lists.

2. Conceptual Literature

According to the Article 12, "The community development associations shall devise community development plans, prepare budgets and actively promote community development according to community characteristics and the needs of residents, in addition to being in line with government policies and projects originally designed by the community." In particular, the most important contents of the Article 12 of the community development associations are its key principles, which are defined as follows: (1) establishing or constructing community development association centers; (2) strengthening the environment and improving sanitation in community development associations; (3) maintaining roads and gutters in community development associations; (4) building and improving infrastructure; (5) greening and beautifying community development associations; (6) the construction of public facilities managed by community development associations; (7) creating funds for community development associations; (8) providing social welfare; (9) constructing nurseries; (10) developing local businesses; (11) creating critical measures to preserve the ethos of Taiwanese society and advocate and promote models of public etiquette; (12) preserving and promoting rural culture and traditional public crafts; (13) building traffic safety infrastructure; (14) building community pact-related infrastructure; (15) organizing local civil defense; (16) funding traditional art clubs and local sports teams; (17) funding community senior citizens' clubs; (18) establishing educational classrooms; (19) creating volunteer service teams; (20) establishing community educational services and local libraries; (21) advancing various community-wide activities; (22) enforcing the community disaster assistance, reporting, and prevention drills; and (23) providing other matters related to the spiritual and ethical development of communities [12–16].

3. Conclusions and Future Direction

Due to the rapid innovative applications of technological education following the swift development and popularity of telecommunication and wireless technologies, ever-more diversified methods of technological education play a critical role in various education and lecturing course, lacking all time and distance restrictions. Currently, in order to bridge the urban–rural development gap, the Taiwanese Government has implemented a lot of policies to encourage teenagers to return to their hometowns.

Therefore, this research cross-utilized the preliminary research on the reduction in urban–rural disparities in community development and found that, based on a series of literature reviews, the geographical condition is the most critical factor for several reasons: (1) Taiwan's mountainous terrain resulted in the greatest infrastructure-building projects focusing on cities; (2) most Taiwanese rural community development associations have to confront economic underdevelopment themselves, despite needing more significant economic development than most Taiwanese urban community development associations because the urban developments are prioritized in budgets; (3) most young people of rural origin prefer to move to cities to make money and achieve life development, with Taiwanese rural community development associations needing more younger people to stay in rural community development associations to drive development and growth and diminish the crisis of ageing populations; (4) most professional lecturers and educators do not want to leave cities without comfortable and convenient public transportation, yet Taiwanese rural community development associations need these professionals to assist their development; and (5) the majority of Taiwanese rural community development associations do not have access to professional information and data services, despite needing this information and data more definitely than Taiwanese urban community development associations.

Significantly, political administrators did not achieve these goals as most Taiwanese community development associations experienced scarcities of professional knowledge without any educational support while carrying out their various activities, such as economic development, environmental protection, cultural heritage, public services, etc., as shown in Figure 1.

As a result, the local development depends on entire community development associations, especially in the most mountain regions of Taiwan. Moreover, the community development associations provide a lot of social responsibilities, such as age-related caring activities; support for local, economic, and industrial development; environmental protection and education; maintaining and developing the traditional cultures and arts, etc. Therefore, due to the digital characteristics of technological education, most Taiwanese rural community development associations are able to obtain professional information and data about courses without space and time restrictions. Thus, after strengthening and satisfying professional demands, local economies will experience growth and younger persons would be encouraged to stay their original area, stimulating the development of rural community development associations. Eventually, the positive development cycle is, predictably, will increase and, as a result, the urban–rural disparities will be directly diminished by the diversified application of technological education in Taiwan.

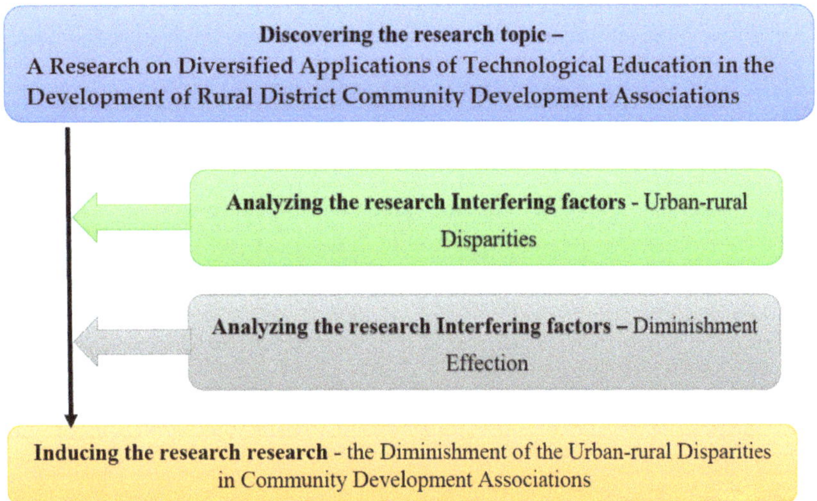

Figure 1. Main research concept.

Author Contributions: Conceptualization, J.-C.M.L. and M.-Y.H.; methodology, M.-Y.H.; validation, J.-C.M.L.; formal analysis, M.-Y.H.; investigation, M.-Y.H.; resources, J.-C.M.L. and C.-L.W.; writing—original draft preparation, M.-Y.H. and C.-L.W.; writing—review and editing, M.-Y.H.; visualization, M.-Y.H.; supervision, M.-Y.H.; project administration, M.-Y.H.; funding acquisition, and C.-L.W. All authors have read and agreed to the published version of the manuscript.

Funding: This research was supported by research project (MOST 110-2420-H-002-003-MY3-Y11017) of the Ministry of Science and Technology and (NTCU110108) of the National Taichung University of Education.

Institutional Review Board Statement: This research did not execute the questionnaires and interview and hence, this research did not require ethical approval.

Informed Consent Statement: Not applicable because this research did not involve any humans.

Data Availability Statement: No new data were created.

Conflicts of Interest: There are no conflict of interest in this research.

References

1. *The Article 12 of the Regulations on Community Development Work*; Ministry of Health and Welfare: Taipei, Taiwan, 2014.
2. Hsieh, Y.M. Interdisciplinarily Exploring the Most Potential IoT Technology Determinants in the Omnichannel E-Commerce Purchasing Decision-Making Processes. *Appl. Sci.* **2020**, *10*, 603. [CrossRef]
3. Hsieh, Y.M. The most sustainable niche principles of social-media education in a higher education contracting era. *Sustainability* **2020**, *12*, 339. [CrossRef]
4. Huang, Y.-M.; Hsieh, Y.M.; Usak, M. A Multi-Criteria Study of Decision-Making Proficiency in Student's Employability for Multidisciplinary Curriculums. *Mathematics* **2020**, *8*, 897. [CrossRef]
5. Chan, Y.-K.; Hsieh, M.-Y.; Usak, M. A Concrete Study on Social-Media Connection of Global Literacy Abilities in MOOCs under the Dual Impacts of Lower Birth-Rate and COVID-19. *Sustainability* **2021**, *13*, 2203. [CrossRef]
6. Usak, M.; Hsieh, M.-Y.; Chan, Y.-K. A Concertizing Research on Making Higher Education Sustainability Count. *Sustainability* **2021**, *13*, 2724. [CrossRef]

7. Duran, M.; Usak, M.; Hsieh, M.-Y.; Uygun, H. A New Perspective on Pedagogical Content Knowledge: Intellectual and Emotional Characteristics of Science Teachers. *Rev. De Cercet. Si Interv. Soc. (RCIS)* **2021**, *72*, 9. Available online: https://www.ceeol.com/search/article-detail?id=959021 (accessed on 23 November 2022). [CrossRef]
8. Wu, T.-L.; Hsieh, M.-Y.; Min, K.-W.; Yu, M.-T.; Ho, C.-T. Use of Sensor Technologies in Online Courses in Post-COVID-19 Era. *Sens. Mater.* **2021**, *33*, 2045–2062. [CrossRef]
9. Huang, C.C.; Chan, Y.-K.; Hsieh, M.Y. The Determinants of ESG for Community LOHASism Sustainable Development Strategy. *Sustainability* **2022**, *14*, 11429. [CrossRef]
10. Chan, Y.-K.; Hsieh, M.Y. An Empirical Study on Higher Education C-ESG Sustainable Development Strategy in Lower-Birth-Rate Era. *Sustainability* **2022**, *14*, 12629. [CrossRef]
11. Hsieh, Y.-M. Online learning era: Exploring the most decisive determinants of MOOCs in Taiwanese higher education. *Eurasia J. Math. Sci. Technol. Educ.* **2016**, *12*, 1163. [CrossRef]
12. Hsieh, Y.-M. Employing MCDM methodology to verify correlationship between social media and service quality in the dynamic m-commerce era. *J. Internet Technol.* **2018**, *19*, 225. Available online: https://jit.ndhu.edu.tw/article/view/1674 (accessed on 20 November 2022).
13. Hsieh, Y.-M.; Usak, M. High Education Radical Transformation Era: How Teachers' Competency can Enhance the Students' Employability. *Rev. De Cercet. Si Interv. Soc. (RCIS)* **2020**, *68*, 95. Available online: https://www.ceeol.com/search/article-detail?id=848474 (accessed on 18 November 2022). [CrossRef]
14. Huang, Y.-M.; Hsieh, Y.M. An Interdisciplinary Research on Students' Employability in Technology Education to Advance Higher Education Enrollment Sustainability. *Sustainability* **2020**, *12*, 1806. [CrossRef]
15. Hsieh, M.Y. The Sustainable Development and Strategic Approaches for Contemporary Higher Education. *Sustainability* **2022**, *14*, 12925. [CrossRef]
16. Huang, C.-C.; Chan, Y.-K.; Hsieh, M.-Y. Preliminary Research on the Sustainable Determinants of Taiwanese Ecotourism with the International Standards. *Int. J. Environ. Res. Public Health* **2022**, *19*, 14489. [CrossRef] [PubMed]

Disclaimer/Publisher's Note: The statements, opinions and data contained in all publications are solely those of the individual author(s) and contributor(s) and not of MDPI and/or the editor(s). MDPI and/or the editor(s) disclaim responsibility for any injury to people or property resulting from any ideas, methods, instructions or products referred to in the content.

engineering proceedings

MDPI

Proceeding Paper

Preliminary Research on the Impact of Bilingual Teaching on Current Technical and Vocational Education in Taiwan †

Chiung-Ling Wang [1], Chien-Yun Dai [1], Jiin-Chyuan Mark Lai [2,*] and Ming-Yuan Hsieh [3,*]

1 Department of Industrial Education, National Taiwan Normal University, Taipei 5671455, Taiwan; Talin8332317@mail.edu.tw (C.-L.W.); dai@ntnu.edu.tw (C.-Y.D.)
2 Department of Applied Foreign Languages, TransWorld University, Douliu 640302, Taiwan
3 Department of International Business, National Taichung University of Education, Taichung 40306, Taiwan
* Correspondence: marklai07@gmail.com (J.-C.M.L.); cpawisely@mail.ntcu.edu.tw (M.-Y.H.)

† Presented at the 3rd IEEE International Conference on Electronic Communications, Internet of Things and Big Data Conference 2023, Taichung, Taiwan, 14–16 April 2023.

Abstract: Based on the blueprint for bilingual education as a national policy, teachers in technical and vocational schools are preparing, in full swing, to teach subjects in English. However, most teachers in the institutions do not have the skills and backgrounds to implement English-mediated instruction (EMI). Therefore, it is necessary to cultivate talents and popularize programs in technical and vocational education and educate talents to have bilingual ability. In addition, international mobility and internship experience in English is also required. The current environment needs to be improved for bilingual education, promoting that bilingual teaching is beneficial to students and teachers. Even though thus puts pressure on them for learning and teaching in technical and vocational schools, bilingual education is the future direction of education in Taiwan.

Keywords: bilingual education; Taiwanese technical and vocational education; higher education

Citation: Wang, C.-L.; Dai, C.-Y.; Lai, J.-C.M.; Hsieh, M.-Y. Preliminary Research on the Impact of Bilingual Teaching on Current Technical and Vocational Education in Taiwan. *Eng. Proc.* **2023**, *38*, 21. https://doi.org/10.3390/engproc2023038021

Academic Editors: Teen-Hang Meen, Hsin-Hung Lin and Cheng-Fu Yang

Published: 21 June 2023

Copyright: © 2023 by the authors. Licensee MDPI, Basel, Switzerland. This article is an open access article distributed under the terms and conditions of the Creative Commons Attribution (CC BY) license (https://creativecommons.org/licenses/by/4.0/).

1. Introduction

In globalization and internationalization [1], communication skills and global perspectives are necessary to enhance a country's competitiveness. Thus, the Taiwanese government instituted and implemented the National Policy for Bilingual Education in 2017. The purpose is to bolster bilingual education, improve the English proficiency of people through demand-driven learning, and eventually enhance the overall national competitiveness. To construct an English–Chinese bilingual country, the following goals are proposed: (1) "Comprehensively strengthen the English proficiency of Chinese people from the demand side", (2) "Use digital technology to shorten the resource gap between urban and rural areas", (3) "Take the bilingual policy and the development of mother tongue culture", and (4) "Create the competitive edge of the younger generation".

The difference between the previous bilingual policy and the new blueprint is to raise the overall competitiveness of the country and to improve the English proficiency of all Taiwanese. To achieve these goals, the educational environment needs to be constructed to promote the atmosphere of learning English for the whole Taiwanese population, especially students [2]. The government considers the demand and supply for bilingual education and realizes that creating high-quality employment opportunities on the demand side and cultivating talents and connecting them to the global society on the supply is critical for the success of the policy. Then, by connecting Taiwan's industries and workforce to the global market, global business opportunities can be created, and global enterprises will invest in Taiwan to provide more employment opportunities [3]. For the policy, the Ministry of Education proposed five strategies and promotion measures, as shown in Figure 1.

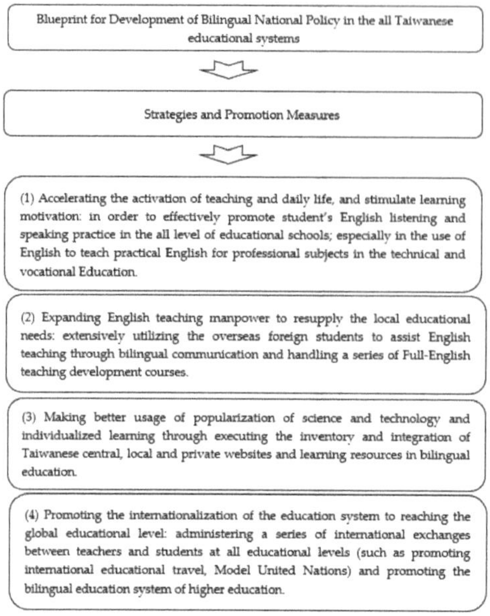

Figure 1. Five strategies and promotion measures for bilingual education of the Taiwanese Ministry of Education.

To diminish the urban–rural gap in English education for teachers and students [4], the Ministry of Education in Taiwan also presented the "Regulations on the Development of School Education in Rural Areas" to stabilize the source of teachers in rural regions, as shown in Figure 2.

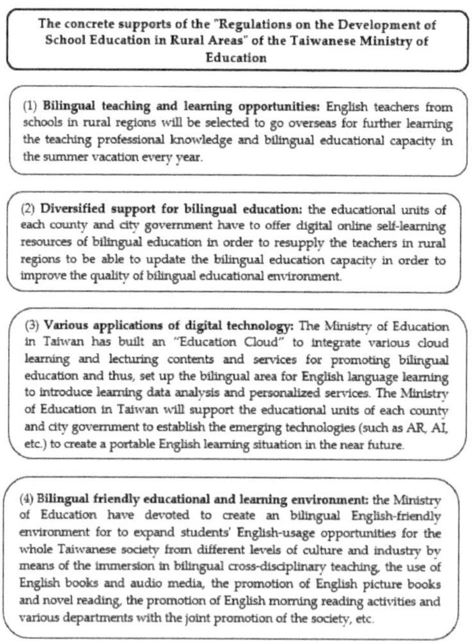

Figure 2. Regulations on the development of school education in rural areas.

Based on the Blueprint for the Development of Bilingual National Policy, the Ministry of Education instituted the "Implementation Plan for All-English Teaching Teacher Training" in 2017 for enhancing the teachers' capability of bilingual education in pre-service and in-service training at all levels of the education system, including elementary school, junior, and senior high schools, technical and vocation, and higher education [5]. By 2020, 21 higher education institutes operated bilingual education and training programs. Up to 2000 students were educated in a bilingual environment in four years. The number will increase to 3000 in eight years and to 5000 in twelve years. In English education, there must be a gap between urban and rural schools. To increase the number of students in bilingual education and narrow the gap between urban and rural schools, the Ministry of Education has put efforts to improve students' abilities and strengthen their English skills by cooperating with ministries and departments so that students can use their bilingual ability in life and the workplace [6].

In Taiwan, technical and vocational education is relatively less supported than other higher education programs, as the majority of people think that a college diploma is mandatory for their children and technicians do not need higher degrees. Since 2021, the Ministry of Education allowed community colleges to award Bachelor's degrees and waive military service for students to study abroad before the age of 30 years old. However, technical and vocational schools are not included in such benefits [7–9]. According to the statistics from the Ministry of Education in 2020, there were around 960,000 technical and vocational students in Taiwan, accounting for 62% of the total number of students. Unfortunately, the resources and funds from the Ministry of Education allocated to each technical and vocational student are only half of that of high school students. The resources for technical and vocational school students are only one-third of that of college or university students. The dominance of the government resource use for higher education is caused by distorted social values, while industries are not willing to support technical and vocational schools. The benefit of higher education drives students to lose interest in going to technical and vocational schools. In addition, only 1924 companies cooperated with institutions in 2020 [10].

Due to the lack of detailed planning, the establishment of standards in technical and vocational education was not strictly regulated, and the "Private School Law" was promulgated and implemented in 1985 to allow donations to start schools. The establishment of private schools is based on the concept of "investing in education", but without enough resources and poor teaching quality and capacity. Under the influence of higher education on students, utilitarianism in educational administrators and the declining birth rate makes the enrollment quota of the technical and vocational schools exceed the total number of applicants since 2016. This situation caused criticism of the establishment of private technical and vocational schools [11–15]. However, the majority of students in the technical and vocational schools cannot enjoy the benefits of the Blueprint for the Development of Bilingual National Policy, as the schools have accepted students with lower academic performance due to academicism and utilitarianism. Nowadays, under the Blueprint for the Development of Bilingual National Policy, technical and vocational schools are facing their biggest challenge, which is bilingual teaching. The policy required teachers and students in the schools to have more English teaching and learning [16–18]. Teachers and students have to have professional knowledge of the original technical and vocational majors and continuously improve their English skills. Therefore, teachers must try bilingual teaching of subjects and English.

However, bilingual education will provide an advantage in job finding and international competitiveness with self-worth [19–22]. Therefore, it is worth exploring the impact of bilingual education in technical and vocational schools. For the exploration, we determined a research motivation and found the research methodology for the evaluation of the research result (Figure 3).

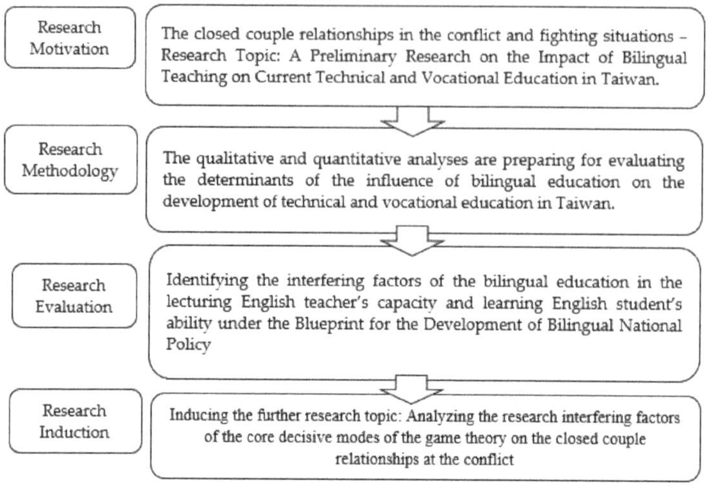

Figure 3. Process of this research.

2. Conclusions and Future Direction

According to the Blueprint for the Development of Bilingual National Policy, all teachers in Taiwan's education system are supposed to teach students in English and Chinese. Since its implementation, teachers in technical and vocational schools have had difficulties because they lack the professional backgrounds to execute EMI. Most students in the schools also have limited English ability, which causes difficulties in learning. Therefore, it is necessary to improve the teaching and learning abilities of teachers and students in technical and vocational schools in the new bilingual education model. Especially, teaching subjects in English requires integrating subject contents and language and continuously adjusting teaching methods to meet the needs of students. The most urgent thing is to focus on talent cultivation and popularize programs. In addition to promoting and cultivating language ability, it is required to increase international mobility and internship experience in English for the competitiveness of teachers and students. The current environment for the schools is not appropriate for bilingual education. In English, students may not learn professional knowledge and skills properly. If there is no appropriate teaching method, bilingual education only brings negative effects on students. However, the final goal of bilingual education in technical and vocational schools is to improve professional work skills and English ability so that students can have better capabilities. Thus, continuous efforts to teach such skills and abilities help overcome the lack of an appropriate environment for bilingual education.

Technical and vocational education in Taiwan has been practiced and implemented for many years. Such education provides the training of practical skills, which have global competitiveness. Numerous high-level technical talents have been raised and contributed to the success and outcome of Taiwan. According to the Blueprint, the implementation of bilingual education pressures teachers and students and affects the quality of the education of technical and vocational schools. However, since bilingual education is especially beneficial for students, the educational system must be improved for appropriate bilingual education.

Author Contributions: Conceptualization, C.-L.W. and M.-Y.H.; methodology, C.-Y.D. and M.-Y.H.; validation, J.-C.M.L. and C.-Y.D.; formal analysis, M.-Y.H.; investigation, C.-Y.D.; resources, J.-C.M.L. and C.-L.W.; writing—original draft preparation, M.-Y.H. and C.-L.W.; writing—review and editing, M.-Y.H.; visualization, C.-Y.D. and C.-Y.D.; supervision, M.-Y.H.; project administration, M.-Y.H.; funding acquisition, and C.-L.W. All authors have read and agreed to the published version of the manuscript.

Funding: This research was supported by the research supporting research projects (MOST 110-2420-H-002-003-MY3-Y11209) of the Ministry of Science and Technology and (NTCU111103) from the National Taichung University of Education.

Institutional Review Board Statement: This research did not execute the questionnaires and interview and hence, this research did not require ethical approval.

Informed Consent Statement: Not applicable because this research did not involve any humans.

Data Availability Statement: No new data were created.

Conflicts of Interest: The authors declare no conflict of interest.

References

1. Azzam, Z. Dubai's private K-12 education sector: In search of bilingual education. *J. Res. Int. Educ.* **2019**, *18*, 227. [CrossRef]
2. Beardsmore, H.B. European models of bilingual education: Practice, theory, and development. *J. Multiling. Multicult. Dev.* **1993**, *14*, 103–120. [CrossRef]
3. De Wilde, V.; Brysbaert, M.; Eyckmans, J. Learning English through Out-of-School Exposure: Which Levels of Language Proficiency are Attained and Which Types of Input are Important? *Biling. Lang. Cogn.* **2020**, *23*, 171–185. [CrossRef]
4. Babino, A.; Stewart, M.A. Remodeling dual language programs: Teachers enact agency as critically conscious language policy makers. *Biling. Res. J.* **2018**, *41*, 272–297. [CrossRef]
5. Grcia, E.E.; Curry-Rodríguez, J.E. The education of limited English proficient students in California schools: An assessment of the influence of Proposition 227 in selected districts and schools. *Biling. Res. J.* **2000**, *24*, 15–35. [CrossRef]
6. Gurney, L.; Liyanage, I. EAL teacher agency: Implications for participation in professional development. *Int. J. Pedagog. Learn.* **2016**, *11*, 49–59. [CrossRef]
7. Haneda, M.; Sherman, B. ESL teachers' acting agentively through job crafting. *J. Lang. Identity Educ.* **2018**, *17*, 402–415. [CrossRef]
8. Katz, S.R. Does NCLB leave the US behind in bilingual teacher education? *Engl. Educ.* **2004**, *36*, 141–152.
9. Leijen, Ä.; Pedaste, M.; Lepp, L. Teacher agency following the ecological model: How it is achieved and how it could be strengthened by different types of reflection. *Br. J. Educ. Stud.* **2000**, *68*, 295–310. [CrossRef]
10. Palmer, D.; Henderson, K.; Wall, D.; Zúñiga, C.E.; Berthelsen, S. Team teaching among mixed messages: Implementing two-way dual language bilingual education at third grade in Texas. *Lang. Policy* **2016**, *15*, 393–413. [CrossRef]
11. Wong, J.W.; Athanases, S.Z.; Banes, L.C. Developing as an agentive bilingual teacher: Self-reflexive and student-learning inquiry as teacher education resources. *Int. J. Biling. Educ. Biling.* **2020**, *23*, 153. [CrossRef]
12. Duran, M.; Usak, M.; Hsieh, M.-Y.; Uygun, H. A New Perspective on Pedagogical Content Knowledge: Intellectual and Emotional Characteristics of Science Teachers. *Rev. Cercet. Interv. Soc.* **2021**, *72*, 9. [CrossRef]
13. Wu, T.-L.; Hsieh, M.-Y.; Min, K.-W.; Yu, M.-T.; Ho, C.-T. Use of Sensor Technologies in Online Courses in Post-COVID-19 Era. *Sens. Mater.* **2021**, *33*, 2045–2062.
14. Usak, M.; Hsieh, Y.M.; Chan, Y.-K. A Concertizing Research on Making Higher Education Sustainability Count. *Sustainability* **2021**, *13*, 2724. [CrossRef]
15. Huang, C.C.; Chan, Y.-K.; Hsieh, M.Y. The Determinants of ESG for Community LOHASism Sustainable Development Strategy. *Sustainability* **2022**, *14*, 11429. [CrossRef]
16. Chan, Y.-K.; Hsieh, M.Y. An Empirical Study on Higher Education C-ESG Sustainable Development Strategy in Lower-Birth-Rate Era. *Sustainability* **2022**, *14*, 12629. [CrossRef]
17. Hsieh, Y.-M. Online learning era: Exploring the most decisive determinants of MOOCs in Taiwanese higher education. *Eurasia J. Math. Sci. Technol. Educ.* **2016**, *12*, 1163. [CrossRef]
18. Hsieh, Y.-M. Employing MCDM methodology to verify correlationship between social media and service quality in the dynamic m-commerce era. *J. Internet Technol.* **2018**, *19*, 225.
19. Hsieh, Y.-M.; Usak, M. High Education Radical Transformation Era: How Teachers' Competency can Enhance the Students' Employability. *Rev. Cercet. Interv. Soc.* **2020**, *68*, 95. [CrossRef]
20. Huang, Y.-M.; Hsieh, Y.M. An Interdisciplinary Research on Students' Employability in Technology Education to Advance Higher Education Enrollment Sustainability. *Sustainability* **2020**, *12*, 1806. [CrossRef]
21. Hsieh, M.Y. The Sustainable Development and Strategic Approaches for Contemporary Higher Education. *Sustainability* **2022**, *14*, 12925. [CrossRef]
22. Huang, C.-C.; Chan, Y.-K.; Hsieh, M.-Y. Preliminary Research on the Sustainable Determinants of Taiwanese Ecotourism with the International Standards. *Int. J. Environ. Res. Public Health* **2022**, *19*, 14489. [CrossRef] [PubMed]

Disclaimer/Publisher's Note: The statements, opinions and data contained in all publications are solely those of the individual author(s) and contributor(s) and not of MDPI and/or the editor(s). MDPI and/or the editor(s) disclaim responsibility for any injury to people or property resulting from any ideas, methods, instructions or products referred to in the content.

engineering proceedings

MDPI

Proceeding Paper

Digital and Traditional Learning: Learning Styles with Music and Technology for Early Childhood Education †

Liza Lee [1] and Chia-Hui Lin [2,*]

1 College of Humanities and Social Sciences, Chaoyang University of Technology, Taichung 413, Taiwan; lylee@gm.cyut.edu.tw
2 Department of Early Childhood Development and Education, Chaoyang University of Technology, Taichung 413, Taiwan
* Correspondence: t2008030@cyut.edu.tw

† Presented at the 3rd IEEE International Conference on Electronic Communications, Internet of Things and Big Data Conference 2023, Taichung, Taiwan, 14–16 April 2023.

Abstract: A non-experimental, a quantitative SPSS research design (independent sample *t*-test, two-way ANOVA, multiple regression analysis) was used to examine the statistically significant difference existing among Taiwan college early childhood education students for the students with blended learning, background demographic characteristics, music emotion, learning styles, technology acceptance, and students' satisfaction. The purpose of this quantitative research study is to examine the statistically significant difference existing among Taiwan college early childhood education students for students with blended learning, background demographic characteristics, music emotion, learning styles, technology acceptance, and students' satisfaction.

Keywords: music emotion; learning styles; technology acceptance; AI; student's satisfaction

Citation: Lee, L.; Lin, C.-H. Digital and Traditional Learning: Learning Styles with Music and Technology for Early Childhood Education. *Eng. Proc.* **2023**, *38*, 19. https://doi.org/10.3390/engproc2023038019

Academic Editors: Teen-Hang Meen, Hsin-Hung Lin and Cheng-Fu Yang

Published: 21 June 2023

Copyright: © 2023 by the authors. Licensee MDPI, Basel, Switzerland. This article is an open access article distributed under the terms and conditions of the Creative Commons Attribution (CC BY) license (https://creativecommons.org/licenses/by/4.0/).

1. Introduction

Early childhood education is defined as any type of educational program that takes care of children birth to five years of age. Ogunnaike [1] proposed that "Early Childhood Education theories provide a framework for understanding the nature, abilities, and how to create learning environments that enhance children's overall development". In 1939, Taiwan Ministry of Education mandated that the kindergarten curriculum had to include music, general knowledge, children's songs, stories, and games [2]. In early childhood education and care (ECEC), music education provides small children with an interest in creating a relationship with music, including both pre-planned and spontaneous activities [3,4]. Basic music elements experiences include listening to music, singing, moving, rhyming, creating music, playing instruments, and body percussion. Music education has a strong positive effect on young children learning; in particular, music enhances spatial perception and auditory skills [5,6]. Music develops creativity, as well as emotional skills [7,8].

In the 21st century, technology has been playing an essential role in our life. However, researchers are still discussing the positive and negative impact it has on early childhood development. Several researchers believe technology may isolate children in their early childhood and make them anti-social. In another viewpoint, some researchers believe technology can support young children to work collaboratively and to become social people [9]. Kirkwood et al. [10] and Kazakoff and Bers [11] proposed that technology can improve pupils' social interactions and benefit their understanding during their learning process. Nowadays, many family favorite pastimes include digital computers, smartphones, and social media [12,13]. In addition, young children are growing up in a digital home environment. Thus, digital musical parenting is a new way for parents to take care of their children [14,15] for young children's exposure to a variety of music learning environments later in life [16].

Anderkin [17] explained that "digital technologies extend possibilities for preschools' participation and engagement in the meaning-making process while simultaneously allowing them to participate in various types of play". Researchers have also indicated that children's interaction with technology allows the development of social and cognitive abilities and student engagement [18,19]. White [20] proposed that early childhood education have the challenge of blending technology into current pedagogy to make sure the technology practices are appropriate. Technology is beneficial for early childhood learners with appropriate development technology practices.

Since the 1960s, there has been an issue with the technology that is appropriate for young children to apply it in early childhood classrooms [16]. However, there are many technology tools allowing pupils to learn from tablet PCs, televisions, digital cameras, laptops, and programmable toys [21]. Even a two-year-old child can naturally touch a screen with a new toy [22]. Digital musical tools, such as smartphones and tablet PCs, have become important parts of young children's musical lives and have been utilized by many parents [23]. "Music can let students experience the expression of emotion" [24]. Music education has a strong positive effect on young children's learning; in particular, music enhances spatial perception and auditory skills [5,25]. Music develops creativity, as well as emotional skills [6]. In addition, Marjanen [26] mentioned that music education provides a strong connection for mothers and babies during their musical interaction and communication.

2. Early Childhood Music and Technology

2.1. Music Emotion

Davidson et al. [27] defined emotion as "a relatively brief episode of the coordinated brain, autonomic, and behavioral changes that facilitate a response to an external or internal event of significance" in an organism. Music is just like a language that communicates emotion to people, plants, and animals with visual perception in our daily life. In addition, music is another rich source of human emotion in which music emotion can be expressed through musical instrument play and lyrics. Huang et al. [28] presented nine basic emotions driven by psycho-emotional adjectives as shown in Figure 1. The whole emotional space is categorized into nine separate emotions: happy, nervous, neutral, delicate, bored, sad, serene, angry, and vigorous.

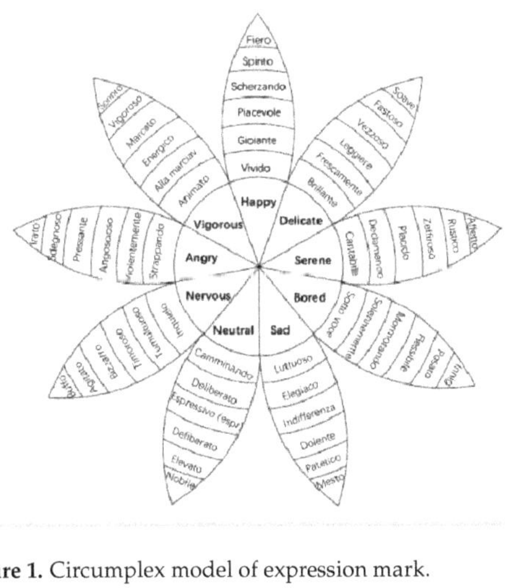

Figure 1. Circumplex model of expression mark.

Fernandez et al. [29] investigated the impact of music that evoked different emotions, such as joy, tenderness, sadness, tension, and so on. Balkwill and Thompson [30] proposed a model of music emotion and discussed that emotion was mediated by psychophysical cues of music or culture-specific cues. The BRECVEMA framework of emotion in music presents the social construction of stereotyping. Music affects emotion. Music has been found to evoke emotions that allow individuals to "value music primarily because the emotion evokes" [31,32]. People use music to change, release, and match their emotions and to release stress [33].

2.2. Blended Learning

Blended learning becomes one of the most accepted learning modes in which learners can learn from digital media and the traditional classroom method [34]. Garrison and Vaughan [35] defined blended learning as "the organic integration of thoughtfully selected and complementary face-to-face and online approaches and technologies". Motteram and Sharma [36] defined blended learning as "the integrated combination of traditional learning with web-based online approaches". Blended learning is characterized by three features: (1) personal contact face-to-face with an instructor; (2) electronic items delivered learning objects; and (3) a blend of traditional and electronic items to achieve learning goals [37]. Besides the physical learning space, learners also can learn from the virtual space with much flexible time and achievable and progressive resources, including Google Meet, Zoom, and Microsoft Teams. In addition, online and face-to-face interactions in the blended course can achieve higher levels of learning and academic achievement [38,39].

Graham [38] depicted the four critical dimensions of interactions in face-to-face and online learning environments in a blended course (Figure 2). According to the dimension of *space*, blended learning mixes the face-to-face and virtual interactions learning methods together. The second *time* dimension for the traditional face-to-face learning environment is live synchronous learning with a limited and shorter time lag than online learning. Additionally, the third *fidelity* dimension of interactions in face-to-face interaction has higher fidelity than online learning. In the final *humanness* dimension, face-to-face interaction has a higher level of humanness than computer interaction learning.

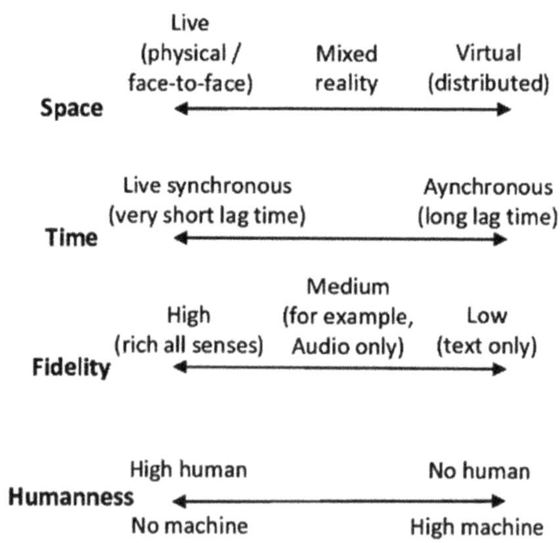

Figure 2. Four dimensions of interaction in face-to face and distributed learning environments.

In the past, people have regarded the two learning methods separately as traditional face-to-face learning and online computer self-paced learning. However, at present, learners have started to mix both methods. In the future, learners are going to apply the major blended learning method (Figure 3).

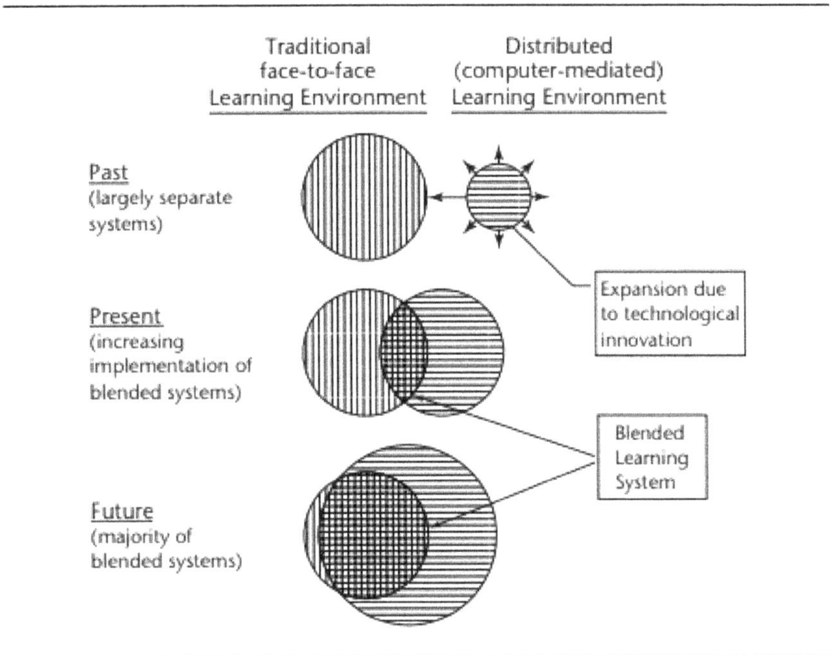

Figure 3. Progressive convergence of traditional F2F and distributed environments allowing development of blended learning systems.

The benefits of blended learning include a personalization/individual learning path, learner-centered approach, self-paced learning, flexibility, enhancement of learners' motivation and engagement improvement in learning outcomes/performance, critical thinking skills, and one-on-one tutoring [40]. In addition, online teaching and learning in Taiwan are both synchronous and asynchronous. In synchronous learning, students can carry out the activities through live video or audio with immediate feedback in real time. Learning can occur individually or in groups or in different places. In asynchronous learning, students can complete their assignments in their own time and speed [41,42]. Online learning occurs in one of two modes, namely asynchronous or synchronous, and is considered to be the safest mode [43]. Students might perceive the transition from virtual learning to remote learning because the lack of in-person learning in the physical classroom may cause a variety of students' emotions.

2.3. Learning Styles (Visual, Auditory, Tactile)

Kolb [44] defined the inherent disposition of a person to the dimensions and continuums of experiential learning, which combined learning preferences. Keefe [45] defined learning styles as affective, psychological characteristics and took a cognitive approach to figure out how learners interacted, saw, and responded to the environment.

Kolb [44] explained the learning style inventory, which includes feeling, watching, thinking, and doing (Figure 4).

Felder and Silverman [46] divided learning styles into four dimensions: perception, input, processing, and understanding. Each dimension has two learning styles. Process (active, reflective) and perception (sensing, intuitive) are from Myers–Briggs, and the Kolb model details input (visual, verbal) and understanding (sequential, global) (Figure 5).

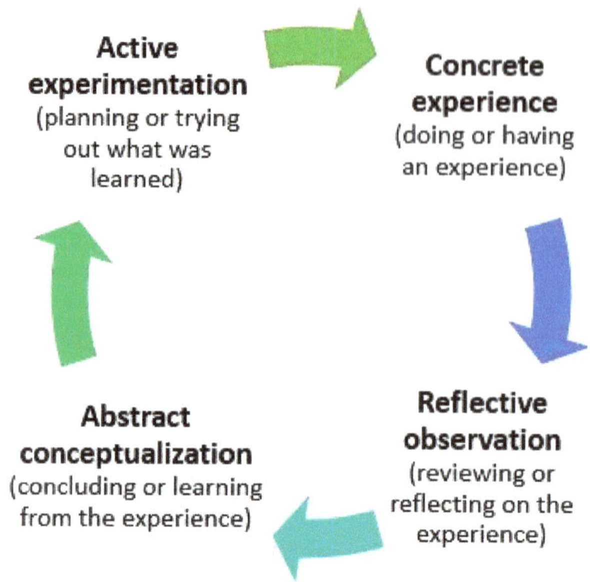

Figure 4. The Experiential Learning Model, Kolb [44].

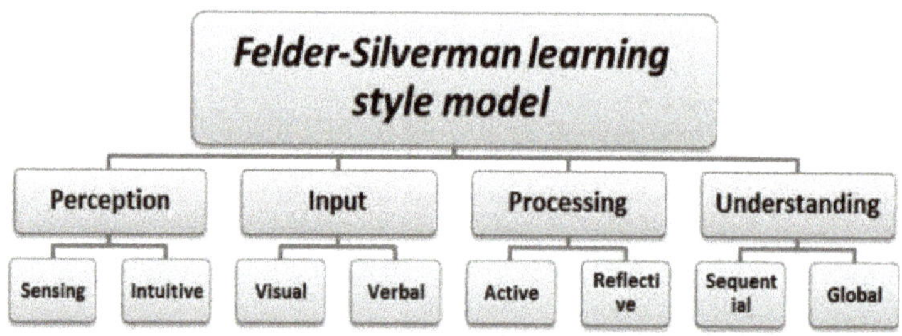

Figure 5. Dimensions of Felder–Silverman's Learning Style Model Dimension Learning.

2.4. Technology in Music Classroom

Technology in the music classroom is popular and important for music teachers to guide young children in learning in the 21st century [47]. Sandy and Murtiyasa [48] have mentioned that multimedia-based learning media improve educational quality and support teaching and learning activities. Burnard [49] defined music technology as the development of music educators using technology that allow students to be independent learners of musical instruments and the creativity of music: "Technology provides ideal media for music education". Students have been influenced by using technology and music in all kinds of settings. Kuzmich [50] stated "With technology and a deliberate plan that you can manage effectively, it is possible to engage students in music making and accomplish your music education goals". In addition, Kuzmich analyzed four factors for music education with technology: achievable goals, methods for immediate personal feedback, expert instruction, and tools focusing on student practice. Music educators have described success in teaching students with the aid of technology. Kuzmich reported the biggest advantage of the technology was the ability to assign different music levels to the students. Owen [51] described technology as bringing confidence to the students to motivate and engage them in new ways. Wang [52] mentioned that music teaching with technology has supported teachers' teaching in the classroom in innovating their teaching methods and enriching learning resources. Music technology can improve teaching and learning efficiency and quality compared to learning music from traditional equipment. Chen [53] and Wang [54] mentioned that music technology teaching improved the shortcoming of traditional music teaching and motivated students' learning enthusiasm, who pay more attention, while also improving the teachers' teaching quality.

According to Fishbein and Ajzen [55], the technology acceptance model (TAM) was developed by Davis in 1985 with the main purpose of providing external variables on the attitude, behavioral intention to use, and actual system use of technology. TAM was used for a multitude of technologies [56,57]. Perceived usefulness (PU) and perceived ease of use (PEOU) are the main two components that influence technology acceptance behavior (Figure 6).

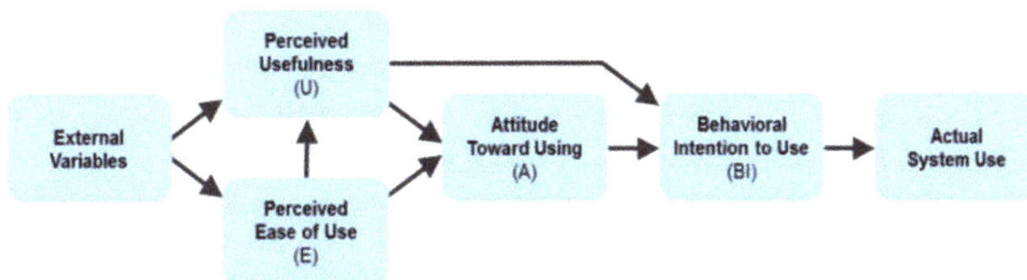

Figure 6. Technology Acceptance Model (TAM) (Davis, 1989).

Artificial intelligence (AI) technology has developed dramatically over the last five years, but AI learning for music still needs to be improved. AI technology can support teachers in preschool music teaching and preschool children in learning in an innovative way, as well as stimulate the students' motivation of learning [58]. AI technology's main purpose is to reflect human beings and technology interaction. Teachers are no longer the only source of knowledge to communicate with the students. AI music technology learning can improve the student's self-learning ability and reduce the time spent directly learning knowledge from the teachers. In addition, it can stimulate students' imagination and appreciation of music [25]. Electronic keyboard instruments are becoming more intelligent recently. According to the student's needs, they can arrange different timbres and melodies at flexible times and locations [59].

3. Hypothesis Development and Methodology

3.1. Research Question and Hypothesis

3.1.1. Research Question

Are there any differences in blended learning, music emotion, learning styles, technology acceptance, and students' satisfaction for being fond of playing instruments or not? (Figure 7).

3.1.2. Hypothesis

H1: *Blended learning and learning styles (visual, auditory, tactile) are statistically significant for students' satisfaction (interaction, structure, support).*

H2: *Music emotion and learning styles (visual, auditory, tactile) are statistically significant for students' satisfaction (interaction, structure, support).*

H3: *Technology acceptance and learning styles (visual, auditory, tactile) are statistically significant for students' satisfaction (interaction, structure, support).*

H4: *Listening to music has significant perceived explanatory variables for blended learning, music emotion, learning styles, technology acceptance, and students' satisfaction.*

H5: *Playing instruments have significant perceived explanatory variables for blended learning, music emotion, learning styles, technology acceptance, and students' satisfaction.*

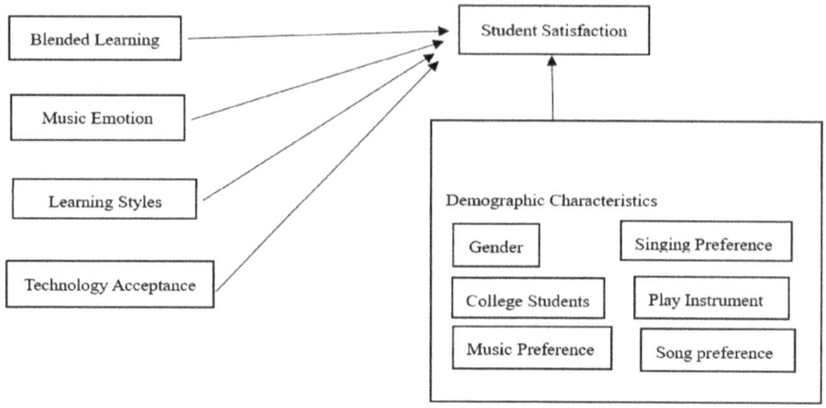

Figure 7. Hypothesized model of background characteristics, blended learning, music emotion, learning styles, technology acceptance, and student satisfaction.

4. Result

4.1. Research Question 1: Independence t-Test

Levene's test for the equality of variances in Table 1 indicates that the variables of 'students who like to play the instruments or not' for blended learning (p = 0.015), music emotion (p = 0.189), learning styles (p = 0.000), technology acceptance (p = 0.036), and students' satisfaction (p = 0.023). There are significant differences in learning styles for students in favor of playing instruments or not. Other variables did not differ significantly.

Table 1. Independent sample *t*-test for all variables of students who like playing instruments or not.

	F	Sig.	t	df	Sig. (Two-Tailed)
Blended Learning	2.624	0.106	2.437 2.377	292 241.601	0.015 0.018
Music Emotion	1.824	0.178	1.316 1.328	292 277.742	0.189 0.185
Learning Styles	2.793	0.096	3.578 3.527	292 253.723	0.000 0.000
Technology Acceptance	0.034	0.853	2.109 2.115	292 272.276	0.036 0.035
Student Satisfaction	1.713	0.192	2.286 2.243	292 248.270	0.023 0.026

4.2. Hypothesis 1: Two-Way ANOVA Analysis

The value of significance (p = 0.000) in Table 2 indicates statistical significance. Therefore, research hypothesis 1 (blended learning and learning styles are statistically significant for students' satisfaction) was supported.

4.3. Hypothesis 2: Two-Way ANOVA Analysis

The value of significance (p = 0.000) in Table 3 indicates statistical significance. Therefore, research hypothesis 2 (music emotion and learning styles are statistically significant for students' satisfaction) was supported.

Table 2. Two-way ANOVA of Blended Learning, Learning Style, and Students' Satisfaction.

Source	Type III Sum of Squares	df	Mean Square	F	Sig.
Corrected Model	41.136	126	0.326	2.284	0.000
Intercept	962.793	1	962.793	6735.689	0.000
Blended Learning	6.061	14	0.433	3.029	0.000
Learning Styles	8.248	23	0.359	2.509	0.000
Blended learning Learning styles	10.789	89	0.121	0.848	0.805

Table 3. Two-way ANOVA of Music Emotion, Learning Style, and Students' Satisfaction.

Source	Type III Sum of Squares	df	Mean Square	F	Sig.
Corrected Model	59.174	235	0.252	2.504	0.000
Intercept	1668.748	1	1668.748	16,595.412	0.000
Music Emotion	17.357	82	0.212	2.105	0.002
Learning Styles	8.478	23	0.369	3.666	0.000
Music Emotion Learning Styles	17.369	130	0.134	1.329	0.112

4.4. *Hypothesis 3: Two-Way ANOVA Analysis*

The value of significance ($p = 0.000$) in Table 4 indicates statistical significance. Research hypothesis 3 (technology acceptance and learning styles are statistically significant for students' satisfaction) was supported.

Table 4. Two-way ANOVA of Technology acceptance, Learning Style, and Students' Satisfaction.

Source	Type III Sum of Squares	df	Mean Square	F	Sig.
Corrected Model	46.145	122	0.378	3.429	0.000
Intercept	983.089	1	983.089	8913.021	0.000
Technology Acceptance	7.094	13	0.546	4.948	0.000
Learning Styles	4.759	23	0.207	1.876	0.013
Technology Acceptance Learning Styles	11.461	86	0.133	1.208	0.149

4.5. *Hypothesis 4: Multiple Regression Analysis*

The value of significance ($p = 0.140$) in Table 5 indicates there is no statistical significance. Therefore, research hypothesis 4 (listening to music has significant perceived explanatory variables for blended learning, music emotion, learning styles, technology acceptance, and students' satisfaction) was not supported.

Table 5. Multiple Regression R-Squared Analyses of Students Who Like to Listen to Music, Blended Learning, Music Emotion, Learning Styles, Technology Acceptance, and Student Satisfaction.

Model	Sum of Squares	df	Mean Square	F	Sig.
Regression	0.166	5	0.033	1.678	0.140
Residual	5.711	288	0.020		
Total	5.878	293			

4.6. *Hypothesis 5: Multiple Regression Analysis*

The value of significance ($p = 0.140$) in Table 6 indicates there is no statistical significance. Therefore, research hypothesis 5 (Playing instruments has significant perceived explanatory variables for blended learning, music emotion, learning styles, technology acceptance and students' satisfaction) was not supported.

Table 6. Multiple Regression R-Squared Analyses of Playing Instruments, Blended Learning, Music Emotion, Learning Styles, Technology Acceptance, and Students' Satisfaction.

Model		Sum of Squares	df	Mean Square	F	Sig.
	Regression	3.461	5	0.692	2.909	0.014
1	Residual	68.539	288	0.238		
	Total	72.000	293			

4.7. *Reliability Anslysis*

Table 7 indicates that Cronbach's alpha for internal consistency of all variables a = 0.876 was an acceptable value of reliability. All of them were more than 0.70; therefore, internal consistency was satisfactory.

Table 7. Reliability Statistics for All Variables.

Cronbach's Alpha	Cronbach's Alpha Based on Standardized Items	N of Items
0.876	0.879	57

4.8. *Factor Analysis and Construct Valisity*

Table 8 shows the results of KMO and Bartlett's test of sphericity. The value of KMO for blended learning was 0.785, and there was a statistical significance of 0.000.

Table 8. KMO and Bartlett's Test Results on Blended Learning, Music Emotion, Learning Styles, Technology Acceptance, and Students' Satisfaction.

Kaiser–Meyer–Olkin Measure of Sampling Adequacy		0.785
Bartlett's Test of Sphericity	Approx. Chi-Squared	550.467
	df	10
	Sig.	0.000

Table 9 shows that factor values were larger than 1 after varimax rotation was extracted, which accounted for approximately 58% of the total variance.

Table 9. Extraction Sums of Squared Loadings on All Variables.

Component	Initial Eigenvalues			Extraction Sums of Squared Loadings		
	Total	% of Variance	Cumulative %	Total	% of Variance	Cumulative %
	1	2.902	58.037	58.037	2.902	58.037
2	0.883	17.657	75.694			58.037
3	0.549	10.971	86.666			
4	0.398	7.957	94.622			
5	0.269	5.378	100.000			

5. Conclusions

Research question 1: The result shows that the value of significance (p = 0.000) for learning styles and other variables did not differ significantly for whether students like to play instruments or not.

'Research Hypothesis 1: Blended learning and learning styles are statistically significant for students' satisfaction' was supported.

'Research Hypothesis 2: Music emotion and learning styles are statistically significant for students' satisfaction' was supported.

'Research Hypothesis 3: Technology acceptance and learning styles are statistically significant for students' satisfaction' was supported.

'Research Hypothesis 4: Listening to music has significant perceived explanatory variables for blended learning, music emotion, learning styles, technology acceptance, and students' satisfaction' was not supported.

'Research Hypothesis 5: Playing instruments has significant perceived explanatory variables for blended learning, music emotion, learning styles, technology acceptance, and students' satisfaction' was not supported.

6. Limitation and Future Study

The participants were only from two colleges. When the researchers were conducting the research and data collection during the pandemic for 2 years, we encountered difficulties in data collection. Colleges would not support completing the questionnaires because many students took remote courses and the return rate was low. Furthermore, the distribution in these groups was irregular. The case was limited to a small portion of educational research. While still valid, the results may be overgeneralized. In addition, we only tested undergraduate students. In future studies, we will compare the difference between public and private students. In addition, we will try to focus on specific musical instruments instead of general musical instruments, such as pianos, tambourines, recorders, ukuleles, and others. We will try to enlarge the number of universities and science and technology universities. Furthermore, we will try to test types of technology. More diverse ways will be tested to conduct the research.

Author Contributions: L.L. was in charge of the project administration, supervision, methodology survey questionnaire, and data collection. Corresponding author C.-H.L. handled the writing, paper structure, and result analysis. All authors have read and agreed to the published version of the manuscript.

Funding: This research received no external funding.

Institutional Review Board Statement: Not applicable.

Informed Consent Statement: Informed consent was obtained from all subjects involved in the study.

Data Availability Statement: Not applicable.

Conflicts of Interest: The authors declare no conflict of interest.

References

1. Ogunnaike, Y.A. Early childhood education and human factor: Connecting theories and perspectives. *Hum. Factor Stud.* **2015**, *21*, 9–26.
2. Son, H.L. *Kindergarten Education*, 2nd ed.; Taipei City Teachers College: Taipei, Taiwan, 1976.
3. FNAE. *National Core Curriculum for Early Childhood Education and Care 2018*; Finnish National Agency for Education: Helsinki, Finland, 2018.
4. Ruokone, I.; Tervaniemi, M.; Reunamo, C.J. The significance of music in early childhood education and care of toddlers in Finland: An extensive observational study. *Music Educ. Res.* **2021**, *23*, 634–646. [CrossRef]
5. Virtala, P. The Neural Basis of Western Music Chord Categorisations—Effects on Development and Music Expertise. Ph.D. Thesis, Institute of Behavioural Sciences, University of Helsinki, Helsinki, Finland, 2015.
6. Gerry, D.; Unrau, A.; Trainor, L.J. Active Music Classes in Infancy Enhance Musical, Communicative and Social Development. *Dev. Sci.* **2012**, *15*, 398–407. [CrossRef] [PubMed]
7. Muhonen, S. Songcrafting: A Teacher's Perspective of Collaborative Inquiry and Creation of Classroom Practice. *Int. J. Music Educ.* **2014**, *32*, 185–202. [CrossRef]
8. Ozel, O. An Exploration of Turkish Kindergarten Early Career Stage Teachers' Technology Beliefs and Practices. Ph.D. Thesis, University of Missouri, Colombia, MO, USA, 2019.
9. Ko, M.Y. YouTube as an Early Childhood Music Education Resource: Parental Attitudes, Beliefs, Usages, and Experiences. Ph.D. Thesis, University of Missouri, Colombia, MO, USA, 2018.
10. Kirkwood, D.; Shulsky, D.; Willis, J. Beyond pinatas, fortune cookies, and wooden shoes: Using the world wide web to help children explore the whole wide world. *Child. Educ.* **2014**, *90*, 11–19. [CrossRef]
11. Kazakoff, E.; Bers, M. Programming in a robotics context in the kindergarten classroom: The impact on sequencing skills. *J. Educ. Multimed. Hypermedia* **2012**, *21*, 371–391.

12. Rideout, V. The Common Sense Media Census: Media Use by Kids Age Zero to Eight. 2017. Available online: http://static1.1.sqspcdn.com/static/f/1083077/27723826/1508433603540/CSM_ZeroToEight_Full+Report.FINAL.2017.1.pdf?token=6XzqNxlxaFwyzQ3IB%2BSUeFKimk8%3D (accessed on 19 October 2017).
13. Wartella, E.; Rideout, V.; Lauricella, A.R.; Connell, S.L. Parenting in the Age of Digital Technology. 2013. Available online: https://cmhd.northwestern.edu/wpcontent/uploads/2015/06/ParentingAgeDigitalTechnology (accessed on 4 June 2013).
14. Palaiologou, I. Children under five and digital technologies: Implications for early years pedagogy. *Eur. Early Child. Educ. Res. J.* **2014**, *24*, 5–24. [CrossRef]
15. Gordon, E. *A Music Learning Theory for Newborn and Young Children*; GIA Publications: Chicago, IL, USA, 2003.
16. Mooney, C.G. *Theories of Childhood: An introduction to Dewey, Montessori, Erikson, Piaget & Vygotsky*, 2nd ed.; Redleaf Press: St Paul, MN, USA, 2013.
17. Anderkin, D.M. An Analysis of Technology Integration within a Play-Based Learning Environment at the Preschool Level. Master's Thesis, Cedarville University, Cedarville, OH, USA, 2015. Available online: http://digitalcommons.cedarville.edu/education_theses/7 (accessed on 15 August 2015).
18. Jorgensen, R.L.; Logan, H. Computer use by preschool children: Rethinking practices as digital natives come to preschool. *Aust. J. Early Child.* **2015**, *33*, 37–44.
19. Cameron, A.L.C. Opening Doors: A Collective Case Study of Integrating Technology in the Preschool through 3rd Grade Classroom in a Developmentally Appropriate Way. Ph.D. Thesis, Pepperdine University, Malibu, CA, USA, 2015.
20. White, H.S. Early Childhood Teacher Perspectives of Developmentally Appropriate Use of Computer Applications. Ph.D. Thesis, Walden University, Minneapolis, MN, USA, 2015.
21. Billington, C. *How Digital Technology Can Support Early Language and Literacy Outcomes in Early Years Settings: A Review of the Literature*; National Literacy Trust: London, UK, 2016.
22. Sharkins, K.A.; Newton, A.B.; Albaiz, N.E.A.; Ernest, J.M. Preschool children's exposure to media, technology, and screen time: Perspectives of caregivers from three early childcare settings. *Early Child. Educ. J.* **2016**, *44*, 437–444. [CrossRef]
23. Marsh, J.; Brooks, G.; Hughes, J.; Ritchie, L.; Roberts, S.; Wright, K. *Digital Beginnings: Young Children's Use of Popular Culture, Media and New Technologies*; University of Sheffield: Sheffield, UK, 2005.
24. Huang, C.; Yu, K. Research on the innovation of college music teaching mode based on Artificial Intelligence. *J. Phys. Conf. Ser.* **2021**, *1915*, 022051. [CrossRef]
25. Putkinen, V.; Tervaniemi, M.; Saarikivi, K.; Ojala, P.; Huotilainen, M. Enhanced development of auditory change detection in musically trained school-aged children: A longitudinal event-related potential study. *Dev. Sci.* **2014**, *17*, 282–297. [CrossRef]
26. Marjanen, K. The Belly-Button Chord: Connections of Pre-and Postnatal Music Education with Early MotherChild Interaction. Ph.D. Thesis, Faculty of Humanities of the Universities of Jyväskylä, Jyväskylä, Finland, 2009.
27. Davidson, R.; Scherer, K.; Goldsmith, H. *The Handbook of Affective Sciences*; Oxford University Press: Oxford, UK, 2003.
28. Huang, S.; Zhou, L.; Liu, Z.; Ni, S.; He, J. Empirical research on a fuzzy model of music emotion classification based on pleasurearousal model. In Proceedings of the 2018 37th Chinese Control Conference, Wuhan, China, 25–27 July 2018; pp. 3239–3244.
29. Fernandez, N.B.; Hars, M.; Trombetti, A.; Vuilleumier, P. Age-related changes in attention control and their relationship with gait performance in older adults with high risk of falls. *Neuroimage* **2019**, *189*, 551–559. [CrossRef]
30. Balkwill, L.L.; Thompson, W.F. A cross-cultural investigation of the perception of emotion in music: Psychophysical and cultural cues. *Music Percept.* **1999**, *17*, 43–64. [CrossRef]
31. Juslin, P.N.; Vastfjall, D. Emotional responses to music. The need to consider underlying mechanism. *Behav. Brain Sci.* **2008**, *31*, 559. [CrossRef]
32. Sloboda, J.A. Music structure and emotional response: Some empirical findings. *Psychol. Music* **1991**, *19*, 110–120. [CrossRef]
33. Behne, K.E. *The Development of "Musikerleben" in Adolescence*; Psychology Press: East Sussex, UK, 1997; pp. 143–159.
34. Bonk, C.J.; Graham, C.R. *The Handbook of Blended Learning: Global Perspectives, Local Designs*; Pfieffer: Aßlar, Germany, 2013.
35. Garrison, D.R.; Vaughans, N.D. *Blended Learning in Higher Education: Framework, Principles, and Guidelines*; John Wiley & Sons Inc.: Hoboken, NJ, USA, 2008.
36. Motteram, G.; Sharma, P. Blending learning in a web 2.0 world. *Int. J. Emerg. Technol. Soc.* **2009**, *7*, 83–96.
37. Hoffman, J.; Miner, N. Real blended learning stands up. *Train. Dev.* **2008**, *62*, 28–31.
38. Graham, C.R. Blended Learning Systems: Definition, Current Trends, and Future Directions. In *Handbook of Blended Learning: Global Perspectives, Local Designs*; Bonk, C.J., Graham, C.R., Eds.; Pfeiffer Publishing: Zurich, Switzerland, 2006.
39. Garrison, D.R.; Vaughan, N. *Blended Learning in Higher Education*; Jossey-Bass: Francisco, CA, USA, 2008.
40. Frederique, B.; Ha, L.; Scott, M.; Clara, N.; Lilibeth, J. Defining student learning experience through blended learning. *Educ. Inf. Technol.* **2020**, *25*, 3049–3069.
41. Hrastinski, S. Asynchronous & Synchronous E-Learning. *Educ. Q.* **2008**, *31*, 51–55.
42. Divakaran, A. Enhancing Math learning using synchronous online teaching. *J. Soc. Sci.* **2015**, *14*, 27–36. [CrossRef]
43. Dada, E.G.; Alkali, A.H. An investigation into the effectiveness of asynchronous and synchronous e-learning mode's on students' academic performance in National Open University (NOUN), Maiduguri Centre. *Int. J. Mod. Educ. Comput. Sci.* **2019**, *5*, 54–64. [CrossRef]
44. Kolb, D. *Experiential Learning: Experience and the Source of Learning and Development*, 2nd ed.; Pearson Education: Harlow, UK, 2015.
45. Keefe, J.W. *Learning Style Theory and Practice*; NASS: Reston, VA, USA, 1987.

46. Felder, R.M.; Silverman, L.K.L. Learning and teaching styles in engineering education. *Eng. Educ.* **1988**, *78*, 674–681.
47. Summers, E.S. The Effects of Parental Involvement and Computer-Based Music Technology on Developing Independent Musicianship. Ph.D. Thesis, Concordia University, Irvine, CA, USA, 2018.
48. Sandy, R.J.; Murtiyasa, B. Developing a multimedia-based learning media for learning matrix transformation. *J. Phys. Conf. Ser.* **2019**, *1265*, 012020. [CrossRef]
49. Burnard, P. Reframing creativity and technology: Promoting pedagogic change in music education. *J. Music Technol. Educ.* **2008**, *1*, 37–55. [CrossRef]
50. Kuzmich, J. Technology is Essential to Accomplished Core Music Standards. *Sch. Band Orch.* **2015**, *6*, 48–50. Available online: http://www.kuzmich4.com/SBO%20Articles/August2015.pdf (accessed on 15 August 2015).
51. Owen, S.L. Student Perceptions of the Efficacy of Smart Music Practice Software. Master's Thesis, California State University, Long Beach, CA, USA, 2015.
52. Wang, H. The application of computer music production technology in college music teaching. *J. Qinghai Norm. Univ.* **2015**, *172*, 132–136.
53. Chen, J. Thoughts on the status quo of computer music in college music teaching. *Contemp. Music* **2016**, *17*, 2.
54. Wang, L. Application of computer music technology in college music teaching. *J. Phys. Conf. Ser.* **2021**, *1915*, 032073. [CrossRef]
55. Fishbein, M.; Ajzen, I. *Belief, Attitude, Intention and Behavior: An Introduction to Theory and Research*; Reading Addison-Wesley Publishing Co.: Boston, MA, USA, 1975.
56. Abrahim, S.; Mir, B.A.; Suhara, H.; Mohamed, F.A.; Sato, M. Structural equation modeling and confirmatory factor analysis of social media use and education. *Int. J. Educ. Technol. High. Educ.* **2019**, *16*, 32. [CrossRef]
57. Dumpit, D.Z.; Fernandez, C.J. Analysis of the use of social media in Higher Education Institutions (HEIs) using the Technology Acceptance Model. *Int. J. Educ. Technol. High. Educ.* **2017**, *14*, 5. [CrossRef]
58. Yu, L.; Ding, J. Application of Music Artificial Intelligence in Preschool Music Education. *IOP Conf. Ser. Mater. Sci. Eng.* **2020**, *750*, 012101. [CrossRef]
59. Lei, D. Research on network teaching of music major based on artificial intelligence technology. *J. Phys. Conf. Ser.* **2020**, *1648*, 042094. [CrossRef]

Disclaimer/Publisher's Note: The statements, opinions and data contained in all publications are solely those of the individual author(s) and contributor(s) and not of MDPI and/or the editor(s). MDPI and/or the editor(s) disclaim responsibility for any injury to people or property resulting from any ideas, methods, instructions or products referred to in the content.

engineering proceedings

MDPI

Proceeding Paper

Web-Based Technology: Trials and Tribulations of Using Online Quizzes in an EFL Course †

Philip Craigie

Department of Applied English, Chaoyang University of Technology, Taichung 413310, Taiwan; philcraigie@cyut.edu.tw

† Presented at the 3rd IEEE International Conference on Electronic Communications, Internet of Things and Big Data Conference 2023, Taichung, Taiwan, 14–16 April 2023.

Abstract: One of the tools to use in the ever-increasing implementation of the flipped classroom is the online quiz to monitor students' formative learning. However, in useful and convenient administering online quizzes, how students engage with them creates a paradox. This case study discusses four main components: the instructor's reasoning for using online quizzes, two different implementation approaches, how the students behaved whilst completing them, and reflecting on how closely the learning results compared to the original intentions. As a result, ways to keep the students on the intended learning path are proposed to limit divergence.

Keywords: web-based technology; online quizzes; flipped classroom; Bloom's Taxonomy; learning styles

Citation: Craigie, P. Web-Based Technology: Trials and Tribulations of Using Online Quizzes in an EFL Course. *Eng. Proc.* **2023**, *38*, 20. https://doi.org/10.3390/engproc2023038020

Academic Editors: Teen-Hang Meen, Hsin-Hung Lin and Cheng-Fu Yang

Published: 20 June 2023

Copyright: © 2023 by the author. Licensee MDPI, Basel, Switzerland. This article is an open access article distributed under the terms and conditions of the Creative Commons Attribution (CC BY) license (https://creativecommons.org/licenses/by/4.0/).

1. Introduction

The flipped classroom has become one of the emerging instructional pedological trends since Bergmann and Sams [1] wrote about how they tried to help their students. Although the initial premise was to provide the information taught in class to students who could not attend, Bergmann and Sams discovered that this information was also consumed by students who were attending class. As students were previewing the class material before starting class, the time spent in class could be spent on more useful activities, such as experiments and projects. Furthermore, teachers can provide more personalized one-on-one instruction.

Whilst other researchers examined the flipped classroom approach, comparisons were drawn to Bloom's Taxonomy, in particular to his cognitive thinking [2–4]. Bloom's Taxonomy, in the revised version, refers to six different levels of cognitive learning, comprising remembering, understanding, applying, analyzing, evaluating, and creating [5,6]. Administering online quizzes typically requires the student to use lower-order thinking, which applies to learning new information before class. In higher-order thinking skills, analyzing, evaluating, and creating are used with critical thinking activities during class. The benefits of this teaching approach are twofold. The first is that while students are learning the basic information, it is more fruitful for them to do so at their own pace and using their own preferred learning style rather than being tied to the teacher's style and pace [7,8]. The second is that during class time, the activities assigned allow students to explore the learning content more deeply and autonomously [9]. Thus, teachers are allowed to focus on those students who need more personalized instruction.

2. Literature Review

One of the key points to successfully implementing a flipped classroom is ensuring that all students have a similar understanding of the preliminary information at the start of class. One way to achieve this is to administer online quizzes, and there have generally been many benefits from this approach [10,11], including the convenience and flexibility when compared to paper quizzes, as well as the ability to provides prompts and immediate feedback to the students. Theoretically, implementing a standard quiz for the topic at hand to all students aims to ensure that a consistent minimum level of knowledge is understood before moving on to the more cognitively challenging tasks to be assigned in class.

Having established that online quizzes are generally an important checkpoint for students to move from lower-order thinking tasks to higher-order, the type of questions posted in online quizzes become essential to ensure that the correct understanding is measured. Clay [12] discussed the many considerations of writing and using online quizzes, including the types of questions and the considerations needed for them. The most common categories of questions used are single and multiple choice, true or false, matching items, fill-in-the-blank, and essay style questions that include short and long answers. With all types of online questions, a great deal of thought is needed for the language used to ensure that questions are not ambiguous and they measure the intended level of knowledge. With the majority of question types, except for long essay answers, each of the question types generally requires the students to use their lower-order thinking skills of remembering, understanding, and applying.

3. Implementation and Reflection of Online Quizzes

For this case study, the researcher's role as a lecturer in the Department of Applied English reflects on the same group of freshmen students taking an oral training course over two semesters. A set course textbook is used within eight units, each having a broad unit question, two listening tracks that discuss issues surrounding the unit question, as well numerous practice activities probing deeper meanings. These practice activities from the book are then used as online quiz questions intended to show that students have the prerequisite understanding of the topic before proceeding with critical thinking and expression activities. To address some of the issues that Clay discussed [12], the reliability and content must be of a high standard. The other attributes of online quizzes that need consideration are rules, including time limits, the ability to retake the quiz, showing incorrectly answered questions, as well as revealing correct answers.

This case study reflects on the freshmen's engagement and observed actions over two semesters for an oral training course. Each semester, the researcher used a different set of quiz-taking rules. For the autumn semester, students were immediately shown a score as well as the incorrect answers. Then, the students could go back into the quiz, revise the incorrect answers and resubmit. In the spring semester, this condition was changed, and students would only be allowed one attempt. Then, after the class finished, the score was revealed along with all answers.

4. Observations of Students' Behavior and Engagement

The following section discusses several issues surrounding each approach for online quizzes. For both semesters, it was observed that very few, if any students prepared by previewing the instructed content before class, despite instructions being provided numerous times including at the start of the semester, as well as closing the previous week's class.

4.1. Autumn Semester

The strategy for implementing this set of online quiz rules was to allow students to see the areas from the day's learning activities from the course textbook that they did not fully understand, allow them to search for and then revise their answers, then resubmit and hopefully see their improvement. The number of retries was unlimited and the quiz would close at the end of the class day. However, after a few weeks, the students quickly adopted their quiz-taking strategies. Rather than reviewing the book's information, the students would immediately attempt the quiz by guessing all of the answers. For the single-answer multiple-choice questions, the students would see which answers were incorrect. Then, they went back into the quiz and randomly guessed another answer before submitting again.

For the more-than-one-answer multiple-choice questions, it was observed that several students selected all choices. Thus, the online automatic marking function was deceived because all the correct answers were selected whilst not checking that all incorrect answers were not selected. Another approach that students adopted was for one of the smarter students to correctly guess all answers and then share the answers via an instant message app with the class. By the end of the semester, the students, unfortunately, appeared to have completed the quizzes by brute force rather than engaging and learning from the course textbook's content.

4.2. Spring Semester

The strategy for implementing this set of online quizzes was aimed to address the issues from the autumn semester while hoping to achieve the learning objectives. In the autumn semester, the first problem mentioned was that students were not engaging with the course textbook's content before attempting the online quiz. The class instruction changed to allocate time for the students to listen to the audio, write their answers in their book, and randomly call students to discuss their answers in front of the class, then the quiz was opened up with only one attempt allowed. The consequences of this approach were that students had to follow the instructor's pace and those students who were not called to speak could passively wait for the answers. From the autumn semester, the second problem mentioned regarding more-than-one answer multiple choice questions was manually reviewed, and if students selected all answers, then their respective scores would be altered to zero and feedback was provided not to do this again. The consequences of this were the loss of the auto-marking benefits, as well as more review work needed. From the autumn semester, the third problem mentioned about students sharing answers appeared. This was addressed by creating a class discussion about the content and online quiz questions before students were instructed to take them. Thus, it was hoped that it would negate the need for students to share answers via instant messaging apps.

4.3. Reflection on Setting the Rules around Online Quizzes

Over these two semesters, two different approaches were taken to administering online quizzes, and as a result, it affected the underlying teaching approach. It is worth noting that very few students, if any, did not preview before class. Hence, strictly adhering to the flipped classroom approach had to be adapted to be more like a blended learning approach. Furthermore, the intended benefits of using online quizzes to foster student-centered learning seemed to be lost, as well as the potential for deeper learning achievements.

One of the main goals of using online quizzes was to allow timely and specific feedback to each student about their understanding of the day's topic. This served two goals. Firstly, an opportunity was provided for students to delve deeper into the course textbook's content and seek the answers to the points where they need improvement. The successful completion of the online quiz was intended to show that the student understood the prerequisite information and would be able to build on this to develop his/her own well-thought-out opinions. Secondly, the successful completion of the online quizzes was to encourage to move onto deeper meaning activities that use higher-order cognitive thinking, thus building up a conversation in a more meaningful and intelligent manner, using relevant points and critical thinking.

However, the students' alternative approach to taking online quizzes resulted that the intended path to the learning objectives was not followed. The original instructional approach intended to be beneficial became a faint idealism. The first concern with the alternative approach was that more time was needed at the start of class to spend on discussing the book's activities before the online quizzes could be completed. This resulted in less time that could be spent on critical thinking activities, which was the original intended use of class time. The second concern with the alternative approach was the benefits of student-centered learning, which included that students could use their preferred learning style, go at their own pace, and choose the order they complete the book's activities that have been lost. Therefore, the best-laid plans fell apart by underestimating the students' desire not to follow them.

5. Strategies to Setup Online Quizzes

Upon reflection on the students' online quiz-taking strategies, a more comprehensive philosophy needs to be thought out. One strategy to make online quizzes more effective is to impose time limits [13,14]. Clay [12] offered some suggestions based on the type of questions being asked. It is recommended that for question types that are true or false questions, students must be allowed 30 s for each question. For multiple-choice questions, 60 s were allowed for each. For short answer questions, students must be allowed 120 s for each. Finally, students must be provided between five to ten minutes to review the work. Ideally, these recommendations need to be adjusted based on the complexity and knowledge level of the students.

The second strategy suggested is to randomly shuffle the answers within each question [12,14]. Many online quiz-setting applications including Google Forms and Microsoft Teams have this option to select. To take this one step further, several LMS applications including TronClass and Moodle can create a unique quiz for each student by randomly drawing questions from an established question bank [14,15]. Ideally with no two online quizzes the same, it hopefully makes it more difficult for students to share answers and forces them to seek answers from their knowledge and learning.

The third strategy is to foster a more positive attitude in students about taking the quizzes. Faculty Focus [16] is a collection of short articles on strategies to make online quizzes more enjoyable, engaging, and beneficial. Some of the strategies include allowing collaboration between students while completing the quiz, using different formats of questions, renaming quizzes to extra-credit exercises, and changing the times when students take the online quiz, just to name a few. Therefore, possibly, a more important issue than the pedagogical reasoning to implement online quizzes is the packaging and presentation of them to encourage students to use them positively.

6. Conclusions

Although the literature has found many positive results of implementing a flipped classroom pedagogy [9,17,18] and using online quizzes as one of its tools [19], this is the tip of the iceberg for an instructor's implementation plan. Managing students' attitudes towards online quizzes, creating a set of rules to minimize students' divergence from the learning objectives, and ensuring that the online quizzes are beneficial, must to be considered to achieve effective results. The overall outcome requires instructors to be watchful of students' behavior and be vigilant with a set of strategies to keep students on the desired learning path.

Funding: This research received no external funding.

Institutional Review Board Statement: This study did not require ethical approval.

Informed Consent Statement: This university does not require any ethics approval at this time for this type of research.

Data Availability Statement: Data sharing is not applicable to this article.

Conflicts of Interest: The authors declare no conflict of interest.

References

1. Bergmann, J.; Sams, A. *Flip Your Classroom: Reach Every Student in Every Class Every Day*; International Society for Technology in Education: Washington, DC, USA, 2012.
2. Alsowat, H. An EFL flipped classroom teaching model: Effects on English language higher-order thinking skills, student engagement and satisfaction. *J. Educ. Pract.* **2016**, *7*, 108–121.
3. Guo, J. The use of an extended flipped classroom model in improving students' learning in an undergraduate course. *J. Comput. High Educ.* **2019**, *31*, 362–390. [CrossRef]

4. Hung, H.T. Flipping the classroom for English language learners to foster active learning. *Comput. Assist. Lang. Learn.* **2015**, *28*, 81–96. [CrossRef]
5. Wilson, L.O. "Bloom's Taxonomy Revised. Understanding the Revised Version of Bloom's Taxonomy", Strategies. 2001. Available online: https://thesecondprinciple.com/essential-teaching-skills/blooms-taxonomy-revised/ (accessed on 3 November 2020).
6. Wilson, L.O. Anderson and Krathwohl Bloom's Taxonomy Revised Understanding the New Version of Bloom's Taxonomy. 2016. Available online: https://quincycollege.edu/content/uploads/Anderson-and-Krathwohl_Revised-Blooms-Taxonomy.pdf (accessed on 22 September 2020).
7. Reid, J.M. The Learning Style Preferences of ESL Students. *TESOL Q.* **1987**, *21*, 87–111. [CrossRef]
8. Reid, J.M. Teachers as Perceptual Learning Styles Researchers. In *Understanding Learning Styles in the Second Language Classroom*; Reid, J., Ed.; Prentice Hall Regents: Upper Saddle, NJ, USA, 1998; pp. 15–25.
9. Wu, W.-C.V.; Hsieh, J.S.C.; Yang, J.C. Creating an online learning community in a flipped classroom to enhance EFL learners' oral proficiency. *Educ. Technol. Soc.* **2017**, *20*, 142–157.
10. Butler, M.; Pyzdrowski, L.; Goodyknoontz, A.; Walker, V. The Effects of Feedback on Online Quizzes. *Int. J. Technol. Math. Educ.* **2008**, *15*, 131–136.
11. Johnson, G.M. Optional Online Quizzes: College Student Use and Relationship to Achievement. Available online: https://cjlt.ca/index.php/cjlt/article/view/26481/19663 (accessed on 26 August 2020).
12. Clay, B. Clay-2001—Is This a Trick Question, Kansas. 2001. Available online: https://kgi.contentdm.oclc.org/digital/api/collection/p16884coll42/id/147/download (accessed on 30 August 2022).
13. Brothen, T.; Wambach, C. The Value of Time Limits on Internet Quizzes. *Teach. Psychol.* **2009**, *31*, 62–64. [CrossRef]
14. Syahid, A. Usability of Moodle Question Types by EFL teachers. In Proceedings of the 2nd INACELT, Palangka Raya City, Indonesia, 16–17 November 2018; pp. 224–237. Available online: http://e-proceedings.iain-palangkaraya.ac.id/index.php/inacelt (accessed on 30 August 2020).
15. Glodowski, K.R.; Thompson, R.H.; Asuncion, E.A. Evidence-Based Recommendations for Programming Quizzes to Improve College Student Behavior in Residential Courses. *J. Behav. Educ.* **2020**, *29*, 543–570. [CrossRef]
16. Faculty Focus. Designing Better Quizzes: Ideas for Rethinking Your Quiz Practices. 2018. Available online: https://www.k-state.edu/assessment/toolkit/measurement/Special-Report-designing-better-quizzes.pdf (accessed on 17 September 2022).
17. Hsieh, J.S.C.; Wu, W.-C.V.; Marek, M.W. Using the flipped classroom to enhance EFL learning. *Comput. Assist. Lang. Learn.* **2017**, *30*, 1–21. [CrossRef]
18. Shih, H.C.J.; Huang, S.H.C. College students' metacognitive strategy use in an EFL flipped classroom. *Comput. Assist. Lang. Learn.* **2019**, *33*, 755–784. [CrossRef]
19. Cohen, D.; Sasson, I. Online quizzes in a virtual learning environment as a tool for formative assessment. *J. Technol. Sci. Educ.* **2016**, *6*, 188–208.

Disclaimer/Publisher's Note: The statements, opinions and data contained in all publications are solely those of the individual author(s) and contributor(s) and not of MDPI and/or the editor(s). MDPI and/or the editor(s) disclaim responsibility for any injury to people or property resulting from any ideas, methods, instructions or products referred to in the content.

engineering proceedings

MDPI

Proceeding Paper

Using YouTube as an Effective Educational Tool to Improve Engineering Mathematics Teaching during the COVID-19 Pandemic †

John C.-C. Lu

Department of Civil Engineering, Chung Hua University, Hsinchu 30012, Taiwan; cclu@chu.edu.tw

† Presented at the 3rd IEEE International Conference on Electronic Communications, Internet of Things and Big Data Conference 2023, Taichung, Taiwan, 14–16 April 2023.

Abstract: An investigation and assessment of virtual learning in engineering mathematics during the COVID-19 pandemic were explored in this study. The result showed that the media platform was an especially useful technology for students to create, share, learn, and interact with others. YouTube, a free media-sharing website, has proved to be an effective educational tool to add a new dimension to education in increasing student engagement, motivation, understanding, and achievement. Thus, students' learning models on the OpenCourseWare YouTube channel were researched to investigate how virtual activities in e-learning of engineering mathematics during coronavirus confinement were implemented and describe how YouTube was used for teaching engineering mathematics by engaging students in mathematical problem-solving.

Keywords: COVID-19; engineering mathematics; YouTube channel; online learning

Citation: Lu, J.C.-C. Using YouTube as an Effective Educational Tool to Improve Engineering Mathematics Teaching during the COVID-19 Pandemic. *Eng. Proc.* **2023**, *38*, 24. https://doi.org/10.3390/engproc2023038024

Academic Editors: Teen-Hang Meen, Hsin-Hung Lin and Cheng-Fu Yang

Published: 21 June 2023

Copyright: © 2023 by the author. Licensee MDPI, Basel, Switzerland. This article is an open access article distributed under the terms and conditions of the Creative Commons Attribution (CC BY) license (https://creativecommons.org/licenses/by/4.0/).

1. Introduction

Improving teaching effectiveness has been tried continuously to help students engage more in learning. Therefore, a series of research related to engineering mathematics has been carried out with continuous funding from the Ministry of Education in Taiwan. The above efforts are aimed at attracting students to study engineering mathematics, and this also helps the author to improve teaching. The efforts include (1) integrating key points of engineering mathematics into multimedia teaching materials, (2) creating an appropriate environment for self-learning, (3) providing learning opportunities for students and enhancing comprehension ability by analogy, (4) constructing unit-themed learning materials, and (5) making an environment suitable for action learning. These efforts are especially critical for the conversion of in-person courses into virtual courses during the COVID-19 pandemic.

Paschal, Pacho, and Adewoyin [1] found effective teaching methods for higher educational institutions during the COVID-19 pandemic in Africa through empirical research. Simamora et al. [2] displayed the lecturer's perspectives during the COVID-19 pandemic in higher education and concluded that there was a need to continue exploring alternative learning environments to ensure learning with an effective, efficient, easy-to-access, and high-quality knowledge dissemination process. Marsudi, Lestari, and Hidayati [3] investigated the improvement in mathematics learning achievements of students after using YouTube as a learning media during the COVID-19 pandemic. They observed the impact of the interactive learning model and orientation of mathematics material on conceptual comprehension ability. Antón-Sancho and Sánchez-Calvo [4] recommended increasing the specific training for professors in the pedagogical usage of information and communication technologies addressing the specific knowledge in each area. Kanetaki et al. [5] identified variables that impacted student performance in the educational process disorientation due to the COVID-19 pandemic and concluded that innovative teaching improved students' spatial conceptions. DeCoito and Estaiteyeh [6] revealed that online teaching was viewed negatively by most teachers in terms of student engagement and outcomes. Febrianto, Mas'udah, and Megasari [7] focused on determining the online learning process and the associated obstacles experienced by students. Their investigation showed the importance of the availability of supporting facilities, infrastructure, and facilitated internet access. Shahroury [8] demonstrated that the use of the flipped classroom strategy helped overcome the challenges associated with e-learning and maintain overall performance. The investigation of Beruin [9] revealed a generally unfavorable and unenthusiastic view of online learning during the COVID-19 pandemic. George [10] presented effective teaching and examination strategies that can be utilized for undergraduate learning courses during COVID-19 restrictions. Karasneh et al. [11] recommended that training programs and interdepartmental communication strategies be implemented and use fewer platforms to provide an efficient online learning experience. Libasin et al. [12] concluded that students and lecturers worked together to ensure similar learning outcomes before the pandemic. The work of Patalinghug and Patalinghug [13] displayed that using YouTube as a web-based instructional tool improved students' sociability, grades, learning motivation, and curriculum delivery through utilizing technology-enabled learning. Revelo-Rosero et al. [14] highlighted digital tools and resources available on the web to improve the teaching–learning process inside and outside the classroom. Abubakar and Muhammed [15] provided a rational literature investigation and analysis of the education teachers' pedagogy and YouTube video technology. There have been many related investigations about effective educational tools to improve teaching during the COVID-19 pandemic.

The series of videos on engineering mathematics on YouTube recorded by the author has helped many students learn well on engineering mathematics. Three representative students sent the appreciated messages from 2020 to 2022 on YouTube as follows.

- "Thank you so much, Teacher. You have helped me a lot. I cannot express how grateful I am right now. Thank you so much. Please stay safe and take care."
- "Thank you for your contribution, since mathematics is a universal language, I will always understand. Good video! Greetings from Mexico."
- "You may not believe me but I'm a Spanish speaker and I only know around 7 words in Chinese but it was so clear that I understand what you explain. Thank you so much!"

In this study, the ideas and keys are suggested to establishing a YouTube engineering mathematics teaching channel for readers' reference.

2. Creating Teaching Materials

There have been 39 teaching research projects implemented by the author since 2003, who had a strong interest in multimedia teaching and the integration of digital technology into teaching. Each project was focused on the research and development of innovative teaching materials. The research projects supported by the Ministry of Education in the past five years are shown in Table 1. All teaching materials have been uploaded on YouTube for interested students to use for free. The established OpenCourseWare on YouTube was applied to teaching and learning engineering mathematics during the COVID-19 pandemic. The goals of the projects are as follows.

- Studying according to their own pace
- Making teaching materials more interesting
- Helping students pass various exams
- Improving students' concentration on the study

- Triggering learning motivation
- Solidifying abstract ideas
- Enhancing memory and impression
- Shortening learning time
- Making teaching activities lively and funny
- Expanding information content can satisfy students' thirst for knowledge

Table 1. Projects supported by the Ministry of Education in 2018–2022.

Year	Project Title
2018	Study on the Establishment and Application of Teaching and Problem-Solving Handouts/Videos of the OpenCourseWare "Engineering Mathematics"
2019	Study on Case Study of the Application of OpenCourseWare "Engineering Mathematics"
2020	Study on the Design of Adding Interactive Tests to the Online Video of the OpenCourseWare "Engineering Mathematics"
2021	Study on the Frequently Asked Questions and Its Answers of OpenCourseWare "Engineering Mathematics"
2022	Evaluation of Key Points of OpenCourseWare "Engineering Mathematics" Using Animation Methods

The current research is focused on the animation presentation of the key knowledge of engineering mathematics. All the efforts are conducive to the implementation of student-centered pedagogy, question-and-answer pedagogy, technology-integrated pedagogy, problem-based pedagogy, design thinking pedagogy, and self-study tutoring. There is much interesting information displayed on YouTube. Instructors can improve teaching based on feedback in the detailed message. Figure 1 presents the welcome page of instructional YouTube developed by the author. Figure 2 shows that there were currently 3627 students subscribed to the educational channel. In 28 days, the YouTube channel had 12,899 accumulated views with 3619 non-repetitive audiences (Figures 3 and 4). Figure 5 presents the ordinary differential equation problem-solving process on YouTube, and the application of ordinary differential equations to structural mechanics is presented in Figure 6. Students could self-assess their learning effectiveness through online quizzes in H5P as displayed in Figure 7. The key points of engineering mathematics are presented by animation as shown in Figure 8. The project facilitated online learning of engineering mathematics while maintaining the quality of teaching.

Figure 1. Welcome page for browsing and Q&A (感謝瀏覽, 歡迎提問): YouTube instructional videos developed by the author at Chung Hua University (中華大學).

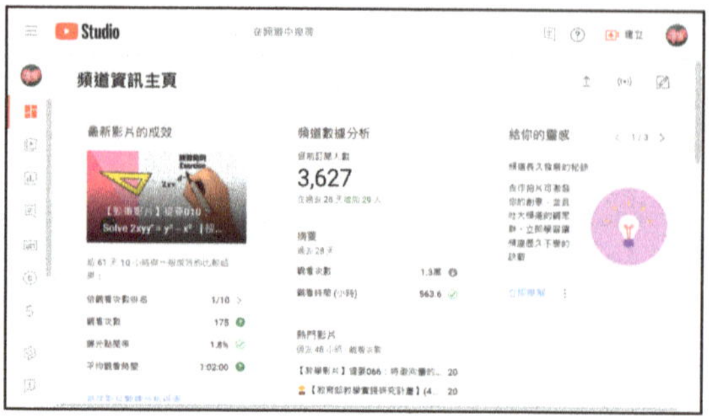

Figure 2. Channel dashboard (頻道資訊主頁): 3627 students subscribed to the engineering mathematics YouTube channel.

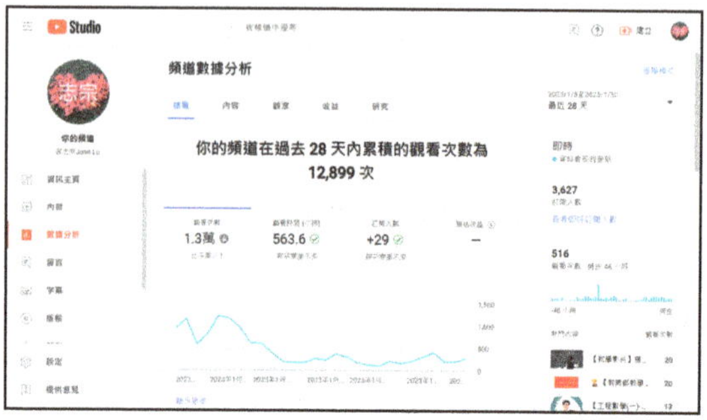

Figure 3. Channel data analysis (頻道數據分析): 12,899 views in the latest 28 days.

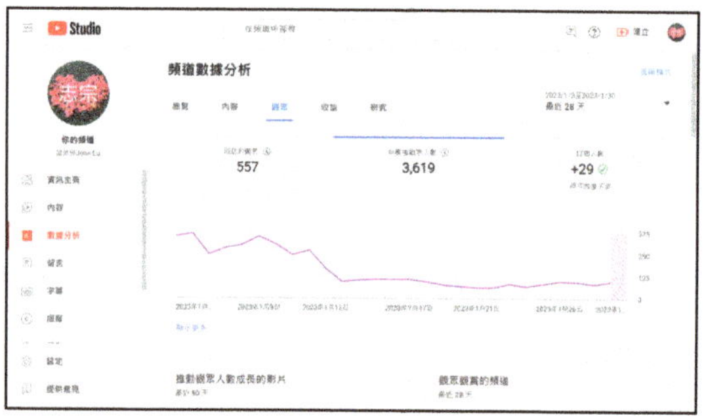

Figure 4. Channel data analysis (頻道數據分析): 3619 non-repetitive audiences in the latest 28 days.

Figure 5. Problem-solving process of ordinary differential equation displaced on YouTube.

Figure 6. Application of ordinary differential equation on structural mechanics.

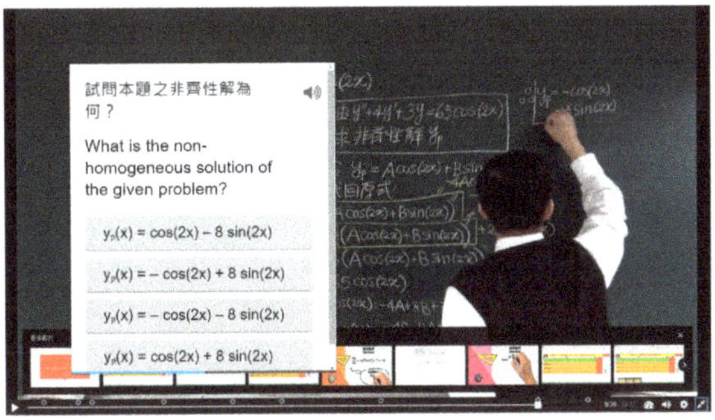

Figure 7. Self-assessment of learning effectiveness through online quizzes.

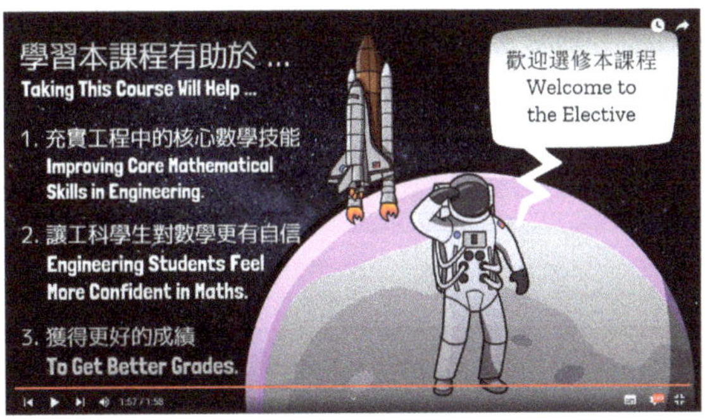

Figure 8. Key points of engineering mathematics are presented by animation.

3. Questionnaire Survey

An online questionnaire survey was conducted on the effectiveness of teaching videos with Google Forms. In total, 77 students from the courses of engineering mathematics in 2020–2022 participated in the survey. The descriptive analysis result showed that planning of learning scored 4.62 on a scale of 5. Thus, the recorded teaching videos on YouTube were effective during the COVID-19 pandemic (Table 2). The scores obtained for each indicator were interpreted based on the criteria of Table 3. The students' responses to the teaching videos are displayed in Table 4. The average score was 4.62, which showed student satisfaction. Most students liked the quality of the engineering mathematics materials in the educational videos, which implies that the quality of the learning videos was satisfactory, and the videos were helpful in online learning during the epidemic.

Table 2. Score on the Likert scale of the questionnaire.

Option	Score
Strongly Disagree	1
Disagree	2
Neither Agree nor Disagree	3
Agree	4
Strongly Agree	5

Table 3. Interpreted students' responses based on scores.

Interval	Students Responses
$1.00 \leq x < 1.80$ [a]	Very Bad
$1.80 \leq x < 2.60$	Bad
$2.60 \leq x < 3.40$	Neutral
$3.40 \leq x < 4.20$	Good
$4.20 \leq x \leq 5.00$	Very Good

[a] x = Scores of each indicator.

Table 4. Students' responses to the developed teaching videos in 2020–2022.

Indicator	Score	Response Category
The teaching materials developed by the teacher can inspire my interest in learning	4.59	Very Good
The recorded digital video provided by the teacher helps me adjust my learning progress	4.65	Very Good
Average	4.62	Very Good

4. Conclusions

Based on the findings in this research, the following conclusions are drawn.

There were 39 teaching research projects implemented by the author since 2003. Each project focused on the development of innovative teaching materials. All of the teaching materials have been uploaded on YouTube for students to use for free. The results of the online survey with 77 students in 2020−2022 showed that students' responses to the developed teaching videos scored 4.62 on a scale of 5, which was very good. Students studied according to their own pace on YouTube. Furthermore, teaching materials made the course more interesting and maintained the quality of instruction. The developed teaching videos integrated the key points of engineering mathematics into multimedia teaching materials and built a suitable environment for active learning. All the efforts were conducive to the implementation of student-centered pedagogy, question-and-answer pedagogy, technology-integrated pedagogy, problem-based pedagogy, design thinking pedagogy, and self-study tutoring.

Funding: This work is supported by the Ministry of Education of Taiwan, Republic of China, through grants PEE1101210 and PEE1110397.

Institutional Review Board Statement: Not applicable.

Informed Consent Statement: Not applicable.

Data Availability Statement: Not available.

Conflicts of Interest: The author declares no conflict of interest.

References

1. Paschal, M.J.; Pacho, T.O.; Adewoyin, O. Teaching methods applied in higher education during COVID-19 pandemic in Africa. *Int. J. Educ. Policy Res. Rev.* **2022**, *9*, 27–40.
2. Simamora, R.M.; de Fretes, D.; Purba, E.D.; Pasaribu, D. Practices, challenges, and prospects of online learning during COVID-19 pandemic in higher education: Lecturer perspectives. *Stud. Learn. Teach.* **2020**, *1*, 185–208.

3. Marsudi, A.S.; Lestari, M.P.; Hidayati, N. The use of YouTube social media in the Covid19 pandemic to improve understanding of mathematical concepts. *Turk. J. Comp. Math. Educ.* **2021**, *12*, 6327–6333.
4. Antón-Sancho, Á.; Sánchez-Calvo, M. Influence of knowledge area on the use of digital tools during the COVID-19 pandemic among Latin American professors. *Educ. Sci.* **2022**, *12*, 635. [CrossRef]
5. Kanetaki, Z.; Stergiou, C.; Bekas, G.; Troussas, C.; Sgouropoulou, C. Analysis of engineering student data in online higher education during the COVID-19 pandemic. *Int. J. Eng. Pedag.* **2021**, *11*, 27–49. [CrossRef]
6. DeCoito, I.; Estaiteyeh, M. Online teaching during the COVID-19 pandemic: Exploring science/STEM teachers' curriculum and assessment practices in Canada. *Discip. Interdiscip. Sci. Educ. Res.* **2022**, *4*, 8. [CrossRef]
7. Febrianto, P.T.; Mas'udah, S.; Megasari, L.A. Implementation of online learning during the COVID-19 pandemic on Madura Island, Indonesia. *Int. J. Learn., Teach. Educ. Res.* **2020**, *19*, 233–254. [CrossRef]
8. Shahroury, F.R. E-learning during COVID-19 epidemic: Experience of a university from Jordan. *Acad. Strateg. Manag. J.* **2022**, *21*, 1–14.
9. Beruin, L.C. STEM students' conceptions of online learning during COVID-19 pandemic: A phenomenographic study. *J. Pedag. Res.* **2022**, *6*, 143–167. [CrossRef]
10. George, M.L. Effective teaching and examination strategies for undergraduate learning during COVID-19 school restrictions. *J. Educ. Technol.* **2020**, *49*, 23–48. [CrossRef]
11. Karasneh, R.; Al-Azzam, S.; Muflih, S.; Hawamdeh, S.; Muflih, M.; Khader, Y. Attitudes and practices of educators towards e-learning during the COVID-19 pandemic. *Electron. J. E-Learn.* **2021**, *19*, 252–261. [CrossRef]
12. Libasin, Z.; Azudin, A.R.; Idris, N.A.; Rahman, M.S.A.; Umar, N. Comparison of students' academic performance in mathematics course with synchronous and asynchronous online learning environments during COVID-19 crisis. *Int. J. Acad. Res. Progress. Educ. Dev.* **2021**, *10*, 492–501. [CrossRef] [PubMed]
13. Patalinghug, H.F.; Patalinghug, M.E. YouTube as a web-based instructional tool in higher education: Technology-enabled learning during the COVID-19 pandemic. *J. Progress. Educ.* **2022**, *12*, 1115–1126. [CrossRef]
14. Revelo-Rosero, J.; Yaguana-Campos, A.; Cadena-Heredia, V.; Andrade-Erazo, C. Web 2.0 as a tool for university teaching in times of pandemic COVID-19. *Rev. Cátedra* **2023**, *6*, 36–54. [CrossRef]
15. Abubakar, H.; Muhammed, H.B. A systematic literature review on teaching teachers pedagogy through YouTube video technology. *J. Digit. Educ. Technol.* **2023**, *3*, ep2301. [CrossRef]

Disclaimer/Publisher's Note: The statements, opinions and data contained in all publications are solely those of the individual author(s) and contributor(s) and not of MDPI and/or the editor(s). MDPI and/or the editor(s) disclaim responsibility for any injury to people or property resulting from any ideas, methods, instructions or products referred to in the content.

engineering proceedings

MDPI

Proceeding Paper

Visual Effects Analysis Based on Computer Processing: Taking Rik Oostenbroek's Works as an Example †

Nan-Hu Lu * and Shun-Hsin Huang

Department of Communication Arts, Chaoyang University of Technology, Taichung 413310, Taiwan; sunlight023@gmail.com
* Correspondence: nhlu@gm.cyut.edu.tw

† Presented at the 3rd IEEE International Conference on Electronic Communications, Internet of Things and Big Data Conference 2023, Taichung, Taiwan, 14–16 April 2023.

Abstract: More artists have come to use computer processing to create works with amazing visual effects. Rik Oostenbroek's works, his creative style, and methods are studied in this research. Born on 25 May 1989, Rik Oostenbroek is a Netherlands-based illustrator and designer. He focuses on digital art creation. His works are recreated by photography, including dynamic images and graphic design with gorgeous colors and dynamic and fluent lines. Currently living in Hilversum, Rik Oostenbroek is a self-taught Dutch freelance artist, designer, and art director working for many brands such as Apple, Nike, Epson, and Viacom. Known for his dynamic forms and unique abstract shapes, Rik is skilled in various design styles, including graphic design, typography, 3D graphics, art direction, surreal/abstract illustration, and a mixture of photo manipulation and retouching. His pioneering style is mainly based on the various patterns created by computers quickly and accurately, which attracts many imitators to follow. As this type of visual style brings greater commercial benefits and suits the visual collocation of various products, Rik's style is of research value. By summarizing Rik's works and comparing them with others, this article demonstrates the creation process and method of his work using computers and illustrates the limitations and challenges arising from those processes. Blender 3.0, a third-party free software, is used in this demonstration.

Keywords: computer processing; Rik Oostenbroek; Blender; 3D graphics; visual effect

Citation: Lu, N.-H.; Huang, S.-H. Visual Effects Analysis Based on Computer Processing: Taking Rik Oostenbroek's Works as an Example. *Eng. Proc.* **2023**, *38*, 25. https://doi.org/10.3390/engproc2023038025

Academic Editors: Teen-Hang Meen, Hsin-Hung Lin and Cheng-Fu Yang

Published: 21 June 2023

Copyright: © 2023 by the authors. Licensee MDPI, Basel, Switzerland. This article is an open access article distributed under the terms and conditions of the Creative Commons Attribution (CC BY) license (https://creativecommons.org/licenses/by/4.0/).

1. Introduction

Rik Oostenbroek has always worked as a freelancer for global brands, including Apple, Mercedes Benz, AON, Adobe, HP, AT&T, Viacom, Nike, ESPN, Wacom, Swatch, BMW, and so on [1]. Being famous for his dynamic gradient form, tons of energy, and sense of speed, Rik has previously held various posts and became proficient in various design styles. In addition to his enthusiasm for traditional photography and printing, he has also been exploring new styles based on computer processing. By facing new challenges brought by computer graphics, he expanded his vision. In this article, Rik's works are reviewed from two dimensions, including (1) the differences between the aesthetics of his works and traditional creation and (2) creation methods. At the same time, his works of the same type are recreated to display the creation process.

2. Background of Creation

According to Adobe Blog, the tireless artist, Rik, was born in 1989 and did not have a creative background until he discovered Photoshop in 2006. After that, Rik focused on jamming and coming up with abstract creations. Then, he decided on his passion and career goals when he became a freelancer in 2009. He spent at least eight hours a day discovering new things to maintain his colorful work and shape his skills, such as 3D, typography, photo manipulation, and vector illustration [2]. As seen in many of his works, he was greatly influenced by the digital computer creation method. He could quickly complete amazing visual effects using computer software and attracted a group of online followers to imitate the same style.

3. Work Style

An important element in the visual form of Rik's work style is generally called "gradient" performance, as shown in Figure 1. It uses color as the basis to show the aesthetics. In nature, all colored lights are composed of three primary lights (red, green, and blue). Among the well-known rainbow colors (including red, orange, yellow, green, blue, indigo, and violet), red, green, and blue are the three primary colors of lights, but each color presents gradations rather than clear boundaries, which is the basic principle of gradational color expression.

Figure 1. Rik Oostenbroek often uses Silent Wave as his work's title and the gradient streamline's expression. Source: Created by Rik Oostenbroek (2022) https://rikoostenbroek.com/wallpaper-of-the-month (accessed on 23 September 2022).

Owing to the integration of gradient performance in various themes, Rik's works are beloved by customers and audiences. However, many imitators are trying to copy his style. The reason is that this type of work creates a large market, and it is not difficult to make as long as you master the characteristics of the software.

3.1. Multicolor and Organic Form

A lot of abstract expressions are mixed with realism in Rik's portfolio. Most of his works are colorful and organic. The organic shape mainly shows irregular and asymmetric objects with a certain radian on the edge. By discarding geometric and sharp expression forms, his works represent a sense of speed and vitality in a circular form of expression, as shown in Figure 2. According to Rik, his works are based on his experience of nature and his dislike of being constrained. That is why he changed from a company employee to a freelancer, which brought him greater flexibility and freedom in his creation. He tried to soak in every single experience he faced in life. It could be a simple sunset or a nice trip abroad. Overall, he said he just got inspired by things he found beautiful. Music influences his work a lot as well [2].

Figure 2. LYRS Rik fully grasped computer technology to show colorful gradient performance and smooth and irregular shapes. Source: Created by Rik Oostenbroek https://rikoostenbroek.com/lyrs, (accessed on 23 September 2022).

3.2. Limitations of Forms and Easiness to Imitate

3.2.1. Style of Rik Oostenbroek

Curved circles and rainbowlike ribbons were adopted in Rik's works to distribute on the screen, and the audience moves their eyes in the direction of the ribbons they appreciate. Rik called the series of works Arcus, and he said it took him about five years at different stages of his creative career. This series can be said to be the beginning of his success and popularity, laying a foundation for his success as a freelancer. It also encourages many young artists to learn from him and start pursuing their creative careers. Although the Arcus series is colorful and bright, it can also deliver a visual sense of motion. It brings a more intuitive feeling to artistic creation, and people focus more on the shape than the content, which is also a frequent problem in abstract painting. Rik claimed that he has been suffering from severe depression since he was 13 years old and has struggled with it, experiencing extreme ups and downs. He needed help finding the balance. However, while making the Arcus series, he gradually found a way to accept depression. From the original fight to find his uniqueness from the creation, he could accept the unique characteristics of depression, no matter how bad the situation was [1].

As revealed by the Arcus series, it is difficult for the audience to distinguish the differences between works with the change of mood. People may express their moods and relieve pressure from the process of artistic creation. At the same time, the visual effects of the final products also bring a pleasant and positive feeling. According to Rik's description, when everyone recognized the finished products and the creation process, he felt great encouragement and confidence. It can be said that the Arcus series has laid the foundation for his unique style in the future.

3.2.2. Connotations of Work to Avoid Imitation

As this series of works is easy to copy, Rik believed there is also a disadvantage to building things people like. Many imitators began to appear. However, Rik viewed these imitations from a more positive perspective. His style often inspires a new generation of creators, and some great Youtubers creatively produced a tutorial on how to make the "Oostenbroek" style.

The Arcus style is easy to imitate due to several factors. First, there is a limitation of the software. Several kinds of software can be used to make similar works, and the principle of 3D production is the same. If no new expression program languages can be developed, everyone uses existing expression production methods to make similar products. Figure 3 shows the author of this paper using the same technology to create a virtual 3D model.

Figure 3. Endless Cycle is a virtual 3D model created by the author of this paper using the same technology and different software.

Secondly, the abstract form has no connotation, and the appearance is easy to be imitated. Those with high requirements for creative technology and realistic form are naturally difficult to imitate. The only way to avoid the imitation of Arcus is to focus on the aesthetics of the shape. At the same time, specific images can be combined with the work to improve the connotations, making the audience feel the emotional resonance in addition to visual stimulation, which cannot be achieved by simple external imitation.

4. Demonstration of Creation Process

As this type of work is made by 3D software, the 3D production process must be strictly followed, from building models, selecting materials and maps, setting animation, camera adjustment, and lighting design, to the final output. Targeting the production process of Rik's Arcus series, how this series is made is revealed in this article.

The circular ring structure is the most important thing for this type of work. The 3D topological modeling function generates the model with the head and tail connected, and then the 3D software is employed to form the procedural map. Finally, the camera is used to capture the complete visual angle of the modeling and render it. The details are as follows.

4.1. Phase One: Topological Modeling

Topology is the study of the invariable properties of specific objects under a specific transformation, called continuous mapping. Mobius bands and clover knots belong to the application of topological modeling. A cup can be transformed into a doughnut after being pulled, deformed, or shrunk. This process is called continuous deformation, as the cup and the doughnut are homeomorphic [3], as shown in Figure 4. A Mobius band is one of the shapes commonly used by sculptors, and 3D software imitates these shapes to show an endless cycle of visual experience. Make Pictures full of feelings of flow and speed. It is not easy to let the vision stay in fixed places on pictures.

First, the 3D software TopMod 2.5 is applied to create the Mobius band (Möbiusband in German), a surface with only one side (surface) and one boundary, as shown in Figure 5. It was discovered independently by the German mathematicians August Ferdinand Möbius and Johann Benedict Listing as an important topological structure. We can only draw on one side if we draw a whole circle along the band on its outside or inside. A common circle can be replaced with the Mobius band due to the rotating visual effect brought by the Mobius band, which is more tensile than that of the general circle.

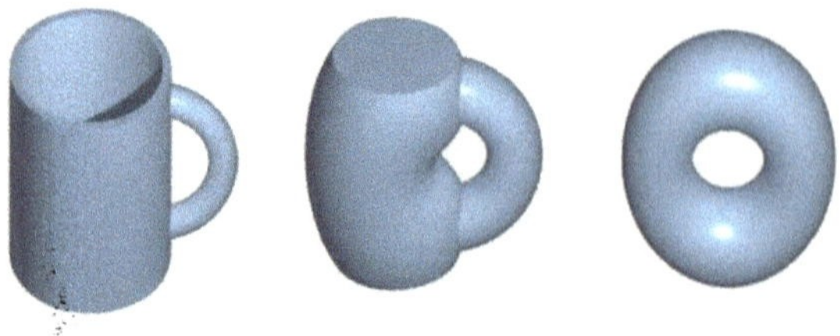

Figure 4. A bottle and a doughnut are the same in topology. Source: National Taiwan Science Education Center, CASE Journal, https://case.ntu.edu.tw/blog/?p=20714, (accessed on 10 September 2022).

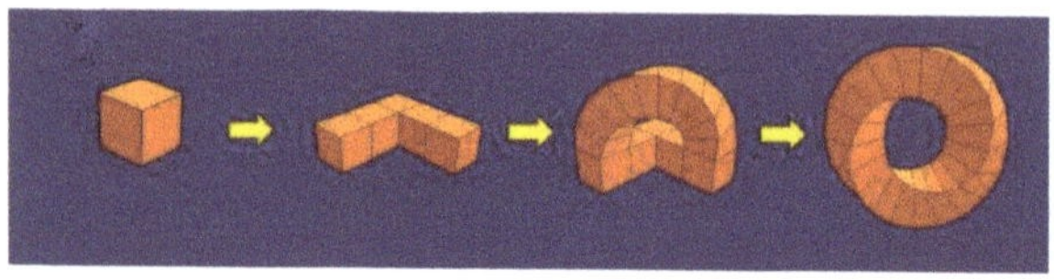

Figure 5. Create and build a Mobius band, demonstrated by the author.

After the completion of creation, the Blender software is used to modify the shape of the objects and determine the right angle of the camera position.

4.2. Phase Two: Procedural Texture

Every 3D special effect software has its own way of producing special visual effects on the surface of products, including the material and textures of objects. Except for using the setting on the Material Panel to establish materials, it creates new visual effects by cabling basic and other materials on a group of nodes after the completion of the modeling. Each node performs operations on materials to form a grid distribution, as shown in Figure 6. One node can be transferred to several nodes, or several nodes can transfer to the next node. In this way, a complicated material appearance can be realized.

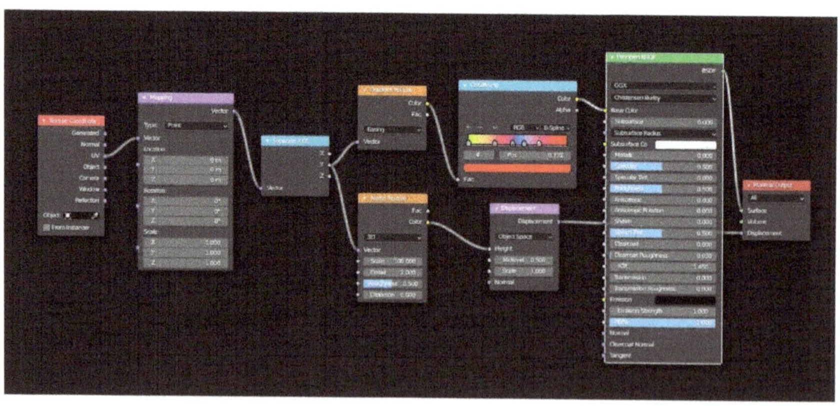

Figure 6. Nodes of E-Generation work, provided by the author.

4.2.1. Texture Node

The node makes it easier to operate textures or materials. With better control, more complex and wonderful materials can be created. When creating a node system, various data processing pipelines are described in which data "originate from" nodes of various sources and "flow through" nodes to represent different processing and filtering stages. Finally, "flow-in" indicates the output or node of the destination. Nodes can be connected in various ways, and their "attributes" or "parameters" can be adjusted to control the behavior of each node. Hence, a free combination of special effects of different materials can be generated. The operation becomes intuitive, roughly dividing the nodes into colored and textured nodes. With these functions, the three-dimensional surface effect of objects becomes more realistic. Generally, the characteristics can be divided into physical and color texture characteristics. The use of nodes facilitates the application effect of procedural texture and improves visual effects. As the operation of traditional software uses single input, the design of nodes breaks this limitation, greatly enhancing the convenience and intuition of vision and operation.

4.2.2. Procedural Texture

The texture is a way of adding details by projecting images and patterns onto the surface. The images and patterns of projections can affect not only color but also specular reflection, transparency, and even forged 3D depth. Typically, images and patterns are projected during rendering, but texture mapping can also be applied in sculpture, painting, and deformed objects. A procedural texture is defined mathematically. It can produce a fairly stereo visual effect on the plane through computer operation, as shown in Figure 7. Compared with traditional texture mapping, procedural texture saves a lot of time. Usually, when mapping, the computer automatically combines different textures. Therefore, a few layers can be combined to produce special effects in procedural texture. However, performance limitations and random uncertainties exist in some operations, which require more tests and adjustments.

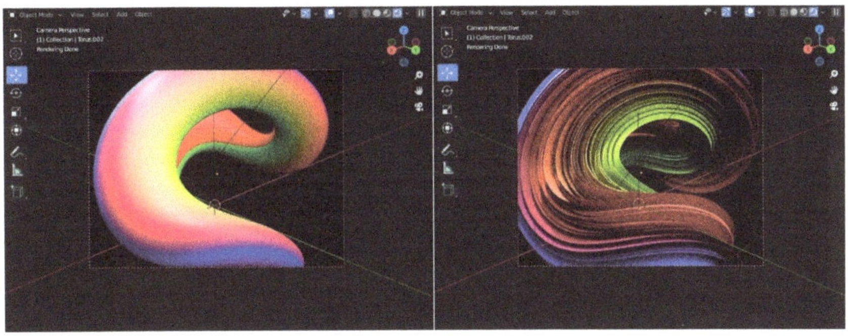

Figure 7. Procedural texture effect of E-Generation work, provided by the author.

Similar effects can also be produced without using procedural textures. However, it is necessary to have relevant shapes when making models. Limited by the changes in shapes, it takes work to form diversified changes. Moreover, it takes considerable time to make relevant shapes. Figure 8 shows traditional modeling used to create a similar effect. From the picture, the slight differences between the two different production methods are observed.

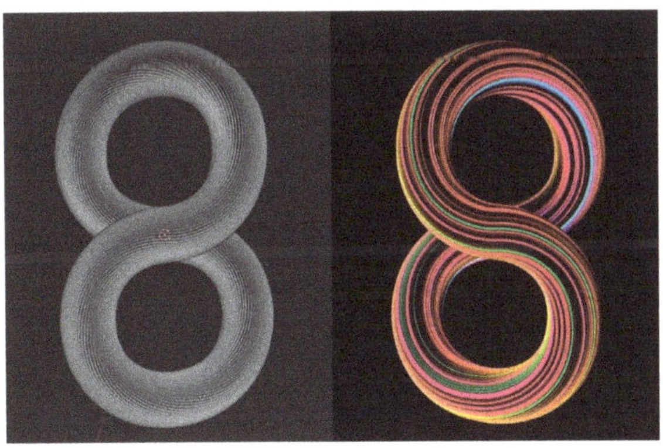

Figure 8. Traditional modeling used to produce similar effects, as demonstrated by the author.

4.3. Phase Three: Camera Position Adjustment and Lighting Design

By imitating all the functions and operation methods of real cameras, Blender Camera expands the usage of cameras. To achieve virtual 3D composite effects, VR, or 360° surround effects, Blender's photography can create a variety of visual possibilities that are not available in traditional photography. This article only demonstrates how to achieve relevant photography functions by targeting the visual effects of the Arcus series. Figure 9 shows how the Blender camera imitates the operation of a real camera.

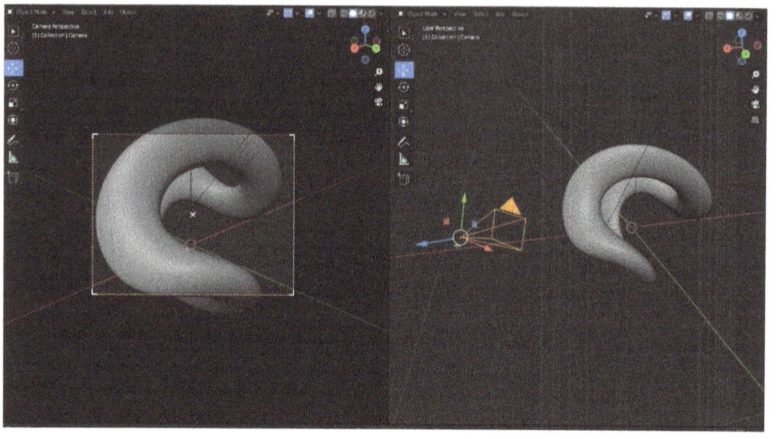

Figure 9. The Blender camera imitates operation of a real camera.

4.3.1. Camera Manipulation

Although Rik's works include dynamic films and graphic pictures, most of the Arcus series is planar. The lighting design and the camera angle adjustment follow the completion of procedural texture. For example, most Arcus-like products are generated by procedural mapping. Procedural mapping produces a three-dimensional effect of the plane, which cannot be fully mastered at the beginning. With subtle changes in the light, the three-dimensional light and shadow of the body change.

Moreover, the parameters of procedural texture are slightly adjusted to show colors. Thus, such works do not start from clear designs. Instead, they can only head towards the broad lines of body and dynamic directions. When the procedural mapping is finished, the final output can be delivered by selecting the perfect angle of the camera.

In all 3D software, a perspective window shows concrete and detailed operations. After modeling and procedural mapping are completed, the most appropriate position is found by relying on the rotation and shifting functions of the perspective windows. At this moment, similar to the camera's viewing window, a perspective window helps determine the most appropriate angle. It is different from the general operation to fine-tune and find the perfect angle with a real camera. To see the results of the final products of this camera, it must be matched with that of a perspective camera to ensure those angles are the same.

4.3.2. Lighting Design

The stereoscopic sense of most important objects is based on lights. Here, 3D lights can produce important visual effects. At the same time, the direction generated by the light source makes the picture look oriented and uniform. Blender software provides four light sources: point light, sunlight, spotlight, and regional light. Most of these works use the regional light design to make the light fall evenly on the object to illuminate every part. As shown in Figure 10, four light sources are set up to distribute the light evenly. The primary light source slants down from the top right, and a large area of refraction lies above the overall atmosphere and three-dimensional feeling. The other two regional lights are auxiliary light sources to make the body look more stereoscopic. Although the primary and secondary light sources can be distinguished, the overall light irradiation is symmetrical. It is worth noting that the shadow generated by light changes with the adjustment of light angle and distance, and the area of shadow should be in harmony with the picture as it will also be the focus.

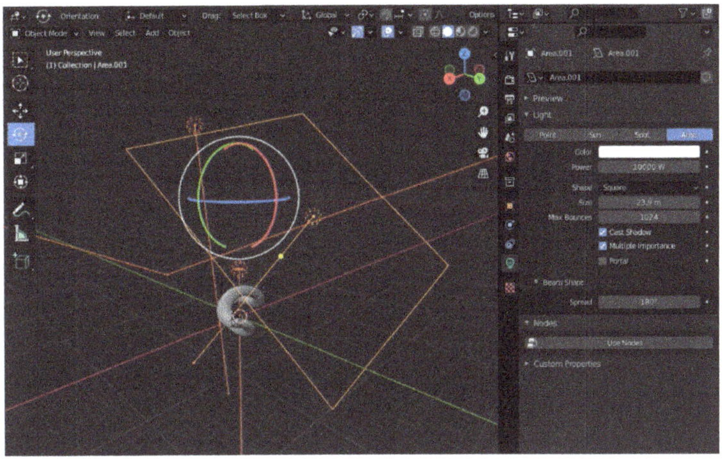

Figure 10. E-Generation set up four light sources to make the light spread evenly.

4.4. Output

Realistic visual effects are created by fully grasping the software and practicing. Several effects cannot be easily shot, so you may need to rely on computers. In addition, computers cannot achieve special effects at once. Instead, they must be tried and corrected constantly to achieve the perfect effect. After the test and adjustment, the most appropriate visual effect is selected with the light shading and camera angle adjustment. These are the details that must be paid attention to. Each step of production deserves repeated inspection. In particular, the pictures in the production process must be rendered first. View the pictures one by one and modify them gradually before you output the film to ensure the quality. Therefore, the operation is time-consuming. An effect is created consistently with Rik Oostenbroek's Arcus series using the method proposed in this article. With the help of computers, pictures that are difficult to be shot are created, such as the E-Generation (Figure 11), which shows the flowing clouds and endless visual effects.

Figure 11. E-Generation creates a virtual spoondrift effect similar to the real one using the method proposed in this article.

5. Conclusions

In the future, computer special effects will bring huge commercial benefits as they fulfill various requirements that cannot be achieved by photographs. The popularization of special effect production will also be significant. The works of Rik Oostenbroek are simultaneously welcomed and imitated by everyone to bring new business opportunities. Although the technology of special effects production is limitedly revealed, personal computers can also make amazing effects. It has become a trend to create innovative effects using computers, which may gradually become a production method to greatly improve the visual effects of images. For the production company or individuals, the most important things are to cultivate a sense of aesthetics, generate more experience, and operate the software well to improve the quality of work.

Author Contributions: Conceptualization, N.-H.L. and S.-H.H.; methodology, N.-H.L. and S.-H.H.; validation, N.-H.L.; formal analysis, N.-H.L.; data curation, N.-H.L.; writing—original draft preparation, N.-H.L. and S.-H.H.; writing—review and editing, N.-H.L. and S.-H.H.; supervision, N.-H.L. All authors have read and agreed to the published version of the manuscript.

Funding: This research received no external funding.

Data Availability Statement: Data will be available on request.

Conflicts of Interest: The authors declare no conflict of interest.

References

1. Aarcus. Available online: https://rikoostenbroek.com/arcus-collection (accessed on 23 September 2022).
2. Artist Spotlight: Rik Oostenbroek. Available online: https://blog.adobe.com/en/publish/2018/04/10/artist-spotlight-rik-oostenbroek (accessed on 23 September 2022).
3. Wu, R.-F. Math brain-Is a Coffee Mug Donut? Available online: https://case.ntu.edu.tw/blog/?p=20714 (accessed on 10 September 2022).

Disclaimer/Publisher's Note: The statements, opinions and data contained in all publications are solely those of the individual author(s) and contributor(s) and not of MDPI and/or the editor(s). MDPI and/or the editor(s) disclaim responsibility for any injury to people or property resulting from any ideas, methods, instructions or products referred to in the content.

engineering proceedings

MDPI

Proceeding Paper

Using Big Data to Analyze Impact of Visual Design on Social Networking †

Tung Wu [1,*], Ying-Ming Su [2] and Ding-Chin Chou [2]

1 Doctoral Program in Design, College of Design, National Taipei University of Technology, Taipei 10608, Taiwan
2 Department of Architecture, College of Design, National Taipei University of Technology, Taipei 10608, Taiwan
* Correspondence: z88930@gmail.com; Tel.: +886-095-679-5199
† Presented at the 3rd IEEE International Conference on Electronic Communications, Internet of Things and Big Data Conference 2023, Taichung, Taiwan, 14–16 April 2023.

Abstract: How to use social media to publicize policies is one of the most critical topics for local governments to operate social networking. Thus, this study is carried out to analyze the relationship between post-presenting and visual design taking the Facebook fan page of the Construction Office of Taitung County Government as an example and using big data technology. The results show that the promotion effect of images and text is higher than that of videos, reposted news, and shared posts. Moreover, the type of image and text is more attention with the position with more than 100 words, and there is no significant difference in the promotion effect of text layout, hashtags, and image forms (graphics or photos). The results of this study provide a reference for future government management of social media, reducing the waste of human resources and ineffective costs.

Keywords: policy promotion; social media; communication; visual design; big data; social networking

1. Introduction

The government has gradually changed from sovereignty to interactivity in its policy promotion. Social media is an excellent communication network between the government and the public. Facebook is still the most widely used social platform in Taiwan in 2022, and its market share is still as high as 60% [1]. Through its algorithm and big data design, it is easier to induce users to participate in communication between social networks. The use of interactive functions can strengthen public participation in policy communication and improve the mechanism of citizen participation [2]. For the government, using social media to improve policy recognition is essential [3]. The primary purpose of this study is to enable the government to effectively promote policies on social networks and make their audiences receive information more effectively. In the past, researchers mainly summarized and analyzed the content of posts on Facebook fan pages. According to the user experience theory, visual perception during the process is found to affect their experience [4]. With the development of information technology (ICT), social media has become the most popular communication channel and formed a unique communication logic different from traditional media. Social media is challenging to define media ecology. In particular, the increasing diversity of built-in social media services makes social media ecology research pertinent. Thus, research on the social ecology of social media has received more attention. Therefore, we selected the "Taitung Construction" Facebook fan page operated by the Taitung County Government in 2015 as the research target of this study.

Citation: Wu, T.; Su, Y.-M.; Chou, D.-C. Using Big Data to Analyze Impact of Visual Design on Social Networking. *Eng. Proc.* **2023**, *38*, 23. https://doi.org/10.3390/engproc2023038023

Academic Editors: Teen-Hang Meen, Hsin-Hung Lin and Cheng-Fu Yang

Published: 21 June 2023

Copyright: © 2023 by the authors. Licensee MDPI, Basel, Switzerland. This article is an open access article distributed under the terms and conditions of the Creative Commons Attribution (CC BY) license (https://creativecommons.org/licenses/by/4.0/).

2. Literature Review

2.1. Digitalization of Government Information

The government has undergone a digital transformation in response to the advent of the information age. Through the effective management of social media, the government can improve the transparency of government implementation of government affairs, and the open online world can make people more willing to express their opinions [5,6]. At the same time, it also breaks the traditional limitations of space and time and promotes the public's enthusiasm to attend to civic issues [2]. Social media is a type of self-media that allows the government to publish information without considering the political stance of other media. The public can directly and openly express their opinions to the government, forming a more effective two-way communication [7]. In recent years, the government has gradually achieved results in the operation of social media. At the same time, they also face the following problems: they do not understand the audience's preferences, resulting in the low promotion of information, spending much money to hire Internet celebrities to promote policies, but they cannot effectively increase the public's recognition of the policy [8], with a lack of human resources to maintain social media, resulting in information omission or multiple people operating social media, making the overall message-style different [9]. These need to be overcome in future policy promotion.

2.2. Related Research on Facebook Fan Pages

Since Facebook opened the fan page function, research exploring its interaction mode has been presented one after another. In the early days of little advertising and few fake accounts, the posting location of the message and the comment's content greatly impacted the post impressions [10]. According to research results in recent years, the content, time, and medium of posts of several types of Facebook fan pages affect post impressions. On Facebook fan pages with English teaching, the pictures in warm colors and the posts posted on working days have higher promotion benefits [11]. Facebook fan pages on commercial brands show that the text content of posts (incentive marketing, brand characteristics, and message vividness) significantly affect the user's interactive behavior. At the same time, the posting time and the post impressions have no significant impact on the interactive behavior [12]. For Facebook fan pages run by the government, the posts' image quality, clarity, and entertainment can significantly affect the degree of policy recognition and support [3]. According to the post's content, entertainment messages and clarification messages have a significant positive impact on fan interaction. Livelihood messages have a significant positive impact on the number of likes and shares, and overall, the post's content trumps the medium [13].

2.3. Research on Visual Design and User Experience

Through the above analysis, the user's feelings are the main factors affecting the interaction of fans. Cognition and emotion have an impact on human thinking. Aesthetic and pleasant things improve efficiency [4]. The typesetting of design posts is used to provide users with aesthetic needs and satisfy their visceral level. Then, the designer needs to consider the behavioral level of the user's operational experience and allow users to enter a reflective level [14], thereby increasing the user engagement of Facebook fan pages. User experience is a perceptual process that results from repeated tests and adds rational information to emotional appeals [15]. Increasing cognitive ease of use, emotional interest, and emotional attachment can enhance the user's sense of identity [16]. Therefore, Facebook fan pages operated by the government must not divide information types by business but consider audience types, use information visualization to enhance user's memory, and effectively link policy issues to make two-way communication more complete [17].

3. Research I: Post-Presenting and Promotion Effect

We take the "Taitung Construction" Facebook fan page operated by the Taitung County Government as the research object. It is explored whether different post-presentation methods affect the effect of policy promotion. The number of Facebook fan page fans was about 7400. About 42.4% of fans set their residence as Taitung County; male users accounted for 51.3%; female users accounted for 48.7%; the majority of users were age 35 to 44 years old, and this fan page did not purchase advertisements.

3.1. Research Methods

We set the research period from January to September 2022. It published 163 posts, an average of 18 posts per month, including 97 policy advocacy posts, 40 policy achievements, 18 related announcements, and 8 relay messages. Posts on the fan page were grouped into five types: video, image and text, reposted Facebook posts, reposted news, and reposted YouTube news according to the presentation. Among them, the number of "repost YouTube news" accounted for 58.9% of all posts. However, its promotion data were relatively low compared to other types. This implied that the policy promotion effect of the fan page was not satisfactory. Therefore, we set the post-presentation as an independent variable, set the promotion data of the backend system of the meta business suite as a dependent variable, and conducted an Analysis of Variance by SPSS.

3.2. Research Results

3.2.1. Descriptive Statistics

According to the descriptive statistics (Table 1), the fan page released 7 videos, 22 images and text, 8 reposted Facebook posts, 30 reposted news, and 96 reposted YouTube news. The presentation of "image and text" obtained the highest promotion benefits [11], followed by "reposted Facebook posts." There was little difference in the post impressions for videos, reposted news, and YouTube news.

Table 1. Descriptive statistics of Research 1.

Post-Presenting	Post Impressions		
	Mean	N	Std. Deviation
Image and text	1945.32	22	1707.784
Videos	592.71	7	181.767
Repost news	581.20	30	290.102
Repost Facebook posts	1056.25	8	528.603
Repost YouTube news	616.61	96	282.803
Total	809.98	163	815.637

3.2.2. Research Analysis and Results

The hypothesis of this study is "post-presenting affects the post impressions on the fan page," and the results of the Analysis of Variance are shown in Table 2. The research results show significant differences in the influence of post-presentation methods on the post impressions, $F(4,158) = 18.47$, $p < 0.001$. Further analysis showed that the post impressions of "image and text" ($M = 1945.32$, $SD = 1707.784$) were higher than "repost Facebook posts" ($M = 1056.25$, $SD = 528.603$), "repost YouTube news" ($M = 616.61$, $SD = 282.803$), "videos" ($M = 592.71$, $SD = 181.767$), and "repost news" ($M = 581.20$, $SD = 290.102$), so overall, the hypothesis is supported.

Table 2. ANOVA result of Research 1.

Post-Presenting	ANOVA				
	Sum of Squares	Do	Mean Square	F	Sig.
Between Groups	34,333,129.704	4	8,583,282.426	18.466	<0.01
Within Groups	73,439,689.241	158	464,808.160		
Total	107,772,818.945	162			

3.2.3. Other Research Findings

According to the presentation method of posts stipulated by Facebook, "image and text" and "repost Facebook posts" are text on top of multiple pictures, and "videos" are played in real time. The layout of "reposted news" and "reposted YouTube news" is presented in a single full-page static image. The interactive interface affects the user's preference [4], so we analyze the impact of visual design on the promotion effect for highly interactive posts.

4. Research II: Visual Design on Promotion Effect

The focus of this study is to summarize the post-editing model with a high promotion effect and low cost for the government. Since the "Taitung Construction" fan page does not have a single administrator, the editing habits are also different.

4.1. Research Methods

In this study, from the original 163 posts, excluding the posts with low interaction, 37 posts were screened out. The factors affecting the user's visual design were used as independent variables, and the promotion data of the backend system of the meta business suite were used as a dependent variable. The independent variables were divided into four categories: number of words (less than 100 words, more than 100 words), typesetting (one line per sentence, no line break), #Hashtag (used, not used), and central visual design (photos, post-production images), and an Independent Sample *t*-test by SPSS was conducted.

4.2. Research Results

4.2.1. Descriptive Statistics

According to the descriptive statistics (Table 3), the fan page had no apparent preference and unique style in the number of posted characters, typesetting, #Hashtag, and the central visual design, verifying that the fan page mentioned in the preceding paragraph is not a single administrator. According to the average post impressions, the post impressions with more than 100 words, no line breaks, the use of #Hashtag, and photos are better than the average. To check whether the above operating factors significantly impact the post impressions, an Independent Sample *t*-test was performed.

Table 3. Descriptive Statistics of Research II.

Independent Variables	Operating Factors	Number of Psst	Average Post Impressions	Std. Deviation
Number of words	Less than 100 words	10	752.40	359.37
	More than 100 words	27	1773.04	1597.32
Typesetting	One line per sentence	23	1391.87	1230.95
	No line break	14	1670.21	1777.89
#Hashtag	Used	13	1746.77	1449.30
	Not used	24	1362.00	1454.36
Central visual design	Photos	18	1869.89	1570.82
	Post-production images	19	1144.11	1253.55
Total		37	1497.19	

4.2.2. Research Analysis and Results

There are four research hypotheses in this study. Hypothesis 1: "The number of words posted will affect the promotion effect, and the post impressions with more than 100 words is higher than the other"; Hypothesis 2: "The type of typesetting will affect the promotion effect, and the post impressions without line breaks is higher than the other"; Hypothesis 3: "#Hashtag will affect the promotion effect, and the post impressions with using #Hashtag is higher than the other"; Hypothesis 4: "The central visual design will affect the promotion effect, and the post impressions presented by photos is higher than the other."

The independent sample t-test results are shown in Table 4. The research results show that the number of posted words significantly differs in the post impressions, $F = 4.402$, $p = 0.043$, $t = -3.114$, $df = 31.873$, p-value = 0.004. Further analysis shows that the number of words is more than 100 words (M = 1773.04, SD = 1597.321), and the post impressions are higher than 100 words or less (M = 752.40, SD = 359.371), so Hypothesis 1 is supported. The research results of Hypothesis 2 ($t = -0.515$, p-value = 0.612), Hypothesis 3 ($t = 0.769$, p-value = 0.447), and Hypothesis 4 ($t = 1.558$, p-value = 0.128) show that they cannot reject the null hypothesis. #Hashtag, typesetting, and central visual design have no significant difference in the impact of the post impressions.

Table 4. *T*-Test results of Hypothesis 1–4.

Post-Presenting	Post Impressions		*T*-Test for Equality of Means		
	F	Sig.	T	Df	Sig. (2-Tailed)
		Hypothesis 1			
Equal variances assumed	4.402	0.043 *	−1.985	35	0.055
Equal variances not assumed			−3.114	31.873	0.004
		Hypothesis 2			
Equal variances assumed	0.044	0.836	−0.563	35	0.577
Equal variances not assumed			−0.515	20.654	0.612
		Hypothesis 3			
Equal variances assumed	0.680	0.415	0.769	35	0.447
Equal variances not assumed			0.770	24.810	0.449
		Hypothesis 4			
Equal variances assumed	0.297	0.589	1.558	35	0.128
Equal variances not assumed			1.548	32.521	0.131

4.3. Other Research Findings

A hashtag is a function similar to tag sharing. In addition to turning the text blue, it increases the attention between the posts. Generally speaking, proper use can increase post impressions, and it has always been a function advocated by marketing units. It is found in this study that using this function has no significant effect on promoting the "Taitung Construction" Facebook Fan Page. It is speculated that the hashtag used on the fan page is of low generality or the hashtag is only used to distinguish between fonts and colors, which makes it impossible to link to popular keywords, so it cannot significantly increase the post impressions.

5. Conclusions

This study aims to explore the impact of visual design on community interaction and to analyze whether the different post-presenting significantly differ in their promotion effect. In the past, most research focused on the cognitive division of the post's content but did not discuss it according to the type of posting. Therefore, with post-presentation for statistical analysis, we verified that posts with images and texts are more likely to arouse users' interest.

Usually, the government does not have enough professional human resources to operate fan pages. It is common for different business units to operate a fan page jointly. According to the business distribution of the government, if there are projects beyond professional such as producing films, posters, and dummies, extra expenses are usually entrusted to external agencies for execution. However, through this research, we found that the way of video and post-graphics does not have a significant impact on users. Instead, the number of texts in the post positively impacts the policy promotion effect.

To save the cost and time of the government, the study result needs to be used as a reference. For the content of future policy promotion posts, clearer texts must be used to inform the public of relevant information, and it does not require gorgeous typesetting and design. Clear information text with photos is the best business strategy for this research subject.

The limitation of this study is that the research object is the construction unit of the local government agency, and the hypotheses are only four. After statistical analysis, only the first hypothesis is supported. The preliminary inference is that the user group of the fan page is primarily middle-aged. Therefore, it belongs to the online communication mode that young people are used to including the typesetting method of one sentence per line, the use of #Hashtag, and the presentation method of memes which have no significant impact on the audience of the fan page. By summarizing the above factors of visual design and policy promotion, it is possible to establish a standard post-editing process for the government to operate the fan profession to avoid problems such as information gaps during the subsequent handover of business personnel.

Author Contributions: Conceptualization, T.W.; methodology, T.W.; validation, T.W.; formal analysis, T.W.; investigation, T.W.; resources, T.W. and T.W.; writing—original draft preparation, T.W. and T.W.; writing—review and editing, T.W.; visualization, T.W.; supervision, Y.-M.S. and D.-C.C.; project administration, T.W. All authors have read and agreed to the published version of the manuscript.

Funding: This research received no external funding.

Institutional Review Board Statement: Not applicable.

Informed Consent Statement: Not applicable.

Data Availability Statement: No new data were created.

Conflicts of Interest: The authors declare no conflict of interest.

References

1. Taiwan Network Information Center. *2022 Taiwan Internet Report*; Taiwan Network Information Center: Taipei, Taiwan, 2022.
2. Hsu, Y.-C. *Public Usage of Government Social Media Websites: The Case of Taiwan Tourism Bureau's Facebook Fan Page*; National Chung Hsing University: Taichung, Taiwan, 2013.
3. Chen, Y.-H.; Hsieh, H.-P. The Key Factors of Enhancing Governments' Policy Support through Social Media. *NTUA GCA* **2021**, *2021*, 44–55.
4. Ma, J.; Chen, C.-C. Examination The Relationship between Emotional Change and Functional Animation. *J. Design.* **2019**, *24*, 1–16.
5. Bonsón, E.; Torres, L.; Royo, S.; Flores, F. Local e-government 2.0: Social media and corporate transparency in municipalities. *Gov. Inf. Q.* **2012**, *29*, 123–132. [CrossRef]
6. Guillamón, M.-D.; Ríos, A.-M.; Gesuele, B.; Metallo, C. Factors influencing social media use in local governments: The case of Italy and Spain. *Gov. Inf. Q.* **2016**, *33*, 460–471. [CrossRef]
7. Gilfoil, D.M.; Aukers, S.M.; Jobs, C.G. Developing and Implementing a Social Media Program While Optimizing Return on Investment—An MBA Program Case Study. *Am. J. Bus. Educ. (AJBE)* **2014**, *8*, 31–48. [CrossRef]
8. Lin, M.-Y. The Research on the Influence of Influencers' Political Endorsement and Policy Marketing Effect: Using Internet Public Opinion Big Data to Analyze Its Impact on Social Media. Master's Thesis, Department of Public Administration and Policy, National Taipei University, New Taipei City, Taiwan, 2021. Available online: https://hdl.handle.net/11296/834k8b (accessed on 8 November 2022).
9. Yeh, M.-K. Research on the Use of Social Media in Policy Marketing and Public Consultation by Chinese Civil Servants—Using the Technology Acceptance Model as an Application. Master's Thesis, Department of Administration and Policy, Tunghai University, Taichung, Taiwan, 2017. Available online: https://hdl.handle.net/11296/w49yc6 (accessed on 8 November 2022).
10. De Vries, L.; Gensler, S.; Leeflang, P.S.H. Popularity of Brand Posts on Brand Fan Pages: An Investigation of the Effects of Social Media Marketing. *J. Interact. Mark.* **2012**, *26*, 83–91. [CrossRef]
11. Hu, S.-J. A Facebook Fanpage Effectiveness Study-the Case of "hopenglish". Master's Thesis, Institute of Political Science, National Taiwan University, Taipei, Taiwan, 2016.
12. Chen, L.-C. A Study on Fan Page Posting Stratege—A Case of Uniqlo Taiwan. Master's Thesis, Department of Business and Management, National United University, Miaoli County, Taiwan, 2018. Available online: https://hdl.handle.net/11296/xn28je (accessed on 8 November 2022).

13. Lu, C.-W. A Study on Social Media Strategy of Financial and Economic Affairs: A Case of Ministry of Finance. Master's Thesis, Department of Agriculture Economics, National Taiwan University, Taipei, Taiwan, 2020.
14. Norman, D.A. *Emotional design: Why We Love (or Hate) Everyday Things*; Civitas Books: New York, NY, USA, 2004.
15. Chen, H.-C. Green Innovation Design Model of Kansei Engineering. Ph.D Thesis, Graduate School of Design, Doctoral Program, National Yunlin University of Science and Technology, Yunlin, Taiwan, 2016. Available online: https://hdl.handle.net/11296/k38qk9 (accessed on 8 November 2022).
16. Tsai, C.-H. A Behavioral Model of Users' Usage of Social Network Website—The Viewpoints of Cognition, Affect, and Behavior. Master's Thesis, Department of Business Administration, National Chiayi University, Chiayi, Taiwan, 2014.
17. Hung, Y.-F. Information Design in the Internet Media Era, New Trends in Government Performance Communication. *Taiwan Econ. Res. Mon.* **2021**, *44*, 39–46. [CrossRef]

Disclaimer/Publisher's Note: The statements, opinions and data contained in all publications are solely those of the individual author(s) and contributor(s) and not of MDPI and/or the editor(s). MDPI and/or the editor(s) disclaim responsibility for any injury to people or property resulting from any ideas, methods, instructions or products referred to in the content.

engineering proceedings

MDPI

Proceeding Paper

Exploring Use Behavior of Self-Service Ordering at Restaurants with Application Unified Theory of Acceptance and Use of Technology Model †

Kuan-Chuan Tao [1], Chi-Yueh Hsu [1,*], Jen-Jen Yang [2,*], Ming-Fu Ke [1] and Chia-Hsuan Chiang [1]

1 Department of Leisure Service Management, Chaoyang University of Technology, Taichung 412, Taiwan; willytao0218@cyut.edu.tw (K.-C.T.); obvious10226878@gmail.com (M.-F.K.); jannies0920@gmail.com (C.-H.C.)
2 Department of Business Administration, Chaoyang University of Technology, Taichung 412, Taiwan
* Correspondence: cyhsu@gm.cyut.edu.tw (C.-Y.H.); jjyang@cyut.edu.tw (J.-J.Y.)
† Presented at the 3rd IEEE International Conference on Electronic Communications, Internet of Things and Big Data Conference 2023, Taichung, Taiwan, 14–16 April 2023.

Abstract: In this study, UTAUT2 is used to explore the use behavior by consumers regarding a self-ordering system. A total of 227 valid questionnaires were collected after being distributed with an effective recovery rate of 90.8%. Confirmatory factor analysis and a structural equation model were utilized via Smart PLS 2.0. The result shows that only several research hypotheses have significant effects. Among the adjustment variables, age and habits have a significant impact on user behavior. Experience and convenience conditions have a significant impact on use intention. Experience and habit have a significant impact on use intention. The overall model fit is excellent with high explanatory power. According to the research results, the public accepts self-service ordering technology systems in restaurants. The higher the level of pleasure after use, the higher the willingness to use the system in the future.

Keywords: self-service technology; UTAUT2; self-service ordering

Citation: Tao, K.-C.; Hsu, C.-Y.; Yang, J.-J.; Ke, M.-F.; Chiang, C.-H. Exploring Use Behavior of Self-Service Ordering at Restaurants with Application Unified Theory of Acceptance and Use of Technology Model. *Eng. Proc.* **2023**, *38*, 26. https://doi.org/10.3390/engproc2023038026

Academic Editors: Teen-Hang Meen, Hsin-Hung Lin and Cheng-Fu Yang

Published: 25 June 2023

Copyright: © 2023 by the authors. Licensee MDPI, Basel, Switzerland. This article is an open access article distributed under the terms and conditions of the Creative Commons Attribution (CC BY) license (https://creativecommons.org/licenses/by/4.0/).

1. Introduction

With the advance of science and technology and the changes in the consumption pattern, technology has gradually been introduced into the workplace, and the mode of operation of the catering industry has gradually been oriented towards customer self-service in Taiwan. Operations through self-service technology (SST) can effectively reduce investment costs and improve quality service [1]. Ho et al. [2] pointed out that introducing self-service innovative technology provides customers with new services or transactions. Self-ordering service has gradually been introduced in the catering industry in Taiwan, and operating through SSTs allows customers to order foods and drinks from touch-screen menus [3]. Restaurants can effectively reduce labor and investment costs, increase revenue, and improve quality services [1,3]. Self-service technology is a service method in which the operators know the consumers' needs through SSTs, and consumers do not need to interact with service staff or wait to access the service [3,4]. Operating SSTs by themselves reduces waiting times. The industry has introduced self-service technology, which is increasingly becoming a part of industry service.

SSTs refer to consumers interacting with the Internet or machines on their own without interacting with service staff [5]. The factors for customers to use self-service technology are ease of use, timesaving, convenience, privacy, correctness, improving order accuracy, reliability, and fun to increase guest satisfaction [3,6,7]. The customer's perceived authenticity has impacted their behavior intentions [8]. The introduction of self-service innovation technology has provided customers with new channels for service or transaction. For shy customers, self-service technology can be used to avoid embarrassment through consumers' new service options [2]. According to Cisco [9], 61% of consumers worldwide are willing to shop in self-service stores and 52% are willing to use self-service checkouts. Therefore, this study was conducted to explore the operation of self-service technology equipment by consumers in restaurants. In this study, self-service technology is defined as the replacement of human interaction services by consumers through technology system equipment. The UTAUT2 model is considered based on the consumer environment and is often used in technology.

2. Methodology

We took the self-service technology users in restaurants as the research objects in order to explore the behaviors of self-service users in restaurants through questionnaire surveys. The research tool was based on the UTAUT2 proposed by Venkatesh et al. [10], and dimension items were taken from Venkatesh et al. [10] for the creation of the questionnaire. The questionnaire was tested online, and snowball sampling was used as a study method. The period of time for the formal questionnaire test was from 6 April to 29 April 2019, and 305 copies were distributed in the form of network questions with 277 valid responses, at a return rate of 90.8%. Smart PLS2.0 was used to analyze the results of the formal questionnaires, including confirmatory factor analysis and structural equation model.

3. Result

The personal background variables of customers who used self-service ordering technology in restaurants included gender, age, and experience. There were 101 males (36.5%) and 176 females (63.5%) among the respondents. Out of 277 respondents, 14.4% were under 20 years old, 45.8% were 21–30 years old, 14.4% were 31–40 years old, 16.6% were 41–50 years old, and 8.7% were over 51 years old. A total of 23.1% had used self-service ordering technology within the last month, 32.5% in the last one to six months, 10.8% in the last six months to a year, and 33.6% had last used it over one year ago.

Smart PLS 2.0 was used to verify the research framework and research hypotheses, including the path coefficient and direct and indirect relationships. The significance t-value > 1.96 was regarded as a critical value. Table 1 shows that hedonic motivation affects use intention. Habit affects use intention and user behavior, while use intention affects use behavior. The estimated impacts were 0.149, 0.596, 0.321, and 0.505 at a significant p-value of <0.05.

Table 1. Path Coefficient of Dimensions.

Variable Paths	Standardization Coefficient	SE	t-Value	p-Value
Performance expectance→Use intention	0.027	0.041	0.646	0.519
Effort expectancy→Use intention	−0.011	0.035	0.311	0.756
Social influence→Use intention	−0.003	0.024	0.100	0.920
Facilitating conditions→Use intention	0.112	0.060	1.851	0.065
Facilitating conditions→Use behavior	0.047	0.039	1.189	0.235
Hedonic motivation→Use intention	0.149	0.050	2.925	0.004 *
Habit→Use intention	0.596	0.045	13.178	0.000 *
Habit→Use behavior	0.321	0.055	5.750	0.000 *
Use Intention→Use behavior	0.505	0.055	9.162	0.000 *

Note: * $p < 0.05$.

The t-value of each adjustment path > 1.96 is critical value, as shown as Table 2. The age adjustment of habit to use behavior, the experience adjustment of facilitating conditions to use intention, and the experience adjustment of habit to use intention reach had a significant level p-value of <0.05, and the adjustment coefficients are 0.108, 0.159, and −0.201, respectively. However, other adjustment paths were not significant.

The paths of this study are directly connected to the influence relationship. Among the path influences of the adjustment variables and external variables, only hedonic motivation and habit have positive effects on use intention. The coefficient values are 0.149 and 0.596

with indirect positive effects on user behavior with the coefficient values of 0.075 and 0.162. In addition, use intention has a direct positive effect on user behavior, and the coefficient values are 0.321 and 0.505. According to the results shown in Tables 2 and 3, the hypotheses H6, H7, H8, H9, H13, H19, and H20 are supported, but the hypotheses H1, H2, H3, H4, H5, H10, H11, H12, H14, H15, H16, H17, H18, H21, and H22 are not supported. The path model diagram is shown in Figure 1.

Table 2. Path Coefficient of Adjustment Variables.

Adjustment Variables	Adjustment Paths	Standardization Coefficient	SE	t-Value	*p*-Value
Gender	Facilitating conditions→Use intention	0.023	0.035	0.645	0.519
	Hedonic motivation→Use intention	0.005	0.030	0.166	0.868
	Habit→Use intention	−0.065	0.041	1.589	0.113
	Habit→Use behavior	−0.028	0.028	1.019	0.309
Age	Facilitating conditions→Use intention	−0.036	0.041	0.889	0.375
	Hedonic motivation→Use intention	0.099	0.035	0.277	0.782
	Habit→Use intention	0.061	0.050	1.257	0.210
	Habit→Use behavior	0.108	0.042	2.552	0.011 *
Experience	Facilitating conditions→Use intention	0.159	0.065	2.464	0.014 *
	Hedonic motivation→Use intention	−0.020	0.032	0.617	0.538
	Habit→Use intention	−0.201	0.064	2.994	0.003 *
	Habit→Use behavior	−0.028	0.038	0.707	0.480
	Use intention→Use behavior	0.024	0.033	0.684	0.495

Note: * $p < 0.5$.

Table 3. Direct Influence Between Various Dimensions.

	Use Intention		Use Behavior		
	Direct	Indirect	Direct	Indirect	Total Effect
Performance expectance	--			--	--
Effort expectancy	--			--	--
Social influence	--			--	--
Facilitating conditions	--		--	--	--
Hedonic motivation	0.149			0.075	0.075
Habit	0.596		0.321	0.162	0.321
Use Intention			0.505		0.505
R^2	0.686		0.715		

According to the results of Table 4, the GOF of the model in this study is 0.758, which means that the model fits well [11]. The R2 of use intention and use behavior are 0.686 and 0.715, which means that the model has high explanatory power.

Table 4. Overall Model Fit.

	AVE	Composite Reliability	R^2	Cronbach's Alpha	Redundancy	GOF
Performance expectance	0.691	0.899		0.852		
Effort expectancy	0.766	0.929		0.898		
Social influence	0.879	0.935		0.864		
Facilitating conditions	0.699	0.874		0.786		
Hedonic motivation	0.781	0.914		0.861		0.732
Habit	0.679	0.894		0.843		
Use Intention	0.822	0.932	0.686	0.892	−0.008	
Use behavior	0.786	0.917	0.715	0.864	0.438	

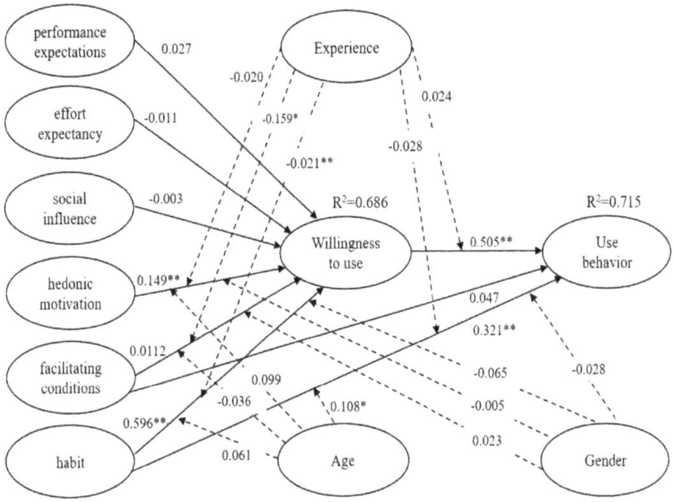

Note: * t > 1.96, $p < 0.05$; ** t > 2.58, $p < 0.01$

Figure 1. Path model diagram.

According to the results of each path in the research model, H6, H7, H8, and H9 are all supported, which is different from the results of Macedo [12]. First of all, performance expectations and effort expectancy have no significant impact on behavioral intentions. Although consumers are satisfied with the self-service ordering technology in restaurants, it is speculated that consumers may not feel that there is a significant difference between ordering through service staff and using self-service ordering technology. The traditional method of ordering meals is the transmission of information. Therefore, this is not affected in this study. In addition, social influence and facilitating conditions have no effects on behavioral intention. It has been speculated that the self-service ordering technology in restaurants has not yet been popularized in the catering industry in Taiwan. At present, the restaurant's service staff still serves. Therefore, this has no effect in this study. Hedonic motivation and habits have impacts on behavioral intentions. Consumers have used 3C products in their daily lives, and the acceptance of using technology systems for self-service ordering in restaurants in the future is high. Using self-service ordering technology during meals can also have entertainment effects. Therefore, hedonic motivation and habits have positive impacts on consumers who use self-service ordering technology.

Of the adjustment variables, only H13, H19, and H20 are supported. The explanation of the adjustment variables is as follows.

(1). H13: The age of self-service technology users in restaurants has a significant effect on H8.

The research results show the influence of age-adjusted habits on user behavior, as the younger age group is more influenced than the older age group. The habits of the younger age group and the older group have positive effects on user behavior, which means young people as opposed to middle aged individuals have positive opinions on using self-service ordering technology in restaurants.

(2). H19: The experience of self-service technology users in restaurants has a significant effect on H4.

The results show the experience-adjusted facilitating conditions and facilitating conditions affect use intention in the less experienced group than the more experienced group. If the customers feel the restaurant operator's ulterior motives for self-service ordering in restaurants, they will use this technology system continually the next time.

(3). H20: The experience of self-service technology users in restaurants has a significant effect on H7.

The results show that in terms of experience-adjusted habits regarding behavioral intention, the less experienced group and the more experienced group have almost the same influence on habit. The less experienced group and the more experienced group have a positive effect on use intention. This means that customers who have a habit of using self-service technology when ordering in restaurants will continue to use this technology system at their next visit.

4. Conclusions and Suggestions

The research result is summarized as follows.

(4). Current situation of self-service ordering technology in restaurants

In the UTAUT2 model, it was found that users consider self-service ordering technology to be convenient and easy to use in restaurants and using this type of system makes users feel that they are keeping up with the social era. However, using self-service ordering technology has not become a habit in modern life, and there is only a medium to high degree of willingness to continue using it in the future.

(5). UTAUT2 model explores the verification of the user intention and user behavior of self-service ordering technology in restaurants

- (a) The hedonic motivation of self-service technology users in restaurants has a significant effect on behavior intention, which indicates that users feel interested and entertained by using self-service technology to order meals during the meal process. They want to continue using this self-service technology to order meals in the future.
- (b) The habits of self-service technology users in restaurants have a significant effect on behavior intention and user behavior, which indicates that customers use self-service technology more. They are willing to use it in the future.
- (c) The behavioral intention of self-service ordering technology users in restaurants has a positive and significant impact on user behavior. The stronger user desire to use self-service order technology in restaurants in the future also affects the use intensity.

(6). UTAUT2 model explores the verification of adjustment variables of self-service ordering technology in restaurants.

The habit of using self-service ordering positively affects the younger group more intensively than the older group. In pursuing convenience, young people accept using self-service ordering technology in restaurants more. Users with more experience have a higher degree of acceptance. Therefore, the UTAUT2 model fits well with exploring self-service ordering technology in restaurants.

The recommendations and future research recommendations of this research are as follows:

- Self-service ordering technology in restaurants is acceptable to the general public. However, the convenience and ease of use of self-service ordering technology does not affect the willingness to use it in the future. If the restaurant industry intends to use these systems in the future, this system helps restaurant operations, and the industry may have to consider the gross profit for future operations. When modern technology has become more advanced, the use of self-service ordering technology in restaurants may also become a norm in the future.
- If users obtain a sense of pleasure from using self-service ordering technology in restaurants, their willingness to use these systems in the future will be high. In the future, if the industry continues to use self-service ordering technology systems, the

industry needs to improve the ability of users to interact with these systems. It was found that young people are more likely to continue using this technology in the future, so the operators should design activities such as minigames, guessing games, and lotteries on the system for customers to interact with.

Author Contributions: Conceptualization, K.-C.T. and C.-Y.H.; methodology, C.-H.C. and M.-F.K.; writing—original draft preparation, K.-C.T. and M.-F.K.; writing—review and editing, J.-J.Y. and C.-Y.H. All authors have read and agreed to the published version of the manuscript.

Funding: This research received no external funding.

Institutional Review Board Statement: Not applicable.

Informed Consent Statement: Dear respondents: Good day! Greatly appreciate that you take time to fill out this questionnaire. This academic study is "Exploring Use Behavior of Self-Service Ordering at Restaurants with Application Unified Theory of Acceptance and Use of Technology Model" The purpose of this study is to explore the factors that impact Switch's users' behavior. This is an anonymous questionnaire; all information is strictly confidential and for research purposes only. Please do not fill out this questionnaire if you disagree with this statement. Thanks for your cooperation with sincere appreciation.

Data Availability Statement: Not applicable.

Conflicts of Interest: The authors declare no conflict of interest.

References

1. Yang, Y.C.; Yang, C.C.; Yang, Y.Y.; Lin, S.C. Reduction in resistance to self-service technology adoption. *Chung Chou J. Manag. Humanit. Sci.* **2012**, *1*, 149–165.
2. Ho, S.H.; Ko, Y.Y.; Huang, C.H. Effects of self-service technology innovations on customer value: The case of Internet banking. *Electron. Commer. Stud.* **2007**, *5*, 463–484.
3. Lord, E. Is Now the Right Time for a Self-Service Ordering System? Lightspeed. Self-Service Technology for Restaurants | Light Speed POS 2017. Available online: http://lightspeedhq.com (accessed on 21 April 2021).
4. Beatson, A.; Coote, L.V.; Rudd, J. Determining consumer satisfaction and commitment through self-service technology and personal service usage. *J. Mark. Manag.* **2006**, *22*, 853–882. [CrossRef]
5. Wang, H.J.; Cheng, C.C.; Hsu, J.L. A study on the acceptance and operating performance of self-service electronic ordering service for catering industry-integrating views of the customers and operators. *J. Perform. Strategy Res.* **2012**, *9*, 63–83.
6. Zhu, F.X.; Wymer, W.; Chen, I. IT-based services and service quality in consumer banking. *Int. J. Serv. Ind. Manag.* **2002**, *13*, 69–90. [CrossRef]
7. Chiu, Y.T.H.; Hofer, K.M. Service innovation and usage intention: A cross-market analysis. *J. Serv. Manag.* **2015**, *26*, 516–538. [CrossRef]
8. Chen, Q.; Huang, R.; Hou, B. Perceived authenticity of traditional branded restaurants (China): Impacts on perceived quality, perceived value, and behavioral intentions. *Curr. Issues Tour.* **2020**, *23*, 2950–2971. [CrossRef]
9. Cisco. Self-Service Shopping Grows in Popularity, According to Cisco Study. 2013. Available online: http://itersnews.com (accessed on 21 April 2010).
10. Venkatesh, V.; Thong, J.Y.; Xu, X. Consumer acceptance and use of information technology: Extending the unified theory of acceptance and use of technology. *MIS Q.* **2012**, *36*, 157–178. [CrossRef]
11. Wetzels, M.; Odekerken-Schroder, G.; van Oppen, C. Using PLS path modeling for assessing hierarchical construct models: Guidelines and empirical illustration. *MIS Q.* **2009**, *33*, 177–196. [CrossRef]
12. Macedo, I.M. Predicting the acceptance and use of information and communication technology by older adults: An empirical examination of the revised UTAUT2. *Comput. Hum. Behav.* **2017**, *75*, 935–948. [CrossRef]

Disclaimer/Publisher's Note: The statements, opinions and data contained in all publications are solely those of the individual author(s) and contributor(s) and not of MDPI and/or the editor(s). MDPI and/or the editor(s) disclaim responsibility for any injury to people or property resulting from any ideas, methods, instructions or products referred to in the content.

engineering proceedings

MDPI

Proceeding Paper

Building a University Department Brand through Brand Association Network Technology †

Kuan-Yin Lee [1], Prasana Kumar Samanta [2] and Serhan Demirci [1,*]

1 Department of Marketing and Logistics Management, Chaoyang University of Technology, Taichung 413310, Taiwan; kylee@cyut.edu.tw
2 Department of Business Administration, Chaoyang University of Technology, Taichung 413310, Taiwan; prasanna.samanta.777@gmail.com
* Correspondence: sdemirci@cyut.edu.tw
† Presented at the 3rd IEEE International Conference on Electronic Communications, Internet of Things and Big Data Conference 2023, Taichung, Taiwan, 14–16 April 2023.

Abstract: Taiwan's birthrate has been declining for nearly a decade, with the newborn population falling from 305,312 in 2000 to 153,820 in 2021. Education has been adversely affected by the low birthrates but also by other industries. Schools must therefore review their brand positioning and develop their brand value. Brand influence, from products to retail stores, from people to organizations, was regarded as the most valuable intangible asset in the past. Schools need to establish brand symbols so that relevant briefings, publicity, stationery, posters, and packaging convey a consistent image of the school. For example, the University of Wisconsin–Madison and the Chaoyang University of Technology have used this method to build a series of brand identities. Of course, the department inherits the brand spirit of the school but is full of its characteristics and positioning. Despite this, few departments establish their brand symbols to achieve consistent brand communication, which is an important research gap in the present study. In this research, the faculty and students of the department were invited to extract the brand knowledge of the Department of Marketing and Logistics Management of Chaoyang University of Science and Technology through the idea of free associations. A total of 7 questionnaires for faculties and 50 questionnaires for students were collected; the questionnaire was composed of five dimensions. As a result, 23 symbols were obtained to represent the Department of Marketing and Logistics Management. The Department of Design's brand symbols can be used for related cultural product design, event posters, and other applications to create a specific brand association and improve brand equity, brand reputation, and memory. These symbols can also be used in related cultural products, event posters, and other applications.

Keywords: department brand symbol; free-association method; networking technology

Citation: Lee, K.-Y.; Samanta, P.K.; Demirci, S. Building a University Department Brand through Brand Association Network Technology. *Eng. Proc.* **2023**, *38*, 30. https://doi.org/10.3390/engproc2023038030

Academic Editors: Teen-Hang Meen, Hsin-Hung Lin and Cheng-Fu Yang

Published: 25 June 2023

Copyright: © 2023 by the authors. Licensee MDPI, Basel, Switzerland. This article is an open access article distributed under the terms and conditions of the Creative Commons Attribution (CC BY) license (https://creativecommons.org/licenses/by/4.0/).

1. Introduction

Aaker pointed out that the benefit of a brand is to provide the company's value proposition to consumers or to maintain the relationship between the organization and its customers, thereby enhancing consumers' trust in the company's other brand products, brand trustworthiness, expertise, and consumer confidence preference. The same applies to the education industry as well [1].

Pinar, Girard, and Basfirinci pointed out that the management of higher education brands and strengthening brand equity through brand-related links is also an important issue for universities [2]. Of course, when managers want to manage brands but fail to build their knowledge sets, brand knowledge cannot be accumulated and transmitted, and consumers' attitudes and behaviors towards brands are unclear. One of the most important aspects of brand knowledge is its identification system. Consumers' cognition can be established through the transmission of identification systems and symbols and is influenced to support a brand. Therefore, a brand manager needs to think about how to transmit the core of brand knowledge through the identification system to make brand management more efficient [3].

When a department's brand wants to convey its vision and spirit through the brand symbol, it is essential to understand the function and role of the university and then set the vision and goals. A university is not only limited to academic research and teaching inheritance but also shoulders the role of social function and student achievement development. Boyer pointed out that the university has four functions: discovery, teaching, application, and integration [4]. Therefore, for the department to convey a brand spirit and core through its brand symbol, we focus on the brand symbol of the design department through academic, positioning characteristics, teaching, and societal aspects.

We employed the interview method to find the most common keywords associated with the department's brand from the previous department's chairman, teachers, and students to design a symbol set for the production of educational materials to shape the department's brand value. At the same time, the symbol provides the stakeholders of society with a stronger sense of identity by using the brand power of the department.

2. Literature Review

A study by Iftach and Orly indicates that schools possess five characteristics [5]: (1) academic characteristics that reduce social inequality by promoting knowledge, research, learning, and achieving excellence; (2) traditional characteristics, which share a common sense of history and reputation accumulation, and respect for religious beliefs; (3) national characteristics, which are attachment, love, and commitment and care for the community; (4) organizational characteristics, reflecting innovation, efficiency, positivity, and development that enable students and teachers to achieve their potential; and (5) social and moral characteristics, emphasizing mutual respect and fairness. A complete brand symbol for a school must be created based on its vision and performance characteristics so that the students can feel the brand's entire value. However, the department continues to maintain the school's values, so brand symbols were also designed in this study aligned with the five characteristics as shown in Figure 1.

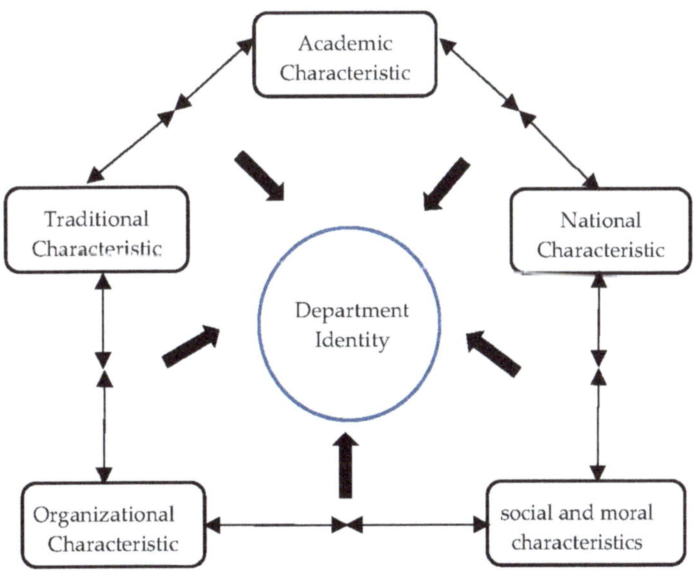

Figure 1. Department identity.

3. Research Methods

This research takes the Department of Marketing and Logistics Management of Chaoyang University of Science and Technology in Taiwan as its object. The department contains students from both the day department and the research institute. It consists of 16 teachers, 4 of whom have served as department heads. Faculty and students were invited to participate in the brand association activity to find a brand vocabulary that matches the department's brand to create a symbol that represents the department.

Faculties and students of the department could use the idea of free association to take the department's brand as the object of the association, collect vocabulary related to the department from five aspects, and associate them with the department's brand. The idea of free association is the simplest and easiest way to describe brand association. By asking consumers "If you think of a certain brand, what will you think?", a brand mental map was created to outline the brand image. Researchers can use free associativity to measure the strength of the initial association based on the order in which the words appear. Furthermore, brand preference can also be assessed based on consumers' associations.

In five dimensions, academic, traditional, national, organizational, and social and moral characteristics, our study examines the following question, "When you think of this department, what do you think of?" We used teachers' free time, an association meeting, and five free student meetings to study the association vocabulary of the department's members to create a brand for the department.

4. Research Results

Eight teachers from the Department of Marketing and Logistics Management at the Chaoyang University of Science and Technology filled out the questionnaires for this study. On average, the teachers had worked in the department for nine years and actively contributed to its development. Therefore, they represented brand vocabulary ideas. In addition, 105 students filled out the questionnaire. During the department's training process, students were divided into freshmen and seniors to understand the associated terminology of the department's brand. As a result of the relatively high integrity of third-year students' participation in department activities, a total of 40 samples were collected, with 28 samples of second-year students and 20 samples of first-year students.

The free-association principle was used to sort 46 words into 5 aspects. There were seven nouns and five adjectives in the traditional, eight nouns and four adjectives in the academic aspect, five nouns and four adjectives in the organizational aspect, four nouns and four adjectives in the social morality aspect, and three nouns and two adjectives in the country-oriented aspect. The words associated with the subjects [6] were reserved for the second-stage questionnaire test when they reached 4% of the total words. Despite not reaching 4% of the total, the data were still discussed and added to the relevant associations, since "organization" was repositioned and strengthened (Table 1).

Table 1. Vocabulary representing the five dimensions of the department.

Characteristics	Noun Vocabulary	Adjective Vocabulary
Traditional orientation	Mark Ting, orientation, chorus competition, purple, early morning activity, competition, departmental society	Lively, teamwork, innovation, enthusiasm, passionate
Academic orientation	Internships, competitions, licenses, multi-disciplinary learning, special topics, five-year consistency, micro-courses, graduate students	Serious, professional, innovative, and resourceful
Organizing	Competitions, off-campus visits, innovative curriculum design, special topics, industry–university cooperation	Teamwork, innovation, positive, professional
Social morality	Exchange of foreign students, special businessmen, achievements of friends, international exchanges	Love, respect, diverse communication, fairness
Country orientation	Public welfare activities, USR local marketing, industry–university cooperation	Caring, loving

5. Conclusions

Based on organizational vision design, this study adopts brand identification and symbolic meaning in brand equity construction and creates the logo of the department. Using the existing department's logo, a consistent image of the department and improve visibility and memory are constructed. Five aspects are used to help facilitate the application of positive word associations to literature and publicity materials. Icons need to be combined with the nature of the activity in each application. To create a stronger relationship between the map and the event theme, the icon combination is used to learn about the link map and the marketing relationship and gain a sense of uniqueness.

Additionally, it is suggested that the designed icon combination be used primarily in the promotion of the department in the future, including the arrangement of related activities, related information documents for freshmen, admission promotional videos, and so on. Icons can be matched in different combinations so that the department icon has become an identification combination with high exposure. Therefore, the students in the department have a greater sense of the importance of the icon combination. In the future, we can also take this opportunity to train marketing students to design their icon combination.

Author Contributions: Conceptualization, K.-Y.L.; Methodology, K.-Y.L.; Investigation, K.-Y.L.; Writing—original draft, P.K.S.; Writing—review and editing, S.D. All authors have read and agreed to the published version of the manuscript.

Funding: This research received no external funding.

Institutional Review Board Statement: Not applicable.

Informed Consent Statement: All participants in the study were students and informed consent was given prior to the questionnaires.

Data Availability Statement: Not applicable.

Conflicts of Interest: The authors declare no conflict of interest.

References

1. Aaker, D.A. Measuring brand equity across products and markets. *Calif. Manag. Rev.* **1996**, *38*, 102–120. [CrossRef]
2. Pinar, M.; Girard, T.; Basfirinci, C. Examining the relationship between brand equity dimensions and university brand equity: An empirical study in Turkey. *Int. J. Educ. Manag.* **2020**, *34*, 1119–1141. [CrossRef]
3. Koll, O.; Von Wallpach, S.; Kreuzer, M. Multi-method research on consumer–brand associations: Comparing free associations, storytelling, and collages. *Psychol. Mark.* **2010**, *27*, 584–602. [CrossRef]
4. Boyer, E. The university of the future: Boyer revisited. *High. Educ.* **1998**, *36*, 253–272.
5. Iftach, G.; Shapira-Lishchinsky, O. Principals' perceptions of school identity: Logo, vision and practice. *Int. J. Educ. Manag.* **2019**, 33, 1170–1184. [CrossRef]
6. John, D.R.; Loken, B.; Kim, K.; Monga, A.B. Brand concept maps: A methodology for identifying brand association networks. *J. Mark. Res.* **2006**, *43*, 549–563. [CrossRef]

Disclaimer/Publisher's Note: The statements, opinions and data contained in all publications are solely those of the individual author(s) and contributor(s) and not of MDPI and/or the editor(s). MDPI and/or the editor(s) disclaim responsibility for any injury to people or property resulting from any ideas, methods, instructions or products referred to in the content.

engineering proceedings

Proceeding Paper

Mystery of Big Data: A Study of Consumer Decision-Making Behavior on E-Commerce Websites †

Chen-Sheng Pai and Shieh-Liang Chen *

Department of Business Administration, Asia University Taichung, Taichung 41354, Taiwan; chansn.pai@gmail.com
* Correspondence: peterchen@asia.edu.tw; Tel.: +886-4-23323456 (ext. 5540)
† Presented at the 3rd IEEE International Conference on Electronic Communications, Internet of Things and Big Data Conference 2023, Taichung, Taiwan, 14–16 April 2023.

Abstract: Using big data analysis, we study the consumer life cycle based on the following four aspects: customer acquisition, participation, profit, and return visit rate. The Google Merchandise store is selected as a case study to collect data during January–December 2022. Thirteen traffic source dimension elements of the four layers were summarized and analyzed, and the following results were obtained. Consumers complete a conversion rate of 57 million. The late contact point affects the conversion rate, which is much higher than that in the early and middle periods. Reducing the number of touchpoints in the conversion increases the revenue. Understanding customers' shopping habits helps improve advertising results. Thus, website managers need to introduce Google Analytics 4 analytics at different stages for site quality and business effectiveness.

Keywords: e-commerce site; consumer behavior; Google Analytics 4

Citation: Pai, C.-S.; Chen, S.-L. Mystery of Big Data: A Study of Consumer Decision-Making Behavior on E-Commerce Websites. *Eng. Proc.* **2023**, *38*, 29. https://doi.org/10.3390/engproc2023038029

Academic Editors: Teen-Hang Meen, Hsin-Hung Lin and Cheng-Fu Yang

Published: 25 June 2023

Copyright: © 2023 by the authors. Licensee MDPI, Basel, Switzerland. This article is an open access article distributed under the terms and conditions of the Creative Commons Attribution (CC BY) license (https://creativecommons.org/licenses/by/4.0/).

1. Introduction

Since 1994, online stores have appeared, and after 1997, online business activities have become popular, replacing onsite stores. According to the UNCTAD report, in 2019, the scale of global e-commerce reached USD 26.7 trillion, equivalent to one-third of the global gross national product. At the beginning of 2020, eMarketer's forecast for the global e-commerce market share of retail sales was 16.1%, but with the COVID-19 pandemic, the ratio increased significantly. Since the advent of the Internet, e-commerce websites have provided a platform for buyers and merchants to transact and facilitate the smooth operation of transactions through the participation and communication of both parties. Scholars in the field of marketing and information decision-making systems believe that high satisfaction rates have a considerable impact on consumers to use or repeat purchases [1–3], which is important for influencing an individual's intention to continue using the product.

However, past literature on e-commerce platforms has focused primarily on community traits [4,5], community values [6–8], and community identity, and community loyalty [9]. There is little research on e-commerce websites and consumer decision-making behavior. Thus, we summarize, compare, and analyze the 13 traffic source dimension elements in the four major components being customer acquisition, participation, profit, and return visit rate, and we explore the connotation of the interaction mechanism of e-commerce websites.

Since the design of the purchase process affects consumer purchasing behavior [10], we used the five stages of consumer decision-making in the EKB core model [11–13] to analyze the need recognition, information search, and program evaluation (alternative evaluation) of purchasing and post-purchasing outcomes. The correlation between website service quality and the five stages of consumer decision-making behavior was considered because there were different connotations of website service quality with the consumption decision-making process. We integrated the intersection of rational views of website service quality and consumers' decision-making emotional views to develop this systematic theoretical framework.

The main objectives of this study are to (1) understand customers' shopping behaviors to improve advertising results, (2) analyze and improve the quality of the website and the business effectiveness, and (3) put forward the management implications.

The second chapter of this article reviews the relevant literature on the service quality and consumer decision-making of e-commerce websites, and the third chapter explains the analysis structure, data collection, and data analysis methods. The fourth chapter presents the research results and findings, and the fifth chapter concludes with recommendations and the direction of follow-up research.

2. Literature Review

Consumer decision-making is the process of purchasing products after considering various service quality factors [14]. The service industry attaches great importance to service process management, and service process design helps to formulate service improvement countermeasures and carry out new service development [15]. In this section, e-commerce website service quality, consumer decision-making mode, and a discussion are presented.

2.1. E-Commerce Website Quality of Service

Website service quality is related to platform quality [16–23], information quality [24–28], relationship quality [29–34], and quality of interaction [32,35]. The Internet has led to the increasing importance of e-commerce [36], and consumers can make purchases according to their needs without time or space constraints [37]. Post-purchase evaluation is an important information exchange for a two-way interaction between sellers and consumers and is also part of customer relationship management activities [38]. A high satisfaction evaluation is beneficial to website performance management, competitive advantage, repurchase behavior, and consumer trust [39], while too many bad reviews lead to low satisfaction, bad corporate image, complaints, low repurchases, negative word-of-mouth communication, and mistrust [40], so the interaction relationship directly affects consumers' willingness to repurchase [16,19,21]. Thus, we take consumer behavior as the main analysis data and then summarize the factors of website service quality that consumers value.

Understanding the connotation of website service quality and performing appropriate operations helps effectively manage customer relations. Consumers prefer a transactional website. The quality of website service is valued even beyond the product price [41], and the quality of website service is valued by e-commerce [42]. Website services are interactive such as e-learning and interaction [43,44], and application interaction designs [45] also convey messages that give customers a sense of trust in the quality of the website's services [35]. Based on the above literature review, no unified view was found to focus on the function of the network information platform. Several studies focused on the characteristics of the information provided on the transaction website, and the relationship maintenance with customers was emphasized. Thus, we explore the service quality of the website from the aspects of customer acquisition, participation, profit, and return visit rate. For consumer decision-making behavior, each component is described below.

2.1.1. Customer Acquisition

Customer acquisition is the default page after logging in and it provides detailed views without revealing personal information. Examples include customer composition (gender, age group, interests, place of residence, and device use) and browsing behavior (frequency of visits, browsing time, participation, and new and existing visitors), and customer origin/medium.

2.1.2. Participation

Participation helps understand what users have executed and what event transitions have been achieved. Websites provide preset events after announcing, and users can also set their events, which can be recorded in the participation reporting area. In addition to events and conversions, the common page views can be seen in "Participation" > "Pages and Screens", with in-depth statistics and insights.

2.1.3. Profit

Profit statements can be used to view the revenue generated by websites or their applications. Through products, advertisements, and subscription programs, these reports are used to understand the number of customer views of each selling good, the number of advertising exposure of the application, or other information on the revenue. The profit statement is divided into profit overview, e-commerce purchases, in-app purchases, and publisher advertising.

2.1.4. Return Visit Rate

The frequency and length of interaction with the website are shown after the user's first visit to the website or APP for the user to understand value according to the additional revenue generated by the user after the first visit to the website.

2.2. *Consumer Decision-Making Model*

Consumer decision-making patterns refer to the process of finding, purchasing, using, evaluating, and disposing of products or services [46]; participating in the acquisition, use, and disposal of products and services; engaging in decision-making processes and actual behavior [11–13]; and making decisions related to people's purchase and use of products or services [14]. It is narrowly defined as the decision made by individuals at various stages to acquire goods or services, while the broad definition includes the consumption behavior of non-profit organizations, industrial organizations, and intermediaries [11–13]. In addition, Yue and Stuart pointed out that virtual community transactions involved a range of factors with different influencing factors at different stages which affected decision-making at each stage, such as target selection, information processing, memory, degree of involvement, attitude, interference, and consumer attribution [11–13,47].

Regarding the consumption decision-making process model, several studies adopted the following three stages: pre-purchase, consumption, and post-purchase. The influencing factors include the brand, the product itself, or the type of service, while the influencing factors in the consumption stage include website services, other customers, store design, and equipment performance. These factors become the basis for satisfaction and service quality evaluation in the post-purchase stage [41]. Other studies used seven stages of consumer decision-making, including demand confirmation, information search, substitute evaluation, purchase, consumption, post-purchase evaluation, and abandonment, which were affected by electronic forums, electronic bulletin boards, and virtual community formation [48]. The previous studies divided consumer decision-making into different stages according to the time of occurrence or divided into three or five stages. The five stages of analysis included need recognition, information search, alternative evaluation, purchase, and post-purchase [11–13,47,49].

2.2.1. Demand Confirmation Stage

The demand confirmation stage is based on consumers' intrinsic demand motivation and external stimulus [11–13]. The attributes of stores affect purchase decisions, and consumers' trust level affected website performance [50]. Website performance is affected by brand identity, product knowledge, and product involvement [10], and intrinsic demand drives actual behavior. Website performance is related to the differentiation and comparison of the form of the desired need and the actual object, the experience of product use, the satisfaction value of expectations, the feeling of satisfaction, the external conditions of consumers to look for better products and designs, and international brands [48].

2.2.2. Information Search Stage

In the information search stage, third-party search service providers use intermediaries to obtain purchase information [32]. Keyword searches affect purchasing behavior and change website sales performance [51] for easy use and for consumers to trust the website more [52]. Online stores are subject to competition from external search processes [39], and they improve their profits by adjusting pricing strategies [53]. Influencing factors at this stage include knowledge [47], product presentation, evaluation of alternatives, communication, word of mouth, knowledge sharing [48], brand loyalty, and other factors, but there is a risk of information authenticity [54]. Simplifying the graphical design allows consumers to better understand whether the composition of the product meets needs such as the expected benefits, price, model, and idea guidance to obtain specific purchase objects [10]. Message response times and prices influence further information search behavior to find alternative options or go directly to the buying behavior stage.

2.2.3. Program Evaluation Stage

The scenario evaluation stage refers to alternatives to choose from [47], distinguishing alternative attributes and differences, and making it easier for buyers to make a decision. When consumers gather the required information, they evaluate various feasible solutions in four parts, as follows: evaluation criteria, beliefs, attitudes, and intentions [51]. Reference [51] pointed out that at this stage, consumers attach importance to the detailed evaluation of products on the website as well as the provision of services such as secure transaction mechanisms, money-back guarantees, and third-party guarantees. Alternatives are evaluated to select the best option by interacting with the seller [48]. If the substitution demand is met, the purchase is made, otherwise the purchase may be stopped or the search may be restarted.

2.2.4. Purchase Stage

The purchase stage refers to the best option chosen by the consumer and taking the action of purchasing [11–13] through a well-established product recommendation system and a mechanism that helps consumers identify products [55,56]. Consumers' purchasing preferences, funds, social status, and shopping preferences influence the purchase decision impact [50]. Brands can enhance the corporate image [54], and consumers increasingly rely on reviews published by social networks to guide their purchase behavior [57], influenced by purchase process design, branding, product recognition [10], product quality assurance, clear refund mechanism, and product delivery specifications. Factors such as certificates issued by trust institutions, adequate security technical descriptions, privacy protection [51], culture, and personality influence consumer purchasing behavior [48], while appropriate payment mechanisms enable consumers to increase their willingness to pay and carry out purchasing actions [58], and then enter the post-purchase behavior stage.

2.2.5. Post-Purchase Behavior Stage

In the post-purchase behavior stage, consumer perception and satisfaction have positive value, and even increase customer stickiness and strengthen competitive advantage [59]. Consumers' purchase experience and their memory influence future purchase decisions [11–13]. Good brand experience [10] provides quality returns and services [59] and brand loyalty, which increases consumers' willingness to buy and pay and recommend to other consumers [60]. Post-purchase reviews are related to satisfaction [50] and assist new consumers in making decisions [10]. Bad reviews lead to low satisfaction, poor corporate image, complaints, low repurchases, and negative word-of-mouth communication [40]. At this stage, the consumer's payment status determines whether to make a profit [20,61]. Buyers' wishful abandonment [39] or evaluation of manipulative behavior [57] are factors that contribute to disputes. Even if it is a good product, there are still cases of it being returned and exchanged. There must be a good contract as a constraint, including refund and return mechanisms, which provide consumer protection. Since consumers are dissatisfied with the goods they have chosen, leading to returns and exchanges or even litigation, mechanisms must be properly developed [59].

In summary, using the five-stage model of consumer decision-making adopted by most scholars as the research index and Google Analytics 4 (GA4), we analyze the logical concepts and influencing factors of consumers when making purchase decisions. The result provides the service quality elements of consumer demand and subsequent comparison of the transaction website at each stage.

3. Research Methods

GA4 was used as a research analysis tool to study four major reports, including customer acquisition, participation, profitability, and return visit rate. Quantitative analysis in different dimensions was carried out to specify the relationship and impact between indicators, interpret key indicators, and improve website performance.

3.1. Research and Analysis

Based on the data compilation and theoretical discussion in Chapter 2, the GA4 traffic analysis tool was applied for website traffic analysis and website performance improvement. Google Merchandise Store was selected as an analysis case, and data during January–December 2022 were collected to determine indicators and their impact on the performance of the e-commerce platform of Google Merchandise Store. Indicator traffic was analyzed and compared for websites, and feedback was gathered to improve it (Figures 1 and 2).

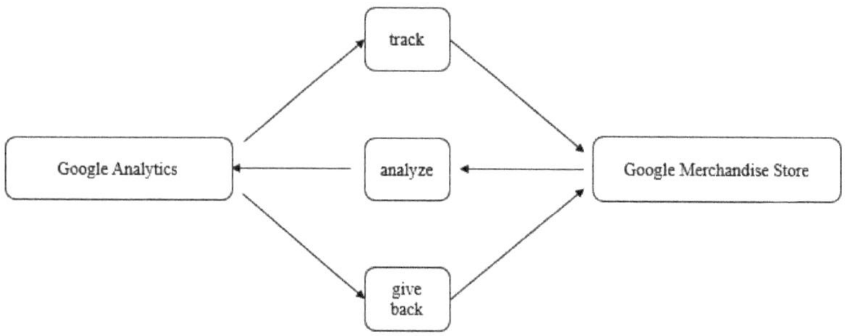

Figure 1. Research methodology.

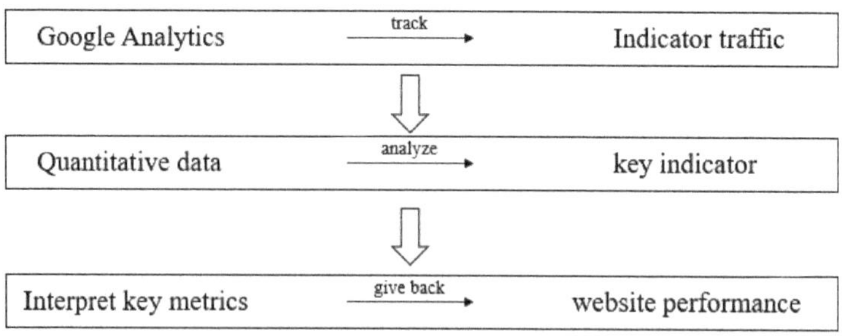

Figure 2. Analysis process.

3.2. Data Analysis

Based on the data, we interpreted the four major reports of GA4 analysis tools: customer acquisition, participation, profit, and return visit rate (Table 1). Different dimensions (Dimension) and their corresponding specific indicators (Metric) are shown in Table 2. The meaning behind the data is statistically summarized as the business significance represented by the analysis to provide a reference and suggestion for optimizing and improving website performance.

Table 1. Four major reports.

Customer acquisition	Customer acquisition is the default page after logging in to GA4, which can provide detailed viewer information without revealing personal information. Examples include customer composition (gender, age group, interests, place of residence, device use), browsing behavior (frequency of visits, browsing time, participation, new and existing visitors), and customer origin/medium (how you came to the website).
Participate	Participation allows us to understand what users have executed and what event transitions have been achieved. GA4 itself provides some preset events that can be started after announcing, and you can also set your own events, bury parameters, and then set events as transitions! Once you've set up an event or conversion, you can do more detailed drill-down tracking in the participation reporting area.In addition to events and conversions, you can also see the common page views of General GA from the past in "Participation"—"Pages and Screens", as well as in-depth statistics and insights on a page-by-page basis!
Profit	Profit reports can be used to view the revenue generated by our own websites or applications, through products, advertisements, and subscription programs. We can use these reports to understand the number of customer views of each selling good, the number of advertising impressions of the application, or other information that can bring revenue to the merchant. In the profit statement, it is divided into profit overview, e-commerce purchases, in-app purchases, and publisher advertising.
Return visit rate	Displays the frequency and length of interaction with the website after the user's first visit to the website or APP, and can be used to understand the user value according to the additional revenue generated by the user after the first visit to the website.

Table 2. Dimensions and indicators.

Dimension	Metric
Traffic sources, platform devices, Geolocation, Life cycle	Number of new users, interactive sessions, participation, interaction sessions per user, average participation time, event count, conversions, total revenue.

We analyzed the traffic sources, sessions, interactive sessions, participation, interactive sessions per user, activity per session, average participation time, event count, and conversions. The interaction between the dimensions and indicators was used to interpret the meaning and the impact on the performance of the website, put forward the best improvement policy, find out the best conversion path for consumers, improve the revenue of the website, and propose a guiding principle of sustainable growth. The list of specific traffic sources and indicator data collected and analyzed in this study are shown in Tables 3 and 4.

Table 3. Traffic sources.

Source of Traffic	Illustrate
Direct	Traffic from the visitor directly to the destination website URL
Organic Search	The traffic of visitors to the target website through search engines
Referral	Traffic from visitors to the target website through links from external websites
Paid Search	Traffic from visitors who enter your website by clicking on keyword ads in search results
Organic Social	Traffic from visitors to the target website via S ocial

Table 4. Indicators.

Indicators	Illustrate
User	Total number of active users
Work Phase	The number of sessions started on the website or application
Interactive session phase	Lasting more than 10 s, transitioning events, or more than 2 screen or web browsing sessions
Average time per session participation	User participation time per session
Each user interaction session	Number of interactive sessions per user (Interactive Sessions/User)
Activity per session	The average number of events per session
Participation	Percentage of interaction sessions ("interactive sessions" divided by "sessions")
Event count	The number of times the user triggered the event
conversion	The number of times a user triggered a conversion event
Total revenue	Sum of revenue from purchases, subscriptions, and advertising ("Purchase Revenue" plus "Subscription Revenue" plus "Advertising Revenue")

3.3. *Research Objectives and Verification*

The interrelationship between internal traffic indicators was analyzed based on the total revenue measurement. The interrelationship of sessions, interactive meeting stages, average single session participation time, interactive sessions per user, activities per session, participation, event count, and conversion was also analyzed to quantify the data, observe and explain their meaning, and find out the advantages and disadvantages. The ultimate goal is to improve the physical fitness of the site and enhance the performance of the website (Figure 3).

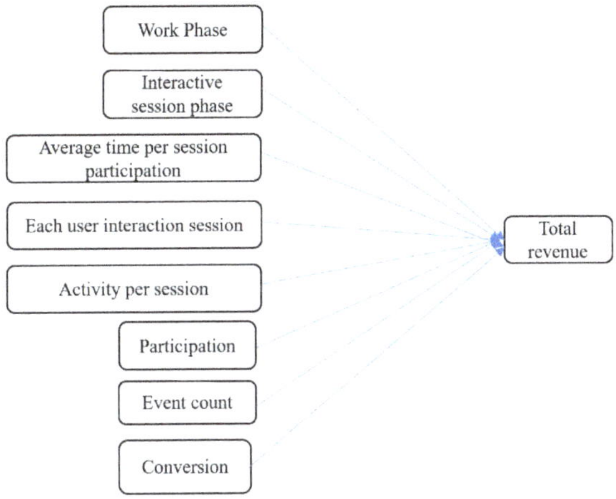

Figure 3. Relationship between internal traffic indicators.

The impact of traffic sources was investigated on total website revenue, as well as the impact of sessions, interactive sessions, average time per session participation, interactive sessions per user, activity per session, participation, event count, and conversions as mediating factors on total revenue (Figure 4).

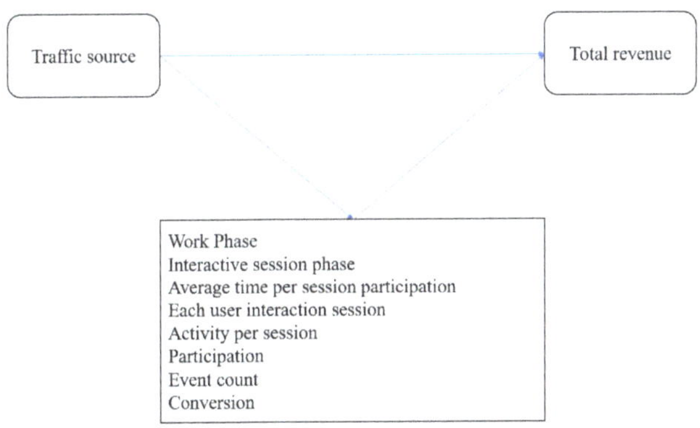

Figure 4. Relationship between traffic sources, indicators, and total revenue.

4. Empirical Analysis

Target websites were investigated based on GA4. The data of 365 days were collected and used for quantitative statistical analysis. The relationship between different dimensions, specific indicators, and the impact on website performance was analyzed to determine website performance indicators. The traffic sources of Google Merchandise Store were quantified and analyzed. The results are presented in Tables 5 and 6. The correlation and impact between external traffic and indicators are discussed later in this article.

Table 5. Basic traffic data of target website.

During the analysis	1 January 2022~31 December 2022
Total number of days	365 days
Site name	Google Merchandise Store
Site type	E-commerce
User (Person)	896,921
New User (Person)	843,322
Participation (%)	63.34
Average time involved	1 min 22 s

Table 6. External traffic data of target website.

During the analysis	1 January 2022~31 December 2022
Total number of days	365 days
Site name	Google Merchandise Store
Site type	E-commerce
The type of traffic from the external source	Flow rate (times)
Total external source traffic	896,921
Direct	336,892
Organic Search	312,536
Unassigned	74,139
Referral	49,625
Paid Search	47,906
Display	44,783
Paid Shopping	19,655
Organic Social	9615
Paid Video	9586
Email	9220
Organic Video	7451
Affiliates	2426
Organic Shopping	1889

4.1. Website Indicator Performance

The traffic metrics of the target website were analyzed with SPSS to find the degree of mutual influence between the indicators (Table 7). The result shows that the meeting stage is positively correlated with the total return. The result of the indicator corresponding to the traffic source is shown in Table 8. There is a positive correlation between interactive work stages and total returns.

Table 7. Sessions and total revenue.

		Total Revenue
Work Phase	Pearson correlation	0.981 **
	Saliency (two-tailed)	0.000
	N	13

** Correlation is significant at the 0.01 level (double-tailed).

Table 8. Relationship between interactive sessions and total revenue.

		Total Revenue
Interactive session phase	Pearson correlation	0.976 **
	Saliency (two-tailed)	0.000
	N	13

** Correlation is significant at the 0.01 level (double-tailed).

The relationship between the participation time in the average single work stage and the traffic source is shown in Table 9. There is no correlation of the participation time in the average single work stage and the traffic source.

Table 9. Average participation time and total benefit per work phase.

		Total Revenue
Average time per session participation	Pearson correlation	0.240
	Saliency (two-tailed)	0.431
	N	13

Each user's interaction stage does not correlate with total revenue (Table 10).

Table 10. Interactive sessions and total revenue per user.

		Total Revenue
Each user interaction session	Pearson correlation	0.172
	Saliency (two-tailed)	0.573
	N	13

The activities at each work stage are not correlated with total returns (Table 11).

Table 11. Activity and total revenue per session.

		Total Revenue
Activity per session	Pearson correlation	0.233
	Saliency (two-tailed)	0.444
	N	13

There is no correlation between participation and total revenue (Table 12).

Table 12. Participation and total revenue.

		Total Revenue
participation	Pearson correlation	0.151
	Saliency (two-tailed)	0.621
	N	13

The event count is positively correlated with the total revenue (Table 13).

Table 13. Event count and total revenue.

		Total Revenue
Event count	Pearson correlation	0.987 **
	Saliency (two-tailed)	0.000
	N	13

** Correlation is significant at the 0.01 level (double-tailed).

Conversions and total revenues are positively correlated (Table 14).

Table 14. Conversions and total revenues.

		Total Revenue
Conversion	Pearson correlation	0.970 **
	Saliency (two-tailed)	0.000
	N	13

** Correlation is significant at the 0.01 level (double-tailed).

4.2. *Comparison of Website Traffic Source*

The analysis and comparison process are shown in Figure 5.

The impact of traffic sources was analyzed on total website revenue, the impact of sessions, interactive sessions, average time per session participation, interaction sessions per user, activity per session, participation, event count, and conversions. The analysis results are shown in Table 15.

The standardized coefficient of traffic source and total revenue was β: 0.979 **** at the significance level of 0.000, showing that traffic source is positively correlated with total revenue. Thus, traffic sources affected total revenue. Interactive sessions are negatively correlated with total revenue as an intermediary factor between traffic sources and total revenue (β: −1.998 ** at the significance level of 0.008). The average participation time of a single work stage is positively correlated with the total revenue as an intermediary factor between the traffic source and total revenue (β: 0.234 * at the significance level of 0.043). Each user's interaction session does not correlate with the total revenue as the intermediary factor between the traffic source and the total revenue (β: 0.150 at the significance level of 0.103). Activity at each work stage is negatively correlated with total revenue as an intermediary factor between traffic sources and total revenue (β: −0 419 * at the significance level of 0.033). Participation is positively correlated with total revenue as an intermediary factor between the traffic source and total revenue (β: 0.155* at the significance level of 0.040). The event count is positively correlated with the total revenue as an intermediary factor between the traffic source and total revenue (β: 2.739 ** at the significance level of 0.010). Conversion does not correlate with total revenue as an intermediary factor between the traffic source and total revenue (β: −0.683 ** at the significance level of 0.399).

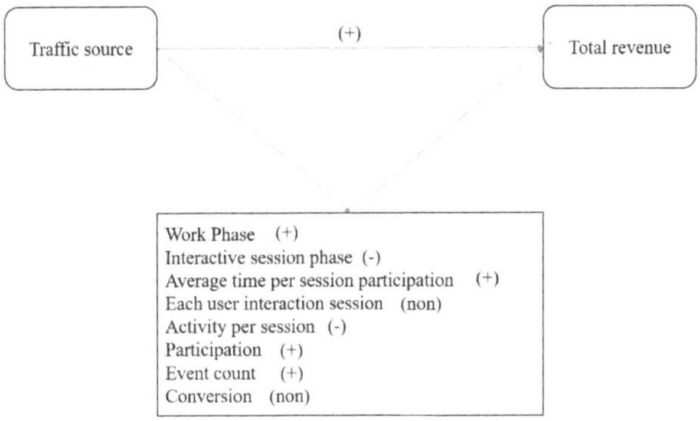

Figure 5. Website traffic source analysis and comparison process.

Table 15. Regression coefficient analysis.

Coefficient a						
Model		Non-Normalized Coefficients B	Non-Normalized Coefficients Standard Error	Normalization Factor B	T	Salience
1	(constant)	−21,056.449	26,123.721		−0.806	0.437
	Traffic sources	3.005	0.190	0.979	15.802	0.000
2	(constant)	−121,831.188	81,298.402		−1.499	0.208
	Traffic sources	2.822	2.388	0.919	1.182	0.303
	Interactive session phase	−6.074	1.220	−1.998	−4.978	0.008
	Average time per session participation	1931.629	662.366	0.234	2.916	0.043
	Each user interaction session	113,799.812	54,065.559	0.150	2.105	0.103
	Activity per session	−15,285.154	4758.270	−0.419	−3.212	0.033
	Participation	296,519.318	98,753.591	0.155	3.003	0.040
	Event count	0.224	0.048	2.739	4.672	0.010
	conversion	−1.328	1.409	−0.683	−0.942	0.399

The analysis result of the traffic source data of the Google Merchandise Store is shown in Table 16. The number of days required for conversion was 0 with 577,899.00 conversions accounting for 37.85%. For Organic Search, after 1 touchpoint, the number of days required to convert was 0.9, and there were 327,165.00 conversions accounting for 21.43%. Paid Search, after 1 touchpoint, showed the number of days required to convert was 0.9 with 71,781.00 conversions accounting for 4.70%. The results showed that Direct, Organic Search, and Paid Search resulted in higher conversion times, shorter changeover time, shorter path length (all contact points), and higher transition.

Table 16. Conversion path.

	Traffic Sources	Conversion	Purchase Proceeds	The Number of Days It Takes for the Conversion to Occur	The Contact Point That Passed before Conversion
		1,526,776	2,572,828.91	1.69	2.78
1	DirectDirect	577,899	1,155,992.7	0	1
2	Organic Search	327,165	81,713.78	0.9	1
3	Paid SearchPaid Search	71,781	26,162.68	0.9	1
4	DisplayDisplay	50,836	161	0.57	1
5	UnassignedUnassigned	45,037	909.44	0.01	1
6	ReferralReferral	41,841	38,973.62	1.27	1
7	Organic Search × 3	38,230	112,658.25	4.67	3
8	Organic Search × 4	32,637	85,832.99	3.54	4
9	Paid Shopping	27,260	11,338.76	0.89	1
10	Organic Search × 2	23,040	126,939.31	4.11	2

5. Conclusions and Recommendations

The purpose of this study is to analyze the traffic data of the Google Merchandise Store with GA4 and explore the relationship between the variables of related traffic indicators. The impact of website performance was also investigated. The results were used to make website operation suggestions, improve website content, optimize website physique, enhance website performance, and achieve the purpose of long-term development of websites. The conclusions of this study are described as follows.

The results of this study show that the correlation between the work stage and the total income is significant. Thus, the working stage is positively correlated with the total return, and the total income is subject to the related variables. The interaction session is significantly correlated with the total revenue, indicating that the total return is affected by the change in the interactive work stage variables. The participation time in the average single work stage is not significantly related to the total return. The correlation between each user's interactive session and total revenue is not significant. The activity of each work stage is not significantly related to the total return. The correlation between participation and total revenue is not significant, while the correlation between event count and total revenue is significant. Event count is positively correlated with total revenue, and the conversion is also significantly correlated with the total return.

The traffic sources are positively correlated with total revenue, and traffic sources affect total revenue, while interactive sessions are negatively correlated with total revenue. When interactive sessions increase, total revenue decreases. The average participation time of a single work stage is positively correlated with the total revenue. Each user's interaction work stage is not related to the total revenue. The activities of each work stage are negatively correlated with the total revenue. Participation is positively correlated with total revenue, and event count is positively correlated with total revenue. Conversion does not correlate with total revenue.

The results of this study can be used as a reference for website managers or digital marketers to find the best conversion path model for the target website at various specific comparison times.

Practical suggestions can be made for website operators engaged in online store operation or shopping websites.

First, the interaction work and the activity of each session needs to be reduced to increase the average time per session, participation, and event count, which can help increase total website revenue. Second, the number of days for a conversion needs to be reduced so that a conversion can increase conversion rates. Website managers or digital marketers must find the best conversion path model for the target website at various specific comparison times. The traffic source is an important variable affecting conversion. Direct, Organic Search, and Paid Search have high conversion rates. The demand identification stage, information search stage, and solution evaluation stage must be designed to meet consumer's needs and to motivate purchasing.

As technology is changing rapidly and information in the online world is updated rapidly, future follow-up research is required. First, it is necessary to collect and analyze data from different dimensions and metrics. We only analyzed the two dimensions of visitor type and traffic source to discuss specific indicators such as bounce rate, page view rate, departure rate, click-through rate, and conversion rate. Other dimensions and indicators need to be analyzed to increase the integrity of the theoretical framework. Other analysis methods need to be adopted with GA4 traffic analysis tools to analyze the relationship between various dimensions and indicators more accurately and propose more reliable and powerful analysis results. Then, the research analysis method is more evolved and strengthened, which is worth exploring for subsequent researchers. The main source of visitors to the Google Merchandise Store, the target website of this study, is the United States. In the future, other variables that affect different cultural backgrounds under different cultural differences also must be explored.

Author Contributions: Conceptualization, C.-S.P. and S.-L.C.; methodology, C.-S.P. and S.-L.C.; software, C.-S.P.; validation, C.-S.P. and S.-L.C.; formal analysis, C.-S.P.; investigation, C.-S.P.; resources, C.-S.P.; data curation, C.-S.P.; writing—original draft preparation, C.-S.P.; writing—review and editing, C.-S.P.; visualization, C.-S.P. All authors have read and agreed to the published version of the manuscript.

Funding: This research received no external funding.

Institutional Review Board Statement: Not applicable.

Informed Consent Statement: Not applicable.

Data Availability Statement: Not applicable.

Conflicts of Interest: The authors declare no conflict of interest.

References

1. Bhattacherjee, A. An empirical analysis of the antecedents of electronic commerce service continuance. *Decis. Support Syst.* **2001**, *32*, 201–214. [CrossRef]
2. Bhattacherjee, A. Understanding information systems continuance: An expectation- confirmation model. *MIS Q.* **2001**, *25*, 351–370. [CrossRef]
3. Gustafsson, A.; Johnson, M.D.; Roos, I. The Effects of Customer Satisfaction, Relationship Commitment Dimensions, and Triggers on Customer Retention. *J. Mark.* **2005**, *69*, 210–218. [CrossRef]
4. McAlexander, J.H.; Schouten, J.W.; Koenig, H.F. Building Brand Community. *J. Mark.* **2002**, *66*, 38–54. [CrossRef]
5. Muniz, A.M., Jr.; O'Guinn, T.C. Brand Community. *J. Consum. Res.* **2001**, *27*, 412–432. [CrossRef]
6. Eggert, A.; Ulaga, W. Customer Perceived Value: A Substitute for Satisfaction in Business Market? *J. Bus. Ind. Mark.* **2002**, *17*, 107–118. [CrossRef]
7. Monroe, K.B. *Pricing: Making Profitable Decision*, 3rd ed.; McGraw-Hill: New York, NY, USA, 1990.
8. Sheth, J.N. *Consumption Values and Market Choice*, 7th ed.; South-Western Publishing: Nashville, TN, USA, 1991.
9. Farquhar, J.; Rowley, J. Relationships and Online Consumer-Communities. *Bus. Process Manag. J.* **2006**, *12*, 162–177. [CrossRef]
10. Teo, T.S.H.; Yeong, Y.D. Assessing the consumer decision process in the Digital marketplace. *Omega* **2003**, *31*, 349–363. [CrossRef]
11. Engel, J.E.; Blackwell, R.D.; Kollat, D.T. *Consumer Behavior*, 1st ed.; Holt Rinehart and Winston: New York, NY, USA, 1968.
12. Engel, J.E.; Blackwell, R.D.; Miniard, P.W. *Consumer Behavior: The Role of the Consumer in Marketing*, 2nd ed.; Holt Rinehart and Winston: New York, NY, USA, 1973.
13. Engel, J.E.; Blackwell, R.D.; Miniard, P.W. *Consumer Behavior*, 6th ed.; The Dryden Press: Chicago, IL, USA, 1990.
14. Walters, C.G.; Paul, G.W. *Consumer Behaviors: An Integrated Framework*, 3rd ed.; Irwin: Homewood, IL, USA, 1970.
15. Korsching, P.F.; El-Ghamrini, S. Rural telephone company adoption of service innovations: A community field theory. *Rural Sociol.* **2003**, *68*, 387–409. [CrossRef]
16. Lu, J.; Wang, L.; Hayes, L.A. How do technology readiness, platform functionality and trust influence C2C user satisfaction? *J. Electron. Commer. Res.* **2012**, *13*, 50–69.

17. Lu, L.; Cooper, C.L.; Kao, S.F.; Zhou, Y. Work Stress, Control Beliefs and Well-being in Greater China: An Exploration of Sub-cultural Differences between the PRC and Taiwan. *J. Manag. Psychol.* **2003**, *18*, 479–510. [CrossRef]
18. Thadathil, P.; Ghosh, A.K.; Sarupria, J.S.; Gopalakrishna, V.V. An interactive graphical system for XBT data quality control and visualization. *Comput. Geosci.* **2001**, *27*, 867–876. [CrossRef]
19. Wang, S.M.; Judy, L.C.C. The effect of social influence on bloggers' usage intention. *Online Inf. Rev.* **2010**, *35*, 50–65. [CrossRef]
20. Wu, C.C.; Liu, Y.F.; Chen, Y.J.; Wang, C.J. Consumer responses to price discrimination: Discriminating bases, inequality status, and information disclosure timing influences. *J. Bus. Res.* **2012**, *65*, 106–116. [CrossRef]
21. Xie, M.; Wang, H.; Goh, T.N. Quality Dimensions of Internet Search Engines. *J. Inf. Sci.* **1998**, *24*, 365–372. [CrossRef]
22. Yen, Y.R.; Tsai, B.Y. Exploring the influential factors toware the continuance intention of on-linr of books purchase. *Int. J. Organ. Innov.* **2011**, *3*, 140–157.
23. Yue, G.; Stuart, B. Virtual item purchase behavior in virtual worlds: An exploratory investigation. *Electron. Commer. Res.* **2009**, *9*, 77–96.
24. Ainin, S.; Bahri, S.; Ahmad, A. Evaluating portal performance: A study of the National Higher Education Fund Corporation (PTPTN) portal. *Telemat. Inform.* **2012**, *29*, 314–323. [CrossRef]
25. Bonner, J.M. Customer interactivity and new product performance: Moderating effects of product newness and product embeddedness. *Ind. Mark. Manag.* **2010**, *39*, 485–492. [CrossRef]
26. Chen, Y.R.R. Exploring environmental scanning for China's government information: A government official's perspective. *Public Relat. Rev.* **2013**, *39*, 581–583. [CrossRef]
27. DeGroote, S.E.G.; Marx, T.G. The impact of IT on supply chain agility and firm performance: Anempirical investigation. *Int. J. Inf. Manag.* **2013**, *33*, 909–916. [CrossRef]
28. Freitas, A.; Knapa, T.; O'Riain, S.; Curry, E. W3P: Building an OPM based provenance model for the Web. *Future Gener. Comput. Syst.* **2011**, *27*, 766–774. [CrossRef]
29. Alejandro, T.B.; Souza, D.V.; Boles, J.S.; Ribeiro, Á.H.P.; Monteiro, P.R.R. The outcome of company and account manager relationship quality on loyalty, relationship value and performance. *Ind. Mark. Manag.* **2011**, *40*, 36–43. [CrossRef]
30. Bloemer, J.; Pluymaekers, M.; Odekerken, A. Trust and affective commitment as energizing forces for export performance. *Int. Bus. Rev.* **2013**, *22*, 363–380. [CrossRef]
31. Čater, T.; Čater, B. Product and relationship quality influence on customer commitment and loyalty in B2B manufacturing relationships. *Ind. Mark. Manag.* **2010**, *39*, 1321–1333. [CrossRef]
32. Gu, B.; Konana, P.; Chen, H.M. Identifying consumer consideration set at the purchase time from aggregate purchase data in online retailing. *Decis. Support Syst.* **2012**, *53*, 625–633. [CrossRef]
33. Lee, Y.K.; Kim, Y.S.; Lee, K.H.; Li, D.X. The impact of CSR on relationship quality and relationship outcomes: A perspective of service employees. *Int. J. Hosp. Manag.* **2012**, *31*, 745–756. [CrossRef]
34. Tian, Q. Social Anxiety, Motivation, Self-Disclosure, and Computer-Mediated Friendship: A Path Analysis of the Social Interaction in the Blogosphere. *Commun. Res.* **2011**, *40*, 237–260. [CrossRef]
35. Loiacono, E.T.; Watson, R.T.; Goodhue, D.L. WebQual (TM): A Web Site Quality Instrument. In Proceedings of the AMA Winter Conference, Austin, TX, USA, 22–25 February 2002.
36. Hagel, J.; Armstrong, A. Net Gain: Expanding markets through virtual communities. *McKinsey Q.* **1997**, *1*, 140–153. [CrossRef]
37. Woo, G.K.; Chang, L.; Stephen, J.H. Effects of an Online Virtual Community on Customer Loyalty and Travel Product Purchases. *Tour. Manag.* **2004**, *25*, 343–355.
38. Kim, J.; Wansink, B. How Retailers' Recommendation and Return Policies Alter Product Evaluations. *J. Retail.* **2012**, *88*, 528–541. [CrossRef]
39. Keaveney, S.M.; Huber, F.; Herrmann, A. A model of buyer regret: Selected prepurchase and post purchase antecedents with consequences for the brand and the channel. *J. Bus. Res.* **2007**, *60*, 1207–1215. [CrossRef]
40. Kalamas, M.; Laroche, M.; Makdessian, L. Reaching the boiling point: Consumers' negative affective reactions to firm-attributed service failures. *J. Bus. Res.* **2008**, *61*, 813–824. [CrossRef]
41. Sun, C.C.; Lin, G.T.R. Using fuzzy TOPSIS method for evaluating the competitive advantages of shopping websites. *Expert Syst. Appl.* **2009**, *36*, 11764–11771. [CrossRef]
42. Ruyter, K.D.; Wetzels, M.; Kleijnen, M. Customer adoption of e-service: An experimental study. *Int. J. Serv. Ind. Manag.* **2001**, *12*, 184–207. [CrossRef]
43. Lanzilottia, R.; Ardito, C.; Costabile, M.F.; DeAngeli, A. Do patterns help novice evaluators: A comparative study. *Int. J. Hum.-Comput. Stud.* **2011**, *69*, 52–69. [CrossRef]
44. Yang, S.C.; Liu, S.F. The study of interactions and attitudes of third-grade students learning information technology via a cooperative approach. *Comput. Hum. Behav.* **2005**, *21*, 45–72. [CrossRef]
45. Arvin, S.A.; House, D.H. Modeling architectural design objectives in physically based space planning. *Autom. Constr.* **2002**, *11*, 213–225. [CrossRef]
46. Schiffman, L.G.; Leslie, L. *Consumer Behavior*, 2nd ed.; Prentice Hall: Englewood Cliffs, NJ, USA, 1993.
47. Puccinelli, N.M.; Goodstein, R.C.; Grewal, D.; Price, R.; Raghubir, P.; Stewart, D. Customer Experience Management in Retailing: Understanding the Buying Process. *J. Retail.* **2009**, *85*, 15–30. [CrossRef]

48. Valck, K.D.; van Bruggen, G.H.; Wierenga, B. Virtual communities: A marketing perspective. *Decis. Support Syst.* **2009**, *47*, 185–203. [CrossRef]
49. Roozmand, O.; Aghaee, N.G.; Hofstede, G.J.; Nematbakhsh, M.A.; Baraani, A.; Verwaart, T. Agent-based modeling of consumer decision making process based on power distance and personality. *Knowl.-Based Syst.* **2011**, *24*, 1075–1095. [CrossRef]
50. Kiang, M.Y.; Ye, Q.; Hao, Y.; Chen, M.; Li, J. A service-oriented analysis of online product classification methods. *Decis. Support Syst.* **2011**, *52*, 28–39. [CrossRef]
51. van Schie, R.J.G.; Donkers, B.; Dellaert, B.G.C. Savings adequacy uncertainty: Driver or obstacle to increased pension contributions? *J. Econ. Psychol.* **2012**, *33*, 882–896. [CrossRef]
52. Chau, P.Y.K.; Hu, P.J.H.; Lee, B.L.P.; Au, A.K.K. Examining customers' trust in online vendors and their dropout decisions: An empirical study. *Electron. Commer. Res. Appl.* **2007**, *6*, 171–182. [CrossRef]
53. Chhabra, M.; Das, S.; Sarne, D. Expert-mediated sequential search. *Eur. J. Oper. Res.* **2014**, *234*, 861–873. [CrossRef]
54. Backhaus, K.; Steiner, M.; Lugger, K. To invest, or not to invest, in brands? Drivers of brand relevance in B2B markets. *Ind. Mark. Management.* **2011**, *40*, 1082–1092. [CrossRef]
55. Lin, C.T.; Hong, W.C.; Chen, Y.F.; Dong, Y. Application of salesman-like recommendation system in 3G mobile phone online shopping decision support. *Expert Syst. Appl.* **2010**, *37*, 8065–8078. [CrossRef]
56. Lin, J.C.; Hsieh, P.L. The influence of technology readiness on satisfaction and behavioral intentions toward self-service technologies. *Comput. Hum. Behav.* **2007**, *23*, 1597–1615. [CrossRef]
57. van Dem, Y.K.; van Trijp, H.C. Cognitive and motivational structure of sustainability. *J. Econ. Psychol.* **2011**, *32*, 726–741. [CrossRef]
58. Li, T.; Meshkova, Z. Examining the impact of rich media on consumer willingness to pay in online stores. *Electron. Commer. Res. Appl.* **2013**, *12*, 449–461. [CrossRef]
59. Kuo, Y.F.; Wub, C.M.; Deng, W.J. The relationships among service quality, perceived value, customer satisfaction, and post-purchase intention in mobile valueadded services. *Comput. Hum. Behav.* **2009**, *25*, 887–896. [CrossRef]
60. Fedorikhin, A.C.; Park, W.; Thomson, M. Beyond fit and attitude: The effect of emotional attachment on consumer responses to brand extensions. *J. Consum. Psychol.* **2008**, *18*, 281–291. [CrossRef]
61. Wu, J.; Wu, Y.; Sun, J.; Yang, Z. User reviews and uncertainty assessment: A two stage model of consumers, willingness-to-pay in online markets. *Decis. Support Syst.* **2013**, *55*, 175–185. [CrossRef]

Disclaimer/Publisher's Note: The statements, opinions and data contained in all publications are solely those of the individual author(s) and contributor(s) and not of MDPI and/or the editor(s). MDPI and/or the editor(s) disclaim responsibility for any injury to people or property resulting from any ideas, methods, instructions or products referred to in the content.

engineering proceedings

MDPI

Proceeding Paper

Analysis and Research on Activity-Based Bottleneck Model †

Hua Li [1], Yuanxia Zhang [1,*], Jie Lv [1], Daoqing Gong [2] and Xinyan Gan [2]

1 School of Computer Science and Engineering, Yulin Normal University, Yulin 537000, China; jgxylh3018@163.com (H.L.); lvjie@ylu.edu.cn (J.L.)
2 School of Public Health and Management, Guangxi University of Chinese Medicine Nanning, Nanning 530200, China; 18775300556@163.com (D.G.); wsx66113969@163.com (X.G.)
* Correspondence: jgxyzyx@126.com
† Presented at the 3rd IEEE International Conference on Electronic Communications, Internet of Things and Big Data Conference 2023, Taichung, Taiwan, China, 14–16 April 2023.

Abstract: Based on Vickrey's bottleneck model, we propose an activity-based bottleneck model to study traffic congestion on freeways during holidays. Based on the activity-based bottleneck model, we study the bottleneck ladder charging problem under the linear marginal activity utility and compare it with the traditional bottleneck model. Compared with the activity-based bottleneck model, the traditional bottleneck model overestimates the queue delay at the bottleneck, the level of step charging, and the start and end time of peak hours. Therefore, the activity-based step charging method disperses the traffic pressure on peak sections to a certain extent, avoiding traffic accidents caused by the transition period between charging and free.

Keywords: bottleneck model; traffic congestion; activity-based approach; step charge

Citation: Li, H.; Zhang, Y.; Lv, J.; Gong, D.; Gan, X. Analysis and Research on Activity-Based Bottleneck Model. *Eng. Proc.* **2023**, *38*, 28. https://doi.org/10.3390/engproc2023038028

Academic Editors: Teen-Hang Meen, Hsin-Hung Lin and Cheng-Fu Yang

Published: 25 June 2023

Copyright: © 2023 by the authors. Licensee MDPI, Basel, Switzerland. This article is an open access article distributed under the terms and conditions of the Creative Commons Attribution (CC BY) license (https://creativecommons.org/licenses/by/4.0/).

1. Introduction

According to the notice of the State Council on Approving and transmitting the implementation plan of the Ministry of Transport and other departments on Exempting small bus tolls on major holidays, Guangxi Zhuang Autonomous Region has implemented the policy of exempting high-speed tolls for small buses with seven seats or less during major holidays since 2012. In recent years, with the continuous increase in vehicle ownership, short-distance and self-driving travel on holidays has increased, and the free traffic congestion and traffic accidents during holidays have also attracted much attention. According to the statistics released by Guangxi Expressway, during May Day this year, the traffic flow of the Expressway in the region is expected to be 3,784,600 vehicles (756,900 vehicles per day), a decrease of 51.2% over the same period in 2021 and an increase of 54.79% over the previous holiday. Among them, the traffic volume of small passenger cars with less than seven seats is 3,338,500 vehicles (667,700 vehicles per day).

From the perspective of expressway operation in the whole region, the traffic congestion sections are mainly concentrated in 29 Expressway sections, including Fujian Quanzhou to Guangxi Nanning Expressway (G72) of Guilin North Sujia interchange to Inner Mongolia Baotou to Guangdong Maoming Expressway (G65) of the bureaucratic Tian interchange section, the Bozhai to Luzhai section of G72, Lowe bridge section of G72, Nanning East toll station of G72, and Liujing toll station of G72. According to the analysis, the causes of congestion are large traffic flow, rear-end collisions, or scraping accidents. There are human behaviors in the transition stage between stopping and going, too. Drivers do not understand the detailed rules of exemption, so they slow down or stay at the entrance of the toll station until 0 o'clock to pass for free, resulting in a large number of queues at the entrance of the highway toll station, causing congestion. In addition, to leave the expressway before the start of charging, drivers are speeding or frequently changing lanes and overtaking, resulting in traffic accidents, which also causes Expressway congestion.

The policy of free highway access on holidays has caused many traffic problems, and there needs to be a perfect system to control congestion and ensure the normal service level of roads. There is considerable research and application on congestion pricing. In terms of theoretical research, in the 1980s, von Platen et al. [1] proposed a static congestion pricing model based on the marginal cost principle, which is also known as the marginal cost pricing model. On the basis of marginal cost pricing theory, Defermos et al. [2] studied the optimal congestion pricing model for general road networks and identified two types of congestion charging: path charging and road section charging, both of which can optimize the system. Yang et al. [3] extended the marginal charging pricing principle to general road networks and established an elastic congestion pricing model under the constraint that the capacity of road sections is limited. In 1993, Smith [4] presented a new dynamic model of peak period traffic flows on congested capacity-constrained urban road networks. While respecting the first in, first out (FIFO) discipline of road traffic queues and the exit capacities of road links, the model determines the (time-varying) costs incurred in traversing the various routes when (time-varying) route inflows are specified. It is proved that if this model is used to find route costs then a dynamic user equilibrium exists.

2. Related Work

2.1. Theoretical Research on Traditional Bottleneck Model

Vickrey [5] first proposed the classical bottleneck model in 1969. He assumed that there is a bottleneck road with limited traffic capacity connecting the residential and working places of residents, and the traffic capacity of other places except the bottleneck road is large enough. That is, only the bottleneck has crowded queues. Assuming that the traffic capacity of the bottleneck road is S vehicles/unit time, queuing occurs when the arrival rate of the bottleneck road exceeds S. Therefore; every traveler faces a trade-off between the cost of congestion time and the cost of planned delay. The basis of travelers' decisions is to choose a departure time to minimize their travel costs. Under the equilibrium state of the bottleneck model, the total travel cost of all travelers is equal.

2.2. No Charge Cost

The travel cost of travelers is determined by the travel time (α), early arrival time for work (β), and late time (γ) and is a linear function of

$$C(t) = \alpha + \beta + \gamma \tag{1}$$

If the travel cost (C) of all travelers is the same during the departure period, it reaches a balance. Assuming that the earliest person (t_0) and the latest person (t_e) do not encounter a queue, and $C(t_0) = C(t_e) = C$. Then, it is expressed as

$$\tau(t)\begin{cases} t_0 = t^* - \dfrac{\gamma}{\beta+\gamma}\left[\dfrac{N}{S}\right] \\ t_e = t^* + \dfrac{'\gamma}{\beta+\gamma}\left[\dfrac{N}{S}\right] \end{cases} \tag{2}$$

where t^* is the ideal arrival time. The cost of travel for each person is

$$C = \frac{\beta\gamma}{\beta+\gamma}\left[\frac{N}{S}\right] = \delta\frac{N}{S} \tag{3}$$

where $\delta = \beta\gamma/(\beta+\gamma)$.

According to the research results of Arnott et al., the Total Travel Cost (TTC), Scheduled Delay Cost (SDC), and Trip Cost (TC) are defined as

$$TTC = SDC = \frac{\delta}{2}\left[\frac{N^2}{S}\right] \tag{4}$$

$$TC = TTC + SDC = \delta\left[\frac{N^2}{S}\right] \tag{5}$$

2.3. Social Optimum

The TC of the system is minimized as the social optimum. Three conditions must be satisfied. (1) Bottleneck roads operate at full capacity during the peak period, (2) the departure rate must not be higher than the capacity, and (3) the earliest departure and the latest departure have the same planned delay cost. Then,

$$TTC = 0,\quad TC^0 = TTC + SDC = \frac{\delta}{2}\left[\frac{N^2}{S}\right] \tag{6}$$

where TC^0 is the socially optimal TC. From the condition, the TC of all travelers during the peak time is $\delta\frac{N}{S}$, and the departure and arrival rates are s. The dynamic charge can be obtained as

$$\tau(t)\begin{cases}\delta\left[\frac{N}{S}\right]-\beta(t^*-t), t\in[t_0,t^*]\\ \delta\left[\frac{N}{S}\right]-\gamma(t-t^*), t\in[t^*,t_e]\end{cases} \tag{7}$$

The socially optimal dynamic toll is composed of two linear functions with different slopes, and t^* attains its maximum value. Each traveler adjusts the departure time accordingly, thus eliminating queues and halving the TC.

3. Method

3.1. Tiered Charging Based on Activity-Based Bottleneck Model

Through the analysis of data over the years, we find that the peak period of the highway during holidays shows a regular distribution, mainly divided into three peak periods with the same pattern. We assume that the peak period of a peak section is from t_p to t_q. According to the classical bottleneck model, Equation (7) shows that dynamic toll is a function of the time change, and it is difficult for travelers to accurately measure their toll amount and thus reasonably arrange their travel time. Therefore, in real life, ladder charging is often used to replace dynamic charging, which has high operability and strong applicability.

In this system, the first traveler and the last traveler do not need to queue, that is $T(t_p) = T(t_q) = 0$. Thus, $\tau(t_p) = \tau(t_q) = 0$. Under the equilibrium condition, the dynamic charge at the bottleneck can be obtained as

$$\tau(t)\begin{cases}\int_{t_p}^{t} u_l(t)dt + \int_{t}^{t_p} u_r(t)dt + \beta(t-t_p), t\in[t_p,t^*]\\ \int_{t_q}^{t} u_l(t)dt + \int_{t}^{t_q} u_r(t)dt + \gamma(t_q-t), t\in[t^*,t_q]\end{cases} \tag{8}$$

where $u_l(t)$ and $u_r(t)$ are marginal functions.

3.2. Optimal Step Charge Model of Linear Function

According to the empirical research of Tseng et al., the marginal utility based on the activity method is a linear function that changes with time. The marginal utility of leaving home decreases with time, and the time utility of returning home increases with time [6,7]. Thus, the following equations are obtained.

$$\begin{cases}u_l(t) = g_0 + g_1 t,\ g_1 < 0\\ u_r(t) = h_0 + h_1 t,\ h_1 > 0\end{cases} \tag{9}$$

Bring the above equations into Equation (8), we obtain

$$\tau(t)\begin{cases} \frac{k_1}{2}t^2 + (k_2+\beta)t - \left[\frac{k_1}{2}(t_p)^2 + (k_2+\beta)t_p\right], t \in [t_p, t^*] \\ \frac{k_1}{2}t^2 + (k_2-\gamma)t - \left[\frac{k_1}{2}(t_q)^2 + (k_2-\gamma)t_q\right], t \in [t^*, t_q] \end{cases} \tag{10}$$

where, $k_1 = g_1 - h_1$, $k_2 = g_0 - h_0$. The piecewise conic is obtained, as shown in Figure 1. At the same time, the peak start and end time can be calculated as follows.

$$\begin{cases} t_p = \dfrac{(\beta+\gamma)t^* + \frac{1}{2}k_1\left(\frac{N}{S}\right)^2 + (k_2-\gamma)\left(\frac{N}{S}\right)}{(\beta+\gamma) - k_1\left(\frac{N}{S}\right)} \\ t_q = t_p + \dfrac{N}{S} \end{cases} \tag{11}$$

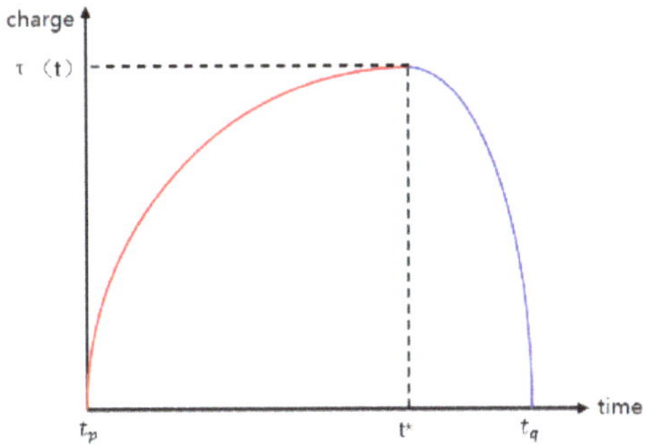

Figure 1. Time-varying tolls under linear marginal activity utility.

3.3. The Optimal Step Charge Based on Laih Model

The Laih model is used for step charging in the relevant bottleneck model, which is widely used in step charging due to its simplicity and real-time nature, such as in congestion charging in Singapore [8,9].

The optimal toll level ρ, toll start time t^+ and end time t^- are determined first to satisfy $\rho = \tau(t^+) = \tau(t^-)$. The core idea of congestion charging is to replace queues at congestion with tolls, and the total revenue from maximizing step tolls must equal the maximum elimination of queues. Therefore, the charge level, the start time, and the end time of the charge should meet the following equations.

$$\begin{cases} \frac{k_1}{2}(t^+)^2 + (k_2+\beta)t^+ - \left[\frac{k_1}{2}(t_p)^2 + (k_2+\beta)t_p + \rho\right] = 0 \\ \frac{k_1}{2}(t^-)^2 + (k_2-\gamma)t^- - \left[\frac{k_1}{2}(t_q)^2 + (k_2-\gamma)t_q + \rho\right] = 0 \end{cases} \tag{12}$$

The start time and the end time of the charge are as follows.

$$\begin{cases} t^+ = \dfrac{-(k_2+\beta) + \sqrt{A}}{k_1} \\ t^- = \dfrac{-(k_2-\gamma) + \sqrt{B}}{k_1} \end{cases} \tag{13}$$

where $A = (k_2 + \beta + k_1 t_p)^2 + 2k_1 p$, $B = (k_2 - \gamma + k_1 t_q)^2 + 2k_1 p$, and according to the optimal conditions, the following equation can be obtained.

$$(\sqrt{A} + \sqrt{B})\left(1 + \frac{\rho k_1}{\sqrt{AB}}\right) - (\beta + \gamma) = 0 \tag{14}$$

By solving the above equation, we can finally determine values of ρ, t^+ and t^-.

4. Experiments and Discussion

This section lists data for numerical analysis of the previous derivation results. The input parameters of the example are shown in Table 1 [10]. The total number of commuters N is 5000 vehicles, and the traffic capacity S of the bottleneck is 2000 vehicles per hour. Therefore, the duration of the whole early peak period is 2.5 h.

Table 1. Example input parameters.

N (Vehicle): 5000	S (vehicle/hour): 2000	t^* (hour): 12:00
α (yuan/hour): 10.0	β (yuan/hour): 6.0	γ (yuan/hour): 19.0
u_l: 11.0	u_r: 8.0	g_0: $-1/2$
g_1: 10.0	h_0: 5/6	h_1: 2.0

The simulation experiment of step charging is carried out in two ways: Vickrey's classic bottleneck model and the activity-based bottleneck model. The results are shown in Table 2.

Table 2. Comparison results of step charges under different bottleneck models.

Model Solution	Travel-Based Bottleneck Model	Bottleneck Model Based on Activity Method
Maximum optimal dynamic charge $\tau(t^*)$ (yuan)	11.40	7.51
Optimal step charge ρ (yuan)	5.70	4.17
Charge start time t^+	08:03	07:48
Charge end time t^-	09:18	09:09
Step charging time length $t^+ - t^-$ (hours)	1.25	1.35

In real life, public travelers are rarely aware of the external diseconomy caused by their travel behavior. Although congestion charging can improve travel efficiency, most of the public is more or less opposed to this policy. In Table 2, the tiered charge based on the activity bottleneck model ρ is far lower than the traditional tiered charging level based on the travel bottleneck model and is easier to be accepted by the public.

Table 2 also shows the comparison of the tiered charging solutions under different bottleneck models. The step charging time length of the first model is 1.25 h, and the step charging time interval of the second model is 1.35 h. Moreover, the start time of the second model is the earliest, and the start time of the first model is the latest. Under the optimal step charge, the traditional trip-based bottleneck model has the highest charge level, indicating that the trip-based bottleneck model overestimates the congestion charge level. Therefore, compared with the traditional trip-based bottleneck model, the activity-based bottleneck model improves the charging efficiency, that is, eliminates the queuing phenomenon when the charge is lower.

5. Conclusions

According to the problems of traffic congestion and frequent accidents on freeways during holidays, we propose a bottleneck model based on the activity method, which charges congestion fees according to the density of road congestion to regulate the travel time, travel mode, and travel path and alleviate traffic congestion to a certain extent.

Although tiered charging cannot eliminate congestion and queue, it can disperse road traffic pressure to the greatest extent. Highway operation managers can timely release information to inform travelers, provide travel advice, travel path planning, and other services for travelers and improve the service level of operation management.

Author Contributions: Conceptualization, H.L.; methodology, H.L.; software, J.L.; investigation, H.L.; writing—original draft preparation, H.L.; writing—review and editing, Y.Z.; supervision, Y.Z.; project administration, D.G.; funding acquisition, X.G. All authors have read and agreed to the published version of the manuscript.

Funding: This work was supported in part by the new generation of information technology innovation project of China University Industry University Research Innovation Fund #2020ITA03039.

Institutional Review Board Statement: Not applicable.

Informed Consent Statement: Not applicable.

Data Availability Statement: The data used in this article involves privacy issues. If necessary, please send us email.

Conflicts of Interest: The authors declare no conflict of interest.

References

1. Von Platen, P.; Hallmann, A.; Pomprapa, A.; Leonhardt, S.; Pickerodt, P.A.; Russ, M.; Taher, M.; Braun, W.; Hinken, L.; Köbrich, R.; et al. System identification of the FiO2 to SpO2 relationship during mechanical ventilation. *Autom. Med. Eng.* **2021**. [CrossRef]
2. Defermos, S.; Sparrow, F.T. Optimal Resource Allocation And Toll Patterns In User-Optimised Transport Networks. *J. Transp. Econ. Policy* **1971**, *5*, 184–200.
3. Yang, H.; Zhang, X. Optimal Toll Design in Second-Best Link-Based Congestion Pricing. *Transp. Res. Rec. J. Transp. Res. Board* **2003**, *1857*, 85–92. [CrossRef]
4. Smith, M.J. A new dynamic traffic model and the existence and calculation of dynamic user equilibria on congested capacity-constrained road networks. *Transp. Res. Part B Methodol.* **1993**, *27*, 49–63. [CrossRef]
5. Vickrey, W.S. Congestion theory and transport investment. *Am. Econ. Rev.* **1969**, *34*, 414–431.
6. Nahmias-Biran, B.H.; Shiftan, Y. Using activity-based models and the capability approach to evaluate equity considerations in transportation projects. *Transportation* **2020**, *47*, 2287–2305. [CrossRef]
7. He, X.; Li, X.; Ji, H. Secure Transmission Based on Marginal Utility in UAV-assisted Cognitive Radio Network. In Proceedings of the 2021 IEEE International Conference on Communications Workshops (ICC Workshops), Montreal, QC, Canada, 14–23 June 2021; IEEE: Piscataway, NJ, USA, 2021.
8. Decorla-Souza, P. Exploratory Evaluation of a Concept Combining Incentivized On-Demand Ridesharing with Congestion Pricing. *Transp. Res. Rec.* **2021**, *2675*, 440–451. [CrossRef]
9. Romero, F.; Gomez, J.; Paez, A. Toll roads vs. Public transportation: A study on the acceptance of congestion-calming measures in Madrid. *Transp. Res. Part A Policy Pract.* **2020**, *142*, 319–342. [CrossRef]
10. Li, Z.C.; Lam, W.; Wong, S.C. Bottleneck model revisited: An activity-based perspective. *Transp. Res. Part B* **2014**, *68*, 262–287. [CrossRef]

Disclaimer/Publisher's Note: The statements, opinions and data contained in all publications are solely those of the individual author(s) and contributor(s) and not of MDPI and/or the editor(s). MDPI and/or the editor(s) disclaim responsibility for any injury to people or property resulting from any ideas, methods, instructions or products referred to in the content.

engineering proceedings

MDPI

Proceeding Paper

Creating Effective Educational Videos on YouTube in Higher Education †

John C.-C. Lu

Department of Civil Engineering, Chung Hua University, Hsinchu 30012, Taiwan; cclu@chu.edu.tw

† Presented at the 3rd IEEE International Conference on Electronic Communications, Internet of Things and Big Data Conference 2023, Taichung, Taiwan, 14–16 April 2023.

Abstract: Educational videos have become an important part of higher education. Thus, it is necessary to discuss practical tips for creating high-yield educational videos. The electronic medium is integrated into traditional courses and serves as the cornerstone of many blended courses. Multiple studies have demonstrated that video is a highly effective educational tool for learning, particularly for hard-to-visualize processes and procedural education. Video allows learners to view content at their own pace and view on-demand materials repetitively. Video on YouTube can be integrated as a supplement to course content to provide targeted information and enrich the learning environment for students. In this article, creating an effective video on YouTube is introduced, providing examples of effective recording and integrating video into engineering mathematics courses on YouTube.

Keywords: educational video; YouTube channel; higher education; engineering mathematics

Citation: Lu, J.C.-C. Creating Effective Educational Videos on YouTube in Higher Education. *Eng. Proc.* **2023**, *38*, 32. https://doi.org/10.3390/engproc2023038032

Academic Editors: Teen-Hang Meen, Hsin-Hung Lin and Cheng-Fu Yang

Published: 20 June 2023

Copyright: © 2023 by the author. Licensee MDPI, Basel, Switzerland. This article is an open access article distributed under the terms and conditions of the Creative Commons Attribution (CC BY) license (https://creativecommons.org/licenses/by/4.0/).

1. Introduction

YouTube videos are commonly used in higher education. D'Aquila et al. [1] concluded that students viewed the videos primarily for exam review, and video usage improves student performance. Jackman [2] used YouTube in three psychology classes at the University of Trinidad and Tobago and found that YouTube can be used in tertiary education pedagogy. Hoa et al. [3] proposed building a healthy and positive YouTube Kids channel suitable for the characteristics of primary school students. Nabayra [4] investigated students' experiences with the use of teacher-created videos in learning mathematics online and revealed that teacher-created videos were helpful for students to self-assess their learning progress. Makruf et al. [5] concluded that Moodle was less optimal in the implementation and evaluation of learning in higher education. Hendriyani et al. [6] developed online learning videos to improve students' creative thinking skills. Bauk [7] discussed the benefits and challenges of international virtual exchange in teaching and learning at higher education institutions. Noetel et al. [8] systematically reviewed the effects of video on learning in higher education and concluded that videos were unlikely to be detrimental and usually improved student learning. Fyfield et al. [9] suggested that videos be accompanied by learning activities rather than watched passively. Miner and Stefaniak [10] compared instructors' and students' perceptions regarding the use of video during instruction. Kohler and Dietrich [11] showed that both social status and individual predispositions influenced the outcome of educational videos. Shoufan [12] tried to use viewers' ratings and participation to measure the quality of educational videos. The findings of Çelik et al. [13] showed that the most used social media tools were also used for educational purposes, and the most significant exception in this subject was TikTok. Boey et al. [14] investigated how the integration of technology contributed to the development of attributes or learner profiles suitable for the workplace and real-world performance while also meeting skill-based learning outcomes.

YouTube can be a powerful tool for education beyond the classroom. Educational videos can be easily uploaded to YouTube, and using these virtual teaching materials has been highly favored by educational organizations. Many studies have shown the application and effectiveness of creating educational videos on YouTube. Therefore, the author has recorded more than 1800 teaching videos on engineering mathematics. Most recorded videos of teaching last less than 15 min and have been uploaded to YouTube for students to view for free. Without advertisements, students can concentrate on self-study. The synchronous class recording is avoided to improve the quality of recording teaching videos. The script writing, directing, and editing of self-recording videos are completed by the author, which saves time for editing. Sometimes, re-recording 3–5 times improves the quality of the teaching video, though most videos are recorded in one take. Each video deals with one theme to construct students' confidence through the thematic teaching materials.

In this article, the experience of recording high-quality teaching videos is shared. A questionnaire analysis was conducted on the effectiveness of teaching videos for 26 students in the engineering mathematics courses. The statistical analysis result showed that planning for learning was evaluated as favorably with a score of 4.81 on a 5-point scale. This shows the effectiveness of teaching videos on YouTube.

2. Effective Educational Videos

The video needs to be recorded using the fundamentals and principles of photography. Furthermore, it is also necessary to understand the golden ratio of the contents and the principles of art. The following are suggestions based on our experience with more than 1800 self-recorded teaching videos.

- **Why do I need to record teaching videos?** Making the classroom joyful and interesting is an important consideration. Attractive teaching videos are required for students to preview and review lessons at their convenience through teaching videos on YouTube. The instructional videos on YouTube are set to public mode, and the interference of advertisements is not allowed;
- **How do I make lesson videos for students?** It is required to be accustomed to recording instructional videos by camera and computer screen capture. Camera skills and computer screen capture software are demanded. The learning process may take several weeks;
- **What should I do if funds are insufficient?** Many teaching units on the campus can provide cameras or teaching software. Teaching equipment can be rented from the academic affairs office. Higher-priced equipment may not be required. Instructional videos can be recorded with a cheap microphone;
- **What should I do if there is no studio room?** The classroom is a suitable studio with a digital camera. A high-resolution digital camera is needed for recording. Synchronous recording with classes must be avoided to improve teaching video quality. In addition, when using an empty classroom, it is necessary to overcome the psychological barrier of no audience;
- **How to improve my digital literacy?** It is necessary to be motivated to learn new digital skills. Even though digital hardware and software are constantly evolving, video production skills have not changed much. There are good ideas for integrating new technologies into teaching by searching for 3C products;
- **Is the artistic effect of teaching video insufficient?** Students want to learn logical thinking and problem-solving from teaching videos rather than appreciate video styles. In addition, if the teacher writes mathematical symbols neatly, it helps students' learning;
- **What should I learn about digital software?** There are many kinds of digital software for making educational videos, such as Power Director, Camtasia, and MS Teams, which are the most commonly used. There are similar applications, and the choice depends on whether it is easy to obtain the licensed software;
- **No time to record instructional videos?** Teachers are usually busy. Therefore, only interest and enthusiasm can provide sufficient motivation. At the right time, in the right place, and with the right people, the teacher needs to complete the recording of

teaching videos. If teachers have enough time to record teaching videos, the quality and quantity of the videos can be gradually improved;

- **It is necessary to have an understanding of copyright law.** Teachers need to understand the copyright law. Although the copyright law has restrictions, it also allows protection. Therefore, teachers need to understand the law before recording teaching videos and collect the materials for recording the teaching video. If necessary, Creative Commons-licensed materials can also be used;
- **Personal capability.** Every teacher has different specialties and abilities. Teachers do not have to record teaching videos to achieve academic success with their abilities. The academic achievements required are multifaceted, and academic efforts in all aspects need to be appreciated;
- **Pre-arrangement of course planning is required.** Usually, teachers can be prepared for their teaching courses in the next few years. Planning the courses is conducive to recording teaching videos. It is convenient for teachers to record the whole educational video of a subject;
- **Awareness of teaching is one of our academic achievements.** Most teachers agree that teaching is an academic achievement equivalent to research accomplishment. Therefore, achievements are critical for teachers in college. Paying attention to the learning effects of students is important for college teachers;
- **Actively participate in teaching workshops or communities.** Through academic information exchanges in communities or workshops, teachers can obtain assistance and the latest information on educational training. It also creates opportunities for academic cooperation with teachers on campus. In the teaching community, information can be obtained, such as the application of the metaverse in education;
- **Sharing the recorded teaching materials.** Sharing and helping each other increase the influence of educational videos. Self-made teaching videos can be provided on free platforms such as YouTube. Virtual interaction with talented students throughout the world is possible;
- **Publishing teaching results on YouTube.** Uploading videos on YouTube is a type of publication, and the influence of YouTube is greater than that of journals or conference papers. The public release provides feedback on teaching achievements through discussions with viewers;
- **The publishing platform for reviews.** Submission of journal or conference papers requires rigorous peer review. Equivalently, teaching videos on YouTube need to pass a strict YouTube review process. Furthermore, students can become reviewers who provide valuable feedback;
- **Digitization preserves teaching records.** Many teachers have accumulated a large amount of teaching output. If the teaching materials can be organized and recorded into digital files, they are more conducive to inheritance, promotion, and preservation. One of the advantages of digitization is that it improves the efficiency of teaching;
- **When the going gets tough, the tough get going.** The production of teaching videos requires time and energy. It is difficult to achieve successful results without extraordinary perseverance. It is criticized sometimes for keeping free instructional videos on YouTube;
- **The traditional teaching mode still needs to be preserved.** Although the teaching mode can be changed through the recording of teaching videos, the traditional face-to-face teaching mode is still needed as communication between teachers and students is important;
- **Try to do all the work independently.** Teaching is a long-term career. It is difficult to maintain a studio team to record teaching videos for a long time. Therefore, teachers need to train themselves to become screenwriters, directors, and video editors. Then, recording teaching videos may last longer;

- **Writing course handouts.** In addition to recording teaching videos, teachers need to provide lecture notes for students' reference. Lecture notes are uploaded next to the corresponding teaching video so that students can easily click on them for reference;
- **Enjoy yourself.** Recording instructional videos is a difficult task. If the teachers enjoy recording teaching videos, they can continue for a long time.

3. Questionnaire Survey

An online questionnaire survey was conducted to collect data to understand how satisfied students were. The questionnaire was distributed to students in the engineering mathematics course. The questionnaire was created on a Likert 5-point scale, as shown in Tables 1 and 2. Questionnaires with full responses were included in the analysis.

Table 1. The score for each option.

Option	Score
Strongly Disagree	1
Disagree	2
Neither Agree nor Disagree	3
Agree	4
Strongly Agree	5

Table 2. Interpreted students' responses based on scores.

Interval	Students Responses
$1.00 \leq x < 1.80$ [a]	Very Bad
$1.80 \leq x < 2.60$	Bad
$2.60 \leq x < 3.40$	Neutral
$3.40 \leq x < 4.20$	Good
$4.20 \leq x \leq 5.00$	Very Good

[a] x = Scores of each indicator.

Based on the results of the 26 students' responses to the questionnaire, the score of each indicator was calculated as follows. The students' satisfaction with teaching videos scored 4.81 on average. Most students favored the quality of the engineering mathematics materials in the educational videos. The quality of the learning videos was evaluated as very good (Table 3).

Table 3. Students' responses to the developed teaching videos.

Indicator	Score	Response Category
The teaching materials developed by the teacher can inspire my interest in learning	4.77	Very Good
The recorded digital video provided by the teacher helps me adjust my learning progress	4.86	Very Good
Average	4.81	Very Good

4. Conclusions

Educational videos allow students to view on-demand materials repeatedly and learn at their own pace. Furthermore, videos on YouTube can be integrated to provide targeted information and enrich the learning environment for students. In this article, tips for creating an effective video on YouTube are provided to help record and integrate a video on YouTube. It is helpful for teachers to be interested in photography and have fundamental photographic concepts. It is recommended to understand the golden ratio of contents and the ten principles of art. Teachers are encouraged to train themselves to become screenwriters, directors, and video editors. Then, recorded teaching videos can last longer. Independent recording is recommended. Creating educational videos can go on as long as teachers enjoy the process. However, it is difficult to obtain successful results without extraordinary perseverance. The questionnaire survey result shows that students' responses to the teaching videos were excellent, with an average score of 4.81. Quality and supplementary materials are important in educational videos.

Funding: This work is supported by the Ministry of Education of Taiwan, Republic of China, through grants PEE1101210 and PEE1110397.

Institutional Review Board Statement: Not applicable.

Informed Consent Statement: Not applicable.

Data Availability Statement: Not available.

Conflicts of Interest: The author declares no conflict of interest.

References

1. D'Aquila, J.M.; Wang, D.; Mattia, A. Are instructor generated YouTube videos effective in accounting classes? A study of student performance, engagement, motivation, and perception. *J. Account. Educ.* **2019**, *47*, 63–74. [CrossRef]
2. Jackman, W.M. YouTube usage in the university classroom: An argument for its pedagogical benefits. *Int. J. Emerg. Technol. Learn.* **2019**, *14*, 157–166. [CrossRef]
3. Hoa, P.D.; Hien, N.T.; Anh, L.K.; Giang, L.H. Creating effective educational video on Youtube Kids for primary student. *Am. J. Educ. Res.* **2021**, *9*, 368–375.
4. Nabayra, J.N. Mathematics learning in the new normal through teacher-created videos: The freshmen university students' experience. *Int. J. Arts. Humanit. Stud.* **2022**, *2*, 22–27. [CrossRef]
5. Makruf, I.; Rifa'i, A.A.; Triana, Y. Moodle-based online learning management in higher education. *Int. J. Instr.* **2022**, *15*, 135–152. [CrossRef]
6. Hendriyani, M.E.; Rifqiawati, I.; Lestari, D. Online learning videos to develop creative thinking skills of students. *Res. Dev. Educ.* **2022**, *2*, 67–75. [CrossRef]
7. Bauk, S. New media in higher education: Sharing some positive practices. *Int. J. Schol. Pap. Media Commun.* **2022**, *15*, 51–61.
8. Noetel, M.; Griffith, S.; Delaney, O.; Sanders, T.; Parker, P.; del Pozo Cruz, B.; Lonsdale, C. Video improves learning in higher education: A systematic review. *Rev. Educ. Res.* **2021**, *91*, 204–236. [CrossRef]
9. Fyfield, M.; Henderson, M.; Heinrich, E.; Redmond, P. Videos in higher education: Making the most of a good thing. *Australas. J. Educ. Technol.* **2019**, *35*, 1–7. [CrossRef]
10. Miner, S.; Stefaniak, J.E. Learning via video in higher education: An exploration of instructor and student perceptions. *J. Univ. Teach. Learn. Pract.* **2018**, *15*, 2. [CrossRef]
11. Kohler, S.; Dietrich, T.C. Potentials and limitations of educational videos on YouTube for science communication. *Front. Commun.* **2021**, *6*, 581302. [CrossRef]
12. Shoufan, A. Estimating the cognitive value of YouTube's educational videos: A learning analytics approach. *Comput. Hum. Behav.* **2019**, *92*, 450–458. [CrossRef]
13. Çelik, B.; Uzunboylu, H.; Demirbaş-Çelik, N. Higher education students' social media platform preferences for educational purposes. *Rev. Educ. Distancia* **2023**, *23*, 72. [CrossRef]
14. Boey, C.K.; Sathish, S.; Koh, S.N.A. Impact of technology-enabled project-based assessments on learner outcomes in higher education. *Int. J. Mob. Learn. Organ.* **2023**, *17*, 131–148. [CrossRef]

Disclaimer/Publisher's Note: The statements, opinions and data contained in all publications are solely those of the individual author(s) and contributor(s) and not of MDPI and/or the editor(s). MDPI and/or the editor(s) disclaim responsibility for any injury to people or property resulting from any ideas, methods, instructions or products referred to in the content.

engineering proceedings

MDPI

Proceeding Paper

A Study on Problems Encountered by Students and Their Causes When Learning About E-Book Production Based on Analytic Hierarchy Process †

Shin Liao [1] and Li-Yi Ma [2,*]

1 Department of Graphic Arts and Communications, National Taiwan Normal University, Taipei 106308, Taiwan; sliao@ntnu.edu.tw
2 Department of Graphic Communications, Shih-Hsin University, Taipei 116002, Taiwan
* Correspondence: liyima@mail.shu.edu.tw
† Presented at the 3rd IEEE International Conference on Electronic Communications, Internet of Things and Big Data Conference 2023, Taichung, Taiwan, 14–16 April 2023.

Abstract: Many colleges' graphic design departments have created curriculums related to digital publishing. While digital reading has become popular and various technologies of e-book production are mature, there are urgent demands to strategically and systematically cultivate professionals for the digital content industries. Thus, this research was to examine the problems associated with e-book production and their causes for future curriculum design. A literature review of college teaching courses in design-related departments was conducted, and an analytic hierarchy process (AHP) was applied with twelve senior digital publishing practitioners. The results showed that "interactive setting" was the most common problem. The next most common problems were "color labeling" and "graphic setting". In-depth interviews with seven experts which were selected from technical and vocational colleges and the digital publishing industry revealed the causes of these problems. The reasons of mistakes were categorized into 54 causes for the seven most common problem categories. Among the causes of "interactive setting" errors, "setting the media playback controller" happened most frequently.

Keywords: digital publishing; e-book production; analytic hierarchy process

Citation: Liao, S.; Ma, L.-Y. A Study on Problems Encountered by Students and Their Causes When Learning About E-Book Production Based on Analytic Hierarchy Process. *Eng. Proc.* **2023**, *38*, 33. https://doi.org/10.3390/engproc2023038033

Academic Editors: Teen-Hang Meen, Hsin-Hung Lin and Cheng-Fu Yang

Published: 25 June 2023

Copyright: © 2023 by the authors. Licensee MDPI, Basel, Switzerland. This article is an open access article distributed under the terms and conditions of the Creative Commons Attribution (CC BY) license (https://creativecommons.org/licenses/by/4.0/).

1. Introduction

The popularity of mobile networks along with the increasing number of e-book readers using large-screen mobile devices promotes various digital information services. This has boosted global digital reading, and the e-book industry is booming. The new network-based media has a significant impact on traditional publishers. The publishing industry is facing rapid changes in production workflows and business operations. With fierce competition and changing reading habits, publishers need to actively cultivate professionals in e-book production to overcome future challenges. For e-book publishers, independent work is an inevitable trend. Digital capability in text, image, video, and voice integration is what e-book publishers need to have. Traditional publishers must think about their market position and transformation. The professional competence and "know-how" in the traditional publishing industry have also changed [1]. Since the production of e-books is related to readability for readers, the correct content layout of e-books and comfort level to read are the key factors for the success of e-books. The hardware used is also important. Flexible materials and touchable user interface (UI) design have been introduced [2,3], and the imposition and distribution of pictures and texts have a significant impact on a reader's comprehension in e-book publication [4].

The purpose of this research is to understand the common problems and reasons that students have for producing e-books based on surveys and interviews. The results provide a reference for future teaching curriculum development to educate students about e-book production based on theory and practice. The research results also recommend the requirements for related student training [5]. The specific purposes of this research are as follows:

- To examine the current teaching status of e-book production courses in related departments at technical and vocational colleges.
- To explore the most common problems that students encounter in e-book production.
- To understand the reasons for the most common problems encountered by students when learning about e-book production.

2. Literature Review

E-Book Production

E-books integrate texts, pictures, sounds, and other elements using multimedia software as the carriers. Through the Internet, information technology, hardware equipment, and copyright management mechanisms, e-books have changed the business operations and marketing channels of traditional publishers. They create new business models in new markets, produce digital knowledge, and develop new distribution and service chains [6]. Since e-books are in digital forms, data processing and dissemination are performed within the network. E-books provide text-based reading services in black-and-white picture books (E-ink carrier) or colored picture books (LCD carrier). Digital readers are provided with various multimedia functions, which add extra value to e-books [7].

Many scholars have studied the digital publishing industry and indicated that those who intend to enter into or transform the digital publishing industry must be familiar with the new online media and technologies and must "create friendly reading formats for different readers". Imagination and cross-media communication are mandatory to produce e-books that are well typographic and friendly to read [8]. From the manual of the Adobe InDesign software, the most preferred by publishers, the functions of InDesign include "workspace", "layout", "text style", "print style", "tables", "long document functions", "drawing", "graphics", "frames and Objects", "transparency effects", "colors", and "interactive documents". These operations are important in publication production. Overall, digital publishing requires more diverse professional abilities than traditional publishing as an e-book is an information medium that is suitable for modern people's lifestyles [9].

From the literature review, it was found that typesetting and page layout are most important in e-book production. Furthermore, considering the contents of digital media, the production of audio-visual materials and the script of e-books must be contained. The traditional printing processes are dismissed. Instead, a new workflow is adopted, including manuscript production, layout, and finished file output for image and text collection, file creation, type and sequence arrangement, master template creation, style setting, layout structure design, image and text integration, interactive effect setting, final preview output, and web release.

Regarding the recent development of digital publishing courses, private institutions offer more related courses than public institutions. Four-year universities and two-year vocational colleges offer digital publishing courses, with four-year universities offering more courses. The number of courses related to digital publishing is increasing every year. Colleges and universities are focusing more on talent cultivation for the digital publishing industry. More than 20,000 students took digital courses in the past five years.

3. Methodology

3.1. Research Method and Objects

The research method included a literature review, an analytic hierarchy process, and expert interviews, as described below.

To understand the "common problems in e-book production", we compiled the "hierarchical questionnaire for common problems in e-book production" based on the literature review and expert opinions for data collection. For the survey, we invited senior digital publishing experts who have "actually engaged in the production of digital publications". Overall, 17 questionnaires were sent, and 12 valid questionnaires were returned. The respondents were all females, with 50% having graduated from junior colleges and universities, 66% having worked in publishing companies for three to five years, and 33% having worked for over ten years.

To understand the most common problems and their causes in e-book production, expert interviews were conducted. Seven technical and vocational college lecturers and digital publishing experts were recruited, including two senior lecturers teaching e-book production courses, two experts in digital publishing, two experts in APP and digital publishing services, and an expert in the digital prepress department. The records of the interviews were compared with the result of the literature review and the questionnaire survey to clarify the differences between the needs of the industry and the current teaching content in colleges.

3.2. Research Tools

To investigate the most common problems related to students' learning e-book production, the "Questionnaire for common problems in e-book production" was created with a nine-point scale. The results were pairwise compared between the experts. The hierarchical structure of the questionnaire contained file management, layout setting, text and table settings, graphic settings, color labeling, interactive setting, and tool panel setting. The hierarchical structure is shown in Figure 1.

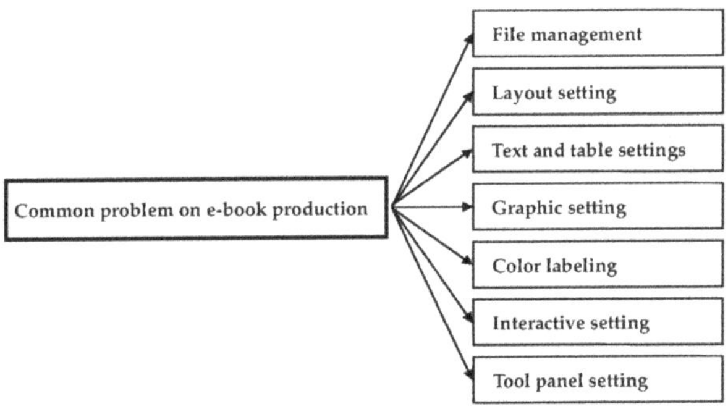

Figure 1. Hierarchical structure of this research.

Based on the literature review, the possible causes for common problems in e-book production were classified into seven categories. This first draft was reviewed by the experts to reduce the number of causes and finalize the topics to be investigated. As the results, 54 causes for the most common problems were identified.

The data were processed using the Expert Choice 11 statistical software for hierarchical analysis. The consistency index (C.I.) was calculated for verification. The verification method is as follows: λ_{max} is the maximum eigenvalue of the pairwise comparison matrix A, and n is the order of the matrix. If $\lambda_{max} = n$, then the pairwise comparison matrix A is consistent.

$$C.I. = (\lambda max - n)/(n - 1) \quad (1)$$

where C.I. = 0 when the before and after judgments are completely consistent. When C.I. > 0.1, judgments are biased and incoherent. When C.I. $\leq$ 0.1, judgments are not consistent but acceptable. When a problem is complicated, more judgments are needed for the pairwise comparison, and the order of the pairwise comparison matrix increases. Then, it becomes difficult to maintain the consistency of judgments. Therefore, a "random index" (R.I.) is proposed to modify the C.I. value to adjust the differences between different orders in order to obtain a "consistency ratio" (C.R.) [10]. A C.I. value can be adjusted by the R.I. value in matrices of different orders to obtain a consistency ratio. When C.R. ≤ 0.1, the degree of matrix consistency is satisfactory. The calculation is carried out as follows:

$$C.R. = C.I./R.I. \quad (2)$$

4. Results and Discussions

The questionnaire contained the seven categories of file management, layout setting, text and table settings, graphic settings, color code, interactive settings, and tool panel settings. Twelve experts in e-book production were invited to finalize the questionnaire. The purpose was to explore the causes of problems in e-book production. By comparing the importance of the causes in pairs, the experts chose the most common causes. During the analytic hierarchical process (AHP) using a nine-point scale of "equally prone to errors", "slightly prone to error", "more prone to error", "frequently prone to error", and "extremely prone to error" (one, three, five, seven, and nine points, respectively), the experts were asked to choose the most appropriate causes by comparing the importance of the causes in pairs. The weight of each cause was obtained. The weights of the common problems in e-book production are shown in Table 1.

Table 1. Weight values of the most common problems in e-book production.

The Most Common Problems	Weight	% Weight	Subhead
interactive setting	0.219	21.9%	1
color labeling	0.155	15.5%	2
graphic setting	0.150	15%	3
tool panel setting	0.133	13.3%	4
text and table settings	0.122	12.2%	5
layout setting	0.115	11.5%	6
file management	0.105	10.5%	7
C.I = 0.010; R.I = 1.32; C.R = 0.007 < 0.1			

The results showed that "interactive setting" had the highest weight of 21.9%, followed by "color labeling" (15.5%), "graphic setting" (15%), "tools panel petting" (13.3%), "text and table settings" (12.2%), "layout setting" (11.5%), and "file management" (10.5%).

The 54 causes of the seven most common problems in e-book production were determined by the experts as follows:

- Interactive settings had 11 causes, including "setting the media playback controller". The other common causes were "setting bookmark", "setting directory", "setting animation function", "setting hyperlink", "use swf viewer", 'setting object status function", "setting page turning effect", "setting page transition effect", "insert multimedia files (music, video)", and "setting button".
- Color labeling had four causes, including "don't notice the color model of the image file", "using different color models", "don't know how to add a color swatch", and "don't know how to add a gradient color".
- Graphic setting had eight causes, including "the wrong way of importing (placing) the picture (missing picture)", "when the image is zoomed, it is often skewed (without proportional scaling)", "don't know how to set the object anchor", and "when importing (placing) the image file, the link disappears". The rest of the causes were "don't know how to set the filter effect", "don't know how to set the text around the image", "don't know the difference between text and graphic boxes", and "when drawing with Bezier curves, the graph is often not closed".
- Tool panel settings had five causes, including "using the wrong operating tool", "screen mode switching error", "don't know which tool to choose", "cannot find the tool panel setting", and "don't know how to reopen the tool panel".
- Text and table settings had 13 causes, including "use of too many fonts", "incorrect text indentation settings", and "errors occur when setting composite fonts". The other

causes were "the text characters do not show properly", "wrong text direction when mix Chinese and English text", "do not know how to substitute the missing font", "don't know how to set the space inside the text box", "do not avoid the orphan or widow", "text overflow does not expand", "wrong text alignment settings", "do not know how to put symbols and pictures in the text box", "don't know the steps to insert the table", and "text in the table is not aligned".

- Layout setting had six causes, including "unclear layer management", "wrong page size", "wrong document type (print, web, digital)", "wrong binding side", "wrong main page setting", and "wrong chapter and page number settings".
- File management had seven causes, including "problems when converting to other formats", "the linked file was not encapsulated", "no naming specification", "don't know how to downgrade to lower versions and formats", "the graphic file size is either too big or not have enough resolution", and "problem on file format setting".

5. Conclusions

Most of the e-book production courses adopt Adobe InDesign as a learning tool as it is the mainstream software for digital publishing, especially for design layout and typesetting. However, its output diveristy and further application are relatively lacking. Today, more e-book editing applications have been released in the market. Although learning InDesign still is the most popular application for e-book production, schools should encourage students to have experience in other digital publishing workflows, such as XML and other web codings, to enhance their competitiveness. Therefore, school lecturers need to introduce and analyze the functions and characteristics of different e-books from a broader perspective and nurture students to have the conceptions and techniques necessary for general e-book production. For different types of e-books, content, and design in different formats, students need to learn how to use professional digital publishing technology. Therefore, this research was carried out to understand the common problems and causes for these problems in e-book production based on surveys and interviews. We defined seven common problems that students encounter in e-book production. The AHP results showed that "interactive setting" had the highest weight, indicating that it was the most common problem in e-book production. The second highest weight was identified for "color labeling", followed by "graphic setting", "tool panel setting", "text and table settings", "layout setting", and "File Management". The causes were also identified for each problem. These results provide a reference for teaching curriculum development with corresponding theory and practice.

Author Contributions: Conceptualization, S.L. and L.-Y.M.; methodology, S.L.; software, S.L.; validation, S.L. and L.-Y.M.; formal analysis, L.-Y.M.; investigation, S.L.; resources, S.L.; data curation, S.L. and L.-Y.M.; writing—original draft preparation, S.L.; writing—review and editing, L.-Y.M.; visualization, L.-Y.M.; supervision, S.L.; project administration, S.L.; funding acquisition, S.L. All authors have read and agreed to the published version of the manuscript.

Funding: This research is funded by the National Science Council TAIWAN (MOST 104-2511-S-003-035-).

Institutional Review Board Statement: Not applicable.

Informed Consent Statement: Informed consent was obtained from all subjects involved in the study.

Data Availability Statement: The data presented in this study are available on request from the corresponding author. The data are not publicly available due to they are part of the research project founded by the National Science Council TAIWAN.

Acknowledgments: This research is part of a larger research project funded by the National Science Council TAIWAN (MOST 104-2511-S-003-035-). We would like to express our gratitude to the experts and scholars for their assistance and valuable advice during the course of this research.

Conflicts of Interest: The authors declare no conflict of interest.

References

1. Liao, S.; Yang, C.-C. The study of the hierarchical competencies required for the digital publishing professionals. In Proceedings of the Hawaii International Conference on Social Sciences, Honolulu, HI, USA, 30 May 2011.
2. Lam, P.; Lam, S.L.; Lam, J.; McNaught, C. Usability and usefulness of eBooks on PPCs: How students' opinions vary over time. *Austr. J. Educ. Technol.* **2009**, *25*, 30–44. [CrossRef]
3. Song, S.-M. The Study on Relativity of Electronic-Reader R&D Strategy and Marketing. Master's Thesis, National Taipei University of Technology, Taipei, Taiwan, 2010.
4. Yueh, H.-P.; Lin, W.-J.; Lee, M.-C.; Huang, J.-Y.; Yeh, S.-T. Manga Reading with E-Readers: A Preliminary Usability Study, *Instr. Technol. Media Learn.* **2011**, *95*, 69–78.
5. Albert, W.L.; Yao, C.H.; Ku, J.B.; Yang, M.S.; Huang, J.T.; Fu, C.; Lin, B.T.; Kuo, W.F. A Study on the Curriculum Development for the Program of Precision Mold and Die Design and Analysis Engineering in University of Science and Technology. *NKNU J.* **2008**, *24*, 81–94.
6. Zeng, J.; Huang, M.; Liu, Y. A study on learning satisfaction of college students on the layout of Studio Classroom publications in different media. *J. Graph. Commun. Arts* **2009**, 15–28. [CrossRef]
7. Zhu, B.-Y. The Present Situation and Past of Digital Publishing in Taiwan. *New Books Recent Publ. Taiwan* **2011**, *145*, 65–68. (In Chinese)
8. Liao, S.; Pan, Y.C. Professional abilities of the employees working in digital publishing industry. In Proceedings of the 3rd International Conference on Information Sciences and Interaction Sciences (ICIS 2010), Chengdu, China, 23–25 June 2010.
9. Heilmann, J.; Linna, H. The technology and applications of the new generation of electronic books. In Proceedings of the 2001 Annual Technical Conference of the Technical Association of the Graphic Arts (TAGA), San Diego, CA, USA, 6–9 May 2001; pp. 581–590.
10. Saaty, T.L. How to make a decision: The analytic hierarchy process, *Eur. J. Oper. Res.* **1990**, *48*, 9–26. [CrossRef]

Disclaimer/Publisher's Note: The statements, opinions and data contained in all publications are solely those of the individual author(s) and contributor(s) and not of MDPI and/or the editor(s). MDPI and/or the editor(s) disclaim responsibility for any injury to people or property resulting from any ideas, methods, instructions or products referred to in the content.

engineering proceedings

Proceeding Paper

An Essential Study on IoT Applications on Community Development Association Development Advancement †

Jiin-Chyuan Mark Lai [1], Chiung-Ling Wang [2] and Ming-Yuan Hsieh [3,*]

1 Department of Applied Foreign Languages, TransWorld University, Douliu 640302, Taiwan; marklai07@gmail.com
2 Department of Industrial Education, National Taiwan Normal University, Taipei 5671455, Taiwan; lin8332317@mail.edu.tw
3 Department of International Business, National Taichung University of Education, Taichung 403454, Taiwan
* Correspondence: cpawisely@mail.ntcu.edu.tw; Tel.: +886-975-118-922

† Presented at the 3rd IEEE International Conference on Electronic Communications, Internet of Things and Big Data Conference 2023, Taichung, Taiwan, 14–16 April 2023.

Abstract: With respect to widely evaluating and elementally assaying the comprehensive possible applications of Internet of Things (IoT) technology for efficiently advancing administrative and operating performances during the implementation of community development missions, this research attempts to initially analyze the various means of employing Internet of Things (IoT) technology by assessing the affairs and activities of 6918 contemporary community associations in Taiwan in order to inductively construct the most effective Internet of Things (IoT) digital system, not only to update their original documentary administrative system to easily and conveniently implement and record administrative missions, activities, and services but also to handle and display the bulk of news and information from most community residents and the Taiwanese government (such as the Council of Agriculture, Executive Yuan). However, after conducting a systematic survey on research related to the Internet of Things (IoT) technology in recent years, there is no one able to directly analyze the shortages and challenges of the diversified applications of Internet of Things (IoT) technology in administrative performance. Specifically, the greatest challenges are the Taiwanese individual information protection regulations and the laws regarding the various applications of digital data (such as the Taiwan Personal Data Protection and Telecommunication Acts), which apparently hinder the development of Internet of Things (IoT) technology in the enhancement of community associations in Taiwan. Therefore, the suggestion contributed by this research is to institute basic digital data using ACT for all 6918 community associations in Taiwan in order to directly force the application of IoT technology to advance the performances and achievements of community development associations.

Keywords: community development association; global contagious disease; digital data applications

Citation: Lai, J.-C.M.; Wang, C.-L.; Hsieh, M.-Y. An Essential Study on IoT Applications on Community Development Association Development Advancement. *Eng. Proc.* **2023**, *38*, 35. https://doi.org/10.3390/engproc2023038035

Academic Editors: Teen-Hang Meen, Hsin-Hung Lin and Cheng-Fu Yang

Published: 26 June 2023

Copyright: © 2023 by the authors. Licensee MDPI, Basel, Switzerland. This article is an open access article distributed under the terms and conditions of the Creative Commons Attribution (CC BY) license (https://creativecommons.org/licenses/by/4.0/).

1. Introduction

The briefest administrative programs of the Executive Yuan, which is the highest administrative government department, include "live and work in peace and contentment", "continuous reproduction breed in an endless succession", and "balanced development in Taiwan". Therefore, in order to improve population loss in the outermost districts of the city, the rapid establishment of community development associations has become the one of local revitalization practices of these administrative programs in an attempt to achieve the "live and work in peace and contentment", "continuous reproduction breed in an endless succession", and "balanced development in Taiwan" goals.

In order to strengthen rural development and improve population aging issues, the Taiwanese government's Ministry of Health and Welfare began to institute and administer the Regulations on Community Development Work on 18 September 2014 to improve current Taiwanese communities, enhance the welfare of residents, and construct a wonderful, aggressive, integrated, and interdependent modern group and entire society. According to Article 12 of the Regulations on Community Development Work: "The community development associations shall devise community development plans, prepare budgets and actively promote community development according to community characteristics and the needs of residents, in addition to being in line with government policies and projects originally designed by the community." Based on the basic purpose of the development of a community development association in Article 12 of the Regulations on Community Development Work [1], projects receiving government funds must respect government policies, as stated above. The 13 principles are listed as:

- The construction of public facilities:
  (1) The construction or repair of community centers.
  (2) Protecting the environment and improving sanitation in communities.
  (3) The maintenance of roads and gutters in communities.
  (4) The collation of and increase in parking facilities.
  (5) Community greening and beautification.
  (6) Other matters relating to the construction of public facilities.
- The development of production and welfare:
  (1) The establishment of community production and construction funds.
  (2) The promotion of social welfare.
  (3) The establishment of community nurseries.
  (4) Promoting the development of community businesses.
  (5) Other matters relating to the development of production and welfare.
- Spiritual and ethical development:
  (1) Important measures for improving the ethos of society as well as advocating and promoting models of public etiquette.
  (2) The maintenance and promotion of rural culture and traditional crafts.
  (3) The establishment of traffic safety in the community.
  (4) The establishment of community pacts.
  (5) The promotion of civil defense.
  (6) The establishment of art and recreation teams.
  (7) The establishment of community senior citizens' clubs.
  (8) The establishment of community growth classrooms.
  (9) The establishment of community volunteer service teams.
  (10) The establishment of community libraries.
  (11) Advocating for community-wide activities.
  (12) Promoting community disaster reporting and prevention drills.
  (13) Other matters relating to spiritual and ethical development.

Therefore, Figure 1 illustrates the 13 principles of Article 12 of the Regulations on Community Development Work as:

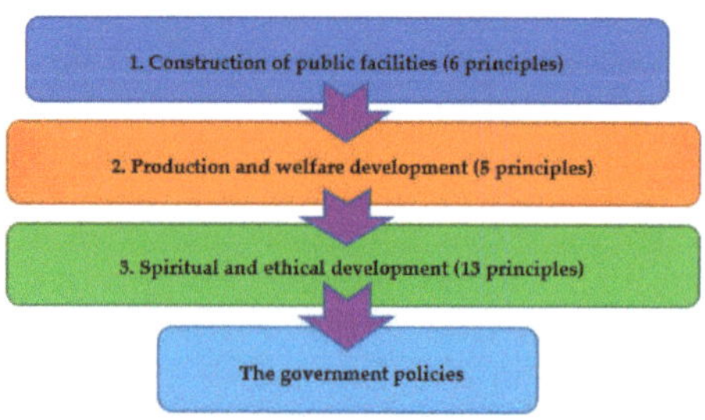

Figure 1. The 13 principles of the government policies.

However, as an urgent and serious issue, aging populations in remote or special needs areas of Taiwan require the most significant services and activities provided from all Taiwanese community development associations, which have had the most severe lack of useful methods, tools, and technologies to support them in advancing their operational effectiveness and implementing efficiency in the majority of Taiwanese community development associations, especially those in remote or special needs regions. Therefore, with the 100 percent coverage rate of fourth-generation mobile networks and the more than 50 percent coverage rate of the fifth-generation mobile networks, the majority of empirical industrialists have recommended considering employing wireless and internet technologies to directly construct unique digital connections for Taiwanese community development associations via the Internet [2–5].

These digital connections between community development associations will not only advance the efficiency of internal services and activities, such as providing the elderly with meals at community care locations (places) and providing entertainment services at community clubs for the elderly (places), etc., but will also strengthen the operational effectiveness of external services and activities, such as the development of tourism, the promotion of local culture, marketing regional agricultural productions, etc. [6].

Based on the rapidly spreading infections of contemporary global contagious diseases, such as severe acute respiratory syndrome (SARS) in 2003 and the coronavirus disease 2019 (COVID-19) in 2019, the majority of educational and on-the-job training courses were implemented though a series of virtual wireless connections using the IoT technology [7–10], as shown in Figure 2:

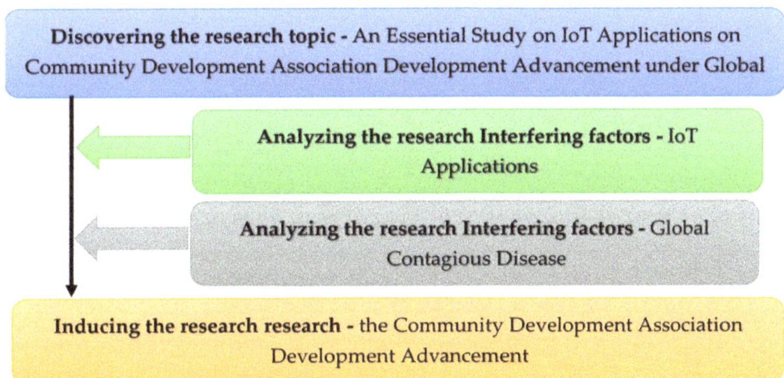

Figure 2. The main research concept.

2. Conceptual Literature

Significantly, in view of the accumulation of outcomes and performances, the number of Taiwanese community development associations has been greater than 6918 since 2020, based on the official statistics from the Ministry of Health and Welfare. Subsequently, since 2019, the number of directors and supervisors (persons) of community development associations was up to 108,524 individuals, including the 6918 director generals, 75,440 directors, and 26,166 supervisors. The number of people serviced was 22,014,889, and they were served by the 145,935 members comprising all Taiwanese community development associations. Momentously, up to 12,612,351 people were attended to through welfare services or activities for some specific communities' services and activities, such as community care locations offering meal services for the elderly (places), elderly entertainment services, activities for the elderly hosted by community clubs (places), meal services, and family entertainment services, such as community common rooms for mothers (classes), the promotion of local tourism, cultural development activities for folk, literature, and art (teams), regional security services from community mutual-help programs (teams), etc.

Eventually, the fundamental aspects developed were directly created and comprehensively instituted, and these developed aspects related to production and development, such as financial and economic development for industries located in community development associations; life aspects, such as cultural and living developments for regional residents in community development associations; and ecological aspect, such as environmental introduction, protection, and education for in-depth circumstantial understanding in community development associations. There were still some various outstanding services and activities in multiple aspects, such as tourism development, including economic living, transportation, and shopping demands, the regional cultural promotion of guidelines for living, and extensive environmental introduction, education, and protection propaganda, etc. [11–14].

In order to directly complete the organizational structure, positively enhance the service functions, and aggressively establish the community activities of community development associations in Taiwan, all Taiwanese community development associations must obtain the strongest and most powerful methods to be able to not only reinforce the current functions and services but also to satisfy a series of accumulated deficiencies. Furthermore, in association with the swift and sustainable development of services and activities in all Taiwanese community development associations, virtual methods, digital tools, and vital systems have been desired for each Taiwanese community development association.

As a result, Internet of Things ("IoT") technology has been considered a supporting digital technology to advance the administrative effectiveness and operating efficiency of the diversified services and activities of the 6918 current Taiwanese community development associations, according to its virtual, digital, and systematical characteristics. This is because the IoT is a type of digital computed connection that can transfer various data in association with a series of 3C ("computer, communication and consumer") electronic diversified devices such as smart phones, individual notebooks, and personal computers. In addition, IoT technology is definitely able to comprehensively connect the diversified devices and humans via integrating the complicated digital data collected from each universally unique identifier ("UUID") from the electronic devices with digital connection functions. Recently, with the rapid development of wireless and internet technologies, the coverage rate of the fourth-generation mobile networks has reached 100 percent since 2018, and the fifth-generation mobile networks have up to 49 percent coverage since 2020, based on the official 2020 report of the National Communication Commission ("NCC") in Taiwan.

3. Conclusions and Future Direction

With respect to widely evaluating and elementally analyzing the various possibilities and diversified applications of IoT technology for strengthening the internal administrative performances and implementation efficiencies of the services and activities provided by the current 6918 community development associations in Taiwan, these Taiwanese community development associations obviously demand a large, powerful, stable, and highly compatible digital system to assist them in handling large volumes of data and information in order to advance the implementation of community services and activities. In particular, the digital system can not only directly advance the efficiency of the community associations but could also assist the Taiwanese government in supporting and understanding the situation and consequences of implementing a digital system for the current 6918 community development associations. Extraordinarily, with respect to the hyper-speed and hyper-connected characteristics of the IoT technology, each 6918 community association could update the efficiency of their administrative performance and their efficiency in implementing services and activities through the diversified applications of the IoT technology by means of sharing more news and information between each Taiwanese community development association.

However, with respect to recent research [15,16] in IoT-technology-related fields in association with the administrative performances of community development associations, the most critical shortages and challenges are obviously the individual digital data protection and security regulations and rules (such as the Taiwan Personal Data Protection and Telecommunication Acts). After the executing a series of preliminary analyses, the empirical suggestion is to institute rules for the application of basic digital data in all 6918 current Taiwanese community development associations in order to easily introduce IoT technology into each community development association in order to successfully confront a series of threats due to serious global contagious diseases.

Author Contributions: Conceptualization, J.-C.M.L. and M.-Y.H.; methodology, M.-Y.H.; validation, J.-C.M.L.; formal analysis, M.-Y.H.; investigation, J.-C.M.L.; resources, J.-C.M.L. and C.-L.W.; writing—original draft preparation, M.-Y.H. and C.-L.W.; writing—review and editing, M.-Y.H.; visualization, J.-C.M.L.; supervision, M.-Y.H.; project administration, M.-Y.H.; funding acquisition, and C.-L.W. All authors have read and agreed to the published version of the manuscript.

Funding: This research was supported by the research supporting research projects (MOST 110-2420-H-002-003-MY3-Y11209) of Ministry of Science and Technology and (NTCU111103) from the National Taichung University of Education.

Institutional Review Board Statement: This research did not execute the questionnaires and interview and hence, this research did not require ethical approval.

Informed Consent Statement: Not applicable because this research did not involve any humans.

Data Availability Statement: No new data was created.

Conflicts of Interest: The authors declare no conflict of interest.

References

1. *The Article 12 of the Regulations on Community Development Work*; Ministry of Health and Welfare: Taipei, Taiwan, 2014.
2. Hsieh, M.-Y. Interdisciplinarily Exploring the Most Potential IoT Technology Determinants in the Omnichannel E-Commerce Purchasing Decision-Making Processes. *Appl. Sci.* **2020**, *10*, 603. [CrossRef]
3. Hsieh, M.Y. The Most sustainable niche principles of social media education in a higher education contracting era. *Sustainability* **2020**, *12*, 399. [CrossRef]
4. Huang, Y.-M.; Hsieh, M.Y.; Usak, M. A Multi-Criteria Study of Decision-Making Proficiency in Student's Employability for Multidisciplinary Curriculums. *Mathematics* **2020**, *8*, 897. [CrossRef]
5. Chan, Y.-K.; Hsieh, M.Y.; Usak, M. A Concrete Study on Social-Media Connection of Global Literacy Abilities in MOOCs under the Dual Impacts of Lower Birth-Rate and COVID-19. *Sustainability* **2021**, *13*, 2203. [CrossRef]
6. Usak, M.; Hsieh, M.Y.; Chan, Y.-K. A Concretizing Research on Making Higher-Education Sustainability Count. *Sustainability* **2021**, *13*, 2724. [CrossRef]
7. Duran, M.; Usak, M.; Hsieh, M.-Y.; Uygun, H. A New Perspective on Pedagogical Content Knowledge: Intellectual and Emotional Characteristics of Science Teachers. *Rev. de Cercet. si Interv. Sociala* **2021**, *72*, 9. [CrossRef]
8. Wu, T.-L.; Hsieh, M.-Y.; Min, K.-W.; Yu, M.-T.; Ho, C.-T. Use of Sensor Technologies in Online Courses in Post-COVID-19 Era. *Sensors Mater.* **2021**, *33*, 2045–2062. [CrossRef]
9. Huang, C.-C.; Chan, Y.-K.; Hsieh, M.Y. The Determinants of ESG for Community LOHASism Sustainable Development Strategy. *Sustainability* **2022**, *14*, 11429. [CrossRef]
10. Chan, Y.-K.; Hsieh, M.-Y. An Empirical Study on Higher Education C-ESG Sustainable Development Strategy in Lower-Birth-Rate Era. *Sustainability* **2022**, *14*, 12629. [CrossRef]
11. Hsieh, M.-Y. Online learning era: Exploring the most decisive determinants of MOOCs in Taiwanese higher education. *Eurasia J. Math. Sci. Technol. Educ.* **2016**, *12*, 1163. [CrossRef]
12. Hsieh, Y.-M. Employing MCDM methodology to verify correlationship between social media and service quality in the dynamic m-commerce era. *J. Internet Technol.* **2018**, *19*, 225. [CrossRef]
13. Hsieh, Y.-M.; Usak, M. High Education Radical Transformation Era: How Teachers' Competency can Enhance the Students' Employability. *Rev. Cercet. Interv. Soc. (RCIS)* **2020**, *68*, 95.
14. Huang, Y.-M.; Hsieh, M.-Y. An Interdisciplinary Research on Students' Employability in Technology Education to Advance Higher Education Enrollment Sustainability. *Sustainability* **2020**, *12*, 1806. [CrossRef]
15. Hsieh, M.-Y. The Sustainable Development and Strategic Approaches for Contemporary Higher Education. *Sustainability* **2022**, *14*, 12925. [CrossRef]
16. Huang, C.-C.; Chan, Y.-K.; Hsieh, M.Y. Preliminary Research on the Sustainable Determinants of Taiwanese Ecotourism with the International Standards. *Int. J. Environ. Res. Public Health* **2022**, *19*, 14489. [CrossRef] [PubMed]

Disclaimer/Publisher's Note: The statements, opinions and data contained in all publications are solely those of the individual author(s) and contributor(s) and not of MDPI and/or the editor(s). MDPI and/or the editor(s) disclaim responsibility for any injury to people or property resulting from any ideas, methods, instructions or products referred to in the content.

engineering proceedings

MDPI

Proceeding Paper

Initial Study of Core Decisive Modes of Game Theory on Closed Couple Relationships in Conflict Situations †

Chun-Ming Shih [1], Jiin-Chyuan Mark Lai [2,*] and Ming-Yuan Hsieh [1,3,*]

1 Graduate School of Human Sexuality, Shu-Te University, Kaohsiung 82445, Taiwan
2 Department of Applied Foreign Languages, TransWorld University, Douliu 640302, Taiwan
3 Department of International Business, National Taichung University of Education, Taichung 40306, Taiwan
* Correspondence: marklai07@gmail.com (J.-C.M.L.); s22871101@stu.edu.tw or capwisely@mail.ntcu.edu.tw (M.-Y.H.)
† Presented at the 3rd IEEE International Conference on Electronic Communications, Internet of Things and Big Data Conference 2023, Taichung, Taiwan, 14–16 April 2023.

Abstract: In life, making a decision is always required. Various situations demanding decision making are encountered in life. We apply the core decisive modes of game theory to closed couple relationships in conflict situations for the following reasons. (1) Life can be explained by the comprehensive decisions in games, as various conflict situations appear in a closed couple relationship. (2) Humans are influenced by diverse factors such as love, sacrifice, intuition, and the benefit of individual personality. (3) Information asymmetry is generated in subjective and objective environments. (4) Human nature is selfish. Based on the results of this research on the three core decisive modes in game theory, there are in-depth considerations and better win-win solutions considering the prisoner's dilemma. Game theory is further assayed in the further research for "Analyzing the research interfering factors of the core decisive modes of the game theory on the closed couple relationships at the conflict".

Keywords: decisive mode; game theory; closed couple relationships

Citation: Shih, C.-M.; Lai, J.-C.M.; Hsieh, M.-Y. Initial Study of Core Decisive Modes of Game Theory on Closed Couple Relationships in Conflict Situations. *Eng. Proc.* **2023**, *38*, 34. https://doi.org/10.3390/engproc2023038034

Academic Editors: Teen-Hang Meen, Hsin-Hung Lin and Cheng-Fu Yang

Published: 26 June 2023

Copyright: © 2023 by the authors. Licensee MDPI, Basel, Switzerland. This article is an open access article distributed under the terms and conditions of the Creative Commons Attribution (CC BY) license (https://creativecommons.org/licenses/by/4.0/).

1. Introduction

Game theory (GT) was induced from the "bad out of the small" theorem of a mathematician, von Neumann, in 1928. GT created the "zero-sum" concept in the individual decision-making process of a game. In the decisive situation of a zero-sum game, the increased remuneration is exactly equal to the increased loss of the opponent, but the sum of the remuneration and loss is always zero. This is the most important theoretical assumption in GT [1]. The concept of "Nash equilibrium (NE)" was the reason for the Nobel Prize in Economics in 1994. NE introduces the best solution in a non-cooperative decisive game [2]. To extend the applicability of GT to economic behavior analysis, von Neumann and Morgenstern addressed the economic behavior for the initial formation of the systematic GT, as GT provides logical mathematic calculations to quantify decision-making activities and offers diversified conditions such as cooperation, decision-making time, and information structure.

In the NE of GT, the basic assumptions are as follows. (1) There are N participants in a decisive game and strategies of (n-1) participants [3]. (2) Each participant in the game pursues the best outcome without considering other participants' maximum profits and potential benefits [4]. If any participant pursues deviating benefits from the current strategies without considering maximized profits or potential benefits, individual decision-making actions or behaviors belong to the non-Nash equilibrium (NNE) in the GT. However, in intimate relationships such as lovers or partners, the economic concepts and mathematic measurements of GT cannot be used. Therefore, we apply the three most popular decisive models to facilitate and promote the development of closed couple relationships and detect subtle changes in the relationships. The decisive models of GT applied in closed-couple relationships are (1) the prisoner's dilemma, (2) the convergent choice, and (3) divergence selection [4].

2. Literature Review

For the definition of a game, a game is classified according to the competitive or confrontational decision-making behaviors with which different goals or interests are used to obtain profits or benefits. For the profits or benefits, each participant considers the possible decision-making of the opponent to choose the most reasonable strategies [5]. GT is designed to consider and analyze the predicted and actual decision-making behaviors and strategies for individual optimization. In these games and strategies, the familiar incentive structures are constructed and expressed in different interactions in the game category. GT is for the analysis of the interaction between decision-makers through mathematical calculations. The biggest difference between the traditional decisive assessment and the GT analysis is that GT incorporates and considers each decision-maker's knowledge and expectation of the opponent as the opponent tries to maximize remuneration for decision-making payoff. Thus, GT is applied to interdisciplinary research such as biology, economics, international relations, computer science, political science, and military strategy by measuring the interaction in games. GT is a mathematical theory to institute the equilibrium point of a two-person zero-sum game and study decisive phenomena under the struggled situations or hostile conditions. GT is one of the greatest achievements in economics to deal with the psychological factors in complex decision-making processes or games in the 20th century.

3. Research Method

To apply the three core decisive models of GT to the analysis of closed couple relationships in conflict situations, the four research steps (research motivation, research methodology, research evaluation, and research induction) were used [6–12] in this research (Figure 1).

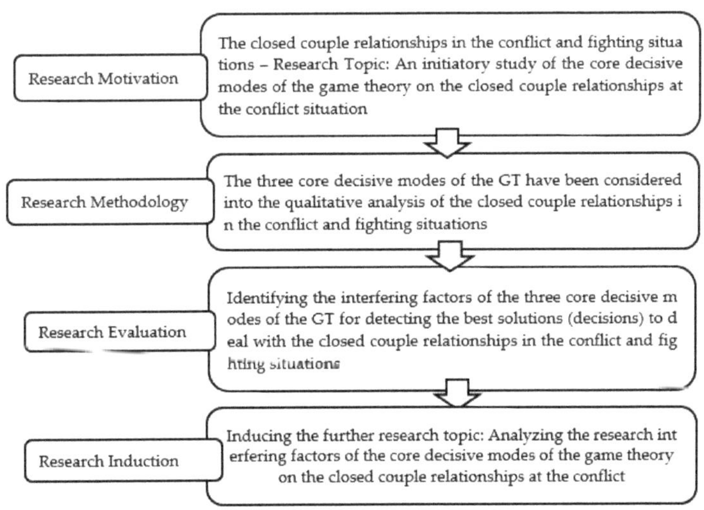

Figure 1. Research method.

4. Results and Discussion

Of the core decisive models, the most famous one is the prisoner's dilemma. The prisoner's dilemma is a typical model for exploring and demonstrating information asymmetry and the best benefits of GT in a dangerous and separate situation. In the beginning, an officer separates two prisoners to construct a situation with information asymmetry between them. Then, the officer implements the prisoner's dilemma to create the prisoner's confession. The best strategy for them is to be in prison for the shortest period. In a closed couple relationship, the "Be frank and lenient; resist and be strict" situation always works for coercing the confession of partners to obtain the truth [13]. Due to separation, two prisoners always have to be in the prisoner's dilemma, in which there are four situations: confession vs. confession, confession vs. no-confession, no-confession vs. confession, and no-confession vs. no-confession as displayed in Table 1.

Table 1. Prisoner's dilemma situation.

Decision	B: Confession	B: No-Confession
A: Confession	5-year sentence	10-year sentence for B Release of A
A: No-Confession	10-year sentence for A Release of B	6-month sentence

Convergent choice is used in daily activities of closed couple relationships because the relationship enforces the couple to choose the same actions for each other even though they are not willing to do so. As a result, there are four situations: favorite activity vs. favorite activity, favorite activity vs. non-favorite activity, non-favorite activity vs. favorite activity, and non-favorite activity vs. non-favorite activity. For example, when a male watches a sports game with his girlfriend, she forces herself to watch the game to make her boyfriend feel happy. The powerful driving force of the convergent choice game is love in the closed couple relationship [14]. The divergent selection (also called the weak chicken or coward game) is employed in conflict situations of the closed couple relationship. The divergence selection is constructed on the two drivers' conflict, too. When two cars speed toward each other on the same street, the one who turns first is a coward. There are four situations, as shown in Table 2. The two drivers do not have enough time to consider each other's choices as the two cars are driving fast and have to choose the best strategy. However, human nature often results in two consequent situations: both turn to be cowards or both go straight to be in a crash. Hence, divergence selection is the opposite of convergent choice.

Table 2. Divergent selection in a two-car crash.

Decision	Driver B: Straight	Driver B: Turn
Driver A: Straight	Crash	A wins B loses
Driver A: Turn	A loses B wins	Tie

In this research, the three decision models in GT were considered to discuss a closed couple relationship in a conflict situation with the main concept of Figure 2.

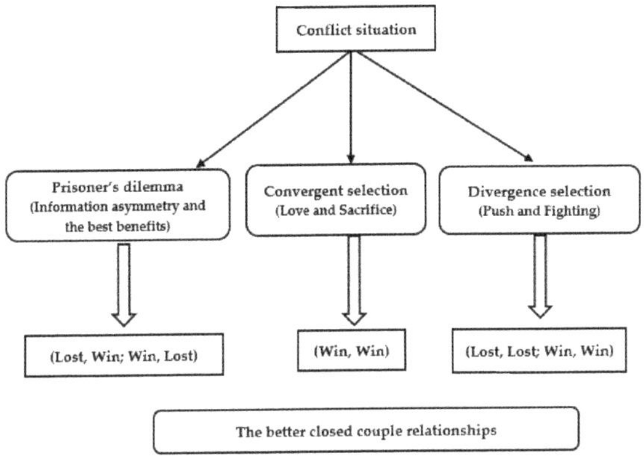

Figure 2. Revaluated method.

5. Conclusions and Recommendations

The core decisive model of GT of a closed couple relationship in a conflict situation is reviewed. Life is full of decisions in games. Various conflicts appear in a closed couple relationship in which love and sacrifice play a dominant role. Information asymmetry is generated by selfish human nature and an objective environment. In the three core decisive models, decisions are based on betrayal and cooperation in the prisoner's dilemma, while decisions are made based on sacrifice in the convergent selection. Life is always to make decisions. Each model explains the various situations for decision making encountered in life. Due to frequent decision-making opportunities, more diversified methods of GT play a critical role in selection and activities. To make the best decision, decision models need to be researched further. To find the best decision, GT can be used as shown in Figure 3 [15–18]. GT is designed and applied to human decision-making processes and the evaluation of a situation to institute the best strategies under the lowest uncertainty and risks. Further discussion is required to research interfering factors of decision-making: information asymmetry, individual traits, emotional impact, love, and sacrifice. To find the best decisions, closed couple relationships can be applied to conflicts.

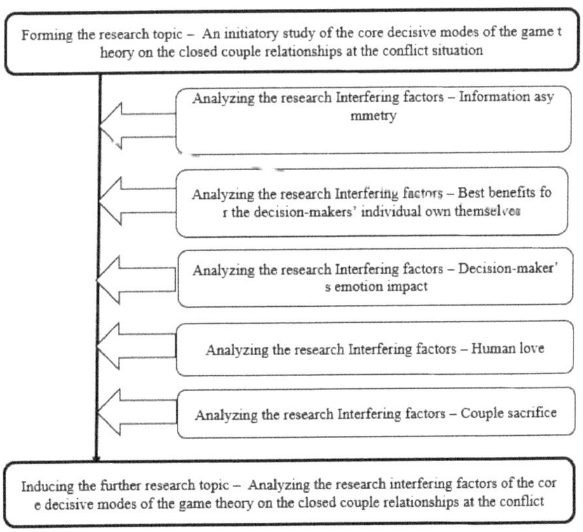

Figure 3. Process of GT application.

Author Contributions: Conceptualization, C.-M.S. and M.-Y.H.; methodology, M.-Y.H.; validation, C.-M.S.; formal analysis, J.-C.M.L.; investigation, J.-C.M.L.; resources, C.-M.S.; data curation, C.-M.S.; writing—original draft preparation, M.-Y.H.; writing—review and editing, M.-Y.H.; visualization, J.-C.M.L.; supervision, C.-M.S.; project administration, M.-Y.H. All authors have read and agreed to the published version of the manuscript.

Funding: This research was supplied by the research supporting research projects (MOST 110-2420-H-002-003-MY3-Y11209) of the Ministry of Science and Technology and (NTCU111103) by the National Taichung University of Education.

Institutional Review Board Statement: This research did not execute the questionnaires and interview and hence, this research did not require ethical approval.

Informed Consent Statement: Not applicable because this research did not involve any humans.

Data Availability Statement: No new data were created.

Conflicts of Interest: There are no conflicts of interest in this research.

References

1. Nash, J.F., Jr. Equilibrium points in n-person games. *Mathematics* **1950**, *36*, 48.
2. Nash, J.F., Jr. The bargaining problem. *Econometria* **1950**, *18*, 155. [CrossRef]
3. Nash, J.F., Jr. Non-cooperative games. *Ann. Math.* **1951**, *54*, 286. [CrossRef]
4. The Concise Encyclopedia of Economics: Prisoners' Dilemma. Available online: http://www.econlib.org/library/Enc/PrisonersDilemma.html (accessed on 12 May 2023).
5. Robinson, J. An iterative method of solving a game *Ann. Math* **1951**, *54*, 296.
6. Huang, C.C.; Chan, Y.-K.; Hsieh, M.Y. The Determinants of ESG for Community LOHASism Sustainable Development Strategy. *Sustainability* **2022**, *14*, 11429. [CrossRef]
7. Chan, Y.-K.; Hsieh, M.Y. An Empirical Study on Higher Education C-ESG Sustainable Development Strategy in Lower-Birth-Rate Era. *Sustainability* **2022**, *14*, 12629. [CrossRef]
8. Hsieh, M.Y. The Sustainable Development and Strategic Approaches for Contemporary Higher Education. *Sustainability* **2022**, *14*, 12925. [CrossRef]
9. Huang, C.-C.; Chan, Y.-K.; Hsieh, M.-Y. Preliminary Research on the Sustainable Determinants of Taiwanese Ecotourism with the International Standards. *Int. J. Environ. Res. Public Health* **2022**, *19*, 14489. [CrossRef]
10. Chan, Y.-K.; Hsieh, M.-Y.; Usak, M.A. Concrete Study on Social-Media Connection of Global Literacy Abilities in MOOCs under the Dual Impacts of Lower Birth-Rate and COVID-19. *Sustainability* **2021**, *13*, 2203. [CrossRef]
11. Usak, M.; Hsieh, M.-Y.; Chan, Y.-K. A Concertizing Research on Making Higher Education Sustainability Count. *Sustainability* **2021**, *12*, 1806–1821.
12. Duran, M.; Usak, M.; Hsieh, M.-Y.; Uygun, H. A New Perspective on Pedagogical Content Knowledge: Intellectual and Emotional Characteristics of Science Teachers. *Rev. De Cercet. Si Interv. Soc.* **2021**, *72*, 9–32. [CrossRef]
13. The End of the Zero-Sum Game. Available online: http://technologizer.com/2011/01/27/the-end-of-the-zero-sum-game/ (accessed on 12 May 2023).
14. Greenwald, A.; Littman, M.L. Introduction to the special issue on learning and computational game theory. *Mach. Learn.* **2007**, *67*, 3. [CrossRef]
15. Wu, T.-L.; Hsieh, M.-Y.; Min, K.-W.; Yu, M.-T.; Ho, C.-T. Use of Sensor Technologies in Online Courses in Post-COVID-19 Era *Sens. Mater.* **2021**, 33, 2045–2062.
16. Hsieh, Y.-M. Online learning era: Exploring the most decisive determinants of MOOCs in Taiwanese higher education. *Eurasia J. Math. Sci. Technol. Educ.* **2016**, *12*, 1163. [CrossRef]
17. Hsieh, Y.-M.; Usak, M. High Education Radical Transformation Era: How Teachers' Competency can Enhance the Students' Employability. *Rev. Cercet. Interv. Soc.* **2020**, *68*, 95. [CrossRef]
18. Hsieh, Y.-M. Employing MCDM methodology to verify correlationship between social media and service quality in the dynamic m-commerce era. *J. Internet Technol.* **2018**, *19*, 225.

Disclaimer/Publisher's Note: The statements, opinions and data contained in all publications are solely those of the individual author(s) and contributor(s) and not of MDPI and/or the editor(s). MDPI and/or the editor(s) disclaim responsibility for any injury to people or property resulting from any ideas, methods, instructions or products referred to in the content.

engineering proceedings

MDPI

Proceeding Paper

Aspect and Sentiment Classification Mechanisms of Student After-Class Self-Evaluated Comments: Investigation on Nonsense Data, Feature Extraction, and Classification Models †

Chih-Yueh Chou * and Tzu-Yi Chuang

Department of Computer Science and Engineering, Yuan Ze University, Taoyuan City 32003, Taiwan; s1096016@mail.yzu.edu.tw
* Correspondence: cychou@saturn.yzu.edu.tw

† Presented at the 3rd IEEE International Conference on Electronic Communications, Internet of Things and Big Data Conference 2023, Taichung, Taiwan, 14–16 April 2023.

Abstract: Students' after-class self-evaluated comments are useful for understanding students' learning and reflecting teacher's teaching. Researchers and engineers have attempted to apply educational data mining techniques, such as text analysis, sentiment analysis, machine learning, and deep learning to develop classification mechanisms of students' self-evaluated comments. This study was carried out to develop aspect and sentiment classification mechanisms to automatically classify students' self-evaluated comments into seven aspect categories and three sentiment categories. We investigated the impact of whether we should exclude nonsense data or not, the impact of different feature extraction methods, and the impact of different classification models on classification accuracy. The results showed that the combination of bidirectional encoder representations from transformers (BERT) word embedding feature extraction and Random Forest classification showed the best accuracy (90.7%) on aspect classification when including nonsense data, whereas the combination of BERT-word embedding feature extraction and Random Forest classification had the best accuracy (93.2%) on aspect classification when excluding nonsense data. Including nonsense data reduced the classification accuracies. In addition, the combination of one-word bag-of-words feature extraction and Random Forest classification presented the best accuracy (99.5%) with regard to sentiment classification.

Keywords: self-evaluated comments; educational data mining; sentiment analysis; machine learning; deep learning

Citation: Chou, C.-Y.; Chuang, T.-Y. Aspect and Sentiment Classification Mechanisms of Student After-Class Self-Evaluated Comments: Investigation on Nonsense Data, Feature Extraction, and Classification Models. *Eng. Proc.* **2023**, *38*, 43. https://doi.org/10.3390/engproc2023038043

Academic Editors: Teen-Hang Meen, Hsin-Hung Lin and Cheng-Fu Yang

Published: 27 June 2023

Copyright: © 2023 by the authors. Licensee MDPI, Basel, Switzerland. This article is an open access article distributed under the terms and conditions of the Creative Commons Attribution (CC BY) license (https://creativecommons.org/licenses/by/4.0/).

1. Introduction

Students' evaluated comments on their learning or teachers' teaching after a class or course help teachers or organizations understand students' learning status, predict students' performance, and reflect teachers' teaching [1]. Students' evaluated comments offer rich information on learning or teaching [2], but it is labor-intensive and time-consuming to read students' comments. Recently, many text classification techniques and applications have been developed, including different feature extraction methods and classification models [3]. Researchers have applied text classification techniques to automatically classify the aspect and sentiment categories of students' comments. For example, Yu et al. have developed a sentiment classification mechanism to classify students' self-evaluated comments into three sentiment categories: positive, neutral, and negative [4]. Sindhu's team has applied deep learning to develop a classification mechanism to classify comments into six aspect categories and three sentiment categories [5]. Onan has developed a sentiment classification mechanism and compared sentiment classification accuracies of different combinations of feature extraction methods and classification models [6].

We developed aspect and sentiment classification mechanisms and compared the aspect and sentiment classification accuracies of combinations of different feature extraction methods and classification models, including conventional machine learning models and deep learning models. In addition, as several students' self-evaluated comments are nonsense or not related to the class, these data may be excluded or classified as being in the "others" category. However, the application of the classification mechanism needs to deal with nonsense data. We compared the aspect classification accuracies of excluding and including (i.e., classified as "others") nonsense data.

2. Method

2.1. Data Collection, Labelling, and Balancing

Students' self-evaluated comments were collected in a programming course at a university. Students were asked to write down their self-evaluated comments in a system after each class. In total, 1640 anonymous comments were collected. These comments were written in Chinese, English, or a mixture of Chinese and English. Two researchers reviewed these comments and classified them into seven aspect categories: interest, gain, positivity, speed and difficulty, teacher, overall, and others. The interest aspect was to know if students were interested in the content of the class, such as "The second example is interesting." or "Today's class is boring.". The gain aspect was to know if students learned from the class or not, such as "I learned how to design a menu.". The positivity aspect was to know if students wanted to make effort to review or practice the content, such as "I need more practice after class.". The speed and difficulty aspect was to know if students' feelings for the teacher's lecture speed or the difficulty of the content, such as "Today's content is easy for me." or "Today's lecture speed is too fast for me.". The Teacher aspect was to know the students' evaluation of the teacher, such as "The teacher is serious in class.". The overall aspect was to know the students' entire evaluation of the class, such as "Good!". Others aspect was to know if the comment is nonsense or not related to the class, such as "123" or "Hello!".

Each comment was also labeled with its sentiment state from three sentiment categories: positive, neutral, or negative. Two researchers independently labeled the aspect and sentiment categories of each comment. If the two researchers' labels were different, a discussion was conducted to determine the final label. Table 1 lists the labeled aspect and sentiment distributions of comments. The distributions of comments are unbalanced. The unbalanced data affect the training of classification models with a bias toward categories with more data. Therefore, comments were balanced by repeating the comments of the categories with fewer comments to the amount of the category with the maximum amount of comments.

Table 1. Labeled aspect and sentiment distributions of comments.

Sentiment	Aspect						
	Interest	Gain	Positivity	Speed and Difficulty	Teacher	Overall	Others
Positive	297	391	117	71	74	205	92
Neutral	0	0	0	2	0	46	138
Negative	4	0	3	145	0	6	49

For measuring the cohesion of comments within each aspect category, the average cosine similarity of comments of each aspect category was calculated (Table 2). The results showed that the "others" category comments had the lowest average cosine similarity. That is, comments in the "others" category were more diverse than other categories. The reason may be that comments in the "others" category were nonsense or not related to the class that was not classified as other aspect categories.

Table 2. Average cosine similarities of different aspect category comments.

	Interest	Gain	Positivity	Speed and Difficulty	Teacher	Overall	Others
Average cosine similarity	0.428	0.394	0.397	0.407	0.592	0.552	0.269

2.2. Feature Extraction of Comments

Features of comments can be extracted by different methods. This study adopted and compared different feature extraction methods. The first feature extraction method was bag-of-words; that is, a feature vector of a comment is generated by checking whether specific feature words exist in the comment or not. The feature words can be a set of one-word segmentations, a set of two-word segmentations, a set of three-word segmentations, or a set of segmentations with the unspecific number of words that were segmented by meaning. Two or more Chinese words together may lead to specific meanings. Thus, segmentations of Chinese sentences are important. We adopted one-word segmentation and unspecific number–word segmentation (k-words; i.e., k-gram, in which a word is a unit) for generating feature vectors of comments and compared their accuracies. These specific feature words were analyzed from all comments. The k-words segmentations were analyzed and segmented by MONPA [7].

The second feature extraction method was sequence-of-words. Words with different sequences may have different meanings. For instance, "John loves Mary." is different from "Mary loves John.". However, the bag-of-words method extracts these two sentences into the same feature vector and cannot distinguish between them. The sequence-of-words method divides sentences into word segmentations (a segmentation may be a word or several words), assigns each specific word segmentation an index, and transforms a comment into a sequence vector as its feature vector. We adopted one-word and k-word segmentations for transforming comments into sequence-of-words feature vectors. The k-word segmentations were also analyzed and segmented by MONPA.

The third feature extraction method was Doc2vec embedding [8]. Several words have semantic relationships. For example, "good" and "excellent" are positive adjectives. The bag-of-words method does not deal with semantic relationships among words. Word2Vec analyzes semantic relationships among words based on their context words from texts and transforms each word into a feature vector [9]. Words with close semantic relationships have similar feature vectors. Doc2Vec is modified from Word2Vec and generates a feature vector as a distributed representation of a text. This study adopted Doc2vec to transform a comment into a feature vector.

The fourth feature extraction method was BERT-word embedding [10]. BERT-word embedding analyzes texts based on BERT pre-trained model. The BERT pre-trained model is trained from analyzing texts in BooksCorpus and Wiki. BERT-word embedding generates feature vectors taking context into account. We adopted BERT-word embedding (paraphrase-xlm-r-multilingual-v1 sentence-transformers model) [11] to transform comments into 768-dimensional feature vectors.

2.3. Classification Models

Classification needs supervised machine learning models. Thus, several supervised classification models were adopted and compared in this study. First, conventional machine learning classification models, including Naïve Bayes (NB), Support Vector Machine (SVM), and Random Forest (RF), were adopted along with bag-of-words, Doc2Vec embedding, and BERT-word embedding extraction methods.

Second, deep learning classification models, including one-layered long short-term memory (1-LSTM), two-layered LSTM (2-LSTM), 1-LSTM with attention mechanism (1-LSTM + attention), and 2-LSTM + attention were adopted, along with the sequence-of-words extraction method. LSTM is a type of recurrent neural network (RNN) for processing sequential data [12]. LSTM uses forget gates to determine which information is preserved.

The attention mechanism defines each output from weighted contributions of inputs and trains these weights [13].

Third, BERT for the Sequence Classification model was also adopted when BERT-word embedding extraction was adopted.

2.4. Model Training and Validation for Calculating Classification Accuracy

Different feature extraction methods and classification models were combined to train models and calculate the classification accuracy. Aspect classification was conducted, both including and excluding the "others" category data (i.e., nonsense data) for comparison. Sentiment classification was conducted, including the "others" category data. In this study, 5-fold cross-validation was used. Data were randomly assigned into 5 datasets. In turn, a dataset was chosen for validation of classification accuracy, and other datasets were used for training classification models. Models were used for classifying the validation dataset and compared the classification results to the labeled classifications to calculate the classification accuracy. The average classification accuracy of five turns was calculated as the classification accuracy of the model.

The classification mechanisms and validations were implemented in Python 3.7, including the Gensim, Keras, Matplotlib, NumPy, Pandas, Scikit-learn, and TensorFlow libraries. The parameters of random state, batch size, and epochs were assigned to 42, 32, and 20, respectively.

3. Result

3.1. Accuracies of Aspect Classifications, including Nonsense Data

Table 3 lists the accuracies of aspect classifications of different feature extraction methods and classification models, including nonsense data. The results were sorted by accuracy from low to high. The results show that the combination of BERT-word embedding feature extraction method and RF classification model have the best accuracy (90.7%). The accuracies of the combinations of the bag-of-words feature extraction method and conventional classification models (NB, SVM, RF) range from 22.1% to 77.2. Compared to bag-of-words, Doc2Vec embedding has better accuracies (from 48.7% to 81.6%) when applied to the same classification models. When applying bag-of-words or Doc2Vec embedding feature extraction methods, RF has the best accuracy, SVM has the second best accuracy, and NB has the worst accuracy.

Table 3. Accuracies of aspect classifications of different extractions and models, including nonsense data.

Feature Extraction	Classification Model	Accuracy
Bag-of-words (k-word)	NB	0.221
Bag-of-words (one-word)	NB	0.223
Bag-of-words (k-word)	SVM	0.450
Doc2Vec embedding	NB	0.487
Bag-of-words (one-word)	SVM	0.496
Doc2Vec embedding	SVM	0.734
Bag-of-words (one-word)	RF	0.772
Bag-of-words (k-word)	RF	0.772
Doc2Vec embedding	RF	0.816
Sequence-of-words (one-word)	1-LSTM	0.830
Sequence-of-words (k-word)	1-LSTM	0.843
BERT-word embedding	BERT for Sequence Classification	0.845
Sequence-of-words (one-word)	1-LSTM + Attection	0.858
Sequence-of-words (one-word)	2-LSTM	0.852
BERT-word embedding	SVM	0.859
Sequence-of-words (one-word)	2-LSTM + Attention	0.866
Sequence-of-words (k-word)	2-LSTM + Attention	0.869
Sequence-of-words (k-word)	2-LSTM	0.870
Sequence-of-words (k-word)	1-LSTM + Attection	0.878
BERT-word embedding	RF	0.907

The accuracies of the combinations of sequence-of-words feature extraction method and deep learning classification models ranged from 83% to 87.8%, which were better than that of the combinations of bag-of-words or Doc2Vec extraction methods and conventional machine learning classification models. The accuracies of the combinations of BERT-word embedding extraction and BERT for Sequence Classification, SVM, and RF were 84.5%, 85.9%, and 90.7%. The results showed that the BERT-word embedding extraction method is better than bag-of-words, Doc2Vec, and sequence-of-words.

3.2. Accuracies of Aspect Classifications, Excluding Nonsense Data

Table 4 shows the accuracies of aspect classifications of different feature extraction methods and classification models, excluding nonsense data. The trend of accuracies excluding nonsense data is similar to the trend of accuracies, including nonsense data, but the accuracies, including nonsense data, are lower than the accuracies, excluding nonsense data. The reason may be that the nonsense data have low cohesion (Table 2).

Table 4. Accuracies of aspect classifications of different extractions and models excluding nonsense data.

Feature Extraction	Classification Model	Accuracy
Bag-of-words (k-word)	NB	0.249
Bag-of-words (one-word)	NB	0.261
Bag-of-words (k-word)	SVM	0.504
Doc2Vec embedding	NB	0.545
Bag-of-words (one-word)	SVM	0.561
Doc2Vec embedding	SVM	0.806
Bag-of-words (k-word)	RF	0.808
Bag-of-words (one-word)	RF	0.830
Doc2Vec embedding	RF	0.845
Sequence-of-words (k-word)	1-LSTM	0.893
Sequence-of-words (one-word)	2-LSTM + Attention	0.895
Sequence-of-words (one-word)	1-LSTM + Attection	0.903
Sequence-of-words (k-word)	2-LSTM + Attention	0.903
Sequence-of-words (one-word)	1-LSTM	0.910
Sequence-of-words (k-word)	1-LSTM + Attection	0.914
Sequence-of-words (k-word)	2-LSTM	0.914
Sequence-of-words (one-word)	2-LSTM	0.919
BERT-word embedding	SVM	0.921
BERT-word embedding	BERT for Sequence Classification	0.923
BERT-word embedding	RF	0.932

The results also show that the combination of the BERT-word embedding feature extraction method and the RF classification model has the best accuracy (93.2%). The combinations of bag-of-words and conventional classification models have low accuracies. Doc2Vec embedding extractions show better accuracies than bag-of-words. The combinations of sequence-of-words and deep learning models have good accuracies (from 89.3% to 91.9%). Lastly, BERT-word embedding extraction method is better than bag-of-words, Doc2Vec, and sequence-of-words.

3.3. Accuracies of Sentiment Classifications

Table 5 presents the accuracies of sentiment classification of different feature extraction methods and classification models. The combinations of bag-of-words and NB and SVM have low accuracies (from 39.4% to 77.8%). The combinations of sequence-of-words and deep learning models have good accuracies (from 96.5% to 98%). Surprising results are that the combinations of bag-of-words and RF have high accuracies (k-words: 97.9% and one-word: 99.5%). The accuracy of sentiment classifications is better than that of aspect classifications. The reason may be that the aspect classifications have six or seven categories, and the sentiment classifications only have three categories. Another possible reason may be that the aspect classifications were easier than aspect classifications.

Table 5. Accuracies of sentiment classifications of different extractions and models.

Feature Extraction	Classification Model	Accuracy
Bag-of-words (k-word)	NB	0.394
Bag-of-words (one-word)	NB	0.417
Bag-of-words (k-word)	SVM	0.733
Bag-of-words (one-word)	SVM	0.778
Sequence-of-words (k-word)	2-LSTM	0.965
Sequence-of-words (one-word)	1-LSTM + Attention	0.969
Sequence-of-words (one-word)	2-LSTM + Attention	0.972
Sequence-of-words (k-word)	1-LSTM	0.973
Sequence-of-words (one-word)	2-LSTM	0.973
Sequence-of-words (k-word)	2-LSTM + Attention	0.976
Sequence-of-words (one-word)	1-LSTM	0.977
Bag-of-words (k-word)	RF	0.979
Sequence-of-words (k-word)	1-LSTM + Attention	0.980
Bag-of-words (one-word)	RF	0.995

4. Conclusions

We developed aspect and sentiment classification mechanisms of student after-class self-evaluated comments and validated their classification accuracies. The results showed that the accuracies of sentiment classifications were better than those of aspect classifications. In addition, the aspect classifications excluding nonsense data have better accuracies than the classifications including nonsense data. We also explored different feature extraction methods, including bag-of-words, sequence-of-words, Doc2Vec embedding, and BERT-word embedding. First, the results showed that Doc2Vec embedding is better than bag-of-words. Second, sequence-of-words was an excellent feature extraction method, along with deep learning classification models. Third, BERT-word embedding is generally an excellent feature extraction method. In addition, classification models were explored, including NB, SVM, RF, 1-LSTM, 2-LSTM, 1-LSTM + attention, 2-LSTM + attention, and BERT for Sequence Classification. The results revealed that RF is generally an excellent classification model.

This study has limitations. First, not all possible combinations of feature extraction methods and classification models were implemented and compared. Second, comments were collected in a programming course at one university. Comments from different courses may diversify and reduce the accuracy of classifications.

Author Contributions: Conceptualization, C.-Y.C. and T.-Y.C.; project administration, C.-Y.C.; supervision, C.-Y.C.; methodology, C.-Y.C. and T.-Y.C.; software, T.-Y.C.; formal analysis, T.-Y.C.; investigation, T.-Y.C.; validation, T.-Y.C.; writing, C.-Y.C. All authors have read and agreed to the published version of the manuscript.

Funding: This research received no external funding.

Institutional Review Board Statement: The study was conducted in accordance with the Declaration of Helsinki, and it was approved by the Research Ethics Committee, National Taiwan Normal University (REC number: 202108ES020, 8 September 2021).

Informed Consent Statement: Informed consent was obtained from all subjects involved in the study.

Data Availability Statement: Data are contained within the article.

Conflicts of Interest: The authors declare no conflict of interest.

References

1. Holmes, L.E.; Smith, L.J. Student evaluations of faculty grading methods. *J. Educ. Bus.* **2003**, *78*, 318–323. [CrossRef]
2. Alhija, F.N.-A.; Fresko, B. Student evaluation of instruction: What can be learned from students' written comments? *Stud. Educ. Eval.* **2009**, *35*, 37–44. [CrossRef]
3. Mirończuk, M.M.; Protasiewicz, J. A recent overview of the state-of-the-art elements of text classification. *Expert Syst. Appl.* **2018**, *106*, 36–54. [CrossRef]
4. Yu, L.-C.; Lee, C.-W.; Pan, H.; Chou, C.-Y.; Chao, P.-Y.; Chen, Z.; Tseng, S.; Chan, C.; Lai, K.R. Improving early prediction of academic failure using sentiment analysis on self-evaluated comments. *J. Comput. Assist. Learn.* **2018**, *34*, 358–365. [CrossRef]
5. Sindhu, I.; Daudpota, S.M.; Badar, K.; Bakhtyar, M.; Baber, J.; Nurunnabi, M. Aspect-based opinion mining on student's feedback for faculty teaching performance evaluation. *IEEE Access* **2019**, *7*, 108729–108741. [CrossRef]
6. Onan, A. Mining opinions from instructor evaluation reviews: A deep learning approach. *Comput. Appl. Eng. Educ.* **2020**, *28*, 117–138. [CrossRef]
7. Yeh, W.-C.; Hsieh, Y.-L.; Chang, Y.-C.; Hsu, W.-L. MONPA: A Multitask Chinese Segmentation, Named-entity and Part-of-speech Annotator. In Proceedings of the 31st Conference on Computational Linguistics and Speech Processing (ROCLING 2019), New Taipei City, Taiwan, 3–5 October 2019; pp. 241–245.
8. Le, Q.; Mikolov, T. Distributed representations of sentences and documents. In Proceedings of the International Conference on Machine Learning, Beijing, China, 21–26 June 2014; pp. 1188–1196.
9. Mikolov, T.; Chen, K.; Corrado, G.; Dean, J. Efficient estimation of word representations in vector space. *arXiv* **2013**, arXiv:1301.3781.
10. Devlin, J.; Chang, M.-W.; Lee, K.; Toutanova, K. Bert: Pre-training of deep bidirectional transformers for language understanding. *arXiv* **2018**, arXiv:1810.04805.
11. Reimers, N.; Gurevych, I. Sentence-bert: Sentence embeddings using siamese bert-networks. *arXiv* **2019**, arXiv:1908.10084.
12. Hochreiter, S.; Schmidhuber, J. Long short-term memory. *Neural Comput.* **1997**, *9*, 1735–1780. [CrossRef]
13. Vaswani, A.; Shazeer, N.; Parmar, N.; Uszkoreit, J.; Jones, L.; Gomez, A.N.; Kaiser, Ł.; Polosukhin, I. Attention is all you need. *Adv. Neural Inf. Process. Syst.* **2017**, *30*, 5999–6009.

Disclaimer/Publisher's Note: The statements, opinions and data contained in all publications are solely those of the individual author(s) and contributor(s) and not of MDPI and/or the editor(s). MDPI and/or the editor(s) disclaim responsibility for any injury to people or property resulting from any ideas, methods, instructions or products referred to in the content.

engineering proceedings

MDPI

Proceeding Paper

Applying Regression Model for Changes in Population Age Structure on Domestic Tourism Demand †

Tzong-Shyuan Chen and Jui-Yuan Chu *

Department of Leisure Services Management, Chaoyang University of Technology, Taichung 413310, Taiwan; jsmike@cyut.edu.tw
* Correspondence: jychu@cyut.edu.tw

† Presented at the 3rd IEEE International Conference on Electronic Communications, Internet of Things and Big Data Conference 2023, Taichung, Taiwan, 14–16 April 2023.

Abstract: Changes in age structure are common in most countries because of aging and lower birth rates. In this study, we constructed a domestic tourism demand model using the concept of the proportion of the prime tourist population to investigate the effect of changes in the age structure of the population. A regression model with the time series data of the number of domestic tourists in Taiwan from 2001 to 2021 was adopted in this study. The results showed that there was a significant positive relationship between the proportion of the prime tourist population and domestic tourism demand, indicating that aging impacted the domestic tourism market negatively.

Keywords: population age structure; aging; domestic tourism

Citation: Chen, T.-S.; Chu, J.-Y. Applying Regression Model for Changes in Population Age Structure on Domestic Tourism Demand. *Eng. Proc.* **2023**, *38*, 47. https://doi.org/10.3390/engproc2023038047

Academic Editors: Teen-Hang Meen, Hsin-Hung Lin and Cheng-Fu Yang

Published: 27 June 2023

Copyright: © 2023 by the authors. Licensee MDPI, Basel, Switzerland. This article is an open access article distributed under the terms and conditions of the Creative Commons Attribution (CC BY) license (https://creativecommons.org/licenses/by/4.0/).

1. Introduction

Aging is a common issue as the population age structure changes in many countries around the world in the twenty-first century. According to the United Nations World Population Prospects [1], the proportion of the global population over the age of 65 will increase to 16.4% in 2050, up from 9.7% in 2022. In terms of Taiwan's elderly population, the proportion of people aged over 65 years old exceeded 14% in 2018, entering an aged society, and will exceed 20% in 2025, becoming a super-aged society. Aging affects the consumption, investment, and savings of the whole economy, as well as annuities, retirement benefits, and government revenues and expenditures. It also affects the labor supply and demand in the labor market, the industrial structure, and, ultimately, long-term economic development [2].

Tourism is an integral part of modern life and plays an important role in the development of the global economy. A country's tourism market is composed of two different markets, inbound tourism and domestic tourism, which is also referred to as internal tourism. Generally speaking, the domestic tourism market gradually expands with the development of the economy, the increase of national income, and the adjustment of the vacation system. According to the World Tourism Organization, the domestic tourism market is ten times larger than the international tourism market [3]. With global tourism deeply impacted by COVID-19, the importance of domestic tourism to the development of the domestic tourism industry has become more evident. Looking at the development of the domestic tourism market in Taiwan, according to the data from the Tourism Bureau, the number of domestic tourists was 190.376 million in 2016, which showed the peak of the number of tourists in recent years (Figure 1). After that, the number of domestic tourists gradually decreased, and the number in 2019 was 169.279 million. At the same time, however, the number of outbound tourists reached a record high. The number of domestic tourists in 2020 was affected by COVID-19, and the number of tourists was 142.970 million, a decrease of 15.5% compared with the number of tourists in 2019.

The domestic tourism market is the core of internal tourism, and its market changes are critical to the future development of the domestic tourism industry. In the face of the recent decline of the domestic tourism market and the different performances of outbound tourism, this study aims to examine the impact of aging on the domestic tourism market through constructing a regression model for domestic tourism demand based on changes in the age structure of the population. The findings provide a reference for strategies of tourism-related industries.

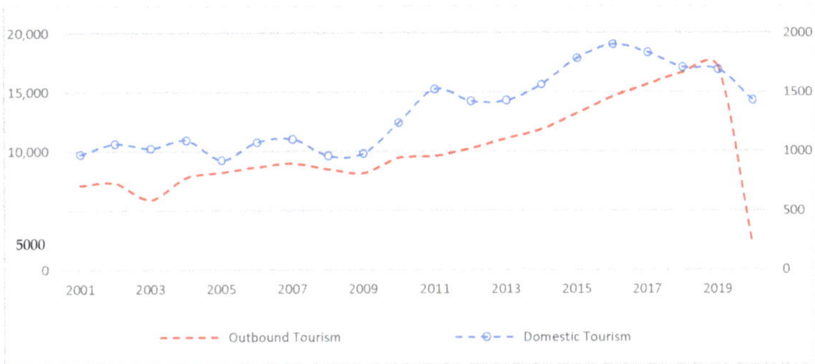

Figure 1. Changes in the number of domestic and outbound tourists from 2001 to 2020.

2. Literature Review

2.1. Domestic Tourism

A review of past studies on tourism demand shows that many related studies mainly have focused on the relationship between tourism demand and its influencing factors, as well as the characteristics and influencing factors of tourism demand. Earlier research on tourism demand mainly focused on international tourism and used aggregate data to analyze the impact of tourism on the economy of a country or region. In recent years, microdata have been used to explore individual or family tourism consumption decisions and influencing factors [4]. However, domestic tourism has received relatively little attention and research among those tourism demand studies.

Using data from 1976 to 1996, the influencing factors of South Korea's tourism expenditure were explored. The empirical results show that working hours, family size, and education time were important factors affecting South Korean domestic tourism expenditure. A two-stage decision budgeting model was used to study the domestic tourism demand of Swedish households. Using monthly data from January 1996 to October 1996, a model of the domestic tourism expenditure system was constructed including accommodation, food and beverage, shopping, transportation, and miscellaneous expenses as variables. Since Swedish households had no consumption records in several domestic tourism expenditure items, the study tried to overcome the sample selection bias caused by zero expenditure. The Hecman model was first used to derive the correction factor Mills λ. Then, this correction factor was put into the Quadratics AIDS system to derive the model coefficients to calculate the income, expenditure, and price elasticities of different household characteristics.

The domestic tourism demand of Chinese urban and rural residents was investigated using a hierarchical linear model. Using data from the National Household Travel Survey from 1996 to 2007, urban residents in 35 major cities and rural residents in 30 provinces were questioned from 2000 to 2007. The study found that absolute personal income was the main factor affecting the demand for domestic tourism. Also, it is observed from the results of the hierarchical linear model that the absolute income effect varied across cities/provinces, showing significant heterogeneity. In addition, relative income has a significant impact on travel demand in certain sub-regions of China. A structural equation model was also used to construct the relationship between tourism motivation, perceived benefits, perceived values, and behavioral intentions in Mongolia, using questionnaire data. The empirical results verified that travel motivation had a significant direct effect on perceived benefits, perceived values, and behavioral intentions.

2.2. Population Age Structure and Consumption

With the coming aging society, the impact of the age structure of the population on consumption has been increasingly emphasized and discussed. The life cycle hypothesis provides a good basis for empirical analysis of the relationship between population age structure and consumption. According to the life cycle hypothesis, an individual's consumption decisions are part of a long-term plan, and the consumption decisions at each age stage in life depend on the allocation of an individual's lifetime resources, rather than on current income. The average propensity to consume in the middle-aged and high-income period is smaller than that in the young and old, which makes the consumption plan formed at an average consumption level. That is, as individuals grow older, their consumption expenditures increase accordingly. In studies [5–7], the life cycle hypothesis was confirmed, showing that as the proportion of the aging population increased, aggregate consumption increased accordingly. However, studies [8,9] showed that population aging had a negative impact on consumption (Table 1).

Table 1. Studies on the relationship between population age structure and consumption.

Author	Age Structure Variables	Results
Attfield and Cannon [5]	Proportion of population by age	Aggregate spending rises as senior population rises
Lefèbvre [10]	Age effect of cohort analysis	Household spending decreases as the head of household age increases
Erlandsen and Nymoen [7]	Proportion of the prime saving population	Aggregate consumption increases as the rate of prime savings decreases
Estrada et al. [6]	Young dependency ratio Old dependency ratio	According to 153 economies worldwide, there is a positive relationship between population aging and aggregate consumption; however, the analysis of 31 Asian developing economies shows a significant negative relationship between population aging and aggregate consumption
Li and Li [9]	Population of age over 65	The increase of elderly population has a negative impact on consumption
Boonyasana and Chinnakum [8]	Young dependency ratio Old dependency ratio	young dependency ratio and old dependency ratio have a negative impact on consumption

The most important key factor in exploring the impact of population age structure on consumption is the setting of population age structure variables. From studies on the relationship between population age structure and consumption in the past two decades, it was found that most of them showed the dynamics of age structure change in terms of the population proportion. Since domestic tourism demand is mainly surveyed for people over 12 years old in Taiwan, the population age structure variables of this study refer to the concept of the proportion of the prime saving population [7] and set the age structure variable of tourism demand according to the proportion of the prime tourist population. Then, the impact of the aging population on the domestic tourism market can be explored.

2.3. Age and Travel

In tourism, the relationship between the age structure of the population and tourism demand has not been discussed in the literature. It is more common to use a cross-sectional age-life cycle to explore its impact on individual or family tourism demand. Henthorne [11], Jang et al. [12], Mehmetoglu [13], and Bernini and Cracolici [14] found a significant positive relationship between age and tourism expenditure, suggesting that population aging had a positive impact on the development of tourism markets in a country or region. However, Dardis et al. [15], Wang et al. [4], and Frleta and Jurdana [16] showed that younger people spent more on tourism than older people and that population aging may have a negative impact on the tourism market in these countries or regions. The quantile regression model was used to classify tourism expenditure into different quantiles of expenditure, analyze the relationship between different components of expenditure and age, and explore the effect of aging on tourism expenditure.

Age-life-cycle variables are usually measured in terms of the age of individuals or household heads or expressed as dummy variables. Such variable settings cannot show the differences in the proportion of the population across age groups. In addition, such a setting makes generational and age effects confounded, so it is not possible to fully clarify the true effect of age on consumption. Therefore, exploring the impact of aging on tourism demand in terms of age-life-cycle variables fails to capture the real impact of demographic age structure changes on consumption and the dynamic process of change.

In summary, we constructed a model of domestic tourism demand, which sets the age structure variables of tourism demand in terms of the proportion of the prime tourist population, to investigate the impact of aging on the domestic tourism market using time-series data.

3. Method

3.1. Variables of Tourism Demand Population

In general, the dependency ratio is often used in age structure variables to represent changes in age structure [7]. The dependency ratio is the ratio of the dependent population to the working population, i.e., the sum of the 0–14-year-old population and the over-65-years-old population to the 15−64-year-old population. Given that domestic tourism demand is surveyed from people over 12 years old in Taiwan, the age structure variables of this study refer to the concept of the prime saving population ratio [7], and the proportion of the prime tourist population is set as

$$\text{AGE} = \frac{(\text{Population } 26 - 65 \text{ years old})}{(\text{Population } 12 - 25 \text{ years old})\text{and } 66 + \text{ years old })} \tag{1}$$

According to Equation (1), the plot of the prime tourist population proportion from 1981 to 2061 is drawn as Figure 2. The proportion of the prime tourist population reached its peak in 2016, and gradually declined after 2017.

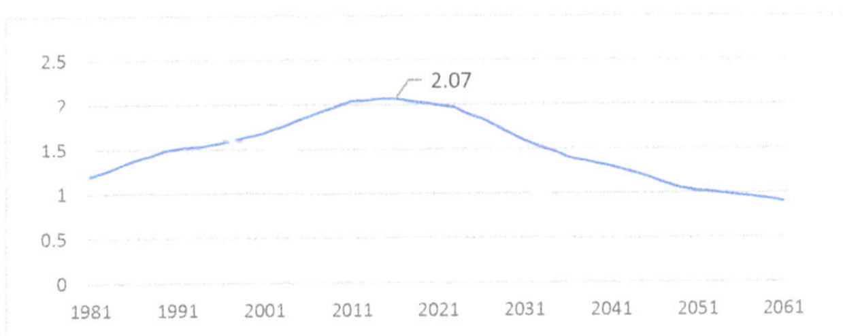

Figure 2. Changes in the proportion of the main tourist population from 1981 to 2061 for Taiwan.

3.2. Model Setting

According to economic theory, the quantity of tourism demand is mainly affected by variables such as income, price, population, and preferences. Therefore, the empirical model of domestic tourist demand in this study is set as follows.

$$TD_i = \alpha + \beta_1 PNI_i + \beta_2 CPI_i + \beta_3 LTD_i + \beta_4 AGE_i + \beta_5 HP_i + \beta_6 COV_i + \varepsilon_i \tag{2}$$

In (2), TD is the number of domestic tourists, PNI is the average national income per person, CPI is the consumer price index, and LTD is the number of domestic tourists in the previous period (representing habit formation) [17,18]. AGE is the proportion of the prime tourist population, HP is the dummy variable of the vacation policy, and COV is the dummy variable of COVID-19; ε is the error term, $i = 1, 2, \ldots, 20$, representing years of 2001 to 2020.

4. Results

Stata statistical software was used to estimate the coefficients of the model, and the empirical results are shown in Table 2. The $\overline{R}^2$ value of the model is 0.896, which shows that the fitness of the model is excellent, and only vacation policy is not significant among the six explanatory variables.

Table 2. Coefficients of estimated demand for domestic tourism.

Variables	β	t
PNI	0.30	2.90 *
CPI	−7249.24	−2.78 *
LDT	0.77	4.15 *
AGE	222,386.90	3.03 *
HP	−9289.80	−0.76
COV	−35,925.38	−2.52 *
Con	121,405.90	1.35
$\overline{R}^2$	0.896	

Note: * $p < 0.05$.

The positive coefficient ($\beta = 0.30, t = 2.90$) of PNI indicates that there is a significant positive relationship between income and domestic travel demand: as the average national income per person increases, domestic travel demand increases. The coefficient ($\beta = -7249.24, t = -2.78$) of CPI is negative, indicating that an increase in prices has a significant negative impact on domestic travel demand. The coefficient of LTD is significantly positive ($\beta = 0.77, t = 4.15$), indicating that domestic tourism has become a part of daily life, and the formation of travel habits has a significant positive effect on domestic tourism demand. The positive estimated coefficient ($\beta = 222386.9, t = 3.03$) of AGE indicates that the age structure of the population has a significant positive effect on the number of domestic tourists. Thus, as the elderly population increases, the proportion of the prime tourist population gradually becomes smaller, and the number of domestic tourists decreases, indicating that aging has a negative impact on the domestic tourism market.

The negative coefficient ($\beta = -9289.80, t = -0.76$) of the HP, although it is not significant, indicates that vacation policy leads to a decline in domestic tourism demand. It is caused by the uncertainty of people's income due to vacation policy. The negative coefficient ($\beta = -35925.38, t = -2.52$) of COV indicates that the COVID-19 epidemic has a significant negative impact on domestic tourism demand. The impact of the epidemic on the domestic tourism market is estimated to be 36 million fewer tourists a year.

Finally, the fitted values of (2) are plotted with the number of domestic tourists in Figure 3. It shows that the tourism demand model constructed in this study has a satisfactory fitness for the number of domestic tourists. Also, it indicates that a domestic travel demand model that considers the age structure of the population is suitable for analysis of changes in domestic tourists and for understanding the impact of aging on the domestic tourism market.

Figure 3. Number of domestic tourists and fitted value.

5. Conclusions

Lower birth rates and aging have become two major features of the changes in the age structure of the population because of the low fertility rate of women of reproductive age and the lengthening of life expectancy in many countries around the world. Aging is crucial to a country's economic or industrial development, so it is important to have information on the impact of aging on the economy or industry and then plan strategies for industrial development.

This study determines the proportion of the prime tourist population, which refers to the concept of the proportion of the prime saving population of Erlandsen and Nymoen [7]. With population age structure variables, a domestic tourism demand model was constructed to explore the impact of aging on the domestic tourism market using time series data from 2001 to 2020 in Taiwan. The results showed that there was a significant positive relationship between the proportion of the prime tourist population and domestic tourism demand, indicating that with the advent of an aging society, the proportion of the prime tourist population gradually declines, and the number of domestic tourists decreases accordingly. Therefore, it is obvious that aging is not conducive to the development of the domestic tourism market. For domestic-tourism-related industries, it is important to actively expand the source of the inbound tourism market to maintain the sustainable development of the industry. As for the domestic tourism senior market, developing new tourism products, increasing their consumption willingness and frequency, and slowing the impact of aging on the tourism industry are issues that the tourism industry must actively face and solve.

Author Contributions: T.-S.C. contributed to conceptualization, methodology, formal analysis, and writing the original draft. J.-Y.C. contributed to formal analysis and writing—reviewing and editing. All authors have read and agreed to the published version of the manuscript.

Funding: This research received no external funding.

Institutional Review Board Statement: Not applicable.

Informed Consent Statement: Not applicable.

Data Availability Statement: Not applicable.

Conflicts of Interest: The authors declare no conflict of interest.

References

1. United Nations. *World Population Prospects 2022*; United Nations: New York, NY, USA, 2022.
2. Fann, G.-J.; Hsu, Y.-H. Socio-Economic Impacts of Population Aging in Taiwan. *Taiwan Geriatr. Gerontol.* **2010**, *5*, 149–168.
3. Page, S.J.; Paul, B.; Graham, B.; Joanne, C. *Tourism: A Modern Synthesis*; Thomson Learning: London, UK, 2001.

4. Wang, Y.; Rompf, P.; Severt, D.; Peerapatdit, N. Examining and identifying the determinants of travel expenditure patterns. *Int. J. Tour. Res.* **2006**, *8*, 333–346. [CrossRef]
5. Attfield, C.L.F.; Cannon, E. *The Impact of Age Distribution Variables on the Long Run Consumption Function*, 3rd ed.; Discussion Paper NO03/546; Clarendon: Oxford, UK, 2003; Volume 2, pp. 68–73.
6. Estrada, G.; Park, D.; Ramayandi, A. Population Aging and Aggregate Consumption in Developing Asia. In *Aging, Economic Growth, and Old-Age Security in Asia*; ADB Economics Working Paper Series No.282; Asian Development Bank: Mandaluyong, Philippines, 2011; pp. 111–129.
7. Erlandsen, S.; Nymoen, R. Consumption and population age structure. *J. Popul. Econ.* **2008**, *21*, 505–520. [CrossRef]
8. Boonyasana, P.; Chinnakum, W. Impact of Population Aging on Consumption and Saving in Thailand: Structural Break Approach. *Thai J. Math.* **2019**, 109–120.
9. Li, Z.L.; Li, H. Analysis of The Impact of Population Aging on Household Consumption. *Int. J. Bus. Soc. Sci.* **2014**, *6*, 102–107.
10. Lefebvre, M. *Population Aging and Consumption Demand in Belgium*; CREFF Working Paper; CREPP: Liege, Belgium, 2006.
11. Henthorne, T.L. An analysis of expenditures by cruise ship passengers in Jamaica. *J. Travel Res.* **2000**, *38*, 246–250. [CrossRef]
12. Jang, S.S.; Bai, B.; Hong, G.-S.; O'Leary, J.T. Understanding travel expenditure patterns: A study of Japanese pleasure travelers to the United States by income level. *Tour. Manag.* **2004**, *25*, 331–341. [CrossRef]
13. Mehmetoglu, M. Nature-based tourists: The relationship between their trip expenditures and activities. *J. Sustain. Tour.* **2007**, *15*, 200–215. [CrossRef]
14. Bernini, C.; Cracolici, M.F. Demographic change, tourism expenditure and life cycle behaviour. *Tour. Manag.* **2015**, *47*, 191–205. [CrossRef]
15. Dardis, R.; Derrick, F.; Lehfeld, A.; Wolfe, K.E. Cross-section studies of recreation expenditures in the United States. *J. Leis. Res.* **1981**, *13*, 181–194. [CrossRef]
16. Frleta, D.S.; Jurdana, D.S. Uderstanding Tourist Spending on Culture and Entertainment. In Proceedings of the 27th International Scientific Conference on Economic and Social Development, Rome, Italy, 1–2 March 2018; pp. 448–460.
17. Kulendran, N. Modeling quarterly tourist flows to Australia using cointegration analysis. *Tour. Econ.* **1996**, *2*, 203–222. [CrossRef]
18. Salman, A.K.; Shukurb, G.; von Bergmann-Winberg, M.-L. Comparison of Econometric Modelling of Demand for Domestic and International Tourism: Swedish Data. *Curr. Issues Tour.* **2007**, *10*, 323–342. [CrossRef]

Disclaimer/Publisher's Note: The statements, opinions and data contained in all publications are solely those of the individual author(s) and contributor(s) and not of MDPI and/or the editor(s). MDPI and/or the editor(s) disclaim responsibility for any injury to people or property resulting from any ideas, methods, instructions or products referred to in the content.

engineering proceedings

Proceeding Paper

Multivariate Statistical Model of College Students' Purchase Decisions for Starbucks in Taiwan †

Jen-Chieh Chung [1] and Ying-Hsiao Lai [2,*]

[1] Department of Marketing and Logistics Management, Chaoyang University of Technology, Taichung 413310, Taiwan; jcchung@cyut.edu.tw
[2] Department of Leisure Service Management, Chaoyang University of Technology, Taichung 413310, Taiwan
* Correspondence: ylai@cyut.edu.tw
† Presented at the 3rd IEEE International Conference on Electronic Communications, Internet of Things and Big Data Conference 2023, Taichung, Taiwan, 14–16 April 2023.

Abstract: The purpose of this study was to explore college students' purchase behaviors using a multivariate statistical model. Because many college students today have part-time jobs, they have increased opportunities to consume higher-quality brands. In Taiwan, Starbucks is regarded as a high-quality coffee brand, costing more than most local coffee brands. To clarify why such students in Taiwan are willing to pay more for Starbucks, data were collected from a sample of 402 college students. The results showed differences in purchase decisions between students majoring in different disciplines and lifestyles. The findings had theoretical and practical implications especially for the marketing strategy of the Taiwanese coffee industry.

Keywords: multivariate statistical model; brand image; lifestyle

Citation: Chung, J.-C.; Lai, Y.-H. Multivariate Statistical Model of College Students' Purchase Decisions for Starbucks in Taiwan. *Eng. Proc.* **2023**, *38*, 38. https://doi.org/10.3390/engproc2023038038

Academic Editors: Teen-Hang Meen, Hsin-Hung Lin and Cheng-Fu Yang

Published: 26 June 2023

Copyright: © 2023 by the authors. Licensee MDPI, Basel, Switzerland. This article is an open access article distributed under the terms and conditions of the Creative Commons Attribution (CC BY) license (https://creativecommons.org/licenses/by/4.0/).

1. Introduction

According to a report from the International Coffee Organization (ICO), Taiwan plays a vital role in the coffee market. In 2021, coffee consumption reached 122 cups per capita in Taiwan. Moreover, the market is still growing [1] (Figure 1). The results of a coffee consumption survey conducted in Taiwan by dailyview.com in 2021 are very interesting. Out of a total of 1121 Taiwanese participants, around 40% drink coffee every day. In the past year, almost 70% of Taiwanese visited Starbucks at least once. Consumers under 30 years old prefer Starbucks [2]. Hence, the coffee market in Taiwan is competitive, especially among young consumers.

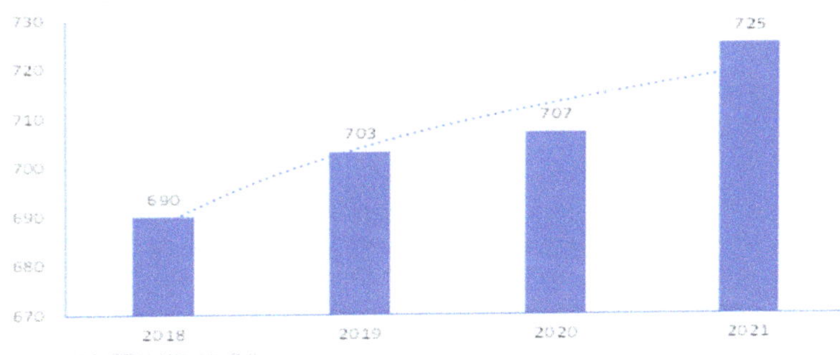

Figure 1. 2021 Taiwan coffee consumption statistics (Source: Trade Statistics Tables, International Coffee Organization).

The EKB Model proposed by Engel, Kollat, and Blackwell [3,4] has been examined by numerous researchers because of its comprehensive structure. The purpose of its use in our study is to identify whether there are any differences among college students. To answer this question, the AIO (activity, interests, and opinion) inventory is an adequate method to apply. Although the AIO measurement was adapted based on segmented markets [5–8], it is also useful for distinguishing young consumers [9].

Since Starbucks is a leading brand of coffee shops in Taiwan now, it is important to explore why it is popular among the young generation. Particularly college students in Taiwan love this brand. According to the definition by the American Marketing Association (AMA) [10], the term brand refers to the combination of name, term, sign, symbol, and design by which we can identify a certain product or its service from its competitors. Chernatory and McWilliam [11] suggested that a brand is a symbolic way to identify the difference between one product and other competitors. Moreover, the brand is a commitment to the consistency of its quality and a tool to project its self-image. Vigeneron and Lester [12] proposed the concept that a prestige brand evolves in three stages. From the lowest degree to the highest degree of prestige, there are "upmarket brands", "premium brands", and "luxury brands". They believed that the major difference between a "prestige brand" and a "nonprestige brand" was that consumers believe that purchasing a prestige brand is a symbol of socioeconomic status and wealth.

Brand awareness is usually considered to be one of the major factors by which to value a product [13]. In addition, Hoyer and Brown [14] indicated that brand awareness is a priority when consumers are choosing a product. Keller [15] proposed that brand knowledge includes "brand awareness" and "brand image". Brand image plays an important role in the marketing field since it includes the response from consumers to the brand name, sign, or image [16]. Hence, the brand image remains in the consumer's mind, and prestige brands have an important influence on purchase decisions. Thus, the purpose of this study was to build a purchase decision model with lifestyle, brand image, and demographic variables for exploring college students' coffee consumption behavior.

2. Methodology

We targeted college students; so, a university in Taiwan was selected to provide the participant pool. Since a stratified random sampling method was used, finding an appropriate sample size for this population was the first step. In 2017, 10,806 students were enrolled in this university. Under a confidence level of 95% and maximum error d = 0.05, and using the following equation

$$n = \frac{N}{N\left(\frac{2d}{Z_{\alpha/2}}\right)^2 + 1} \quad (1)$$

the minimum sample size was determined to be 371. Meanwhile, the controlled variables were gender, college, and age. Data were collected in the chosen university between 1 October 2017 and 31 December 2017. A total of 430 questionnaires were distributed to students in the university, and 402 (93.5%) were completed.

3. Results

Table 1 shows the descriptive statistics results. Since a stratified random sampling method had been adopted, a chi-square (goodness-of-fit) test was used to examine if the sampling structure was suitable for the population. All variables were fit for the population structure.

Among the respondents, 26.1% (n = 105) were freshmen, 25.4% (n = 102) were sophomores, 29.4% (n = 118) were juniors, and 19.2% (n = 77) were seniors. In addition, 36.3% (n = 146) of respondents were male and 63.7% (n = 256) were female. A total of 40.0% (n = 161) of the students majored in Management, 14.7% (n = 59) in Science and Engineering, 14.7% (n = 59) in Design, 18.2% (n = 73) in the Humanities and Social Sciences, and

12.4% (n = 50) in Informatics. Regarding monthly disposable income, 25.1% of respondents (n = 101) spent under 5000 NTD, 38.8% (n = 156) spent between 5001 and 8000 NTD, 25.1% (n = 101) spent between 8001 and 12,000 NTD, and 10.9% (n = 44) of respondents spent over 12,000 NTD.

Table 1. Demographic Characteristics of Sample.

		Sample		Population	
Variable	**Categories**	**Frequency**	**Percent (%)**	**Frequency**	**Percent (%)**
Gender	Male	146	36.3%	4421	40.9%
	Female	256	63.7%	6385	59.1%
College	Management	161	40.0%	4047	37%
	Science and Engineering	59	14.7%	1577	15%
	Design	59	14.7%	1729	16%
	Humanities and Social Sciences	73	18.2%	1945	18%
	Informatics	50	12.4%	1508	14%
Age	Freshmen	105	26.1%	2633	24%
	Sophomores	102	25.4%	2795	26%
	Juniors	118	29.4%	2705	25%
	Seniors	77	19.2%	2673	25%
Monthly Disposable Income (NTD)	3000–4999	101	25.1%		
	5001–7999	156	38.8%		
	8001–11,999	101	25.1%		
	12,000 and above	44	10.9%		

Table 2 shows that among the respondents, 65.9% (n = 265) purchased a coffee drink fewer than 5 times per month, 22.6% (n = 91) purchased a coffee drink between 6 and 10 times per month, 7.7% (n = 31) purchased a coffee drink between 11 and 20 times per month, and 3.7% (n = 15) purchased a coffee drink 21 or more times per month. Regarding the money spent on coffee, around 70% of the respondents spent under 150 NTD per month. Moreover, around 70% of the respondents visited Starbucks less than once a month. However, 47.8% (n = 192) of the respondents said they spent under 150 NTD when they visited Starbucks, and 47.5% (n = 191) of the respondents spent between 151 and 250 NTD. Around half of the respondents went to Starbucks with friends.

The purpose was to identify the brand image effect among college students. Therefore, exploratory factor analysis (EFA) was adopted in this study. The first step revealed that three variables had adequate reliability: consumer behaviors (Cronbach's α = 0.82), decision-making (Cronbach's α = 0.78), and AIO (Cronbach's α = 0.83). According to Kaiser and Rice [17], the value of the KMO measure of sampling adequacy must be greater than 0.8. The result of this study was 0.82, which demonstrated that the EFA was appropriate. Another index used was the Bartlett test, the value of which was 1191.32, with a p-value of 0.00. Thus, both indices showed that the EFA was appropriate.

We used an EFA with an orthogonal varimax rotation. An exploratory principal component factor analysis was performed to assess the validity of the lifestyle constructs and determine the potential groupings of lifestyle groups. An initial analysis of 15 brand image items found that 3 items had low factor loading (less than 0.5). Thus, the total of the final items was 12. The results indicated that six items were related to the staff's performance, one was about the Starbucks' environment, and five items were related to the brand image.

Table 2. Consumers Preferences.

Variable	Categories	Frequency	Percent (%)
Average times drinking coffee per month	Under 5 times	265	65.9%
	6–10	91	22.6%
	11–20	31	7.7%
	21+ times	15	3.7%
Average of each coffee consumption spend (NTD)	Under 150 NTD	273	67.9%
	151–250	104	25.9%
	251–400	15	3.7%
	Over 400 NTD	10	2.5%
Average of monthly Starbucks consumption	Under 1 time	261	64.9%
	2	74	18.4%
	3	32	8.0%
	4	15	3.7%
	5	8	2.0%
	6	2	0.5%
	7+	10	2.5%
How much spent on average at Starbucks	Under 150 NTD	192	47.8%
	151–250	191	47.5%
	251–400	16	4.0%
	Over 400 NTD	3	0.7%
With company?	No	61	15.2%
	Classmates	75	18.7%
	Colleagues	6	1.5%
	Friends	178	44.3%
	Family and relatives	69	17.2%
	Others	13	3.1%

Based on these findings, F1 was named " Service Quality", and F2 was named "Brand Value". Table 3 presents the results of the factor analysis. Based on the factor analysis results, there were two factors in this study. Then, a cluster analysis was used to determine the different groups. Since AIO measurements were adopted to examine college students' lifestyles, the results showed that college students could be divided into four groups: "prefer the brand", "coffee is a kind of gift", "coffee is a refreshing drink", and "drinking coffee is a social activity".

Table 3. Factor Analysis Results for Brand Image.

Items	F1	F2
Factor 1: Service Quality		
Staff have a good attitude.	0.874	
Staff members are professional.	0.865	
Staff uniforms are clean.	0.840	
Starbucks has high service quality.	0.838	
Staff are efficient.	0.832	
Environment and design style attract me.	0.645	
Staff can quickly answer my questions.	0.565	
Factor 2: Brand Value		
Starbucks' logo makes me feel stylish.		0.836
I care about the brand if it brings a sense of pride.		0.829
The Starbucks brand usually represents innovative ideas.		0.762
Starbucks coffee and snacks are high quality.		0.593
The quality of the coffee appeals to me.		0.546

4. Discussion

College students in Taiwan, especially those who live in urban areas, spend the majority of their free time with classmates and friends playing virtual games, going to KTV (karaoke), and shopping. Due to geographic limitations and cultural differences, Taiwanese students do not have many opportunities to participate in outdoor activities. Therefore, having dinner together or drinking coffee together represent important events that provide an emotional connection with other students. One of the reasons why Starbucks is a top choice for college students is because its brand image represents "high quality". Another reason is "self-identity", that is, having a Starbucks cup in their hands has a special meaning. Furthermore, it draws college students' attention. Compared to other luxury products, Starbucks' products are much easier to obtain.

5. Conclusions

College students are willing to pay more to buy Starbucks coffee even though not all of them can afford it. Students drink coffee as a habit, and they go to Starbucks coffee shops with friends. The facts show that Starbucks' prices are two to three times more than other brands. However, students treat "going to Starbucks with friends" as a kind of social activity. Since Starbucks' marketing position in Taiwan is "high quality", it is considered a "luxury" product. Hence, it is a welcome gift to college students. In sum, Starbucks' coffee is not only a drink for college students in Taiwan but also an "identity" for them.

Author Contributions: Conceptualization, J.-C.C. and Y.-H.L.; methodology, J.-C.C.; validation, J.-C.C. and Y.-H.L.; formal analysis, J.-C.C. and Y.-H.L.; resources, J.-C.C. and Y.-H.L.; data curation, J.-C.C.; writing—original draft preparation, J.-C.C.; writing—review and editing, J.-C.C. and Y.-H.L.; supervision, J.-C.C. and Y.-H.L.; funding acquisition, none. All authors reviewed and edited the manuscript. All authors have read and agreed to the published version of the manuscript.

Funding: This research received no external funding.

Institutional Review Board Statement: Not applicable.

Informed Consent Statement: Not applicable.

Data Availability Statement: No new data were created or analyzed in this study. Data sharing is not applicable to this article.

Conflicts of Interest: The authors declare no conflict of interest.

References

1. International Coffee Organizaion. Available online: ico.org (accessed on 21 April 2021).
2. Dailyview Survey. Available online: dailyview.tw (accessed on 15 May 2022).
3. Kollat, D.T.; Engel, J.F.; Blackwell, R.D. Current Problems in Consumer Behavior Research. *J. Mark. Res.* **1970**, *7*, 327–332. [CrossRef]
4. Engel, J.F.; Blackwell, R.D.; Miniard, P.W. *Consumer Behavior*, 8th ed.; Dryden Press: New York, NY, USA, 1995.
5. Plummer, J.T. The Concept and Application of Life Segmentation. *J. Mark.* **1974**, *38*, 33–37. [CrossRef]
6. Wind, Y. Issues and Advances in Segmentation Research. *J. Mark. Res.* **1978**, *15*, 317–337. [CrossRef]
7. Yankelovich, D.; Meer, D. Rediscovering market segmentation. *Harv. Bus. Rev.* **2006**, *84*, 122. [PubMed]
8. Darden, W.R.; Reynolds, F.D. Backward Profiling of Male Innovators. *J. Mark. Res.* **1974**, *11*, 79–85. [CrossRef]
9. Kotler, P.; Keller, K.L. *Marketing Management*, 14th ed.; Pearson: New York, NY, USA, 2011.
10. American Marketing Association. *Marketing Definitions: A Glossary of Marketing Terms*; American Marketing Association: Chicago, IL, USA, 1960.
11. Chernatory, L.D.; McWilliam, G. Branding Terminology the Real Debate. Marketing Intelligence and Planning. *Mark. Intell. Plan.* **1989**, *7*, 29–32.
12. Vigneron, F.; Lester, W.J. A Review and a Conceptual Framework of Prestige-Seeking Consumer Behavior. *Acad. Mark. Sci. Rev.* **1999**, *2*, 1–15.
13. Aaker, D.A. *Managing Brand Equity: Capitalizing on the Value of a Brand Name*; The Free Press: New York, NY, USA, 1991. [CrossRef]
14. Hoyer, W.D.; Brown, S.P. Effects of Brand Awareness on Choice for a Common, Repeat-Purchase Product. *J. Consum. Res.* **1990**, *17*, 141–148. [CrossRef]
15. Keller, K.L. Conceptualizing Measuring and Managing Customer-Based Brand Equity. *J. Mark.* **1993**, *57*, 1–22. [CrossRef]

16. Magid, J.M.; Cox, A.D.; Cox, D.S. Quantifying Brand Image: Empirical Evidence of Trademark Dilution. *Am. Bus. Law J.* **2006**, *43*, 1–42. [CrossRef]
17. Kaiser, H.F.; Rice, J. Little Jiffy, Mark Iv. *Educ. Psychol. Meas.* **1974**, *34*, 111–117. [CrossRef]

Disclaimer/Publisher's Note: The statements, opinions and data contained in all publications are solely those of the individual author(s) and contributor(s) and not of MDPI and/or the editor(s). MDPI and/or the editor(s) disclaim responsibility for any injury to people or property resulting from any ideas, methods, instructions or products referred to in the content.

engineering proceedings

MDPI

Proceeding Paper

Power of Networking: Study on Mutual Cameos †

Yu-Cheng Tu, Jhih-Ling Jia and Kuan-Yin Lee *

Department of Marketing and Logistics Management, Chaoyang University of Technology, 168, Jifeng E. Rd., Wufeng District, Taichung 413310, Taiwan; tuyc@cyut.edu.tw (Y.-C.T.); chuya861102@gmail.com (J.-L.J.)
* Correspondence: kylee@cyut.edu.tw; Tel.: +886-4-23323000 (ext. 7070)
† Presented at the 3rd IEEE International Conference on Electronic Communications, Internet of Things and Big Data Conference 2023, Taichung, Taiwan, 14–16 April 2023.

Abstract: YouTube is Google's most successful video streaming platform. As a medium, it creates a low-cost and highly flexible personal channel. YouTubers are part of a popular emerging industry. Each YouTuber's channel shows its characteristics. YouTubers occasionally collaborate with other YouTubers to shoot. Such a commercial cooperation is called a "Cameo (Feat)". In this study, we explored the cameo effect of guest appearances using the promoter score (likes plus dislikes) and net promoter score (likes minus dislikes). We gathered data in the form of videos from the top 100 YouTubers. The results show that the number of views increased by 63.4%, the number of likes increased by 62.2%, the number of dislikes increased by 67.8%, and the number of comments increased by 36.0%. Overall, the promoter score increased by 62.4%, and the net promoter score increased by 61.9%.

Keywords: cross over; feat; promotional effect; YouTuber; cameo effect

1. Introduction

Founded in 2005, YouTube is a video-sharing website from the United States that allows users to upload, watch, share, and comment on videos. Its founding concept was originally to facilitate the sharing of video clips among friends, and it later evolved into a place for users to publish and share their works. Since 2011, YouTube has approached the channel of mainstream media and has become an emerging industry. YouTubers are like celebrities and have their own fans and followers. When the number of fans and followers increases, these YouTubers become public figures, and fans also change their consumption, behaviors, and habits because of idol worship [1].

In this study, the Internet thermometer of YouTuber rankings is used. The YouTuber category is divided into the top 100 YouTubers were selected based on the magnitude of internet traffic they generate and their significant positive impact. We selected the top 100 YouTubers based on the volume of Internet traffic as the research object. With their visibility, the influence of the volume of Internet traffic is determined.

A brand alliance is a marketing strategy in which two or more established brands are jointly presented to consumers [2]. Cameos have become prevalent phenomena within the current landscape of YouTubers. The commercial term is a "joint name", which brings different influences, such as a mutual influence between brands or the influence of popularity. Commercial co-branding cooperation is also explored in this study from the perspective of the YouTuber, applying it to the interaction effect of inter-YouTuber collaboration, this study proposes the following two research questions for investigation:

(1) Whether the guest's appearance affects the YouTuber's various indicators and promotion.
(2) What is the impact of the cameo appearance on the promoter score?

Citation: Tu, Y.-C.; Jia, J.-L.; Lee, K.-Y. Power of Networking: Study on Mutual Cameos. *Eng. Proc.* **2023**, *38*, 44. https://doi.org/10.3390/engproc2023038044

Academic Editors: Teen-Hang Meen, Hsin-Hung Lin and Cheng-Fu Yang

Published: 27 June 2023

Copyright: © 2023 by the authors. Licensee MDPI, Basel, Switzerland. This article is an open access article distributed under the terms and conditions of the Creative Commons Attribution (CC BY) license (https://creativecommons.org/licenses/by/4.0/).

2. Literature Review

2.1. Development Status and Trend of YouTube

YouTube is the world's first online video and audio platform, and it allows users to upload, watch, share, and comment on videos [3]. The YouTube platform was created to facilitate the sharing of video clips among friends, and it later evolved into a place for users to publish and share their works. On the YouTube platform, uploading videos has become a profession. From a single instance of sharing life with friends, a variety of video have developed to be shared with the world, transcending country-to-country restrictions, narrowing the distance between people from all over the world, and bringing different business opportunities. So far, YouTube is still the most used online video and audio platform in the world, with an average of 720,000 h of video uploads per day and 250 million views per day. This only counts the number of views on a TV screen, excluding mobile phones, computers, and other electronic products. Judging from the above data, global users rely on the YouTube platform to a high degree. In 2018, YouTube crashed for an hour, and some people reported the case. YouTube has become inseparable from people.

2.2. Types of YouTubers

According to the KOL Radar [1], YouTuber types are classified into 29 categories: life, unpacking, dressing, beauty, 3C, games, sports, parent–child, pets, feelings, constellations, travel, food, imitation, funny, magic, dance, music, photography, film, calligraphy, illustration, cosplay, medical care, finance and economics, law, education, language teaching, and image management. Compared with traditional TV stations, Taiwanese YouTubers are popular among audiences, and more groups have shifted their traditional cable-TV-viewing habits to YouTube in the Internet age. Every YouTuber has a set of business models. The most important factor for the operation of a successful channel is the number of audiences. Similar to the STP strategy in marketing analysis, market segmentation is necessary to determine the target group. The most important factor is positioning the channel. We divide YouTube's operating mode into the following aspects [4]:

- Relevancy to people: It is very important to have a constant and fixed frequency of uploading videos. This is also a tacit understanding of subscribers. With this tacit understanding, fans habitually reply to videos. Such an interaction is an excellent way to maintain fans.
- Set of Keywords: Keywords make it easier for consumers to search for videos.
- Therefore, it is imperative to comprehend the discourse employed by the primary target demographic, time-critical subjects, or undisclosed phenomena, and incorporate high-ranking search outcomes. The volume of hashtags in videos increases the probability of the video being searched.
- Partners: When the popularity of a YouTuber increases, their guest invitations increase. At this time, encountering "pig teammates" must be avoided. To make an appealing video, fans must feel the "accompanying text" but maintain a high-quality channel image. In videos, rules must be set so that the audience can pay the bill.

2.3. Video Contents

- Length of the video: When a video is longer than six or seven minutes, or if the video duration exceeds ten minutes while containing abundant content, it becomes exceedingly challenging to sustain the audience's attention, resulting in eventual video skipping. Video planning: Although the YouTuber's "intent" cannot be seen or touched by the audience, it cannot be neglected. From script design and shooting props and scenes to the final editing process, YouTubers need to prepare carefully. There must be a bitter process behind success. Only by doing an excellent job on a video can it truly impress the hearts of the audience and become a draw to attract potential fans.
- Improve interaction: Youtubers' revenues come from data such as the numbers of views and clicks. In short, it is a variable income. At this time, resources must be

integrated. When a YouTuber becomes a commodity, the most important step for them is to increase the exposure rate of their product. In terms of online platforms, Facebook fans, Instagram, or Snapchat are used. Through these platforms, different customer groups can be reached to receive different responses. At the same time, it is important to improve the quality of the channel to increase the number of followers.

- Image thumbnails: Each video has a small thumbnail. This thumbnail is one of the audience's first impressions of the video. It is necessary to spend more time on the thumbnail, color matching, layout, and so on.

2.4. How to Improve Videos

Establishing a good word-of-mouth effect leads to positive communication. We conduct an analysis, taking into account the following three viewpoints:

- Partners must be carefully chosen to reach different YouTubers categories and cooperate, triggering different degrees of discussion and creating new topics.
- Channel thumbnails must be eye-catching so that the audience clicks on them, and the title must be topical and explosive to increase exposure.
- The video must not be too long. The beginning is the most important, and a concise and powerful film can help the audience stay focused.

To build a reputation and foster positive engagement, it is essential to generate topics that encompass novel and intriguing subjects or reference contemporary events. As long as the content is sufficiently appealing, it will resonate with the audience, resulting in high traffic volume and aiding the individual in attaining influential status as a YouTuber [5].

2.5. Cameo Effect

A brand alliance is a commercial marketing method through which two or more brands conduct cross-industry cooperation, combine the advantages of both parties, and jointly launch products, thereby attracting consumer groups from different fields. Through this cooperation model, customers of different brands and consumer groups in different fields can reach another unfamiliar brand. The use of the synergistic effect of co-branding creates greater benefits. This is similar to YouTube's cameo videos. They cooperating with each other, thus reaching greater audiences, bringing a novel sense of freshness, increasing the audience's attention and attracting fans [2].

3. Research Results and Analysis

The top 100 YouTubers collected are classified, and their cameo effects are discussed. The classification method is as follows: the top 100 YouTubers are divided into two groups. The groups distinguish those with cameos and those without cameos. The non-cameo objects serve as the reference value for comparing the effects between the two. The number of likes and the number of no-likes in each combination are added and subtracted to further judge the promotion efficiency. By utilizing the non-cameo object data as a baseline, a comparative analysis is conducted to assess the effects of the cameo object data.

With cameos, the number of views increased by 63.4%, and the number of likes increased by 62.2% (Table 1). The number of downvotes increased by 67.8%, and the number of comments increased by 36%. Mutual guest appearances improved all YouTuber indicators with positive effects. At the same time, the number of downvotes also increased, and the increase was greater than the number of likes. Therefore, the theme of the video affected the response in addition to with cameos.

Table 1. Guest appearances for top 100 YouTubers.

Feat or Not	Views	Likes	Dislikes	Comments
No feat (Base)	226,321	3971	177	550
Feat	369,720 (+63.4%)	6439 (+62.2%)	297 (+67.8%)	748 (+36%)

YouTubers with cameo objects had a positive impact on various indicators. The number of views increased due to the cameos, indicating that the cameos successfully attracted the attention of the audience and expanded the original audience group. Topicality also increased, and the number of comments increased. Not only did the original audience group leave messages but the audience of the guest audience also left messages because of the guest. The effect of guest appearances was important, but the numbers of likes and downvotes fluctuated greatly, changing the benefits of cooperation.

Table 2 shows the effect without cameos. The promoter score increased by 62.4%, and the net promoter score increased by 61.9%. The comparison results show that both the promoter score and the net promoter score increased significantly. Cameos increased the discussion of the video. A positive correlation exists between the relative attention and audience interest, indicating that higher relative attention levels have a greater impact.

Table 2. Degree of promotion of cameo effect.

Feat or Not	Promoter Score (Likes Plus Dislikes)	Net Promoter Score (Likes Minus Dislikes)
No feat (Base)	226,321	3971
Feat	369,720 (+63.4%)	6439 (+62.2%)

Combining the comparison results of Tables 1 and 2, it is confirmed that guest appearances had a multiplicative effect on all data, creating a synergistic effect. Thus, YouTubers can use guest appearances to increase their audiences. Cameos affect the increase in audience but have a negative effect. The most critical factor is the choice of the topic.

4. Conclusions

We found that the cameos have a significant effect. YouTubers use each other's unique styles to create unconventional video content to reach out to each other's fans and attract more people to subscribe and watch and for increased success. The audience's attention can be drawn for a positive impact and to improve various indicators, although the number of downvotes also increases due to the influence of the cameo. However, it increases the promoter score. The cameo effect is the benefit brought about by co-branding. By combining one's brand with an unfamiliar brand, more novel and eye-catching products can be created by stimulating the consumers' interest in co-branded products. With a brand loyalist, this cooperation model creates a synergy.

This research shows that the data can objectively reflect the general preferences of the masses but cannot subjectively express the personality traits of YouTubers. No matter how the data changes, the "cameo" is an indispensable tool for YouTubers. Cameos are one of the trends in YouTube video creation these days as YouTubers influence each other's channels and popularity. At the same time, the interaction between YouTubers and the audience is relatively important, and everyone wants the audience to continue watching the next video. In the present context, audiences seek novelty and change, making any form of collaboration and modification beneficial in attracting viewers. Cameos are generally helpful for YouTubers, but cameos must be chosen according to the different goals pursued by individuals rather than applying the same method to everyone. To maximize the benefits, it is necessary to understand positioning and reality and to tailor the content after evaluation.

Author Contributions: Conceptualization, Y.-C.T. and K.-Y.L.; methodology, Y.-C.T.; validation, K.-Y.L. and J.-L.J.; formal analysis, Y.-C.T.; data curation, J.-L.J.; writing—original draft preparation, Y.-C.T.; writing—review and editing, Y.-C.T. and J.-L.J.; supervision, K.-Y.L.; project administration, Y.-C.T. All authors have read and agreed to the published version of the manuscript.

Funding: This research was funded by Chaoyang University of Technology.

Institutional Review Board Statement: Not applicable.

Informed Consent Statement: Not applicable.

Data Availability Statement: Not applicable.

Conflicts of Interest: The authors declare no conflict of interest.

References

1. KOL Radar. Insight Report. 2022. Available online: https://app.kolradar.com/search (accessed on 11 August 2022).
2. Ruekert, R.W.; Rao, A. Brand alliances as signals of product quality. *Sloan Manag. Rev.* **1994**, *36*, 87–97.
3. Schouten, A.P.; Janssen, L.; Verspaget, M. Celebrity vs. Influencer endorsements in advertising: The role of identification, credibility, and Product-Endorser fit. *Int. J. Advert.* **2020**, *39*, 258–281. [CrossRef]
4. Chetioui, Y.; Benlafqih, H.; Lebdaoui, H. How fashion influencers contribute to consumers' purchase intention. *J. Fash. Mark. Manag.* **2020**, *24*, 361–380. [CrossRef]
5. De Veirman, M.; Cauberghe, V.; Hudders, L. Marketing through Instagram influencers: The impact of number of followers and product divergence on brand attitude. *Int. J. Advert.* **2017**, *36*, 798–828. [CrossRef]

Disclaimer/Publisher's Note: The statements, opinions and data contained in all publications are solely those of the individual author(s) and contributor(s) and not of MDPI and/or the editor(s). MDPI and/or the editor(s) disclaim responsibility for any injury to people or property resulting from any ideas, methods, instructions or products referred to in the content.

engineering proceedings

MDPI

Proceeding Paper

Assessing Impact of Tablet-Based Digital Games on Mathematics Learning Performance †

Sarocha Tepho and Niwat Srisawasdi *

Faculty of Education, Khon Kaen University, Khon Kaen 40000, Thailand; sarocha_tepho@kkumail.com
* Correspondence: niwsri@kku.ac.th

† Presented at the 3rd IEEE International Conference on Electronic Communications, Internet of Things and Big Data Conference 2023, Taichung, Taiwan, 14–16 April 2023.

Abstract: The popularity of digital games offers new-generation students an effective mathematics learning approach. In this study, a digital game intervention is used to enhance the learning of school mathematics for primary school students. The participants were first-grade children identified by the school as having difficulties in mathematics. The students were assigned to interact with the game implemented as a parallel part of normally scheduled class activities for one semester. The results revealed that using digital games for cognitive engagement with the regular mathematical content being taught in a classroom increased students' learning potential.

Keywords: digital game; cognitive training; mathematical disabilities; mathematics learning; numeracy training

Citation: Tepho, S.; Srisawasdi, N. Assessing Impact of Tablet-Based Digital Games on Mathematics Learning Performance. *Eng. Proc.* **2023**, *38*, 41. https://doi.org/10.3390/engproc2023038041

Academic Editors: Teen-Hang Meen, Hsin-Hung Lin and Cheng-Fu Yang

Published: 27 June 2023

Copyright: © 2023 by the authors. Licensee MDPI, Basel, Switzerland. This article is an open access article distributed under the terms and conditions of the Creative Commons Attribution (CC BY) license (https://creativecommons.org/licenses/by/4.0/).

1. Introduction

Sources contributing to mathematical achievement include numerical knowledge and underlying general cognitive processing. Mathematical skills also are an integral component of basic literacy. Improving students' proficiency in mathematics plays a key role in many countries' education strategies [1,2]. Despite the importance of mathematics, the majority of students in K-12 education regard mathematics unfavorably and recognize it as a frustrating and difficult subject that causes learning fatigue, pressure, and anxiety [3–7]. Therefore, mathematics shows the highest student failure rate [8]. Scholars have pointed out that the traditional teaching method deserves criticism and leads to learning problems. For example, students' exposure to complex problems is very limited. Hence, the traditional method does not advance the development of students' problem-solving skills, conceptual understanding, or critical analysis [9,10].

Due to substantial technological developments during the past decade, many researchers and educators are incorporating technology into education [11,12]. Among the various technology-supported learning methods, digital game-based learning (DGBL) is perceived as an effective way to learn mathematics because digital games are powerful in presenting complex mathematical concepts as they provide an alternative media where the interaction and exploration of the content are inherent. For instance, it was indicated that using DGBL applications in mathematics promotes students' perseverance and improves their engagement in learning [13,14]. Thus, gaming could contribute to young learners' cognitive development [15]. Researchers also found that numerous technology-based games improved students' numerical magnitude knowledge by training domain-specific and domain-general skills and improving mathematical learning performance. For example, playing adaptive and computerized number games involves numerical magnitude comparisons using dots, numbers, or arithmetic problems and improves preschool- and kindergarten-age children's numerical magnitude knowledge [16,17]. However, most research has been focused on improving mathematics learning of children with a disease that causes learning difficulty in mathematics. Mathematics learning disabilities (MLD) are used broadly to describe difficulty with a variety of math skills that cannot be explained by a lack of intelligence, neurological abnormalities, or insufficient teaching [18]. Currently, studies on mathematics learning examine working memory and numeracy training programs for children with math learning difficulties.

Early numeracy skills training may impact children's math achievement more than other programs for domain-general mathematics skills. However, there are still gaps in the literature for a better understanding of the phenomena, which needs further study. Based on the literature review, we hypothesized that playing a mathematics-based game would improve the numerical fluency, mathematical fluency, and mathematics learning achievement of children with mathematical learning disabilities. Thus, we investigated the effectiveness of a tablet-based game intervention in mathematics learning to answer the following research questions.

- Do MLD students who use tablet-based digital game intervention demonstrate greater mathematical fluency?
- Do MLD students who use tablet-based digital game intervention demonstrate greater mathematics learning achievement?

2. Literature Review

2.1. DGBL

Learning during playing games on a digital platform is called DGBL learning. DGBL transforms digital games into educational tools [19]. Multiple studies have indicated that game-based learning has numerous benefits, including allowing students to enjoy learning [20], improving their engagement, and motivating them to explore the unknown [21,22]. Digital game-based learning is an interactive experience that integrates educational content into gaming activities and is relevant to almost all subjects [12,23]. Digital game-based learning attracts students' attention, fosters a positive attitude toward learning, heightens their interest and engagement, and contributes to the growth of problem-solving-related critical thinking skills [20,22].

2.2. DGBL for Mathematics Learning

Mathematics is fundamental for developing human cognition and advancing many other disciplines [12]. Children in the 21st century must be capable of logical, creative, critical, and rational thoughts and ideas [23]. Nevertheless, many elementary and high school students find mathematics discouraging. Specifically, students often dislike mathematics because they believe it to be tedious, difficult, and irrelevant [6]. Therefore, it is essential to establish effective strategies for arousing students' interest in mathematics, enhancing their conceptual understanding, and supporting them in acquiring arithmetic skills [24].

In mathematics education, DGBL is regarded as a learning tool that helps students acquire conceptual knowledge [25], practice arithmetic skills [26], and increase classroom engagement [25]. For several decades, the pedagogical potential of DGBL has been widely recognized [26,27]. Students will learn mathematics effectively when they can independently and productively generate mathematical concepts. Therefore, using DGBL to teach and study mathematics helps students develop self-awareness [28]. In addition to other fundamental skills such as reading and problem-solving, children can master fundamental mathematics concepts and skills while they play the game and complete the tasks. These interactions assist in learning and acquiring abilities [29], suggesting that DGBL assists pupils in enhancing their mathematical performance.

3. Tablet-Based Digital Game Intervention

In Ref. [18], a brain training program was created based on computer games (Figure 1). We translated the game content into the Thai language version. This tablet game provides narratives and themes to simulate a video game environment. Each game features twelve minigames that vary in game design, topic, necessary action, and types of inputs. Within each minigame, the level of difficulty increases proportionally to the number of stimuli that were recalled depending on the working memory capacity and numerical factors. After completing each assignment, the player receives a gift, and their development through each mission was awarded with badges that increased their learning status.

Figure 1. Screenshots of the tablet-based game intervention.

4. Research Methodology

4.1. Participants

In total, 10 first graders (5 females and 5 males) from two classes in a municipal primary school in the northeastern area of Thailand were recruited, considering mathematical fluency and prior mathematics knowledge assessments. The students had a limited capacity for working memory and scored at or below the 20th percentile on the tests in the two classes. Their abilities were different in participating in the game in the experiment.

4.2. Training Intervention

Participants were exposed to the gaming intervention for 16 weeks, one regular semester. The training games were played on 7-inch touchscreen tablets, which recorded the children's reactions, too. During the semester, we offered the intervention twice or thrice per week for 20 min every session in a total of 800 min of training.

4.3. Research Instruments

A multiple-choice questionnaire with 25 items was used as a research tool for evaluating the mathematics learning achievement of the students. Surveys were conducted before and after the game intervention. The maximum possible score on this assessment was 25. A correct response scored one point, while an incorrect one scored zero. Mathematical fluency was evaluated using a "Math Fluency" examination. Participants were given one minute to execute as many arithmetic operations as possible in each set. The raw score was determined by how many additional problems were completed successfully within the allocated time.

4.4. Data Collection and Analysis

All the students experienced the game intervention for 800 min for 16 weeks. They spent 31 min completing the mathematics learning achievement questionnaire as a post-test. The obtained data from questionnaire surveys were analyzed for descriptive statistics such as arithmetic mean, standard deviation, frequency, and ratio. Individual actual gains were calculated as the percentage of the absolute gain.

$$\text{Gain (individual)} = (\text{post-test score} - \text{pre-test score})/\text{pre-test score} \times 100\ (\%) \quad (1)$$

The use of the single student's gain and its related calculations was used as an empirical justification of course effectiveness.

5. Results and Discussions

5.1. Research Question 1

For research question 1, descriptive statistics and individual normalized learning gain were analyzed. The descriptive statistics of the students' pre- and post-test mathematical fluency scores are presented in Table 1.

Table 1. Descriptive statistics on students' mathematical fluency.

Measurement	Mathematical Fluency			
	Mean	Standard Deviation (SD)	Maximum Score	Minimum Score
Pre-test	1.50	1.43	5	0
Post-test	16.60	5.23	26	11

Table 1 shows that the students' mathematical fluency scores after receiving the tablet-based game intervention (mean = 16.60, SD = 5.23) were higher than those before the intervention (mean = 1.50, SD = 1.43). Figure 2 shows the normalized gains ranked from 0.20 to 0.45 and from low to medium levels.

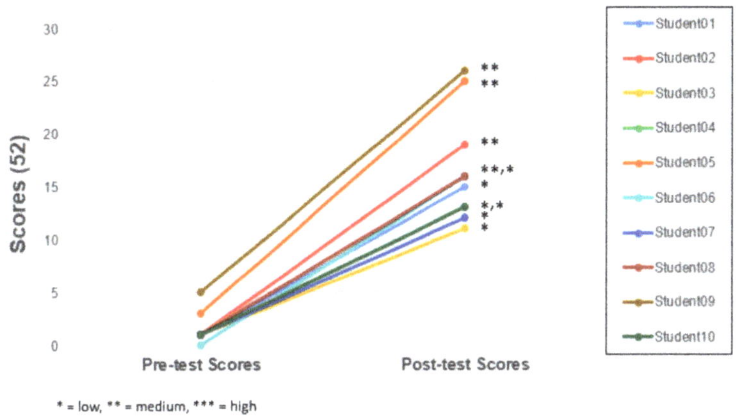

Figure 2. Results on mathematical fluency for individual students.

These results indicate that students improved mathematical fluency with the proposed tablet-based digital game intervention. These findings correspond to the results of Refs. [18,30] and show that numeric training benefits children at risk of mathematical learning disabilities to a larger extent.

5.2. Research Question 2

For research question 2, individual normalized learning gain and descriptive statistics were analyzed. Table 2 shows the descriptive statistics for the students' pre-and post-test scores on mathematics learning achievement.

Table 2. Descriptive statistics on students' mathematics learning achievement.

Measurement	Mathematical Learning Achievement			
	Mean	Standard Deviation (SD)	Maximum Score	Minimum Score
Pre-test	12.50	2.59	17	7
Post-test	17.40	2.88	23	14

Table 2 shows that the students' mathematics learning achievement scores after receiving the game intervention (mean = 17.40, SD = 2.88) were higher than those before the intervention (mean = 12.50, SD = 2.59). Figure 3 shows the ranking of the single-student normalized gains from 0.08 to 0.83, with the ranking of their size effects from low to high levels.

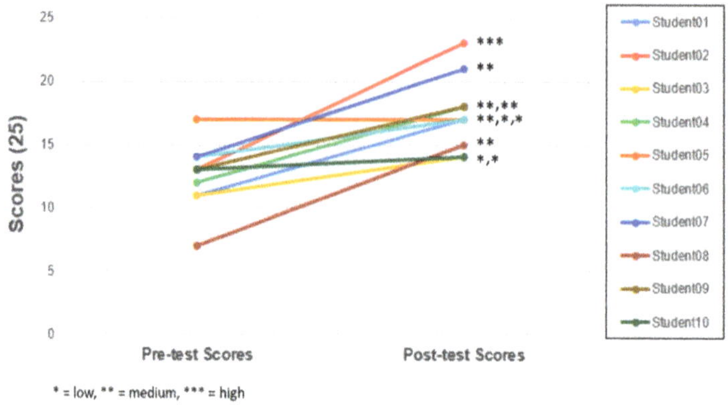

Figure 3. Results on mathematics learning achievement for individual students.

Intervention effects were observed in other mathematics performance training for students' mathematics development, as those who finished the numerical training enhanced mathematics learning performance [16].

6. Conclusions

Even though the number of studies on working memory and numeracy training has increased significantly over the past decade, it is still necessary to determine whether children with math learning impairments can benefit from them [18]. Thus, we researched if tablet-based digital game interventions allowed mathematics learning gains by measuring learning achievement and mathematical fluency for primary school students with mathematical challenges. The results revealed that the tablet-based digital game intervention had significantly favorable effects on the enhancement of students' mathematical fluency. In addition, primary school students with mathematical difficulties enhanced their mathematics learning performance considerably.

This study has limitations, such as a small sample size and short intervention periods. These necessitate a study with a larger number of students and a longer period to explore a long-term impact. The short period could alter the students' interaction with the games and the extent of the observed effects. In addition, future research with game interventions and training studies with other teaching aids and strategies are also required to find a way to overcome the inherent limits in actual classroom settings.

Author Contributions: Conceptualization, N.S.; methodology, N.S.; software, N.S.; validation, N.S.; formal analysis, N.S. and S.T.; investigation, N.S. and S.T.; resources, N.S.; writing, original draft preparation, N.S. and S.T.; writing, review and editing, N.S.; visualization, N.S. and S.T.; supervision, N.S.; project administration, N.S. All authors have read and agreed to the published version of the manuscript.

Funding: This research received no external funding.

Institutional Review Board Statement: Not applicable.

Informed Consent Statement: Not applicable.

Data Availability Statement: The data used in this research are available upon request.

Acknowledgments: The authors would like to express their gratitude to Kerry Lee, Department of Early Childhood Education, Faculty of Education and Human Development, Education University of Hong Kong, for providing the digital game.

Conflicts of Interest: The authors declare no conflict of interest.

References

1. Boris, G. *Science, Technology, Engineering, and Mathematics (STEM) Education: An Overview*; Congressional Research Service: Washington, DC, USA, 2018; Version 4; pp. 1–30.
2. European Schoolnet. *European Schoolnet's 2018 Annual Report*; European Schoolnet: Brussels, Belgium, 2018; pp. 1–20.
3. Deng, L.; Wu, S.; Chen, Y.; Peng, Z. Digital game-based learning in a shanghai primary school mathematics class: A case study. *J. Comput. Assist. Learn.* **2020**, *36*, 709–717. [CrossRef]
4. Geist, E. The anti-anxiety curriculum: Combating math anxiety in the classroom. *J. Instr. Psychol.* **2010**, *37*, 24–29.
5. Ofer, J.; Bos, B. The design and application of technology-based courses in the mathematics classroom. *Comput. Educ.* **2009**, *53*, 1133–1137. [CrossRef]
6. Luhan, J.; Novotna, V.; Kriz, J. ICT support for creative teaching of mathematic disciplines. *Interdiscip. Stud. J.* **2013**, *2*, 89.
7. Sun, L.; Ruokamo, H.; Siklander, P.; Li, B.; Devlin, K. Primary school students' perceptions of scafolding in digital game-based learning in mathematics. *Learn. Cult. Soc. Interact.* **2021**, *2*, 100457. [CrossRef]
8. Huang, Y.M.; Huang, S.H.; Wu, T.T. Embedding diagnostic mechanisms in a digital game for learning mathematics. *Educ. Technol. Res. Dev.* **2014**, *62*, 187–207. [CrossRef]
9. Bikić, N.; Maričić, S.M.; Pikula, M. The efects of diferentiation of content in problemsolving in learning geometry in secondary school. *Eur. J. Math. Sci. Technol. Educ.* **2016**, *12*, 2783–2795.
10. Kinard, J.T.; Kozulin, A. *Rigorous Mathematical Thinking: Conceptual Formation in the Mathematics Classroom*; Cambridge University Press: Cambidge, UK, 2008.
11. Hussein, M.H.; Ow, S.H.; Cheong, L.S.; Thong, M.K. A digital game-based learning method to improve students' critical thinking skills in elementary science. *IEEE Access* **2019**, *7*, 96309–96318. [CrossRef]
12. Hwa, S.P. Pedagogical change in mathematics learning: Harnessing the power of digital game-based learning. *Educ. Technol. Soc.* **2018**, *21*, 259–276.
13. Ke, F. A case study of computer gaming for math: Engaged learning from gameplay? *Comp. Educ.* **2008**, *51*, 1609–1620. [CrossRef]
14. Tsai, F.H.; Yu, K.C.; Hsiao, H.S. Exploring the factors infuencing learning efectiveness in digital gamebased learning. *Educ. Technol. Soc.* **2012**, *15*, 240–250.
15. Martinovic, D.; Ezeife, C.I.; Whent, R.; Reed, J.; Burgess, G.H.; Pomerleau, C.M. "Criticproofng" of the cognitive aspects of simple games. *Comput. Educ.* **2014**, *72*, 132–144. [CrossRef]
16. Francesco, S.; Patrizio, T.; Daniela, L.; Marco, Z. *Training Numerical Skills with the Adaptive Videogame "The Number Race": A Randomized Controlled Trial on Preschoolers*; Elsevier: Amsterdam, The Netherlands, 2016; Volume 5, pp. 20–29.
17. Wilson, A.J.; Dehaene, S.; Dubois, O.; Fayol, M. Effects of an adaptive game intervention on accessing number sense in low-socioeconomic-status kindergarten children. *Mind Brain Educ.* **2009**, *3*, 224–234. [CrossRef]
18. Muñez, D.; Lee, K.; Bull, R.; Khng, K.H.; Cheam, F.; Rahim, R.A. Working memory and numeracy training for children with math learning difficulties: Evidence from a largescale implementation in the classroom. *J. Educ. Psychol.* **2022**, *114*, 1866. [CrossRef]
19. Prensky, M. Digital game-based learning. *Comput. Entertain.* **2003**, *1*, 21. [CrossRef]
20. Kyriakides, A.O.; Meletiou-Mavrotheris, M.; Prodromou, T. Mobile technologies in the service of students' learning of mathematics: The example of game application A.L.E.X. in the context of a primary school in Cyprus. *Math. Educ. Res. J.* **2016**, *28*, 53–78. [CrossRef]
21. O'Rourke, J.; Main, S.; Hill, S.M. Commercially available digital game technology in the classroom: Improving automaticity in mental-maths in primary-aged students. *Aust. J. of Teach. Educ.* **2017**, *42*, 50–70. [CrossRef]
22. Lotherington, H.; Ronda, N.S. Gaming geography: Educational games and literacy development in the Grade 4 classroom. *Can. J. Technol.* **2003**, *35*, n3. [CrossRef]
23. Nahdi, D.S. Keterampilan matematika di a bad 21. *J. Cakrawala Pendas* **2019**, *5*, 133–140.
24. Kiili, K.; Devlin, K.; Multisilta, J. Editorial: Is game-based math learning finally coming of age? *Int. J. Serious Games* **2015**, *2*, 1–4. [CrossRef]
25. Meletiou-Mavrotheris, M.; Prodromou, T. Pre-service teacher training on game-enhanced mathematics teaching and learning. *Technol. Knowl. Learn.* **2016**, *21*, 379–399. [CrossRef]
26. Chang, M.; Evans, M.A.; Kim, S.; Norton, A.; Deater-Deckard, K.; Samur, Y. The effects of an educational video game on mathematical engagement. *Educ. Inform. Technol.* **2016**, *21*, 1283–1297. [CrossRef]
27. Iten, N.; Petko, D. Learning with serious games: Is fun playing the game a predictor of learning success? *Br. J. Educ. Technol.* **2014**, *47*, 151–163. [CrossRef]
28. Giannakas, F.; Kambourakis, G.; Papasalouros, A.; Gritzalis, S. A critical review of 13 years of mobile game-based learning. *Educ. Technol. Res. Dev.* **2018**, *66*, 341–384. [CrossRef]

29. Byun, J.; Joung, E. Digital game-based learning for K-12 mathematics education: A meta-analysis. *Sch. Sci. Math.* **2018**, *118*, 3–4. [CrossRef]
30. Nemmi, F.; Nymberg, C.; Helander, E.; Klingberg, T. Grit is associated with structure of nucleus accumbens and gains in cognitive training. *J. Cogn. Neurosci.* **2016**, *28*, 1688–1699. [CrossRef]

Disclaimer/Publisher's Note: The statements, opinions and data contained in all publications are solely those of the individual author(s) and contributor(s) and not of MDPI and/or the editor(s). MDPI and/or the editor(s) disclaim responsibility for any injury to people or property resulting from any ideas, methods, instructions or products referred to in the content.

engineering proceedings

MDPI

Proceeding Paper

Teaching Knowledge in the Logic and Engineering Method through Board Games †

Kuan-Yin Lee [1] and Prasana Kumar Samanta [2,]*

[1] Department of Marketing and Logistics Management, Chaoyang University of Technology, Taichung 413310, Taiwan; kylee@cyut.edu.tw
[2] Department of Business Administration, Chaoyang University of Technology, Taichung 413310, Taiwan
* Correspondence: prasanna.samanta.777@gmail.com
† Presented at the 3rd IEEE International Conference on Electronic Communications, Internet of Things and Big Data Conference 2023, Taichung, Taiwan, 14–16 April 2023.

Abstract: Today, students are accustomed to visual information and need engaging, stimulating, and fun teaching in logic and engineering methods for learning purposes. Traditional presentations and the transfer of memory information cannot arouse students' interest. By implementing logic and engineering methods in their teaching, teachers strive to promote active learning and deepen the learning effect by motivating students and engendering interest in learning. Problem-based learning (PBL) allows students to clarify or solve problems by identifying gaps in their knowledge, constructing clear educational topics, and integrating relevant information (PBL). This study aims to motivate students to learn through problem-oriented game-based approaches which help to solve chain board game development problems that increase learning efficiency. Third-year students have already mastered the basics of theoretical knowledge and have gained relevant work experience in industry–academy internships, class cadre work, and community service. Learning effectiveness and methods of learning, social interaction, and subject engagement are significantly different, according to an analytical study of SPSS data. The use of methods and the organization of notes are significantly more likely to be observed in students with high learning outcomes than in students with low learning outcomes. Learning outcomes do not differ significantly between students with high and poor peer ratings of board games. Students with high peer ratings of board games perform better than their poor peers in terms of learning methods. Students with poor peer scores in board games have significantly more body input than students with good scores. The results of this study can be helpful to teachers in subsequent curriculum design to improve students' learning effectiveness.

Keywords: game learning method; logic; engineering method; learning effectiveness

Citation: Lee, K.-Y.; Samanta, P.K. Teaching Knowledge in the Logic and Engineering Method through Board Games. *Eng. Proc.* **2023**, *38*, 46. https://doi.org/10.3390/engproc2023038046

Academic Editors: Teen-Hang Meen, Hsin-Hung Lin and Cheng-Fu Yang

Published: 27 June 2023

Copyright: © 2023 by the authors. Licensee MDPI, Basel, Switzerland. This article is an open access article distributed under the terms and conditions of the Creative Commons Attribution (CC BY) license (https://creativecommons.org/licenses/by/4.0/).

1. Introduction

Learning outcomes need to be improved by effective teaching strategies in response to educational changes [1]. Razon defined play as a voluntary and freely performed activity that provides happiness [2]. Games can stimulate growth, improve skills, and improve mood. By enhancing experience and fun, board games stimulate students' enthusiasm for learning [3]. Additionally, teachers are incorporating learner-centered principles into their courses and modifying traditional teaching methods [4]. Playing games allows the present generation to learn about complex interconnected knowledge bases while having fun. Games are a good way for educators to convey knowledge to learners and engage them in the course material.

The learner-centered teaching method is becoming increasingly popular among teachers. They try to arouse students' motivation and interest in learning through teaching activities design, thereby promoting active learning and deepening the learning effect. Learning topics that provide students with insight into their knowledge gaps can help them to identify and correct the knowledge gaps using problem-based learning (PBL).

Games play a role in our everyday lives when they become part of our lives, bringing us entertainment and learning, as well as uniting people's feelings and sometimes serving as educational tools. The board game industry has grown in popularity over the past few years. By playing board games over the internet, people can interact in "face-to-face" contact and with 3C products in order to enjoy games in "face-to-face" contact.

In reviewing the teaching site, several problems were found.

- People are always carrying their mobile phones in the 3C era.
- Concentration time is shortened year by year.
- Despite cross-domain learning, multiculturalism, and youth influence, there are small groups in the classroom that are even more difficult to cross.
- The learners seem to have unfulfilled ambitions and talents, and always have the feeling that the teacher does not understand what students want.

Therefore, the following is possible.

- Through board games, the interaction is brought back to the human world, and at the same time, it assists teachers in teaching.
- Sharing and learning from the results with peers to enhance human-to-human interaction.
- Developing related board games using chain knowledge and enhancing students' learning abilities through the board game development program.

Therefore, this study developed a complete and exclusive board game for students studying chain enterprise management. Students can learn the benefits of board games by using their brains, activating their brains, and increasing interest in life through board game activities.

Teachers teach students the basic knowledge and guide students to design a "chain enterprise theme board game" through problem-based learning (PBL) and learn the content of chain enterprise management and build team soft power. Furthermore, lecturers from the board game industry will teach students how to develop board games based on theoretical foundations.

Students should learn relevant knowledge and set goals in order to develop chain board games based on the dominance of learning. In addition to integrating knowledge, students must also ensure that play and learning are enjoyable for peers by focusing on factors such as entertainment, ease of play, and comprehension.

PBL is a teaching method that can improve students' ability to solve problems and manage related decisions. Furthermore, it facilitates active learning, strengthens memory, promotes teamwork, and fosters active learning attitudes [5]. The research results use problem-oriented and game-based learning strategies which are used to help students learn through action, which enhances their interest in learning and results in better learning outcomes.

2. Literature Review

2.1. Gamification Design

Gamification means that through the design of systems, services, organizations, and activities, learners have the same experience and motivation as games to influence user behavior [6]. Games are entertaining, enhance interpersonal relationships, and integrate elements of education and learning, allowing gamers to cultivate creativity, and emotional management, and improve learning stability [7].

Since games are an effective learning method, many theories become the basis for arguments. ARCS motivation theory mentions that games are a process of enhancing learning motivation, including attention, relevance, confidence, and satisfaction [8]. Learning environments and contexts play a key role in the authenticity of learning activities and knowledge [9]. As a result, using games to motivate students to learn and allowing them to play within a game context will transform the current educational scene and attract learners.

2.2. Learning Input

Reeve and Tseng pointed out that student input was important, and students were not only passive recipients of information but also active learners [10]. Increasing students' engagement helps achieve better learning outcomes, and teachers' teaching practices have a positive outcome [11].

Most studies examined students' learning engagement through multiple facets which were distinguished from behavioral and emotional engagement. Emotional engagement can refer to school identification, school belonging, liking for school, or being bored at school [12], while behavioral engagement includes participating in activities in and out of the classroom [13]. Game development is the focus of this research, specifically game development for learning, as opposed to game development for commercial value. This study intends to apply the knowledge gained. Therefore, when developing games, the behavioral side needs to be able to understand the knowledge taught, and the emotional side can interact well with teachers and peers to achieve the purpose of the game learning curriculum. Therefore, we use study skills, emotional engagement, and interaction engagement as aspects of studying students' learning engagement.

2.3. Learning Effectiveness

Student academic performance is the most basic definition of learning effectiveness (midterm and final grades) [14]. Learning outcomes include cognitive learning and perceptual learning. Cognitive learning includes changes in personal psychology [15] and perceptual learning is defined as the changes in learners' perceptions of skills and knowledge levels before and after learning experience. In this study, we define learning outcomes as the subjective outcomes obtained from one's learning gains after participating in the learning process of board games.

3. Research Method

The junior students already have a basic understanding of theory and accumulated relevant learning achievements. Competitor awards, industry–academy experience, class cadre experience, and community involvement are among these achievements. The research involved 125 students from two classes.

3.1. Curriculum Design

The weekly content and teaching activities are shown in Table 1.

Table 1. Weekly content and teaching activities.

<table>
<tr><th>Weekly</th><th>Content</th><th>Teaching Activities</th></tr>
<tr><td>Week 01</td><td>Course Introduction</td><td>1. The teacher explains the course objectives, teaching methods, and assessment methods
2. Introduction to PBL
3. Small group (group of 5–6 people)</td></tr>
<tr><td>Week 02</td><td>1. History and Introduction of Chain Enterprise Development—Chain Store</td><td rowspan="7">1. The theoretical knowledge of chain enterprises is taught by teachers
2. The case is read and discussed within the groups
3. The results of the discussion are shared
4. Board game experience</td></tr>
<tr><td>Week 03</td><td>2. Chain Enterprise Market Opportunities</td></tr>
<tr><td>Week 04</td><td>3. Chain Headquarters Management, Organization, and Development Conditions</td></tr>
<tr><td>Week 05</td><td>4. Chain Headquarters Management—Franchisor</td></tr>
<tr><td>Week 06</td><td>5. Affiliate Entrepreneurial Opportunity Assessment</td></tr>
<tr><td>Week 07</td><td>6. Franchisee Store Operation Strategy—Franchisee</td></tr>
<tr><td>Week 08</td><td>7. The Relationship between Franchisees of the Chain Headquarters</td></tr>
</table>

Table 1. *Cont.*

Weekly	Content	Teaching Activities
Week 09	Board Game Development and Design Speech	1. The development process of board games is shared 2. Board game setting mechanism
Week 10	Semi-Structured Interview and Board Game Questionnaire	
Week 11	Interim Report: Publication of PBL Works	The results of board games are shown
Week 11	Interim Report: Publication of PBL Works	Other group members are encouraged to play board games designed and developed by themselves

3.2. *Board Game Experience Process Program*

Board game experience and development flow chart is shown in Figure 1.

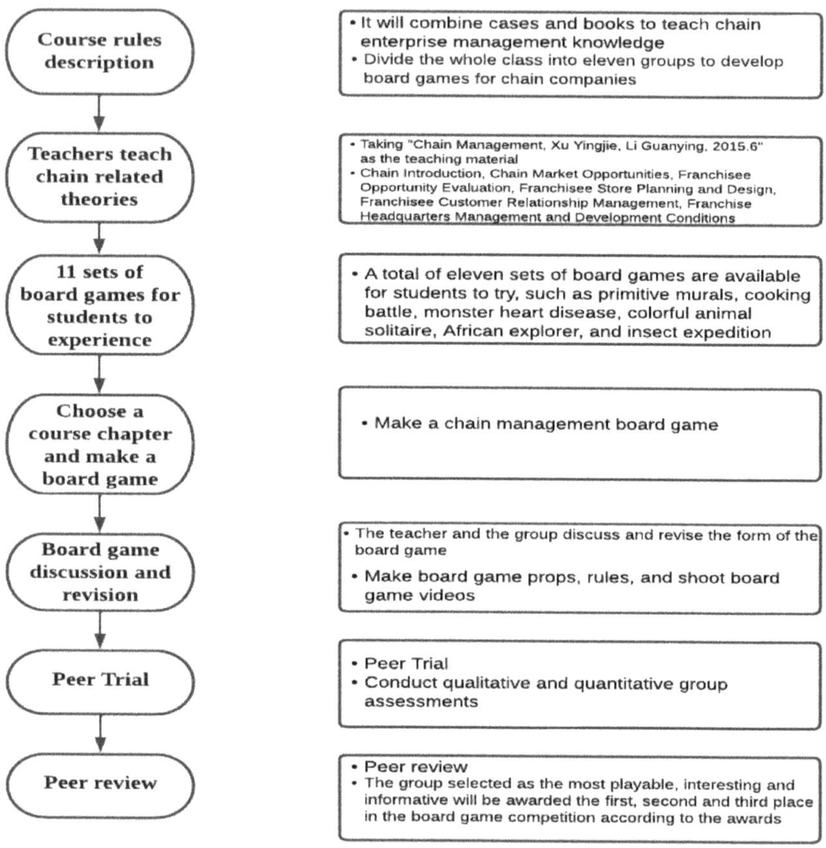

Figure 1. Board game experience and development flow chart.

3.3. *Board Game Content Design*

This research plan review process is conducted by teachers with more than ten years of experience in teaching chain courses and in practicing chain courses to ensure that the content is valid. Table 2 shows the basic content that the group should develop for developing board games based on at least one key concept from the chain chapter.

Table 2. Chain chapters and foci.

Chain Chapters	Focus	Chain Chapters	Focus
1. Development History and Introduction of Chain Enterprises	1-1 Chain definition 1-2 Types of chain operations 1-3 Chain fee (three gold) 1-4 Chain enterprise value chain	5. Affiliate Entrepreneurial Opportunity Assessment	5-1 Startup funding rule of thirds 5-2 Affiliate survey
2. Chain Enterprise Market Opportunities	2-1 Advantages and disadvantages of chain headquarters 2-2 Advantages and disadvantages of franchisee 2-3 Chain development trend	6. Franchise Store Operation Strategy	6-1 Window classification 6-2 Magnet theory 6-3 Display method 6-4 VP PP IP
3. Chain Headquarters Management, Organization and Development Conditions	3-1 Chain organization chart 3-2 Affiliate type 3-3 Chain headquarters and their conditions 3-4 Member's conditions	7. The Relationship Between Franchisees of the Chain Headquarters	7-1 Forms of communication 7-2 Chain headquarters training content 7-3 The basic content of the franchise company

4. Conclusions

Since "Management of Chain Enterprises" is a compulsory course of the department, we asked the students on the course to develop a board game of chain chapter knowledge and let the group monitor the developed board game and rate the learning gains of chain knowledge. After the game was over, a questionnaire on learning engagement and learning effectiveness was issued and tested, and suggestions for future course design and arrangement were put forward as a reference for future related issues and follow-up research.

Author Contributions: Conceptualization, K.-Y.L.; methodology, K.-Y.L.; validation, K.-Y.L.; investigation, P.K.S.; writing—original draft, P.K.S.; writing—review and editing, P.K.S. All authors have read and agreed to the published version of the manuscript.

Funding: This research received no external funding.

Institutional Review Board Statement: The study did not require ethical approval.

Informed Consent Statement: All participants in the study were students and informed consent was given prior to the questionnaires being administered.

Data Availability Statement: Not applicable.

Conflicts of Interest: The authors declare that research was conducted in the absence of any commercial or financial relationships that could be construed as a potential conflict of interest.

References

1. Richardson, C.; Percy, M.; Hughes, J. Nursing therapeutics: Teaching student nurses care, compassion and empathy. *Nurse Educ. Today* **2015**, *35*, e1–e5. [CrossRef] [PubMed]
2. Razon, N. *Okul Öncesi Eğitimde Oyunun, Oyunda Yetişkinin Işlevi*; Ya-Pa Okulöncesi Eğitimi ve Yaygınlaştırılması Semineri: İstanbul, Turkey, 1985.
3. Garris, R.; Ahlers, R.; Driskell, J.E. Games, motivation, and learning: A research and practice model. In *Simulation in Aviation Training*; Routledge: England, UK, 2017; pp. 475–501.
4. Schmidt, C.K.; Davis, J.M.; Sanders, J.L.; Chapman, L.A.; Cisco, M.C.; Hady, A.R. Exploring nursing students' level of preparedness for disaster response. *Nurs. Educ. Perspect* **2011**, *32*, 380–383. [CrossRef] [PubMed]
5. Saarinen-Rahiika, H.; Binkley, J.M. Problem-based learning in physical therapy: A review of the literature and overview of the McMaster University experience. *Phys. Ther.* **1998**, *78*, 195–207. [CrossRef] [PubMed]
6. Hamari, J.; Koivisto, J.; Sarsa, H. Does gamification work?—A literature review of empirical studies on gamification. In Proceedings of the 2014 47th Hawaii International Conference on System Sciences, Waikoloa, HI, USA, 6–9 January 2014; pp. 3025–3034.
7. Ozorio, B.; Fong, D.K. Chinese casino gambling behaviors: Risk taking in casinos vs. abstract investments. *UNLV Gaming Res. Rev. J.* **2004**, *8*, 3.

8. Keller, J.M.; Kopp, T. Application of the ARCS model of motivational design. In *Instructional Theories in Action: Lessons Illustrating Selected Theories and models*; Reigeluth, C.M., Ed.; Lawrence Erlbaum: Hillsdale, NJ, USA, 1987.
9. Brown, J.S.; Collins, A.; Duguid, P. Situated cognition and the culture of learning. *Educ. Res.* **1989**, *18*, 32–42. [CrossRef]
10. Reeve, J.; Tseng, C.M. Agency as a fourth aspect of students' engagement during learning activities. *Contemp. Educ. Psychol.* **2011**, *36*, 257–267. [CrossRef]
11. Ogunmokun, O.A.; Unverdi-Creig, G.I.; Said, H.; Avci, T.; Eluwole, K.K. Consumer well-being through engagement and innovation in higher education: A conceptual model and research propositions. *J. Public Aff.* **2021**, *21*, e2100. [CrossRef]
12. Leithwood, K.; Jantzi, D. Transformational school leadership effects: A replication. *Sch. Eff. Sch. Improv.* **1999**, *10*, 451–479. [CrossRef]
13. Lipps, G.; Norris, C.; Pignal, J. Measuring school engagement. *Educ. Q. Rev. Online* **2003**, *9*, 25–34.
14. Piccoli, G.; Ahmad, R.; Ives, B. Web-based virtual learning environments: A research framework and a preliminary assessment of effectiveness in basic IT skills training. *MIS Q.* **2001**, *25*, 401–426. [CrossRef]
15. Siegler, R.S.; Shipley, C. Variation, selection, and cognitive change. In *Developing Cognitive Competence: New Approaches to Process Modeling*; Lawrence Erlbaum Associates, Inc.: Mahwah, NJ, USA, 1995; pp. 31–76.

Disclaimer/Publisher's Note: The statements, opinions and data contained in all publications are solely those of the individual author(s) and contributor(s) and not of MDPI and/or the editor(s). MDPI and/or the editor(s) disclaim responsibility for any injury to people or property resulting from any ideas, methods, instructions or products referred to in the content.

engineering proceedings

MDPI

Proceeding Paper

Applying a Multivariate Simultaneous Linear Regression Model to Analyze the Mediating Effect of English Leisure Activities on English Learning Motivation and Learning Effectiveness †

Chih-Wei Lin [1,2], Yun-Tsan Lin [3], Shao-Leon Lian [1], Su-Shiang Lee [1], Pei-Jun Zen [1] and Wei-Hsun Hsu [4,*]

1 Department of Leisure Service Management, Chaoyang University of Technology, Taichung 413310, Taiwan; cwlin@cyut.edu.tw (C.-W.L.); shaoleonlian95@gmail.com (S.-L.L.); sulee@cyut.edu.tw (S.-S.L.); jz801108@gmail.com (P.-J.Z.)
2 Department of Business Administration, Chaoyang University of Technology, Taichung 413310, Taiwan
3 Department of Leisure Industry Management, National Chin-Yi University of Technology, Taichung 411030, Taiwan; yuntsan@ncut.edu.tw
4 Department of Marketing and Logistics Management, Chaoyang University of Technology, Taichung 413310, Taiwan
* Correspondence: s11126617@gm.cyut.edu.tw; Tel.: +886-4-23323000 #7558
† Presented at the 3rd IEEE International Conference on Electronic Communications, Internet of Things and Big Data Conference 2023, Taichung, Taiwan, 14–16 April 2023.

Abstract: The purpose of this study was to investigate the relationships between English learning motivation, English leisure activities, and the English learning effect on elementary school children by using a linear regression model. The collected data were analyzed with descriptive statistics, multivariate simultaneous linear regression, and mediation tests. The result showed that elementary students lack parental involvement in English learning. Regular parental support leads to better learning outcomes. English leisure activities are crucial for improving motivation and effectiveness. Schools and parents should prioritize English skills through daily learning and leisure activities to incrementally improve English ability and positively impact learning.

Keywords: English leisure activities; English learning strategy; English teaching; linear regression model

Citation: Lin, C.-W.; Lin, Y.-T.; Lian, S.-L.; Lee, S.-S.; Zen, P.-J.; Hsu, W.-H. Applying a Multivariate Simultaneous Linear Regression Model to Analyze the Mediating Effect of English Leisure Activities on English Learning Motivation and Learning Effectiveness. *Eng. Proc.* **2023**, *38*, 48. https://doi.org/10.3390/engproc2023038048

Academic Editors: Teen-Hang Meen, Hsin-Hung Lin and Cheng-Fu Yang

Published: 27 June 2023

Copyright: © 2023 by the authors. Licensee MDPI, Basel, Switzerland. This article is an open access article distributed under the terms and conditions of the Creative Commons Attribution (CC BY) license (https://creativecommons.org/licenses/by/4.0/).

1. Introduction

Under globalization, English has become the figurative language of internationalization. It is the driving force for every country to pursue economic growth and leap to the international stage. Mastering fluent English is similar to holding a golden ticket to enter the village. Therefore, the direction of talent training for international students becomes internationalized, and the development direction of Chinese students is presented as a compulsory subject for examinations. It tends to move international education towards internationalization [1]. Using the linear regression model, the course of individual leisure behaviors is verified as it is a commonly used form of empirical science. The model has been used for the study of consumer mobile payment behavior [2], cultural and creative hotel accommodation behavior [3], tourist revisit behavior [4], and travel website ordering behavior [5]. Taiwanese teachers use English activities to integrate English in their teaching to make the curriculum diverse and educational. Teachers try to incorporate the concept of leisure education into the curriculum through the use of English activities to help students develop in a variety of ways. Such teaching activities have good results in the short term. However, most researchers suggest extending the research period or offering opportunities for students to develop an intrinsic and spontaneous interest in English learning to examine the maintenance of students' learning effectiveness [6–10].

Chen [11] believed that teaching with passion inspired motivation to learn English, and an ideal English learning environment is the key to learning. The English learning outcomes of Taiwanese students are evaluated by four English competencies: listening, speaking, reading, and writing. Although the short-term results of paper tests are mostly positive, long-term and practical applications have not shown satisfactory results [6–10]. Therefore, English leisure activities that integrate leisure education and English activities may become mainstream education in the future. Therefore, this study aims to understand the relationship between primary school children's English leisure activities and learning effectiveness through various English activities and assessment methods.

2. Literature Review

2.1. English Learning Motivation

Brophy [12] believes that motivation refers to students' subjective experiences, especially what made students willing to participate in classroom activities and their reasons for doing so. Based on the literature review and discussion, he defines "English learning motivation" as any behavior and thinking that promotes acquiring, retaining, and retrieving knowledge to learn English. It is determined by students' internal and external interactive environmental factors [13].

2.2. English Leisure Activities

Kelly [14] defines leisure as a free and selective activity that is mainly based on leisure. Leisure is often included in formal education, whether formal or informal. Leisure is a learning activity that balances learners' physical and mental learning [15]. Based on the literature review and discussion, we define "English leisure activities" as English activities that students are exposed to daily. These activities help students achieve the goals of leisure, relaxation, and learning [16].

2.3. English Learning Effectiveness

Wang [17] pointed out that students are a principal part of education, and the success or failure of education depends on students' learning effectiveness. Therefore, before analyzing the term "learning effectiveness", we must take into account the direct and indirect learning effectiveness of students and pay attention to the learning effectiveness dimensions of students' cognitive, emotional, and motor skills. Based on the literature review and discussion, we define "English learning effectiveness" as the performance of students' cognition, affection, and skills improvement in English listening, speaking, reading, and writing after English-related activities and courses in or out of school.

2.4. Learning Motivation, Leisure Activities, and Learning Effectiveness of English

According to the research of Hung [13], English learning motivation and learning strategies of elementary school children were tested through multivariate stepwise regression. It was found that English learning motivation and strategies were positively correlated, and the higher the motivation, the more English learning strategies were used.

Hypothesis 1: *Elementary school children's English learning motivation significantly impacts English leisure activities.*

Wang [18] found that English learning motivation has a significant positive correlation with English learning effectiveness. Lin and Chen [19] pointed out that students' learning motivation through e-learning positively impacts learning outcomes.

Hypothesis 2: *Elementary school children's English learning motivation significantly impacts English learning effectiveness.*

An experimental study on secondary school students pointed out that using multimedia English songs to integrate secondary school English grammar into teaching improved students' learning effectiveness [7]. An experimental study on elementary school children pointed out that after receiving English song learning, the students made significant progress in oral reading flow, and most students had a positive attitude towards English song practice and singing. Listening to the melody builds vital interest rather than entering teaching at the beginning, and the selection of songs should be in line with the student's level [8].

Hypothesis 3: *Elementary school children's English leisure activities significantly impact English learning effectiveness.*

An experimental study on elementary school children pointed out that teaching English picture books had a significant effect on improving the overall performance of students' English ability and the overall performance of children's learning motivation with a sense of identity and a positive learning attitude [9].

Hypothesis 4: *Elementary school children's English leisure activities have a mediating relationship effect on learning motivation and English learning effectiveness.*

3. Research Method and Procedure

We adopted a cluster sampling method to conduct a questionnaire survey as the primary research method to collect relevant information. The research subjects were fifth and sixth-grade elementary school students in Taiwan in 2016. The schools were classified according to the school type of the Ministry of Education: large, medium, and small schools. The number of classes was calculated according to the proportion of the school type; then, the number of students was selected according to the number of classes.

The pre-test questionnaires were collected from 1 March 2016 to 15 March 2016. The questionnaires were collected after expert validity confirmed the questionnaire's item summarization and question clarity. The number of questionnaires issued was 245, the number of valid samples recovered was 234, and the recovery rate of valid samples was 95.5%. The formal questionnaire of this study was implemented from 1 April 2016 to 30 April 2016, and the estimated number of official questionnaires was 960. The number of questionnaires issued was 817, the number of valid samples recovered was 757, and the recovery rate of valid samples was 92.7%.

The questionnaire contained four parts. The first part was the personal information of elementary school children. The second part was the English learning motivation scale which was subdivided into four dimensions: "value", "expectancy", "affective", and "executive volition". The third part—the English leisure activity scale—was divided into two types: "static" and "dynamic", and the fourth part—the English learning effectiveness scale—was divided into three dimensions: "cognition", "affective", and "skill". Except for the first part, the basic personal information, the rest of the scales were calculated by using a Likert five-point scale. Five professional scholars tested the validity of this research questionnaire. After the items were revised according to expert opinions, it became the pre-test questionnaire for this research. In this study, SPSS was used for the statistical analyses.

Through factor analysis, the Bartlett's sphericity test results showed that the scale of this study was significant. The KMO value of each scale was 0.923 (English learning motivation), 0.794 (English leisure activities), and 0.908 (English learning effectiveness). In addition, after analysis of the "English learning motivation" scale, the CR value of the 20th question did not meet the testing standards and was deleted from the official questionnaire. All data were analyzed for reliability and validity, descriptive statistics, multivariate simultaneous regression analysis, and mediating effect.

4. Result and Discussion

4.1. Descriptive Statistics

The subjects were females (48.5%) and males (51.5%) in the fifth grade (50.9%) and the sixth grade (49.1%). Most students had no person to speak English to (69%) in their families, while other students had other people to speak English to (20.5%) or their mother (11.4%) or father (8.1%). Most of the students started to learn English in kindergarten (67.1%) and at home (9.1%), and 27.5% had not learned English. Most students had no English tutoring after school (36.1%), while other students had three to four years (21.9%), five years and above (21.5%), and one to two years (20.5%). The majority were male in the fifth grade, had no person to speak English to, and started to learn English in kindergarten.

4.2. Multivariate Simultaneous Regression

In the regression analyses, the factors of English learning motivation were used as independent variables while English leisure activities was used as a dependent variable. The results of multiple simultaneous regression analyses of elementary school children's English learning motivation on English leisure activities showed four factors of English learning motivation as "value", "expectancy", "affective", and "executive volition". Moreover, the explanatory power of the four factors of the learning activity value was 30.7%, and the simplified VIF of the four factors were 3.499, 3.201, 2.344, and 1.784, in order. The VIF value was less than 5, indicating that there was no apparent linear relationship between the four factors (Table 1).

$$\text{English Leisure Activities} = 0.177 \times \text{value} + 0.180 \times \text{expectancy} + 0.346 \times \text{affective} + (-0.139) \times \text{executive volition} \quad (1)$$

Table 1. Regression analysis result of English leaning motivation on English learning effectiveness.

Factor	B	S.E.	T	β	VIF
Value	0.176	0.056	3.127 *	0.177	3.499
Expectancy	0.189	0.057	3.313 *	0.180	3.201
Affective	0.332	0.045	7.450 *	0.346	2.344
Executive volition	−0.14	0.041	−3.425 *	−0.139	1.784
		R^2 0.307			

* $p < 0.05$.

With English learning motivation as the independent variable and English learning effectiveness as the dependent variable, the multivariate simultaneous regression analysis results are shown in Table 2. "Value", "expectancy", and "affective" reached a significant level. Furthermore, the three factors of English learning motivation explained the overall explanatory power of English leisure activities up to 69.2%, and the VIF values of the three factors were 3.499, 3.201, and 2.344, in order. The VIF value of less than 5 indicated no apparent linear relationship between the three factors. The standardized regression equation is as follows.

$$\text{English learning effectiveness} = 0.248 \times \text{value} + 0.274 \times \text{expectancy} + 0.385 \times \text{affective} \quad (2)$$

With English leisure activities as independent variables and English learning effectiveness as dependent variables, the results of multiple simultaneous regression analyses are shown in Table 3. The "static" relationship between English leisure activities and "dynamic" two factors reached a significant level. Furthermore, the two factors explained the overall explanatory power of English learning affect up to 39.9%, and the VIF value of the two factors was 1.908. The VIF value was less than 5, indicating no apparent linear relationship between the two factors. The standardized regression equation is as follows.

$$\text{English learning effectiveness} = 0.344 \times \text{static} + 0.343 \times \text{dynamic} \quad (3)$$

Table 2. Regression analysis of English learning motivation on English learning effectiveness.

Factor	*B*	S.E.	*T*	*β*	VIF
Value	0.262	0.040	6.550 *	0.248	3.499
Expectancy	0.306	0.040	7.562 *	0.274	3.201
Affective	0.392	0.032	12.439 *	0.385	2.344
Executive volition	0.21	0.021	0.716	0.019	1.784
		R^2 0.692			

* $p < 0.05$.

Table 3. Regression on Analysis of English Leisure Activities on The English Learning Effectiveness.

Factor	*B*	S.E.	*T*	*β*	VIF
Static	0.314	0.036	8.824 *	0.344	1.908
Dynamic	0.359	0.041	8.782 *	0.343	1.908
		R^2 0.399			

* $p < 0.05$.

4.3. Mediating Effect

According to the concept and test procedure of the mediating effect [20], we analyzed the mediating effect on English leisure activities, learning motivation, and learning effect. The mediating effect test is shown in Table 4.

Table 4. Mediating effect of English leisure activities on learning motivation and learning effectiveness.

Condition	IV	DV	Coef.	*β*	T
Condition 1	English learning motivation	English learning effectiveness	b1	0.821	39.447 *
Condition 2	English learning motivation	English leisure activities	b2	0.515	16.525 *
Condition 3	English leisure activities	English learning effectiveness	b3	0.284	12.895 *
Condition 4	English learning motivation	English learning effectiveness	b′1	0.674	30.676 *

* $p < 0.05$.

Baron and Kenny (1986) [20] proposed that the concept of a mediating effect needed to be confirmed through three regression equations. The equation was tested for whether there was a mediating effect between the independent variable and the dependent variable under the following four conditions.

1. English learning motivation has a significant impact on English learning effectiveness.
2. English learning motivation has a significant impact on English leisure activities.
3. When considering English learning motivation and the influence of English leisure activities on the English learning effect, English leisure activities significantly impact the English learning effect.
4. When considering the influence of English learning motivation and leisure activities on the English learning effect, after controlling for the variable of participation in leisure activities, the influence of the original English learning motivation on the English learning effect is significantly reduced.

If the estimated value of the English learning motivation coefficient does not reach a statistically significant level, the influence disappears, and English leisure activities become a complete mediating variable. If not, English leisure activities are a part of the mediating variables. The above four conditional transformation regression equations are as follows.

(1) In Y = b1X + a1, b1, estimates must be statistically significant.

(2) In Z = b2X + a2, b2, estimates must be statistically significant.
(3) In Y = b′1X+ b3Z +a3, b3 estimates must be statistically significant.
(4) In Y = b′1X+ b3Z +a3, b′1 estimates must be statistically significant.

According to the analysis results, English learning motivation greatly influenced the effectiveness of English learning (b1 = 0.821, t = 39.447, $p < 0.05$), English learning motivation to English leisure activities (b2 = 0.515, t = 16.525, $p < 0.05$), and English leisure activities to English learning effectiveness (b3 = 0.284, t = 12.895, $p < 0.05$). Therefore, the first three conditions were met, but English learning motivation significantly affected the English learning effect (b′1 = 0.674, t = 30.676, $p < 0.05$), Thus, the complete mediation effect was not accepted. However, b1 to b′1 decreased significantly by 0.147. English leisure activities partially mediated between English learning motivation and the learning effect. The estimated values of the standardized coefficients among the variables were aggregated, and the path coefficient diagram is shown in Table 4.

As shown in the path coefficient diagram in Figure 1, the total effect value of English learning motivation on English learning effect is 0.821, the direct effect value of English learning motivation on English leisure activities is 0.515, and the direct effect value of English leisure activities on the English learning effect is 0.515. Therefore, English learning motivation forms an indirect mediating effect on the English learning effect through English leisure activities. The total effect of English learning motivation on English learning effectiveness is 0.821, indicating that each standard deviation unit of English learning motivation has a unit change of 0.821 on English learning effectiveness.

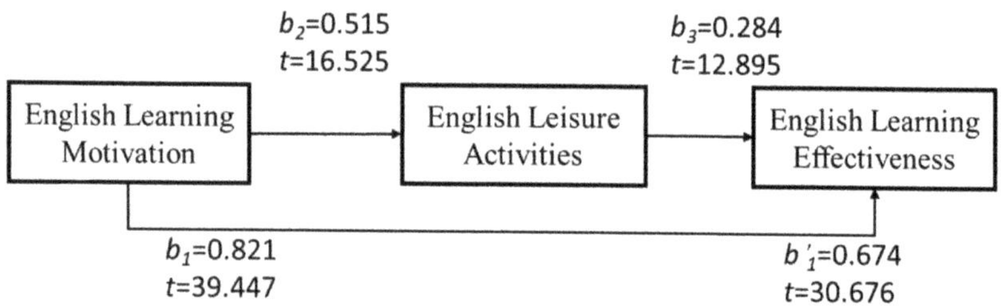

Figure 1. English learning motivation, English learning effectiveness, and English leisure activity path map.

5. Conclusions

Based on the research results, the following conclusions are drawn.

5.1. The Correlation between English Learning Motivation, Leisure Activities, and Learning Effectiveness of Elementary School Children

English learning motivation significantly impacts the English learning effect, and English leisure activities significantly impact the English learning effect. The executive volition factor of English learning motivation harms participation in English leisure activities. The executive volition factor in the item design is more specific and passive for elementary school children to achieve goals. Therefore, if students have more things to complete in their studies, they reduce their participation in English leisure activities. However, whether students have learning motivation is extremely important for participation in English leisure activities and the performance of learning effectiveness. Learning motivation is divided into intrinsic and extrinsic learning motivation. The motivation of students to learn is not biased. Regarding intrinsic or extrinsic motivations, students' English learning does not require external environmental factors and assistance but needs their expectations and willingness for English learning to have better English learning results.

5.2. Correlation between English Leisure Activities and the English Learning Effectiveness of Elementary School Children

The English leisure activities of elementary school children significantly impact English learning effectiveness, and the relationship between the two variables is positive. If primary school children participate in English-related leisure activities more, there will be a more positive output in learning effectiveness. However, they tend to participate in certain types of activities in the middle grades, and they are less likely to extend their range and participate in many ways. It is believed that the primary purpose of English leisure activities is to improve school children's learning motivation and increase their interest in English learning from activity participation. Therefore, the learning effect becomes better if students try English leisure activities in many ways. At the same time, the English learning of school children is not limited to learning in the classroom. When participating in relevant English leisure activities, they must participate in groups or be led by parents and teachers. Therefore, it is also essential to have an environment for English leisure activities in the company of others. In this case, although parents have little time to accompany and participate, they need to also carefully choose the destination of their children's after-school learning or activities to confirm that the student's English learning has a lasting effect.

5.3. The Mediating Effect of English Leisure Activities on Elementary School Children's English Learning Motivation and Effectiveness

There is a partial intermediary between the English leisure activities of elementary school children and English learning motivation and the learning effect. English learning motivation directly affects the English learning effect. If there is an intervention of English leisure activities, English leisure activities impact the effectiveness of English learning. Therefore, if students' English learning motivation is low, by promoting students' participation in English-related activities, students' learning effectiveness can also be improved. Although English learning in Taiwan is mainly based on books and branches of learning, many institutions have made various changes to make students more interested in English. Schools have also followed this trend. English singing, drama, reading, and other activities are integrated into English teaching, adding a variety of styles to English learning. These are relatively common English learning methods. School children could do various types of activities that schools could provide to find out whether students engage in English-related activities in their daily life. The results also show that students' participation in English leisure activities is lower than expected. The concept of multi-party learning English has yet to be popularized. No matter which unit or individual, those who want students to advance their English skills should understand that learning is not a matter of learning English. It requires long-term efforts and training. Contact through English leisure activities in daily life allows students to have more opportunities to learn English in their spare time so that students can develop their interest in learning. Therefore, English leisure activities are an essential factor for enhancing learning motivation and promoting good learning effectiveness.

6. Recommendation

Based on the research results, suggestions for future research are as follows.

1. It is suggested that follow-up researchers gradually expand the research scope through industry–university cooperation or assist the government with expanding the research object to understand the differences in the English learning status of elementary school children in Taiwan. In the research object, future researchers may try to add subjects from different grades to see whether shorter English learning years impact school children's learning effectiveness.
2. The variables in this study are English learning motivation, English leisure activities, and English learning effectiveness. The verification results showed that English leisure activities had a mediating effect on English learning motivation and learning effectiveness, but the influence coefficient was not high. There are two possible

reasons. First, to cause more than one mediating factor between English learning motivation and English learning effect, researchers can find out the variables that may affect both motivation and effect through a literature review. Second, there are few items in English leisure activities, and there needs to be previous literature to follow the research on English leisure activities. The composition of the items is determined by reference to relevant English strategic activities. To conduct in-depth investigations on this variable through on-the-spot observation of elementary school children's participation in English leisure activities and to help relevant units have more information on the English learning situation of school children when making policies, more detailed content of the items is demanded.

Author Contributions: Conceptualization, P.-J.Z., C.-W.L. and S.-S.L.; methodology, C.-W.L.; writing—original draft preparation, Y.-T.L., S.-L.L. and P.-J.Z.; writing—review and editing, W.-H.H. and Y.-T.L. All authors have read and agreed to the published version of the manuscript.

Funding: This research received no external funding.

Institutional Review Board Statement: Not applicable.

Informed Consent Statement: Not applicable.

Data Availability Statement: Not applicable.

Acknowledgments: All subjects' enthusiastic participation is greatly appreciated.

Conflicts of Interest: The authors declare no conflict of interest.

References

1. Li, C.L.; Chuang, H.Y.; Fu, M.F. The study of developing English learning motivation scale for elementary school students in Taiwan. *J. Kaohsiung Appl. Inst. Technol.* **2006**, *3*, 1–33.
2. Lin, C.W.; Lee, S.S.; Tang, K.Y.; Kang, Y.X.; Lin, C.C.; Lin, Y.S. Exploring the users behavior intention on mobile payment by using TAM and IRT. In Proceedings of the 2019 3rd International Conference on E-Society, E-Education and E-Technology, Madrid, Spain, 2–4 August 2019; pp. 11–15.
3. Lin, C.W.; Yang, C.P.O.; Huang, C.H.; Wu, P.C. Analyze the Behavioral Process of Consumers Choosing Cultural and Creative Hotels by Using Multiple Regression Analysis: From the Perspective of AIDA Theory. In Proceedings of the 2022 International Conference on Electrical, Computer and Energy Technologies (ICECET), Prague, Czech Republic, 20–22 July 2022; pp. 1–5.
4. Nugraha, P.A.; Mao, T.Y.; Lin, C.W.; Lin, Y.S. The Impact of Food Image, Customer Satisfaction, and Food Quality on Revisit Intention Under Multiple Regression Model. In Proceedings of the 2021 IEEE International Conference on Social Sciences and Intelligent Management (SSIM), Taichung, Taiwan, 29–31 August 2021; pp. 1–4.
5. Lesnussa, Y.R.A.; Mao, T.Y.; Lin, C.W.; Shih, C.H. The Effect of Service Quality, Price and Promotion on Customer Satisfaction in PegiPegi.com. In Proceedings of the 2021 IEEE International Conference on Social Sciences and Intelligent Management (SSIM), Taichung, Taiwan, 29–31 August 2021; pp. 1–7.
6. Chen, H.W. Effects of Multimedia English Teaching Program on Promoting Learning Achievement and Learning Motivation for English Low Achievers. Master's Thesis, National Taipei University of Education, Taipei City, Taiwan, 2011.
7. Chen, J.Y. The Integration of Multimedia English Songs into the English Grammar Instruction to Enhance Junior High School Students' English Learning Achievement and Learning Motivation. Master's Thesis, University of Kang Ning, Taipei City, Taiwan, 2014.
8. Cheng, C.H. Effects of English Song Singing on Reading Fluency and Learning Attitude of Taiwanese Fifth Graders. Master's Thesis, National Taipei University of Education, Taipei City, Taiwan, 2015.
9. Tsai, M.J. The effects of English picture books of predictable stories teaching to improve English ability and learning motivation of elementary students. *Shu-Te Online Stud. Humanit. Soc. Sci.* **2017**, *12*, 128–155.
10. Chang, H.S.; Lin, M.H. Using Consciousness-Raising Grammar Teaching in EFL Elementary English Class: Effects on English Grammar Learning. *Intergrams* **2013**, *14*, 1–32.
11. Chen, W.R. Integrating Intrinsical Motivating Activities into Vocabulary Instruction. Master's Thesis, National Taipei University of Education, Taipei City, Taiwan, 2010.
12. Brophy, J. *Motivating Students to Learn*; Routledge: London, UK, 2004.
13. Hung, M.Y. The Relationships among Elementary School Students' Foreign Language Anxiety, EFL Learning Motivation and Strategy. *Forum Educ. Res.* **2012**, *3*, 49–67.
14. Kelly, J.R. *Leisure*; Allyn & Bacon: Needham Heights, MA, USA, 1996.
15. Lin, Q.-R. Discussion on the Implementation Strategy of Leisure Education for Primary School Children. Master's Thesis, Pingtung University Department of Physical Education, Pingtung County, Taiwan, 2009.

16. Lin, J.A. Be the Angel for Kids during their English Learning Journey. *Educ. Bimon.* **2009**, *509*, 84–87.
17. Wang, R.Z. Analysis of Student Learning Effectiveness. *Eval. Bimon.* **2010**, *27*, 62.
18. Wang, T.W. A Study of the Relationship among Family Social Capital, English Learning Motivation and Achievement: Sixth Graders in Banciao City in New Taipei County. Master's Thesis, National Taipei University of Education, Taipei City, Taiwan, 2011.
19. Lin, M.H.; Chen, H.C.; Liu, K.S. A study of the effects of digital learning on learning motivation and learning outcome. *Eurasia J. Math. Sci. Technol. Educ.* **2017**, *13*, 3553–3564. [CrossRef]
20. Baron, R.M.; Kenny, D.A. The moderator–mediator variable distinction in social psychological research: Conceptual, strategic, and statistical considerations. *J. Personal. Soc. Psychol.* **1986**, *51*, 1173. [CrossRef] [PubMed]

Disclaimer/Publisher's Note: The statements, opinions and data contained in all publications are solely those of the individual author(s) and contributor(s) and not of MDPI and/or the editor(s). MDPI and/or the editor(s) disclaim responsibility for any injury to people or property resulting from any ideas, methods, instructions or products referred to in the content.

engineering proceedings

MDPI

Proceeding Paper

Research on Innovation of Agricultural Product Logistics Circulation System under the Background of Big Data †

Deming Lei [1], Hongzhen Lin [2,*] and Yawen Tai [2]

[1] College of Arts and Law, Wuhan University of Science and Technology, Wuhan 430065, China; linjingziran@126.com
[2] School of Management, Wuhan University of Science and Technology, Wuhan 430065, China; tai20011224@126.com
* Correspondence: linhongzhen@126.com

† Presented at the 3rd IEEE International Conference on Electronic Communications, Internet of Things and Big Data Conference 2023, Taichung, Taiwan, 14–16 April 2023.

Abstract: Digital rural agricultural engineering is the most widely representative innovative application of digital technology in the field of modern China's new agricultural technology and economic development and the practice of China's new rural economy and society under the background of the process of modern social knowledge networking, intelligent information society, and the digital sharing of social information resources. Using the method of literature research, this paper analyzes the problems existing in the current logistics circulation system of agricultural products. Through the co-occurrence analysis of technical keywords, this paper studies the current situation in the field of agricultural product logistics circulation systems and puts forward countermeasures and measures to promote the innovation and development of agricultural product logistics circulation systems. Finally, it is concluded that we should explore and innovate to build a fresh circulation supply chain system of new smart agricultural products in the new era of agriculture driven by rural big data.

Keywords: agriculture products; artificial intelligence; logistics circulation system; big data

Citation: Lei, D.; Lin, H.; Tai, Y. Research on Innovation of Agricultural Product Logistics Circulation System under the Background of Big Data. *Eng. Proc.* **2023**, *38*, 54. https://doi.org/10.3390/engproc2023038054

Academic Editors: Teen-Hang Meen, Hsin-Hung Lin and Cheng-Fu Yang

Published: 28 June 2023

Copyright: © 2023 by the authors. Licensee MDPI, Basel, Switzerland. This article is an open access article distributed under the terms and conditions of the Creative Commons Attribution (CC BY) license (https://creativecommons.org/licenses/by/4.0/).

1. Introduction

Digital village is the application of networks, informatization, and digitalization in agricultural development and rural economy and society. It is an endogenous modernization transformation development path used to improve farmers' modern information skills. The implementation of the digital village strategy will fundamentally change market factors, such as agricultural scale, ecological environmental protection, fresh food e-commerce, self-media public participation, and user consumption experience by presenting higher requirements and new challenges. The innovation of the modern agricultural product circulation system involves multiple dimensions, such as the main body of the circulation market, the circulation business model, the circulation infrastructure, the quality traceability supervision, and the circulation technical means, and is an important part of the implementation of the digital village strategy. China's existing agricultural product circulation system is based on a decentralized-centralized-decentralized model with a large agricultural product wholesale market as the center, mainly offline face-to-face bulk transactions, and storage and transportation at room temperature with extremely low cold chain rates. The disadvantages of low price, high cost, high corruption, and information asymmetry have led to the contradiction between the low price of fresh and high-quality agricultural products in the hands of farmers and the high and poor quality of residents' purchases.

2. Literature Review

The current research on agricultural product circulation systems in China and abroad mainly covers several aspects: (i) The theoretical framework, system, and strategy of agricultural product logistics, including organization and system, logistics technology and infrastructure, supply chain strategy, just-in-time strategy, continuous replenishment strategy, value-added service strategy, guided production strategy, etc.; (ii) The rural e-commerce circulation system, which mainly includes the agricultural product supply chain system, information system, circulation system, organizational system, service system, safety system, agricultural product cold chain logistics system, etc., and its existing problems. These existing problems mainly include the widening of the gap between market size and market circulation, backward circulation methods and business formats, insufficient circulation infrastructure, weak strength of circulation entities, serious waste of circulation links, low efficiency, and failure to form a supply chain of agricultural products; (iii) The design of the information system, including the agricultural product e-commerce system, e-commerce application, logistics information technology, fresh food e-commerce ecosystem, and the fourth-party agricultural product e-commerce service platform [1]; (iv) Research on the distribution channels of rural e-commerce, including circulation links, cost structure, profit distribution, channel operation performance, channel power changes, participants, circulation formats, government policy support systems, channel strategy, structure, function, and terminal [2]; (v) Research on the efficiency of rural e-commerce circulation, which mainly includes the evaluation index system of agricultural product circulation efficiency, agricultural product circulation mode, circulation structure, distribution network layout, agricultural product circulation efficiency framework, and the influences and mechanisms of the circulation efficiency of agricultural products [3]; and (vi) Research on the rural e-commerce model, which mainly includes the online direct sales of agricultural products of origin, cross-border agricultural product e-commerce, fresh agricultural product e-commerce, the exploration of new standards for non-standardized agricultural products, and county-level e-commerce [4].

The research on rural e-commerce big data in China and abroad mostly adopts technologies and methods such as cluster analysis, recommendation algorithms, and association rules.

Law, Li Rongrong (2017) built a Hadoop-based big data mining framework for e-commerce platforms based on the characteristics of big data in e-commerce platforms and analyzed the mining process in detail [5]. Peng, Zhen Long, and Huang and You Lan (2014), taking the e-commerce data of tea set companies as an example, used the FP-grow algorithm to obtain frequent itemsets and mined user behaviors through association rules, which further realized user classification and precise marketing [6]. Reference [7] crawled the review data of water purifiers, conducted cluster analysis research on it, and identified the advantages and disadvantages of the water purifiers. Zhang Yumei and Jin Yibo (2019) used web crawler technology to analyze the characteristics of Wuwei agricultural products on Taobao.com. They collected and analyzed data such as the number and names of payers and finally put forward measures and suggestions on the brand building, quality control, and industrial extension of Wuwei's characteristic agricultural products [8]. Jiao Huiying et al. (2019) used machine learning algorithms to mine the data of residents' and enterprises' behavior, electricity consumption, and payment on the State Grid e-commerce platform to realize user portrait analysis and provide new ideas on the comprehensive management of business processes, which is the business model of the State Grid e-commerce platform [9]. Qian Dandan and Zhou Jinhai (2019) took the e-commerce data of Chinese herbal medicine companies as an example, designed a business intelligence architecture platform, and used the K-Means clustering analysis algorithm to group consumers to achieve personalized marketing for different consumers [10]. Other scholars have studied the big data processing process of rural e-commerce logistics systems. The McKinsey Global Institute defines big data as a collection of data that is so large that its acquisition, storage, management, and analysis far exceed the capabilities of traditional database software tools. Big data has rich resource types and diverse application processing methods, but the basic process of information processing is roughly similar, including four stages of data collection, processing and integration, analysis, and data interpretation [11]. The processing flow is shown in Figure 1 and includes obtaining the required data from the data source, processing the data of different structures in a unique way, standardizing it into a unified data type, processing and analyzing it through appropriate data analysis methods and tools, and finally obtaining a reasonable result before using visualization technology to show the data to users [12].

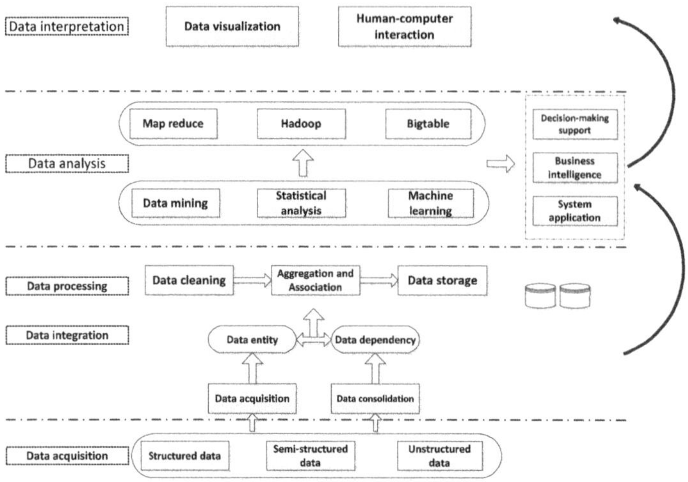

Figure 1. Big data processing process.

Some scholars have conducted research on the data mining of rural e-commerce logistics system. Data mining, also known as data collection and data exploration, is a technology that uses machine learning, mathematical statistics, data visualization, and other multi-field technologies to sort out valuable laws and models from massive amounts of noisy and incomplete information. The basic tasks include classification and prediction, cluster analysis, association rules, time series patterns, deviation detection, intelligent recommendation, etc. [13]. Data mining methods are generally used to determine the characteristics of target data sets or to summarize current information to further predict future situations [14], as shown in Figure 2.

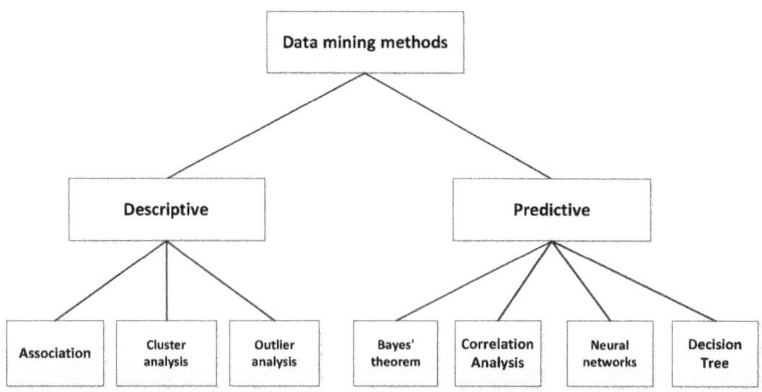

Figure 2. Data mining methods.

3. Analysis on the Problems Existing in the Logistics Circulation System of Agricultural Products

3.1. The Service Is Backward, and the Resource Integration Ability Is Not Strong

How can we develop a middle class in the current rural e-commerce environment in China? The transaction data system of agricultural product networks in various parts of the country and the data of agricultural and sideline products market supply and demand information platform data are not necessarily mature and perfect. E-commerce online transaction information and data in large production areas, price trends, and market supply and demand of major agricultural products in various agricultural markets are also relatively incomplete, comprehensive, and transparent. The constraints also restrict the large-scale sellers and traditional networks of rural e-commerce in China. Fast, timely, and accurate connections between sellers, origin e-commerce platforms, traditional Chinese rural sales network terminal e-commerce platforms, and large-scale agricultural products e-commerce buyers information services are needed, and because the current agricultural product e-commerce is not enough to improve its own relatively open and effective industrial operation supporting system and the combination of industrial policy support and supporting support, it is not in line with the current domestic market innovations in various agricultural fields known as "Internet + agriculture". The separation and independence of elements and rural entrepreneurial talent space resources limit the potential and development of the future rural e-commerce market to bring huge investment opportunities potential space and potential huge growth market space.

The grade of cold chain logistics equipment is not high. Most of the facilities and equipment in the cold chain logistics of agricultural products are outdated, and some even use cold storage built in the 1980s. The existing cold storage refrigeration equipment and refrigeration principle are backward, the technology update speed is slow, and the investment starting point is generally low. The cold chain logistics of agricultural products have the characteristics of heavy assets, high investment, and a long return period. Considering the cost, most investors generally adopt low-cost business strategies, and the planning and design costs of procurement, refrigeration equipment, and construction concepts are generally low.

3.2. The Rural E-Commerce Model Is Unclear, and the Blind Duplication of Construction Is Serious

Rural e-commerce is a systematic project, and its elements mainly include e-commerce platforms, operation teams, county and rural logistics distribution systems, cold chain facilities, training and incubation, etc. The operation mode of e-commerce platforms can be divided into B2C, B2B, 020, micro businesses, etc. The county area is the main battlefield of rural e-commerce, with the county-level operation center being the main carrier. In the process of rural e-commerce construction, the planning and guidance of the county government play a key role. However, due to the extreme lack of top-level design teams and technical operation teams proficient in rural e-commerce in the county, the model is unclear and blind duplication of construction is serious. At present, it is very difficult for domestic counties to cooperate with platforms such as Village Tao, Ganjie.com, and Juchao.com, and they are on the verge of being eliminated.

The construction of rural e-commerce is a complex systematic project, which not only requires the support of government policies and the guarantee of special financial expenditures to improve the rural logistics network and infrastructure but also requires departments such as agriculture, supply and marketing, transportation, postal services, e-commerce, express delivery, and other departments. Cooperation and collaboration among other enterprises. In addition to the e-commerce giants (Alibaba, JD.com, Suning, etc.) deployed in the rural market, traditional supply and marketing enterprises have also accelerated the development of rural e-commerce through deepening comprehensive reforms. At the same time, provincial-level rural e-commerce website models have emerged, including the docking model of housing, the resource docking model of agricultural enterprises and cooperatives, the direct sales model of e-commerce companies' self-built platforms, the county-level e-commerce model, the "local government + e-commerce platform + local operator" tripartite operation platform, and other models. Despite this, the development of rural e-commerce still faces problems such as imperfect infrastructure and supporting conditions, poor logistics, and distribution paths, homogeneous competition among enterprises, lack of e-commerce professionals, and blindly repeated platform construction in various places, which seriously hinder the rural economy's development and the realization of the rural revitalization strategy.

3.3. Weak Innovation and Lack of Core Competitiveness

At present, the overall construction mode of the national rural e-commerce platform is mostly dominated by large private enterprises whose own strength is still relatively single and weak and has not yet been able to form a national rural e-commerce platform with regional representative characteristics. The business platform system is its underlying structure, which fails to meet the actual support needs of the national "agriculture, rural areas and farmers" policy. The government's policy orientation mostly focuses on how to enter the e-commerce platform, carry out talent training, and decorate and upgrade store management, etc., which has led to the general similarity of how to build rural e-commerce in various places, without any obvious regional differences, and the lack of continuous innovation. The investment in support policies such as technology and R&D is less than that in local areas, which makes it difficult for local areas to promote innovation with independent intellectual property rights, develop and operate rural e-commerce platforms, and explore new business models for industrial development models. Another problem of rural e-commerce is the lack of brand effect. In the current online business environment, whether it is Tmall, Taobao, JD.com, or other platforms, there are many miscellaneous products, and the competitiveness of stores is insufficient. Creating a rural e-commerce atmosphere requires not only guidance and support from local county-level governments but also more and more need to rely on a group of industry benchmarking and leading enterprises to demonstrate, drive or guide, and give play to their industry leaders. It plays a leading role in the market leading and establishes its industry benchmark. At present, although the number of leaders in the field of rural e-commerce in China is still very small, and many small and medium-sized traditional e-commerce companies have also started their rural e-commerce business in the country, in the end, most companies will still be based on the industry. Due to the policy orientation problem, it eventually fell into a chaotic situation of the same development model and no actual replicable rural e-commerce performance growth model. It is urgent to cultivate two-three rural e-commerce leaders who can become unicorn enterprises and, proceeding from reality, cultivate professional operation teams, enhance R&D and innovation capabilities, establish industry benchmarks, and drive an industry development atmosphere.

4. Countermeasures and Measures for Promoting the Innovative Development of Agricultural Products Logistics Circulation Systems

4.1. Adjust the Direction of Guidance, Focus on Research and Development, and Promote Innovation

To promote the innovative development of agricultural products logistics circulation systems, we suggest (i) adjusting the direction of government policy support and guidance from the policy orientation of e-commerce training, outlet decoration, and e-commerce base construction to focusing on research and development and encouraging technological innovation; (ii) strictly formulating policy support fund standards, increasing the proportion of enterprise R&D investment, technical patents, team structure, talent education, etc. in the evaluation indicators, selecting enterprises that focus on R&D, and promoting the innovation ability of local enterprises; and (iii) implementing the unicorn leader plan, focusing on fresh food e-commerce, agricultural materials e-commerce, rural distribution, agricultural cold chain processing, county-level new retail and other fields. The project team of the rural e-commerce unicorn will provide strong and continuous support to its team leader.

4.2. Improve the Support Mechanism and Change Subsidies into Equity

To improve the support mechanism and change subsidies into equity, we suggest transforming the financial subsidy funds related to rural e-commerce into a entrusted equity investment + entrusted loan model, in accordance with the "Company Law" and the modern corporate governance structure, vigorously supporting mixed-ownership enterprises to develop rural e-commerce, and vigorously promoting a 4:3:3 equity structure, that is, 40% of state-owned capital, 30% of social capital, and the implementation of a governance model of 30% of management. In this model, state-owned shareholders do not directly participate in business operations and operate in full accordance with market rules, give full play to the guiding role of financial funds, mobilize the enthusiasm of social capital to participate, and ensure that the executive team continues to be cohesive.

4.3. Strengthen the Construction of Cold Chain Logistics Infrastructure

In the entire agricultural product circulation system, the overall improvement of the cold chain logistics infrastructure is the fundamental and key point to truly solve the problem of agricultural product circulation quality. Guided by the promotion of policy implementation from the perspective of policy top-level design and industry overall planning, based on the innovation of fourth-party logistics and rural e-commerce business models, we suggest taking Jiangxi as an example and relying on supply and marketing cooperatives to build the province's cold chain backbone network, trying to improve and solve practical problems, such as the extremely low circulation rate of the cold chain, the backward circulation method, unsound systems, weak cold chain logistics infrastructure, small scale, and the lack of a network of cold chain logistics, so as to guarantee the realization of high-quality circulation of agricultural products, improve the circulation quality of agricultural products, shorten the circulation cycle of agricultural products, and promote the construction and implementation of the cold chain circulation system in terms of policy and practice.

5. Conclusions

At present, the modern circulation mode system of bulk agricultural products in my country still adopts the traditional decentralized-semi-centralized-quasi-decentralized mode with the traditional large-scale regional agricultural product wholesale market center as the logistics center. There are serious disadvantages, such as an obvious asymmetry of information, and the traditional offline market is mainly based on the face-to-face trading mode of bulk products. The successful implementation of the digital village strategic plan has brought about fundamental changes in the structure of various market factors such as the scale of traditional agricultural operations, ecological environmental protection, and the development of fresh food e-commerce, which has prompted the existing extensive circulation price system of Chinese agricultural products. Once again, fundamental change is urgently needed. Taking the reform and innovation experiment of the new rural e-commerce model as the starting point for exploring the efficiency of improving the circulation of agricultural and sideline products and taking the innovative experiment of constructing the cold chain logistics system of agricultural and sideline products as the breakthrough point to innovate and improve the quality of the food circulation system, the big data method of agricultural products and the dynamic modeling of agricultural products are used. Analysis tools are used to complete the design analysis and operation practice of the integrated e-commerce platform for fresh agricultural products and explore and innovate the construction of a new smart fresh agricultural product circulation supply chain system in the new era of agriculture driven by the background of rural big data.

Author Contributions: Conceptualization, D.L.; methodology, H.L.; formal analysis, Y.T.; resources, D.L.; writing—original draft preparation, H.L.; funding acquisition, D.L. All authors have read and agreed to the published version of the manuscript.

Funding: This research was funded by Hubei SME Research Center: Research on the development of rural e-commerce SMEs from the perspective of government support, grant number [HB-SME2021C07].

Institutional Review Board Statement: Not applicable.

Informed Consent Statement: Not applicable.

Data Availability Statement: No new data were created.

Conflicts of Interest: The authors declare no conflict of interest.

References

1. Wang, S.; Ding, Z. Agricultural products e-commerce ecosystem—A theoretical analysis framework. *China Rural. Obs.* **2015**, *4*, 39–48.
2. Zhao, X.; Li, C. Transformation of agricultural product circulation channels: Evolution law, dynamic mechanism and development trend? *Manag. World* **2012**, *3*, 81–95.
3. Ouyang, X.; Huang, F. The influence of my country's entry into WTO on the circulation efficiency of rural agricultural products in China. *Agric. Technol. Econ.* **2013**, *1*, 68–76.
4. Chen, L. Looking at the development trend of agricultural products e-commerce from the Ali platform. *China Circ. Econ.* **2015**, *6*, 58–64.
5. Li, R. Research on big data mining of e-commerce platform based on Hadoop. *Nanfang Agric. Mach.* **2017**, *48*, 111+169.
6. Peng, Z.L.; Huang, Y.L. Research on E-Commerce Intelligence Based on IOT and Big Data. *Appl. Mech. Mater.* **2014**, *496–500*, 1889–1894.
7. Ding, Y. *Design and Research of Big Data Technology Based on E-Commerce Platform*; Wuhan Engineering University: Wuhan, China, 2018.
8. Zhang, Y.; Jin, Y. Data analysis of Wuwei's characteristic agricultural products based on Web data mining—Taking Taobao e-commerce platform as an example. *Enterp. Technol. Dev.* **2019**, *9*, 161–163+165.
9. Jiao, H.; Xin, C.; Liu, J. Application of Big Data Technology in State Grid E-commerce Platform. *Software* **2019**, *40*, 82–84.
10. Qian, D.; Zhou, J. The application of business intelligence based on big data in the analysis of e-commerce data. *Electron. Commer. Serv.* **2019**, *4*, 29–30+96.
11. Cheng, B.; Li, X. Discussion on the data processing flow and quality control of scientific research achievements in colleges and universities. *Libr. J.* **2020**, *39*, 48–55.
12. Meng, X.; Ci, X. Big Data Management: Concepts, Technologies and Challenges. *Comput. Res. Dev.* **2013**, *50*, 146–169.
13. Wang, B.; Peng, W. Research on Intelligent Collection Model of Service Enterprise Diagnostic Information Intelligence Based on Data Mining. *Sci. Technol. Manag. Res.* **2020**, *40*, 184–192.
14. Luo, P.; Yang, G.; Ning, Z. Research on Data Mining Methods and Knowledge Discovery. *Libr. Inf. Work.* **2011**, *55*, 217–219+228.

Disclaimer/Publisher's Note: The statements, opinions and data contained in all publications are solely those of the individual author(s) and contributor(s) and not of MDPI and/or the editor(s). MDPI and/or the editor(s) disclaim responsibility for any injury to people or property resulting from any ideas, methods, instructions or products referred to in the content.

engineering proceedings

MDPI

Proceeding Paper

Using The KANO Model and Importance-Performance Analysis (IPA) to Explore Student Needs in Synchronous Distance Learning Combined with The Gamification Element †

Shih-Chieh Chen * and Jun-Cheng Liu *

Department of Industrial Design, National Yunlin University of Science and Technology, Yunlin 640111, Taiwan
* Correspondence: chenshih@yuntech.edu.tw (S.-C.C.); therock5116@gmail.com (J.-C.L.)

† Presented at the 3rd IEEE International Conference on Electronic Communications, Internet of Things and Big Data Conference 2023, Taichung, Taiwan, 14–16 April 2023.

Abstract: Since the COVID-19 pandemic, synchronous distance learning has become a common teaching method. However, there are problems in synchronous distance learning, such as distractions and cheating. Many studies have shown that gamification has been used as one way to improve student's learning motivation. Thus, we explored the functional requirements of synchronous distance learning software and different distance learning acceptance with gamification elements by students using KANO and an importance-performance analysis. The results of the study showed that most students thought it important to integrate games into synchronous distance learning, and gamification brought more fun to learning while achieving the purpose of learning.

Keywords: synchronous distance learning; gamification elements; technology acceptance mode; KANO model; importance-performance analysis

Citation: Chen, S.-C.; Liu, J.-C. Using The KANO Model and Importance-Performance Analysis (IPA) to Explore Student Needs in Synchronous Distance Learning Combined with The Gamification Element. *Eng. Proc.* **2023**, *38*, 56. https://doi.org/10.3390/engproc2023038056

Academic Editors: Teen-Hang Meen, Hsin-Hung Lin and Cheng-Fu Yang

Published: 28 June 2023

Copyright: © 2023 by the authors. Licensee MDPI, Basel, Switzerland. This article is an open access article distributed under the terms and conditions of the Creative Commons Attribution (CC BY) license (https://creativecommons.org/licenses/by/4.0/).

1. Introduction

The COVID-19 pandemic caused school closures globally in 2020, with 107 countries shutting down schools, as reported by UNESCO. While synchronous distance learning became the new norm, it faced challenges such as decreased interaction between students and teachers [1]. To overcome these challenges, digital game-based learning was used to enhance such interactions. Studies have shown that a competitive gaming environment can increase students' motivation, academic performance, and gamification elements in synchronous distance learning, improving student focus and interaction [2,3]. By analyzing 15 basic functions of synchronous distance learning software from Google Meet and Microsoft Teams and 11 gamification elements from digital game learning software such as Blooket and Duolingo, we assessed students' attitudes toward the different functions of these elements. For this research, the KANO and importance-performance analysis (IPA) questionnaires were used. The results of this study serve as a basis for future research.

2. Research Background

2.1. Gamification

Deterding, Dixon, Khaled, and Nacke defined gamification as the "integration of game design elements into non-game contexts" [4]. Using gamification aims to increase user engagement and immersion by incorporating elements of game design such as mechanics and game flow. Gamification also changes the way an event or process operates, making it more captivating and enjoyable for the user.

2.2. Synchronous Distance Learning

Synchronous distance learning means that teachers and students conduct online teaching and learning in different places through media, such as the Internet and video software [5]. Due to the COVID-19 pandemic, Microsoft Teams and Google Meet were widely used in teaching.

2.3. KANO Model

The KANO two-dimensional quality model is a tool that can evaluate the quality of functions or services. It was developed by Noriaki Kano in 1984 to provide a way to help organizations understand the needs and preferences of their customers [6]. The KANO model categorizes customer requirements into five distinct categories based on feedback from existing and potential customers. The model identifies what features are important to their customers and makes informed decisions on product or service development. These five categories are as follows (Figure 1).

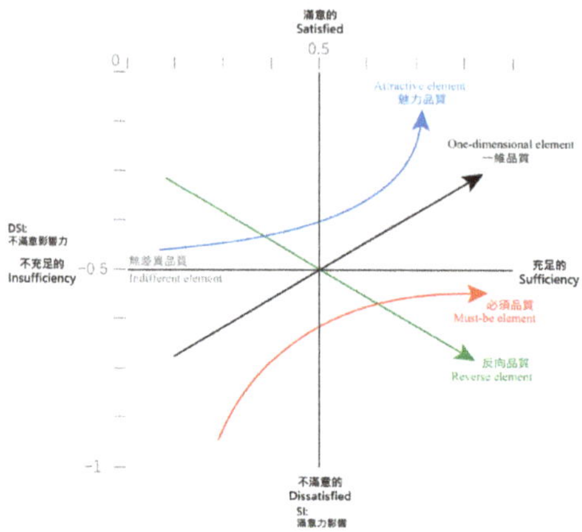

Figure 1. KANO two-dimensional quality model, adapted with permission from Ref. [6]. Copyright 1984 Kano.

- Attractive element (A): Adequate presence enhances satisfaction, while an inadequate presence is still acceptable.
- One-dimensional element (O): More of this quality leads to greater satisfaction, while fewer leads to less satisfaction.
- Must-be element (M): Adequate presence is expected, while inadequate presence leads to dissatisfaction.
- Indifferent element (I): The presence or absence of this quality does not impact satisfaction.
- Reverse element (R): Adequate presence causes dissatisfaction, while absence leads to satisfaction.

2.4. Classification of Kano 2-Dimensional Quality Elements

Matzler and Hinterhubery improved the Kano model by introducing a "Two-Dimensional Quality Element Classification Table" [7]. This table can categorize relative quality elements based on responses to forward and reverse questions (Table 1).

Table 1. Two-dimensional quality element classification table of the KANO model, adapted with permission from Ref. [7]. Copyright 1998 Matzler and Hinterhuber.

Functional Requirement	Dysfunction				
	Like	Must-have	No-Comment	Bearable	Dislike
Like	Q	A	A	A	O
Must-have	R	I	I	I	M
No-Comment	R	I	I	I	M
Bearable	R	I	I	I	M
Dislike	R	R	R	R	Q

M—Must be; O—One-dimensional; A—Attractive R—Reverse; I—Indifferent; Q—Questionable.

2.5. Quality Improvement Index

Matzler and Hinterhube introduced the Quality Improvement Index (QII) [7] to prioritize the quality elements that require improvement. QII comprised two components: the "Satisfaction Index (SI)" and the "Dissatisfaction Index (DSI)". They formulated the calculation of the QII as follows.

$$\mathrm{SI} = (\mathrm{A} + \mathrm{O})/(\mathrm{A} + \mathrm{O} + \mathrm{M} + \mathrm{I}) \quad (1)$$

$$\mathrm{DSI} = (\mathrm{O} + \mathrm{M})/(\mathrm{A} + \mathrm{O} + \mathrm{M} + \mathrm{I}) \times -1 \quad (2)$$

QII can be used to determine which quality elements need improvement. The higher the SI value is toward one, the greater the impact it has on customer satisfaction. On the other hand, the closer the DSI value is to -1, the stronger the negative impact it has on customer satisfaction when quality is insufficient.

2.6. IPA

Martilla and James introduced the IPA analysis [8], which is a valuable and practical set of analytical methods to understand consumers' perceptions of product and service quality. This result provides a foundation for evaluating and improving perception. By using satisfaction and significance as evaluation criteria, a two-dimensional matrix graph was created and divided into four quadrants (Figure 2). Quadrant 1 was for "Keep Up the Good Work", which signified customers' high satisfaction and the high importance of the project, making it a strength of the company. Investment should be sustained to maintain its competitive edge. Quadrant 2 was for "Concentrate Here", highlighting low satisfaction but the high importance of the service, making it crucial to improve. Quadrant 3 is for "Low Priority", depicting the low satisfaction and low importance of the service, making it a low priority for improvement. Quadrant 4 is for "Possible Overkill", suggesting the high satisfaction but low importance of the service, indicating the potential to reallocate resources to more critical areas.

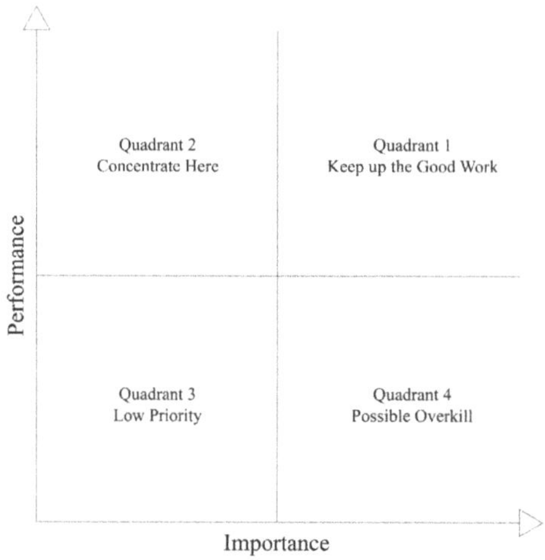

Figure 2. IPA model, adapted with permission from Ref. [8]. Copyright 1997 Martilla and James.

3. Research Methods

We distributed questionnaires to students and graduate students (19 to 25 years old) of the Yunlin University of Science and Technology in Yunlin County, Taiwan, through Google Forms. A total of 60 valid questionnaires were collected. The research questionnaire consisted of three parts on a 5-point Likert scale ("Strongly Agree", "Agree", "Neutral", "Disagree", and "Strongly Disagree"). The first part was to collect basic information and ask about technology acceptance with an explanation of "trying to understand students' acceptance of distance learning software." The second part was created based on the IPA model on a 7-point Likert scale ("Very Strongly Agree", "Strongly Agree", "Agree", "Neutral", "Disagree", "Strongly Disagree", and "Very Strongly Disagree") to understand students' subjective feelings about the function. The third part reflected the KANO two-dimensional quality model. According to Matzler and Hinterhuber [7], five evaluation indicators were included for the answers to positive and reverse questions on a 5-point Likert scale ("Like", "Must-have", "No-Comment", "Bearable", and "Dislike"). SI and DSI were calculated with Equations (1) and (2) as quality improvement indicators.

4. Results

In total, 40 females and 20 males took part in the survey. Thirty-one had a high acceptance of distance learning software, while 29 had a low acceptance of technology. Students with a high acceptance of distance learning software were classified according to the degree of satisfaction and dissatisfaction. The satisfied group showed a high SI for A1, A5, A9, A16, A19, A22, and A24, while the dissatisfied group had a low DSI for A1, A4, A6, and A13. Students with a low acceptance of distance learning showed a high SI for A4, A9, A16, and A24 and a low DSI for A1, A4, and A6 (Table 2).

Table 2. KANO quality improvement index.

	ITEM	High Acceptance		Low Acceptance	
		SI	DSI	SI	DSI
Synchronous Distance Learning software	A1. Link Invitation	0.61	−0.68	0.41	−0.66
	A2. Host	0.23	−0.16	0.21	−0.21
	A3. Automatic Roll Call	0.53	−0.13	0.42	−0.15
	A4. Share The Screen	0.57	−0.80	0.48	−0.72
	A5. Multiple Share The Screen	0.63	−0.17	0.25	−0.14
	A6. Meeting Messages	0.39	−0.58	0.38	−0.52
	A7. Raise Hands	0.39	−0.42	0.41	−0.41
	A8. Emoji	0.39	−0.13	0.41	−0.21
	A9. Group	0.63	−0.23	0.45	−0.28
	A10. Homework	0.57	−0.33	0.21	−0.21
	A11. Vote	0.47	−0.27	0.39	−0.21
	A12. Whiteboard	0.42	−0.16	0.31	−0.17
	A13. Background Effects	0.58	−0.58	0.55	−0.31
	A14. Schedule	0.32	−0.10	0.25	−0.21
	A15. Screenshot	0.42	−0.23	0.36	−0.32
gamification elements	A16. Game	0.74	−0.06	0.48	−0.07
	A17. Game Rewards	0.48	−0.06	0.38	−0.10
	A18. Game Loss Penalty	0.39	−0.06	0.11	−0.07
	A19. Achievement	0.61	−0.10	0.39	−0.07
	A20. Points or Coins	0.52	−0.10	0.22	−0.15
	A21. Exchange	0.52	−0.13	0.26	−0.11
	A22. Custom Roles	0.65	−0.10	0.41	−0.10
	A23. Collect	0.24	−0.07	0.15	−0.15
	A24. Friends	0.68	−0.19	0.48	−0.11
	A25. Leaderboards	0.33	−0.07	0.26	−0.07
	A26. Gamified Data Dashboards	0.48	−0.16	0.36	−0.14

4.1. Kano Two-Dimensional Quality Elements

With the average value of SI and DSI, a quadrant diagram was drawn for the participants of this survey (Figure 3).

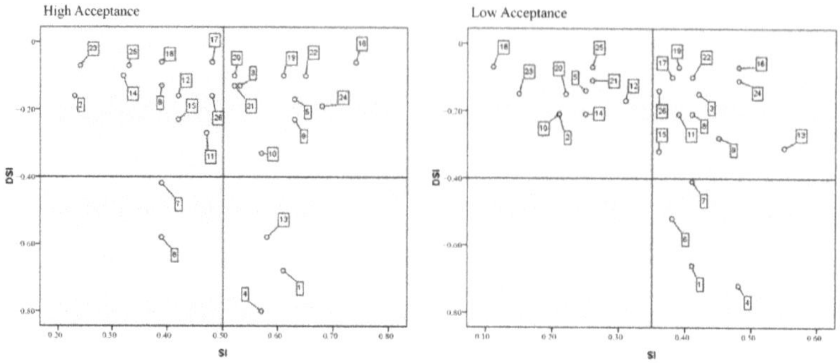

Figure 3. KANO Quality Improvement Index.

The attractive element in Quadrant 1 showed that the participants were satisfied with the software. Even without these elements, they would not feel bad. For students with a high acceptance of distance learning software, A3, A5, A9, A10, A16, A19, A20, A21, A22, and A24 were positioned in Quadrant 1. For students with a low acceptance of distance learning software, A3, A8, A9, A11, A13, A15, A16, A17, A19, A22, A24, and A26 were found in Quadrant 1.

The indifferent element was found in Quadrant 2. The participants were not satisfied or dissatisfied regardless of whether the service or function was available or not. For students with a high acceptance of distance learning software, A2, A8, A11, A12, A14, A15, A17, A18, A23, A25, and A26 were plotted in Quadrant 2, while for students with a low acceptance of distance learning software, A2, A5, A10, A12, A14, A18, A20, A21, A23, and A25 were found.

Quadrant 3 showed the must-be elements. Without this element, the participants were dissatisfied. However, even with this element, the participants were not particularly satisfied as the element was considered a basic service. For students with a high acceptance of distance learning software, A6 and A7 were found in Quadrant 3. One-dimensional elements were positioned on Quadrant 4. The fewer the elements, the more dissatisfied the participants were. However, with this element, their satisfaction increased. For students with a high acceptance of distance learning software, A1, A4, and A13 were plotted on Quadrant 4, while A1, A4, and A6 were found in Quadrant 4 for students with a low acceptance of distance learning software.

4.2. IPA Analysis

According to the IPA results, the quadrant diagram of the high acceptance and low acceptance of distance education software is drawn in Figure 4. For the students with a high acceptance of distance education software, most functions were found in Quadrants 1 and 3. In Quadrant 1, A1, A3, A4, A5, A6, A7, A10, A11, A13, A16, A19, A22, and A26 were found, while A8 and A21 were on Quadrant 2, A2, A9, A12, A14, A15, A17, A18, A23, and A25 were on Quadrant 3, and A10 and A20 were on Quadrant 4.

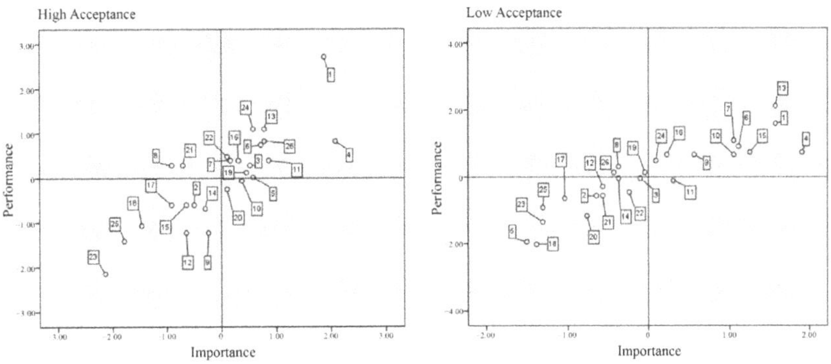

Figure 4. Quality improvement index from IPA analysis.

For the students with a low acceptance of distance learning, A1, A4, A7, A6, A9, A10, A13, A15, A16, and A24 were plotted on Quadrant 1, A8, A19, and A26 were on Quadrant 2, A2, A3, A5, A12, A14, A17, A18, A20, A21, A22, A23, and A25 were on Quadrant 3. In Quadrant 4, there was only A11.

5. Conclusions

The KANO model and IPA analysis results revealed that the participants in this study thought that integrating games into synchronous distance learning software was required even though gamification elements were not the main focus for the development of the software. They regarded that gamification enhanced learning experiences by making them more enjoyable and fulfilling and the goal of fun learning. Without gamification, they might feel dissatisfaction. The IPA result revealed that emojis were significant for the participants to use the software. These small yet powerful symbols serve as a crucial medium of communication, and diverse emojis were an element of excitement for the students, thereby elevating their engagement in synchronous distance learning. Additionally, other crucial gamification elements, such as achievements, exchange mechanisms, and gamified dashboards, must also be prioritized for development.

Author Contributions: All authors contributed equally to this work. All authors have read and agreed to the published version of the manuscript.

Funding: This research was funded by [National Science and Technology Council, R.O.C], grant number [NSTC 111-2410-H-224-021].

Institutional Review Board Statement: Not applicable.

Informed Consent Statement: Not applicable.

Data Availability Statement: Not applicable.

Conflicts of Interest: The authors declare no conflict of interest.

References

1. Leontyeva, I.A. Modern Distance Learning Technologies in higher education: Introduction problems. *Eur. J. Math. Sci. Technol. Educ.* **2018**, *14*, 1–8. [CrossRef] [PubMed]
2. Cagiltay, N.E.; Ozcelik, E.; Ozcelik, N.S. The Effect Of Competition On Learning In Games. *Comput. Educ.* **2015**, *87*, 35–41. [CrossRef]
3. Süme, M.; Aydın, C.H. Design Principles for Integrating Gamification into Distance Learning Programs in Higher Education: A Mixed Method Study. *Int. J. Serious Games* **2022**, *9*, 79–91. [CrossRef]
4. Deterding, S.; Dixon, D.; Khaled, R.; Nacke, L. From game design elements to gamefulness: defining gamification. In Proceedings of the 15th International Academic MindTrek Conference: Envisioning Future Media, New York, NY, USA, 28 September 2011.
5. Anderson, R.; Beavers, J.; Vandegrift, T.; Videon, F. Videoconferencing and presentation support for synchronous distance learning. In Proceedings of the 33rd Annual Frontiers in Education, Westminster, CO, USA, 5–8 November 2003.
6. Kano, N. Attractive quality and must-be quality. *J. Japan Soc. Qual. Control* **1984**, *31*, 147–156.

7. Matzler, K.; Hinterhuber, H. How to make product development projects more successful by integrating Kano's model of customer satisfaction into quality function deployment. *Technovation* **1998**, *18*, 25–38. [CrossRef]
8. Martilla, J.A.; James, J.C. Importance-Performance Analysis. *J. Mark.* **1997**, *41*, 77–79. [CrossRef]

Disclaimer/Publisher's Note: The statements, opinions and data contained in all publications are solely those of the individual author(s) and contributor(s) and not of MDPI and/or the editor(s). MDPI and/or the editor(s) disclaim responsibility for any injury to people or property resulting from any ideas, methods, instructions or products referred to in the content.

engineering proceedings

MDPI

Proceeding Paper

Need of Design Knowledge and Use of Novice Designer for Product Sketching †

Chen Li and Song Xu *

School of Design and Art, Beijing Institute of Technology, Zhuhai 519088, China
* Correspondence: xusong@bitzh.edu.cn

† Presented at the 3rd IEEE International Conference on Electronic Communications, Internet of Things and Big Data Conference 2023, Taichung, Taiwan, 14–16 April 2023.

Abstract: Design knowledge is the basis for designers to solve design problems. In product design, sketches assist designers in thinking as an important step. Novice designers need to use different types of design knowledge in sketching to solve design problems and find product solutions. In this study, we determined the novice designers' knowledge needs and processes in sketching and discovered their sketch ideas to understand how they solve design problems.

Keywords: design knowledge; sketching; design process; design education

Citation: Li, C.; Xu, S. Need of Design Knowledge and Use of Novice Designer for Product Sketching. *Eng. Proc.* **2023**, *38*, 58. https://doi.org/10.3390/engproc2023038058

Academic Editors: Teen-Hang Meen, Hsin-Hung Lin and Cheng-Fu Yang

Published: 28 June 2023

Copyright: © 2023 by the authors. Licensee MDPI, Basel, Switzerland. This article is an open access article distributed under the terms and conditions of the Creative Commons Attribution (CC BY) license (https://creativecommons.org/licenses/by/4.0/).

1. Introduction

Design knowledge refers to the information on product processes including multi-level information such as user communities, product attributes, organizational strategies, and technical principles that promote design innovation [1,2]. In the innovation design of products, the early design is usually vague and undefined and is difficult to disintegrate under fixed standards. Designers solve such design problems with personal experience in capturing, storing, and reusing various knowledge by analyzing and visualizing the possible effects of the solution. Designers have flexibility in conceptual design according to design requirements to show different results. Manolya and Gero studied the differences between experienced and novice designers in the use of knowledge in the conceptual design process and found that their knowledge supported the thinking process [3]. They also found that expert designers were more structured in cognitive activities and used knowledge more strategically, resulting in more than three times higher productivity than novice designers [3]. Thus, the challenge for novice designers is how to strategically access and filter effective knowledge for innovation in any given situation [4].

In the conceptual design of products, designers rarely use a computer-aided system (CAD) because the exact geometric information is not defined yet. Instead, they sketch to stimulate their design thinking [5]. Sketches show the core idea of designs and play an irreplaceable role in the positioning, development, expression, deduction, and formation of product concepts [6,7]. The analysis of design knowledge in the sketching process is a dynamic process involving knowledge acquisition, processing, and application [8]. Previous studies revealed that the design knowledge in the sketching process could be effectively and explicitly described by design protocols [9]. However, in this study, we try to discover how novice designers use their knowledge to stimulate a variety of product ideas during the sketching process. For this study, cognitive activities, sketching behaviors, and sketching outcomes of novice designers are recorded in sketching by Protocol Analysis. Based on the function behavior strategy (FBS) ontology and the theory of creative segmentation, we select and code the design knowledge in sketching and graphically describe the use of knowledge in different stages of sketching to describe how knowledge is used in the creative activities of novice designers. The findings are useful to find the type of support that design education needs to provide for novice designers.

2. Theoretical Backgrounds

2.1. FBS Ontology

FBS ontology is used to identify knowledge in the sketching process and describe knowledge use patterns. In the FBS model, Gero used two categories of variables to describe different aspects of the design: function variables to describe the teleology of the object and behavior variables referring to what the object does including expected behavior variables and structural behavior variables [10]. The structure variable refers to the objects' components and their relationships. The FBS ontology interprets the design process based on the design requirement (R) with which designers construct the connection among function, behavior, and structure according to their experiences [10]. Specifically, the designer ascribes the function (F) to the expected behavior (Be), at the same time derives the structural behavior (Bs) from the feasible structure (S), compares the expected behavior with the structural behavior, and represents the design description (D). Additionally, the FBS framework proposes eight fundamental knowledge processes for all design activities (Figure 1).

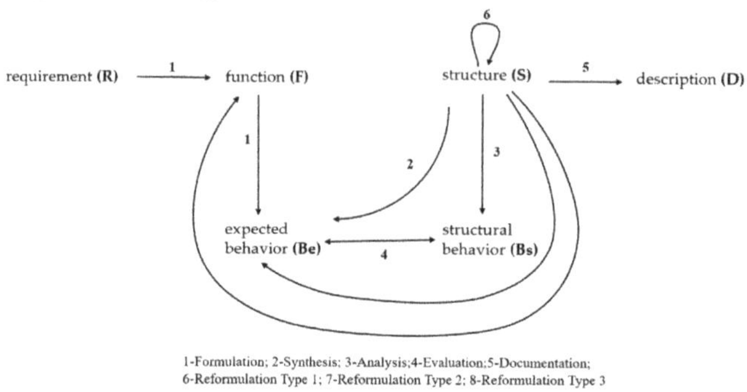

Figure 1. FBS Frameframework and fundamental knowledge processes.

2.2. Creative Segment Theory

In the sketching process, the designer constantly generates inspiration from multiple perspectives. Each acquisition of inspiration is regarded as a knowledge process and provides possibilities for new design solutions. Thus, the process of sketching consists of multiple segments. In the Creative Segment Theory, Creative Segments (CS) are used as nodes to describe the sketching process, and a tree structure is adopted to describe the relationship between creative activities [11]. Usually, a designer's sketching process contains more than one creative segment, alternating between finding ideas and expressing them. During sketching, designers have different knowledge processes and applications and involve various design information. When finding new ideas, designers constantly stimulate their minds to explore the influence of design knowledge. When expressing ideas, designers elaborate and evaluate newly generated ideas [8]. After the creative segments are divided, it is better understood how knowledge supports designers' creative activities and builds knowledge patterns.

3. Research Method

To understand the needs and use of design knowledge by novice designers, we studied the knowledge process of product sketching through an experimental design. We recruited two sophomore students majoring in product design. Their average age was 18.5 years old. They received one year of fundamental training in product design but had not participated in product practice projects. They were regarded as novice designers. We refer to them as Designer 1 and Designer 2 hereafter.

3.1. Experiment Design

The experiment design required the two designers to design a "Better Reading Lamp" as their topic. A product idea was provided. The maximum time for design was limited to 45 min. Designers used paper and pens for their sketching. In this experiment, the sketching processes of the two designers did not interfere with each other. They found the development direction of the product concept according to their points of view and recorded the derivation process of the product concept until they finally came up with the most satisfactory one (Figure 2).

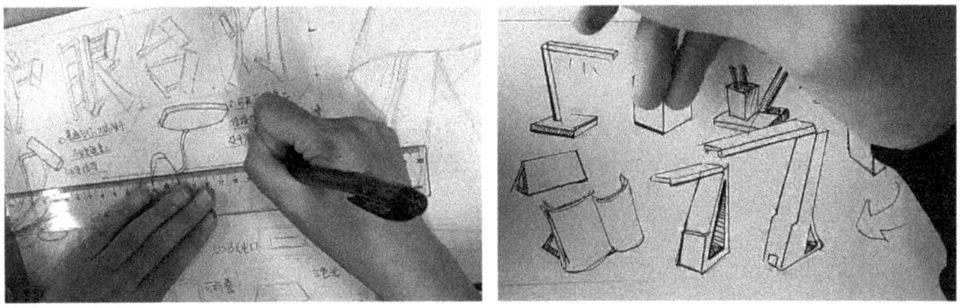

Figure 2. Sketching process of Designers 1 and 2 (the Chinese in the Figure has no effect in understanding the experiment).

3.2. Data Collection and Coding

The two designers were required to report their ideas during the sketching process. Thinking Aloud was used for protocol analysis. This method required designers to verbally report their thinking while completing the design task. The design protocols reflected what the designer saw, thought, and did at the cognitive level and could be objectively restored in the time series of the designer's thinking [12]. The video equipment was used to record the audio-visual data of the designers. After the experiment, we processed the data by segmenting and coding and discussed the results to obtain consistent results.

4. Results and Discussion

4.1. Creative Segment of Sketching Process

Through manual identification, nine Creative Segments were identified in the sketch data of the two designers. The Creative Segment Trees displayed the path from the conventional to the most satisfactory lamp (Figure 3).

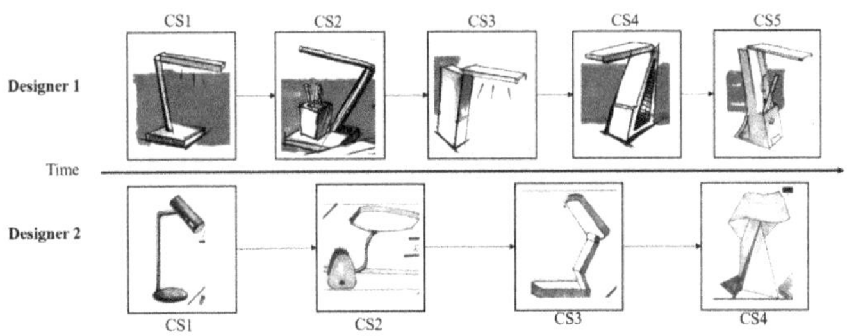

Figure 3. Creative Segment Tree of sketching process.

Under the design requirement, Designer 1 first thought of a table lamp. Since most of the reading behavior occurs at the desk, the table lamp needs to be stably placed on the desk. CS 1 consisted of a head, a body, and a base. The lamp head could be folded to adjust the range of light irradiation up and down. Designer 1 compared this lamp with the conventional table lamp used in daily life and thought a multi-function table lamp would be better. CS2 had a large volume pen holder on the lamp base, and its body could be adjusted. Considering that the pen holder would take up too much space on the desktop and would easily clash with the body, Designer 1 changed the overall shape of the lamp into CS3 which had a foldable lamp head and the pen holder below the lamp head. Designer 1 found that the shape of CS3 was not stable when the lamp head was unfolded, so the square-shaped support was changed into a more stable triangle one, and the storage space of the pen holder was reduced (CS4). Designer 1 reduced the volume of the supporting structure, making CS5 a more unique and stable Z-shape, and added buttons and other details.

At the beginning of sketching, Designer 2 also thought that the most common object suitable for reading was a table lamp, whose light source could be adjusted (CS1). By replacing the lamp base with a clamp structure, Designer 2 proposed CS2, which could be clamped on the table. Designer 2 thought that reading would not only take place at home, so a portable lamp was better. By associating with the M-shaped folding structure seen in life, Designer 2 puts forward a small-size CS3 which was portable but not light enough. CS4 simulated the structure of a paper airplane. The body of CS4 was changed from a block to a surface and folded to make the lamp portable and stable.

4.2. Analysis of Design Knowledge Needs

Based on the FBS ontology, 251 codes were extracted from the design protocols: 119 from Designer 1 and 132 from Designer 2. The data distribution is shown in Table 1. In terms of the proportion of the number of codes, F was mentioned the least. Given R, both novice designers relied on their life experiences to find out potential product functions. Behavior variables in the FBS model were classified as Be and Bs. In the sketching process, designers usually compare the two variables to evaluate the design solutions. Among all the FBS codes, knowledge of product structure was the most mentioned in the sketching process, accounting for nearly half of the data of Designer 1 and about one-third of the data of Designer 2. The two novice designers spent more time in the later stage of sketching to use the accumulated structural knowledge and frequently hesitated over which structure could be used for a more satisfactory solution. Therefore, novice designers needed to be supported in accessing and filtering effective knowledge on product structure.

Table 1. FBS codes of Designers 1 and 2.

Code	Designer 1	Designer 2
F	5 (4.2%)	9 (6.8%)
Be	19 (16.0%)	26 (19.6%)
Bs	23 (19.3%)	27 (20.5%)
S	62 (52.1%)	47 (35.6%)
D	10 (8.4%)	23 (17.4%)
Total	119 (100%)	132 (100%)

4.3. Analysis of Knowledge Processes

After being given the design task, the two designers first understood the design requirement, then searched for existing cases in life to determine the functions of the first product, then developed design ideas, and enhanced the product details. To further understand the knowledge processes based on different creative segments, we presented the relationships between F, Be, Bs, and S in time series according to the FBS framework, as shown in Figures 4 and 5.

During Designer 1's sketching process, analysis (S→Bs) was present in every Creative Segment. Structure, as core knowledge, mostly appeared between adjacent segments from CS2 to CS5. Designer 1 explored the reformulation of the lamp body structure to the next one many times (Reformulation Type 1, S→S′). Formulation (F→Be) appeared in the early CS1, CS2, and CS5, which represented a product function breakthrough in the designer's sketching process. Evaluation (Be↔Bs) appeared mainly in the last three Creative Segments, and the interaction of behavioral knowledge and structural knowledge finally led to the most satisfactory lamp for Designer 1.

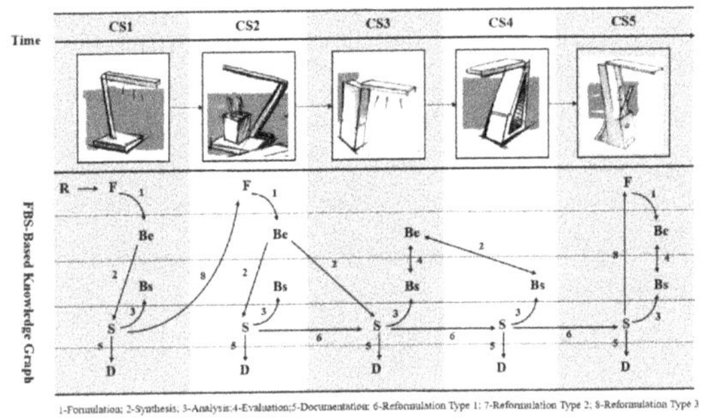

Figure 4. Knowledge graph of Designer 1's sketching.

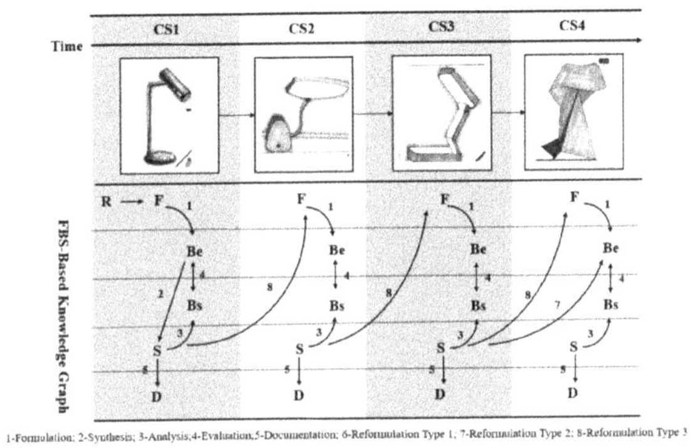

Figure 5. Knowledge graph of Designer 2's sketching.

Designer 2 conducted frequent Formulation (F→Be), Analysis (S→Bs), and Evaluation (Be↔Bs) in the four Creative Segments. This pattern of using knowledge was different from that of Designer 1. Designer 2 used Reformulation Type 3 (S→F) between Creative Segments, leading to four significant breakthroughs. The lamp that Designer 2 is most satisfied with is proposed according to life experience. In CS4, Designer 2 evaluated the expected behavior and structural behavior (Be↔Bs), while enriching the functional details of a product concept.

5. Conclusions

In the process of product conceptual design, knowledge supports the thinking of designers which is important in design education. Knowledge in the design process is usually the information stored, organized, and integrated by the designer in memory. We used the FBS ontology and the Creative Segment Theory to analyze design knowledge needs and the sketching process. In the early stages of product design, sketches aided designers in thinking and focusing a large amount of knowledge. Novice designers lack strategic design knowledge and design practices for the sketching process. We found that novice designers had common knowledge needs in the sketching process. They could quickly access the functional knowledge of each solution according to the design requirements. However, to promote the innovation of the solution, they needed to master the appropriate structural knowledge, which is a major concern in design education. Novice designers could develop the habit of using knowledge during the sketching process. The knowledge process with the connection point between creative segments would promote the inheritance relationship of product features. The innovation process of product development with the core knowledge leads to significantly different product features. However, both approaches allowed novice designers to find the ultimate design solution.

Author Contributions: Conceptualization, C.L. and S.X.; methodology, C.L.; software, S.X.; validation, C.L. and S.X.; formal analysis, C.L.; investigation, C.L.; resources, C.L. and S.X.; data curation, C.L.; writing—original draft preparation, C.L. and S.X; writing—review and editing, C.L. and S.X; visualization, C.L. and S.X; supervision, S.X.; project administration, C.L.; funding acquisition, C.L. and S.X. All authors have read and agreed to the published version of the manuscript.

Funding: This research was supported by the School of Design and Art, Beijing Institute of Technology, Zhuhai (ZX-2020-031).

Institutional Review Board Statement: Not applicable.

Informed Consent Statement: Not applicable.

Data Availability Statement: Not available due to privacy restrictions.

Acknowledgments: We would like to extend our gratitude to the School of Design for their generosity to lend us the experimental venue and to the two subjects who took time out of their busy schedule to participate in the experiment, without whom the research would not have been completed successfully.

Conflicts of Interest: The authors declare no conflict of interest.

References

1. Bertola, P.; Teixeira, J.C. Design as a knowledge agent: How design as a knowledge process is embedded into organizations to foster innovation. *Des. Stud.* **2003**, *24*, 181–194. [CrossRef]
2. Tu, J.W.; Li, Y.; Li, W.Q.; Xiong, Y. Knowledge retrieval model and implementation for product innovative design. *Comput. Integr. Manuf. Syst.* **2013**, *19*, 300–308.
3. Kavakli, M.; Gero, J.S. Strategic knowledge differences between an expert and a novice designer. In *Human Behaviour in Design*; Springer: Berlin/Heidelberg, Germany, 2003; pp. 42–52.
4. Van Aken, J.E. Valid knowledge for the professional design of large and complex design processes. *Des. Stud.* **2005**, *26*, 379–404. [CrossRef]
5. Ponn, J.; Lindemann, U.; Diehl, H.; Müller, F. Sketching in Early Conceptual Phases of Product Design: Guidelines and Tools. In Proceedings of the Eurographics Workshop on Sketch-Based Interfaces and Modeling, Grenoble, France, 30–31 August 2004.
6. Taborda, E.; Chandrasegaran, S.K.; Kisselburgh, L.; Reid, T.; Ramani, K. Enhancing visual thinking in a toy design course using freehand sketching. In Proceedings of the ASME 2012 International Design Engineering Technical Conferences & Computers and Information in Engineering Conference, Chicago, IL, USA, 12–15 August 2012.
7. Tang, H.H.; Lee, Y.Y.; Gero, J.S. Comparing collaborative co-located and distributed design processes in digital and traditional sketching environments: A protocol study using the function–behaviour–structure coding scheme. *Des. Stud.* **2011**, *32*, 1–29. [CrossRef]
8. Chen, S.; Yang, Z.Y.; Sun, L.Y.; Lou, Y. Research on design knowledge analytical method during sketching-combining acoustic energy feature and creative segment theory. *J. Zhejiang Univ. (Eng. Sci.)* **2015**, *49*, 2073–2082.
9. Gero, J.S.; Kan, J.; Purmohamadi, M. Analysing Design Protocols: Development of Methods and Tools. In Proceedings of the ICoRD'11 International Conference on Research into Design, Bangalore, India, 10–12 January 2011.
10. Gero, J.S.; Kannengiesser, U. The function-behaviour-structure ontology of design. In *An Anthology of Theories and Models of Design: Philosophy, Approaches and Empirical Explorations*; Chakrabarti, A., Blessing, L.T.M., Eds.; Springer: New York, NY, USA, 2014; pp. 263–283.
11. Sun, L.; Xiang, W.; Chai, C.; Wang, C.; Huang, Q. Creative segment: A descriptive theory applied to computer-aided sketching. *Des. Stud.* **2014**, *35*, 54–79. [CrossRef]
12. Sarkar, P.; Chakrabarti, A. A support for protocol analysis for design research. *Des. Issues* **2013**, *29*, 70–81. [CrossRef]

Disclaimer/Publisher's Note: The statements, opinions and data contained in all publications are solely those of the individual author(s) and contributor(s) and not of MDPI and/or the editor(s). MDPI and/or the editor(s) disclaim responsibility for any injury to people or property resulting from any ideas, methods, instructions or products referred to in the content.

engineering proceedings

Proceeding Paper

Applying Linear Regression Analysis to Identify Willingness of Using Environment-Friendly Electric Motorcycle-Sharing for Tourism Activities: A Case Study of GoShare †

You-Jie Huang [1], Sing-Yu Jhuang [1], Yi-Shin Lin [2] and Ho-Yi Chan [1,*]

1 Department of Leisure Service Management, Chaoyang University of Technology, Taichung 413, Taiwan; a2470712@cyut.edu.tw (Y.-J.H.); s10926603@gm.cyut.edu.tw (S.-Y.J.)
2 Department of Living Sciences, National Open University, Nantou 247, Taiwan; yishin10106@gmail.com
* Correspondence: s11126601@gm.cyut.edu.tw
† Presented at the 3rd IEEE International Conference on Electronic Communications, Internet of Things and Big Data Conference 2023, Taichung, Taiwan, 14–16 April 2023.

Abstract: This study aimed to evaluate tourist characteristics, rental factors, non-rental factors, and use intention of shared motorcycles in tourism activities. Linear regression analysis was used to compare the differences and influences of variables. A convenience sampling survey method was adopted to investigate GoShare motorcycle-sharing service. Questionnaires were distributed to 271 respondents aged 20–29 years old who used motorcycles in tourism. With data descriptive statistics, *t*-test, one-way analysis of variance, and regression analysis were conducted, and four main results were obtained: (1) The respondents tended to follow a "Freestyle travel" type. (2) As a rental factor, "Environmental Efficiency" was the most important. (3) "Renting is not easy" was the most important reason not to rent a motorcycle. (4) Tourist characteristics and rental factors impacted use intention significantly. Therefore, the following suggestions were made in this study: it is necessary to (1) strengthen the promotion of the GoShare motorcycle-sharing service, (2) enhance the quality of the rental service, (3) improve the mobile application, and (4) focus on in-depth tourism and expand the services at scenic spots.

Keywords: motorcycle sharing; tourist characteristics; rental factor; use intention; linear regression analysis

Citation: Huang, Y.-J.; Jhuang, S.-Y.; Lin, Y.-S.; Chan, H.-Y. Applying Linear Regression Analysis to Identify Willingness of Using Environment-Friendly Electric Motorcycle-Sharing for Tourism Activities: A Case Study of GoShare. *Eng. Proc.* **2023**, *38*, 57. https://doi.org/10.3390/engproc2023038057

Academic Editors: Teen-Hang Meen, Hsin-Hung Lin and Cheng-Fu Yang

Published: 28 June 2023

Copyright: © 2023 by the authors. Licensee MDPI, Basel, Switzerland. This article is an open access article distributed under the terms and conditions of the Creative Commons Attribution (CC BY) license (https://creativecommons.org/licenses/by/4.0/).

1. Introduction

According to the Survey of Travel by R.O.C Citizens of the Tourism Bureau 2020 of the Ministry of Transportation and Communications, the majority (71.0%) of the young adults had a road trip in their automobiles, with riding motorcycles ranked fifth (7.5%) compared to the MRT (9.2%) and the Taiwan Railway (8.4%). The number of people aged 18–30 years old using motorcycles during their trips was significantly higher than that of the other age groups, as young people commonly rode motorcycles for tourism activities [1]. Most people use their automobiles and motorcycles for tourism activities. Using motorcycle-sharing services for tourism activities reduces carbon emissions and avoids the waste of resources. However, tourists' habits of using transportation are not easily changed. Studies have found that tourism characteristics, payment methods, energy saving, carbon reduction, and comfort level were all important factors in choosing transportation methods [2,3]. *Commonwealth Magazine* [4] wrote that people between 20–30 years old had the second highest use intention of motorcycle-sharing services. The statistics of the Ministry of Transportation and Communications have shown that the rate of owning automobiles and motorcycles by people aged between 20–30 years old was lower than that of people aged 31–39 years old [5]. Thus, people aged between 20–30 years old are more likely to use motorcycle-sharing services. GoShare has tried to develop its business in tourism among the three major motorcycle brands in

Taiwan. In this study, the use intention of GoShare for tourism activities was investigated to estimate the potential market for the age group of 20–30 years olds based on tourism characteristics and rental factors using linear regression analysis.

2. Literature Review

2.1. Sharing Economy

In the original sharing economy system, idle resources are rented and returned. To make sharing economy profit-making, companies try to make a service specifically for the public to share and use [6].

2.2. Tourist Characteristics

The World Tourism Organization (1991) defines a tourist as a person who travels within his or her residence or anywhere else, regardless of the purpose and means of transport. Schneider explored the motivation of adventure tourists and found that there were significant differences between active adventure tourists and moderate adventure tourists [7]. Wu studied mutual relations among personality traits, travel motivations, travel constraints, and travel intentions for senior citizens in Kaohsiung City. Personality traits were categorized as affinity, conscientiousness, extroversion, neuroticism, and open-mindedness. The study found that personality traits significantly impacted travel motivation and behavior intention [8].

This study aims to explore different travelers' current situations and characteristics in using the GoShare motorcycle-sharing service in tourism activities. Referring to Hsieh et al., 10 questions such as "Self-planning", "Degree of Innovation", "Experience Depth", "Travel Pace", "Time Concept", "Tourism Risk", "Sense of Direction", "Saving", "Physical Load", and "Energy Conservation and Carbon Reduction" were included in the questionnaire [2].

2.3. Motorcycle-Sharing Service Rental Factor

Hsu and Chang investigated the Taiwan motorcycle-sharing service market to find out the relationship between consumers' product involvement and value proposition recognition. Their results showed that consumers paid more attention to convenience and battery life, especially female consumers [9]. Chio et al. studied and analyzed the heterogeneity of the selection behavior of travel carriers, and they discovered that the place of living affected the use of the carrier [3]. For rental factors for the GoShare motorcycle-sharing service participating in tourism activities, current research mostly focuses on the purchase of electric locomotives. Based on the research of Chang, Lin, Hsu, and Lu on the factors of renting a shared scooter, we added the elements of tourism activities to make appropriate adjustments, dividing the GoShare motorcycle-sharing service into four factors: environmental efficiency, product attribute, convenience, and perceived value for participating tourism activities [10].

2.4. Use Intention

Fishbein and Ajen defined use intention as the subjective likelihood that one performs, and found a correlation between use intention and actual behavior [11]. Hsieh et al. found that travelers with a higher degree of innovation, depth of experience, time planning, and physical load had a higher use intention [2]. Chuang et al. found that playfulness, aesthetics, and excellence effectively increased purchase intention [12]. We explored the use intention for GoShare motorcycle-sharing service in tourism activities, defining use intention as tourists' willingness based on Fishbein and Ajzen's definition of the use of intentions, adjusted and measured as appropriate for this study.

3. Research Methodology

3.1. Research Subjects

Questionnaires were distributed to people aged 20–29 who had used motorcycles for tourism activities. An online questionnaire survey was conducted using a convenient sampling method on LINE and FACEBOOK. The survey was carried out from 6 April to 4 July 2019. A total of 403 questionnaires were distributed, and 132 incomplete answers were deducted, while 271 valid questionnaires were recovered with an effective recovery rate of 67.20%.

3.2. Research Method

GoShare motorcycle-sharing service is commonly used for daily transportation or emergency vehicles. To explore consumer's adoption behavior and use intention for GoShare motorcycle-sharing service during tourism, we used the characteristics of tourists and rental factors to interpret and predict consumer intentions. Research Framework (Figure 1) is listed below.

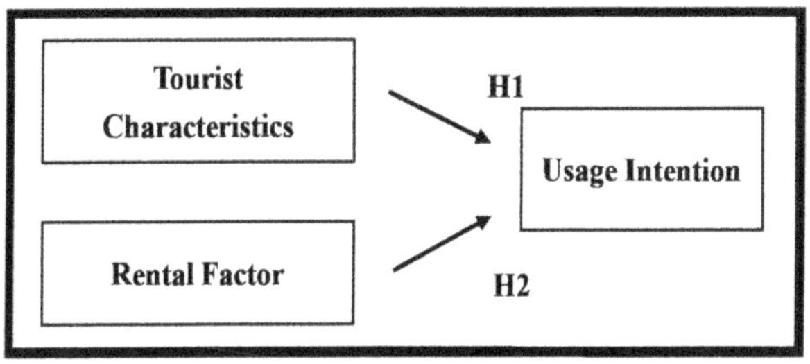

Figure 1. Research framework.

According to the above research framework, two research assumptions were proposed.

H1: *Tourist characteristics has a significant impact on use intention.*

H2: *Rental factor has a significant impact on use intention.*

4. Results

4.1. Tourist Characteristics on Use Intention

Tourist characteristics had a significant effect on use intention (F = 30.749, $p = 0.000 < 0.001$). Freestyle travel ($\beta = 0.289$, $p = 0.000 < 0.01$) and enthusiasm ($\beta = 0.289$, $p = 0.000 < 0.001$) had a significant positive effect on use intention. Freestyle travel was more influential than enthusiasm, which means that travelers were more likely to use the GoShare motorcycle-sharing service in tourism activities. Planners did not have any significant effect as GoShare motorcycle-sharing service existed anywhere and it took time to locate motorcycles, which delayed the trip. Freestyle travel and enthusiasm did not force travelers to follow the planned schedule and handle unexpected situations. Therefore, such travelers were more accepting and tried the GoShare motorcycle-sharing service. The explained variance for tourist characteristics on use intention was 0.285, which explained use intention by 28.5%. Regression analysis of tourist characteristics each dimension on use intention (Table 1) is listed below.

Table 1. Regression analysis of tourist characteristics each dimension on use intention.

	Unstandardized Coefficient β	Standardized Coefficient β	T	VIF
(constant)	2.516		9.693	0.000
Freestyle	0.252	0.289	4.173 ***	1.553
Planner	0.057	0.071	1.119	1.288
Enthusiasm	0.220	0.271	3.838 ***	1.616
	F = 30.749 ***		$R^2 = 0.285$	

*** Standardized loadings are significant at $p < 0.001$.

4.2. Rental Factors on Use Intention

Rental factors had a significant effect on use intention (F = 76.073, $p = 0.000 < 0.001$). Environmental efficiency ($\beta = 0.219$, $p = 0.000 < 0.001$) and product attribute ($\beta = 0.482$, $p = 0.000 < 0.01$) had a significant positive effect on use intention. Product attribute had a greater impact than environmental efficiency. Travelers were more likely to use the GoShare motorcycle-sharing service because of them. Product attribute showed the lowest recognition in the rental factor, but the regression analysis result showed that its impact was the largest. This indicated that the respondents were concerned about the battery life of rented motorcycles. However, the recognition rate was relatively low. Therefore, improving the product attribute effectively increased use intention. Convenience and perceived value did not have any significant effect. The explained variance for rental factors on use intention was 0.570. Regression analysis of rental factors each dimension on use intention (Table 2) is listed below.

Table 2. Regression analysis of rental factors each dimension on use intention.

	Unstandardized Coefficient β	Standardized Coefficient β	T	VIF
(constant)	1.118		4.717	0.000
Environmental Efficiency	0.214	0.219	3.596 ***	1.984
Product Attribute	0.441	0.482	6.450 ***	2.987
Convenience	0.055	0.058	0.825	2.658
Perceived Value	0.074	0.082	1.356	1.952
	F = 76.073 ***		$R^2 = 0.570$	

*** Standardized loadings are significant at $p < 0.001$.

4.3. Tourist Characteristics and Rental Factors on Use Intention

Using regression analysis, the relationship between tourist characteristics, rental factors, and use intention was analyzed. They had a significant impact (F = 147.748, $p = 0.000 < 0.001$). Tourist characteristics ($\beta = 0.159$, $p = 0.003 < 0.001$) and rental factors ($\beta = 0.712$, $p = 0.000 < 0.001$) had a significant impact on use intention. The influence of rental factor was greater than tourist characteristics. Rental factor showed a greater impact on using the motorcycle-sharing service in tourism activities. Tourist characteristics had less influence than the rental factor. By improving the quality of the GoShare motorcycle-sharing service, use intention will increase. Regression Analysis of Tourist Characteristics and Rental Factors on use intention (Table 3) is listed below.

Table 3. Regression analysis of tourist characteristics and rental factors on use intention.

	Unstandardized Coefficient β	Standardized Coefficient β	T	VIF
(constant)	0.733		2.946	0.000
Tourist Characteristics	0.159	0.155	3.022 **	1.388
Rental Factor	0.712	0.655	12.763 **	1.388
	F = 147.748 ***		$R^2 = 0.550$	

Standardized loadings are significant at ** $p < 0.01$, *** $p < 0.001$.

5. Conclusions

Based on the research results, young adults' perception of tourist characteristics, rental factors, and use intention of the GoShare motorcycle-sharing service in tourism activities was summarized as follows.

5.1. Summary

5.1.1. Tourist Characteristics

The characteristics of tourists are grouped into three types: freestyle travel, planning, and enthusiasm. The respondents who led a companion during the trip thought they could plan the trip and handle unexpected incidents. The conservative respondents showed no intention to invest in tourism and cared more about the process than the destination.

5.1.2. Rental Factor

There are four types of rental factors: environmental efficiency, product attribute, convenience, and perceived value. Environmental efficiency was the major rental factor of the motorcycle-sharing service. As the awareness of environmental protection is prevalent, using electric energy is regarded as a power source without emitting gases and saves energy in tourism. However, product attributes did not have much recognition as rental factors. The respondents believed that the appearance and services of the GoShare motorcycles were not important factors for rent.

5.1.3. Tourist Characteristics on Use Intention

Users with freestyle travel and enthusiasm were more willing to use the GoShare motorcycle-sharing service. The selling point of the GoShare motorcycle-sharing service is renting and returning anywhere. Thus, users were not afraid of unexpected situations and could enjoy themselves without worrying about times for arrival and departure at destinations.

5.1.4. Rental Factors on Use Intention

Rental factors of environmental efficiency and product attributes had a significant impact on the use intention of the GoShare motorcycle-sharing service. Product attributes affected use intention more than environmental efficiency. The GoShare motorcycles had an attractive appearance with low noise, energy saving, and carbon reduction, which effectively increase use intention.

5.2. Relevant Recommendations

Relevant recommendations for the industry were suggested as follows based on the research results.

5.2.1. Promotion of Environmental Significance

The respondents were willing to use the GoShare motorcycle-sharing service in tourism activities to save energy and reduce carbon. Thus, the industry must strengthen the promotion of environmental protection significance.

5.2.2. Improvement of Product and Rental Services

The results of the study showed that the GoShare motorcycle-sharing service had a greater impact on use intention, but fewer respondents used it before. The appearance, battery life, and rental service need to be improved.

5.2.3. Improvement of Mobile Application

The respondents indicated that it was difficult to use due to the limited number of motorcycles. The GoShare motorcycle-sharing service mobile application needs to locate motorcycles with GPS information for increasing the intention of use of the GoShare motorcycle-sharing service.

5.2.4. In-Depth Tourism and Service at Scenic Spots

The respondents who are willing to use pursued freestyle travel and had enthusiasm. Although the GoShare motorcycle-sharing service promotes urban use, it must include attractions in each city accessible by motorcycles in order to make relevant tourist maps or activities. This will attract tourists in traveling. In addition, the expansion of rental services in tourism can also solve the problem of transportation in and to tourist destinations where there are not many cars or charging stations.

Author Contributions: Conceptualization, Y.-J.H. and S.-Y.J.; methodology, Y.-J.H.; validation, S.-Y.J., H.-Y.C. and Y.-S.L.; formal analysis, Y.-J.H., H.-Y.C. and S.-Y.J.; investigation, S.-Y.J. and H.-Y.C.; writing—original draft preparation, S.-Y.J. and Y.-S.L.; writing—review and editing, Y.-J.H. and H.-Y.C.; visualization, H.-Y.C.; supervision, Y.-J.H.; project administration, Y.-J.H. All authors have read and agreed to the published version of the manuscript.

Funding: This research received no external funding.

Institutional Review Board Statement: Not applicable.

Informed Consent Statement: Not applicable.

Data Availability Statement: The data presented in this study are available on request from the corresponding author.

Conflicts of Interest: The authors claim no conflict of interest.

References

1. Tourism Bureau, Ministry of Transportation and Communications. 2020 Survey of Travel by R.O.C Citizens. Available online: https://srda.sinica.edu.tw/datasearch_detail.php?id=3342 (accessed on 17 November 2021).
2. Hsieh, C.H.; Feng, C.M.; Lin, W.C. Modal Choice Preference in Scenic Areas-A Case of Sun Moon Lake Route, Taiwan Tourist Shuttle Bus. *J. Chin. Inst. Transp.* **2014**, *26*, 35–62.
3. Chiou, Y.C.; Wen, C.H.; Chen, W.Y.; Wang, M.H.; Fu, C.; Tseng, H.M.; Lin, Y.S. Modeling Mode Choice Behaviors by Using Mixed Logit Models. *Transp. Plan. J.* **2014**, *43*, 143–172.
4. CommonWealth Magazine. Analysis of Sharing Locomotive Using Behavior. Available online: https://magazine.feg.com.tw/magazine/en/magazine_detail.aspx?id=12141 (accessed on 23 October 2021).
5. Department of Statistics, Ministry of Transportation and Communications. *Commonly Used Transportation Statistics*; Department of Statistics, Ministry of Transportation and Communications: Taipei City, Taiwan, 2014.
6. McKercher, B.; Prideaux, R.; Cheung, C.; Law, R. Achieving voluntary reductions in the carbon footprint of tourism and climate change. *J. Sustain. Tour.* **2010**, *18*, 297–317. [CrossRef]
7. Schneider, P.P. Exploring the motivation and personality traits of 85 adventure travelers: A hierarchical model approach. *Eur. J. Tour. Res.* **2012**, *5*, 72–75. [CrossRef]
8. Wu, S.T. The Study of Mutual Relations among Personality Traits, Travel Motivations, Travel Constraints, and Travel Intentions for Senior Citizens in Kaohsiung City. *J. Hosp. Tour.* **2019**, *16*, 55–79.
9. Hsu, C.C.; Chang, W.C. Discussion on the Relationship between Consumers' Product Involvement and Value Proposition Recognition: Taking Taiwan's electric locomotive market as an example. *Ind. Des.* **2017**, *125*, 51–56.
10. Chang, C.I.; Lin, Y.H.; Hsu, Y.L.; Lu, C.C. Research on the Factors of Renting Shared Scooter. *Manag. Inf. Comput.* **2021**, *10*, 105–113.
11. Fishbein, M.; Ajzen, I. Belief, attitude, intention and behavior: An introduction to theory and research. *Philos. Rhetor.* **1975**, *10*, 130–132.
12. Chuang, P.T.; Chen, K.Y.; Lo, W.J. Impact of Customer Experience Value on Purchase Intention of Smart Motorcycle. *Manag. Inf. Comput.* **2018**, *7*, 149–158.

Disclaimer/Publisher's Note: The statements, opinions and data contained in all publications are solely those of the individual author(s) and contributor(s) and not of MDPI and/or the editor(s). MDPI and/or the editor(s) disclaim responsibility for any injury to people or property resulting from any ideas, methods, instructions or products referred to in the content.

engineering proceedings

Proceeding Paper

Marketing Education through PBL: A Case of Social Media Influencer Marketing †

Serhan Demirci, Kuan-Yin Lee * and Ahmad Muzakki Binamak

Department of Marketing and Logistics Management, Chaoyang University of Technology, 168, Jifeng E. Rd., Wufeng District, Taichung 413310, Taiwan
* Correspondence: kylee@cyut.edu.tw

† Presented at the 3rd IEEE International Conference on Electronic Communications, Internet of Things and Big Data Conference 2023, Taichung, Taiwan, 14–16 April 2023.

Abstract: Project-based learning (PBL) has been widely implemented in higher education under various disciplines. In this study, sophomore-year marketing students' perception of PBL-based course outcomes was analyzed to explore implications of PBL for better student learning and marketing education practices. Data were collected through students' written narratives at the end of the semester. The results suggest that the PBL methodology significantly increases student motivation and engagement, along with the successful application of learned skills in building social media influencer brands. However, teamwork and participation with group mates were found to be significant challenges for students. Recommendations for future work and practical implications are proposed in this article.

Keywords: PBL; social media; marketing; education

Citation: Demirci, S.; Lee, K.-Y.; Binamak, A.M. Marketing Education through PBL: A Case of Social Media Influencer Marketing. *Eng. Proc.* **2023**, *38*, 59. https://doi.org/10.3390/engproc2023038059

Academic Editors: Teen-Hang Meen, Hsin-Hung Lin and Cheng-Fu Yang

Published: 28 June 2023

Copyright: © 2023 by the authors. Licensee MDPI, Basel, Switzerland. This article is an open access article distributed under the terms and conditions of the Creative Commons Attribution (CC BY) license (https://creativecommons.org/licenses/by/4.0/).

1. Introduction

Recently, project-based learning (PBL) has become popular among various educational institutions, ranging from elementary schools to higher education [1]. Especially within higher education, designing and implementing a learning program based on a student's active participation and understanding has become mainstream [2]. PBL is described as a learning method that allows students to engage in a sustained, collaborative fashion and to focus on a specific project that is often organized around a driving question [3]. PBL not only helps students in building core competencies in subjects but also in developing generic skills such as communication, teamwork, and leadership [4]. Previous research showed that PBL is effective in increasing learning outcomes and cognitive growth in higher education [5]. Significant results were obtained in the vocational training of nursing students [6]. Various applications of the PBL method have been widely observed in engineering education as project-based learning has shown to be highly effective in subjects where practical training and application are essential [7,8].

This study aims to describe and summarize the educational activities performed in the course "Internet and Social Media Marketing" in the 2021 academic year at the Department of Marketing and Logistics Management, Chaoyang University of Technology, Taiwan. The course is for sophomores to introduce various social media platforms, encourage students to learn various marketing methods and strategies on these platforms, and increase the effectiveness of marketing activities. Using the social media influencer model, students are encouraged to establish their social media brands and practice various methods. The activities implemented are (1) building a social media brand, (2) developing a marketing plan, (3) planning a content calendar, (4) creating social media content, and (5) data analysis based on feedback.

The course, Internet and Social Media Marketing, provides students with practical experience in social media marketing. In building a social media brand, a social media influencer or KOL (key opinion leader) is invited for hands-on training in creating and managing social media accounts is the main axis of this course. Due to its nature, practical learning, creative and analytical thinking, and problem-solving are part of this process. For that reason, the PBL methodology is highly suitable in this context.

The motivation behind the project was to deliver a learning environment where predefined problems and situations were given. Students were encouraged to come up with solutions, ideas, and strategies. Through PBL methodology, the instructor resumed the role of a facilitator and a consultant in helping and guiding students throughout the semester as they built their social media brands. Thus, students took the role of the self-directed learner, which is a main characteristic of the PBL learning methodology [9]. Based on these goals, this study aimed to answer the following research questions: (1) based on the PBL method, how to develop a course curriculum for internet and social media influencer marketing course, (2) what the learning outcomes of PBL methodology are in marketing education, and (3) whether PBL method has a positive effect on student motivation and participation.

The article is organized as follows. Section 2 covers the current literature on PBL methodology and Section 3 summarizes the PBL methodology and learning activities implemented in this course. The results are presented in Section 4 and discussed in Section 5.

2. Literature Review

Problem-Based Learning

Problem-based learning refers to an instructional methodology where learners build knowledge through various inquiry-based activities, solving problems to accomplish a meaningful real-world result [10,11]. For that reason, it is also called project-based learning because learners go through a process of developing and delivering a project. Such an approach has shown to be valuable in providing hard and soft skills at the same time, which is rather challenging in higher education. Hard skills refer to cognitive knowledge, professional skills, theories, and methods directly related to the subject of the course [1]. On the other hand, soft skills refer to supplementary skills such as problem-solving and teamwork [12]. In a traditional learning environment where teachers are the transmitters of knowledge while students act as the receptors of knowledge, such skills are difficult to build [13].

The main features of PBL include a driving question, focus on deliverable outcomes, participation in educational activities, teamwork among students, using scaffolding techniques, and delivering tangible results [11]. Among these main features, building deliverable results is regarded as the most crucial one [14]. The creation of tangible results requires learners to find solutions, come up with ideas, creative-thinking, and application of knowledge during the process.

3. Materials and Methods

3.1. PBL Course Design

PBL is based on organizing learning activities around a project. In this methodology, there are five main features: (1) projects are central, not supplementary, (2) projects are based on problems or questions, (3) students are involved in constructive exploration, (4) projects are student-driven but teacher supervised, and (5) projects are based on real-life situations not academic.

Concerning these five axes, the PBL methodology also involves certain principles. First, it emphasizes long-term goals. For example, in this study, the course curriculum of the whole semester (i.e., 18 weeks) is allocated to PBL activities. Second, all activities are student-centered. Other than stating the main points, an instructor does not assume the role of lecturer but rather provides students with marginal exploration. Third, all learning activities are carried out in collaboration as a team where students are required to cooperate. Fourth, students are implicitly asked to have soft skills such as time management, communication, leadership, and problem-solving. Such soft skills are not required directly for the activities but are implicitly required to complete the given tasks. Fifth, students are required to deliver real-life outcomes, such as building a social media brand and creating and publishing social media posts regularly based on a content plan. Finally, projects involved the use of technology-based tools, which is also inevitably the main feature of a course about the internet and social media platforms.

Evaluation of student performance is one of the challenging aspects of PBL methodology as it involves teamwork and difficult to identify the effort of each student [15].

3.2. Curriculum Development

One of the goals of this study is to develop a new course curriculum based on the PBL methodology. The need for a new curriculum is rooted in the fact that PBL requires a teacher and learner to play different roles than in the traditional learning environment [16]. The PBL methodology requires a new approach as a whole, including course activities, course material, feedback mechanism, and evaluation. Thus, it is necessary to develop a new course curriculum.

First, the instructor redesigns the course outline to include new subjects (i.e., trending topics such as TikTok, podcasting, and NFT). As the topic of the course is based on the internet and social media, students are expected to be more interested and engaged. However, it is necessary to provide enough flexibility, opportunities, and space for the learners to have more opportunities to engage with the course subject.

Students enrolled in the course were organized into groups of 6–8 members to build a social media influencer brand from scratch. Each student had a different role and responsibility. Students needed to communicate and share the tasks that require management and leadership skills. Each group was asked to deliver social media content that was monitored and evaluated by the course instructor where the instructor only had to supervise and provide feedback and comments.

4. Results

The course curriculum was composed of 18 weeks of different activities where each group was required to create and develop social media influencer accounts. As a part of marketing education, students were asked to devise a marketing plan. Each group was asked to discuss and devise a business plan and a social media plan with measurable goals and metrics. After the instructor provided foundation knowledge, each group developed their marketing plan by discussing it in groups.

Later, students were asked to discuss in groups to explore what type of social media content could be used in building their brand. They were asked to research current trends and popular influencers in their domain and curate ideas that could be used in their social media marketing plan. In this way, each group formulated their channel's unique marketing mix. This activity was followed by another discussion about building social media content plan. Each group was asked to prepare a content calendar (Figure 1) to experience the important role of social media editor and to practice the role of social media account manager.

As the semester continued, each group established their social media influencer brand. Several groups chose to start YouTube channels (Figure 2), while others preferred Instagram or TikTok. Students had total control over and responsibility for their social media channels. The instructor provided feedback and guidance when needed.

The evaluation method applied in the course included a midterm oral presentation where each group shared the current state of their social media channels. A final written group report was written to create a "media kit", which was an executive summary of the brand commonly used by social media influencers. Other than these assignments, students were asked to fill out a survey to answer questions about their perception of the course, providing cognitive and emotional feedback. Meanwhile, students' participation in group discussions and attendance in class were recorded and monitored carefully by the instructor. Participation and teamwork are essential in the PBL methodology. Thus, students were frequently reminded and encouraged to come to the classroom on time and join group discussions as much as possible.

Figure 1. Social media content calendar of a group. Note: The Chinese in the Figure has no effect in understanding the outcome of this learning activity.

Figure 2. YouTube channel of a group. Note: The Chinese in the Figure has no effect in understanding the outcome of this learning activity.

A total number of 56 responses were collected ($n = 56$). A majority of students ($n = 48$) shared that "they believe this course helped them achieve their learn goals". Similarly, a significant group of students ($n = 37$) shared that "they feel motivated about this course". An interesting finding of the results was that many students ($n = 22$) expressed that they felt uncomfortable and had negative emotions due to problems in teamwork and cooperation with group mates. Teamwork is an important aspect of PBL and encourages students to work in groups. Based on our analysis of this study, we found that teamwork was the most challenging part of the course.

5. Discussion

PBL has been a popular learning method in higher education teaching different subjects. Most related studies showed the significant benefits of PBL. In this study, we explored how PBL methodology could be used in marketing education, specifically social media marketing. The results coincided with previous literature, suggesting PBL showed significantly higher motivation in students. Students were highly active and engaged all through the semester. This also evidenced from the attendance and participation being high throughout the semester. An interesting finding of this study is about students' perception of teamwork. As mentioned, many students complained about non-participating group members and were emotionally affected by the challenges brought by the necessity to work in groups. Teamwork is a core axis of the PBL methodology, so it is inevitably embedded in the course design. For many students, such a necessity might be a new experience that they did not have in previous courses. This finding underlines the benefit of the PBL methodology as it enables students to experience teamwork and face challenges to build soft skills.

6. Conclusions

The PBL methodology provides new opportunities in higher education. This study helps to understand how to employ PBL methodology in marketing education. First, our results presented that the course curriculum and group activities included in the course showed satisfactory results. Student participation and grades coincided with this result. Overall learning outcomes were satisfactory as expected, whereas the PBL methodology proved to be encouraging student participation and motivation.

Author Contributions: Conceptualization, S.D. and K.-Y.L.; methodology, S.D.; validation, K.-Y.L. and A.M.B.; formal analysis, S.D.; data curation, A.M.B.; writing—original draft preparation, S.D.; writing—review and editing, S.D. and A.M.B.; supervision, K.-Y.L.; project administration, S.D. All authors have read and agreed to the published version of the manuscript.

Funding: This research was funded by Chaoyang University of Technology.

Institutional Review Board Statement: Not applicable.

Informed Consent Statement: Not applicable.

Data Availability Statement: Not applicable.

Conflicts of Interest: The authors declare no conflict of interest.

References

1. Vogler, J.S.; Thompson, P.; Davis, D.W.; Mayfield, B.E.; Finley, P.M.; Yasseri, D. The hard work of soft skills: Augmenting the project-based learning experience with interdisciplinary teamwork. *Instr. Sci.* **2018**, *46*, 457–488. [CrossRef]
2. Prince, M. Does active learning work? A review of the research. *J. Eng. Educ.* **2004**, *93*, 223–231. [CrossRef]
3. Fincher, S.; Knox, D. The Porous Classroom: Professional Practices in the Computing Curriculum. *Computer* **2013**, *46*, 44–51. [CrossRef]
4. Rodríguez, J.; Laverón-Simavilla, A.; del Cura, J.M.; Ezquerro, J.M.; Lapuerta, V.; Cordero-Gracia, M. Project Based Learning experiences in the space engineering education at Technical University of Madrid. *Adv. Space Res.* **2015**, *56*, 1319–1330. [CrossRef]
5. Downing, K. Problem-Based Learning and the Development of Metacognition. *High. Educ.* **2009**, *57*, 609–621. [CrossRef]
6. Tiwari, A.; Lai, P.; So, M.; Yurn, K. Comparison of the Effects of Problem-based Learning and Lecturing on the Development of Students' Critical Thinking. *Med. Educ.* **2006**, *40*, 547–554. [CrossRef] [PubMed]
7. Dym, C.L.; Agogino, A.M.; Eris, O.; Frey, D.D.; Leifer, L.J. Engineering design thinking, teaching and learning. *J. Eng. Educ.* **2005**, *94*, 103–120. [CrossRef]
8. Mills, J.E.; Treagust, D.F. Engineering education—Is problem based or project-based learning the answer? *Australas. J. Eng. Educ.* **2003**, *3*, 2–16.
9. Allen, D.E.; Donham, R.S.; Bernhardt, S.A. Problem-based learning. *New Dir. Teach. Learn.* **2011**, *128*, 21–29. [CrossRef]
10. Brundiers, K.; Wiek, A. Do we teach what we preach? An international comparison of problem- and project-based learning courses in sustainability. *Sustainability* **2013**, *5*, 1725–1746. [CrossRef]
11. Krajcik, J.S.; Shin, N. Project-based learning. In *The Cambridge Handbook of the Learning Sciences*, 2nd ed.; Sawyer, R.K., Ed.; Cambridge University Press: Cambridge, UK, 2014.
12. Casner-Lotto, J.; Barrington, L. *Are They really Ready to Work? Employers' Perspectives on the Basic Knowledge and Applied Skills of New Entrants to the 21st Century U.S. Workforce*; Partnership for 21st Century Skills; 1 Massachusetts Avenue NW Suite 700E: Washington, DC, USA, 2006.
13. Alorda, B.; Suenaga, K.; Pons, P. Design and evaluation of a microprocessor course combining three cooperative methods: SDLA, PBL and CnBL. *Comput. Educ.* **2011**, *57*, 1876–1884. [CrossRef]

14. Helle, L.; Tynjälä, P.; Olkinuora, E. Project-based learning in post-secondary education—Theory, practice and rubber sling shots. *High. Educ.* **2006**, *51*, 287–314. [CrossRef]
15. Dutson, A.J.; Todd, R.H.; Magleby, S.P.; Sorensen, C.D. A review of literature on teaching engineering design through project-oriented capstone courses. *J. Eng. Educ.* **1997**, *86*, 17–27. [CrossRef]
16. Saunders-Smits, G.N.; de Graaff, E. The development of integrated professional skills in AE through problem-based learning in design projects. In Proceedings of the 2003 American Society for Engineering Education Annual Conference & Exposition, Nashville, TN, USA, 22–25 June 2003.

Disclaimer/Publisher's Note: The statements, opinions and data contained in all publications are solely those of the individual author(s) and contributor(s) and not of MDPI and/or the editor(s). MDPI and/or the editor(s) disclaim responsibility for any injury to people or property resulting from any ideas, methods, instructions or products referred to in the content.

engineering proceedings

MDPI

Proceeding Paper

Using Text Mining to Explore Acceptance of Zero-Touch Fintech Service Innovation: Internet Word-of-Mouth of R Digital Bank †

Shu-Chin Huang, Ching-Mei Zhu and Maw-Liann Shyu *

Department of Business Administration, Chaoyang University of Technology, Taichung 413310, Taiwan; schuang2@gm.cyut.edu.tw (S.-C.H.); a419002@gmail.com (C.-M.Z.)

* Correspondence: mlshyu@cyut.edu.tw; Tel.: +886-4-23323000 (ext. 4438)

† Presented at the 3rd IEEE International Conference on Electronic Communications, Internet of Things and Big Data Conference 2023, Taichung, Taiwan, 14–16 April 2023.

Abstract: By using the texts of Internet word-of-mouth, the most popular digital banking services and its users' perceptions of the usefulness and ease of use of these services were explored. Python was used to capture 2026 valid texts on the Industrial Square Banking Service Edition, covering the period of 2016−2020. The results show that (1) the digital banking Fintech provided digital and traditional services, (2) positive comments were seen more than negative comments for confirmation and perceived usefulness while negative comments were more than positive comments for satisfaction and continuance intention, and (3) satisfaction via Internet word-of-mouth was seen the most frequently.

Keywords: Internet community; Internet word-of-mouth; text mining; digital banking; fintech

Citation: Huang, S.-C.; Zhu, C.-M.; Shyu, M.-L. Using Text Mining to Explore Acceptance of Zero-Touch Fintech Service Innovation: Internet Word-of-Mouth of R Digital Bank. *Eng. Proc.* **2023**, *38*, 65. https://doi.org/10.3390/engproc2023038065

Academic Editors: Teen-Hang Meen, Hsin-Hung Lin and Cheng-Fu Yang

Published: 29 June 2023

Copyright: © 2023 by the authors. Licensee MDPI, Basel, Switzerland. This article is an open access article distributed under the terms and conditions of the Creative Commons Attribution (CC BY) license (https://creativecommons.org/licenses/by/4.0/).

1. Introduction

The pandemic drove people to change the way they used financial services. On the other hand, it encouraged the financial industry to launch new financial services in response to the current situation so that the various developmental trends in progress could accelerate its evolution and magnify its effects, thus bringing a mindset to reshape future development [1]. In other words, the epidemic not only promoted the financial industry to accelerate digital transformation but also expanded the scope of digital services. People were eager for zero contact and gradually developed the habit of using digital financial services. Technological advances spurred financial innovation and brought new opportunities. Unlike traditional banks based on branches, digital banks are fundamentally different in terms of their architecture, which is based on the Internet Protocol (IP) that facilitates the application of Buyology. Therefore, traditional banks must transform themselves into digital banks through Business Process Re-engineering (BPR). The core function of Fintech services is to help businesses and consumers better manage their financial operations and living consumption expenditures.

In addition, many empirical studies support the importance of Internet Word of Mouth (IWOM) in influencing consumer behavior or decision-making [2–4]. Both positive and negative word-of-mouth have different impacts. Positive word-of-mouth can attract consumers to discover new service providers and reinforce the perception that consumers have made the right choice. On the other hand, positive word-of-mouth helps companies reduce marketing expenses, attract new customers, increase revenue, and promote products and services [5].

The existing studies on the marketing value and importance of IWOM has attracted many scholars' interests. The main research issues fall into three categories: (1) exploring how consumers process and receive messages and the factors that influence consumers' processing of messages, (2) identifying the characteristics of word-of-mouth communicators and their motivations, and (3) focusing on how to measure the effectiveness of word-of-mouth [6]. However, the specific content of IWOM has yet to be discussed and clarified.

Clarifying the content of IWOM helps industry players and scholars better understand the information processing that influences consumers' decision-making and guide industry players to improve their services or products. Hence, this study's objective is to determine the positive and negative word-of-mouth on the Internet.

The factors that affect users' willingness to adopt new technology and continue to use it are even more important in the current industry environment, where emerging smart businesses are becoming increasingly advanced. Many scholars have proposed theoretical models and related empirical evidence to understand whether users continue to use the new technology after adoption. Among them, the theoretical model proposed by Bhattacherjee [7] has been cited or modified by many scholars and is regarded as a significant theoretical basis for the continuous adoption of new technology research.

Since in the information system (IS), the user's decision of continuous usage is similar to the consumer decision of repurchase, Bhattacherjee [7] proposed a Post-Acceptance Model of IS Continuance. The model emphasizes the discussion of the Post-Acceptance concept and argues that confirmation and satisfaction can already cover the effects of Pre-Acceptance. Additionally, it asserts that (1) confirmation has a positive effect on perceived usefulness after system acceptance, (2) user satisfaction is influenced by the degree of post-acceptance confirmation and perceived usefulness, and (3) user intention of IS continuance intention is influenced by post-acceptance satisfaction and perceived usefulness. In other words, confirmation, perceived usefulness, and satisfaction are important factors that influence IS continuance intention.

Unlike the previous quantitative validation method of the Post-Acceptance Model of IS Continuance which used questionnaires to examine the causal relationships among the concepts of confirmation, perceived usefulness, satisfaction, and continuance intention of using digital banking services, we focused more on the contents and meanings of the aforementioned research concepts in IWOM to identify the factors that influence users' continued use of digital banking services.

Based on the above, the objectives of this study are (1) to examine what digital banking services are the most popular among IWOM, (2) to explore what content issues are covered by positive and negative IWOM, and (3) to identify the importance of the factors that influence users' continuous use of digital banking systems (confirmation, perceived usefulness, satisfaction, and other concepts). The results provide the basic information on customer experience design and customer relationship maintenance to increase customer life-cycle values. The research process in this study is shown in Figure 1.

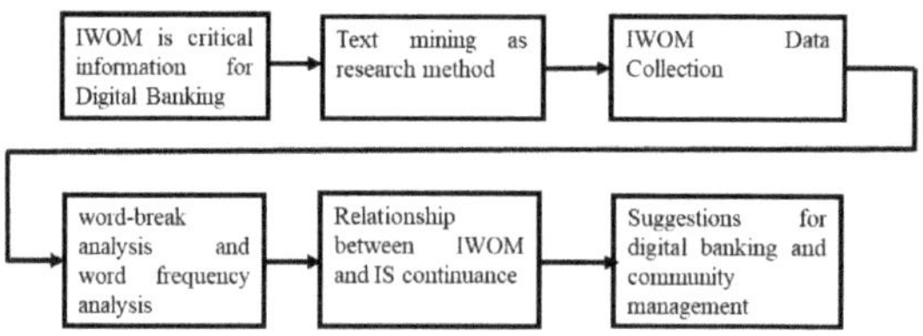

Figure 1. Research process.

2. Research Methodology

2.1. Data Collection

The texts collected in this study included 2026 texts related to R digital banking obtained from the banking service section of "PTT Bulletin Board System". The texts included article titles, posters' IDs, times, replies, pictures, and so on. The exclusion criteria of the articles included content not related to R Digital Bank (advertisement) and comments with non-textual content (such as trendy cyber words and emoticons).

2.2. Operational Definitions

The operational definitions of the research concepts (confirmation, perceived usefulness, satisfaction, and continuance intention) are summarized, and examples of positive and negative descriptions for each research concept are presented in Table 1.

Table 1. Operationalizing definitions of research concepts.

Concepts	Operationalizing Definitions
Confirmation	User perception of consistency between expectations and actual performance of digital banking information systems.
Perceived Usefulness	Users' perception that the use of digital banking systems contributes to the completion of specific tasks or brings benefits.
Satisfaction	The mental state of users after using the digital banking system
Continuance Intention	The user's intention to continue to use the digital banking system in the future

Source: Adapted from Ref. [8].

2.3. Analysis Process

This study was conducted to understand the experience of digital-banking users in using digital services in online word-of-mouth and to analyze the text by using the Chinese word-breaking system (CRTC) for the concepts of Confirmation, Perceived Usefulness, Satisfaction, Continuance Intention, and Service Items. By deleting unrelated conjunctions, pronouns, and other superfluous words, the number of occurrences of those concepts was calculated to determine the frequency of various concepts and their importance. In the actual analysis, all the texts were analyzed by word-break analysis, and then the word frequency analysis of each service item was estimated. Secondly, each type of word break was divided into positive and negative comments, and the idea attribution and frequency analysis were conducted concerning their meanings. Finally, the word frequency analysis of each research idea was carried out to understand the important ideas disseminated in the R digital bank's IWOM that influenced the IS continuance.

3. Results

3.1. Term Frequency Analysis of Popular Digital Banking Services in IWOM

To understand what services were more frequently discussed by the community in IWOM, we calculated their term frequency (TF). The results are summarized in Table 2. The results show that the most mentioned digital banking services in order are digital services, traditional banking services, savings/foreign exchange, marketing activities, apps, cards, interest rates, and currency.

Table 2. Term frequency analysis of various services of digital banking.

Service Items	Frequency	TF
Digital Services	5564	0.2266
Traditional Banking Services	4622	0.1882
Savings/Foreign Exchange	3524	0.1435
Marketing Activities	2998	0.1221
APP	2835	0.1155
Cards	1636	0.0666
Interest Rates	1332	0.0542
Currency	561	0.0228
Subtotal	23,072	0.9396

3.2. Term Frequency Analysis of Positive and Negative Comments in IWOM

To understand the relative importance of positive and negative comments on each research concept in IWOM, each research concept was classified into positive and negative comments, and the frequency and TF-IDF of the top three descriptors of positive and negative comments was calculated.

The analysis results showed that the descriptions of perceived usefulness appear most frequently in the positive comments, followed by satisfaction, confirmation, and continuance intention (Table 3). Satisfaction was the most frequent description in the negative comments, followed by confirmation, perceived usefulness, and continuance intention (Table 4). On the other hand, among the concepts, positive comments appeared more than negative comments regarding confirmation and perceived usefulness, while negative comments appeared more regarding satisfaction and continuance intention.

Table 3. Term frequency analysis of positive comments in IWOM.

Concepts	Rank	Descriptions	Frequency	TF
Confirmation	1	Easy	26	0.0011
	2	Okay	26	0.0011
	3	Not bad	19	0.0008
	Total		325	0.0132
Perceived Usefulness	1	Convenient	30	0.0012
	2	Novel	30	0.0012
	3	Enhancing	29	0.0012
	Total		325	0.0132
Satisfaction	1	Attitude	24	0.0010
	2	Happy	18	0.0007
	3	Relaxed	18	0.0007
	Total		227	0.0092
Continuance Intention	1	Looking forward to	18	0.0007
	2	Love it	5	0.0002
	2	Like it	5	0.0002
	Total		45	0.0018

The total represents the number of occurrences of the positive descriptions of each concept.

Table 4. Term frequency analysis of negative comments in IWOM.

Concepts	Rank	Descriptions	Frequency	TF
Confirmation	1	Error	105	0.0043
	2	Drawback	17	0.0007
	3	Overdue	16	0.0007
	Total		194	0.0079
Perceived Usefulness	1	Get no reply	25	0.0010
	2	Credit card fraud	17	0.0007
	3	Too little	7	0.0003
	Total		85	0.0035
Satisfaction	1	Worried	40	0.0016
	2	Whatever	24	0.0010
	3	Dumbfounded	23	0.0009
	Total		353	0.0144
Continuance Intention	1	Useless	19	0.0008
	2	Termination	16	0.0007
	2	Substitution	8	0.0003
	Total		51	0.0021

The total represents the total number of occurrences of the negative descriptions of each concept.

3.3. Term Frequency Analysis of Each Research Concept

To understand the relative importance of each research concept in IWOM, the frequency and TF-IDF of confirmation, perceived usefulness, satisfaction, and continuance intention are calculated in the study through word analysis (Table 5). The results show that the frequency and TF-IDF of satisfaction were the highest, followed by perceived usefulness, confirmation, and continuance intention.

Table 5. Term frequency analysis of research concepts.

Concepts	Frequency	TF
Satisfaction	580	0.0236
Perceived Usefulness	410	0.0167
Confirmation	398	0.0162
Continuance Intention	96	0.0039

Term frequency analysis (TF) represents the number of occurrences of the positive descriptions of each concept.

4. Conclusions

The research results show that the descriptions of perceived usefulness appeared most frequently in the positive comments, followed by satisfaction, confirmation, and continuance intention. Satisfaction was the most frequent description in the negative comments, followed by confirmation, perceived usefulness, and continuance intention. Moreover, positive comments appeared more than negative comments regarding confirmation and perceived usefulness, while negative comments appeared more than positive comments regarding satisfaction and continuance intention. Finally, satisfaction was the most frequent among the IWOM, followed by perceived usefulness, confirmation, and continuance intention.

This result is different from that of the previous research model of Post-Acceptance of IS Continuance. Most existing studies used questionnaires to collect data, and the research questions focused on the factors influencing the adoption of new technologies and the causal relationships among them to gain a deeper understanding of users' technology-continuance behavior. However, we used text mining as the research method to explore the content of IWOM from the text obtained. IWOM was one reason that consumers adopted or continued to use new technology services. Therefore, the method of this study contributes to finding out users' continued adoption of new technology and systems as well as the influence of the adoption or continuing use of new technology and systems by other users. Most survey methods used positive questions to explore the continuance experience of the respondents, while we included negative comments (i.e., negative experience). Thus, a complete description of the continuance experience was provided in this study.

Based on the study's results, two suggestions were made, including redesigning the Frequently Asked Questions and Answers and developing a user-oriented digital financial platform system re-optimization.

Discussion services of digital banking need to be provided to explore the needs of consumers from a large amount of user data, thus optimizing digital banking services. For example, bankers can use it to redesign Frequently Asked Questions and Answers to reduce the cost of personalized customer service and improve the efficiency of customer service. Negative comments give negative experiences for users of digital-banking services to continuously use the services and for services to be improved in digital banking. The negative comments for perceived usefulness and confirmation help explore the problems or difficulties encountered by service users during the operation of the service system. Therefore, it is suggested that digital bankers use the comments to develop a user-oriented platform and for its re-optimization. In addition, sharing positive comments may help digital bankers attract potential users to become actual customers and, therefore, to have reference information for various communication strategies to improve their positive images.

Author Contributions: Conceptualization, S.-C.H., M.-L.S. and C.-M.Z.; methodology, S.-C.H.; writing—original draft preparation, S.-C.H., C.-M.Z. and M.-L.S.; writing—review and editing, S.-C.H. and M.-L.S. All authors have read and agreed to the published version of the manuscript.

Funding: This research received no external funding.

Institutional Review Board Statement: Not applicable.

Informed Consent Statement: Not applicable.

Data Availability Statement: Not applicable.

Conflicts of Interest: The authors declare no conflict of interest.

References

1. Zhang, K.; Huang, C.Y. Trends in the Development of Financial Institutions in the Post-Epidemic Era. *Financ. Inf. Q.* **2022**, *3*, 154–196.
2. Lin, Y.-T.; Chen, Y.-C. The Impact of the Positive and Negative E-WOM on Consumers' Purchase Intention: The Moderation Effect of the Product Type and Product Involvement. *Electron. Commer. Stud.* **2015**, *13*, 167–193.
3. Chevalier, J.; Mayzlin, D. The effect of word of mouth on sales: Online book reviews. *J. Mark. Res.* **2006**, *43*, 345–354. [CrossRef]
4. Clemons, E.; Gao, G.; Hitt, L. When Online Reviews Meet Hyper Differentiation: A Study of the Craft Beer Industry. *J. Manag. Inf. Syst.* **2006**, *23*, 149–171. [CrossRef]
5. Reichheld, F.F.; Sasser, W.E. Zero defections: Quality comes to services. *Harv. Bus. Rev.* **1990**, *6*, 105–111.
6. Tsai, P.J.; Wu, Y.F.; Chuang, Y.H. Effect of internet interactivity on E-Word-of-Mouth: Comparison of Facebook and Youtube. *Commer. Manag. Q.* **2016**, *17*, 81–111.
7. Bhattacherjee, A. Understanding Information Systems Continuance: An Expectation-Confirmation Model. *MIS Q.* **2001**, *25*, 351–370. [CrossRef]
8. Mishra, A.; Shukla, A.; Rana, N.P.; Currie, W.L.; Dwivedi, Y.K. Re-examining post-acceptance model of information systems continuance: A revised theoretical model using MASEM approach. *Int. J. Inf. Manag.* **2023**, *68*, 102571. [CrossRef]

Disclaimer/Publisher's Note: The statements, opinions and data contained in all publications are solely those of the individual author(s) and contributor(s) and not of MDPI and/or the editor(s). MDPI and/or the editor(s) disclaim responsibility for any injury to people or property resulting from any ideas, methods, instructions or products referred to in the content.

engineering proceedings

Proceeding Paper

Applying Regression Analysis to Work Stress, Leisure Constraint and Leisure Participation of Nantou County Junior High School Teachers †

You-Jie Huang [1], Ming-Kuo Chen [2,*], Yu-Yu Lo [1] and Ho-Yi Chan [1]

1 Department of Leisure Service Management, Chaoyang University of Technology, Taichung City 413310, Taiwan; a2470712@cyut.edu.tw (Y.-J.H.); fish7295@yahoo.com.tw (Y.-Y.L.); s11126601@gm.cyut.edu.tw (H.-Y.C.)
2 General Education Center, Chaoyang University of Technology, Taichung City 413310, Taiwan
* Correspondence: mgchen@cyut.edu.tw

† Presented at the 3rd IEEE International Conference on Electronic Communications, Internet of Things and Big Data Conference 2023, Taichung, Taiwan, 14–16 April 2023.

Abstract: This research aims to discover the existing state of public junior high school teachers' work stress, leisure constraint, and leisure participation. A questionnaire investigation approach was adopted to collect data from public junior high school teachers in Nantou County. A total of 300 questionnaires were distributed, and 295 valid questionnaires were returned. Descriptive statistics and multiple regression analysis were carried out with the data, and the following results were obtained. Nantou County public junior high school teachers' work stress was at an upper-medium level. Student behavior had the most significant impact on work stress. Teachers' leisure constraint was at a lower-medium level. The highest leisure constraint was from the structural constraint. Teachers' leisure participation was at a lower-medium level, and the most frequent participation was entertainment activities. Work stress and leisure constraints had a significant effect on leisure participation. Conclusions and suggestions are provided for the reference of the junior high school teachers, junior high school, and educational institutions for future studies.

Keywords: junior high school teachers; work stress; leisure constraint; leisure participation

Citation: Huang, Y.-J.; Chen, M.-K.; Lo, Y.-Y.; Chan, H.-Y. Applying Regression Analysis to Work Stress, Leisure Constraint and Leisure Participation of Nantou County Junior High School Teachers. *Eng. Proc.* **2023**, *38*, 70. https://doi.org/10.3390/engproc2023038070

Academic Editors: Teen-Hang Meen, Hsin-Hung Lin and Cheng-Fu Yang

Published: 29 June 2023

Copyright: © 2023 by the authors. Licensee MDPI, Basel, Switzerland. This article is an open access article distributed under the terms and conditions of the Creative Commons Attribution (CC BY) license (https://creativecommons.org/licenses/by/4.0/).

1. Introduction

Since the 1990s, Taiwan has undergone major changes in education including class-size reduction, open education, Grade 1–9 curriculum changes, 12-year basic education, teacher evaluations for professional development, and positive discipline for 108 national curriculum guidelines. The role of a teacher is no longer simply to impart knowledge. Today's teachers not only have to face complicated teaching or administrative work but deal with the increasingly complex behavioral problems of students and communication between parents and teachers. In addition, adapting to the volatile education policy reform and the impact of a declining birthrate are causing work and mental pressure on teachers.

Chi [1] points out that teachers use leisure participation to relieve stress, improve interpersonal relationships, increase work efficiency, and achieve self-realization when faced with work stress. Wu and Liu [2] found that the overall work stress of primary school teachers in the greater Taipei area had different correlations for different leisure activities. Hsu et al. [3] found a positive correlation between work stress and leisure participation. The previous results showed that work stress and leisure participation affect each other. Huang et al. [4] found that the barriers to participation in leisure activities of junior high school teachers have intrapersonal and structural constraints. Ho [5] also found that the leisure constraint of Taiwanese teachers was positively correlated with leisure participation. Thus, work stress, leisure participation, and leisure constraints interact with each other.

Nantou County is located in the center of Taiwan and is a landlocked county including 13 rural areas. Rural schools were founded in every area. This makes Nantou County have the most rural schools in Taiwan [6]. Therefore, to understand the current situation of work stress, leisure participation, and leisure constraints among junior high school teachers in Nantou County, we analyzed the impact of teachers' work stress and leisure constraints on leisure participation.

2. Literature Review

2.1. Teacher Work Stress

Teaching is one of the professions with the highest amount of work stress [7]. The relevant research on teachers' work stress was first proposed [8]. Teachers' work stress comes from many aspects [9–16]. When teachers are engaged in teaching or administrative-related work, they are faced with expectations, requirements, and responsibilities assigned to their work beyond their tolerance. When individuals are unable to adapt, negative emotions such as anxiety, tension, anger, and frustration arise. Many influencing factors cause teachers' work stress. A teacher work stress scale was developed according to the sources of teachers' work stress: student behavior, parent–teacher communication, working engagement, interpersonal relationship, and professional knowledge [16,17].

2.2. Leisure Constraint

We adopted the views from refs. [18–23] to define teacher leisure constraints. Teachers engage in preferred leisure activities in their leisure time but are affected by internal or external factors, resulting in a reduction or inability to participate in leisure activities. This is defined as teacher leisure constraints. Leisure constraints are sorted out and summarized into three types of factors that hinder individuals' participation in leisure activities [18]: intrapersonal constraints, interpersonal constraints, and structural constraints [24–28].

2.3. Leisure Participation

Referring to previous studies [23,29–33], we define teacher leisure participation as the free choice and participation of teachers in their leisure time for the activities of interest. In this study, we used subjectivity analysis, factor analysis, and multidimensional scaling [27,34–36]. In recent years, subjectivity analysis has been used for the leisure participation of teachers. In this study, subjectivity analysis is used to divide leisure activity participation into six categories: outdoor, sports, social, entertainment, hobby, and knowledge [23,37,38].

2.4. Relationship between Work Stress, Leisure Constraint, and Leisure Participation

References [1,39,40] revealed that the leisure constraints of junior high school teachers were significantly and negatively correlated with leisure participation. Tseng [37] found that the external and interpersonal obstacles of secondary school teachers in leisure constraints were negatively correlated with sports leisure activities. Although the subjects were not teachers, leisure constraints had a significant negative impact on leisure participation [21,23,41,42]. Based on the above, we assumed that the leisure constraints of junior high school teachers have a significant negative impact on leisure participation.

Teachers' work stress showed a small positive correlation with leisure participation [43]. The working engagement of work stress was significantly and negatively correlated with the sports and hobbies of leisure participation [1]. Professional knowledge and family leisure participation had a significant negative correlation. Instructional coaching was significantly negatively correlated with participation in sports. The increased pressure of interpersonal relationships encouraged part-time administrative teachers to find opportunities to communicate and coordinate with others, which increased participation. Teachers' work stress had a significant negative impact on the degree of leisure participation [40,41]. Based on the above, it was assumed that the work stress of junior high school teachers had a significant impact on leisure participation [44].

3. Research Methodology

3.1. Research Subjects

The qualified public junior high school teachers serving in Nantou County in the 110th academic year were recruited as the research subjects, including part-time administrative teachers, full-time teachers, and class teachers. Teacher trainees, acting teachers, and substitute teachers were excluded. Stratified sampling was used for the sampling of a questionnaire survey. Based on the proportion of teachers in remote areas and general areas [45], 135 teachers from rural schools and 165 from schools in general areas were invited.

3.2. Research Method

The questionnaire consisted of four parts. The first part was for basic personal information. The second part contained a modified teacher work stress scale [16,17], and the third part included a modified leisure constraint scale [46,47]. The fourth part contained a leisure participation scale [39–41]. The research structure is shown in Figure 1.

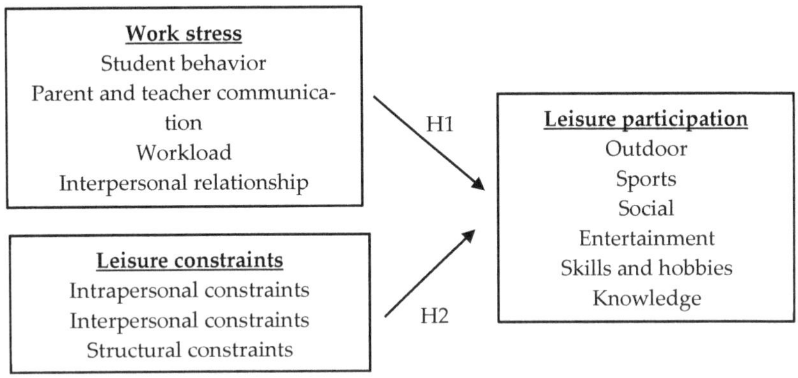

Figure 1. Research framework.

According to the above research framework, two research assumptions were proposed.

H1. *Work stress has a significant impact on leisure participation.*

H2. *Leisure constraints have a significant impact on leisure participation.*

4. Results

The snowball method was used to conduct online surveys. A total of 300 questionnaires were connected between June and July 2022. After excluding invalid questionnaires, 295 valid questionnaires were collected with an effective recovery rate of 98.33%. Cronbach's α reliability analysis was performed on the questionnaires [46], and their reliability was validated with all scores over 0.75.

4.1. Basic Information of Subjects

The basic information of the subjects in this study is as follows. There were more female teachers (182), accounting for 61.7% of teachers, and 113 male teachers (38.3%). A total of 112 teachers were 41–50 years old (38%), and 102 were 31–40 years old (34.6%), while 46 (15.6%) were over 51 years old. A total of 35 were 21–30 years old (11.9%). An amount of 180 were married (61%), while 115 were not (39%). A total of 84 teachers had a teaching experience of 16–20 years (28.5%), 65 with over 20 years (22%), 63 with 11–15 years (21.4%), 49 with 6–10 years (16.6%), and 34 with less than 5 years (11.5%). The largest number (144) had the title of a mentor (48.8%), 75 had the title of full-time teacher (25.4%), 51 had the title of teacher and group leader (17.3%), and 25 had the title of directors (8.5%).

4.2. Work Stress, Leisure Constraints, and Leisure Participation

The average score of the overall work stress of junior high school teachers in Nantou County was 3.02, indicating that their work stress was moderately high. Teachers' perception of work stress was the highest for student behavior (3.44), followed by professional knowledge (3.28). Work stress was less in interpersonal relationships (2.56), parent–teacher communication (2.76), and working engagement (2.98).

The average overall score of leisure constraints was 2.93, indicating that the teachers did not feel leisure constraints considerably. The constraints were caused by structural constraints (3.12), while intrapersonal constraints (2.77) and interpersonal constraints (2.90) were less pronounced.

The frequency of leisure participation showed the highest average score (3.71). The average score of leisure participation for knowledge, outdoor leisure, sports leisure, social leisure, and hobby leisure was 2.93, 2.68, 2.53, 2.37, and 2.00, respectively.

4.3. Work Stress on Leisure Participation

Work stress had a significant positive impact on leisure participation. Working engagement had a significant positive effect on outdoor leisure participation (t = 1.985; $p < 0.05$; β = 0.164). Parent–teacher communication had a significant positive effect on sports leisure participation (t = 2.284; $p < 0.05$; β = 0.220). Interpersonal relationship to social leisure participation (t = 3.276; $p < 0.01$; β = 0.207), to hobby leisure participation (t = 2.908; $p < 0.01$; β = 0.184), and knowledge leisure participation (t = 4.022; $p < 0.001$; β = 0.255) had a significant positive impact, and student behavior (t = 2.258; $p < 0.05$; β = 0.215) had a significant positive impact on entertainment leisure participation.

Each cause of work stress had a significant negative impact on leisure participation. Student behavior on outdoor leisure participation (t = −2.915; $p < 0.01$; β = −0.281) and sports leisure participation (t = −2.850; $p < 0.01$; β = −0.269) had a significant negative effect. Professional knowledge had a significant negative effect on sports leisure participation (t = −3.808; $p < 0.001$; β = −0.264). Professional knowledge had a significant negative effect on hobby leisure participation (t = −4.277; $p < 0.001$; β = −0.296). The results are detailed in Table 1.

Table 1. Impact of work stress on leisure participation.

Dependent Variable	Influence Dimensions	B	β	SS
Outdoor	Student Behavior **	−0.332	−0.281	3.1%
	Working Engagement *	0.193	0.164	
Sports	Student Behavior **	−0.457	−0.269	7.2%
	Parent–Teacher Communication *	0.373	0.220	
	Professional Knowledge ***	−0.418	−0.264	
Social	Interpersonal Relationship ***	0.207	0.206	7.2%
Entertainment	Student Behavior *	0.289	0.215	5.4%
Hobby	Interpersonal Relationship **	0.212	0.184	7.5%
	Professional Knowledge ***	−0.382	−0.296	
Knowledge	Interpersonal Relationship ***	0.328	0.255	6.6%

*** Standardized loadings are significant at $p < 0.001$. ** Standardized loadings are significant at $p < 0.01$. * Standardized loadings are significant at $p < 0.05$.

4.4. Work Stress on Leisure Participation

Each cause of work pressure had a significant positive impact on leisure participation. Structural constraints showed significant positive effects on social leisure participation (t = 5.373; $p < 0.001$; β = 0.265), entertainment leisure participation (t = 4.776; $p < 0.001$; β = 0.312), hobby leisure participation (t = 7.947; $p < 0.001$; β = 0.475), and knowledge leisure participation (t = 4.500; $p < 0.001$; β = 0.277). In addition, intrapersonal constraints had a significant positive impact on knowledge leisure participation (t = 3.122; $p < 0.01$; β = 0.259).

Each cause of leisure constraints had a significant negative impact on leisure participation. Intrapersonal constraints had a significant negative impact on outdoor leisure participation (t = −2.634; $p < 0.01$; β = −0.218), sports leisure participation (t = −4.713; $p < 0.001$; β = −0.359), and entertainment leisure participation (t = −2.567; $p < 0.05$; β = −0.227). Interpersonal constraints had a significant negative impact on outdoor leisure participation (t = −3.201; $p < 0.01$; β = −0.246), sports leisure participation (t = −3.582; $p < 0.001$; β = −0.254), social leisure participation (t = −3.392; $p < 0.001$; β = −0.272), and hobby leisure participation (t = −4.756; $p < 0.001$; β = −0.357).

Such results showed that interpersonal constraints impacted leisure participation negatively but structural constraints had a positive impact. The results are detailed in Table 2.

Table 2. Impact of work pressure on leisure participation.

Dependent Variable	Influence Dimensions	B	β	SS
Outdoor	Intrapersonal constraints	−0.288	−0.218	17.9%
	Interpersonal constraints	−0.375	−0.246	
Sports	Intrapersonal constraints	−0.684	−0.359	30.5%
	Interpersonal constraints	−0.556	−0.254	
Social	Interpersonal constraints	−0.417	−0.272	11.2%
	Structural constraints	0.434	0.343	
Entertainment	Intrapersonal constraints	−0.341	−0.227	6.6%
	Structural constraints	0.445	0.312	
Hobby	Interpersonal constraints	−0.636	−0.357	22.1%
	Structural constraints	0.697	0.475	
Knowledge	Intrapersonal constraints	0.449	0.259	17.2%
	Structural constraints	0.454	0.277	

5. Conclusions

Based on the results of this study, the following conclusions are drawn.

Junior high school teachers in Nantou County felt work stress at a moderately high level. This result is similar to the findings of refs. [43,47–49]. Student behavior was the most important cause of work stress followed by professional knowledge, which is also similar to the results of refs. [43,47,49,50]. In many rural schools in Nantou County, students are less willing to learn. Teachers feel pressure and have to follow many educational policies and institutional changes. Teachers have to continue studying and increasing their professional knowledge; so their workload is increasing day by day with work stress. However, the teachers did not feel leisure constraints as much. They mainly had structural constraints and had the least interpersonal constraints. Intrapersonal constraints significantly impacted leisure constraints, which is similar to the findings of refs. [27,51]. The leisure constraints of the teachers in Nantou County were only affected by external factors. Since the spread of COVID-19 in 2019, many restrictions have been formulated to reduce the movement of people. When people went out for leisure activities, they needed to consider whether the leisure environment was safe and disinfected. This made structural leisure constraints relatively important.

For the teachers' leisure participation, entertainment was the most important followed by knowledge, outdoor, sports, social, and hobby constraints. This result is similar to the findings of refs. [27,51,52]. The teachers' leisure participation frequency was reduced since 2019 due to COVID-19. The teachers' work stress from student behavior significantly impacted entertainment leisure participation positively and outdoor leisure participation and sports leisure participation negatively. Knowledge had a significant negative impact on sports and hobby leisure participation. High pressure from these factors impacted leisure participation in general, which supported the first hypothesis. Structural constraints were affected by leisure constraints, which also impacted social, entertainment, hobby, and knowledge leisure participation. This supported the second hypothesis of this study.

Based on these results, the following suggestions are made. The teachers need to have co-study time through teaching and research meetings each semester. Teachers can solve problems encountered in teaching through discussion, and brainstorm to find better teaching methods and improve students' learning motivation and effectiveness. The teachers can also participate in diversified leisure activities and the cultivation of their hobbies to relax their body and mind, and relieve work stress. As the work stress of the teachers mainly came from student behavior and professional knowledge, it is recommended that the school administration support and assist teachers in dealing with student and parent issues. In addition, as the travel time for Nantou County rural schoolteachers to participate in the off-campus teacher study course is longer, school administrators need to organize such courses and stress relief-related courses on campus. Then, teachers can use their spare time to participate in learning, enrich their professional knowledge, and reduce work stress.

The Ministry of Education is currently actively promoting students having tablets, financially supporting schools to purchase tablets, and providing learning opportunities for teachers to enhance their professional knowledge using tablets for classes and facilities to avoid network problems in learning. Education administrators must listen to the opinions of all parties and evaluate the appropriateness and feasibility of the policy carefully. Public resources can relieve teachers' work stress by understanding and respecting the teachers.

Author Contributions: Conceptualization, Y.-J.H. and M.-K.C.; methodology, Y.-J.H. and Y.-Y.L.; writing—original draft preparation, Y.-J.H., Y.-Y.L. and H.-Y.C.; writing—review and editing, M.-K.C. and Y.-J.H. All authors have read and agreed to the published version of the manuscript.

Funding: This research received no external funding.

Institutional Review Board Statement: Not applicable.

Informed Consent Statement: Not applicable.

Data Availability Statement: The data of this study was used to support the findings of this study are available from the corresponding author upon request.

Conflicts of Interest: The authors declare no conflict of interest.

References

1. Chi, C.C. A Study on Working Stress, Leisure Participation and Leisure Constraints for Teachers with Administrative Duties in Junior High Schools, Taichung City, Taiwan. Master's Thesis, Feng Chia University, Taichung City, Taiwan, 2011.
2. Wu, S.P.; Lui, P.S. A Study on Relationship between Teacher's Job Stress and leisure participation of Elementary School Teacher in Greater Taipei Area. *J. Work. Leis.* **2008**, *1*, 49–61.
3. Hsu, Y.W.; Chung, C.C.; Tsai, W.C.; Huang, M.L. A Study Analyzes Relationships among Job Stress, leisure participation and Health Condition Based on Kindergarten Teachers in Yunlin and Chiayi County. *J. Sport Leis. Hosp. Res.* **2010**, *51*, 80–105.
4. Huang, J.Z.; Kao, W.Y.; Shen, Y.K. Study on the Factors of leisure Barriers for Junior High School Teachers in Hsinchu City and County. *J. Sport Leis. Hosp. Res.* **2010**, *1*, 53–70.
5. Ho, Y.P. Discussion of the Literature Review of the Teachers' leisure participation and leisure Barriers in Taiwan. *J. NPUST Phys. Educ.* **2012**, *1*, 77–95.
6. Ruan, Y.X. [Rural Education 2] Myths in Rural Areas Actual Distribution and Analysis of Regions. Available online: https://reurl.cc/GAp1ox (accessed on 17 August 2021).
7. Kyriacou, C. Teacher stress directions for future research. *Educ. Review. Ducational Rev.* **2001**, *53*, 27–35. [CrossRef]
8. Kyriacou, C.; Sutcliffe, J. Teacher stress: Prevalence, sources and symptoms. *Br. J. Educ. Psychol.* **1978**, *48*, 159–167. [CrossRef] [PubMed]
9. Litt, J.M.; Turk, G.H. Sources of stress and dissatisfaction in experienced high school teachers. *J. Educ. Res.* **1985**, *78*, 178–185. [CrossRef]
10. Liu, Y.H. Study of work stress of Secondary and Elementary School Teachers. *Sch. Adm.* **2011**, *72*, 77–98.
11. Kuo, Y.H. A Study on the Relationship between Job Stress and Working Burnout for Tainan City Junior High School Teachers--Moderate Effect of School Support and Resilience. Master's Thesis, National University of Tainan, Tainan, Taiwan, 2012.
12. Chang, Y.C. A Study of the Relationships Between Job Stress and Emotional Labor for Junior High School Teachers in Taichung City. Master's Thesis, Central Taiwan University of Science and Technology, Taichung City, Taiwan, 2013.
13. Chen, H.H.; Hsu, C.H.; Huang, Y.T. A Study on the Relationship between work Pressure and leisure Coping of Elementary School Teachers. *J. Sport Recreat. Res.* **2013**, *7*, 138–159.

14. Yeh, R.C.; Lu, M.S.; Chen, Y.C.; Lin, Y.C. Effects of Teachers' Emotional Intelligence and Job Stress on Job Satisfaction: An Empirical Study on Junior-High School Teachers of Pingtung County. *J. Meiho Univ.* **2014**, *33*, 177–194.
15. Chen, Y.C. A Study of the Relationship between Elementary School Teacher's Working Pressure and Leisure Participation in Taichung City. Master's Thesis, Central Taiwan University of Science and Technology, Taichung City, Taiwan, 2014.
16. Tan, H.P. A Study on the Relationship of Job Stress, Leisure Participation and Physical–Mental Health among Teachers-A Case Study of Keelung Municipal Elementary and Junior High Schools. Master's Thesis, Ching Kuo Institute of Management and Health, Keelung City, Taiwan, 2019.
17. Fu, Y.C. The Relationships of Job Stress, Leisure Participation, and Physical-Mental Health among Elementary School Teachers in Taitung County. Master's Thesis, National Taitung University, Taichung City, Taiwan, 2016.
18. Crawford, D.W.; Jackson, E.L.; Godbey, G.C. Reconceptualizing barrier to family leisure. *Leis. Sci.* **1987**, *9*, 119–127. [CrossRef]
19. Lin, S.F. A Study of the Association Among Leisure Motivation- Leisure Constraints and Well-Being for Middle-Aged and Elderly Females Participating in Chorus. Master's Thesis, National Kaohsiung Normal University, Kaohsiung City, Taiwan, 2017.
20. Chen, W.L. A Study on the Leisure Participation and Leisure Barriers of Older—A Case Study of a Community Care Station in Taoyuan City. Master's Thesis, Kainan University, Taoyuan City, Taiwan, 2018.
21. Chien, M.K. The Relationship among leisure Motivation, leisure participation and leisure constraint of senior--a case study in southern. *Leis. Ind. Res.* **2019**, *17*, 31–46.
22. Chen, H.F. The Leisure Participation and Leisure Constraints of Tainan Senior Citizens. Master's Thesis, University of Kang Ning, Tainan City, Taiwan, 2020.
23. Kan, C.I. A Study on the Relationship between Leisure Constraints and Participation for Junior High School Teachers in Pingtung County. Master's Thesis, Tajen University, Pingtung County, Taiwan, 2021.
24. Chen, N.C. The Relationships of Leisure Boredom to Leisure Constraints, Leisure Participation, and Leisure Satisfaction among Adolescents. Master's Thesis, National Taiwan Sport University, Taoyuan City, Taiwan, 2000.
25. Chang, H.M.; Kao, C.H. The Related Study of Between leisure Needs and leisure Constraint-Take Example by Changhua Residents. *Phys. Educ. J.* **2001**, *30*, 143–152.
26. Lai, C.S. The Construction of Leisure Constraint Scale–An Example of College Students in Taipei. Master's Thesis, National Taiwan Sport University, Taoyuan City, Taiwan, 2002.
27. Liang, Y.F. Research on Leisure Activity Particpation and Leisure Constraints: Analysis of Junior High School Teachers in Chang-Hua County. Master's Thesis, Dayeh University, Changhua City, Taiwan, 2004.
28. Yuan, H.H. A Study on Leisure Activities Participation and Constraints of Teachers in Kinmen. Master's Thesis, Ming Chuan University, Taipei City, Taiwan, 2008.
29. Dumazedier, J. *Sociology of Leisure*; Elsevien: New York, NY, USA, 1974.
30. Chen, S.P.; Wan, C.H.; Su, P.H.; Cheng, H.C. A Study on the Effect Job Stress of Kindergarten Preschool Educators in Chiagi County. *Sport. Tour. Stud.* **2019**, *8*, 1–10.
31. Liao, Y.J. Research on Middle Age and Elderly Leisure Participation and Health. Master's Thesis, China University of Science and Technology, Taipei City, Taiwan, 2017.
32. Lin, Y.J. TA study on the behavior of recreation areas chosen by tourists. *City Plan.* **1984**, *10*, 33–49.
33. Teng, C.C. The Research of Comprehensive High School Teachers' Leisure Needs and Leisure Participation. Master's Thesis, Chaoyang University of Technology, Taichung City, Taiwan, 2002.
34. Wang, M.S. The Relationships between the Patterns of Leisure Participation and Deviant Behaviors among Adolescents in Taipei City. Master's Thesis, University of Taipei, Taipei City, Taiwan, 2003.
35. Li, K.Y. A Study on the Leisure Behavior of Administrative Staff in National High School Vocational Schools—Tunghai University. Master's Thesis, University of Taipei, Taipei City, Taiwan, 2005.
36. Chang, C.C. The Influence of Job Stress and Leisure Participation on Physical and Mental Health of Elementary School Teachers in Ping-Tung County. Master's Thesis, Asia University, Taichung City, Taiwan, 2008.
37. Tseng, Y.H. A Study on Leisure Participation and Leisure Constraints: Exemplify of Public and Private Junior High School Teachers in Kaohsiung. Master's Thesis, Shu-Te University, Kaohsiung City, Taiwan, 2015.
38. Lai, C.J. The Study of Teachers' Job Stress, Leisure Participation and Well-Being for Junior High School Teachers in Taichung City. Master's Thesis, National Changhua University of Education, Changhua City, Taiwan, 2017.
39. Chuang, S.Y. A Study of Leisure Participation Motivation and Leisure Constraints for Junior High School Teachers in Chiayi County. Master's Thesis, National Taiwan University of Sport, Taichung City, Taiwan, 2013.
40. Chen, T.Y. The Study of Teachers' Leisure Participation Motivation, Leisure Satisfaction and Leisure Constraints in Miaoli. Master's Thesis, Chaoyang University of Technology, Taichung City, Taiwan, 2014.
41. Cheng, S.C. leisure participation, Constraints, and Academic Pressure of National Chiayi Senior High School Students. *J. Taiwan Soc. Sport Manag.* **2010**, *10*, 55–74.
42. Ou, C.H. The Study on the Influence of Firefighter's Leisure Obstruction and Leisure Activity Participation on Life Satisfaction- A Case Study of Kaohsiung and Pingtung Area. Master's Thesis, Meiho University, Pingtung County, Taiwan, 2014.
43. Lou, M.H. The Study of Leisure Participation, Job Stressand Leisure Satisfaction with the reference to Junior High School Teachers. Master's Thesis, Chaoyang University of Technology, Taichung City, Taiwan, 2011.

44. Hsu, W.L. A Study of Relationship among Job Stress, Leisure Participation and Leisure Constraints of Elementary School Teachers in Tainan City. Master's Thesis, Southern Taiwan University of Technology, Tainan City, Taiwan, 2014.
45. Department of Education, Nantou County Government. School Information at All Levels in the 11th School Year [Raw Data]. Available online: https://www.ntct.edu.tw/p/404-1000-304.php (accessed on 13 August 2022).
46. Wu, M.L. *SPSS Operation and Application: The Practice of Quantitative Analysis of Questionnaire Data*; WuNan: Taipei, Taiwan, 2009.
47. Lee, T.J. A Study on Work Stress and Sports Participation of Secondary School Teachers in Greater Kaohsiung Area. Master's Thesis, National Kaohsiung Normal University, Kaohsiung City, Taiwan, 2011.
48. Ouyang, L.Y. A Study on the Relationship between Job Stress and Job Burnout of Junior High School Teachers in Pingtung County. Master's Thesis, National Pingtung University of Science and Technology, Pingtung City, Taiwan, 2012.
49. Chen, Y.H. The Impact of Leisure Participation, Leisure Benefits and Leisure Obstacles on Happiness for Junior High School Teachers in Kaohsiung. Master's Thesis, Shu-Te University, Kaohsiung City, Taiwan, 2016.
50. Chang, C.F. A Study of Junior High School Teachers' Job Stress and Coping Strategies in Fengyuan, Taichung city. Master's Thesis, Nanhua University, Chiayi County, Taiwan, 2013.
51. Crawford, D.W.; Jackson, E.L.; Godbey, G.C. A hierarchical model of leisure constraints. *Leis. Sci.* **1991**, *13*, 309–320. [CrossRef]
52. Lin, C.C. A Research on the Relation between Leisure Participation and Leisure Constraints: Analysis of Junior High School Teachers in Gangshan Area. Master's Thesis, Shu-Te University, Kaohsiung City, Taiwan, 2014.

Disclaimer/Publisher's Note: The statements, opinions and data contained in all publications are solely those of the individual author(s) and contributor(s) and not of MDPI and/or the editor(s). MDPI and/or the editor(s) disclaim responsibility for any injury to people or property resulting from any ideas, methods, instructions or products referred to in the content.

engineering proceedings

MDPI

Proceeding Paper

DEA Approach to Evaluate Research Efficiency of Departments in University †

Shu-Mei Chao [1,2,*] and Mu-Jin Chen [3]

1 Department of Education, National Chengchi University, Taipei 11605, Taiwan
2 Research and Development Office, National Chengchi University, Taipei 11605, Taiwan
3 Center of Teacher Education, Chaoyang University of Technology, Taichung 413310, Taiwan; mujinc@gmail.com
* Correspondence: smjau@nccu.edu.tw
† Presented at the 3rd IEEE International Conference on Electronic Communications, Internet of Things and Big Data Conference 2023, Taichung, Taiwan, 14–16 April 2023.

Abstract: The purpose of this study is to evaluate the research efficiency of 40 departments in a university from 2015 to 2017. In this study, data envelopment analysis (DEA) is used with a non-parametric mathematical linear planning approach, and an appropriate model is proposed for evaluating the research efficiency of the university's departments. The analyzed items are selected based on the relevant literature on research efficiency. The result of this study helps research policymakers and motivates faculty and researchers to take the initiative for better-quality research with limited resources and international competitiveness.

Keywords: higher education institution; departments' research efficiency; data envelope analysis

1. Introduction

1.1. Research Motivation

The research object of this study is a university specializing in the humanities and social sciences (case university hereinafter). Compared with natural science, humanities and social sciences are mostly restricted by language, and their research is normally on regional or local issues. Regarding industry–university cooperation and budget allocation, the case university is in an inferior position. Although the case university has many disciplines within 200 in the QS ranking, the school's overall ranking falls between 601–650, and its efforts are not easily highlighted in the world university rankings. Even its performance is underestimated. Therefore, to improve research efficiency, the case university established a subsidy and reward system and provided professors and researchers with appropriate assistance and incentives to improve academic research standards and stimulate research energy. Therefore, under a limited budget and investment, the case university must rationally evaluate each unit's research efficiency to clarify the problems of inefficient units, identify the benchmark units as the objects of learning, and use the resources effectively.

The case university comprises several colleges, including liberal arts, science, law, commerce, social sciences, foreign languages and literature, communication, international affairs, and education. Initially, the colleges had a classification system to group similar departments. The colleges did not have a substantive administrative structure. Instead, resources were managed by individual departments. Over time, departments became segregated, and resources became limited. The case university recently proposed the substantiation of its colleges to integrate its resources. Currently, the College of Law, College of Communication, College of International Affairs, and College of Education have been substantiated as physical colleges.

The case university has four physical and five non-physical colleges. According to a study by Chao and Chen [1], the research efficiency of the physical colleges of the

Citation: Chao, S.-M.; Chen, M.-J. DEA Approach to Evaluate Research Efficiency of Departments in University. *Eng. Proc.* **2023**, *38*, 71. https://doi.org/10.3390/engproc2023038071

Academic Editors: Teen-Hang Meen, Hsin-Hung Lin and Cheng-Fu Yang

Published: 29 June 2023

Copyright: © 2023 by the authors. Licensee MDPI, Basel, Switzerland. This article is an open access article distributed under the terms and conditions of the Creative Commons Attribution (CC BY) license (https://creativecommons.org/licenses/by/4.0/).

case university was better than non-physical colleges between 2015 and 2017. Therefore, we analyzed the research efficiency performance of 40 departments in the non-physical colleges of the case university using a data envelopment analysis and a non-parametric mathematical linear planning approach.

1.2. Research Objectives

Based on the above, the research objectives of this study are as follows:

1. To evaluate the research efficiency of the case universities' departments using data envelopment analysis (DEA);
2. To identify the issues of research inefficiency in the departments of the case university through this study;
3. To identify benchmarks as a reference for departments to learn from;
4. To make recommendations to the departments and management to solve the management problems associated with research inefficiency.

1.3. Research Questions

Therefore, the questions to be studied in this research are defined as follows:

1. What is the research efficiency of the departments in the case university?
2. What are the issues of inefficiency in some departments of the case university?
3. What are the priorities for improvement in the relatively inefficient departments of the case university?
4. Which departments are regarded as benchmarks for the case university's departments to follow?
5. How are the problems of inefficient research management in the case university solved?

2. Literature Review

In this section, the methods for assessing research efficiency are introduced to explain the theory, main models, and characteristics of the research method (data envelopment analysis) adopted in this study.

Universities assess the efficiency of academic research of different disciplines scientifically based on standardization and consistency of management. However, the estimation of university-research efficiency is a controversial issue, and the measurement of the so-called efficiency is under debate.

In the UK, the quality of research and allocation of research funds is assessed by the Research Excellence Framework (REF), which focuses on defining what research excellence means and identifying what constitutes "research excellence". The key features of the REF are related to the evaluation of three elements: (1) Output Quality, (2) Research Impact, and (3) Research Environment. The results of the REF evaluation are the product of an expert review or examination based on appropriate indicators [2]. Other countries such as the United States, Germany, Australia, New Zealand, and Norway also attach great importance to the evaluation of university research efficiency. University research assessments are conducted to determine how to allocate research grants and motivate universities to pursue excellent research [3].

Scientific methods to evaluate research efficiency are important for research management. Commonly used methods to evaluate research efficiency include data envelopment analysis, peer review method, Delphi method, bibliometric method, hierarchical analysis, gray correlation analysis, and fuzzy integrated evaluation method [3,4].

DEA is derived from Farrell's method of evaluating the relative efficiency of multiple inputs and outputs [5]. Later, Charnes, Cooper, and Rhodes improved it into the CCR model [6] based on DEA. DEA uses a mathematical model to obtain the production frontier for measuring efficiency without a preset production function model. This non-parametric quantitative technique assesses the relative efficiency of evaluated units, commonly known as decision-making units (DMUs) [7–10]. The input and output data of DMU are used to find the production frontier through the mathematical model by comparing the actual data of each DMU with the production frontier. Each DMU's relative efficiency and relative inefficiency are measured to achieve the goal of relative efficiency improvement [11]. Rhodes defined DEA as "a nonlinear (nonconvex) programming model providing a new definition of efficiency for use in evaluating activities of not-for-profit entities participating in public programs" (p. 429) [6].

The DEA is addressed as a linear programming problem as stated by Charnes et al. [6], and the mathematical model to calculate the efficiency is expressed as follows [12].

$$Min\ E_0 = \sum_{i=1}^{m} v_i x_{i0}$$

$$\sum_{i=0}^{m} v_i x_{ij} - \sum_{r=1}^{s} u_r y_{rj} \geq 0$$

$$\sum_{r=1}^{m} u_r y_{r0} = 1$$

$$v_i\ , u_i \geq 0, i = 1, \ldots, m;\ r = 1, \ldots, s$$

where each DMU uses m inputs to obtain s outputs. Hence, the j-th DMU uses x_{ij} units of input i to produce y_{rj} units of output r. Additionally, u_r represents the weighted outputs and v_i the weighted inputs.

The result is categorized into an efficient group (efficiency value equal to 1) and an inefficient group (efficiency value less than 1). A completely inefficient DMU assumes a value of 0. Data envelopment analysis has been used for years to study the efficiency of university research [12–15].

According to Hsu [16], the main methods for selecting input and output items for efficiency assessment are (1) literature analysis, (2) evaluation or review of indicators, and (3) the Delphi method. Since most studies to evaluate the efficiency of educational institutions used the literature analysis method to select the input and output items, we summarized and analyzed the input and output items by referring to the past literature to evaluate the research efficiency of the teaching unit of the university. The research output items are evaluated by the research-oriented indicators assessed by the university department, and the research output items are regulated by the evaluation of the faculty performance assessment. In the output, special reference is made so that research output items are evaluated by the research-oriented indicators of the case university departments. The research output items are regulated by the research evaluation of the faculty performance assessment to make the selected items more objective and applicable to the case university.

3. Research Methodology

3.1. Research Framework

The framework of this study is based on the research motivation, research objectives, and related literature. The research framework in Figure 1 is used to empirically analyze the "research efficiency of departments of the case university".

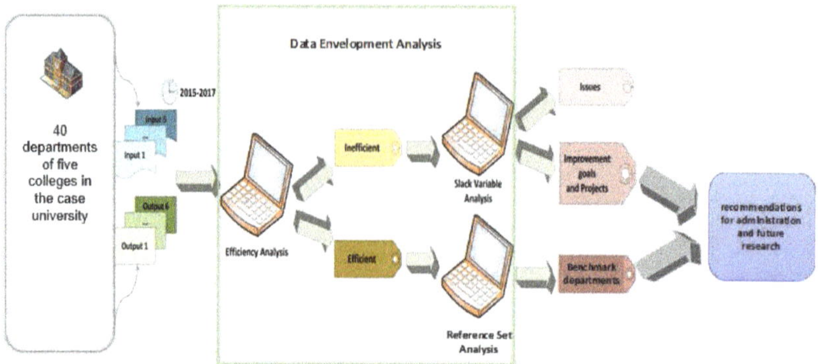

Figure 1. Research framework.

3.2. Research Subjects and Decision Making Units

The samples used in this study are the departments of a national university in Taiwan, including 40 departments of the five non-physical colleges as the decision-making units. The research efficiency of 40 departments from 2015 to 2017 was analyzed. In the data analysis, each college was de-identified and coded as A, B, C, and so on, and the N department of the A college was coded as A- N.

3.3. Input and Output Items

In this study, the inputs and outputs selected were summarized and analyzed by considering the attributes of the case university and referring to the literature. There were five inputs (FTE professors, FTE associate professors or higher, academic research awards attained, doctorates, and research funding) and six outputs (approved budget, approved projects, citations in international databases, number of publications under peer review, professionally reviewed monographs (chapters) and patents, and winners of iconic external academic awards).

If we limited the calculation to the number of journal articles, it would not be able to include the overall research performance and would likely be influenced by the characteristics of the discipline and hide the performance of certain disciplines such as humanities and social sciences. This is often represented by an important monograph [17]. In addition, the Ministry of Education promulgated the "Regulations Governing Accreditation of Teacher Qualifications at Junior Colleges and Institutions of Higher Education", which includes professors' patents as the research results for promotion and examination. thus establishing patents as an important indicator of the research output of university professors. In addition, the university also recognizes patents as the result of evaluating professors' research performance to encourage them to engage in industry–academia collaboration. In practice, it is common to combine the number of books and patents as one of the research output items [17,18]. We added the number of books and chapters to include "the number of professionally reviewed monographs (chapters) and patents" as one of the research output items.

3.4. Data Collection and Processing

The data were collected from 2015 to 2017, and the sources of data for this study were official databases and website information from the following four sections: (1) the Higher Education Database of the Ministry of Education, (2) the Teacher's Publication Catalogue Database, (3) the case university's official website, and (4) the Scopus database.

In the DEA model for efficiency assessment, the number of inputs and outputs increase when the number of inputs increases, which is known as isotonicity. To examine this relationship, we used Pearson Product-moment Correlation Analysis. In this study, the CCR and BCC models for data envelopment analysis were used to calculate the research efficiency values of each department, including the overall efficiency, pure technical efficiency, and scale efficiency values for each year. The DEAP2.1-XP software developed by the Centre for Efficiency and Productivity Analysis at the University of Queensland, Australia, was used to perform the calculations.

According to Lovell [19], if the DMU is a profit-making organization, the market demand is random and uncontrollable, and the use of inputs can be freely adjusted. Thus, the input-oriented model was used to measure it. To measure the achievement of each output with the same level of input, we adopted an output-oriented model that emphasizes the measurement of research efficiency from an output perspective.

Finally, Slack Variable Analysis (SVA) was used to provide targets and magnitudes of improvement for relatively inefficient units so that relatively inefficient units could further be understood, whether they had too much input or not enough output, and obtain specific and quantitative data. The quantitative data were obtained to achieve relative efficiency and provide management with references and guidelines. In addition, through Reference Set Analysis, benchmark analysis was conducted to identify benchmark units and examine how often the relatively efficient units were used as reference objects for improvement by the inefficient units. The units referred to more often were listed as learning benchmarks.

4. Results and Recommendations

In this study, data were collected from 40 departments from 2015 to 2017 and were analyzed by using data envelopment analysis to analyze the research efficiency of each department. The results are as follows.

4.1. Stable Overall Research Efficiency

In this study, we analyzed the research efficiency of the departments in the non-physical colleges of the case university. The overall efficiency of the departments only showed a few differences in the three years, with the worst annual average in 2016 (0.841), the second best in 2015 (0.880), and the best in 2017 (0.888). The overall research efficiency was stable (Table 1).

Table 1. Annual department efficiency value—CCR model.

Item	Department Code	CCR -Overall Efficiency				
		2015	2016	2017	Mean	SD
1	A-1	0.818	1.000	0.943	0.920 *	0.076
2	A-2	0.985	0.921	0.518	0.808	0.179
3	A-3	0.493	0.945	0.859	0.766	0.170
4	A-4	1.000	1.000	1.000	1.000 ***	0.000
5	A-5	0.658	1.000	1.000	0.886 **	0.161
6	A-6	1.000	0.845	1.000	0.948 **	0.073
7	A-7	0.914	0.972	1.000	0.962 *	0.036
8	B-1	0.742	0.438	0.735	0.638	0.123
9	B-2	1.000	0.944	1.000	0.981 **	0.026
10	B-3	1.000	1.000	1.000	1.000 ***	0.000
11	B-4	1.000	1.000	1.000	1.000 ***	0.000
12	B-5	1.000	1.000	1.000	1.000 ***	0.000
13	C-1	1.000	1.000	1.000	1.000 ***	0.000
14	C-2	0.866	1.000	0.912	0.926 *	0.056
15	C-3	1.000	0.996	0.769	0.922 *	0.108
16	C-4	1.000	0.712	1.000	0.904 **	0.136

Table 1. *Cont.*

Item	Department Code	CCR -Overall Efficiency				
		2015	2016	2017	Mean	SD
17	C-5	1.000	0.989	0.833	0.941 *	0.076
18	C-6	1.000	0.727	1.000	0.909 **	0.129
19	C-7	1.000	0.817	1.000	0.939 **	0.086
20	C-8	1.000	1.000	0.832	0.944 **	0.079
21	C-9	0.167	0.091	1.000	0.419 *	0.412
22	C-10	1.000	0.342	1.000	0.781 **	0.310
23	E-1	1.000	0.735	1.000	0.912 **	0.125
24	E-2	0.948	1.000	1.000	0.983 **	0.025
25	E-3	0.793	0.929	0.628	0.783	0.107
26	E-4	0.828	0.484	0.530	0.614	0.132
27	E-5	0.635	1.000	0.753	0.796 *	0.152
28	E-6	1.000	1.000	0.913	0.971 **	0.041
29	E-7	1.000	0.874	1.000	0.958 **	0.059
30	E-8	1.000	1.000	1.000	1.000 ***	0.000
31	E-9	1.000	0.937	1.000	0.979 **	0.030
32	F-1	0.468	0.691	1.000	0.720 *	0.218
33	F-2	1.000	1.000	1.000	1.000 ***	0.000
34	F-3	0.428	0.755	0.395	0.526	0.141
35	F-4	0.682	0.975	0.652	0.662	0.012
36	F-5	0.783	0.722	1.000	0.835 *	0.119
37	F-6	1.000	0.571	0.250	0.607 *	0.307
38	F-7	1.000	1.000	1.000	1.000 ***	0.000
39	F-8	1.000	0.556	1.000	0.852 **	0.209
40	F-9	1.000	1.000	1.000	1.000 ***	0.000
Mean		0.880	0.841	0.888	0.870	

Note: The average * represents the number of research units that achieved a relative efficiency of 1.000 over the years. A * indicates that the unit has an efficiency value of 1 for one year, ** indicates an efficiency value of 1 for two years, and *** indicates that the unit has an efficiency value of 1 for three years.

4.2. Inefficiency Due to Scale Inefficiency and Technical Inefficiency

Among the 40 departments of the case university in this study from 2015 to 2017 with a total of 120 units (40 units/year × 3 years), 28 units (23%) were inefficient due to scale inefficiency; 5 units (4%) were inefficient due to pure technical inefficiency; and 22 units (18%) were inefficient due to both pure technical inefficiency and scale inefficiency. The overall inefficiency of 22 units (18%) was due to both pure technical inefficiency and scale inefficiency. In particular, 15 (63%) of the overall inefficient units (24) in 2016 were inefficient, indicating that departments were largely invested in resources but did not fulfill the optimal scale of output. (Table 2)

Table 2. The overall efficiency of the departments in 2015–2017 unit classification—BCC model.

Category/Period (Number of Units)	2015 (16 Departments)	2016 (24 Departments)	2017 (15 Departments)
Scale inefficiency (Pure technical efficiency = 1)	A-1,A-2,A-5 E-3 F-4,F-5 (6 departments)	A-2,A-3,A-6 B-2 C-3,C-5,C-7 E-1,E-3,E-4,E-7 F-1,F-3,F-4,F-8 (15 departments)	A-1,A-3 C-2,C-5 E-4,E-5,E-6 (7 departments)
Pure technical inefficiency (Scale efficiency = 1)	A-3 C-9 (2 departments)	C-9,C-10 (2 departments)	F-6 (1 department)
Pure technical inefficiency and Scale inefficiency	A-7 B-1 C-2 E-2,E-4,E-5 F-1,F-3 (8 departments)	A-7 B-1 C-4,C-6 E-9 F-5,F-6 (7 departments)	A-2 B-1 C-3,C-8 E-3 F-3,F-4 (7 departments)

4.3. Importance of Research Funding and External Academic Awards and Citations

According to the slack variable analysis, among the inefficient departmental inputs, research funding needed to be improved in 2015, and the improvement rate was 44%. In 2016, research funding especially needed to be improved, and the improvement rate was 37%. In 2017, research funding and the number of doctorates were the items to be improved, and the improvement rate was 52%. Among the inefficient departmental inputs from 2015 to 2017, the most important items for improvement were reductions in research funding, which means that more inefficient resources were invested in research (Figure 2).

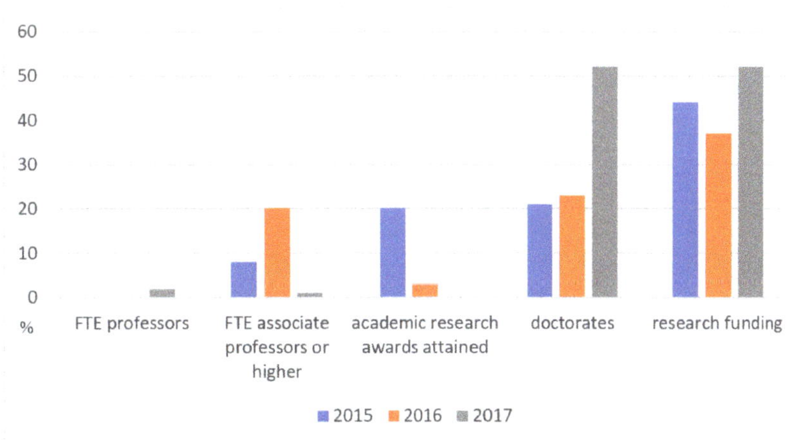

Figure 2. Input improvement for inefficient DMUs—CCR model.

Among the overall inefficient departmental output items, the number of external academic awards needed to be improved in 2015, and the improvement rate was 700% (the actual was 2, and the improvement target was 16), followed by an increase in the number of international citation database articles, and the improvement rate was 153%. The number of external academic awards needed to be improved in 2017 with an improvement rate was

125% (the actual value was 4, and the improvement target was 9), followed by an increase in the number of approved research projects with an improvement rate was 102%. Overall, the most important item for improvement from 2015 to 2017 was the increase in the number of extramural academic award recipients, followed by increasing the number of articles in the international citation database (Figure 3).

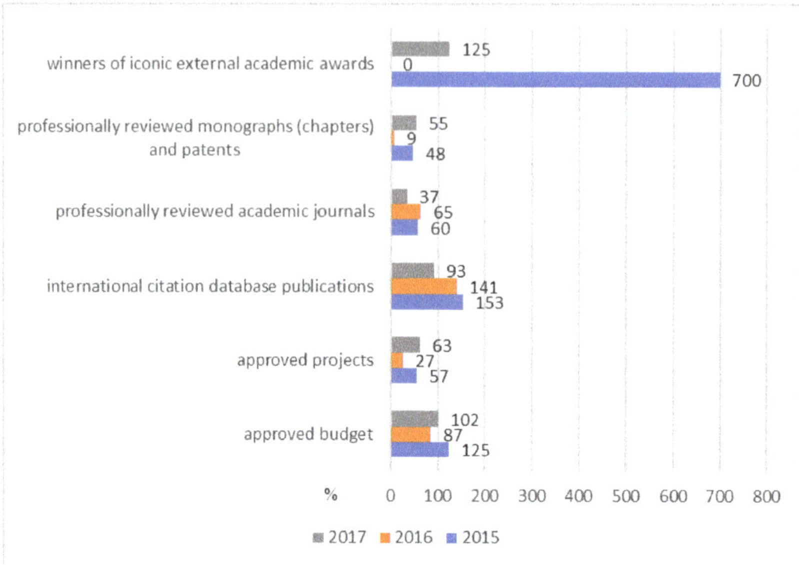

Figure 3. Output improvement for inefficient DMUs—CCR mode.

4.4. *Twenty-Five Benchmark Units from 2015–2017*

The departments with a high number of references in the reference set analysis (a strong efficiency unit, referred to more than three times) had high overall efficiency and were used as a learning benchmark for other units (Table 3).

Table 3. CCR model—reference set analysis table.

Item	Department Code	2015	2016	2017
1	A-1	-	0	-
2	A-2	-	-	-
3	A-3	-	-	-
4	A-4	0	1	0
5	1-5	-	1	2
6	A-6	2	-	5
7	A-7	-	-	0
8	B-1	-	-	-
9	B-2	0	-	0
10	B-3	9	14	7
11	B-4	6	15	9
12	B-5	4	9	4

Table 3. *Cont.*

Item	Department Code	2015	2016	2017
13	C-1	0	9	2
14	C-2	-	1	-
15	C-3	1	-	-
16	C-4	5	-	1
17	C-5	2	-	-
18	C-6	0	-	0
19	C-7	1	-	0
20	C -8	7	0	-
21	C -9	-	-	4
22	C -10	0	-	0
23	E-1	0	-	1
24	E-2	-	0	5
25	E-3	-	-	-
26	E-4	-	-	-
27	E-5	-	0	-
28	E-6	0	1	-
29	E-7	2	-	4
30	E-8	1	2	8
31	E-9	3	-	3
32	F-1	-	-	0
33	F-2	1	11	3
34	F-3	-	-	-
35	F-4	-	-	-
36	F-5	-	-	0
37	F-6	0	-	-
38	F-7	4	4	2
39	F-8	5	-	7
40	F-9	0	1	0

Note: More than 3 times are benchmark units, and "-" refers to not relative efficiency units.

5. Conclusions

In this research, we adopted the DEA methodology to evaluate the research efficiency of the case. We analyzed the research efficiency of departments of non-physical colleges from 2015 to 2017 to identify the departments that must be improved and the departments that could be the benchmarks and improve research efficiency. In summary, the study result provides specific research results and recommendations for further reference by the research policymakers.

Author Contributions: Conceptualization, S.-M.C. and M.-J.C.; methodology, S.-M.C. and M.-J.C.; validation, S.-M.C.; formal analysis, S.-M.C.; investigation, S.-M.C.; data curation, S.-M.C.; writing—original draft preparation, S.-M.C.; writing—review and editing, S.-M.C. and M.-J.C.; supervision, S.-M.C. and M.-J.C. All authors have read and agreed to the published version of the manuscript.

Funding: This research received no external funding.

Institutional Review Board Statement: Not applicable.

Informed Consent Statement: Not applicable.

Data Availability Statement: Not applicable.

Conflicts of Interest: The authors declare no conflict of interest.

References

1. Chao, S.-M.; Chen, M.-J. An Evaluation of the Research Efficiency of Colleges in the University—Case Study of the C University. In Proceeding of the 2021 IEEE International Conference on Social Sciences and Intelligent Management (SSIM), Taichung, Taiwan, 29–31 August 2021. [CrossRef]
2. Yang, Y. The latest reform of research assessment in UK universities—Planning and implementation of the Research Excellence Framework (REF). *Eval. Bimon.* **2014**, *52*, 41–47.
3. Wang, R.Z. *International Evaluation of University Research Performance*, 1st ed.; Higher Education Evaluation & Accreditation Council of Taiwan: Taipei, Taiwan, 2008.
4. Zhang, L. Research and Development of Research Performance Evaluation Methods in Universities. Master's Thesis, Nanjing University of Science and Technology, Nanjing, China, 2012.
5. Farrell, M.J. The measurement of productive efficiency. *J. R. Stat. Soc.* **1957**, *120*, 253–290. [CrossRef]
6. Charnes, A.; Cooper, W.W.; Rhodes, E. Measuring the efficiency of decision making units. *Eur. J. Oper. Res.* **1978**, *2*, 429–444. [CrossRef]
7. Ji, Y.-b.; Lee, C. Data envelopment analysis. *Stata J.* **2010**, *10*, 267–280. [CrossRef]
8. Lin, T.T.; Lee, C.-C.; Chiu, T.-F. Application of DEA in analyzing a bank's operating performance. *Expert Syst. Appl.* **2009**, *36*, 8883–8891. [CrossRef]
9. Samoilenko, S.; Osei-Bryson, K.-M. Using data envelopment analysis (DEA) for monitoring efficiency-based performance of productivity-driven organizations: Design and implementation of a decision support system. *Omega* **2013**, *41*, 131–142. [CrossRef]
10. Arshadi, H.; Goudarzi, R.; Okhovati, M. A data envelopment analysis (DEA) approach to evaluate the research efficiency of Iranian universities. *Int. J. Inf. Sci. Manag.* **2022**, *20*, 95–108.
11. Sun, S. *Data Envelopment Analysis: Theory and Application*, 1st ed.; Yangzhi Culture: Taipei, Taiwan, 2004.
12. Munoz, D.A. Assessing the research efficiency of higher education institutions in Chile: A data envelopment analysis approach. *Int. J. Educ. Manag.* **2016**, *30*, 809–825. [CrossRef]
13. Johnes, G.; Johnes, J. Measuring the research performance of UK economics departments: Application of data envelopment analysis. *Oxf. Econ. Pap.* **1993**, *45*, 332–347. [CrossRef]
14. Lee, B.L.; Worthington, A.C. A network DEA quantity and quality-orientated production model: An application to Australian university research services. *Omega* **2016**, *60*, 26–33. [CrossRef]
15. Rhaiem, M. Measurement and determinants of academic research efficiency: A systematic review of the evidence. *Scientometrics* **2017**, *110*, 581–615. [CrossRef]
16. Hsu, F.-Y. Operating Efficiency Evaluation Study of Faculties in Private University—A Case Study of Certain Medical University. Master's Thesis, National Sun Yat-sen University, Kaohsiung, Taiwan, 2010.
17. Fu, Y.C. Evaluating the Development Strategy of Our University: An Example of Data Envelopment Analysis. Master's Thesis, National Chengchi University, Taipei, Taiwan, 2007.
18. Liao, H.-M. Assessing the Research Performance of Private Universities in Taiwan Using Data Envelopment Analysis and Clustering. Master's Thesis, Fu Jen Catholic University, Taipei, Taiwan, 2008.
19. Lovell, C.A.K. Production frontiers and productive efficiency. In *The Measurement of Productive Efficiency: Techniques and Applications*, 1st ed.; Oxford University Press: New York, NY, USA, 1993; pp. 3–67.

Disclaimer/Publisher's Note: The statements, opinions and data contained in all publications are solely those of the individual author(s) and contributor(s) and not of MDPI and/or the editor(s). MDPI and/or the editor(s) disclaim responsibility for any injury to people or property resulting from any ideas, methods, instructions or products referred to in the content.

engineering proceedings

MDPI

Proceeding Paper

Patent Statistics and Analysis of Development Trends of Technology-Assisted Instruction †

Shang-Pin Li [1], Wen-Shin Hsu [2,3] and Wen-Ling Hsu [4,*]

1 Department of Early Childhood Development and Education, Chaoyang University of Technology, Taichung 413310, Taiwan; spli@cyut.edu.tw
2 Department of Medical Information, Chung Shan Medical University, Taichung 402201, Taiwan
3 Informatics Office Technology, Chung Shan Medical University Hospital, Taichung 402201, Taiwan
4 Department of Marketing and Logistics Management, Chaoyang University of Technology, Taichung 413310, Taiwan
* Correspondence: wenling801@gmail.com
† Presented at the 3rd IEEE International Conference on Electronic Communications, Internet of Things and Big Data Conference 2023, Taichung, Taiwan, 14–16 April 2023.

Abstract: Patents and related statistics are the indicators of technological progress. There have been controversial discussions about the value of patents as an indicator of technological advancement. Patent research has been conducted for research and development to develop the core framework of the United States Patent and Trademark Office. This study was conducted to classify the patents in terms of technology-assisted instruction using the Patent Co-citation Analysis (PCA) method and factor analysis. For the analysis, education, demonstration, rendition, instructional aids, instructional equipment, teaching aids, and didactic materials were chosen as keywords to construct a citation relationship network of patents and to classify core patent issues. The study results showed that 225 patents were cited more than 25 times. They were classified into 11 categories. The result provided information on the development and application of technology-assisted education to develop teaching tools further.

Keywords: patent statistics; citation analysis; patent co-citation analysis; technology-assisted instruction

Citation: Li, S.-P.; Hsu, W.-S.; Hsu, W.-L. Patent Statistics and Analysis of Development Trends of Technology-Assisted Instruction. *Eng. Proc.* **2023**, *38*, 72. https://doi.org/10.3390/engproc2023038072

Academic Editors: Teen-Hang Meen, Hsin-Hung Lin and Cheng-Fu Yang

Published: 4 July 2023

Copyright: © 2023 by the authors. Licensee MDPI, Basel, Switzerland. This article is an open access article distributed under the terms and conditions of the Creative Commons Attribution (CC BY) license (https://creativecommons.org/licenses/by/4.0/).

1. Introduction

Teaching aids, classroom digitization equipment, and technology have continued to be improved over the past two decades. Recently, educational technology, including auxiliary teaching of textbooks and teaching aids, has become an issue for educational changes and the integration of classroom teaching. Combining traditional education with teaching aids improves the interest and quality of students in classroom learning [1]. In traditional teaching methods, adding devices to teaching materials devices assists classroom instruction and supports classroom learning and teaching process with tools, technologies, equipment, software environment, and information-based resources. It helps students overcome their learning difficulties with interesting topics and comfortable feelings. With the development of teaching aids, researchers pay attention to intellectual property rights which are key in many fields of business. Learning for development is conclusive for the education system in the era of modern technology. Due to the increasing importance of knowledge, private companies, research institutes, and colleges have found that protecting intellectual property rights is critical. These previous works led us to find the best way for intellectual property rights such as patents [2].

The OECD Patent Statistics Manual (OECD, 2009) gives a detailed list of patents as statistical indicators of inventive activity and covers the advantages and disadvantages of the indicators of patent statistics in depth. Patent analysis has been regarded as a tool for the techno-economic analysis of R&D management and productivity of enterprises, as well as international innovation performance. Therefore, patents are considered a sufficient source of technical and commercial knowledge about the progress of technology, market trends, and ownership [3]. It is used as bibliometric data with various techniques to manipulate and analyze them. Patent citation analysis is the most widely used [4,5]. The patent citation analysis provides technical indicators such as patent citations, the cycle time of technology, and the impact index of technology. These indices have been used as indicators of the quality of technological assets, the economic value of innovation output in the market value equation, and technological coupling and knowledge flows within borders. Patent documents contain important research findings for educational, industrial, commercial, legal, and policy-makers [6]. Therefore, this research aims to study the development trends in technologies using patent information.

2. Literature Review

2.1. Teaching Aids

Teachers are presenters and players to encourage students to participate in learning and keep them vigilant and efficient in class [7]. In traditional teaching methods, adding devices to teaching materials assists classroom instruction and supports learning and teaching with tools, equipment, software, and information-based resources. It helps students overcome learning difficulties, makes the textbook interesting, and makes students feel competent [8].

2.2. Patent Analysis

The patent analysis requires bibliometric data with various techniques to manipulate and analyze it. Among them, patent citation analysis is the most adopted tool. It has been used to evaluate the competitiveness of firms [9], develop technology plans [10], prioritize R&D investment [11], or monitor technological change in firms [12]. Patent citation analysis is related to the bibliometric analysis of patent documents. Essentially, the methodology is citation-based to integrate patents precisely from the scientific paper databases [13]. Co-citation refers to different scientific mappings involving two processes: the cluster structure of co-cited documents and co-citation analysis. The result of co-citation clustering is to assign research papers to a co-citation cluster [14]. Recent studies have compared five citation-based approaches, including cross-reference, bibliographic coupling, co-citation, and text-based methods [15].

The co-citation analysis calculates the frequency of co-referenced documents to prove their similarity. The number of times co-referenced is not limited because new documents may reference A and B simultaneously. Therefore, the frequency at which documents are commonly cited is used effectively to evaluate their similarity and determine the literature and its evolution. The co-cited situation is presented in Figure 1.

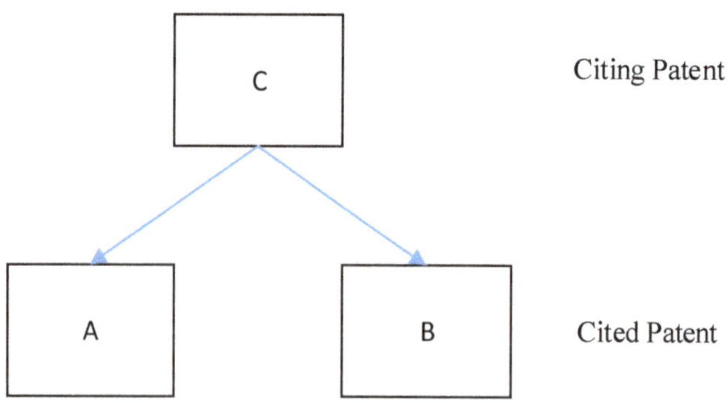

Figure 1. Relationship of co-citation.

3. Methodology

Co-cited analysis was originally used to measure the relationship between two publications. A common citation model can be constructed using co-cited analysis to determine the similarity between patents. We examined the development of technology-assisted instruction by employing the concept of co-citation and established citation relationships of technology-assisted instructions. We classified the technology-assisted instruction to identify the issues involved in patents.

3.1. Research Flow

3.1.1. Phase One: Establishing a Patent Citation Matrix

Confirming keywords for patent data retrieval, creating technology-assisted instruction patent and cited patent database, and establishing a patent citation matrix.

3.1.2. Phase Two: Technology-Assisted Instruction Clustering

Patent co-citation approach, factor analysis, and naming of specification factors.

3.2. Confirming Keywords

To retrieve technology-assisted instruction patents effectively, "education", "demonstration", "rendition", "Instructional Aids", "Instructional Equipment", "teaching aids", and "didactic materials" were set as keywords for subsequent search.

3.3. Sample and Data Collection

This study aimed to investigate the major trends of technology-assisted instruction technologies and to develop the framework using USPTO patent information. The search yielded 2225 technology-assisted instructions issued by the USPTO.

3.4. Measurement

The concept of co-citation in bibliometrics was employed to classify the patent specifications. The design concept was to select the most frequently cited specifications and use them as the specifications for classification. Subsequently, co-citation frequencies were used to evaluate the similarities between the patent specifications. Finally, the patent specifications were classified based on their similarities.

3.5. Similarities between Cited Specifications

Pearson's correlation coefficient was employed to investigate the similarities between pairs of cited specifications. This process consisted of three steps. In Step 1, we calculated the frequency with which the cited specification pairs were cited. In Step 2, the link strength within the cited specification pairs was calculated, and in Step 3, Pearson's correlation coefficients were calculated.

3.6. Factor Analysis for Specification Classification

In bibliometrics, the three most commonly used methods for co-citation analysis are factor analysis, cluster analysis, and multidimensional scaling analysis. This study employed factor analysis to obtain reduced and induction variables.

4. Results

The subjects of this study were 2225 technology-assisted instructions issued by the USPTO. The concept of co-citation classification in citation analysis was employed to develop a patent specification co-citation method for exploring the relationship between citing and cited specifications. In addition, factor analysis was used for specification classification, following which the categories were named based on their characteristics. This allowed us to determine whether the specification categories that were identified and summarized using bibliometrics resemble. Consequently, the correctness of this concept was established.

4.1. *Specification Collection*

The patent database was used as the source for obtaining specifications and selecting those issued by the USPTO. Full-text searches were conducted using "education, demonstration, instructional aids, instructional equipment, teaching aids, didactic materials" as keywords. From the 2225 specifications used as the citing specifications in this study, we obtained 76,298 cited specifications.

4.2. *Selection of Cited Specifications*

After combining repetitive citations and removing cited specifications with few citations, we obtained 206 cited specifications cited more than 30 times (including 1154 citations). In this study, we defined c as 30. Therefore, the citation relationship between cited specifications and citing specifications resulted in a new citation relationship matrix.

4.3. *Similarity Evaluation of the Cited Specifications*

To obtain Pearson's correlation coefficients for the cited specifications, three steps were taken between cited specifications and citing specifications.

4.3.1. Step 1: Calculating the Co-Citation Frequencies of Cited Specification Pairs

After obtaining the co-citation matrix consisting of cited specifications and cited specification pairs, the cited relationship matrix was integrated. The relationship matrix for cited specification pairs was transposed and multiplied to yield a symmetric cited specification co-citation matrix. An examination was conducted to check whether the co-citation matrix contained cited specification pairs with an excessively low co-citation frequency. However, no cited specifications were found to have been co-cited only once or not at all. Therefore, all of the cited specifications were retained.

4.3.2. Step 2: Calculating the Link Strength of Cited Specification Pairs

The co-citation matrix for cited specification pairs was integrated to yield a link strength matrix for cited specification pairs.

4.3.3. Step 3: Calculating Pearson's Correlation Coefficients

The link strength matrix for cited specification pairs was used to create a Pearson's correlation coefficient matrix for cited specification pairs using SPSS.

4.4. *Specification Factors*

In factor analysis, the specifications were classified into 14 categories. However, Categories 12 to 14 were removed because they contained a comparatively smaller number of specifications. Subsequently, we extracted the most frequently cited specifications in each category, identified the commonalities of the specification claims, and named each category based on their claims. The names of the categories were as follows: test system, test generating and formatting system, blended learning educational system, remote teaching system, computer-aided instruction, game-aided instruction, training system and method, internet-based education support, early childhood education aids system, technology-assisted learning, and cognitive ability training system. Table 1 shows the detailed bases for the naming and the commonalities.

Table 1. Bases for Factor Naming and Results.

	Basis for Naming	**Name**
Factor 1	US5321611A US4978305A US5466159A	test system

Table 1. *Cont.*

	Basis for Naming	Name
Factor 2	US6370355B1 US6470171B1 US6162060A	blended learning educational system
Factor 3	US5303042A US5437555A US6064856A	remote teaching system
Factor 4	108 F.3d 1361 134 F.3d 1473 927 F.2d 1200	test generating and formatting system
Factor 5	US5987443A US5974446A US5441415A	computer-aided instruction
Factor 6	US5286036A US5306154A US5035625A	game-aided instruction
Factor 7	US5035625A US4931018A US4680014A	training system and method
Factor 8	US6155840A US6688889B2 US6988138B1	internet-based education support
Factor 9	US5275567A US4968255A US5823782A	early childhood education aids the system
Factor 10	US6118973A US5779486A US6077085A	technology-assisted learning
Factor 11	US5692906A US5957699A US5813862A	cognitive ability training system

5. Conclusions

In teaching, teachers often need to prepare teaching plans through teaching aids according to the characteristics of students. With the advancement of technology, more teaching aids are developed to incorporate science and technology. Teachers can learn about the development of patent-related issues and enhance their understanding of patents and their application in teaching.

This study aimed to develop a co-citation classification method for technology-assisted instruction by examining the characteristics of the specifications and applying the co-citation classification method. Bibliometrics was used for the classification of technology-assisted instruction. This method enabled us to identify the relationships among specification citations. Subsequently, factor analysis was employed to classify specifications with close citation relationships and name the categories.

As a result, technology-assisted instructions were classified and named for categories to assist teachers in conducting patent specification searches. Each specification resembles a small database that contains a variety of data. In the future, the application of this method can be extended to develop a specification citation database, where helpful information such as patent numbers and relevant rulings can be accessed. The proposed specification can be used as basic information for a particular patent.

Author Contributions: S.-P.L.: Conceptualization, Writing—original draft; W.-S.H.: Data curation, Formal analysis; W.-L.H.: Methodology, Writing—review & editing. All authors have read and agreed to the published version of the manuscript.

Funding: This research received no external funding.

Institutional Review Board Statement: Not applicable.

Informed Consent Statement: Not applicable.

Data Availability Statement: Not applicable.

Conflicts of Interest: The authors declare no conflict of interest.

References

1. Wang, J. The Role of Information Technology-assisted Instruction and its Implementation Strategies. *Sci. Insights* **2022**, *40*, 545–547. [CrossRef]
2. Mok, M.S.; Sohn, S.Y.; Ju, Y.H. Conjoint analysis for intellectual property education. *World Pat. Inf.* **2010**, *32*, 129–134. [CrossRef]
3. Paci, R.; Sassu, A.; Usai, S. International patenting and national technological specialization. *Technovation* **1997**, *17*, 25–38. [CrossRef]
4. Narin, F. Patent bibliometrics. *Scientometics* **1994**, *30*, 147–155. [CrossRef]
5. Gupta, V.; Pangannaya, N. Carbon nanotubes: Bibliometric analysis of patents. *World Pat. Inf.* **2000**, *22*, 185–189. [CrossRef]
6. Yoon, B.; Park, Y. A text-mining-based patent network: Analytical tool for high-technology trend. *J. High Technol. Manag. Res.* **2004**, *15*, 37–50. [CrossRef]
7. Alshatri, S.H.H.; Wakil, K.; Jamal, K.; Bakhtyar, R. Teaching Aids Effectiveness in Learning Mathematics. *Int. J. Educ. Res. Rev.* **2019**, *4*, 448–453. [CrossRef]
8. Tonks, D. *Teaching Aids*; Routledge: New York, NY, USA, 2012.
9. Narin, F.; Noma, E.; Perry, R. Patents as indicators of corporate technological strength. *Res. Policy* **1987**, *16*, 143–155. [CrossRef]
10. Mogee, M.E. Using Patent Data for Technology Analysis and Planning. *Res. Manag.* **1991**, *34*, 43–49. [CrossRef]
11. Hirschey, M.; Richardson, V.J. Valuation effects of patent quality: A comparison for Japanese and U.S. firms. *Pacific-Basin Financ. J.* **2001**, *9*, 65–82. [CrossRef]
12. Archibugi, D.; Pianta, M. Measuring technological change through patents and innovation surveys. *Technovation* **1996**, *16*, 146–451. [CrossRef]
13. Karki, M. Patent citation analysis: A policy analysis tool. *World Pat. Inf.* **1997**, *19*, 269–272. [CrossRef]
14. Boyack, K.W.; Klavans, R. Co-citation analysis, bibliographic coupling, and direct citation: Which citation approach represents the research front most accurately? *J. Am. Soc. Inf. Sci. Technol.* **2010**, *61*, 2389–2404. [CrossRef]
15. Liu, X.; Yu, S.; Janssens, F.; Glänzel, W.; Moreau, Y.; De Moor, B. Weighted hybrid clustering by combining text mining and bibliometrics on a large-scale journal database. *J. Am. Soc. Inf. Sci. Technol.* **2010**, *61*, 1105–1119. [CrossRef]

Disclaimer/Publisher's Note: The statements, opinions and data contained in all publications are solely those of the individual author(s) and contributor(s) and not of MDPI and/or the editor(s). MDPI and/or the editor(s) disclaim responsibility for any injury to people or property resulting from any ideas, methods, instructions or products referred to in the content.

engineering proceedings

MDPI

Proceeding Paper

Scientific Empirical Study on Analyzing Learning Outcomes of Cadres in College Student Association Using Hierarchical Linear Regression Model †

Ming-Kuo Chen [1], Ruo-Lan Liu [2] and Hsin-Nan Chien [2,3,*]

1 General Education Center, Chaoyang University of Technology, Taichung 413310, Taiwan; mgchen@cyut.edu.tw
2 Holistic Education Center, National Taiwan Normal University, Taipei 244014, Taiwan; rlliu@ntnu.edu.tw
3 Student Affairs Office, Feng Chia University, Taichung 407102, Taiwan
* Correspondence: hnchien@fcu.edu.tw; Tel.: +886-9-28368564 or +886-9-37430670

† Presented at the 3rd IEEE International Conference on Electronic Communications, Internet of Things and Big Data Conference 2023, Taichung, Taiwan, 14–16 April 2023.

Abstract: We investigated the relationship between the personal backgrounds, learning experiences, and learning outcomes of college students serving as student association cadres by applying the indicators of the Council for the Advancement of Standards in Higher Education (CAS) in the United States. The participants of this study were students at the university student associations in Taiwan, and the sampling was carried out through purposeful sampling. We distributed 1850 questionnaires, and 1761 valid questionnaires were received with a recovery rate of 95%. The self-assessed learning outcomes of the student association cadres were good. Among them, "integration and speculation of cognition" was the highest, followed by "integration and application of knowledge", "inner self-development", "interpersonal interaction ability", "practice ability", and "social care and civic literacy". The self-assessment of the student association cadres in their learning experience was also positive. Among them, "experience in engagement" was the highest, followed by "experience in management", "experience in handling activities", and "experience in training courses"; while "experience in participating in activities" was the lowest. Through a hierarchical linear regression analysis of the American CAS learning outcome indicator, the student association cadres' personal background and learning experience had significant explanatory power. The R-squared value of each aspect ranged from 39.6% to 61.1%. It was obvious that the American CAS indicator could be used to examine the learning outcomes of higher education student associations in Taiwan. The results confirm that studying student associations could become an essential approach for cultivating students' learning outcomes and become an essential reference for Taiwan's universities and colleges to promote community education.

Keywords: student association cadres; student associations learning experience; learning outcomes

Citation: Chen, M.-K.; Liu, R.-L.; Chien, H.-N. Scientific Empirical Study on Analyzing Learning Outcomes of Cadres in College Student Association Using Hierarchical Linear Regression Model. *Eng. Proc.* **2023**, *38*, 73. https://doi.org/10.3390/engproc2023038073

Academic Editors: Teen-Hang Meen, Hsin-Hung Lin and Cheng-Fu Yang

Published: 4 July 2023

Copyright: © 2023 by the authors. Licensee MDPI, Basel, Switzerland. This article is an open access article distributed under the terms and conditions of the Creative Commons Attribution (CC BY) license (https://creativecommons.org/licenses/by/4.0/).

1. Introduction

Universities provide professional and extracurricular courses. Therefore, the learning approach for college students can be divided into in-class and out-class. Professors teach in-class courses through various methods such as teaching, practice, and experiments. Out-class courses depend on various course designs that can guide college students to participate, experience, self-examine, and obtain the effect of learning. Whether it is in-class learning or out-class learning, both should be reviewed through evaluation.

Reference [1] believes that learning outcomes refer to the abilities acquired from the learning process. Marsh [2] points out that the student learning outcome should include knowledge, skills, and personality traits. Learning outcomes can be known as the knowledge, technology, behavior, and value brought to students through the teaching courses provided by the school [3]. Yang [4] points out that many countries often consider learning outcomes in terms of competence; the degree of competence is considered a reflection of the learning outcomes. Marsh [2] believes evaluating learning outcomes in the educational environment will help the students understand the learning objectives, and teachers will master the teaching methods and lesson plans. It has always been an interesting topic for researchers to explore what college students have learned in college, especially in extracurricular activities. Chang [5] contends that the classroom imparts professional knowledge; however, the possibility of students developing talents relies on the learning outcomes that are obtained by students from extracurricular activities. Chen [6] believes that the learning outcomes obtained by students from extracurricular learning can be evaluated at many levels. For example, participation in the association is an important learning activity and is essential for schools to cultivate students' ability to have core literacy. In addition, reference [7] also points out that the experience gained from participating in the association is a critical background condition for entering the workplace in the future, especially for those with association cadre experience, which are the enterprises' targets. Previous studies verify that association cadres are the key to the operation of the association [8–10]. Therefore, it is necessary to use mathematical techniques to establish an evaluation mechanism for the learning experience and achievements of student association cadres.

However, many past studies on the topics of college students in the field of extracurricular activities did not systematically use science and engineering to explore the learning outcomes and related influencing factors. For example, when exploring the issue of college students involved in campus experience, only the variables of community participation were considered [11–13]. After reviewing relevant research, Ref. [7] summarized the learning experience of association cadres into five indicators, including (1) experience in management, (2) experience in handling activities, (3) experience in engagement, (4) experience in training courses, and (5) experience in participating in activities, and compiled a questionnaire for experts and scholars to measure the learning experience of the association cadres.

In addition, Ref. [14] integrated the literature and proposed the following nine core competencies of learning outcomes obtained from extracurricular activities: (1) communication and expression ability, (2) all-life learning ability, (3) team cooperation ability, (4) innovation and creativity ability, (5) problem-solving ability, (6) positive attitude ability, (7) interpersonal ability, (8) critical thinking ability, and (9) international perspective ability. In Ref. [6], the authors distributed 1033 questionnaires according to these nine indicators and concluded that students' extracurricular learning outcomes could achieve the abilities mentioned earlier. The American Council for the Advancement of Standards in Higher Education (CAS) is an important organization that promotes the standards of student affairs, services, and development programs. CAS [15] concluded that the evaluation of learning outcomes includes the following six orientations: (1) integration and application of knowledge, (2) integration and speculation of cognition, (3) inner self-development, (4) interpersonal interaction ability, (5) social care and civic literacy, and (6) practice ability. However, in Taiwan, there is no empirical research on the ability indicators in these six orientations.

Based on the above, this study was carried out to integrate relevant research and the CAS assessment of students' learning outcomes through interviews with experts and scholars and a questionnaire survey. The learning achievement indicators and the connotations of student association cadres were summarized to explore the learning outcome of students' association cadres. Based on a review of the literature, the objectives of this study are defined as follows:

(1) To understand the general situation of the learning experience and learning outcomes of the university association cadres;
(2) To explore the differences in personal background variables on the learning experience and outcomes of the university association cadres;

(3) To explore the relationship between personal background variables, learning experience, and learning outcomes of the university association cadres from a scientific point of view.

2. Research Method

We adopted a questionnaire survey [7] to understand the association cadres' learning experiences and learning outcomes. We used stratified sampling to filter the 155 universities in Taiwan. As a result, 15 universities were selected, including 5 each from northern, central, and southern Taiwan, respectively. A total of 1850 questionnaires were distributed, and 1761 valid responses were returned, with an effective rate of 95%. The questionnaire was divided into the following three parts: personal information, association learning experience, and association learning outcomes. All questions were designed based on a six-point Likert scale. A T-test and variance analysis were conducted to understand the differences in the learning experiences and learning outcomes of university association cadres with different personal backgrounds. Finally, a hierarchical regression analysis was used to explore the relationship between personal background, learning experiences, and learning outcomes of university association cadres.

The Bartlett test, the pre-test through the exploratory factor analysis, showed that a significance level was reached ($p < 0.05$). The KMO values of the students' association learning experience and the learning outcome were 0.932 and 0.963, indicating that the factor analysis was appropriate. Therefore, the formal questionnaire was continued to test the reliability and validity. The KMO values of the student association learning experience and the learning outcome were 0.956 and 0.975. The learning experience was extracted into the following five aspects: experience in management, experience in engagement, experience in training courses, experience in handling activities, and experience in participating in activities. The cumulative variance was 70.13%. The Cronbach's alpha was 0.95, 0.92, 0.91, 0.87, and 0.77 for the five aspects, and the overall reliability was 0.96. The learning outcome was extracted into the following six aspects: integration and application of knowledge, integration and speculation of cognition, inner self-development, interpersonal interaction ability, social care, civic literacy, and practice ability. The cumulative variance was 73.92%. The Cronbach's alpha was 0.95, 0.92, 0.92, 0.93, 0.87, and 0.90 for the six aspects, and the overall reliability was 0.97. Therefore, the questionnaires on the learning experiences and learning outcomes showed a high reliability and validity, and the items of the questionnaire had acceptable internal consistency and stability.

3. Research Findings

The results show that in the learning experiences among the association cadres, the perception of "Experience in engagement" (M = 5.09) was the highest, followed by "experience in organization management" (M = 4.98), "Experience in management" and "Experience in training courses" (M = 4.91), and the "Experience in participating in activities" (M = 4.62). Overall, the perception of the learning experience among the association cadres was significant. In the learning outcomes among the association cadres, perception in "Integration and speculation of cognition" was the highest (M = 5.23), followed by "Integration and application of knowledge" (M = 5.18), "Inner self-development" (M = 5.16), "Interpersonal interaction ability" (M = 5.12), "practice ability" (M = 5.03), and the perception of "social care and civic literacy" (M = 4.74). Overall, the perception of learning outcomes among the association cadres was acceptable.

Regarding the perception of a personal background variable on the learning experience, the *t*-test result showed significant differences in the gender, school attributes, school education system, faculty, grades, association attributes, number of cadres, seniority, positions held, and weekly time devoted to the associations. The results show that the males had a higher perception than the females, private universities had a higher perception than national universities, general universities had a higher perception than the institutes of technology, seniors had a higher perception than juniors, presidents had a higher perception than general cadres, and the self-assessment of the learning experience of those with more engagement time had higher perceptions than those with less engagement time.

A multiple regression analysis was performed to explore the relationship between the personal background variables of the association cadres and their learning experiences. The results show that the R^2 of "Experience in management" was 0.267, that is, the explanatory power was 26.7%. The R^2 of "Experience in handling activities", "Experience in engagement", "Experience in training courses", and "Experience in participating in activities" was 0.134, 0.096, 0.206, and 0.163 respectively. The personal background factor had a significant difference compared with the learning experience (Table 1).

Table 1. Regression Analysis of Background Variable on Learning Experience.

	Organization Management Experience	Event Management Experience	Association Investment Experience	Participation in Training Course Experience	Participation in Activity Experience
R^2	0.267	0.134	0.206	0.163	0.096
F	28.822 ***	13.374 ***	20.509 ***	15.366 ***	8.411 ***
p	0.000	0.000	0.000	0.000	0.000

*** $p < 0.0001$.

The personal background and learning experience were used as predictors, and the learning outcome factor was used as a criterion to proceed with the hierarchical regression analysis to explore the relationship. The results show that the personal background and learning experience significantly impacted the learning outcomes. "Inner self-development" had the highest explanatory power of 61.1%, followed by "Integration and application of knowledge", 59.3%; "interpersonal interaction ability", 52.2%; "Integration and speculation of cognition", 42.8%; and "practice ability", 40.9%. The "social care and civic literacy" had the lowest explanatory power of 39.6%. Overall, the learning experience of the student association cadres had a significant positive impact on the learning outcomes. The results are shown in Table 2.

Table 2. Explanatory Power of the Aspects.

Aspects	Explanatory Power
Experience in management	26.7%
Experience in handling activities	13.4%
Experience in engagement	20.6%
Experience in training courses	16.3%
Experience in participating in activities	9.6%
Integration and application of knowledge	59.3%
Inner self-development	61.1%
Interpersonal interaction ability	52.2%
Integration and speculation of cognition	42.8%
Social care and civic literacy	39.6%
Practice ability	40.9%

4. Conclusions and Suggestions

In terms of learning experience, "Experience in engagement" was the most important, while "Experience in participating in activities" was the least important. Overall, the perception of the learning experience of the student association cadres was significant. In the perception of learning outcomes, the perception of "Integration and speculation of cognition" was the most important, while the perception of "social care and civic literacy" was the least important. Overall, the perception of the learning outcomes of the student association cadres was significant. Student association cadres with different personal backgrounds, such as gender, school attributes, school education system, colleges, grades, association attributes, number of cadres, seniority, positions, and weekly time devoted to associations, had different perceptions of the learning experience and learning outcomes.

The personal background of the student association cadres was significant in the learning experience. "Experience in management" had the highest impact, followed by "Experience in engagement", "Experience in training courses", "Experience in handling activities", and "Experience in participating in activities". As for the results of the learning experiences on the learning outcomes, "Experience in engagement" had the highest impact, followed by "Experience in handling activities", "Experience in training courses", "Experience in management", and "Experience in participating in activities". Overall, the learning experience had a significant positive impact on the learning outcomes.

The study result showed that the accumulation of learning experiences was essential. Higher education practitioners can increase learning opportunities for the student association cadres to participate in and accumulate the learning experiences. According to the results of the learning outcomes, the self-evaluation for students in the "social care and civic literacy" learning outcome was the least important. Therefore, schools must strengthen their engagement in "social care and civic literacy" when providing learning experiences such as contests regarding global issues, international perspective, and civic literacy.

According to the results, the student association cadres who served as the "president", and who had a longer "seniority" and a longer "weekly devotion to the association", had a positive relationship with the learning experience. Therefore, the student association cadres need to undertake the position of "President", devote more time to associate and accumulate seniority and the learning experiences of "Experience in management", "Experience in handling activities", "Experience in engagement", "Experience in training courses", and "Experience in participating in activities". Furthermore, the personal learning experience needs to be strengthened to obtain high-quality learning outcomes, including "Integration and application of knowledge", "Inner self-development", "Interpersonal interaction ability", "Integration and speculation of cognition", and "Practice ability".

Author Contributions: Conceptualization, M.-K.C. and H.-N.C.; methodology, H.-N.C. and R.-L.L.; writing—original draft preparation, M.-K.C., R.-L.L. and H.-N.C.; writing—review and editing, M.-K.C. and H.-N.C. All authors have read and agreed to the published version of the manuscript.

Funding: This research received no external funding.

Institutional Review Board Statement: Not applicable.

Informed Consent Statement: Not applicable.

Data Availability Statement: The data of this study was used to support the findings of this study are available from the corresponding author upon request.

Conflicts of Interest: The authors declare no conflict of interest.

References

1. Spady, W.G. Outcome-Based Education: Critical Issues and Answers. Virginia: American Association of School Administrators. 1994. Available online: https://files.eric.ed.gov/fulltext/ED380910.pdf (accessed on 15 November 2022).
2. Marsh, C.J. A critical analysis of the use of formative assessment in schools. *Educ. Res. Policy Pract.* **2007**, *6*, 25–29. [CrossRef]
3. Peng, S.M. *Institutional Research in Higher Education: Concepts and Applications*; Higher Education Press: Taipei, Taiwan, 2019. [CrossRef]
4. Yang, Y. EU higher education quality assurance policy focusing on student learning outcomes. *Eval. Bimon.* **2011**, *30*, 27–34. [CrossRef]
5. Chang, H.M. *Intervening or Letting go—Problems and Counseling Countermeasures for College Students to Participate in Extracurricular Activities*; Lin, Z.S., Ed.; Student Affairs and Community Counseling Series; Soochow University: Taipei, Taiwan, 2003; Volume 4, pp. 350–366.

6. Chen, M.K. College Students' Co-Curricular Learning Outcomes and Influencing Factors Research–From the Strategic Perspective of Institutional Research. Ph.D. Thesis, National Taiwan Normal University, Taipei, Taiwan, 2018.
7. Chien, H.N. A Study on the Construction of Learning Experience and Outcome Indicators and Investigation for Higher Education Student Organization Cadres. Ph.D. Thesis, National Taiwan Normal University, Taipei, Taiwan, 2020.
8. Huang, Y. Theory and application of university student affairs. In *Student Affairs and Community Counseling*; Lin, C.S., Ed.; Soochow University: Taipei, Taiwan, 2001; pp. 35–80.
9. Chang, T.M. A Preliminary Study on the Relationship between Personal Traits and Leadership Ability—Taking the Leaders of Student Associations in Southern Colleges as an Example. *J. Gen. Educ. Tainan Univ. Technol.* **2007**, *6*, 165–194.
10. Tsai, C.H. The Research on Higher Education Extracurricular Learning Core Competencies and Measuring Indicators. Ph.D. Thesis, National Taiwan Normal University, Taipei, Taiwan, 2012.
11. Astin, A.W. *Assessment for Excellence: The Philosophy and Practice of Assessment and Evaluation in Higher Education*; America Council on Education Macmillan: New York, NY, USA, 1991.
12. Terenzini, P.T.; Springer, L.; Pascarella, E.T.; Nora, A. Influences affecting the development of students' critical thinking skills. *Res. High. Educ.* **1995**, *36*, 23–39. [CrossRef]
13. Kuh, G.D. What we're learning about student engagement from NSSE. *Change Mag. High. Learn.* **2003**, *35*, 24–32. [CrossRef]
14. Tsai, C.H. *Student Association Management and Guidance Theory, Practice and Application*; Soochow University: Taipei, Taiwan, 2016.
15. Wells, J. *Professional Standards for Higher Education*, 9th ed.; Council for the Advancement of Standards in Higher Education: Washington, DC, USA, 2015.

Disclaimer/Publisher's Note: The statements, opinions and data contained in all publications are solely those of the individual author(s) and contributor(s) and not of MDPI and/or the editor(s). MDPI and/or the editor(s) disclaim responsibility for any injury to people or property resulting from any ideas, methods, instructions or products referred to in the content.

engineering proceedings

Proceeding Paper

Pairs Trading Strategies in Cryptocurrency Markets: A Comparative Study between Statistical Methods and Evolutionary Algorithms †

Po-Chang Ko [1,2], Ping-Chen Lin [2,3], Hoang-Thu Do [1,*], Yuan-Heng Kuo [1], You-Fu Huang [2] and Wen-Hsien Chen [3]

1 Department of Intelligent Commerce, National Kaohsiung University of Science and Technology, Kaohsiung 82445, Taiwan; cobol@nkust.edu.tw (P.-C.K.); f110156133@nkust.edu.tw (Y.-H.K.)
2 AI Fintech Center, National Kaohsiung University of Science and Technology, Kaohsiung 82445, Taiwan; lety@nkust.edu.tw (P.-C.L.); hyoufu@nkust.edu.tw (Y.-F.H.)
3 Department of Finance and Information, National Kaohsiung University of Science and Technology, Kaohsiung 82445, Taiwan; wehs71280@nkust.edu.tw
* Correspondence: i109167104@nkust.edu.tw
† Presented at the 3rd IEEE International Conference on Electronic Communications, Internet of Things and Big Data Conference 2023, Taichung, Taiwan, 14–16 April 2023.

Abstract: Pairs trading is a popular quantitative trading strategy with the advantage of a similarity in price movement to financial assets. Assuming that the price spreads of trading pairs are mean-reverting, this strategy exploits the disequilibrium in financial markets to find arbitrage investment opportunities. Pairs trading has been widely applied to stock, ETF, and commodity markets. However, the effectiveness of this method for cryptocurrency markets has yet to be properly explored. Therefore, we examine the profitability of pairs trading for 26 cryptocurrencies traded on the Binance exchange at high frequencies of 1, 5, and 60 min. In addition to the traditional statistical methods of distance, correlation, cointegration, and stochastic differential residual (SDR), we focus on two evolutionary algorithms: genetic algorithm (GA) and non-dominated sorting genetic algorithm II (NSGA-II). During the 79-trading-day period from 11 January to 31 March 2018, NSGA-II showed the best results at all frequencies, with an average return of 2.84%. Among the statistical models, SDR ranks first, whereas Correlation ranks last, with average returns of 1.63% and −0.48%, respectively. The z-test results show that the models are statistically significantly different. We propose NSGA-II as the best candidate for use in pairs trading strategies in cryptocurrency markets.

Keywords: pairs trading; cryptocurrency markets; NSGA-II

Citation: Ko, P.-C.; Lin, P.-C.; Do, H.-T.; Kuo, Y.-H.; Huang, Y.-F.; Chen, W.-H. Pairs Trading Strategies in Cryptocurrency Markets: A Comparative Study between Statistical Methods and Evolutionary Algorithms. *Eng. Proc.* **2023**, *38*, 74. https://doi.org/10.3390/engproc2023038074

Academic Editors: Teen-Hang Meen, Hsin-Hung Lin and Cheng-Fu Yang

Published: 6 July 2023

Copyright: © 2023 by the authors. Licensee MDPI, Basel, Switzerland. This article is an open access article distributed under the terms and conditions of the Creative Commons Attribution (CC BY) license (https://creativecommons.org/licenses/by/4.0/).

1. Introduction

Pairs trading is a strategy in which an investor simultaneously buys undervalued assets and sells overvalued assets when a disequilibrium condition is detected in a specific financial market. This method has proven to be effective in stock markets, commodity markets, and ETF markets [1]. Recently, the cryptocurrency market has emerged as a profitable and risky investment channel for investors. Therefore, this study is carried out to suggest a suitable pairs trading strategy for the cryptocurrency market by answering two questions: (1) Is pairs trading an effective trading strategy in the cryptocurrency market, especially with high trading frequencies? (2) Among the current most popular techniques for pairs selection, which ones are the best?

Most research on pairs trading applies traditional pairs selection methods such as distance, cointegration, or correlation. We explore two more evolutionary algorithms for selecting trading pairs: genetic algorithm (GA) and non-dominated sorting genetic algorithm II (NSGA-II). Moreover, we also conduct rigid tests to examine the significant differences among the methods.

The rest of this article is organized as follows. Section 2 describes the methodologies of pairs selection methods, trading strategies as well as data and experimental design, Section 3 analyzes the trading results and give discussion.

2. Methodology

The research flow chart is depicted in Figure 1 with five steps: processing the data obtained from the Binance Exchange API, selecting trading pairs based on four traditional statistical methods and two evolutionary algorithms, proposing trading strategies, analyzing trading results based on six different criteria, and finally conducting z-tests to confirm the significant differences between the methods.

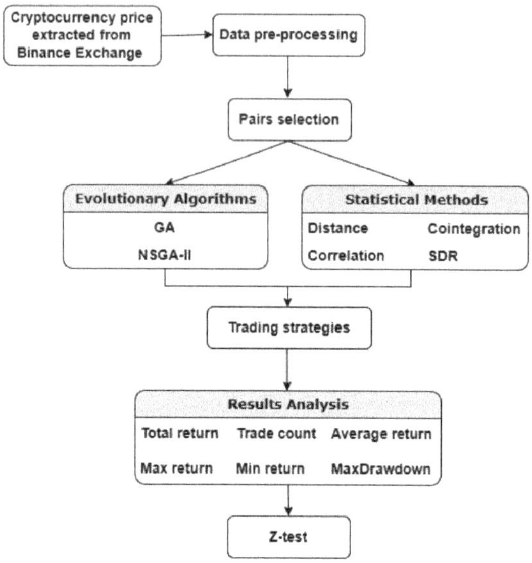

Figure 1. Research flow chart.

2.1. Pairs Selection

We deploy six techniques to select trading pairs, including four statistical methods and two evolutionary algorithms.

2.1.1. Euclidean Distance Method

The Euclidean distance method is still prevalent in choosing trading pairs thanks to its simplicity. The basic concept is that pairs with the smallest distance are selected based on the application of Euclidean squared distance to normalized price series.

2.1.2. Cointegration

Cointegration is a term coined by Granger [2] and Engle and Granger [3] to describe the long-run relationship between two assets. Granger was awarded the Nobel Prize in Economics in 2003 for his contributions to this concept. When two assets are cointegrated, their prices follow a similar pattern or, in financial terms, the two assets have similar risk exposure so that their prices move alongside. Cointegration techniques have been utilized to investigate the co-movement of a wide range of financial assets, including equity shares, commodities, exchange rates, and, more recently, cryptocurrencies. In this study, the augmented Dickey–Fuller test (ADF) [4] is used to determine whether or not a given time series is stationary. The null hypothesis of the ADF test is that there is a unit root (or a stochastic trend) in the residual series, with the alternative being that there is no unit root. If the null hypothesis is rejected, the two assets are cointegrated and selected to form a pair.

2.1.3. Correlation

The correlation is quantified using the correlation coefficient ρ, with values ranging from −1 to +1. A value of +1 means that a perfect positive correlation exists between the two variables, i.e., when one is moving in a certain direction, the other is also moving in the same direction with the same magnitude. A value of −1 means that there is a perfect negative correlation and 0 means there is no correlation at all. If the correlation between the two assets is high, the trader can choose that pair. This value represents a strong relationship between the two stocks. Correlation does not have a well-defined relationship with cointegration. Cointegrated series may have low correlation, and highly correlated series may not be cointegrated.

2.1.4. Stochastic Differential Residual (SDR)

SDR was developed by Do, Faff, and Hamza [5], which applies CAPM and APT theories to specify the residual spread function G_t as follows.

$$G_t = G\left(P_t^A, P_t^B, U_t\right) = R_t^A - R_t^B - \Gamma r_t^m \quad (1)$$

where R_t^A, R_t^B are the expected return on assets A and B, respectively, Γ is the vector of exposure differentials or the sensitivity of the asset price to macroeconomic factors, and r_t^m is the risk premiums.

2.1.5. Genetic Algorithm (GA)

Genetic algorithm (GA) is a computer science technique for solving combinatorial optimization problems. GA simulates evolutionary adaptations of biological populations based on Darwin's theory. To apply GA, we need to determine the coding of chromosomes, the fitness function, the chromosomes selection method, the crossover method, the mutation method, and the termination condition.

2.1.6. Non-Dominated Sorting Genetic Algorithm II (NSGA-II)

NSGA-II is another genetic algorithm with three unique features: a fast non-dominated sorting approach, a fast crowded distance estimation procedure, and a simple crowded comparison operator. NSGA-II follows the steps of population initialization, non-dominated sort, crowding distance, selection, genetic operators, recombination, and selection [6].

2.2. Trading Strategies

After the pairs are generated from the algorithms mentioned above, we set the trading strategies. We apply the Bollinger bands of the spread between security A and security B, in which the upper limit of the Bollinger bands at time t is $SMA_t + 2STD_t$, and the lower limit is SMA_t—$2STD_t$, where SMA is the simple moving average value and STD is the standard deviation value. If the spread exceeds the upper limit, we short security A and long security B. On the contrary, if the spread is smaller than the lower limit, we long security A and short security B. For the timing of closing the position, if the spread value goes back to the SMA value at a certain point in time, we close the position. Additionally, when the position is held over the trading period, the position is forced to close.

The trading strategies are described in Figure 2.

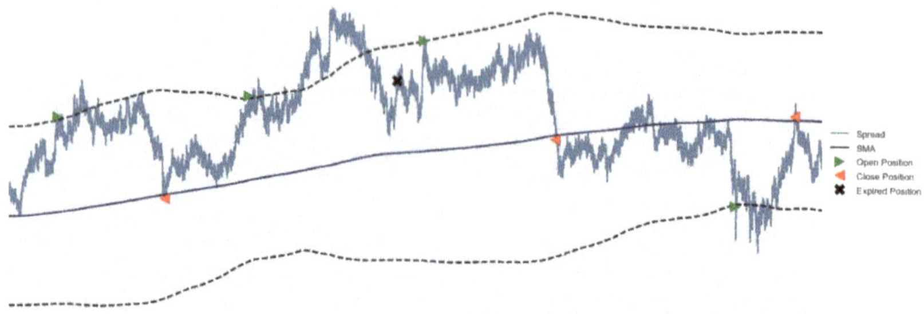

Figure 2. Trading strategies.

2.3. *Data and Experimental Design*

2.3.1. Data

The closing prices are taken from the global cryptocurrency exchange—Binance—for three months, from 3 January 2018 to 31 March 2018, with three different frequencies, 1, 5, and 60 min (1 h). Based on Ref. [7], we start with 183 cryptocurrencies on Binance, and then weed out those that have been inactive in the market for the considered period. Next, we keep 30% of the cryptocurrencies with the top trading volume to ensure liquidity conditions. Finally, the remaining 26 cryptocurrencies continue to the next steps. The currency of the original raw data on Binance is BTC. To make it easier to read and track the experimental results, we have converted the currency intSo USDT (Tether (USDT), a unit of cryptocurrency designed so that a base USD represents each token.

2.3.2. Experimental Design

The experimental setup is described as follows.

- Initial principle for each trade: US $1000.
- The SMA and the STD are calculated based on the sliding window with a forming period of five time units and a trading period of one time unit.
- We select five pairs for every sliding window, except for GA and NSGA-II, because these two methods may not select out five pairs.

3. Result and Discussion

Table 1 describes the trading results of six selection methods based on different criteria.

Table 1. Trading results.

Algorithm	TotalReturn			TradeCount			AVGReturn			MaxReturn			MinReturn			MaxDrawdown		
	1 min	5 min	60 min	1 min	5 min	60 min	1 min	5 min	60 min	1 min	5 min	60 min	1 min	5 min	60 min	1 min	5 min	60 min
Distance	1.66	1.51	1.10	234	215	172	0.01	0.01	0.01	31.31	31.31	19.68	−10.21	−10.24	−14.86	0.08	0.05	0.32
Cointegration	1.39	0.98	0.31	212	204	155	0.01	0.00	0.00	40.91	40.09	33.70	−32.06	−34.10	−32.00	0.38	0.17	0.00
Correlation	−0.52	−0.42	−0.49	201	195	172	0.00	0.00	0.00	36.85	40.52	45.93	−29.94	−30.53	−32.64	0.64	0.00	0.00
SDR	2.11	1.40	1.39	215	208	183	0.01	0.01	0.01	61.71	67.28	65.41	−64.18	−64.48	−57.61	0.52	0.72	0.51
GA	0.48	0.99	1.12	183	171	119	0.00	0.01	0.01	60.88	66.37	56.38	−60.91	−64.48	−57.59	0.01	0.46	0.00
NSGA-II	3.59	2.68	2.24	196	178	113	0.02	0.02	0.02	31.04	27.86	26.08	−28.17	−55.66	−33.31	1.40	1.51	0.50

The experimental results show that NSGA-II has the highest return rates at each data frequency of 1, 5, and 60 min, with an average return rate of 2.84% from 11 January 2018 to 31 March 2018—79 days in total. The average annualized rate of return is about 13.8%. Among the statistical models, SDR performs the best, whereas correlation performs the worst, with average returns of 1.63% and −0.48%, respectively. The average rate of return of GA is 0.86%. Despite a positive return, it is lower than we expected. The possible reason for this is that GA is a single-objective optimization technique, so it does not consider other risk factors, resulting in a lower-than-expected performance. It is also noted that the min return of GA is among the lowest. Table 2 shows that NSGA-II significantly differs from other algorithms at the 99% confidence level in all frequencies, except for GA and SDR.

Table 2. z-test results.

Frequency	1 min	5 min	60 min
NSGA-II	1.0000	1.0000	1.0000
Distance	0.0000 ***	0.0031 ***	0.0000 ***
Cointegration	0.0000 ***	0.0000 ***	0.0000 ***
Correlation	0.0000 ***	0.0000 ***	0.0000 ***
SDR	0.1771	0.2173	0.1961
GA	0.0000 ***	0.0241 **	0.0561 *

* Significant at the 10% level; ** significant at the 5% level; *** significant at the 1% level.

Author Contributions: Conceptualization, P.-C.K., P.-C.L., H.-T.D. and W.-H.C.; methodology, P.-C.K., P.-C.L. and Y.-H.K.; software, Y.-F.H. and Y.-H.K.; validation, P.-C.K. and Y.-F.H.; formal analysis, H.-T.D. and Y.-H.K.; investigation, P.-C.K. and P.-C.L.; data curation, P.-C.L. and Y.-H.K.; writing—original draft preparation, H.-T.D.; writing—review and editing, P.-C.L. and H.-T.D.; visualization, H.-T.D. and Y.-H.K.; supervision, P.-C.K., P.-C.L. and W.-H.C. All authors have read and agreed to the published version of the manuscript.

Funding: This research received no external funding.

Institutional Review Board Statement: Not applicable.

Informed Consent Statement: Not applicable.

Data Availability Statement: The datasets for this study are available upon request from the corresponding author.

Conflicts of Interest: The authors declare no conflict of interest.

References

1. Krauss, C. Statistical arbitrage pairs trading strategies: Review and outlook. *J. Econ. Surv.* **2017**, *31*, 513–545. [CrossRef]
2. Granger, C.W.J. Some properties of time series data and their use in econometric model specification. *J. Econom.* **1981**, *16*, 121–130. [CrossRef]
3. Engle, R.F.; Granger, C.W.J. Co-integration and Error Correction: Representation, Estimation and Testing. *J. Econom. Soc.* **1987**, *55*, 251–276. [CrossRef]
4. Dickey, D.A.; Fuller, W.A. Distribution of the estimators for autoregressive time series with a unit root. *J. Am. Stat. Assoc.* **1979**, *74*, 427–431.
5. Do, B.; Faff, R.; Hamza, K. A new approach to modeling and estimation for pairs trading. In Proceedings of the 2006 Financial Management Association European Conference, Durham, UK, 20–22 April 2006.
6. Yusoff, Y.; Ngadiman, M.S.; Zain, A.M. Overview of NSGA-II for optimizing machining process parameters. *Procedia Eng.* **2011**, *15*, 3978–3983. [CrossRef]
7. Fil, M.; Kristoufek, L. Pairs trading in cryptocurrency markets. *IEEE Access* **2020**, *8*, 172644–172651. [CrossRef]

Disclaimer/Publisher's Note: The statements, opinions and data contained in all publications are solely those of the individual author(s) and contributor(s) and not of MDPI and/or the editor(s). MDPI and/or the editor(s) disclaim responsibility for any injury to people or property resulting from any ideas, methods, instructions or products referred to in the content.

engineering proceedings

MDPI

Proceeding Paper

A Comparison between Digital-Game-Based and Paper-Based Learning for EFL Undergraduate Students' Vocabulary Learning †

Alex Dharmawan Sianturi and Ruei-Tang Hung *

Applied English Department, College of Humanities and Social Sciences, Chaoyang University of Technology, Taichung 413310, Taiwan; alexdharmawan1997@gmail.com
* Correspondence: hb865777@gmail.com

† Presented at the 3rd IEEE International Conference on Electronic Communications, Internet of Things and Big Data Conference 2023, Taichung, Taiwan, 14–16 April 2023.

Abstract: This research aimed to compare two strategies for vocabulary learning (digital-game-based learning and paper-based learning). The research was conducted during the first semester of the academic year 2022/2023. A total of 40 EFL undergraduate students within the Applied English Program of a private university located in the middle part of Taiwan were selected and divided into two groups: digital-based (n = 20) and paper-based (n = 20). The instrument developed by the researcher was pre- and post-vocabulary tests for both groups. The pre-vocabulary test was implemented to identify the level of students' prior knowledge of vocabulary mastery. For the intervention, Kahoot! quiz exercises were conducted weekly for the digital-game-based group, while the paper-based group received the same quiz on paper every week. The post-vocabulary tests showed no significant difference between the students using digital-game-based quizzes and paper-based quizzes during the six-week intervention.

Keywords: vocabulary learning; digital-game-based learning; paper-based learning; Kahoot!

Citation: Sianturi, A.D.; Hung, R.-T. A Comparison between Digital-Game-Based and Paper-Based Learning for EFL Undergraduate Students' Vocabulary Learning. *Eng. Proc.* **2023**, *38*, 78. https://doi.org/10.3390/engproc2023038078

Academic Editors: Teen-Hang Meen, Hsin-Hung Lin and Cheng-Fu Yang

Published: 11 July 2023

Copyright: © 2023 by the authors. Licensee MDPI, Basel, Switzerland. This article is an open access article distributed under the terms and conditions of the Creative Commons Attribution (CC BY) license (https://creativecommons.org/licenses/by/4.0/).

1. Introduction

With the prevalence of mobile devices in modern educational settings, the use of mobile devices has provided teachers with more options to incorporate digital games into their teaching methods. The swift advancement of technology has made it easier to share information and knowledge on global scale, greatly influencing the way the learners and educators learn and teach language at schools. In the classroom, technology includes everything from low-tech items, such as pencils, paper, and chalkboards, to high-tech resources, such as digital learning tools, gadgets, and computers with numerous functions and applications [1]. Students are empowered by technology, since it allows them to decide how they learn, makes education relevant to their digital-orientated lives, and prepares them long term. Students are encouraged to think critically, solve problems, collaborate, and innovate.

M-learning encourages learning and teaching via the use of wireless and mobile technology [2]. Laptops, tablet PCs, PDAs, cell phones, and other mobile devices are all considered aspects of mobile learning. There are many chances for educators to adopt digital learning tools and create an increased usage of mobile devices or mobile learning [3]. Digital-game-based learning (DGBL) is a learning strategy that blends educational information or learning concepts into video games to engage learners [4]. Game-based learning is frequently related to studying the effects of GBL, implementing educational games, and creating game-based educational techniques [5]. The arrival of digital game-based learning occurred during a global technological breakthrough in the late 20th century.

DGBL possesses the capability to capture students' interest and motivation, provide customized learning experiences, while enhancing long-term memory and offering practical opportunities [6].

Vocabulary is central to language for language learners, and a critical element of language pedagogy [7]. Teachers employ a variety of ways and techniques to teach vocabulary to ESL/EFL students, and practitioners must familiarize students with different strategies and approaches [8–10]. Synonyms, antonyms, self-defining context, dramatization, illustrations and pictures, realia, metaphorical sentences, and other strategies were used, respectively, by teachers in vocabulary teaching [11,12]. In addition, vocabulary is the center of language and the structure block of speech, and learning new words is a crucial element of language education [13–15].

There are several approaches to language acquisition in the technology era. Computer-based vocabulary learning software, blogs, cell phones, digital portfolios, online learning, electronic texting, pen pal, electronic displays, electronic reader corpora, mobile-phone-based apps such as Twitter and text messages, social networks, podcasts, and wiki websites are examples of such technologies. The availability of online applications facilitates the creation of interactive materials for students by teachers. Teachers can make learning sessions more fascinating for their students. They can make games or quizzes competitive through a leaderboard [16].

Kahoot! is one of the applicable interactive quizzes implemented in learning vocabulary. Moreover, Kahoot! is complimentary and superficial for the instructor to use for learning in the classroom. Kahoot! also supports creative energy and increases performance to achieve a stronger motivation of the student [17,18]. Several studies proved that the application creates some advantages. Kahoot! is a favorable instrument as it is beneficial for formative assessment, and enhances understanding [19]. The application of gamification improved learners' attention and drive for success [20]. Despite the advantage, some studies found that Kahoot! has some difficulties. First-time users of m-learning were dissatisfied with the functionality and features of the available tools, including the slow internet connection [21]. It showed that a slow internet connection causes the biggest problem for implementing Kahoot! within learning [22–24]. Therefore, it is essential to offer suitable facilities to help new users.

In Taiwan, a majority of university students use advanced technology such as mobile devices, laptops, and computers. Based on prior research, we focused on the use of mobile devices and investigated a vocabulary learning strategy that involved weekly quizzes. The students used Kahoot! as a tool to complete the quizzes. The aim of this research was to compare two different strategies: digital-game-based learning using Kahoot! quizzes, and paper-based learning using paper-based quizzes, and their impact on the vocabulary learning of EFL (English as a foreign language) students. As a result, the following research question was formulated for this research.

1. Is there any significant difference between the mean scores of the students who used digital-game-based learning (Kahoot! quiz) and the mean scores of the students who used paper-based quizzes?
2. What are student's perceptions toward the two strategies for their vocabulary learning?

2. Method

2.1. Sample and Population

The sample consisted of 38 students who were enrolled in an intermediate English course. The participants were purposefully selected among second-year students at Chaoyang University of Technology during the fall semester of the academic year 2022/2023. The participants were separated into two distinct groups: namely the digital-game-based group and the paper-based group, each consisting of 19 students.

2.2. Procedure

The researcher distributed six vocabulary quizzes to both groups to review the material covered in the course. Only the digital-game-based group received the quiz exercise through Kahoot!, whereas the paper-based group received the quiz exercise through paper and pen. Students in the Kahoot! class could see their scores right after the session finished. Contrastingly, the paper-based group took a similar quiz on paper, but received their results the following week. The goal of the quiz exercise for both groups was to see how well students comprehended the vocabulary. The researcher informed the students that the results of those quizzes had no relevance to their grades, but the scores from the post-test would be collected and calculated for grading purposes within the current course. The post-test was given to both groups after six weeks of intervention. A pre-test was also given to both groups to ensure that their vocabulary levels were similar. The same post-test was given to both groups to determine which learning strategy was more effective.

2.3. Instruments

A total of 30 questions were created for both the pre- and post-vocabulary tests, which were administered six weeks apart. To ensure the reliability and validity of the tests [25], they were reviewed and modified by two experts from the Applied English Department at Chaoyang University of Technology. The validity and reliability of the tests were calculated using the KR-21 formula, and the results indicated a score of 0.78.

The questionnaire was distributed to both groups of students after the intervention. The questionnaire included 11 items on a 5-point Likert scale, and the students were requested to complete it. Cronbach's coefficient alpha was determined to evaluate the reliability of the questionnaire. The reliability value in the digital-game-based group was 0.92 and 0.84 in the paper-based group, which were excellent and acceptable for a measure of the 11 items.

3. Results

3.1. Comparison of Gain Score

Standard deviation and mean were calculated for students' results of pre-and post-tests in the two groups. Independent sample *t*-tests and analysis of variance were used to compare the digital game-based and paper-based groups. The *t*-test was used to demonstrate statistical differences between the mean scores of the two groups. The result is presented in Table 1.

Table 1. Comparison of the pre-test between the digital-game-based group and paper-based group.

Group	N	Mean	SD	t. Value	f	Sig. (Two-Tailed)
A	19	45.26	9.949	0.135	1.352	0.978
B	19	45.00	13.170			

Table 1 demonstrates that the digital-game-based group and the paper-based group had a similar mean score on the pre-test (45.26 and 45.00, respectively). The results indicate that the value of significance (two-tailed) was 0.978, exceeding the significance level of 0.05, implying that there was no statistical significance between the two groups. Therefore, the results of both groups were statistically the same. After six weeks of implementing the two strategies, the average score of the students who used digital-game-based learning was 63.11, while the average score of the students who used paper-based learning was 61.16. The significance value (two-tailed) was 0.662. The mean scores for both groups significantly improved; however, there was still no significant difference between the two groups, as shown in Table 2.

Table 2. Comparison of the post-test between the digital-game-based group and paper-based group.

Group	N	Mean	SD	t. Value	f	Sig. (Two-Tailed)
A	19	63.11	12.849	0.441	0.655	0.662
B	19	61.16	14.338			

Table 2 shows that there was no significant difference between the students who took digital-game-based quizzes and paper-based quizzes during the six weeks of intervention.

3.2. Student Perception toward the Digital-Game-Based Quiz (Kahoot!)

Among the items evaluated (see Appendix A), the students' highest perception was for Item 1 ("I find the Kahoot! quiz exciting, interesting, motivating and funny") and Item 3 ("Getting a good grade in every quiz in the class is the most satisfying thing for me"), with a mean score of 4.21. Conversely, the lowest mean score was for Item 6 ("The Kahoot! quiz does not give any chance to cheat"), with a mean of 3.47. Table 3 presents the percentage and mean score of each item assessed.

Table 3. Student perception results for the digital-game-based quiz (Kahoot!).

Item	Strongly Agree		Agree		Neutral		Disagree		Strongly Disagree		Mean
	%	N	%	N	%	N	%	N	%	N	
Q1	37%	7	58%	11	-	-	-	-	5%	1	4.21
Q2	26%	5	68%	13	-	-	-	-	5%	1	4.11
Q3	37%	7	53%	10	5%	1	5%	1	-	-	4.21
Q4	26%	5	58%	11	5%	1	11%	2	-	-	4.00
Q5	11%	2	58%	11	16%	3	16%	3	-	-	3.63
Q6	16%	3	26%	5	47%	9	11%	2	-	-	3.47
Q7	11%	2	58%	11	16%	3	16%	3	-	-	3.63
Q8	21%	4	58%	11	11%	2	5%	1	5%	1	3.84
Q9	16%	3	58%	11	16%	3	5%	1	5%	1	3.74
Q10	16%	3	58%	11	16%	3	11%	2	-	-	3.79
Q11	16%	3	68%	13	5%	1	11%	2	-	-	3.89

3.3. Student Perception toward the Paper-Based Quiz

Student perception was the highest for Item 3 ("Getting a good grade in every quiz in the class is the most satisfying thing for me"), with a mean of 4.42, and the lowest mean was 3.47 for Item 11 ("Getting late results and feedback from the paper-based quiz unmotivated me"). The percentage and mean of each item are presented in Table 4. All the items for paper-based quiz is presented in Appendix B.

Table 4. Student perception results for the paper-based quiz (Kahoot!).

Item	Strongly Agree		Agree		Neutral		Disagree		Strongly Disagree		Mean
	%	N	%	N	%	N	%	N	%	N	
Q1	16%	3	68%	13	16%	3	-	-	-	-	4.00
Q2	11%	2	68%	13	21%	4	-	-	-	-	3.89
Q3	47%	9	47%	9	5%	1	-	-	-	-	4.42
Q4	32%	6	37%	7	26%	5	5%	1	-	-	3.95
Q5	5%	1	74%	14	16%	3	5%	1	-	-	3.79
Q6	16%	3	58%	11	16%	3	11%	2	-	-	3.79
Q7	11%	2	84%	16	5%	1	-	-	-	-	4.05
Q8	5%	1	63%	12	32%	6	-	-	-	-	3.74
Q9	5%	1	47%	9	47%	9	-	-	-	-	3.58
Q10	16%	3	37%	7	37%	7	5%	1	5%	1	3.53
Q11	5%	1	42%	8	47%	9	5%	1	-	-	3.47

4. Conclusions

There was no significant difference between the scores of the students who used digital-game-based learning (Kahoot! Quiz) and paper-based learning, but both groups showed an improvement in the post-test compared to the pre-test. The digital-game-based group showed more improvement after the intervention. The students were motivated to achieve better grades on every quiz based on the questionnaire results. In the digital-game-based group, the students believed that cheating was not possible in Kahoot! quizzes (Item 3), whereas the students in the paper-based learning group felt demotivated waiting for the results each week. It is challenging to determine which strategy was better based on the responses and results of both groups. Further research is necessary to evaluate the perception of these strategies, as this 6-week intervention could not demonstrate significant differences, indicating the need for a further extended intervention.

Author Contributions: Conceptualization, A.D.S. and R.-T.H.; methodology, R.-T.H.; software, A.D.S.; validation, R.-T.H.; formal analysis, A.D.S. and R.-T.H.; investigation, A.D.S. and R.-T.H.; resources, R.-T.H.; writing—original draft preparation, R.-T.H.; writing—review and editing, A.D.S. and R.-T.H.; visualization, A.D.S. and R.-T.H.; supervision, R.-T.H.; project administration, R.-T.H. All authors have read and agreed to the published version of the manuscript.

Funding: This research received no external funding.

Institutional Review Board Statement: Not applicable.

Informed Consent Statement: Not applicable.

Data Availability Statement: The data are unavailable.

Conflicts of Interest: The authors declare no conflict of interest.

Appendix A

1. I find the Kahoot! quiz exciting, interesting, motivating and fun.
2. I feel positive when using the Kahoot! quiz.
3. Getting a good grade in every quizzes in the class is the most satisfying thing for me.
4. The Kahoot! quiz is more challenging than the paper-based quiz.
5. I feel that the Kahoot! quiz was familiar and simple to complete.
6. The Kahoot! quiz does not give any chance to cheat.
7. The Kahoot! quiz technique enables me to learn better.
8. I want this technique to be used in other courses.
9. I prefer taking the quiz online through Kahoot! to a paper-based quiz.
10. The Kahoot! quiz environment is appropriate and convenient for test-taking.
11. Getting immediate results and feedback from the Kahoot! quiz system motivates me.

Appendix B

1. I find the paper-based quiz exciting, interesting, motivating and fun.
2. I feel positive when using the paper-based quiz.
3. Getting a good grade in every quiz in the class is the most satisfying thing for me.
4. The paper-based quiz is more challenging than the online quiz.
5. I feel that the paper-based quiz was familiar and simple to do.
6. The paper-based quiz does not give any chance to cheat.
7. The paper-based quiz technique enables me to learn better.
8. I want this technique to be used in other courses.
9. I prefer taking the paper-based quiz to an online quiz.
10. The paper-based quiz environment is appropriate and convenient for test-taking.
11. Getting late results and feedback from the paper-based quiz unmotivated me.

References

1. Susanti, S. Fun Activities in Teaching Enlish by Using Kahoot! In Proceedings of the 2nd International Seminar on Education Fakultas Tarbiyah Dan Ilmu Keguruan Institut Agama Islam Negeri Batusangkar, Batusangkar, Indonesia, 5–6 September 2017; pp. 453–458.
2. Traxler, J. Defining mobile learning. In Proceedings of the ADIS International Conference Mobile Learning 2005, Qwara, Malta, 28–30 June 2005.
3. Sarrab, M. Mobile Learning (M-Learning) and Educational Environments. *Int. J. Distrib. Parallel Syst.* **2012**, *3*, 31–38. [CrossRef]
4. Coffey, H. Digital Game-Based Learning. *Learn NC* **2009**, 1–3. Available online: https://www.medicina.univr.it/documenti/Avviso/all/all036209.pdf (accessed on 14 June 2023).
5. Sadler, T.; Romine, W.L.; Stuart, P.E. Game-Based Curricula in Biology Classes: Differential Effects Among Varying Academic Levels. *J. Res. Sci. Teachnol.* **2013**, *50*, 479–499. [CrossRef]
6. Deubel, P. Game on. *J. Technol. Horiz. Educ.* **2006**, *33*, 30–35.
7. Saeidi, M.; Mozaheb, M.A. Comparing vocabulary learning of EFL learners by using two different strategies (mobile learning vs. flashcards). *Int. J. Mob. Learn. Organ.* **2012**, *6*, 303–315. [CrossRef]
8. Baleghizadeh, A.S.; Ashoori, A. The Impact of Two Instructional Techniques on EFL Learners' Vocabulary Knowledge: Flash Cards versus Word Lists 1. *Mextesol. J.* **2011**, *35*, 1–9.
9. Hulstijn, H. Intentional and incidental second language vocabulary learning: A reappraisal of elaboration, rehearsal and automaticity. In *Cognition and Second Language Instruction*; Cambridge University Press: Cambridge, UK, 2012. [CrossRef]
10. Larsen-Freeman, D.; Anderson, M. *Techniques and Principles in Language Teaching*, 3rd ed.; Oxford handbooks for language teachers; Oxford University Press: Oxford, UK, 2013.
11. Savignon, S.J. Communicative Language Teaching: Linguistic Theory and Classroom Practice (Chp 1). In *Interpreting Communicative Language Teaching*; Yale University Press: New Haven, CT, USA, 2002.
12. Mehta, N.K. Vocabulary teaching: Effective methodologies. *Internet TESL J.* **2009**, *15*, 14–23.
13. Ghazal, L. Learning vocabulary in EFL contexts through vocabulary learning strategies. *Novitas-ROYAL* **2007**, *1*, 84–91.
14. Schmitt, N. Review article: Instructed second language vocabulary learning. *Lang. Teach. Res.* **2008**, *12*, 329–363. [CrossRef]
15. Malone, J.E. How Vocabulary is Learned, by S. Webb and P. Nation (2017). *Instr. Second Lang. Acquis.* **2019**, *3*, 115–118. [CrossRef]
16. Ayumi; Chan, A.M. Improving Students English Vocabulary Using Kahoot! In Proceedings of the 3rd International Conference on Educational Development and Quality Assurance (ICED-QA 2020), Online, 27–28 October 2020; Volume 506. [CrossRef]
17. Burguillo, J.C. Using game theory and Competition-based Learning to stimulate student motivation and performance. *Comput. Educ.* **2010**, *55*, 566–575. [CrossRef]
18. Plump, C.M.; LaRosa, J. Using Kahoot! in the Classroom to Create Engagement and Active Learning: A Game-Based Technology Solution for eLearning Novices. *Manag. Teach. Rev.* **2017**, *2*, 151–158. [CrossRef]
19. Ismail, M.A.; Mohammad, J.A.-M. Kahoot: A Promising Tool for Formative Assessment in Medical Education. *Educ. Med. J.* **2017**, *9*, 19–26. [CrossRef]
20. Bicen, H.; Kocakoyun, S. Perceptions of students for gamification approach: Kahoot as a case study. *Int. J. Emerg. Technol. Learn.* **2018**, *13*, 72–93. [CrossRef]
21. Pandolfini, V. Online tutoring roles: Italian teachers' professional development experience. In *Online Tutor 2.0: Methodologies and Case Studies for Successful Learning*; García-Peñalvo, F., Seoane Pardo, A., Eds.; IGI Global: Hershey, PA, USA, 2014; pp. 119–143. [CrossRef]
22. Patil, Y.M.; Kumbhar, P.D. Learning by gamification: An effective active learning tool in engineering education. *J. Eng. Educ. Transform.* **2021**, *34*, 447. [CrossRef]
23. Wirani, Y.; Nabarian, T.; Romadhon, M.S. Evaluation of continued use on Kahoot! As a gamification-based learning platform from the perspective of Indonesia students. *Procedia Comput. Sci.* **2021**, *197*, 545–556. [CrossRef]
24. Göksün, D.O.; Gürsoy, G. Comparing success and engagement in gamified learning experiences via Kahoot and Quizizz. *Comput. Educ.* **2019**, *135*, 15–29. [CrossRef]
25. Neuman, W. *Social Research Methods: Qualitative and Quantitative Approaches*, 6th ed.; Pearson: London, UK, 2013; Volume 30. [CrossRef]

Disclaimer/Publisher's Note: The statements, opinions and data contained in all publications are solely those of the individual author(s) and contributor(s) and not of MDPI and/or the editor(s). MDPI and/or the editor(s) disclaim responsibility for any injury to people or property resulting from any ideas, methods, instructions or products referred to in the content.

engineering proceedings

MDPI

Proceeding Paper

Images of E-Sports in Chinese Newspapers †

Jianfeng Wei and Yufeng Huang *

School of Journalism and Communication, Xiamen University, Xiamen 361005, China; jfwei1124@yahoo.com.tw
* Correspondence: biznewss@xmu.edu.cn

† Presented at the 3rd IEEE International Conference on Electronic Communications, Internet of Things and Big Data Conference 2023, Taichung, Taiwan, 14–16 April 2023.

Abstract: We analyze the reports and comments in newspapers on e-sports using T.A. van Dijk's news discourse and discuss how Chinese newspapers present the configuration and the construction of e-sports. The method of reporting e-sports follows the viewpoints of cyber games. Such viewpoints regarding e-sports as cyber games in media-meditated communication are socially accepted. Through the presentation of these viewpoints, a model of the coverage of e-sports is constructed. E-sports have gained government support, and the media constructs an image of e-sports as an information technology industry because of their technological nature.

Keywords: e-sports; cyber game; image study

Citation: Wei, J.; Huang, Y. Images of E-Sports in Chinese Newspapers. *Eng. Proc.* **2023**, *38*, 80. https://doi.org/10.3390/engproc2023038080

Academic Editors: Teen-Hang Meen, Hsin-Hung Lin and Cheng-Fu Yang

Published: 13 July 2023

Copyright: © 2023 by the authors. Licensee MDPI, Basel, Switzerland. This article is an open access article distributed under the terms and conditions of the Creative Commons Attribution (CC BY) license (https://creativecommons.org/licenses/by/4.0/).

1. Introduction

E-sports have emerged and developed rapidly on a global scale. With internet-based games as the core, competitions in e-sports have become a global industry. For example, the World Cyber Games (WCG) represent one of the world-class e-sports competitions, organized by the South Korean government and Samsung with worldwide recognition. E-sports have been developed in China for more than ten years. Using electronic software, the game sets an environment for competitive battle, emphasizing fairness. The games in e-sports require sophisticated skills that require mental and physical abilities for self-fulfillment. These abilities include hand–eye coordination, operational accuracy, multitasking under time limits, and others. Psychological abilities such as rapid response capability, precise analysis, low anxiety under stress, and tactics are also essential in e-sports. In 2003, the Chinese State General Administration of Sports listed e-sports as one of the 99 official sports. However, the e-sports movement in China does not measure up with other sports. The players lack systematic support and usually seek sponsorship independently. Players who wish to compete in international competitions sometimes give up due to visa issues. This occurs because a number of players participate in illegal cyber games for prize money on behalf of their country. Thus, China has a peculiar administrative system. Athletes of e-sports belong to the state and their club. Although the players compete abroad on behalf of themselves or sponsors, they compete for the country as well. Since this is a new field, many provisions have not yet been introduced, compared to other sports. Moreover, there is not much support, a factor which is related to traditional Chinese social values.

Chinese parents consider e-sports and gaming as improper professions and think winning championships does not ensure a future. Traditional values still lead people to believe that the right path for their children is to study and that playing games is just for entertainment. The media's attitude toward e-sports also follows parents' stereotypes. They tend to regard playing cyber games as an "Internet addiction" that endangers physical and mental health. In 2004, China's State Administration of Radio issued the "Prohibition on broadcasts on cyber games notice", restricting the broadcasting of cyber game programs. This has caused the general public to perceive e-sports as negative. Although Chinese society is gradually diversifying, the atmosphere has changed since ten years ago. Parents are beginning to understand e-sports owing to their innovative campaign and being a profitable industry. Thus, we attempt to answer the question 'How can the Chinese media shape the image and social values of e-sports?' in this study.

2. Literature Review

2.1. Categories of Cyber Games

There are many types of cyber game platforms. The first includes console games that need additional monitors and machines or devices such as a cassette, CD, or other storage media to play arcade games, street fights, Dance Revolution, Taiko, and many other games. The consoles include Nintendo's NES and Wii, Sony Play Station, and Microsoft's Xbox. There are handheld game consoles that are compact in shape and lightweight, such as Nintendo's Game Boy and Sony's Play Station Portable (PSP). Personal computers are also a type of console with a keyboard, mouse, and other external devices used mainly to play online games via the internet. Online games are classified into three types. The first one is the web game, with a web-style interface. Plants vs. Zombies, Happy Farm, and Angry Birds are web games. The second one is multiplayer role-playing games that accommodate numerous online players at the same time with real-time interactions through conversations. Due to the sense of belonging to the community and online dating features, this type of game has gained much popularity. The third is LAN games, for which a computer is set up as a temporary game server while other computers involved in the game connect to the host. The server computer acts as a client and a server while in charge of handling the game data. Such games are restricted to a given number of players at a time. The players cooperate or compete with each other, and the e-sports events are mostly carried out through such games.

2.2. Concept of e-sports

There is no academic definition for e-sports thus far. The only difference between e-sports and playing online games is the inclusion of competition. Thus, the concept must be elaborated to differentiate e-sports from simple game playing. E-sports are intellectual competitions using software and hardware. Through exercise, players are trained to improve cognitive ability, responsiveness, hand–eye coordination, determination, and teamwork. In fact, similar to other sports, e-sports are focused on professionalism with an emphasis on physical and mental exercise and teamwork. There are obvious differences between e-sports and cyber games.

Online games are essentially for leisure and entertainment. In a virtual environment, users play virtual characters to pursue what cannot be done in reality and to experience the enjoyment of playing other roles. In the early stage of the game, the amounts of money and time spent in each level of the game are pre-designed. Thus, regardless of the player's technical ability, the only requirement is to comply with the rules of the game. If players invest more money in the game, they can be upgraded to the next level faster in order to gain advantages over general players. In e-sports, there are no such shortcuts. Players only rely on their skills, because e-sports involve fierce competition. Similar to traditional sports, e-sports players must improve their speed, response, and handling of the electronic equipment through hard work performed daily and drills to obtain good results. Since the opponent is a human, there is no way to predict what the opponent will do next.

The current global e-sports tournament events consist of four main categories: FPS (First-Person Shooter), RAC (Race Game), RTS (Real-Time Strategy), and MOBA (Multi-player Online Battle Arena). In this study, online games are also included, considering the historical development of e-sports.

2.3. Image of e-Sports in Media

An *Image* is interpreted as a mental representation or attitude. When we see an object, its image is retained in our memory even after this object is out of sight. This memory of an image is recalled when a similar object appears. Lippmann stated that people cannot come into contact with every person or thing [1]. Boorstin explained that an image of any object (especially a person) is an artificial impression or reproduction of an external form [2].To cope with this situation, people need to develop a simplified cognitive process to group things with the same or similar qualities together and create a set of images. Then, people use this image to assess the memory of the group. Lippmann believed that images are pictures in minds, similar to a map. An image influences perception of people and things and their roles and positions in society. For people who do not experience things directly, images that the media shapes and forms become critical. Many empirical studies have found that the mass media have the power to shape an image for the public. The relationship between real sport and its digital adaptation is significantly influenced by technological advancements [3]. E-Sports is not only about the computer games and playing itself, but it can also serve as a means of satisfying various needs [4]. From the communication point of view, the media is an important tool for the construction of reality. When reproducing social events or issues, the media select, edit, and dramatize them based on their standards and rules.

3. Study Design

To analyze the image of e-sports in newspapers, we use the topic derivation method of T.A. Van Dijk's news discourse qualitative analysis of materials. T.A. Van Dijk analyzed the three macro-rules of news discourse: deletion, generalization, and construction. The study design for this research is as follows.

The research object in this study was the "China Core Newspapers Full-text Database". e-Sports were recognized by the Chinese State General Administration of Sports as one of the 99 official sports in 2003, and the "Prohibition on broadcasts on cyber games notice" was issued by China's State Administration of Radio in 2004. Search results confirmed that e-sports coverage by the media has increased significantly since 2004. In 2001, there was only 1 report, but 14, 42, and 102 reports were found in 2002, 2003, and 2004, respectively. Thus, data for this research were retrieved from the reports published from 1 January 2005 to 31 December 2014. We searched the database with the keyword "e-sports" and excluded any unrelated e-sports reports. In total, 575 reports were collected as a result.

The collected reports were encoded with dates, press numbers, newspaper companies, and layouts. The Chinese media is different from those of other countries because of the party newspaper. In previous studies, party newspapers were rarely taken into account. Referring to the previous literature, we calculated the frequency of adjectives used to describe the image of e-sports in the media. The adjectives were classified into positive and negative ones to code for the analysis. Before analysis, we randomly selected and coded 50 reports, accounting for 8.7% of all reports. The pilot coding result showed 86% credibility based on mutual consent among experts.

4. Results and Discussion

4.1. Numbers of Reports

The numbers of reports on e-sports published in the media from 2005 to 2014 are 85 (in 2005), 110 (2006), 87 (2007), 82 (2008), 47 (2009), 46 (2010), 37 (2011), 8 (2012), 36 (2013), and 37 (2014). The number of reports increased from 2005 to 2008 but has decreased since then. In 2008, Beijing hosted the Olympic Games. Since e-sports are included in the official competitions in the Olympic Games, the media has not paid much attention to them since 2008. In 2006, "People's Daily" had four reports on e-sports, which was the highest number of reports by the newspaper since 2003, when e-sports became an official sport. "People's Daily" usually speaks for the government's attitude on certain topics. Reports in 2006 reflected the government's policy for e-sports. The largest number (14) of reports were published in the "Chengdu Daily", as the world's largest e-sports platform, Global Gaming League (GGL), was held in Chengdu in 2008. In 2012, due to instability and uncertainty with respect to clubs, well-known players retired. Even though the Association of China e-Sports was founded by professional e-sports players to promote the development and support of e-sports clubs and professional players, clubs disappeared due to reduced sponsorship. With their dual nature of games and sports, e-sports again faced the challenges of social values.

4.2. Content of Reports

A total of 40 newspapers published 575 reports on e-sports content. "Chengdu Daily" published 6.6% of the total number of reports, "China Information World" 4.7%, "China Culture Daily" 4.2%, "Science and Technology Daily" 4.2%, "Beijing business daily" 3.7%, "China Press and Publishing Journal" 3.1%, "China Sports Daily" 3.1%, "Computer World" 2.8%, "Popular Science news" 2.6%, "Popular Computer Weekly", 2.4%, "Beijing daily" 2%, "Guang Ming Daily" 1.7%, and "Computer Partner World" 1.7%. Based on the content analysis, the following trends were found: (1) Within a decade, words describing e-sports in the newspapers became neutral, while negative adjectives were used more than positive ones. This shows that the media coverage of e-sports was one of misunderstanding with prejudice, which caused the public to have no exact understanding of e-sports. (2) The reports revealed the lack of awareness and professionalism in e-sports. (3) The government did not support the development of the e-sports industry, but the local governments of Chengdu and Sichuan Provinces did for their local events. (4) The number of reports in online newspapers was greater than that in sports newspapers. Compared to sports newspapers, online newspapers paid more attention to e-sports and considered e-sports as an extended form of the IT industry rather than a sport.

5. Conclusions

The stereotype positing that "Learning is the noblest of human pursuit" has been accepted, as traditional scholars have influenced people for a long time. Such a philosophy is engrained in the education system and media campaigns. Chinese society has become muted to such a philosophy. As a result, family and society limit themselves in choosing a new way of life, which hinders the development of e-sports. The definition of Chinese newspapers and social media differs from that of Western newspapers. To some extent, the Chinese media represents the official governmental views. In this situation, people involved in organizing e-sports games have been excluded from official mainstream media coverage, and it has not been easy for them to draw the attention of the media and gain government support. For the Chinese Government to develop e-sports, the media needs to overcome the stereotypes of e-sports and have a more positive outlook on this issue. Otherwise, permanent negative stereotypes will remain, further affecting the perception and values of e-sports in the future and consigning it to be an underdeveloped sport.

Similar to track and field sports during the agricultural era and motorsports during the industrial era, the world can foresee the potential of e-sports and is investing in the development of this high-value sport. e-Sports have a positive impact on the development of the technology industry. Therefore, the technology industry needs to improve the performance of hardware and organize sound competition. In China, the media portrays online games with a negative image. If a member of a Chinese family plays e-sports as a lifelong career, the family and society will oppose it. It is necessary to advertise the fact that e-sports are not just for entertainment but represent a new industry and sport which need to be campaigned for by the mass media.

Author Contributions: Both of the two authors contributed to the study conception and design. Material preparation, data collection and analysis were performed by J.W. and Y.H. The first draft of the manuscript was written by J.W. and both authors commented on previous versions of the manuscript. Both authors read and approved the final manuscript. All authors have read and agreed to the published version of the manuscript.

Funding: This research received no external funding.

Institutional Review Board Statement: Not applicable.

Informed Consent Statement: Not applicable.

Data Availability Statement: Not applicable.

Conflicts of Interest: The authors declare no conflict of interest.

References

1. Herbst, S. Walter lippmann's public opinion, revisited. *Harv. Int. J. Press/Politics* **1999**, *4*, 88–93. [CrossRef]
2. Nimmo, D.D.; Savage, R.L. *Candidates and Their Images: Concepts, Methods, and Findings*; Goodyear Publishing Company: Santa Monica, CA, USA, 1976.
3. Hebbel-Seeger, A. The relationship between real sports and digital adaptation in e-sport gaming. *Int. J. Sports Mark. Spons.* **2012**, *13*, 132. [CrossRef]
4. Martončik, M. e-Sports: Playing just for fun or playing to satisfy life goals? *Comput. Hum. Behav.* **2015**, *48*, 208–211. [CrossRef]

Disclaimer/Publisher's Note: The statements, opinions and data contained in all publications are solely those of the individual author(s) and contributor(s) and not of MDPI and/or the editor(s). MDPI and/or the editor(s) disclaim responsibility for any injury to people or property resulting from any ideas, methods, instructions or products referred to in the content.

engineering proceedings

MDPI

Proceeding Paper

Discussion on Diversity of Animation Teaching Methods in Universities †

Yuanxian Chen and Xing Zheng *

Department of Animation, Xiamen University Tan Kah Kee College, Zhangzhou 363121, China; chenyx@xujc.com
* Correspondence: zhengxing@xujc.com
† Presented at the 3rd IEEE International Conference on Electronic Communications, Internet of Things and Big Data Conference 2023, Taichung, Taiwan, 14–16 April 2023.

Abstract: In modern education, it has been generally believed that the most important responsibility of educators is to awaken the ability of students. It is a general goal to develop the students' comprehensive knowledge and practical skills in majoring in animation at universities. Thus, we investigated the effects of learning to understand the social significance of the academic ability and technical skills of graduates. Then, we reviewed the teaching methods in animation and analyzed the uniqueness and complexity of teaching to understand the evolutionary communication method and their complementary resources. We propose a creative method to gradually promote teaching animation. Through the experiments used in the course, the teaching effects were compared. The result of this research provides educators with a reference for developing a method of teaching animation.

Keywords: animation teaching method; complementary resources; creative ability training

Citation: Chen, Y.; Zheng, X. Discussion on Diversity of Animation Teaching Methods in Universities. *Eng. Proc.* **2023**, *38*, 81. https://doi.org/10.3390/engproc2023038081

Academic Editors: Teen Hang Meen, Hsin-Hung Lin and Cheng-Fu Yang

Published: 13 July 2023

Copyright: © 2023 by the authors. Licensee MDPI, Basel, Switzerland. This article is an open access article distributed under the terms and conditions of the Creative Commons Attribution (CC BY) license (https://creativecommons.org/licenses/by/4.0/).

1. Introduction

The purpose of this research was to analyze the problems of animation education in universities and propose an education program. According to the Undergraduate Specialty Catalogue of Higher Institutions, animation is a major offered by the Department of Drama, Film, and Television under the discipline of art. Recently, economic and technical factors have been introduced into the animation major, which has a great impact on education in both colleges and universities. Teaching methods, training of animation skills, and graduates' works have been used as corresponding measures for the future development of animation education.

Modern college education advocates for teachers to awaken the power of students and the society needs college students to have the proper knowledge and skills in physical and mental health [1]. Animation education in colleges and universities must consider the importance of professional training, and at the same time, help students to develop their self-confidence, self-esteem, independence, honesty, and enthusiasm quality, and thus, students can develop both professional skills and independent thinking. Standards for teaching quality have been established with a normative and referential significance for animation teaching. In the *National Standards for Teaching Quality of Animation, Digital Media Arts, and Digital Media Technology*, animation education has been defined as a cross-disciplinary specialty that serves social development and cultural construction, and embodies the deep integration of science and art. This document reveals the requirements for backbone courses, class hours, credits, graduation works criteria, and others, which objectively guides educators to teach the comprehensive knowledge and practical skills in the direction of talent training in both colleges and universities. At the same time, animation education in universities is different compared to those in specialized art colleges. Therefore, it is necessary to understand the situation of animation education in universities to implement these teaching methods and help students develop and use their practical skills and knowledge across different fields.

2. The Situation of Current Students

At present, many colleges and universities offer animation majors in addition to specialized art colleges with their resources. For example, at Tianjin University, the animation major has been set in the Intelligence and Computing Department of the faculty of engineering. In the Communication University of China and Tongji University, an animation major has been set in the faculty of information and media. At Nanjing Normal University and Northeast Normal University, an animation major is offered in the faculty of normal art education. Southeast University, Jinan University, Wuhan University of Technology, and Tan Kah Kee College, Xiamen University all offer animation majors in the faculty of art and design.

Students majoring in animation in universities are enrolled with fine art skills and graduate with a comprehensive knowledge of art, engineering, cultural communication, performance, and film with a Bachelor of Arts. Students are required to undertake an examination for drawing, color representation, sketching, and on-site creation in the animation major course. In universities, the comprehensive quality is focused on admissions with academic scores and art exam scores. The admission of students in animation majors is based on the total score calculation. Students' final score is calculated with different proportions, which differ across the universities. For example, the sum of 40% of the academic subject score and 60% of the art exam score are considered or vice versa. Several universities attach importance to the personality of candidates from different provinces and subsequently adjust the calculation proportion according to it. Different calculations by universities are related to different needs or understanding of the learning ability of science, artistic expression, culture, and others required for animation majors. This admission method may add the requirements and expectations of practical training and graduation works, which has also been reflected in the curriculum structure as shown in Table 1.

Table 1. The curriculum structure of Tan Kah Kee College, Xiamen University.

Course Category		Required Hours	Credit	Hour Ratio	Credit Ratio
Skills education	Required course	986	34	29%	22%
	Elective course	102	6	3%	4%
General education	Required course	442	22	13%	14%
	Elective course	225	12	7%	8%
Professional education	Required course	1121	58	33%	37%
	Elective course	474	23	14%	15%
Total by category	Required ourse	2549	114	76%	74%
	Elective course	801	41	24%	26%
	Specialized course	1595	81	48%	52%
	Non-specialized course	1755	74	52%	48%
	Theory section	1207	72	36%	46%
	Practice section	2113	83	64%	54%
	Total	3320	155	100%	100%

3. Diversity of Animation Teaching Methods

3.1. Inspirational Teaching and Active Learning

Inspirational teaching and emphasis on active learning must be the focus of education methods for animation majors in universities. Most high-school art students learn realistic drawing, color representation, and other related skills through traditional art training.

Repetitive art training has a significant effect on drawing skills in terms of muscle memory and visual imitation, through which students can produce stereotyped and modeled works in a short period of time. However, this repetitive and imitative learning makes training arise without interest and thinking and makes students lazy and passive in learning as a result, which cannot cultivate personality and style. Thus, few art students are interested in art theories and appreciate different artworks or stories outside the classroom, and after they enroll in the animation major at university, they lack creativity and interesting thinking. Confusion and difficulties from students can be felt, as it is hard for them to continue the simple and imitative drawing habits that they gained during their past training. They need to understand that creating animation requires script writing, planning, directing, performing, and sound editing, and they must have an interest in these topics. Animation majors in universities integrate different resources and make use of elective courses for obtaining different knowledge, such as computing, music, communication, management, and so on. Based on the knowledge of different disciplines across the university setting, animation students can gain comprehensive skills, ability, and creative thinking.

3.2. Computer-Assisted Instruction for Creation

Computer-assisted instruction (CAI) was developed in the USA in the 1960s', which has had a positive effect on animation teaching today. CAI improves guiding students' creation. As the core of animation education, teachers need to provide students with a good environment for creativity and understanding [2].Teachers need to teach and guide the creation using communication, interaction, and discussion. This teaching process includes lectures, tutorials, course exercises, technical training, and other activities. Multimedia, computer technology, hypertext, and design platforms help overcome the shortcomings of traditional indoctrination and one-sided teaching. CAI shortens the learning time and improves the teaching quality with an effective strategy. With this system, teachers can pay close attention to students' characteristics and learning habits and monitor the creation process and the final work. During tutoring, teachers can make individual customization and provide adaptive guidance based on students' cognition and ability.

3.3. Technical Teaching and Ability Training

For practical teaching, it is better to have 20 students in each class. In technical teaching, the content and progress for the average level of the class are too easy for excellent students and conversely too difficult for poor students. Therefore, it is better to instruct students to learn actively through practice. At the beginning of the teaching process, the teacher briefly demonstrates the basic steps of the technique on the computer and explains how it is used in work producing. Then, the teacher guides the students to practice their skills through creation. By this process, the teacher gradually understands the abilities and levels of different students through observation and then adjusts the way of technology demonstration. At the same time, technical teaching must emphasize the combination of skills and creations to stimulate students' interest and overcome difficulties. After the beginning stage of teaching, students can master the basic operation of technical skills through personalized learning, and then teamwork can be introduced to conduct in-depth teaching. Teachers organize complementary groups based on student characteristics so that they can learn from each other in the group and improve their creativity and ability through cooperation.

4. Relationship between the Teaching Method and Ability Training

Teaching methods and cultivating students' abilities needs to be unified as a whole and carried out for students to acquire their knowledge and skills. In ability training, which is the core of animation teaching, the year one and two courses are offered as the visual study and professional basis, while the year three and four courses are provided for professional training and the development of creative thinking, respectively. Visual representation is taught to shape characters and scenes and express ideas and thinking through images. The professional basis is learned for story-telling and the concept of motion. Junior students need to understand the characteristics of different materials and media through the practice of artistic thinking for the creation of animation. Senior students need to develop their ability to conceive and create by finding problems, observing life, forming personal views, and creating valuable works, such as commercial projects [3].

Courses in years one and two include Animation Modeling to develop observation and analysis and Animation Representation to develop visual expression. Courses in years three and four offer Materials and Three-Dimensional practice, Principle of Motion, Performance and Sound, Scriptwriting and Storyboard, Video Shooting and Editing, CG software study, and Narrative illustration to practice the ability to observe and investigate social problems, analyze data, find solutions, propose and produce virtual projects to comprehensively organize and cooperate with others, and ultimately to achieve the ability to create animated work

Educators in animation majors must pay close attention to the trends of the creative industry, grasp and analyze the information related to animation, and discuss the dynamics of the animation education system. Teachers must keep communicating with each other and discuss the direction of teaching in courses to maintain the identity of animation education and effectively convey knowledge and new trends to students. For the effectiveness of teaching, professors in universities need to practice and carry out projects and research for guiding and training students for their professions. Furthermore, teachers must understand the difficulties at each stage of creation and evaluate different creation levels to find the optimal conditions for the equipment and management systems in teaching.

The integration of animation art with different fields of knowledge is advantageous to develop students' capabilities. The professional development model of animation majors is successful in developed countries, which involves VR design, computer-aided technology, three-dimensional technology, media art, communication design, video packaging, comics, electronic music, interaction design, and so on. Such an interdisciplinary education saves resources, and benefits universities in terms of management and the academic atmosphere of international, cutting-edge, and scientific disciplines. A flexible and diverse knowledge framework corresponds to the changes in the market demand for talent. Teachers and students across different majors need to interact and cooperate flexibly to ultimately introduce new technologies and create new ideas. On the other hand, today's We Media and other social platforms emphasize experimentation and creativity, highlighting the commercial value of personal creation into the market. Therefore, it is necessary to discuss the new value of synthesis, experiments, and creativity in animation education. It is of great significance to guide students to understand the concepts of art, culture, tradition, history, and others and react to innovative industries. We can obtain the advantages of artistic creativity and technology innovation from the university environment, which objectively urges the diversity of animation teaching.

5. Conclusions

Animation majors in universities need to make full use of the resources available. The curriculum must integrate different disciplines of knowledge so that students can develop their professionalism in a broader vision. The teaching methods must be more flexible and diverse, using new media and tools, emphasizing the communication and interaction between teachers and students, and ultimately cultivating comprehensive practical ability and creative thinking of students. On the education of animation in universities, the teaching process and methods have been proposed in this study. These proposed teaching methods need to be constantly evaluated and adjusted by educators considering both possible criticisms and suggestions.

Author Contributions: Writing—original draft preparation, X.Z.; Writing—review and editing, Y.C. All authors have read and agreed to the published version of the manuscript.

Funding: This research received no external funding.

Institutional Review Board Statement: Not applicable.

Informed Consent Statement: Not applicable.

Data Availability Statement: Data sharing not applicable.

Conflicts of Interest: The authors declare no conflict of interest.

References

1. Hotta, A. The Direction of Design Education in Universities of Art. *Artes: Bulletin of Takarazuka University of Art and Design*, March 1995; 123–132.
2. Mayer, R.E. *The Cambridge Handbook of Multimedia Learning*, 2nd ed.; Cambridge Uiversity Press: Cambridge, UK, 2014.
3. Benshahi, T.C. *Anime and Manga*; Nakamura Gakuen University: Fukuoka, Japan, 2019; pp. 58–64.

Disclaimer/Publisher's Note: The statements, opinions and data contained in all publications are solely those of the individual author(s) and contributor(s) and not of MDPI and/or the editor(s). MDPI and/or the editor(s) disclaim responsibility for any injury to people or property resulting from any ideas, methods, instructions or products referred to in the content.

engineering proceedings

MDPI

Proceeding Paper

A Negative Binomial Regression Model of Student Absenteeism in the Principles of Microeconomics †

Chen Chen [1], Shih-Shen Chen [1,*] and Chin W. Yang [2]

1 College of Economics and Management, Fuzhou University of International Studies and Trade, Fuzhou 350200, China; chenchen@fzfu.edu.cn
2 Department of Economics, Clarion University of Pennsylvania, Clarion, PA 16214, USA; cwyang@clarion.edu
* Correspondence: chensishen@fzfu.edu.cn

† Presented at the 3rd IEEE International Conference on Electronic Communications, Internet of Things and Big Data Conference 2023, Taichung, Taiwan, 14–16 April 2023.

Abstract: We investigated the determinants of absenteeism in the Principles of Microeconomics at a rural state university in Pennsylvania. We apply Poisson statistics and a negative binomial of both Type 1 and 2 models to estimate absenteeism behavior. All cases have GPA as a significant factor. A higher GPA reduces the absence rate by 3.2574 times in the NB1 model. It was found that business majors cut classes significantly less (26.24%) than education majors.

Keywords: attendance; economics education; learning performance factors; academic achievement; university outcome assessment; regression (statistics)

1. Introduction

Absenteeism is a serious problem for universities and often leads to lower grades and hence, a high dropout rate. In a state-supported university, increasing rates of class absenteeism have been observed. Smaller classes may not be a problem for a typical freshman when the grades of popular education courses are good to obtain. A student cruises through less challenging courses without much effort in the wake of ubiquitous grade inflation due to pressure from the administration. One of the challenging courses for a sophomore business major is the Principles of Microeconomics, in which absenteeism is detrimental to a student's grade. Those who excessively decrease classes end up obtaining a letter grade of D or E. The subsequent "repeat of the course" or "take it at a community college during summer" attitudes give rise to an uneasy feeling in the Department of Economics. Before implementing any policy regarding absenteeism, it is important to analyze student characteristics concerning class attendance. Studies on college student absenteeism are scarce, as class attendance is generally not mandatory. Despite the proliferation of the applications of count regression (e.g., ship accidents by [1]; doctor visits by [2]), there seems to be a lack of application to economic and business education. An interesting application is found in the study who employed a Probit model in relation to Principles of Finance classes [3]. Another found a significant but small impact of class attendance on the performance of an introductory statistics class [4].

The most cited study is reported an average absenteeism rate at three major US universities of 1/3, which was comparable to that reported by Rogers and Rogers (2003) in Intermediate Microeconomics at an Australian University [5]. Romer found a positive and significant attendance–performance relationship. Before this, Schmidt (1983) identified a positive correlation between exam performance and time spent in lectures of an Information Processing class [6]. Park and Kerr (1990) also detected an inverse relationship between absenteeism and course grade in a Money and Banking course [7]. In a similar vein, Durden and Ellis (1995) indicated that attendance had a large and significant effect on performance when absenteeism is excessive [8]. Devadoss and Foltz (1996) employed a large sample

Citation: Chen, C.; Chen, S.-S.; Yang, C.W. A Negative Binomial Regression Model of Student Absenteeism in the Principles of Microeconomics. *Eng. Proc.* **2023**, *38*, 82. https://doi.org/10.3390/engproc2023038082

Academic Editors: Teen-Hang Meen, Hsin-Hung Lin and Cheng-Fu Yang

Published: 14 July 2023

Copyright: © 2023 by the authors. Licensee MDPI, Basel, Switzerland. This article is an open access article distributed under the terms and conditions of the Creative Commons Attribution (CC BY) license (https://creativecommons.org/licenses/by/4.0/).

(n = 400) in an Agricultural Economics class and found that perfect attendees had higher average scores than those who cut classes by more than 50% [9].

Marburger (2001) identified a significant correspondence between incorrectly answered questions and missed class on a given date [10]. Kirby and McElroy (2003) examined the determinants of class attendance (n = 368) and found that working hours and travel time had a positive but diminishing marginal effect on course grade. In his detailed model [11], Stanca (2006) found neither a proxy variable nor instrument variable approach that offered a satisfactory solution to the omitted variable in terms of estimating the effect of class attendance on course performance [12]. He proposed a panel data approach (fixed-effect model) to control for the unobservable characteristics of class attendance. Even though class attendance impacted course performance positively and significantly, the magnitude was small.

Chen and Lin (2008) applied a randomized experiment approach to a Public Finance course and found between a 9.4 and 18% improvement in exam performance on average due to attending lectures [13]. The randomized design has the advantage of providing treatments (cause) and responses (effect) understood by social scientists. It also has non-response and selection bias for the experiment. Arulampalam et al. (2012) provided a quantile regression model in an attempt to differentiate the unobserved characteristics relating to class attendance [14]. They found that students in upper quantiles prevailed in terms of the negative absenteeism–performance relationship. When students in UK universities cannot understand the lecture, the difference in learning outcome, regardless of attending class or not, is small.

The above-mentioned result has a common problem. Absenteeism is rarely exogenous despite the efforts to alleviate the instrument, proxy variable approach, panel data, or quantile regression. To solve the problem, Self (2012) applied a negative binomial model to a Principles of Macroeconomics course at a Midwestern university [15]. The results indicated certain factors, including seniors, morning classes, and male students. An attendance policy has a significant impact on the absence rate. However, the model specification test did not directly estimate the impact of GPA or other variables on the absence rate. In this paper, we employ three types of count regressions: Poisson, negative binomial Type I, and Type II to focus on the impact of different majors and GPA on the absence rate.

2. Data and Methodology

To reduce the variations due to faculty- and text-specific factors, we employed a data set from one of the authors who had taken attendance in Principles of Microeconomics (or ECON 212) and used the same text as that of Clarion University, one of fourteen state-supported university systems in Pennsylvania. The school of business has been accredited by the AACSB since 1998, with an enrollment of approximately 650, about 1/10 of the university's enrollment. It serves the educational needs of residents who are largely from the region. We used the data of 604 students, computerized student transcript records of Clarion University from a course in the sample period to minimize exogenous factors consistently. Only those students who received final grades in ECON 212 were included in the sample.

For the number of absences in ECON 212, count regressions are the ideal analysis method. That is, the rate of absence is no longer assumed to be exogenous. The most commonly used count regression models are Poisson and negative binomial types [7]. We first considered the Poisson regression model in which y_i (the number of absences) is drawn from a Poisson distribution.

$$P(Y_i = y_i/\mathbf{x_i}) = e^{-\lambda_i}\lambda_i^{y_i}/y_i! \quad (1)$$

where the expected number of events (absence) given $\mathbf{x_i}$ is λ_i. The most common form for λ_i is as follows:

$$E(y_i/\mathbf{x_i}) = \lambda_i = e^{\mathbf{x_i'}\beta} \quad (2)$$

$$\text{or } \ln \lambda_i = \mathbf{x}_i' \boldsymbol{\beta} \tag{3}$$

Using Equation (3), we formulate the Poisson model as below.

$$\ln E(y_i/\mathbf{x}_i) = \ln \lambda_i = \mathbf{x}_i' \boldsymbol{\beta} \tag{4}$$

where $\mathbf{x}_i$ contains GPA, GENDER, MAJOR1, MAJOR2, and MAJOR3. GPA is a continuous variable on a 4-0 scale. GENDER = 0 for male students, and it is 1 for female students. MAJOR1 = 1 for business majors, and it is zero otherwise. MAJOR2 = 1 for arts and science majors, and it is zero otherwise. MAJOR3 = 1 for business information majors, and it is zero otherwise. The reference group for the three MAJOR dummies is education majors.

Given that the log-likelihood function is concave and that rapid convergence is achieved, the Poisson regression model is computationally efficient with the asymptotic covariance for parameter estimates of

$$\Big[\sum_{i} \hat{\mu}_i \mathbf{x}_i \mathbf{x}_i' \Big]^{-1} \tag{5}$$

where $\hat{\mu}_i$ is the expected value of the Poisson distribution.

Thus, it has become a popular choice in modeling count regressions in accident analysis and the number of times a patient visits a doctor. A Poisson model has a restrictive assumption in which the average and variance of y_i must be the same. When the variance exceeds the average, a case of overdispersion occurs. Cameron and Trivedi (1998) suggested the use of negative binomial (NB) models for robust estimators and bootstrap approaches. Table 1 suggests the following results based on the Poisson regression model. First, GPA is negatively related to absenteeism, with a p-value of 0.000. Second, different from Self (2012), gender is not a significant determinant in explaining absenteeism (p = 0.275) [15]. In general, female students tend to cut fewer classes than male students ($\hat{\beta}_2 = -0.078$). Third, the students of a business major cut classes significantly less than those of education (p = 0.000) and arts and sciences majors (p = 0.107).

Table 1. Estimated results of the Poisson regression model.

Variable	Estimate	Standard Error	t-Statistic	p-Value
C	3.54371	0.158480	22.3606	[0.000]
GPA	−0.639844	0.057290	−11.1684	[0.000]
GENDER	−0.077989	0.071515	−1.09052	[0.275]
MAJOR1	−0.333574	0.087194	−3.82565	[0.000]
MAJOR2	−0.135093	0.083797	−1.61215	[0.107]
MAJOR3	0.069834	0.121339	0.575528	[0.565]

n = 604, mean of dependent variable $\hat{\lambda}$ = 5.09768; standard deviation of dependent variable = 4.66226; Log-likelihood = −487.896.

The negative binomial model type 1 (NB1) assumes the variance w_i is a linear function or a multiple of the mean λ_i or

$$w_i = (1 + \alpha)\lambda_i \tag{6}$$

The negative binomial model type 2 (NB2) assumes the following quadratic relation.

$$w_i = \lambda_i + \alpha\, \lambda_i^2 \tag{7}$$

As both NB1 and NB2 are considered to be a generalization of the Poisson model and are available in many econometrics, a common practice is to run the Poisson, NB1, and NB2 models before choosing the optimal one. Tables 1–3 report the estimated results for the three models.

Table 2. Estimated results of the NB1 regression model.

Variable	Estimate	Standard Error	t-Statistic	p-Value
C	3.57134	0.162156	22.0241	[0.000]
GPA	−0.650996	0.056559	−11.5100	[0.000]
GENDER	−0.084488	0.067383	−1.25385	[0.210]
MAJOR1	−0.304356	0.083130	−3.66121	[0.000]
MAJOR2	−0.134290	0.082046	−1.63677	[0.102]
MAJOR3	0.042738	0.111408	0.383616	[0.701]
ALPHA α	2.75437	0.250689	10.9872	[0.000]

n = 604, $\hat{\lambda} = 5.09768$, standard deviation of dependent variable = 4.66226, Log-likelihood = −1561.95.

Table 3. Estimated results of the NB2 regression model.

Variable	Estimate	Standard Error	t-Statistic	p-Value
C	3.72674	0.196303	18.9846	[0.000]
GPA	−0.710946	0.066194	−10.7403	[0.000]
GENDER	−0.080952	0.073221	−1.10558	[0.269]
MAJOR1	−0.336069	0.089712	−3.74609	[0.000]
MAJOR2	−0.090736	0.091299	−0.993831	[0.320]
MAJOR3	0.090447	0.127622	0.708713	[0.479]
ALPHA α	0.542145	0.046782	11.5888	[0.000]

n = 604, $\hat{\lambda} = 5.09768$, standard deviation of dependent variable = 4.66226, Log-likelihood = −1571.54.

3. Results and Discussions

The results in Table 1 suggest that the average and variance of the dependent variable (numbers of absence in ECON 212) differ significantly. The average of $\hat{\lambda}$ = 5.09768 is much less than variance $\hat{w}$ = 21.7367. The larger variance suggests that several students in the Economics Department reduce classes excessively. Grade inflation allows a student to maintain a GPA of 2.0 to stay in university. The overdispersion is analogous to heteroscedasticity in the OLS model. For a known variance specification, we correct the problem by using the NB1 model, where the variance of the dependent variable is a multiple of its average. Table 2 shows that the results are similar to those of the Poisson model. GPA remains significant in the NB1 model (p = 0.000). The negative coefficient of −0.651 indicates that GPA and the number of absences is significantly and inversely related. That is, the higher the GPA, the less class absence. The magnitude of the relationship for an average student can be derived from Equation (2) or

$$\frac{\partial E(y_i/\mathbf{x_i})}{\partial x_i} = \lambda_i \hat{\beta}_i = 5.09768 * (-0.639) = -3.2574 \quad (8)$$

If GPA is increased by one unit, the expected number of skipped classes is reduced by 3.2574. A student with GPA = 2 cuts 3.2574 classes more than a student with GPA = 3. Missing classes for over one week and a half have a noticeable impact on performance in Principles of Microeconomics.

Female students have a decreased tendency to reduce classes (p = 0.21). The same trend is also observed in the Poisson model (p = 0.275). We surmise that female students at the university are more conscientious in the Principles of Microeconomics course than male students. In the Poisson model, MAJOR1 (business major) is significant in the NB1 model. A business major has less tendency to reduce classes than the reference group (education majors) (p = 0.00). It means that students of an education major feel difficulty and lose interest in the Principles of Microeconomics. The same trend is observed for the students of arts and science majors. The students of business information science seem to have a similar trend as those of education majors in terms of absenteeism.

The difference in the expected rate of absence between majors is evaluated by using Equations (2) and (6).

$$\frac{\lambda_i(\text{MAJOR1} = 1)}{\lambda_i(\text{MAJOR1} = 0)} = \frac{e^{x'\beta(\text{MAJOR1}=1)}}{e^{x'\beta(\text{MAJOR1}=0)}} = \frac{e^{x'\beta - 0.304356}}{e^{x'\beta}} = \frac{1}{e^{0.304356}} = 0.7376 \quad (9)$$

where −0.304356 (Table 2 of the NB1 model) is the estimated coefficient on the dummy variable MAJOR1. In total, 73.76% of the students of business majors did not take Principles of Microeconomics, while those of education majors take 26.24%.

Similar for arts and sciences majors, the ratio of the absence rate is as follows:

$$\frac{\lambda_i(\text{MAJOR2} = 1)}{\lambda_i(\text{MAJOR2} = 0)} = \frac{1}{e^{0.13429}} = 0.8743 \quad (10)$$

The absenteeism rate for the students of arts and sciences majors is 87.43% of those of education majors.

The overdispersion is tested with the estimated coefficient on α (Wald test). In the case of the NB1 model, a p-value of 0.000 leads to the rejection of $H_0 : \alpha = 0$. In addition, the value of α is 2.75437 implies considerable overdispersion (Cameron & Trivedi, 1998, pp 78–79). It is tested by using the LR test, and the result shows the difference in the fitted log-likelihoods of the Poisson and NB1 or $-2 * (-1561.95 + 487.896) = 2148.1$, which far exceeds the critical value of x^2. Both test results indicate that the overdispersion problem precludes the Poisson regression model due to a small group of outliners (excessively cutting classes).

Table 3 presents that GPA is a factor in explaining the absenteeism problem but remains an insignificant predictor. MAJOR1 remains the same as in NB1. MAJOR2 becomes significant (p-value of 32%), and MAJOR3 remains insignificant. The Wald test α indicates the NB2 is appropriate for the Poisson model in terms of the overdispersion problem. The choice between the two negative binomial models is important for the LR test as the Wald test gives indistinguishable results. Since the log-likelihood function of the NB1 model gives a slightly larger value than that of the NB2 model (−1561.95 > −1571.54), we opt for the NB1 model in analyzing the student absenteeism problem at this university.

4. Concluding Remark

Decreasing the attrition rate in public universities is of primary concern to their survival. Declining enrollment, especially in rural areas, causes many problems. A proper policy cannot be effectively implemented without understanding the factors that affect absenteeism. As a majority of college professors do not take attendance, the issue remains unresolved. We applied the Poisson, NB1, and NB2 regression models to analyze the problem. All of the models indicate that GPA is inversely related to absenteeism in Principles of Microeconomics classes. In particular, one unit of GPA is expected to reduce absences by 3.2574 in the NB1 model. In the NB1 model, the students of business majors reduce 26.24% more classes than those of education majors. Due to an overdispersion problem, the Poisson model is inappropriate. A large variance in the number of absences exceeds its corresponding average (21.7367 > 5.09768). The overdispersion problem leads to the question of whether one should implement a mandatory attendance policy. Low-students become disruptive and hence jeopardize the overall welfare of the entire class. The students ought to have the freedom to weigh the positives and negatives between the marginal gain in course performance and attending the class. The overdispersion (excessive cut) problem is found in the NB1 and NB2 models. The LR test indicates that the NB1 model has a slight advantage over the NB2 model. The explanatory variables that determine the number of cut classes are GPA, MAJOR1, and MAJOR2 (marginally significant). Female students tend to cancel the Principles of Microeconomics noticeably. The efforts should thus be focused on those with low GPAs in business information science and education majors.

Author Contributions: Conceptualization, S.-S.C. and C.W.Y.; methodology, C.C.; software, C.C.; validation, S.-S.C.; formal analysis, S.-S.C.; investigation, C.C.; resources, C.W.Y.; data curation, C.W.Y.; writing—original draft preparation, C.C.; writing—review and editing, S.-S.C.; All authors have read and agreed to the published version of the manuscript.

Funding: This research received no external funding.

Institutional Review Board Statement: Not applicable.

Informed Consent Statement: Not applicable.

Data Availability Statement: Data is unavailable due to privacy or ethical restrictions.

Conflicts of Interest: The authors declare no conflict of interest.

References

1. McCullagh, P.; Nelder, J. *Generalized Linear Models*; Chapman and Hall: New York, NY, USA, 1983.
2. Cameron, A.C.; Trivedi, P.K. *Regression Analysis of Count Data*; Cambridge University Press: Cambridge, UK, 1998. [CrossRef]
3. Chan, K.; Shum, C.; Wright, J. Class attendance and student performance in Principles of Finance. *Financ. Pract. Educ.* **1997**, *7*, 58–65.
4. Rodgers, J.R. A Panel-Data Study of the Effect of Student Attendance on University Performance. *Aust. J. Educ.* **2001**, *45*, 284–295. [CrossRef]
5. Romer, D. Do students go to class? Should they? *J. Econ. Perspect.* **1993**, *7*, 167–174. [CrossRef]
6. Schmidt, R.M. Who maximizes what? A study in student time allocation. *Am. Econ. Rev.* **1983**, *73*, 23–28.
7. Park, H.; Kerr, P. Determinants of academic performance: A multinomial logit approach. *J. Econ. Educ.* **1990**, *21*, 101–111. [CrossRef]
8. Durden, G.C.; Ellis, L.V. The effect of attendance on student learning in Principles of Economics. *Am. Econ. Rev.* **1995**, *85*, 343–346.
9. Devadoss, S.; Foltz, J. Evaluation of Factors Influencing Student Class Attendance and Performance. *Am. J. Agric. Econ.* **1996**, *78*, 499–507. [CrossRef]
10. Marburger, D.R. Absenteeism and undergraduate exam performance. *J. Econ. Educ.* **2001**, *32*, 99–109. [CrossRef]
11. Kirby, A.; McElroy, B. The effect of attendance on grade for first year Economics students in University College Cork. *Econ. Soc. Rev.* **2003**, *34*, 311–326.
12. Stanca, L. The Effects of Attendance on Academic Performance: Panel Data Evidence for Introductory Microeconomics. *J. Econ. Educ.* **2006**, *37*, 251–266. [CrossRef]
13. Chen, J.; Lin, T.-F. Class Attendance and Exam Performance: A Randomized Experiment. *J. Econ. Educ.* **2008**, *39*, 213–227. [CrossRef]
14. Arulampalam, W.; Naylor, R.A.; Smith, J. Am I missing something? The effects of absence from class on student performance. *Econ. Educ. Rev.* **2012**, *31*, 363–375. [CrossRef]
15. Self, S. Studying absenteeism in Principles of Macroeconomics: Do attendance policies make a difference? *J. Econ. Educ.* **2012**, *43*, 223–234. [CrossRef]

Disclaimer/Publisher's Note: The statements, opinions and data contained in all publications are solely those of the individual author(s) and contributor(s) and not of MDPI and/or the editor(s). MDPI and/or the editor(s) disclaim responsibility for any injury to people or property resulting from any ideas, methods, instructions or products referred to in the content.

engineering proceedings

Proceeding Paper

Practical Research on Mixed-PPGIS Promoting Public Participation in Urban Micro-Public Space Renewal †

Yinglu Huang [1,*], Xiaojun Huang [2], Junhui Zhao [2] and Yilin Chen [2]

[1] College of Public Affairs, National Taipei University, Taipei 237303, Taiwan
[2] School of Architecture, Huaqiao University, Xiamen 361021, China; whiugwan@163.com (X.H.); zeonfai_chiu@163.com (J.Z.); tatiana2023@163.com (Y.C.)
* Correspondence: archyl1998@outlook.com
† Presented at the 3rd IEEE International Conference on Electronic Communications, Internet of Things and Big Data Conference 2023, Taichung, Taiwan, 14–16 April 2023.

Abstract: Based on the public participation geographic information system (PPGIS), the research platform is designed to facilitate and provide a basis for public participation. The platform includes the application of three core links: information acquisition, planning participation, and decision-making participation. To solve the problems in the planning and design of the micro-public space in China's urban and rural planning system, the renewal practice of the micro-public space in Jimei College Village, Xiamen, China was used as a study case to explore how the mixed-PPGIS can be applied to micro-public space renewal by promoting the realization of "public participation". The research results show that the mixed-PPGIS is a combination of technical methods that leads to the improvement of bottom-up participatory planning. At the same time, the importance of "people" in the mixed-PPGIS construction was emphasized.

Keywords: mixed-PPGIS; micro-public space; Jimei College Village; renewal

Citation: Huang, Y.; Huang, X.; Zhao, J.; Chen, Y. Practical Research on Mixed-PPGIS Promoting Public Participation in Urban Micro-Public Space Renewal. *Eng. Proc.* **2023**, *38*, 84. https://doi.org/10.3390/engproc2023038084

Academic Editors: Teen-Hang Meen, Hsin-Hung Lin and Cheng-Fu Yang

Published: 20 July 2023

Copyright: © 2023 by the authors. Licensee MDPI, Basel, Switzerland. This article is an open access article distributed under the terms and conditions of the Creative Commons Attribution (CC BY) license (https://creativecommons.org/licenses/by/4.0/).

1. Introduction

A micro-space refers to a specific space available in daily life from the perspective of geography. The concept of the vest-pocket park, the neighborhood park, and the pocket park was subsequently applied to the micro-public space [1]. Marcus and Francis described humanized space in which public spaces serve users at different times, ages, and habits [2]. Definitions of an urban micro-public space vary according to its size. The scale of urban micro-public space is usually small, ranging between 300 and 5000 m^2 with several types including street parks, small-sized squares, community parks, stadiums, and others. Its main function is to be present in the residents' daily lives. Micro-public spaces with frequent use are mostly located in areas with excellent accessibility and large demands and requirements of residents [3].

Various problems exist in the planning and design, construction, use, and management of micro-public space due to the lack of responsibility of the personnel working in the current urban and rural planning system. Dong et al. described the planning and design principles of urban micro-public space as "embedding, fixing, and public participation" [3]. Comprehensive research and in-depth analysis of the major issues for micro-public space have been conducted only by a limited number of professionals. Many scholars have proposed renewed practices of micro-public spaces, combining digital participation platforms.

In the Zhonggulou District of Beijing, China, the Zhonggulou community public participation platform was combined with WebGIS and Baidu Map for the process of the reconstruction of the district to provide interactive channels for diversified social cooperation and community participation. Xiang proposed the planning of a micro-public space for Xiling community, Yichang, Hubei, China, through data collection, sorting, and analysis with a PPGIS platform based on public participation in the new era [4]. Peng et al.

designed the micro-public space based on a micro-climate environment with the measured and simulated mobile big data and WeChat data to represent public participation [5]. The innovative use of digital technology was considered to enhance public participation. The construction of multiple opinion feedback channels based on the Internet and multimedia is believed to be a necessary action for the introduction of group wisdom and the realization of "public participation" [6].

Previous studies showed that the establishment of a digital participation platform is important in the preliminary survey, process, and planning for decision making in planning public space through public participation. However, how to participate in the whole process of planning and design has not been described in detail. Hence, the present study designs the participation platform using the mixed public participation geographic information system (PPGIS) and applies it to the renewal of the micro-public space of Jimei School Village, Xiamen. In the design, the public participates in the whole process of information acquisition, project survey, project visualization, planning, and decision making, acquiring public requirements, visual simulation of design solutions, and online interaction.

2. Research Method

PPGIS is a combination of GIS and societal participation. The traditional top-down planning by experts since the 1960s has not been applied to the current social development anymore, so decision making based on the concept of participation has gradually developed. Participation is a process to condense the will of local people, improve their understanding of events to reach a consensus on decision making, and achieve a goal with residents' empowerment. Based on the research on PPGIS, the present study designs a mixed-PPGIS platform characterized by integration which provides convenient channels for the public to participate in planning according to their interest. Three core links are involved in the platform: message acquisition, planning participation, and decision making.

Mixed PPGIS is an integrated PPGIS that is embodied in the following three aspects:

- Combination of community local knowledge with scientific knowledge;
- Combination of the traditional PPGIS workshop with network technology;
- Combination of diverse participatory groups.

The architecture of the mixed-PPGIS platform includes the user layer, display layer, interlayer, and data layer, as mixed-PPGIS integrates the PPGIS workshop and network technology. The network participation is supported by corresponding system software, including Local Space Viewer 4 developed by Suzhou Map New Media Co., Ltd. (Suzhou, China) and WeChat platform developed by China Tencent (Shenzhen, China).

2.1. *Local Space Viewer (LSV4)*

Local Space Viewer 4 is a new, professional, and three-dimensional (3D) digital software program that quickly browses, measures, analyzes, and labels 3D geographic information data. Researchers and residents open maps on computers or mobile phones to browse with the downloaded LSV. Meanwhile, plans and designs are implemented based on the tools provided by LSV based on the questionnaire survey.

2.2. *Community Map: WeChat Official Account*

A WeChat official account of the community map was developed in the present study based on the geographic information data acquired by LSV. Relevant planning information and a thematic map are presented in the WeChat public address through GIS analysis and are shown to the public. The user obtains the thematic information to participate in planning, discussion, and decision making using a simple interaction design.

3. Renewal of Micro-Public Space in Jimei School Village

3.1. Research Design

Jimei School Village in the Jimei District of Xiamen City is located in Jimei Peninsula, Xiamen. The district is a 20 min drive away from the center of Xiamen City, and convenient transportation is available in a total area of 1.35 km^2. Jimei School Village has many school institutions and historical buildings, being a well-known historical and cultural tourism area. According to the master planning, Jimei School Village has developed its historical buildings for an educational function and introduced commercial and tourism activities for sustainable development with the goal of having a "school area + scenic spot".

We formed a research team including 97 undergraduates from several universities and initiated a public space renovation plan from March to June 2021 based on a general survey for this study. Firstly, we conducted participatory mapping through eight PPGIS workshops. The interview was conducted with residents, merchants, enrolled students, and tourists. Meanwhile, we acquired 35 micro-public spaces which were required to be improved and transformed. In the interview, we recorded the basic situation of the micro-public spaces (area, current facilities, functions, and others) and the evaluation of the current situation. An evaluation was carried out for safety, accessibility, comfort, recognizability, and inclusiveness (Figure 1). The micro-public spaces were classified into "community-based", "street-based", and "public space-based" spaces based on the interview and evaluation results. As a result, special drawings using LSV were drawn (Figure 2). Public opinions were collected and sorted through the PPGIS workshop using LSV4 for research and design.

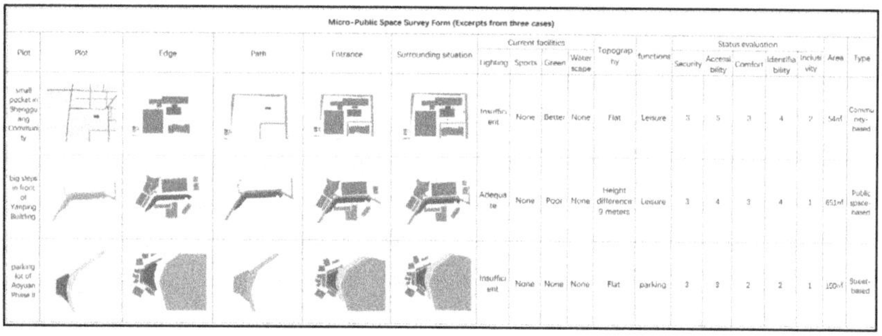

Micro-Public Space Survey Form (Excerpts from three cases)

Plot	Plot	Edge	Path	Entrance	Surrounding situation	Current facilities				Topography	functions	Status evaluation					Area	Type
						Lighting	Sports	Green	Water scape			Security	Accessibility	Comfort	Identifiability	Inclusivity		
small pocket in Shenggu ang Community						Insufficient	None	Better	None	Flat	Leisure	3	5	3	4	2	54㎡	Community-based
big steps in front of Yanping Building						Adequate	None	Poor	None	Height difference 9 meters	Leisure	3	4	3	4	1	651㎡	Public space-based
parking lot of Aoyuan Phase II						Insufficient	None	None	None	Flat	parking	2	3	2	2	1	100㎡	Street-based

Figure 1. Micro-public space design (excerpts from three cases).

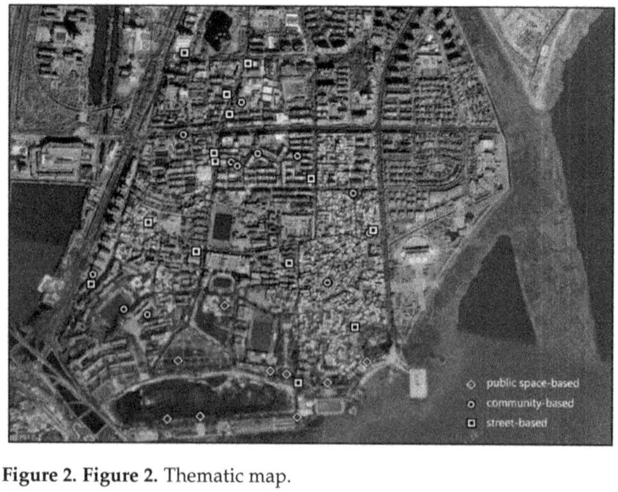

Figure 2. Figure 2. Thematic map.

In the second stage of the research and design, the design for transforming micro-public spaces was constructed based on public demands. Interviews with government departments and relevant construction organizations were performed for a feasibility study. The project design for the first batch of nine micro-public spaces was then confirmed. The project design was conducted by university professors. The three-dimensional drawings were uploaded to LSV (Figure 3) to facilitate the public's browsing and inquiring. Simultaneously, the evaluation of comments and the selection of plans were carried out via WeChat official accounts to promote the participation of the public. Positive responses from the community were received. In online voting at the end of five days, we received 252,238 responses.

Figure 3. Figure 3. 3D drawings based on LSV4.

The project was disclosed with the traditional Chinese planning process in the third stage of the research. Experts from local governments, construction organizations, industry, universities, and on-site viewers were invited to participate in the final voting. Finally, the results of the selected projects were submitted to the government department for implementation.

3.2. Analysis of Research Results

Mixed-PPGIS was integrated into the micro-public space renewal in the present study with public participation. Paul's multidimensional analysis method was referred to for analyzing the planning results. Paul combined the empirical experience of the World Bank regarding community participation for the planning of urban housing, health, and agriculture sectors and proposed a multidimensional analysis framework for community participation based on the "objectives", "intensity", and "instruments" (Figure 4) [7]. In the micro-public space renewal of Jimei School Village, the whole process of "public participation" was realized in the mixed-PPGIS public participation platform.

- We held several PPGIS workshops in the planning stage with LSV4 for participatory mapping and information acquisition of micro-public spaces. GIS thematic map data were produced and analyzed, too. University teams, residents, and tourists participated in information sharing and decisions through agreement, cooperation, and interaction. The planning results were visualized in an LSV thematic map to encourage public participation through WeChat. Other than the participants, administrative workers and experts were also involved. Participation became more active in the negotiation stage, and the goal of the renewal was achieved through understanding and agreement. The final projects were selected via voting through a WeChat public platform. The public's opinion was shared with experts and government workers. With such participation, the objectives of the renewal were achieved based on the consensus reached in planning and decision making.

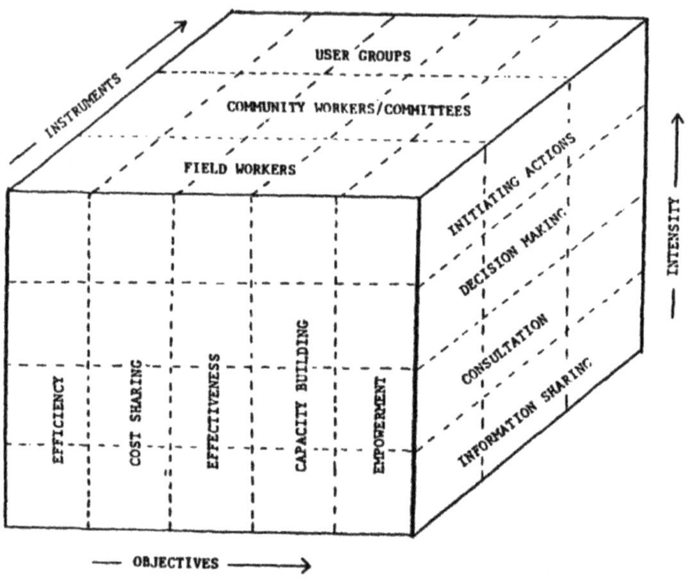

Figure 4. Figure 4. Multidimensional analysis framework for community participation [7].

4. Conclusions and Recommendations

According to the summary of the experience of 50 empirical cases in the World Bank, community participation was not successful in several projects due to technological gaps, weak promotion and supervision, lack of a set of comprehensive services, and failure to implement key policies for the project. Mixed-PPGIS allows technical methods to be integrated into bottom-up participatory planning. Three characteristics such as multiple participation, phase partitioning processing, and mixed development mode were reflected in the present study [8]. The participation platform of mixed-PPGIS played a critical role in the renewal of urban micro-public spaces with "people" as the key factor. The timely and appropriate participation of administrative staff is necessary to promote participatory planning and plays different roles in different stages of the whole micro-public space renewal process. The results show the importance of the university team which plays a role of organization and guidance in the whole process.

Author Contributions: Conceptualization, Y.H.; methodology, Y.H.; software, X.H.; validation, Y.H., Y.C.; formal analysis, Y.H.; investigation, X.H.; resources, J.Z.; data curation, Y.H.; writing—original draft preparation, X.H.; writing—review and editing, Y.H.; visualization, Y.H.; supervision, Y.H.; project administration, Y.H.; All authors have read and agreed to the published version of the manuscript.

Funding: This research was supported by "the 2021 Fujian Social Science Foundation Project", No. FJ2021B194.

Institutional Review Board Statement: Not applicable.

Informed Consent Statement: Informed consent was obtained from all subjects involved in the study.

Data Availability Statement: The data and materials include primary data collected during the teaching practice process, including pictures and texts. The datasets used or analysed during the current study are available from the corresponding author upon reasonable request.

Conflicts of Interest: The authors declare no conflict of interest.

References

1. Danford, R.S.; Strohbach, M.W.; Warren, P.S. Active Greening or Re-wilding the City: How does the intention behind small pockets of urban green affect use? *Urban For. Urban Green.* **2018**, *29*, 377–383. [CrossRef]
2. Marcus, C.C.; Francis, C. *People Places-Design Guidelines for Urban Open Space*, 2nd ed.; John Wiley & Sons: Hoboken, NJ, USA, 1997.
3. Xu, H.; Zhu, J.; Wang, Z.; Fang, W. Exploring the Spatial Pattern of Urban Block Development Based on POI Analysis: A Case Study in Wuhan, China. *Sustainability* **2019**, *11*, 6961. [CrossRef]
4. Fei, X.; Mao, B. Study on the Strategy of Urban Small Public Space Planning Based on the Concept of Park City—Take the Old Town of Jiangling County in Jingzhou as an Example. *Curr. Urban Stud.* **2020**, *19*, 107–114. [CrossRef]
5. Yu, J.; Meng, Y.; Zhou, S.; Zeng, H.; Li, M.; Chen, Z.; Nie, Y. Research on Spatial Delineation Method of Urban-Rural Fringe Combining POI and Nighttime Light Data—Taking Wuhan City as an Example. *Int. J. Environ. Res. Public Health* **2023**, *20*, 4395. [CrossRef] [PubMed]
6. Yun, J.; Yu, W.; Wang, H. Exploring the Distribution of Gardens in Suzhou City in the Qianlong Period through a Space Syntax Approach. *Land* **2021**, *10*, 659. [CrossRef]
7. Paul, S. *Community Participation in Development Projects: The World Bank Experience*; World Bank: Washington, DC, USA, 1987; pp. 8–9.
8. Kacaw, L.; Tsai, B.-W. The Application of PPGIS to Telecoupling Research: A Case Study of the Agricultural Landscape Transformation in an Indigenous Village in Taiwan. *Sustainability* **2023**, *15*, 1577. [CrossRef]

Disclaimer/Publisher's Note: The statements, opinions and data contained in all publications are solely those of the individual author(s) and contributor(s) and not of MDPI and/or the editor(s). MDPI and/or the editor(s) disclaim responsibility for any injury to people or property resulting from any ideas, methods, instructions or products referred to in the content.

engineering proceedings

MDPI

Proceeding Paper

Design of the Speech Emotion Recognition Model †

Hanping Ke *, Feng Luo and Manyin Shi

School of Information, Mechanical and Electrical Engineering, Normal University, Ningde 352100, China; t0407@ndnu.edu.cn (F.L.); t0708@ndnu.edu.cn (M.S.)

* Correspondence: t1502@ndnu.edu.cn

† Presented at the 3rd IEEE International Conference on Electronic Communications, Internet of Things and Big Data Conference 2023, Taichung, Taiwan, 14–16 April 2023.

Abstract: Existing emotional feature methods only represent the limited information on the emotional state and lack the mining and utilization of the correlation between emotional features. Therefore, a new design scheme is proposed based on the psychological acoustic model of the speech spectrum to investigate the characteristics of the spectrum distribution of emotion. The proposed model for speech emotion recognition improves the accuracy of the recognition and provides the basis for the development and application of further developed models for speech emotion recognition.

Keywords: speech emotion; recognition model; design scheme; spectrum feature

1. Introduction

Speech is the most important way for people to communicate. Voice signals contain rich semantic information and carry emotional status effectively. The recognition of the emotional status in speech with the machine learning method is used in virtual reality, driving safety, medicine, customer service quality, and many other applications. Recently, the rapid development of artificial intelligence (AI) and virtual reality (VR) has promoted the publication of various studies on speech emotion recognition [1]. In human–computer interactions and VR immersion, speech emotion recognition plays an important role in the transportation industries, including the automobile, aircraft, and shipbuilding industries. By analyzing the emotional changes that occur during the user purchase process, sales methods and strategies can be adjusted accordingly to increase the quality of the sales. In retail businesses, a better experience can be provided through the analysis of the users' emotional status. Therefore, researching emotion recognition in speech is a universal demand for the development of both e-commerce and retail businesses. However, the technology of the recognition of emotion in speech is not yet mature, which thereby motivates this study to develop and propose a new technology to recognize emotion in speech.

Citation: Ke, H.; Luo, F.; Shi, M. Design of the Speech Emotion Recognition Model. *Eng. Proc.* **2023**, *38*, 86. https://doi.org/10.3390/engproc2023038086

Academic Editors: Teen-Hang Meen, Hsin-Hung Lin and Cheng-Fu Yang

Published: 21 July 2023

Copyright: © 2023 by the authors. Licensee MDPI, Basel, Switzerland. This article is an open access article distributed under the terms and conditions of the Creative Commons Attribution (CC BY) license (https://creativecommons.org/licenses/by/4.0/).

2. Previous Research

In the design of a speech emotion recognition system, the analysis and extraction of emotion features are challenging due to the rich variability of human emotion. In real life, people perceive and recognize the semantic information and emotional states in a speech. Therefore, choosing the appropriate features of speech decides the performance of emotion recognition. As different classifiers have different applicable scenarios, the selection of emotional features depends on which classifiers are used. Thus, researchers have studied the characteristics of emotion together with the classifiers in emotion recognition. According to the study conducted by Koolagudi [2], speech emotion features can be classified into excitation source features, prosodic features, vocal channel features (spectral features), and joint features.

2.1. Excitation Source Features

The excitation source feature arises from the excitation part of speech. According to the speech generation model, the excitation source signal of speech is obtained with the sound channel information. In this information, the linear prediction coefficient (LPC) can be calculated using the linear prediction analysis of speech, and then the excitation source signal is obtained with the LP filtering, which is usually expressed as the LP residual energy. In 1976, the Wakita first used the LP residual energy in speaker recognition, which indicated that the LP residual energy characterizes the paralinguistic information of speech [3]. In 2006, the features of the LP residual energy were used in the identification of speakers [4]. Since then, LP residuals have been applied to speech-emotion recognition. In 2015, Yegnanarayana and Gangashetty added echo features to recognize the differences in the LP residual energy in speech emotion and found that echo features were conducive to the characterization of emotion in segmented speech and were beneficial to improve the recognition rate [5]. In 2017, Gangamohan et al. achieved identification rates of 76 and 69% on the IITH-H and EMO-DB databases, respectively, by calculating the Kullback–Leibler (KL) distance of the excitation source signals [6]. In 2019, Pravena and Govind determined the intensity of the excitation source and the base frequency of the speech signal and calculated its statistical properties using the Gaussian Mixture Model (GMM), which further improved the efficiency of the identification of these excitation source features [7].

2.2. Prosodic Features

In speech, rhythmic information is included in the duration, intensity, and tone of the sound. Base sound, energy, duration, and others in the rhythmic information reflect the emotional state. By extracting the rhythmic information and conducting statistical analyses on it, the characteristics of emotion can thereby be determined. In 2015, Han constructed a multiple Elman network model based on prosody features to identify different emotions based on sensitive rhythmic segments. Recognizing the multi-classifier of emotion, the emotion recognition of the human ear was simulated with a recognition rate of 67.9% [8]. In 2018, Zhang et al. applied a non-linear dynamic model to analyze the emotional speech signals using the chaotic characteristics in the speech sound process and extracted the non-linear features of the emotional speech signal and the commonly used acoustic features (rhythmic features and the Mayer inversion coefficient (MFCC)), following which they characterized the chaotic properties of the emotional speech signals [9].

2.3. Spectral Features

Vocal channel features are also termed as spectral features or segmented features. In speech emotion recognition, speech is divided into segments in 20–30 ms, and the resonant peak and sub-band spectrum energy are analyzed in the frequency domain. The frequency transform usually adopts the discrete Fourier transform. To further enhance the recognition ability, the characteristic parameters are transformed into the inverted spectrum domain to characterize the emotional state. The MFCC, perceptual linear prediction coefficient (PLPC), and linear prediction inversion coefficient (LPCC) in the spectral domain are all used in speech recognition as spectral features for speech recognition. In 2017, Lotfian et al. investigated the emotion recognition of synthetic speech through MFCC analysis, and proposed a novel research scenario, namely, the emotion recognition of robot sounds [10]. In 2019, Jing et al. improved the recognition performance by 6% in the Chinese corpus of the Chinese document-level extractive summarization dataset (CDESD) [11].

2.4. Joint Features

The study of speech and emotion recognition includes feature extraction and emotion classification. Emotion classification models include the hidden Markov model (HMM), GMM, artificial neural network (ANN), and support vector machine (SVM), each of which possesses their advantages and disadvantages that are related to feature selection. Currently, feature extraction is the most concerned area of research. The excitation feature comes from the speech signal source, which is related to the speech excitation source by suppressing the sound channel excitation. According to the principle of digital speech generation, excitation sources are related to the semantic content of speech. These features are used in the recognition and classification of speech. Speech affective states are determined by the tones and semantic contents. Therefore, the excitation source features are not used in extracting the speech subsidiary information, such as tone and intonation and in emotion recognition. Prosodic features are derived from pronunciation characteristics, such as the duration, intensity, and tone of speech. According to the principle of linguistics, the prosodic features present the pronunciation characteristics of speech, and allow better recognition performances compared to the excitation source features. However, the classification of excitation source features, prosodic features, spectral features, and joint features may overestimate the characteristics of language pronunciations. Different languages have differences in pronunciation due to cultural differences. Thus, prosodic features are deemed to not be robust enough to recognize emotion using libraries or linguistic scenarios.

3. Design of the Speech Emotion Recognition Model

Emotion recognition is conducted through the analysis of characteristics in classified emotions. Common emotional features are used to determine a recognition rate and obtain the influencing factors on the features. Then, the spectrum of the emotional features is investigated to discriminate the features and find the distribution law of frequency under different emotions.

3.1. Multi-Scale Spectral Feature Extraction Model

According to the phonological psychoacoustic model, the human ear has different perceptions of speech in different frequencies, as emotional status can be presented in different frequencies. The signal enhancement method improves the representation of different emotional statuses at different frequencies. The spectrum transformation method is used to discriminate the different emotional statuses in the frequency domain. In the psychological acoustic model, the characteristics of the different features according to the emotional status were designed with a multi-scale spectral feature extraction algorithm.

3.2. Speech Emotion Recognition System

In speech emotion recognition, emotional features are closely related to emotion classifiers. Thus, model training and testing with multiple features are performed to improve the emotion recognition rate. Joint features are used for the multi-scale feature recognition in cross-language, and for the improvement of the feature extraction method and classifier design. In this study, we conducted theoretical research and experimental verification. Based on the investigation and analysis of the latest research in the related fields, the emotion features with the speech psychoacoustic model and mathematical statistical method were defined to discover the distribution pattern of the emotion features in the frequency domain. The spectrum features were extracted with speech digital processing to obtain the vectors of the features and design a machine learning method to establish a speech emotion recognition model. In this experiment, the model was verified for its performance through data collection, statistical analysis, and feedback.

3.3. Feature Analysis

Common features, spectrum features, and the emotion feature distribution law were all investigated in this study. For the analysis of the common features, openSMILE was used to extract the features. Here, openSMILE is Munich's Open Source Media Large Feature Space Extraction (openSMILE) Toolkit. These features were then inputted to determine the classifier. The classifier was obtained with the SVM multi-classifier in the Interspeech2009 emotion recognition system to rank the contribution rate of the individual emotional features and obtain the common features affecting the emotional status according to the ranking. In the spectrum feature analysis, it has been assumed that the speech emotion signal contains information in languages (intonation, tone, and so on), and these features were acquired from the speech emotion status. Since the phonetic spectrogram represents the signal properties, the spectrum information of emotion was obtained in the frequency domain in the phonetic spectrogram. These features were marked on their spectra for a base period, MFCC envelope, and frame energy. Statistical analyses for the different frequency components was performed for feature extraction. The threshold was represented by the subband frequency range of T:

$$T = \{t_i \in [f_{i0}, f_{i1}], i = 1, 2, \ldots, m\} \tag{1}$$

where t_i represents the ith subband, and f_{i0} and f_{i1} indicate the start and end frequency of the ith subband, respectively.

3.4. Feature Extraction

For the enhancement of these emotional features, the speech signal was generated using the linear filtering of the excitation source features with sound channel filtering. If the lattice-type excitation source attenuated by 12 dB in the signal amplitude, the labial radiation subsequently increases by 6 dB. Therefore, when the speech is spoken, a 6dB decay occurs in the entire amplitude, leading to an increased frequency by about 1000 Hz, and thereby cause the channel information to decay in the high-frequency region. Therefore, to prevent the loss of emotional information, high-frequency channel information can be enhanced by pre-weighted filtering. Pre-aggravating filtering aggravates the spectrum information in high-frequency regions.

For speech signals, pre-aggravation is performed using a first-order difference equation (Equation (2)):

$$y(n) = s(n) - \alpha \cdot s(n-1) \tag{2}$$

where α is a constraint parameter with a value of 0 to 1. The specific value of α needs to be obtained through experiments for the pre-aggravating $y(n)$.

As the emotional status in speech is mainly reflected in the frequency difference, it is therefore necessary to convert the signal in the time domain into the frequency domain. In this research, the common discrete Fourier transform was adopted to obtain the speech signal in the frequency domain. As the speech signal exhibits long-time and non-stationary characteristics, it is not conducive to feature analysis. However, as the voice has short-time stationary characteristics, frame segmentation is required. We used a frame length of 20 ms and a frame displacement of 10 ms, respectively. In order to obtain the frame level signal in each frame, the Hamming window function is used, as shown Equation (3):

$$w(n) = \begin{cases} 0.54 - 0.45cos(2\pi n/(L-1)), 0 \le n \le L \\ 0, other \end{cases} \tag{3}$$

where L indicates the window length.

The window to the pre-increased signal is expressed as

$$y_w(n) = y(n)w(n) \tag{4}$$

The Discrete Fourier transform was then used to obtain the signal in the frequency domain, as shown in Equation (5).

$$Y_F(k) = \sum_{n=0}^{N-1} y_w(n) e^{-j\frac{2\pi}{N}kn} \tag{5}$$

The frequency signal Y was obtained using the Fourier transform in the whole frequency bandwidth. According to the previous analysis, a different emotional status was presented in the different frequency bands. To highlight and characterize these emotional features, molecular bands were used, while the distribution law in the sub-bands was derived from the threshold T of the feature analysis. According to the distribution law, $Y_F(i)$ was obtained with TFDivided. The subband frequency signal Y was obtained by performing the Fourier transformation $F(i)$ to transform the signal into the temporal subband signal $y_t(i)$. For a detailed representation of the signal frequency, the sine wave S in the subband was used with the sinusoidal modeling of the time-domain signal $s(i, j)$ to extract the amplitude of each sine wave and obtain the final amplitude feature $A(i, j)$. The detailed algorithm calculation process is shown in Figure 1.

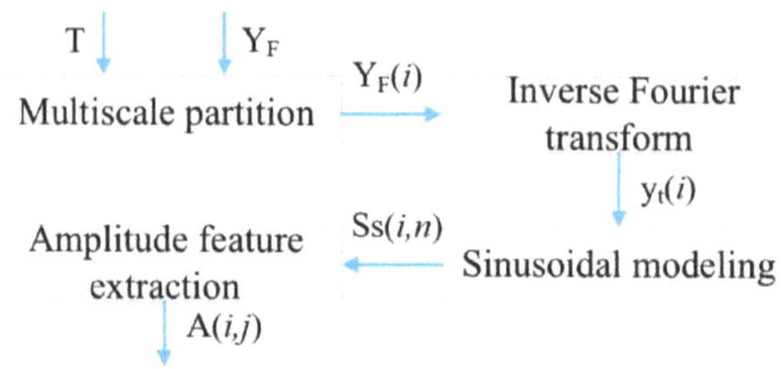

Figure 1. Route of the extract of the amplitude of each sine wave.

The frequency signal Y was determined based on the statistical threshold of each subband obtained from the feature analysis $T = \{t_i \in [f_{i0}, f_{i1}], i = 1, 2, \ldots, m\}$, and was divided into the frequency subband signal $Y_F(i)$. For the frequency subband signal $Y_F(i)$, the Fourier transform was again used to obtain the temporal subband signal $y_t(i, n)$. The inverse Fourier transform was performed with Equation (6).

$$y_t(i,n) = \sum_{k=0}^{N-1} Y_F(i,k)e^{-j\frac{2\pi}{N}kn} \tag{6}$$

Sinusoidal modeling of the subband signal in the time domain was used to obtain the sine signal $S_s(i, j)$, whose model is expressed in Equation (7):

$$s_s(i,n) = \sum_{j=1}^{L} A_j \cos(2\pi f_j \frac{n}{f_s} + \theta_j) \tag{7}$$

Where L represents the number of sinusoidal components, and f_s represents the signal sampling rate, respectively.

The amplitude of each sine wave component was extracted to create a feature matrix $A(i, j)$, with the j-th magnitude feature of the i-th subband in a dimension of $m \times L$, where m represents the number of subbands, and L represents the number of sine wave components of each subband, respectively.

3.5. Emotional Recognition

In the classifier design, the emotion recognition corpus is generally small, and the SVM can thereby be used to obtain a better recognition with a small amount of data. Thus, the multiclass classifier design of the SVM was adopted in this study, and the kernel function was chosen as the radial basis function. The SVM model was trained by inputting features and was subsequently assessed for emotion classification. For training and testing, the training set, validation set, and test set were all selected at a ratio of 6:2:2. The final test results from the fuzzy matrix method were evaluated for the system performance based on the recognition rate.

4. Conclusions

In the context of AI, a new method for speech emotion recognition was proposed using spectrum feature analysis and a multi-scale spectrum feature extraction. The proposed method was validated to be practical and provided an optimized solution for the research and development of speech emotion recognition. In the proposed method, the overall information on the emotional status and characteristics was used to improve the accuracy of emotion recognition. In emotion recognition, the characteristics of emotional statuses were analyzed with different frequency components to define the distribution law of the emotional features on the frequency spectrum. The differences in the frequency components were also found for different emotions. Thus, multi-scale features could be distinguished between different emotions more effectively in the future.

Author Contributions: Conceptualization and methodology of this manuscript were proposed by H.K. Software and experiment were conducted by F.L. M.S. worked out the writing—original draft preparation and writing—review. All authors have read and agreed to the published version of the manuscript.

Funding: This research was supported by the Hainan Provincial Natural Science Foundation of China (No. 2021J011169, 2022J011224, 2020J01435) and the Ningde Normal University for the service local action special plan (No. 2020ZX505).

Institutional Review Board Statement: Not applicable.

Informed Consent Statement: Not applicable.

Data Availability Statement: No new data were created.

Conflicts of Interest: The authors declare no conflict of interest.

References

1. Swain, M.; Routray, A.; Kabisatpathy, P. Databases, features and classifiers for speech emotion recognition: A review. *Int. J. Speech Technol.* **2018**, *21*, 93–120. [CrossRef]
2. Koolagudi, S.G.; Rao, K.S. Emotion recognition from speech: A review. *Int. J. Speech Technol.* **2012**, *15*, 99–117. [CrossRef]
3. Wakita, H. Residual energy of linear prediction to vowel and speaker recognition. *IEEE Trans. Audio Speech Signal Process.* **1976**, *24*, 270–271. [CrossRef]
4. Prasanna, S.R.M.; Gupta, C.S.; Yegnanarayana, B. Extraction of speaker-specific excitation information from linear prediction residual of speech. *Speech Commun.* **2006**, *48*, 1243–1261. [CrossRef]
5. Yegnanarayana, B.; Gangashetty, S.V. Epoch-based analysis of speech signals. *Sadhana* **2015**, *36*, 651–697. [CrossRef]
6. Perez-Espinosa, H.; Gutierrez-Serafin, B.; Martinez-Miranda, J.; Espinosa-Curiel, I.E. Automatic children's personality assessment from emotional speech. *Expert Syst. Appl.* **2022**, *187*, 115885. [CrossRef]
7. Pravena, D.; Govind, D. Significance of incorporating excitation source parameters for improved emotion recognition from speech and electroglottographic signals. *Int. J. Speech Technol.* **2019**, *4*, 787–797. [CrossRef]
8. Wenjing, H.; Haifeng, L. Research on Speech and Emotion Recognition Method Based on Prosodic Paragraph. *J. Tsinghua Univ. (Nat. Sci. Ed.)* **2015**, *s1*, 1363–1368.
9. Ying, S.; Hui, Y.; Xueying, Z. Based on Chaos Characteristics. *J. Tianjin Univ. (Nat. Sci. Eng. Technol. Ed.)* **2018**, *48*, 681–685.
10. Lotfian, R.; Busso, C. Emotion recognition using synthetic speech as neutral reference. In Proceedings of the IEEE International Conference on Acoustics, Speech and Signal Processing, New Orleans, LA, USA, 5–9 March 2017.
11. Jing, S.; Mao, X.; Chen, L. Prominence features: Effective emotional features for speech emotion recognition. *Digit. Signal Process.* **2019**, *72*, 216–231. [CrossRef]

Disclaimer/Publisher's Note: The statements, opinions and data contained in all publications are solely those of the individual author(s) and contributor(s) and not of MDPI and/or the editor(s). MDPI and/or the editor(s) disclaim responsibility for any injury to people or property resulting from any ideas, methods, instructions or products referred to in the content.

engineering proceedings

MDPI

Proceeding Paper

Applying the Unified Theory of Acceptance and Use of Technology Model on the Behavior of Home Buyers Using Housing Apps †

Jui-Chang Chiu, Chi-Yueh Hsu, Chun-Yu Chien *, Yu-Qui Fan and Yu-Wen Cheng

Department of Leisure Service Management, Chaoyang University of Technology, Taichung 413310, Taiwan; jacky@cyut.edu.tw (J.-C.C.); cyhsu@gm.cyut.edu.tw (C.-Y.H.); fang860222@gmail.com (Y.-Q.F.); a0989222936@gmail.com (Y.-W.C.)

* Correspondence: chien_ivy@hotmail.com

† Presented at the 3rd IEEE International Conference on Electronic Communications, Internet of Things and Big Data Conference 2023, Taichung, Taiwan, 14–16 April 2023.

Abstract: It is convenient to have advances in science and technology so that people obtain information without going out. We explore the intention of the use of housing apps based on the unified theory of acceptance and the use of technology. A total of 365 questionnaires were collected with 8 incomplete answers discarded. The snowball sampling method was used for confirmatory factor analysis and SEM structural equation model analysis. The research results show the following. (1) Housing app users can quickly obtain knowledge and information about houses, and it is more convenient. (2) Both effort expectancy and social influence have a direct and positive effect on behavior intention when using the housing app. (3) There is no significant impact after adding moderator variables of gender, age, and income in Unified Theory of Acceptance and Use of Technology, (UTAUT). It is convenient and helpful to use the housing app. Therefore, the use of the housing app will be an indispensable trend.

Keywords: real estate; UTAUT; housing app

Citation: Chiu, J.-C.; Hsu, C.-Y.; Chien, C.-Y.; Fan, Y.-Q.; Cheng, Y.-W. Applying the Unified Theory of Acceptance and Use of Technology Model on the Behavior of Home Buyers Using Housing Apps. *Eng. Proc.* **2023**, *38*, 85. https://doi.org/10.3390/engproc2023038085

Academic Editors: Teen-Hang Meen, Hsin-Hung Lin and Cheng-Fu Yang

Published: 21 July 2023

Copyright: © 2023 by the authors. Licensee MDPI, Basel, Switzerland. This article is an open access article distributed under the terms and conditions of the Creative Commons Attribution (CC BY) license (https://creativecommons.org/licenses/by/4.0/).

1. Introduction

In the past, the traditional real estate industry always relied on real estate agents taking customers to see their dream houses. Now, with the assistance of technology, many people read an overview at the "online house inspection", and then make an appointment with the real estate agents or the landlord to see the houses. At present, a housing app function has been combined with the GPS satellite positioning function to find out the route planning between the user's location and the object. After the outbreak of COVID-19 in 2019, the frequency of using apps and official website information tools by business personnel have increased by 20–30% [1]. Therefore, the impacts and factors of using housing apps are valuable topics for research.

This study is based on the unified theory of acceptance and use of technology (UTAUT) to revise and propose a more complete model and explain the users' behavior in using new technology information. According to the research from Chen [2] and Ke [3], the moderator variables of gender, age, and income are added in this research. Therefore, the intention of use and user behavior of real estate transaction platforms is researched, too. According to the research motivation, the three research objectives of this study are (1) to explore the behavior of housing app users by UTAUT; (2) to verify housing app users' use behavior by adding moderator variables; and (3) to compare two research models to verify the housing app users fit.

Through a literature review and analysis, two research models are used to explore the behavior of housing app users, including research model 1 to verify the behavior of housing app users without adding moderating variables and research model 2 to verify the behavior of housing app users with moderating variables. The frameworks of this study include six dimensions of UTAUT: performance expectancy, effort expectancy, social influence, facilitating condition, behavior intention, and user behavior, and the three moderating variables of gender, age, and income.

2. Data Analysis

The snowball sampling method was used in this research, and the research objects were the users who have downloaded the housing apps or have used the housing apps to search for houses. Google online questionnaires were issued to Line groups and major social platforms. The questionnaire survey was performed as a pre-test and a formal survey. The formal questionnaire test period was from 1 April 2021 to 1 May 2021. A total of 365 questionnaires were distributed with 8 incomplete questionnaires, and the effective response rate was 98%.

The research tool is based on the UTAUT proposed by Venkatesh et al. [4] and refers to the Chinese scales translated by Kao [5] from Venkatesh et al. [4] and Chiang [6]. Referring to Ke [3], the items of this questionnaire were revised according to the purposes of this study. This research defined user behavior as the frequency of using housing apps to search for ideal objects every day. The questionnaire was divided into two parts. The first part was to know the influencing factors and intentions of housing app users' use behavior, including performance expectancy, effort expectancy, social influence, facilitating conditions, and intention of use. The second part was to know the personal background variables, including gender, age, and income. A 7-point Likert scale was used in this research, with 7-points scales for an agreement include 1 "strongly disagree" to 7 "strongly agree".

Smart PLS 2.0 was used to analyze the paths of model 1, UTAUT without moderator variables, and model 2, UTAUT with moderator variables. The results and the fits of the two models were compared. Model 1 is UTAUT without moderator variables, and the results of the relationship estimates for each potential variable are shown in Table 1. The significance test of each path takes a t-value > 1.96 as the critical value, which indicates whether the relationship between the variables reaches significant levels. Table 1 shows that all the paths are significant. The estimated effect of performance expectancy on behavioral intentions is −0.059, which is not significant. The estimated effect of effort expectancy on behavioral intention is 0.158, which reaches a significant level, a p-value < 0.05. The estimated effect of social influence on behavioral intention is 0.417, which reaches a significant level, a p-value < 0.05. The estimated effect of facilitating conditions on user behavior is 0.342, which reaches a significant level, a p-value < 0.05. The estimated effect of behavior intention on user behavior is 0.845, which reaches a significant level, a p-value < 0.05.

Table 1. Path coefficient of dimensions of model 1.

Variable Paths	Standardization Coefficient	SE	t Value	p Value
Performance expectancy→Behavior intention	−0.059	0.039	1.511	0.132
Effort expectancy→Behavior intention	0.158	0.058	2.734 *	0.007
Social influence→Behavior intention	0.417	0.07	5.997 *	0.000
Facilitating conditions→Use behavior	0.342	0.056	6.158 *	0.000
Behavior intention→Use behavior	0.845	0.021	41.047 *	0.000

* $p < 0.05$.

According to the results in Table 2, the path of research model 1 affects the relationships. Effort expectancy and social influence have a direct and positive effect on behavior intention. The coefficient values are 0.158 and 0.417. Effort expectancy, social influence, and behavior intention have an indirect and positive effect on user behavior. The coefficient values are 0.133, 0.352, and 0.845. Facilitating conditions have a direct and positive effect on user behavior. The coefficient value is 0.342. According to the results, H1-2, the effort expectancy of housing APP users have significant effect on behavior intention, H1-3, the social influence of housing APP users have significant effect on behavior intention, H1-4, the facilitating conditions of housing APP users have significant effect on use behavior, and H1-5, the behavioral intention of housing APP users have significant effect on use behavior, are supported. Effort expectancy, social influence, and facilitating condition have positive impacts on behavioral intention, which indicating that with the prevalence of the Internet, information acquisition methods have diversified. The promotion of mass media, Internet media and related industries can increase the number of buyers. The behavioral intention of using housing APP means that App users believe that they can obtain effective housing information from the APP on their smartphones, and improve the users' in finding housing. The path model diagram is shown in Figure 1.

Table 2. Direct and indirect influence between various dimensions of model 1.

Dimensions	Behavior Intention		Use Behavior		
	Directly	Indirectly	Directly	Indirectly	Total Effect
Performance expectancy					
Effort expectancy	0.158			0.133	0.133
Social influence	0.417			0.352	0.352
Facilitating conditions			0.342		0.342
Behavior intention				0.845	0.845
R^2	0.748		0.714		

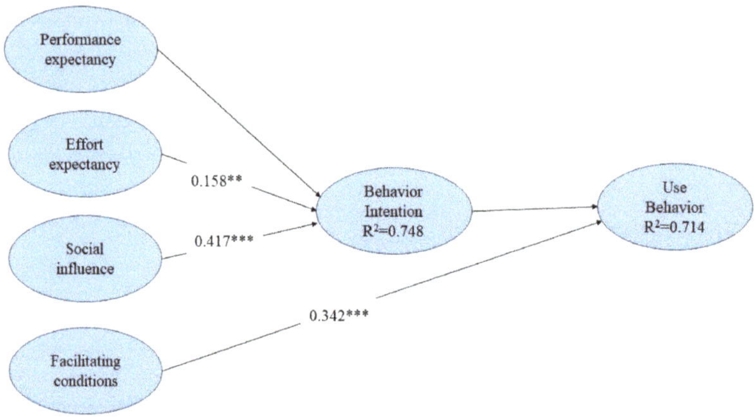

Figure 1. Path diagram of model 1. Note: ** t > 2.58, $p < 0.01$; *** t > 3.29, $p < 0.001$.

PLS-SEM uses GOF value as an indicator of overall model fit. Cronbach's α measures the internal consistency and stability of the dimension, and R^2 represents the explained variation of the model's internal factors. According to the results of Table 3, the GOF of the model in this study is 0.422, which means the model fits well in the model. Cronbach's α is above 0.8, which has good reliability and R^2 of behavioral intention and use behavior are, respectively, 0.748 and 0.714, which means that the model has middle and high explanatory power.

Model 2 shows the results of the UTAUT with moderator variables (Table 4). In the significance test of each path, a t-value > 1.96 is regarded as the critical value, which indicates whether the relationships between the variables reach significant levels. Table 4 shows that social influence affects behavior intention and behavior intention affects user behavior significantly. The estimated effect of social influence on behavior intention is 0.821, the estimated effect of facilitating conditions on user behavior is 0.214, and the estimated effect of behavior intention on user behavior is 0.661 at a significant level, a p-value < 0.05.

Table 3. Goodness-of-fit of model 1.

	Average Variance Extracted, AVE	Composite Reliability	R^2	Cronbach's Alpha	Redundancy	Good of Fitness, GOF
Performance expectancy	0.630	0.931		0.914		
Effort expectancy	0.604	0.924		0.906		
Social influence	0.700	0.933		0.914		
Facilitating conditions	0.818	0.931		0.889		0.422
Behavior intention	0.831	0.936	0.748	0.898	0.172	
Use behavior	0.794	0.920	0.714	0.870	0.564	
Performance expectancy	0.630	0.931		0.914		
Effort expectancy	0.604	0.924		0.906		

Table 4. Path coefficient of dimensions of model 2.

Variable Paths	Standardization Coefficient	SE	T Value	*p* Value
Performance expectancy→Behavior intention	−0.931	0.203	0.458	0.648
Effort expectancy→Behavior intention	−0.170	0.284	0.596	0.551
Social influence→Behavior intention	0.821	0.233	3.524 *	0.000
Facilitating conditions→Use behavior	0.214	0.114	1.879	0.061
Behavior intention→Use behavior	0.661	0.060	11.082 *	0.000

* $p < 0.5$.

The results of the moderator estimates of each moderator variable in model 2 are shown in Table 5. The significance test of each path takes a t-value > 1.96 as the critical value, which indicates whether the relationships between the variables reach a significant level.

Table 5. Path coefficient of moderator variables of model 2.

Moderating Variables	Moderating Paths	Standardization Coefficient	SE	t Value	*p* Value
Gender	Performance expectancy→Behavior intention	0.257	0.287	0.898	0.370
	Effort expectancy→Behavior intention	0.305	0.426	0.716	0.474
	Social influence→Behavior intention	0.053	0.287	0.184	0.854
Age	Performance expectancy→Behavior intention	−0.250	0.319	0.780	0.436
	Effort expectancy→Behavior intention	0.608	0.361	1.683	0.093
	Social influence→Behavior intention	−0.260	0.302	0.860	0.391
	Facilitating conditions→Use behavior	0.148	0.175	0.846	0.398
Income	Effort expectancy→Behavior intention	0.407	0.343	1.188	0.236
	Social influence→Behavior intention	−0.321	0.280	1.146	0.253
	Facilitating conditions→Use behavior	−0.183	0.156	1.177	0.240

According to the results in Table 6, the paths of research model 1 directly and indirectly affect the relationships with moderator variables. Social influence has a direct and positive effect on behavior intention with a coefficient value of 0.821. Social influence has an indirect and positive effect on user behavior with a coefficient value of 0.543. Behavior intention has a direct and positive effect on user behavior with a coefficient value of 0.661. According to the results of hypothesis 2, only H2-3 and H2-5 are supported. The path model diagram is shown in Figure 2.

Table 6. Direct and indirect influence between various dimensions of model 2.

Dimensions	Behavior Intention Direct	Behavior Intention Indirect	Use Behavior Direct	Use Behavior Indirect	Total Effect
Performance expectancy					
Effort expectancy					
Social influence	0.821			0.543	0.543
Facilitating conditions					
Behavior intention			0.661		
R^2	0.714		0.735		

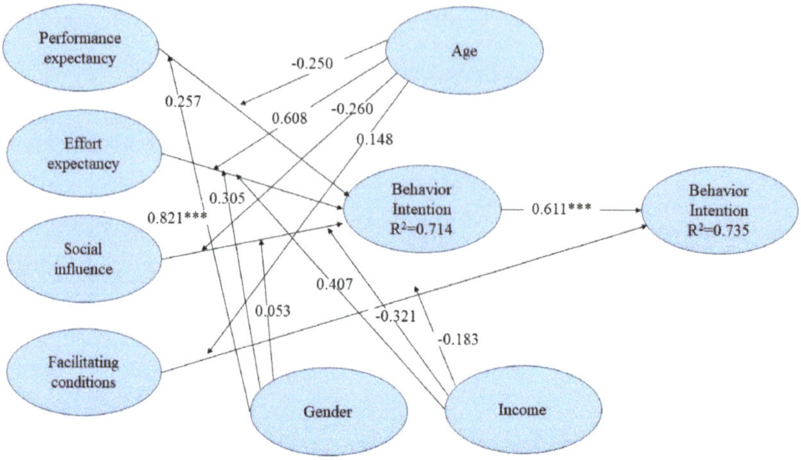

Figure 2. Path diagram of model 2. Note: *** t > 3.29, $p < 0.001$.

According to Table 7, the GOF is 0.420, which presents the model fits well in this study. Cronbach's α is above 0.8, which has good reliability and R^2 of behavioral intention and use behavior are, respectively, 0.714 and 0.735, which means that the model has middle and high explanatory power.

Table 7. Path diagram of model 2.

Dimensions	AVE	Composite Reliability	R^2	Cronbach's α	Redundancy	GOF
Performance expectancy	0.629	0.931		0.914		
Effort expectancy	0.605	0.924		0.906		
Social influence	0.699	0.933		0.914		
Facilitating conditions	0.818	0.931		0.889		0.420
Behavior intention	0.831	0.936	0.714	0.898	−0.231	
Use behavior	0.794	0.920	0.735	0.870	−0.014	

3. Discussion

3.1. Verifying Research Hypotheses

3.1.1. H1: Housing App Users Have a Significant Effect on UTAUT

According to Table VIII, the research model 1, H1-2, H1-3, H1-4, and H1-5 are established, which is consistent with the research from Chen [2]. Performance expectancy, effort expectancy, social influence, and facilitating conditions have a significant positive impact on behavior intention, indicating that with the prevalence of the Internet, the methods of acquiring information have become more diversified. It promotes through the media and online media in related industries to increase the behavior intention of buyers to use the housing apps. It also presents that app users think they can obtain effective housing information from the apps and find a house. In model 1, performance expectancy has no significant impact on behavior intention. It is speculated that although housing app users think the app is easy to use, they do not completely trust all the information in the app because the users still need to confirm the house condition and judge the house before buying it or not.

3.1.2. H2: Housing APP User Moderator Variables Have a Significant Moderating Effect on UTAUT

According to Table 8, in model 2, H2-3, and H2-5 are supported. However, in the moderator path, gender, age, and income do not meet the significant level, which is consistent with the result of Chang [7] and Weng and Huang [8]. Currently, not only do men want to buy a house and use the searching tools to find houses but women can make money independently. Therefore, the results have no significant effect by gender. When age and income variables are not significant factors, it is inferred that younger users may want to see houses through the housing app, but they are less likely to buy a house because of budget. Relatively, older people use housing apps less than young people, and some of them still use a traditional way to see houses.

Table 8. Comparison of the fit of research models.

Indicators	Research Model 1	Research Model 2
GOF value	0.422	0.420
Behavior intention R^2	0.748	0.682
Use behavior R^2	0.714	0.763

3.1.3. Comparison of the Fit of Research Models

According to the analysis results of models 1 and 2, the results and details are as follows.

According to the results in Table VIII, the GOF value of model 1 with external variables is higher, and the R2 of research model 1 is higher. According to Wetzels et al. [9], a GOF value around 0.1 represents a weak model fit, a GOF value greater than 0.25 represents a moderate model fit, and a GOF value greater than 0.36 represents a good model fit. R2 is a measure of the explanatory power of the dependent variables in the model. Hair and Sarstedt [10] pointed out that R2 was divided into weak, moderate, and high explanatory power at 0.25, 0.50, and 0.75. According to the results in Table VIII, the GOF values of the two research models are 0.420 and 0.422 and the R2 is between 0.682 and 0.763, which indicates that both models have a good model fit and a moderate to a high level of explanatory power.

4. Conclusions and Suggestions

4.1. Conclusions

4.1.1. Validation of UTAUT without Moderator Variables

According to the results of model 1, "behavioral intention" has significant explanatory power, affects user behavior, and also increases the user's loyalty, because the users of housing apps can save time and obtain effective housing information without spending too much effort. In recent years, app platforms have brought convenience and dependence to people. Therefore, the use of housing apps has an impact on users.

4.1.2. Validation of UTAUT with Moderator Variables

According to the results of model 2, housing apps are undoubtedly one of the indispensable social tools for young users. However, their income is not equivalent to older people, so the rate of using housing apps for the younger group is not higher than that for older people. Because of gender equality, the concept of males being in charge of housing no longer exists. Both men and women can freely want to buy a house and see a house. Women may also be the head of the family and choose to buy a house. Power is no longer a matter of one-sided choice.

4.1.3. Suggestions

(a) The research results show that users can obtain housing information on the housing apps, which means that the apps have practical value. Therefore, we suggest that app users are more likely to know accurate information about houses through an app that is constantly updated. In addition, the successive launch of apps specially made for housing transaction information are tools that the public can consider using, which also saves the time and energy of looking for houses.

(b) The public generally likes to watch short videos on the Internet, with YouTube becoming a popular platform in recent years. Most of the information included in the housing apps is text and pictures. Although there are photos for users to observe the condition of the houses, it is suggested that real estate agents use small video clips to introduce various housing conditions and introduce a personal grid so that the public can better understand the actual situation and users do not feel that the process is monotonous and instead has a sense of fun.

Author Contributions: Conceptualization, J.-C.C. and C.-Y.H.; methodology, C.-Y.C. and Y.-Q.F.; writing—original draft preparation, Y.-W.C. and C.-Y.H.; writing—review and editing, J.-C.C. and C.-Y.C. All authors have read and agreed to the published version of the manuscript.

Funding: This research received no external funding.

Institutional Review Board Statement: Not applicable.

Informed Consent Statement: Informed consent was obtained from all subjects involved in the study.

Data Availability Statement: Data are contained in the paper.

Conflicts of Interest: The authors declare no conflict of interest.

References

1. Digital Times, "Generate 2D and 3D Layouts Automatically! In the Post-Epidemic Era, How Does Sinyi Realty Inc. Use AI Modeling to Improve the Transaction Rate". 2020. Available online: https://www.bnext.com.tw/article/58337/sinyi-2020 (accessed on 21 April 2021).
2. Chen, S.L. Investigating Consumer's Acceptance Behavior on Pet Application by UTAUT Model. Master's Thesis, National Taipei University of Technology, Taipei, Taiwan, 2015.
3. Ke, M.F. Applying the Unified Theory of Acceptance and Use of Technology Model to the Use Behavior of Self-Service Ordering at Restaurants. Master's Thesis, Chaoyang University of Technology, Taichung, Taiwan, 2019.
4. Venkatesh, V.; Morris, M.G.; Davis, G.B.; Davis, F.D. User acceptance of information technology: Toward a unified view. *MIS Q.* **2003**, *27*, 425–478. [CrossRef]
5. Kao, C.C. Using UTAUT to Explore Vocational-High-School Teacher's Behavioral Intention of Virtual Professional Communities. Master's Thesis, National Taiwan Normal University, Taipei, Taiwan, 2012.
6. Chiang, H.H. A Study on the User Behavior of Online Live Streaming of Fitness. Master's Thesis, Chaoyang University of Technology, Taichung, Taiwan, 2018.
7. Chang, T.J. An Investigation of App Users' Behavior Patterns by Combination of UTAUT and Flow Theory—A Case of App Mobile Community Groups. Master's Thesis, National Taichung University of Education, Taichung, Taiwan, 2012.
8. Weng, C.Y.; Huang, H.P. A Study of LINE TV Usage Based on the UTAUT2 Model and Digital Lifestyle Theory. *J. Cyber Cult. Inf. Soc.* **2017**, 33, 17–63.
9. Wetzels, C.; Odekerken-Schröder, G.; Van Oppen, C. Using PLS path modeling for assessing hierarchical construct models: Guidelines and empirical illustration. *MIS Q.* **2009**, *33*, 177–195. [CrossRef]
10. Hair, J.F.; Ringle, C.M.; Sarstedt, M. PLS-SEM: Indeed a silver bullet. *J. Mark. Theory Pract.* **2011**, *19*, 139–152. [CrossRef]

Disclaimer/Publisher's Note: The statements, opinions and data contained in all publications are solely those of the individual author(s) and contributor(s) and not of MDPI and/or the editor(s). MDPI and/or the editor(s) disclaim responsibility for any injury to people or property resulting from any ideas, methods, instructions or products referred to in the content.

engineering proceedings

MDPI

Proceeding Paper

Sum of Exponential Model for Fitting Data †

Ting-Cheng Chang *, Min-Hao Wu and Ying Lin

Department of Computer Science, College of Information Engineering, Guangzhou Panyu Polytechnic, Guangzhou 511483, China; minhao1616@126.com (M.-H.W.); ying.lins@outlook.com (Y.L.)
* Correspondence: tcchang0615@gmail.com
† Presented at the 3rd IEEE International Conference on Electronic Communications, Internet of Things and Big Data Conference 2023, Taichung, Taiwan, 14–16 April 2023.

Abstract: As an approach to feature estimation, exponential fitting has attracted research interests in mathematical modeling. Semantic networks are used for numerous applications in computers, physics, and biology. However, such applications may have fitting troubles with various mathematical tools. Therefore, we present a novel method of fitting $2n$ data points of a signal to a sum of n exponential functions. The experiments proved that the proposed method operated well for linear and nonlinear functions, as its algorithm was straightforward, practical, and easy to determine. At the same time, the computational intricacy was considerably low, which has specific worth in use.

Keywords: fitting $2n$ data points; a sum of n; exponential function

Citation: Chang, T.-C.; Wu, M.-H.; Lin, Y. Sum of Exponential Model for Fitting Data. *Eng. Proc.* **2023**, *38*, 87. https://doi.org/10.3390/engproc2023038087

Academic Editors: Teen-Hang Meen, Hsin-Hung Lin and Cheng-Fu Yang

Published: 25 July 2023

Copyright: © 2023 by the authors. Licensee MDPI, Basel, Switzerland. This article is an open access article distributed under the terms and conditions of the Creative Commons Attribution (CC BY) license (https://creativecommons.org/licenses/by/4.0/).

1. Introduction

Relevant information with the linear combinations of real and complex exponentials is pervasive in science and engineering applications. Given that Gaspard Riche de Prony developed an approach [1] to resolve the problem for equally spaced samples, numerous advancements, and applications have been proposed. We surveyed the most effective ones to explain their applications and experiences and to allow their application in various fields. A linear combination of exponentials was used in regular differential equations to explain the different physical processes. After being modeled by the remedy of a formula, a combination of exponentials provided valuable information such as decay rates or product residential or commercial properties in a physical system. Likewise, the exponential fitting had an excellent approximation on the compact of the domain with Fourier transformation in complex exponentials [2–6].

The purpose of this study was to present a method of fitting real signal data sampled at a period T in a set of $2n$ data points $\{x(0), x(T), x(2T), \cdots, x([2n-1]T)\}$. The data points to the s curve were composed of n exponential functions with unknown weights and exponents. Mathematically, this involved the solution of the following Equations (1)–(5).

$$x(kT) = \sum_{i=1}^{n} c_i e^{p_i(kT)} \tag{1}$$

For the unknown C_i and P_i in the complex conjugate pairs (P_i is an imaginary number), Equation (1) represents a sum of sinusoids. This curve fitting can have many applications. For example, if $x(t)$ represents the impulse response of a linear time-invariant system, and the Laplace transform of Equation (1) yields the transfer function of an nth-order model of the system.

2. Curve Fitting Method

We let ϕ_i denote $e^{p_i(kT)}$ and x_i denote $x(kT)$. Then, Equation (1) could be rewritten as:

$$\begin{array}{llll} c_1 & +c_2 & +\cdots+c_n & =x_0 \\ c_1\phi_1 & +c_2\phi_2 & +\cdots+c_n\phi_n & =x_1 \\ c_1\phi_1^2 & +c_2\phi_2^2 & +\cdots+c_n\phi_n^2 & =x_2 \\ & & \vdots & \\ c_1\phi_1^{2n-1} & +c_2\phi_2^{2n-1} & +\cdots+c_n\phi_n^{2n-1} & =x_{2n-1} \end{array} \tag{2}$$

Equation (2) is explained by the following theorem.

Theorem 1. *The nonlinear equation, such as Equation (2), possesses a unique solution* $\{c_k, N_k\}$ $(k = 1, 2, \ldots, n)$ *if and only with the following* $n \times n$ *matrix, which is nonsingular.*

$$A \equiv \begin{bmatrix} x_0 & x_1 & \cdots & x_{n-1} \\ x_1 & x_2 & \cdots & x_n \\ \vdots & \vdots & & \vdots \\ x_{n-1} & x_n & \cdots & x_{2n-2} \end{bmatrix} \tag{3}$$

The solution for N_k *is given by the n distinct roots of the* n^{th} *degree polynomial equation.*

$$\det \begin{bmatrix} 1 & x_0 & x_1 & \cdots & x_{n-1} \\ \phi_k & x_1 & x_2 & \cdots & x_n \\ \phi_k^2 & x_2 & x_3 & \cdots & x_{n+1} \\ \vdots & \vdots & \vdots & & \vdots \\ \phi_k^n & x_n & x_{n+1} & \cdots & x_{2n-1} \end{bmatrix} = 0 \tag{4}$$

The solution for c_k *can then be given by:*

$$A = \begin{bmatrix} 1 & 1 & 1 & \cdots & 1 \\ \phi_k & \phi_2 & \phi_3 & \cdots & \phi_n \\ \phi_k^2 & \phi_2^2 & \phi_3^2 & \cdots & \phi_n^2 \\ \vdots & \vdots & \vdots & & \vdots \\ \phi_k^{2n-1} & \phi_2^{2n-1} & \phi_3^{2n-1} & \cdots & \phi_n^{2n-1} \end{bmatrix}^{-1} \tag{5}$$

$$\begin{bmatrix} c_1 \\ c_2 \\ c_3 \\ \vdots \\ c_n \end{bmatrix} = A \begin{bmatrix} x_0 \\ x_1 \\ x_2 \\ \vdots \\ x_{n-1} \end{bmatrix} \tag{6}$$

3. Proof

3.1. Sufficiency Part

It could be supposed that A was nonsingular; the first $n1$ equation of Equation (2) could be arranged as:

$$B = \begin{bmatrix} 1 & 1 & 1 & \cdots & 1 \\ \phi_k & \phi_2 & \phi_3 & \cdots & \phi_n \\ \phi_k^2 & \phi_2^2 & \phi_3^2 & \cdots & \phi_n^2 \\ \vdots & \vdots & \vdots & & \vdots \\ \phi_k^{2n-1} & \phi_2^{2n-1} & \phi_3^{2n-1} & \cdots & \phi_n^{2n-1} \end{bmatrix} \tag{7}$$

$$B\begin{bmatrix} c_1 \\ c_2 \\ c_3 \\ \vdots \\ c_n \end{bmatrix} = \begin{bmatrix} x_0 \\ x_1 \\ x_2 \\ \vdots \\ x_{n-1} \end{bmatrix} \tag{8}$$

We let the columns of the left-most matrix in Equation (8) be denoted as:

$$v_k \equiv \begin{bmatrix} 1 & \phi_k & \phi_k^2 & \cdots & \phi_k^n \end{bmatrix}^T, k = 1,\ 2,\ \cdots, n \tag{9}$$

and let $x_0 \equiv \begin{bmatrix} x_0 & x_1 & \cdots & x_n \end{bmatrix}^T$. Then, Equation (8) showed that x_0 was a linear combination of $\{v_1, v_2, \ldots, v_n\}$.

Next, if the set of $n + 1$ consecutive equations in Equation (2) was considered as the starting point with the second equation, they could be rearranged as:

$$\begin{bmatrix} 1 & 1 & 1 & \cdots & 1 \\ \phi_1 & \phi_2 & \phi_3 & \cdots & \phi_n \\ \phi_1^2 & \phi_2^2 & \phi_3^2 & \cdots & \phi_n^2 \\ \vdots & \vdots & \vdots & & \vdots \\ \phi_1^n & \phi_2^n & \phi_3^n & \cdots & \phi_n^n \end{bmatrix} \begin{bmatrix} c_1\phi_1 \\ c_2\phi_2 \\ c_3\phi_3 \\ \vdots \\ c_n\phi_n \end{bmatrix} = \begin{bmatrix} x_1 \\ x_2 \\ x_3 \\ \vdots \\ x_{n+1} \end{bmatrix} \tag{10}$$

Equation (10) shows that $x_1 \equiv [x_1,\ x_2,\ \ldots,\ x_{n+1}]^T$ was a linear combination of $\{v_1, v_2, \ldots, v_n\}$.

Similarly, if we considered the set of $n + 1$ consecutive equations in Equation (2) starting with the third equation, we could see that $x_2 \equiv [x_2,\ x_3,\ \ldots,\ x_{n+2}]^T$ was a linear combination of $\{v_1, v_2, \ldots, v_n\}$.

This continued until the last $n + 1$ of Equation (2) was taken, from which it was shown that $x_{n-1} \equiv [x_{n-1}, x_n, \ldots x_{2n+1}]^T$ was a linear combination of $\{v_1, v_2, \ldots, v_n\}$.

Equation (3) of the Theorem implies that the vectors $\{x_0, x_1, \ldots x_{n-1}\}$ are linear and independent of each other. Hence, they span an n-dimensional subspace in an $(n + 1)$ dimensional Euclidean space. This subspace must be the same as the one spanned by the vectors $\{v_1, v_2, \ldots, v_n\}$ since each x_i, $i = 0, 1, \ldots, n-1$ is a linear combination of the set $\{v_1, v_2, \ldots, v_n\}$. It follows that the vectors $\{v_1, v_2, \ldots, v_n\}$ are linearly independent and that, from Equation (7) of v_k $(k = 0, 1, \ldots, n)$, they must be distinct.

Moreover, each v_k is a linear combination of $\{x_0, x_1, \ldots x_{n+1}\}$. This implies:

$$v_k = \begin{bmatrix} 1 \\ \phi_k \\ \phi_k^2 \\ \vdots \\ \phi_k^n \end{bmatrix} = d_1\begin{bmatrix} x_0 \\ x_1 \\ x_2 \\ \vdots \\ x_n \end{bmatrix} + d_2\begin{bmatrix} x_1 \\ x_2 \\ x_3 \\ \vdots \\ x_{n+1} \end{bmatrix} + \cdots + d_n\begin{bmatrix} x_{n-1} \\ x_n \\ x_{n+1} \\ \vdots \\ x_{2n-1} \end{bmatrix} \tag{11}$$

Equation (11) could be rearranged as an $(n + 1) \times (n + 1)$ equation system.

$$\begin{bmatrix} 1 & x_0 & x_1 & \cdots & x_{n-1} \\ \phi_k & x_1 & x_2 & \cdots & x_n \\ \phi_k^2 & x_2 & x_3 & \cdots & x_{n+1} \\ \vdots & \vdots & \vdots & & \vdots \\ \phi_k^n & x_n & x_{n+1} & \cdots & x_{2n-1} \end{bmatrix} \begin{bmatrix} -1 \\ d_1 \\ d_2 \\ \vdots \\ d_n \end{bmatrix} = 0 \tag{12}$$

Since the solution of Equation (12) was nontrivial, the determinant of the square matrix had to vanish, leading to Equation (4) which was an nth-degree polynomial equation in N_k because the coefficient of the nth power term of N_k could be seen from Equations (3) and (12) to be $(-1)^n$ det A, which was assumed to be nonzero. The n roots N_k of Equation (4) must be distinct because each N_k had to satisfy Equation (4) and be distinct. Having obtained the distinct values of N_k, $k = 0, 1, \ldots, n$, c_k could be given by the first n equations (Equation (2)), which led to Equation (6) in the Theorem.

As for the uniqueness of the solution, since every solution $\{c_k, N_k\}$ ($k = 1, 2, \ldots, n$) had to satisfy Equation (4), according to the above arguments, Equation (4) produced exactly n distinct values for N_k, and the solution of Equation (2) was unique.

3.2. Necessity Part

Suppose Equation (2) has a unique solution $\{c_k, N_k\}$ ($k = 1, 2, \ldots, n$). This implies the following.

(i) N_k must be distinct from each other; otherwise, non-unique combinations of c_k in Equation (2) exist and are fulfilled.

(ii) None of c_k vanishes; otherwise, the value of N_k associated with a vanishing c_k becomes non-unique.

The first n of Equation (2) gave:

$$C = \begin{bmatrix} 1 & 1 & 1 & \cdots & 1 \\ \phi_1 & \phi_2 & \phi_3 & \cdots & \phi_n \\ \phi_1^2 & \phi_2^2 & \phi_3^2 & \cdots & \phi_n^2 \\ \vdots & \vdots & \vdots & & \vdots \\ \phi_1^{n-1} & \phi_2^{n-1} & \phi_3^{n-1} & \cdots & \phi_n^{n-1} \end{bmatrix}^{-1} \tag{13}$$

$$\begin{bmatrix} c_1 \\ c_2 \\ c_3 \\ \vdots \\ c_n \end{bmatrix} = C \begin{bmatrix} x_0 \\ x_1 \\ x_2 \\ \vdots \\ x_{n-1} \end{bmatrix} \tag{14}$$

The next n of Equation (2), starting with the second equation, gave:

$$D = \begin{bmatrix} 1 & 1 & 1 & \cdots & 1 \\ \phi_1 & \phi_2 & \phi_3 & \cdots & \phi_n \\ \phi_1^2 & \phi_2^2 & \phi_3^2 & \cdots & \phi_n^2 \\ \vdots & \vdots & \vdots & & \vdots \\ \phi_1^{n-1} & \phi_2^{n-1} & \phi_3^{n-1} & \cdots & \phi_n^{n-1} \end{bmatrix}^{-1} \tag{15}$$

$$\begin{bmatrix} c_1\phi_1 \\ c_2\phi_2 \\ c_3\phi_3 \\ \vdots \\ c_n\phi_n \end{bmatrix} = D \begin{bmatrix} x_1 \\ x_2 \\ x_3 \\ \vdots \\ x_n \end{bmatrix} \tag{16}$$

This proceeded until the set of n in the consecutive Equation (2), starting with the nth equation, was reached.

$$E = \begin{bmatrix} 1 & 1 & 1 & \cdots & 1 \\ \phi_1 & \phi_2 & \phi_3 & \cdots & \phi_n \\ \phi_1^2 & \phi_2^2 & \phi_3^2 & \cdots & \phi_n^2 \\ \vdots & \vdots & \vdots & & \vdots \\ \phi_1^{n-1} & \phi_2^{n-1} & \phi_3^{n-1} & \cdots & \phi_n^{n-1} \end{bmatrix}^{-1} \tag{17}$$

$$\begin{bmatrix} c_1\phi_1^{n-1} \\ c_2\phi_2^{n-1} \\ c_3\phi_3^{n-1} \\ \vdots \\ c_n\phi_n^{n-1} \end{bmatrix} = E \begin{bmatrix} x_{n-1} \\ x_2 \\ x_{n+1} \\ \vdots \\ x_{2n-2} \end{bmatrix} \tag{18}$$

Combining Equations (14) to (18) yielded:

$$F = \begin{bmatrix} 1 & 1 & 1 & \cdots & 1 \\ \phi_1 & \phi_2 & \phi_3 & \cdots & \phi_n \\ \phi_1^2 & \phi_2^2 & \phi_3^2 & \cdots & \phi_n^2 \\ \vdots & \vdots & \vdots & & \vdots \\ \phi_1^{n-1} & \phi_2^{n-1} & \phi_3^{n-1} & \cdots & \phi_n^{n-1} \end{bmatrix}^{-1} \quad (19)$$

$$G = \begin{bmatrix} x_0 & x_1 & x_2 & \cdots & x_{n-1} \\ x_1 & x_2 & x_3 & \cdots & x_n \\ x_2 & x_3 & x_4 & \cdots & x_{n+1} \\ \vdots & \vdots & \vdots & \cdots & \vdots \\ x_{n-1} & x_n & x_{n+1} & \cdots & x_{2n-2} \end{bmatrix} \quad (20)$$

$$\begin{bmatrix} c_1 & c_1\phi_1 & \cdots & c_1\phi_1^{n-1} \\ c_2 & c_2\phi_2 & \cdots & c_2\phi_2^{n-1} \\ c_3 & c_3\phi_3 & \cdots & c_3\phi_3^{n-1} \\ \vdots & \vdots & \cdots & \vdots \\ c_n & c_n\phi_n & \cdots & c_n\phi_n^{n-1} \end{bmatrix} = FG \quad (21)$$

4. Examples

Consider the signal:

$$x(t) = 2e^{-2t} - 3e^{t} \quad (22)$$

Sampling this signal at a sampling period $T = 1$ yielded Equation (2) with $c_1 = 2$, $c_2 = -3$, $N_1 = e^{-2} = 0.13533283$, $N_2 = e^{1} = 2.718281828$, $x_0 = -1$, $x_1 = -7.88417491$, $x_2 = -22.130537$, and $x_3 = -60.2516532$. Let us reverse this process. After sampling four consecutive points of the signal $x(t)$ at a uniform sampling period $T = 1$, we obtained the values of $\{x_0, x_1, x_2, x_3\}$, as indicated above, which could be solved for $\{c_1, c_2, N_1, N_2\}$. From Equation (2), we obtained:

$$c_1 + c_2 = -1 \quad (23)$$

$$c_1\phi_1 + c_2\phi_2 = -7.88417491 \quad (24)$$

$$c_1\phi_1^2 + c_2\phi_2^2 = -22.130537 \quad (25)$$

$$c_1\phi_1^3 + c_2\phi_2^3 = -60.2516532 \quad (26)$$

First, we could see that:

$$A = \begin{bmatrix} x_0 & x_1 \\ x_1 & x_2 \end{bmatrix} = \begin{bmatrix} -1 & -7.88417491 \\ -7.88417491 & -22.130537 \end{bmatrix} \quad (27)$$

was nonsingular. N_k was obtained as the root of Equation (4), which, in this case, led to:

$$\begin{aligned} &\det\begin{bmatrix} 1 & -1 & -7.88417491 \\ \phi_k & -7.88417491 & -22.130537 \\ \phi_k^2 & -22.130.537 & -60.2516532 \end{bmatrix} \\ &= -40.029677\phi_k^2 + 114.2293714\phi_k - 14.7260955 \\ &= 0 \end{aligned} \quad (28)$$

This yielded $N_1 = 0.135335283$ and $N_2 = 2.718281828$. Substituting these values into Equation (6) provided:

$$\begin{bmatrix} 1 & 1 \\ 0.13533528 & 2.71828182 \end{bmatrix} \begin{bmatrix} c_1 \\ c_2 \end{bmatrix} = \begin{bmatrix} -1 \\ -7.88417491 \end{bmatrix} \quad (29)$$

from which we obtained $c_1 = 2$ and $c_2 = -3$. From known N_1 and N_2, $p_1 = -2$ and $p_2 = 1$ were obtained according to $\phi_i = e^{piT}$.

5. Conclusions

The problem of fitting $2n$ data points to a curve consisting of n exponential functions was solved. The exponential functions were complex in general, with sinusoids being a special case. The curve-fitting problem was solved by a system of nonlinear algebraic equations. An example has been given to illustrate the procedure of this method.

Author Contributions: Conceptualization, methodology, T.-C.C.; writing—original draft preparation, M.-H.W.; validation, writing—review and editing, Y.L. All authors have read and agreed to the published version of the manuscript.

Funding: This research was funded by the Guangdong College Research Platform and Research Project grant number 2021ZDZX1137 and 2019GKTSCX069, the Panyu Polytechnic Innovation Team grant number 2020CXTD003 (2011/210113263), the Panyu Polytechnic Research Project grant number 2021KJ04 (2011/210113263), and the Department of Education of Guangdong Province, China, grant number 2020KQNCX192.

Institutional Review Board Statement: Not applicable.

Informed Consent Statement: Not applicable.

Data Availability Statement: Not applicable.

Conflicts of Interest: The authors declare no conflict of interest.

References

1. Prony, R. Essai experimental et analytique sur les lois de la Dilatabilit_e des uideselastiques et sur celles de la Force expansive de la vapeur de l'eau et de la vapeur de l'alkool, a differentes temperatures. *J. L'ecole Polytech. Floreal Prairial* **1795**, 3, 24–76.
2. Istratovand, A.A.; Vyvenko, O.F. Exponential analysis in physical phenomena. *Rev. Sci. Instrum.* **1999**, *70*, 1233–1257. [CrossRef]
3. Naus, H.; van Stokkum, I.H.M.; Hogervorst, W.; Ubachs, W. Quantitative analysis of decay transients applied to a multimode pulsed cavity ringdown experiment. *Appl. Opt.* **2001**, *40*, 4416–4426. [CrossRef] [PubMed]
4. Stry, S.; Hering, P.; Mürtz, M. Portable difference-frequency laser-based cavity leak-out spectrometer for trace-gas analysis. *Appl. Phys. B* **2002**, *75*, 297–303. [CrossRef]
5. von Basum, G.; Dahnke, H.; Halmer, D.; Hering, P.; Mürtz, M. Online recording of ethane traces in human breath via infrared laser spectroscopy. *J. Appl. Physiol.* **2003**, *95*, 2583–2590. [CrossRef] [PubMed]
6. Booth, R.S. Fitting Data Containing Multiple Exponentials Using Functionals Derived from Finite Laplace Transforms of Time Moments of the Data. *Nucl. Technol.* **2017**, *198*, 217–227. [CrossRef]

Disclaimer/Publisher's Note: The statements, opinions and data contained in all publications are solely those of the individual author(s) and contributor(s) and not of MDPI and/or the editor(s). MDPI and/or the editor(s) disclaim responsibility for any injury to people or property resulting from any ideas, methods, instructions or products referred to in the content.

engineering proceedings

Proceeding Paper

Applying the Technology Acceptance Model to Understand Financial Practitioners' Intentions to Use the Digital Innovation Learning Platform †

Mei-Su Chen and Wen-Tsung Huang *

Department of Insurance and Finance Management, Chaoyang University of Technology, Taichung 413310, Taiwan; chmeisu@cyut.edu.tw
* Correspondence: wthuang@cyut.edu.tw
† Presented at the 3rd IEEE International Conference on Electronic Communications, Internet of Things and Big Data Conference 2023, Taichung, Taiwan, 14–16 April 2023.

Abstract: Under the 2030 bilingual national policy in Taiwan, the financial industry encourages employees to use a digital innovation learning platform (English e-learning platform, EELP) to strengthen their language skills and improve service capabilities and improve the language skills of financial practitioners. In this study, we used the Technology Acceptance Model (TAM) to understand financial practitioners' intentions to use the digital innovation learning platform. TAM is the one of most influential models for technology acceptance. In the study, a total of 528 questionnaires were collected from financial practitioners with 457 valid ones. The collected data were analyzed by descriptive statistics and regression analysis. From demographic variables, more female financial practitioners used the digital innovation learning platform than males. Most practitioners using e-learning were 50–59 years old in the position of a business manager with 11–15 years of experience. In the technology acceptance model, perceived ease of use and perceived usefulness had a significant positive impact on attitude to use. Convenience, safety, and attitude to use had significant positive effects on willingness. All the hypotheses for the technology acceptance model were supported.

Keywords: digital innovation learning platform; English e-learning platform (EELP); technology acceptance model

Citation: Chen, M.-S.; Huang, W.-T. Applying the Technology Acceptance Model to Understand Financial Practitioners' Intentions to Use the Digital Innovation Learning Platform. *Eng. Proc.* **2023**, *38*, 62. https://doi.org/10.3390/engproc2023038062

Academic Editors: Teen-Hang Meen, Hsin-Hung Lin and Cheng-Fu Yang

Published: 29 June 2023

Copyright: © 2023 by the authors. Licensee MDPI, Basel, Switzerland. This article is an open access article distributed under the terms and conditions of the Creative Commons Attribution (CC BY) license (https://creativecommons.org/licenses/by/4.0/).

1. Introduction

Digital learning (e-learning) is considered to be a learning method for web-based learning or online learning [1]. The popularization of network infrastructure has formed an emerging information circulation channel, and digital learning is characterized by the environment, immediacy, and not being limited by space. The concept of learning is also oriented with interactivity, self-pacing, and tailored repetitions.

While the financial industry has moved towards informatization and technology, the learning system for financial practitioners is also moving toward digitalization. International laws and security risks are all important English education and training courses. With the financial industry adopting the "digital innovation learning system" (also known as the English e-learning platform, EELP) for training courses becoming a trend, it is even more necessary to deeply explore the effectiveness and development of digital learning in enterprises.

Due to the 2030 bilingual national policy in Taiwan, the financial industry encourages employees to use the digital innovation learning platform to improve and strengthen their language skills to improve service capabilities. In this study, we used the Technology Acceptance Model [2] to understand financial practitioners' intentions to use the digital innovation learning platform. The purpose of this study was to understand the influencing factors of financial practitioners' willingness to use digital learning systems and to explore the impact of financial practitioners' perceived ease of use and usefulness of the digital innovation learning platform.

2. Literature Review

2.1. Digital Learning

Experts in education, learning, and information technology give different definitions for digital learning (e-learning), but they are all based on the same foundation—the use of technological products or techniques to learn [3]. The use of digital learning systems allows employees to learn what the company requires of them to improve skills regardless of time and place and understand the latest information related to their industry. Therefore, digital learning has become one of the most widely discussed topics in the IT field. Although digital learning can increase the staff's IT programming, design, system operation, enterprise, and soft skills [4], the employees of the financial industry may not be willing to use it.

2.2. Technology Acceptance Model (TAM)

Many scholars use the technology acceptance model for their research. TAM [2] is one of the most influential models in technology acceptance research. The primary influencing factors on an individual's intention to use new technology are perceived ease of use and perceived usefulness. For an older adult, if they perceive that playing digital games is difficult and a waste of time, then they are unlikely to play. If they perceives that digital games can provide needed mental stimulation and is easy to learn, they are more likely to learn how to play. Although TAM has been criticized, it is still a useful general framework. The research result is consistent with other investigations about older adults' intentions to use new technologies [2].

3. Research Method and Procedure

According to the research using TAM, we found the use of digital innovation learning platforms is positively correlated to users' acceptance of the platform. Thus, we constructed the research model and hypotheses.

H1. *The effect of perceived ease of use and perceived usefulness of financial practitioners using digital innovation learning platforms is positive.*

H2. *The effect of the perceived usefulness of financial practitioners using digital innovation learning platforms on the attitude toward its use is positive.*

H3. *The effect of the perceived ease of financial practitioners using digital innovation learning platforms on the attitude to use is positive.*

H4. *The effect of financial practitioners' attitude toward using digital innovation learning platforms on their willingness to use them is positively correlated.*

H5. *The effect of the convenience of financial practitioners using the digital innovation learning platform on their willingness to use them is positive.*

H6. *The effect of the security of financial practitioners using the digital innovation learning platform on their willingness to use it is positive.*

This study adopted the cluster sampling method using a questionnaire survey to collect relevant information. A total of 528 questionnaires were collected from financial practitioners, and 457 valid questionnaires were used in this study. The return rate was 86.6%. The collected data were analyzed using descriptive statistics and regression analysis. The pre-test questionnaires of this study were collected from 1 March 2022, to 31 March 2022. An expert confirmed the validity of the questionnaire and question clarity.

4. Result and Discussions

4.1. Descriptive Statistics

A total of 164 respondents (35.9%) were aged 50–59, and 129 people (28.2%) were aged 30–39. A total of 192 respondents (42.0%) had a university degree, and 149 (21.7%) had a master's degree. This showed that most of the respondents were highly educated. A total of 98 respondents (21.5%) had a working experience of 11–15 years, and 96 people (21.0%) had a working experience of 7–9 years. A total of 185 financial practitioners (40.5%) were high-level managers, while 111 financial practitioners (24.3%) were middle managers. All details are presented in Table 1.

Table 1. The profile of respondents.

Items	Category	Sample	%
Gender	Male	253	55.4
	Female	204	44.6
Age (years)	20–29	48	10.5
	30–39	129	28.2
	40–49	98	21.4
	50–59	164	35.9
	60 or above	18	4.0
Education background	Non-college graduate	116	25.4
	College	192	42.0
	Graduate school	149	21.7
Working experience (years)	1–3	92	20.1
	4–8	84	18.4
	7–9	96	21.0
	11–15	98	21.5
	16 or above	87	19.0
Position	Low-level manager	65	14.2
	Middle manager	111	24.3
	Manager	96	21.0
	High-level manager	185	40.5

4.2. Reliability

The Cronbach's α was 0.829 to 0.934, showing a high reliability for the questionnaire survey as it was above 0.7 [5]. The adoption of digital innovation learning platforms was strongly linked to each characteristic. All criteria had correlation coefficients higher than 0.5. All criteria showed a beneficial effect on financial practitioners' attitudes and behaviors surrounding e-learning. The details are presented in Table 2.

4.3. Cross Analysis

From the demographic variables, more female financial practitioners used the digital innovation learning platform than males. Most practitioners using e-learning were 50–59 years old and in the position of a business manager with 11–15 years of experience.

Table 2. Summary of reliability analysis.

Construct	Number of Items	Cronbach's α Value
Perceived ease of use	5	0.890
Perceived usefulness	5	0.895
Attitude to use	5	0.879
Convenience	5	0.922
Security	4	0.829
Attitude to use	5	0.899
Willingness to use	6	0.934

4.4. Regression and Hypothesis Test

Regression analysis was used to examine the factors that affected the attitude and intention of financial practitioners to use the digital innovation learning platform. The purpose was to verify the hypothesis established by the technology acceptance model as presented in Table 3.

Table 3. Regression and hypothesis test.

Independent Variable	Dependent Variable		
	Perceived Usefulness β Value	Attitude to Use β Value	Willingness to Use β Value
Perceived ease of use	0.426 ***	0.313 ***	
Perceived usefulness		0.564 ***	
Attitude to use			0.660 ***
Convenience			0.233 ***
Safety			0.224 ***
R^2	0.332	0.576	0.640
Adjusted-R^2	0.359	0.563	0.749
Hypotheses Supported	H1	H2 H3	H4 H5 H6

*** $p < 0.001$.

The following findings were found from the results:

- The perceived ease of use of financial practitioners using the digital innovation learning platform had a positive and significant effect on the perceived usefulness ($\beta = 0.426$, $p < 0.001$), with an explanatory power of 33.2%. Therefore, the research hypothesis H1 was supported. When financial practitioners thought that the digital innovation learning platform was easier to operate, they used the digital innovation learning platform more often.
- The perceived usefulness of financial practitioners using the digital innovation learning platform had a positive and significant effect on the attitude to use ($\beta = 0.564$, $p < 0.001$), with an explanatory power of 57.6%. Therefore, the research hypothesis H2 was supported. When financial practitioners thought that using the digital innovation learning platform was easier and helped them complete their studies, they had a more positive attitude towards using the digital innovation learning platform.
- The perceived ease of use of financial practitioners using the digital innovation learning platform had a positive and significant effect on the attitude to use ($\beta = 0.313$, $p < 0.001$), with an explanatory power of 57.6%. Therefore, the research hypothesis H3 was supported. When financial practitioners thought that using the digital innovation learning platform was easy, they were more confident in the system and had positive attitudes toward using it.

- The attitude of financial practitioners toward using the digital innovation learning platform had a positive and significant effect on their willingness to use ($\beta = 0.660$, $p < 0.001$), with an explanatory power of 64.0%. Therefore, the research hypothesis H4 was supported. When financial practitioners had positive attitudes and feelings toward the use of the digital innovation learning platform, they had a higher willingness to use it.
- The perceived convenience of financial practitioners using the digital innovation learning platform had a positive and significant effect on the willingness to use ($\beta = 0.233$, $p < 0.001$), with an explanatory power of 64.0%. Therefore, the research hypothesis H5 was supported. When financial practitioners used the digital innovation learning platform, the higher the convenience of the learning platform, the higher their willingness to use it was.
- The perceived security of financial practitioners using the digital innovation learning platform had a positive and significant effect on the willingness to use ($\beta = 0.224$, $p < 0.001$), with an explanatory power of 64.0%. Therefore, the research hypothesis H6 was supported. When financial practitioners used the digital innovation learning platform, the higher the security of the learning platform, the higher their willingness to use it was.

5. Conclusions and Recommendations

Due to the 2030 bilingual national policy in Taiwan, the financial industry encourages employees to use the digital innovation learning platform to improve and strengthen the language skills of financial practitioners and strengthen their service capabilities. In this study, TAM was used to understand financial practitioners' intentions to use the digital innovation learning platform.

A total of 457 valid questionnaires were collected out of 528 distributed questionnaires. The results showed that more female financial practitioners used the digital innovation learning platform than male practitioners. A majority of the practitioners were 50–59 years old and business managers with 11–15 years of working experience. The perceived ease of use and perceived usefulness had a significant positive effect on the attitude to use. Convenience, safety, and attitude to use had significant positive effects on willingness. All hypotheses in TAM were supported.

Based on the research results, the following suggestions are proposed:

- It is suggested that follow-up researchers expand the research scope and add some other external variables.
- As the financial industry includes different service items, such as banking, insurance, bonds, and funds, future research needs to include differential analysis among the different financial industries.

Author Contributions: Conceptualization, M.-S.C. and W.-T.H.; methodology, M.-S.C. and W.-T.H.; validation, W.-T.H.; formal analysis, M.-S.C.; investigation, M.-S.C.; resources, M.-S.C.; data curation, M.-S.C.; writing—original draft preparation, M.-S.C.; writing—review and editing, W.-T.H., visualization, W.-T.H.; supervision, M.-S.C. and W.-T.H.; project administration, W.-T.H. All authors have read and agreed to the published version of the manuscript.

Funding: This research received no external funding.

Institutional Review Board Statement: Not applicable.

Informed Consent Statement: Not applicable.

Data Availability Statement: The data that support the findings of this study are available from the corresponding author upon reasonable request.

Conflicts of Interest: The authors declare no conflict of interest.

References

1. Trombley, B.K.; Lee, R. Web-based learning in corporations: Who is using it and why, who is not and why not? *J. Educ. Media* **2002**, *27*, 137–146.
2. Davis, F.D. Perceived usefulness, perceived ease of use and user acceptance of information technology. *MIS Q.* **1989**, *13*, 319–340. [CrossRef]
3. Pollard, E.; Hillage, J. *Exploring e-Learning*; Institute for Employment Studies: Brighton, UK, 2001; Volume 376, p. 65.
4. Bhattacherjee, A.; Hikmet, N. Reconceptualizing Organizational Support and Its Effects on Information Technology Usage: Evidence from the Healthcare Sector. *J. Comput. Inf. Syst.* **2008**, *48*, 69–76.
5. DeVellis, R.F. *Scale Development: Theory and Applications*; Sage Publications, Inc.: Thousand Oaks, CA, USA, 1991.

Disclaimer/Publisher's Note: The statements, opinions and data contained in all publications are solely those of the individual author(s) and contributor(s) and not of MDPI and/or the editor(s). MDPI and/or the editor(s) disclaim responsibility for any injury to people or property resulting from any ideas, methods, instructions or products referred to in the content.

engineering proceedings

MDPI

Proceeding Paper

Literature Review on the Development of Visualization Studies (2012–2022) †

Tianyin Jiang [1,*], Yaxin Hou [2] and Jaebum Yang [3]

1 Academy of Art Design, Fujian Business University, Fuzhou 350012, China
2 School of Art & Design, Hubei University of Technology, Wuhan 430074, China; yxhou2021@foxmail.com
3 College of Art, Design and Sport Science, Department of Art & Design, Dong-Eui University, Busan 47340, Republic of Korea; jbyang@deu.ac.kr
* Correspondence: jiangtianyin2019@naver.com; Tel.: +86-13140695922
† Presented at the 3rd IEEE International Conference on Electronic Communications, Internet of Things and Big Data Conference 2023, Taichung, Taiwan, 14–16 April 2023.

Abstract: In the past decade, the visualization and transformation of data and information have attracted lots of research interest, while visualization has gradually extended to all industries. Based on the retrieval of core literature in a Web of Science search from 2012 to 2022, this study finds that these developments mainly discussed the change of visualization and its related concepts, current research hotspots, and influential journal papers. Consequently, it aims to explore research gaps and provide directional guidance for future research.

Keywords: data and information visualization; visualized analysis; literature review; visualization studies

Citation: Jiang, T.; Hou, Y.; Yang, J. Literature Review on the Development of Visualization Studies (2012–2022). *Eng. Proc.* 2023, *38*, 89. https://doi.org/10.3390/engproc2023038089

Academic Editors: Teen-Hang Meen, Hsin-Hung Lin and Cheng-Fu Yang

Published: 31 July 2023

Copyright: © 2023 by the authors. Licensee MDPI, Basel, Switzerland. This article is an open access article distributed under the terms and conditions of the Creative Commons Attribution (CC BY) license (https://creativecommons.org/licenses/by/4.0/).

1. Introduction

1.1. Research Background

Most industry and business data are newly evaluated with the rapid development of the 'Internet +' in recent years. Big data technology extracts relevant information from a large number of unstructured or multi-structured datasets. As the McKinsey Global Research Institute (MGI) released in their report '*Big Data: The next frontier for Innovation, competition, and productivity*', the development of big data has been observed as the foreword and future trend of contemporary academic development. The process of forming images of objects through visual observation is a sort of mental processing procedure. Visualization improves people's ability to observe things and the formation of the overall concept for them. It displays data or information from different dimensions in graphical ways and promotes decision-makers or stakeholders to quickly and effectively understand the data or information.

Whether it is data visualization or information visualization, it has gradually been put on the frontier of research in this century and has developed prosperously in the past decade. In the vigorous development of big data analysis, InfoVis has been widely used in various data analysis applications. There are many published articles relating to big data analysis, which involves many researchers in diverse disciplines [1–3]. Data visualization has become the premise as an effective method of research and decision making.

1.2. Purpose and Significance

The study selected 'core' articles published from 2012 to 2021 and classified them. The main contents were categorized according to the following aspects. The first is the development of visualization-related concepts and the relationship between them: what are the directions for the evolution of visualization-related concepts based on these concepts? Secondly, according to the collection of the related literature, the specific result of the research influence of visualization technology over the past decade is described with the current research trends. These research results provide directional guidance for subsequent researchers and explore the research gap between studies and applications.

2. Methods and Steps

Literature research based on the retrieval system of the Web of Science (web of knowledge.com (accessed on 1 March 2023)) allows the effective retrieval of the following keywords: 'data visualization' and 'information visualization'. Web of Science is a well-known database that is most commonly used to retrieve literature [4,5]. The information is selected by evaluation of readers, and its credibility and influence are favored by scholars. The content covers journals, books, reports, seminars, and other academic information. It also includes three influential databases, such as Science Citation Index Expanded (SCI-Expanded), Social Sciences Citation Index (SSCI), and Arts & Humanities Citation Index (A & HCI). According to the retrieval of the keywords 'data visualization' and 'information visualization', these two are combined for analysis on the basis of the research objectives.

3. Analysis and Discovery

3.1. Development and Transformation of Visualization-Related Concepts

The development of visualization-related concepts changes over time, and these related concepts are evolving constantly where a contextual relationship is created. The essence is the visual interaction with computers. The conceptual change is mainly reflected in the development of visualization from 'technical implementation' to 'consideration of the recipient's cognition', and from scientific research suitability for professionals to mass communication. Visualization is gradually broadening the field of such influence.

Starting from several terms including data, information, data visualization, scientific visualization, information visualization, and knowledge visualization, the paper sorts out the relationship and common characteristics between different concepts and perceptions.

In information generation, the concept of data has been changed over a long history. Before the appearance of pictographs, people used ropes to record events, and later, stone walls were used. Such records are the essence and an original way for information delivery and representation. By the 10th century AD, people recorded the positions and changes of celestial bodies at different times with simple geometric figures and coordinate concepts together with longitude and latitude [6]. Nowadays, data means digital data [7] collected from various sources and composed of symbols and facts that are discrete and unexplained. When it has nothing to do with other data, it may have no meaning [8]. Information is data that has been interpreted or processed, and therefore contains certain meanings [9]. Usually, information tends to spread more visually than data.

The definition of data visualization is related to a set of techniques designed to extract relevant information from a large amount of unstructured or multi-structured data [10]. It usually refers to the data extraction method and belongs to the technical category. With the continuous evolution of the concept of data visualization, people are accepting new concepts with both scientific visualization as well as information visualization. With the impact of modern computer technology, the connotation of data visualization has been greatly expanded from scientific computing visualization to information visualization and knowledge visualization. Thus, the related research objects include spatial or non-spatial data and human knowledge [11]. As a result, the scope of data visualization is expanded.

Scientific visualization was proposed first in 1987. It refers to the use of computer graphics and image processing to create visual images, replace large and complex digital presentation forms, and help people better understand the concepts of science, technology, and results. Scientific visualization is a further extension of data visualization, which extends the real sensibility of graphic images [12], and its purpose is to help people better understand the meaning of data. Scientific methods are enriched by focusing on guiding data sets and seeing invisible information [13]. Therefore, scientific visualization is more inclined to reveal the information of data in specific application disciplines for professional researchers. Complex geographic data, measurement data, and scientific data are obtained in a process with computers. Re-analysis and combination sometimes need three-dimensional space with light source rendering to understand the real effect, which is regarded as three-dimensional real-world visualization. From the perspective of conceptual development, variants and sub-types based on data and scientific visualization (cartographic visualization) or knowledge domain (statistical visualization) are also created [12].

Information visualization (usually referred to as InfoViz) is one of the tools to allow and/or improve big data analysis. The prosperity of big data analysis has led the related technology to the wide application of information visualization in various fields from finance to sports and politics [10]. Its research and development started in the 1990s and began as a disciplined development. The purpose was to help users explore, understand, and analyze data through progressive, iterative visual exploration [14]. By using computer graphics and interaction to help mankind solve problems [15], and mimicking human's five senses, an immediate understanding of information helps researchers explore and understand deeper information and meanings [1–3]. Therefore, information visualization becomes important in applying abstracted data with interactive visual interfaces for various purposes [16]. Information visualization is closely related to the user's cognition as it pays more attention to people's acceptance, which is more associated with psychology, visual design, and human-computer interaction in addition to computer graphics for business methods. Information visualization is more inclined to visual communication than scientific visualization.

The development of knowledge visualization was carried out in this century. In the working document '*Knowledge Visualization-Towards a New Discipline and its Fields of Application compile*' by Eppler and Burkhard, knowledge visualization is considered to be a milestone and a new research category. Knowledge visualization stresses guiding the spread of knowledge among the public more than information visualization, and it expands the scope of viewing objects by sharing [17]. Knowledge visualization was developed based on information visualization and scientific visualization, and uses visual description to promote the transmission of group knowledge and creative creation. The purpose is to study the use of visual expression to improve knowledge transfer and creation among people [18]. Knowledge visualization is applied to represent knowledge in a visual format. It aims to support cognitive processes; to generate, represent, and build retrieval sharing; and to use knowledge [19]. Based on the above concepts, the difference between various visualization concepts is mainly reflected in the evolution of the definition of visualization over time and their relationships. There are also common features between concepts that are an interactive visualization of data (Figure 1).

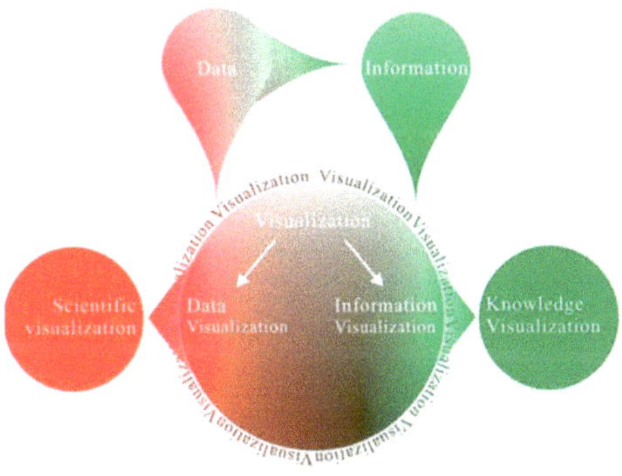

Figure 1. Development of concepts.

3.2. Influence of Visualization Research

This study explores the influence of visualization in three aspects: the influence of research, the influence of journals, and the influence of co-citation. The influence of research largely depends on the time for the development of the discipline and the number of people involved in the research. The influence of journals is not only related to the number of published articles but also depends on the depth and breadth of the research and a universal value. The influence of co-citation is based on professionalism [19]. The impact of journals is judged by the number of publications to prove the journal's attention on data visualization and information visualization.

In line with the retrieval of keywords of 'data visualization' and 'information visualization' and the core attributes, a total of 2229 research articles are merged, and then the top 10 articles published in 853 journals are selected.

The journals in descending order are as follows.

IEEE TRANSACTIONS ON VISUALIZATION AND COMPUTER GRAPHICS, 110 published, BIOINFORMATICS, 92 published, COMPUTER GRAPHICS FORUM, 53 published, BMC BIOINFORMATICS, 48 published, IEEE COMPUTER GRAPHICS AND APPLICATIONS, 34 published, PLOS ONE, 32 published. IEEE ACCESS, 30 published, ISPRS INTERNATIONAL JOURNAL OF GEO-INFORMATION, 29 published, JOURNAL OF VISUALIZATION, 27 published, NUCLEIC ACIDS RESEARCH, 25 published.

To a certain extent, the number of publications is not proportional to the impact factor, and the impact factor is not completely dependent on the number of publications but on the rate of their citations. However, the abuse of impact factor has affected the influence of journals [19]. The quality of publication does not depend on the journal impact factor. The impact factor alone does not evaluate the quality of journals precisely. Hence, the impact factor of a journal has an asymmetric relationship with the impact of journals, as shown (Tables 1 and 2).

Table 1. Journal influence.

Rank	Journal	Quantity
1	IEEE T VIS COMPUT GR	110
2	BIOINFORMATICS	92
3	COMPUTER GRAPHICS FORUM	53
4	BMC BIOINFORMATICS	48
5	IEEE COMPUT GRAPH	34
6	PLOS ONE	32
7	IEEE ACCESS	30
8	ISPRS INT J GEO-INF	29
9	JOURNAL OF VISUALIZATION	27
10	NUCLEIC ACIDS RESEARCH	25

Table 2. Impact factor (2022).

Rank	Journals	Impact Factor
1	NUCLEIC ACIDS RESEARCH	16.971
2	BIOINFORMATICS	6.937
3	IEEE T VIS COMPUT GR	4.597
4	IEEE ACCESS	3.367
5	PLOS ONE	3.240
6	BMC BIOINFORMATICS	3.169
7	ISPRS INT J GEO-INF	2.899
8	IEEE COMPUT GRAPH	2.088
9	COMPUTER GRAPHICS FORUM	2.078
10	JOURNAL OF VISUALIZATION	1.331

In terms of co-citation of articles in data visualization and information visualization, the top 10 co-cited articles comprehensively show the change in the content of professional fields. For example, the D-3 methods mentioned in the data-driven article are not for building a single framework but for providing an efficient operation of data-based documents. This shows that proprietary representation and extraordinary flexibility are necessary for analyzing the visualization [20]. For example, as a tool for the unification of biology, gene ontology mentioned in articles promotes the transformation of the concept of the naming system on the premise of a unified understanding of biology and its interoperability [21].

There is a high degree of correlation between co-cited articles to be manifested in the frequently referred articles of a particular field. For example, in the study of the top 10 journals, 'Empirical Studies in Information Visualization: Seven Scenarios', 'VisDesigner: Expressive Interactive Design of Information Visualizations', and 'D-3: data-driven documents' are influential in information visualization with high frequencies of citation (Table 3).

Table 3. TOP10 co-citation literature distribution based on keywords 'data visualization' and 'information visualization'.

Rank	Title	Cited Times
1	D-3:Data-Driven Documents	84
2	Raphical perception-theory, experimentation, and application to the development of graphical methods	35
3	A multi-level typology of abstract visualization tasks	28
4	Low-level components of analytic activity in information visualization	27
5	Readings in Information visualization: using vision to think	26
6	Protovis: agraphical toolkit for visualization	23
7	Gene ontology: tool for the unification of biology	22
8	The eyes have it: a task by data type taxonomy for information visualizations	22
9	viSNE enables visualization of high dimensional single-cell data and reveals phenotypic heterogeneity of leukemia	17
10	An integrated encyclopedia of DNA elements in the human genome	17

In addition to the citation rate, researchers' academic influence can also be defined based on the number of their publications. First of all, a limited number of researchers concentrate on a certain field. For example, Lee, B. published 34 articles from 2012 to 2021, of which 14 were selected through core screening. Collaborations among influential researchers with him are also found (Table 4).

Table 4. TOP 10 influential researchers ranked by keywords 'data visualization' and 'information visualization'.

Rank	Author	Times	Percentage
1	Ma, Kwan-Liu	9	0.09%
2	Olovira, OsvaldoN, Jr	9	0.09%
3	Chen, Min	8	0.08%
4	Isenberg, Petra	8	0.08%
5	Yuan, xiaoru	8	0.08%
6	Chen, Wei	7	0.07%
7	Jamroz, Dariusz	7	0.07%
8	Lee, Bongshin	7	0.07%
9	Paulovich, Fernado V	7	0.07%
10	Qu, Huamin	7	0.07%

3.3. Research Trends and Imbalances

The overall trend is based on the retrieval of keywords. The research trend is observed in two aspects: (i) the number of published articles each year and the overall trend, and

(ii) the analysis of popular research areas. From 2012 to 2021, the overall trend showed an upward increase with the total number of articles increasing year by year, among which data visualization contributed greatly. The research imbalance is mainly reflected in the ratio of articles that have been highly cited. The research on data visualization has gradually increased, and the total number of articles has increased from 138 in 2012 to 306 in 2021. Over the 10 years, data visualization has become the focus of research. However, the research on information visualization has fluctuated and become stable. The total number of articles is 299. The average is 29.9 articles per year, with the largest number of 46 in 2019 and the least of 16 in 2015 (Figure 2).

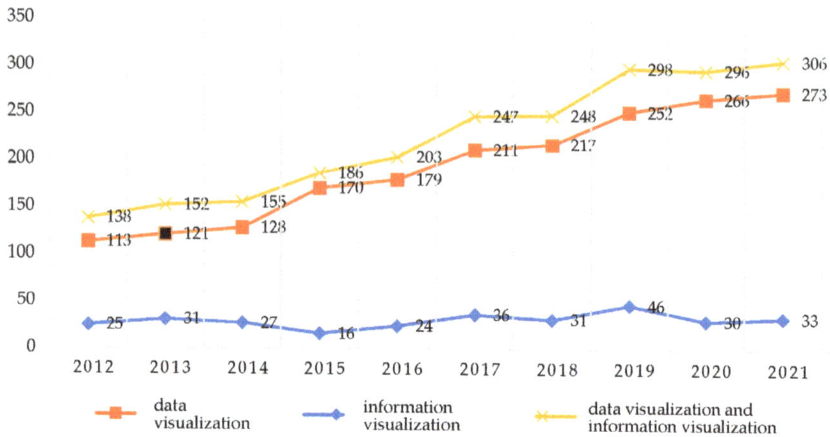

Figure 2. Publication trends of keyword retrieval ('data visualization' and 'information visualization').

The paper retrieves keywords within the research field of these articles, calculates them according to the frequency of occurrence, and then eliminates similar keywords by using cluster analysis. The top 10 applicable research fields are then selected in terms of wide applications, particularly in biology, geography, safety, health, commerce, transportation, energy, and personal applications. In the screening and sorting of keywords, it is found that data visualization and information visualization belong to the category of computer science. Most of the related research is about human-computer interaction and the realization of technology. The main research on visualization is mainly related to computer science. However, bio-related disciplines, biotechnology, biochemistry, biological information, and other applications have gained popularity in the past decade. With the wide application and rapid increase of computing power, the visualization of multi-faceted scientific data is becoming more important than before in engineering, medicine, or climate research (Table 5).

Table 5. Research field distribution based on keywords 'data visualization' and 'information visualization'.

Rank	Research Fields	Quantity	Percentage
1	Computer science	893	40.06%
2	Engineering	366	16.41%
3	Biochemistry molecular biology	294	13.18%
4	Mathematical computational biology	195	8.74%
5	Biotechnology applied microbiology	174	7.80%
6	Chemistry	165	7.40%
7	Mathematics	162	7.26%
8	Science technology other topics	111	4.97%
9	Environment sciences ecology	93	4.17%
10	Medical informatics	77	3.45%

From a trend analysis, the imbalance of research is found in two different aspects. First, the number of articles is different. Among 2229 retrieved articles, 1930 articles (86.6%) deal with data visualization, while 299 articles (13.4%) deal with information visualization. The number of articles on data visualization is 6.74 times higher than that on information visualization (Table 6).

Table 6. Proportion of the number of articles and researchers in the combined retrieval based on keywords 'data visualization' and 'information visualization'.

Keywords	Numbers of Researchers	Percentage
Data visualization	7915	87.6%
Information visualization	1120	12.4%
Data visualization and information visualization	9035	100%

Data visualization requires a large amount of unstructured or multi-structured data compared to information visualization [10]. Data visualization was coined in the 1950s with the advent of computer graphics [22]. Information visualization was introduced in 1999. The development of data visualization needs a longer time and is more mature than information visualization. Due to the required time for development, data visualization needs more research than information visualization. Thus, among 9035 researchers, 7915 (87.6%) are related to data visualization, while 1120 (12.4%) are related to information visualization. The ratio of the researchers for each research area on the visualization is similar to that for articles.

There is a significant difference in the number of highly cited articles. The top 10 articles have a citation rate of more than 500, and the citation rate of the top three articles is higher than 1000 times (Table 7).

Table 7. Information related to TOP10 articles retrieved by 'data visualization and information visualization'.

Rank	Title	Cited Times
1	Integrative Genomics Viewer (IGV): high-performance genomics data visualization and exploration	4098
2	qgraph: Network Visualizations of Relationships in Psychometric Data	1130
3	GGTREE: an R package for visualization and annotation of phylogenetic trees with their covariates and other associated data	1095
4	viSNE enables visualization of high dimensional single-cell data andreveals phenotypic heterogeneity of leukemia	968
5	Mantid-Data analysis and visualization package for neutron scattering and mu SR experiments	871
6	Pathview: an R/Bioconductor package for pathway-based data integration and visualization	698
7	ETE 3: Reconstruction, Analysis, and Visualization of Phylogenomic Data	669
8	Anvi'o: an advanced analysis and visualization plattormfor 'omics data	594
9	Protter: interactive protein feature visualization and integration with experimental proteomic data	586
10	Artemis: an integrated platform for visualization and analysis of high-throughput sequence-based experimental data	565

The research goals of highly cited articles are diverse. When an article is cited by others at a certain time, the article generates attention for its research outcome [23]. For example, the article 'Integrative Genomics Viewer (IGV): high-performance genomics data visualization and exploration' is cited 4098 times. Its research outcome allows a wide range of applications and provides a basis for other research.

4. Conclusions

Based on the keywords of 'data visualization' and 'information visualization', this study finds the development and transformation of concepts, the influence of research, the trend, and the imbalance of research.

4.1. Concept Development and Its Transformation

The concept of data visualization has evolved continuously, covering scientific visualization and information visualization. Scientific visualization is a further extension of data visualization to the real perceptual image. Information visualization is more inclined toward visual communication than scientific visualization, and knowledge visualization is more inclined toward guiding knowledge among the public than information visualization that expands the scope of the viewing object with the characteristics of sharing.

In conceptual transformation, visualization is more influential on the application of scientific research from professionals to mass communication. For example, knowledge visualization promotes visual information that only professionals can present as an explanation that the public understands.

4.2. Research Influence

This article mainly describes and analyzes three bases, namely, the influence of journals, the influence of researchers, and the influence of co-citation articles. In terms of the influence of journals, more influential journals are concentrated in software engineering and biologically related fields, but the relationship between the number of published journals and the impact factor is not proportional. In the aspect of co-citation, the analysis result of co-citation articles shows that the top 10 co-cited articles have high versatility and a high degree of correlation between co-cited articles. This correlation in a particular field implies frequent citations or widely cited topics.

4.3. Research Trends and Imbalance

The research trend is analyzed in two aspects: the number of published articles each year and the overall trend. The total amount of articles is increasing year by year, especially the research on data visualization, but the research on information visualization fluctuates less and becomes stable. This also reflects the imbalance of research. The imbalance is reflected in the analysis of popular research areas that are widely distributed, including biological disciplines, biotechnology, biochemistry, biological information, and other applications. In other fields, engineering and climate research are becoming more important.

5. Research Limitations and Prospects

The result of data visualization and information visualization provides researchers with preliminary insights into their research direction. The paper only employs the Web of Science core database, so the overall number of topics collected from the literature may be relatively biased. Thus, future research needs more literature retrieval pipelines such as Scopus, CSCD, and KCI to obtain comprehensive information. In terms of the objectives, this study is descriptive and mainly discusses the related characterizations. Subsequent research is necessary to explain more about the profound connotation and reasons for the presented results in this study and to conduct a summary analysis.

There are various methods such as cluster analysis [24] and semi-structured interviews [25,26] that are applicable for implementing the related research. Recent studies have focused on empirical methods and applications. With empirical methods, more information research is deployed to the application [27] to know in the practical field.

In the future, the research on visualization will become popular regardless of the number of researchers and articles in development. As a consequence, research on information visualization is still in great need. The reason is that information visualization is mainly for understanding characteristics of multidisciplinary cross-cutting nature, and it is difficult to carry out related research on cognitive science and visual design. These two disciplines have great differences, so the integration of the two disciplines is not easy. In addition to this, the concepts of information visualization reflect the popular trend of visualization technology, though, at present, it does not have much demand.

Author Contributions: J.Y. devised the overall framework, T.J. wrote and summarized the content, and Y.H. collected information and summarized the content. All authors have read and agreed to this version of the manuscript.

Funding: This research received no external funding.

Institutional Review Board Statement: Not applicable.

Informed Consent Statement: Informed consent was obtained from all subjects involved in the study.

Data Availability Statement: The data presented in this study are available in Section 3 of the manuscript.

Conflicts of Interest: The authors declare no conflict of interest.

References

1. Cao, N.; Lin, Y.R.; Sun, X.; Lazer, D.; Liu, S. Whisper, HQu: Tracing the spatiotemporal process of information diffusion in real time. *IEEE Trans. Vis. Comput. Graph.* **2012**, *18*, 2649–2658. [PubMed]
2. Cui, W.; Liu, S.; Tan, L.; Shi, C.; Song, Y.; Gao, Z.; Qu, H.; Tong, X. Textflow: Towards better understanding of evolving topics in text. *IEEE Trans. Vis. Comput. Graph.* **2011**, *17*, 2412–2421. [CrossRef]
3. Ma, J.; Liao, I.; Ma, K.-L.; Frazier, J. Living liquid: Design and evaluation of an exploratory visualization tool for museum visitors. *IEEE Trans. Vis. Comput. Graph.* **2012**, *18*, 2799–2808. [CrossRef] [PubMed]
4. Wamba, S.F.; Akter, S.; Edwards, A.; Chopin, G.; Gnanzou, D. How 'big data' can make big impact: Findings from a systematic review and a longitudinal case study. *Int. J. Prod. Econ.* **2015**, *165*, 234–246. [CrossRef]
5. Aghaei, C.A.; Salehi, H.; Yunus, M.; Farhadi, H.; Fooladi, M.; Farhadi, M.; Ale Ebrahim, N. A comparison between two main academic literature collections: Web of Science and Scopus databases. *Asian Soc. Sci.* **2013**, *9*, 18–26.
6. Ansari, B.; Barati, M.; Martin, E.G. Enhancing the usability and usefulness of open government data: A comprehensive review of the state of open government data visualization research. *Gov. Inf. Q.* **2022**, *39*, 101657.
7. Binder, H.; Blettner, M. Big data in medical science-A biostatistical view. *Dtsch. Arztebl. Int.* **2015**, *112*, 137–142. [CrossRef] [PubMed]
8. Meyer, R. Knowledge visualization. In *Trends in Information Visualization*; University of Munich: Munich, Germany, 2010.
9. Keller, T.; Tergan, S.O. Visualizing knowledge and information: An introduction. In *Knowledge and Information Visualization*; Sigmar, O.T., Tanja, K., Eds.; Springer: Berlin/Heidelberg, Germany, 2005; pp. 1–23.
10. Plaisant, C.; Shneiderman, B. Lightning and Thunder: The Early Days of Interactive Information Visualization at the University of Maryland. *IEEE Comput. Graph. Appl.* **2022**, *42*, 103–113. [CrossRef] [PubMed]
11. Hong, W.; Wang, J. Survey on Visualization and Visual Analytics. *J. Yanshan Univ.* **2010**, *34*, 95–99.
12. Bačić, D.; Fadlalla, A. Business information visualization intellectual contributions: An integrative framework of visualization capabilities and dimensions of visual intelligence. *Decis. Support Syst.* **2016**, *89*, 77–86. [CrossRef]
13. Ping, Z. Business information Visualization: Guidance for Research and Practice. In *Encyclopedia of Microcomputers*; Allen, K., James, G.W., Rosalind, K., Eds.; CRC Press: Boca Raton, FL, USA, 2001; Volume 27, pp. 61–77.
14. Shiravi, H.; Shiravi, A.; Ghorbani, A.A. A survey of visualization.systems for network security. *IEEE Trans. Vis. Comput. Graph.* **2012**, *18*, 1313–1329. [CrossRef] [PubMed]
15. Purchase, H.C.; Andrienko, N.; Jankun Kelly, T.J. Theoretical Foundations of Information Visualization. In *Information Visualization*; Kerren, A., Stasko, J., Fekete, J.D., North, C., Eds.; Springer: Berlin/Heidelberg, Germany, 2008; pp. 46–64.
16. Keim, D.A.; Mansmann, F.; Schneidewind, J.; Ziegler, H. Challenges in Visual Data Analysis. In Proceedings of the Information Visualization (IV 2006), London, UK, 5–7 July 2006.
17. Tergan, S.O.; Keller, T.; Burkhard, R.A. Integrating knowledge and information: Digital concept maps as a bridging technology. *Inf. Vis.* **2006**, *5*, 167–174.
18. Burkhard, R.A. Learning from architects: The difference between knowledge visualization and information visualization. In Proceedings of the Eighth International Conference on Information Visualisation, 2004, IV 2004, London, UK, 16 July 2004.
19. Zupanc, G.K.H. Impact beyond the impact factor. *J. Comp. Physiol. A* **2014**, *200*, 113–116. [CrossRef] [PubMed]
20. Bostock, M.; Ogievetsky, V.; Heer, J. D^3 data-driven documents. *IEEE Trans. Vis. Comput. Graph.* **2011**, *17*, 2301–2309. [CrossRef] [PubMed]
21. Ashburner, M.; Ball, C.A.; Blake, J.A.; Botstein, D.; Butler, H.; Cherry, J.M.; Davis, A.P.; Dolinski, K.; Dwight, S.S.; Eppig, J.T.; et al. Gene ontology: Tool for the unification of biology. *Nat. Genet.* **2000**, *25*, 25–29. [CrossRef] [PubMed]
22. Post, F.H.; Nielson, G.M.; Bonneau, G. *Data Visualization: The State of The Art*; Kluwer Academic Publishers: Boston, MA, USA; Dordrecht, The Netherlands; London, UK, 2003.

23. Keim, D.; Andrienko, G.; Fekete, J.D.; Görg, C.; Kohlhammer, J.; Melançon, G. Visual analytics: Definition, process, and challenges. In *Information Visualization*; Kerren, A., Stasko, J., Fekete, J.D., North, C., Eds.; Springer: Berlin/Heidelberg, Germany, 2008; pp. 154–175.
24. McNabb, L.; Laramee, R.S. Survey of surveys (SoS)-Mapping the landscape of survey articles in information visualization. *Comput. Graph. Forum* **2017**, *36*, 589–617.
25. Lee, S.; Kim, S.H.; Hung, Y.H.; Lam, H.; Kang, Y.A.; Yi, J.S. How do people make sense of unfamiliar visualizations? A grounded model of novice's information visualization sensemaking. *IEEE Trans. Vis. Comput. Graph.* **2015**, *22*, 499–508. [CrossRef] [PubMed]
26. Kandel, S.; Paepcke, A.; Hellerstein, J.M.; Heer, J. Enterprise data analysis and visualization: An interview study. *IEEE Trans. Vis. Comput. Graph.* **2012**, *18*, 2917–2926. [PubMed]
27. Liu, S.; Cui, W.; Wu, Y.; Liu, M. A survey on information visualization: Recent advances and challenges. *Vis. Comput.* **2014**, *30*, 1373–1393. [CrossRef]

Disclaimer/Publisher's Note: The statements, opinions and data contained in all publications are solely those of the individual author(s) and contributor(s) and not of MDPI and/or the editor(s). MDPI and/or the editor(s) disclaim responsibility for any injury to people or property resulting from any ideas, methods, instructions or products referred to in the content.

engineering proceedings

MDPI

Proceeding Paper

Design and Analysis of Education Personalized Recommendation System under Vision of System Science Communication †

Manyin Shi *, Fang Luo, Hanping Ke and Shiliang Zhang

School of Information, Mechanical and Electrical Engineering, Ningde Normal University, Ningde 352100, China; t0407@ndnu.edu.cn (F.L.); t1502@ndnu.edu.cn (H.K.); t8159@ndnu.edu.cn (S.Z.)

* Correspondence: t0708@ndnu.edu.cn

† Presented at the 3rd IEEE International Conference on Electronic Communications, Internet of Things and Big Data Conference 2023, Taichung, Taiwan, 14–16 April 2023.

Abstract: Guided by system science, we propose a cognitive model based on graph theory and explore personalized recommendation algorithms based on a deep knowledge point tracking model by integrating the learning characteristics, prior knowledge, and learning ability of learners. Recommendation of the knowledge point is provided by combining the deep knowledge point tracking model and cognitive model, and personalized curriculum recommendation is provided by combining a knowledge point tracking model and graph theory. A dynamic personalized learning path is recommended by combining the knowledge point network and a student model. Then, teaching resources are recommended, and learning efficiency is improved.

Keywords: system science; personalized education; recommendation system design scheme

Citation: Shi, M.; Luo, F.; Ke, H.; Zhang, S. Design and Analysis of Education Personalized Recommendation System under Vision of System Science Communication. *Eng. Proc.* **2023**, *38*, 91. https://doi.org/10.3390/engproc2023038091

Academic Editors: Teen-Hang Meen, Hsin-Hung Lin and Cheng-Fu Yang

Published: 21 August 2023

Copyright: © 2023 by the authors. Licensee MDPI, Basel, Switzerland. This article is an open access article distributed under the terms and conditions of the Creative Commons Attribution (CC BY) license (https://creativecommons.org/licenses/by/4.0/).

1. Introduction

To solve the problem of "information overload and maze" in education, personalized recommendations of teaching resources are introduced. Personalized recommendation of teaching resources refers to educational information and teaching resources that are provided based on personal information. Different from the "one-to-many" services provided by educational resources, the recommendation meets personal needs, so users do not need to go through many recommendation processes. Thus, the cost of finding the required educational resources can be significantly reduced [1,2]. The personalized recommendation service has changed from the simple mode of "resources found by users" to the intelligent mode of "resources for uses". The service is for learner-centered adaptive learning [3–5]. Adaptive learning refers to the construction of learners' interests and abilities based on online learning behavior data. Learner behavior is predicted to intelligently provide personalized learning paths, resources, or courses according to personal interests and abilities. The personalized recommendation of teaching resources effectively solves the problem of an "information maze" brought on by a massive amount of data. Therefore, the development of a personalized recommendation system is an important topic for research in educational information.

2. Research Status of Related System

Personalized recommendations have been extensively studied. In terms of learning course recommendation, references [6,7] propose a personalized course generation system based on the hierarchical recommendation algorithm. The system provided personalized courses in two stages according to the learning process of users. Personalized courses are recommended based on the analysis of the overall teaching plan and curriculum-based knowledge extracted from the massive knowledge base. Then, target learners can acquire the necessary knowledge before learning the courses through the pre-test module and a hierarchical recommendation algorithm. In learning, a dynamic update of the learner's characteristics is performed with the genetic algorithm that generates personalized course contents according to the learning objectives of the course. Based on knowledge points and knowledge management theory, the 5-layer structure of knowledge and the 4-layer structure of using knowledge are recommended. The knowledge points and learning strategy are designed and developed based on recommended personalized courses [8]. A personalized course learning model for the program of the certificate in English language teaching (CELT) was developed based on the self-organized community [9,10]. The model used the similarity between peers and lecturers, and the pertinence of the algorithm to guide learners by recommending learning objects, pathways, and strategies was verified. The proposed model reduced the workload of the lecturers, and the experimental results showed that the proposed model effectively improved the learning quality and interest of the learners.

The current personalized recommendation system for teaching resources was proposed in 2015 with a depth knowledge tracking (DKT) model based on deep learning theory and knowledge tracking theory. With the big data of learning behaviors, the prediction accuracy of the system was higher than with Bayesian knowledge tracing (BKT) and item response theory (IRT). Without experts' help, teachers could find the necessary knowledge, and favorable conditions for learners' knowledge discovery were determined. In learning, knowledge takes a relatively small part as in the learning process, a cognitive law of knowledge, cognitive theory, and cognitive model are combined in the internal connection for mining knowledge. Based on the knowledge and its interconnection, the characteristics of learning courses can be recognized, which are used for learning courses and path recommendations. However, an appropriate knowledge-tracking model has not yet been developed. With higher accuracy of the deep knowledge-tracking model, the accuracy and satisfaction with recommended knowledge points and learning courses and paths will be improved.

3. Goals and Ideas of System Design

This study aims to develop a personalized recommendation algorithm based on a DKT model that considers the learning characteristics, prior knowledge, and learning ability of learners. Introducing the cognitive model and figure theory, the teaching resources are optimized, and the accuracy of the recommendation of teaching resources is improved. The wider coverage, better recall, and improved recommended rate allow learners to improve learning efficiency, enhance learning interest, optimize the learning effect, and relieve "cognitive overload". Such advantages can be achieved with the proposed algorithm.

In the algorithm, the knowledge point recommendation algorithm, the deep knowledge point tracking model, and the cognitive model are integrated for accurate and enhanced interpretability of the prediction process. The Ebbinghaus forgetting curve is used to simulate learning status and recommend knowledge points that are forgotten. The deterministic inputs, noisy, and gate (DINA) model is also introduced to evaluate the guess and error rates so that the feature vector of the learners' knowledge point can be closer to their real state. The graph theory is also used to explore the internal relationship between the knowledge points. Dependencies, reference relationships, parallel relationships, and pre- and post-relationships are determined to create the personalized knowledge graph of learners. The similarity or subordination between the features found and the feature vector in the knowledge graph of learning courses is tested to validate the algorithm's performance. Based on the inherent relationship between knowledge points and learning courses, a personalized learning path is recommended.

4. System Design

According to the objectives of this study, the following processes are carried out.

- Constructing an algorithm for recommending personalized knowledge based on the DKT model and the cognitive model

- Constructing an algorithm for a personalized course recommendation based on the DKT model and graph theory
- Constructing an algorithm for personalized learning path recommendation based on the combination of knowledge point networks and student models

The DKT model has advantages over BKT and IRT in that it can train with the data parameters and study the algorithm mechanism. The long- and short-term memory model (LSTM) is used to predict the required knowledge points and test the prediction accuracy and operation rate. With the Ebbinghaus forgetting curve, the DINA model and LSTM are used to compare the prediction accuracy, recall rate, test time, and training time. Based on the prediction results, the algorithms recommend knowledge points that learners have not mastered well. In LSTM, the hidden Markov model (HMM) and K-means algorithm are adopted to explore the accuracy of the predicted knowledge relationship graph. The feature vector of the learning course is extracted to create the learning course graph. Then, the subordination relationship between the learners' knowledge map and the learning course is investigated to recommend the relevant learning courses based on the knowledge points that the learners have mastered poorly. The learning path is recommended based on the directed knowledge point network graph and a personalized knowledge status map, which show cognition, ability, goals, and other characteristics of learners. The learning path is optimized for better effectiveness of personalized learning, and the most appropriate learning path is recommended for learners.

The algorithms are developed to improve the accuracy of the prediction and lay the foundation for the subsequent recommendation. For the algorithms, it is necessary to find the characteristic vector of the learner's knowledge points and create a knowledge map. Through similarity or relationship tests of the knowledge graph and learning course characteristics, a personalized learning course recommendation can be provided. The learning path is planned with the group's wisdom and the prior knowledge of the learner. To develop the algorithms, the public data is preprocessed. Then, combining the long and short-term model (LSTM), DINA model, and student model and based on knowledge point map theory and cognitive model theory, the learners' knowledge points are predicted to create a personalized knowledge point map (Figure 1).

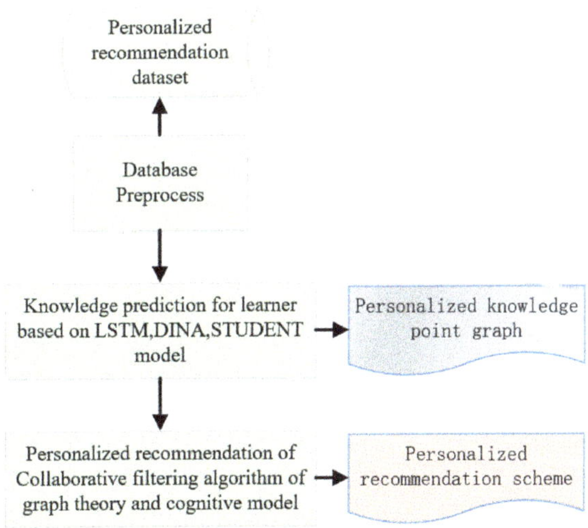

Figure 1. System design route.

5. Experiments

5.1. Dataset

For the experiment, data from 2009–2010 is used. It contains 525,535 sets of the exercise data of 4217 students with the student number, question number, correct answer result, associated knowledge points, answer order, answer type, and others. The data on students' lifestyles and studies is collected from the 10-week cell phone use of 48 students at Dartmouth College in the USA) including learners' mental state, academic performance, and behavioral trends. The size of the data is 53 Gb, containing 32,000 self-reports and pre-post surveys. The Ted dataset, the Edx dataset, and the Datashop data are also used in the experiment.

5.2. Comparison and Evaluation

As comparison algorithms, we use the collaborative filtering algorithm, collaborative filtering recommendation algorithm, single-layer DKT model, two-layer DKT model algorithm, and BKT algorithm.

To evaluate the performance of the algorithms, the following index is calculated.

Mean absolute error:

$$\text{MAE} = \frac{1}{|E^p|} \sum_{(u,a \in E^p)} \left| r_{ua} - r'_{ua} \right| \quad (1)$$

Root-mean-square error:

$$RMSE = \sqrt{\frac{1}{|E^p|} \sum_{(u,a) \in E^p} \left(r_{ua} - r'_{ua} \right)^2} \quad (2)$$

Pearson Correlation factor:

$$\text{PCC} = \frac{\sum_a (r'_a - \overline{r'})(r_a - \overline{r'})}{\sqrt{\sum_a (r'_a - \overline{r'})^2} \sqrt{\sum_a (r_a - \overline{r'})^2}} \quad (3)$$

Precision:

$$\text{P}_u(L) = \frac{N_{tp}}{N_{tp} + N_{fp}} \quad (4)$$

Recommended coverage:

$$\text{COV}_R(L) = \frac{N_d(L)}{N} \quad (5)$$

6. Discussion, and Conclusions

We propose an algorithm for the personalized recommendation of knowledge points, learning paths, resources, and courses for learners. We use the DKT model to determine learning style characteristics, prior knowledge, and learning ability, and the cognitive model and graph theory to optimize personalized teaching resources. The developed algorithm improves the accuracy, coverage, recall, and rate of teaching resource recommendations, enhances learning efficiency, interest, and effect, and overcomes "overload" and "learning flight". The algorithm also makes the knowledge point feature vector more accurate and improves the interpretability of the prediction. Learning courses are recommended to offer more comprehensive and accurate courses according to the learner's needs. The algorithm can be used to develop further personalized educational recommendation systems.

Author Contributions: M.S. proposed conceptualization and methodology of this manuscript. The figures and partial writing were conducted by F.L. H.K. and S.Z. both are responsible for the formal analysis and investigation. All authors have read and agreed to the published version of the manuscript.

Funding: This research was supported by the Hainan Provincial Natural Science Foundation of China (No. 2020J01435, No. 2021J011169, 2022J011224). Ningde Normal University for service local action special plan (No. 2020ZX505).

Institutional Review Board Statement: Not applicable.

Informed Consent Statement: Not applicable.

Data Availability Statement: No data were created.

Conflicts of Interest: The authors declare no conflict of interest.

References

1. Montoya, M.M.; Massey, A.; Lockwood, P.N. 3D collaborative virtual environments: Exploring the link between collaborative behaviors and team performance. *Decis. Sci.* **2018**, *42*, 451–476. [CrossRef]
2. Kevin, C.; Carsten, S.L.; Tae, K.L. Knowledge Tracing to Model Learning in Online Citizen Science Projects. *IEEE Trans. Learn. Technol.* **2019**, *36*, 2253–2278.
3. Piech, C.; Spencer, J.; Huang, J.; Gang, H.G.; Sahami, M. Deep Knowledge Tracing. *Comput. Sci.* **2018**, *7*, 19–23.
4. Melis, E.; Goguadze, G.; Homik, M.; Libbrecht, P. Semantic-aware components and services of Active Math. *Br. J. Educ. Technol.* **2011**, *37*, 405–423. [CrossRef]
5. Dario, N.; Méndez, D.; Ramírez, C.J.; Alberto, J.I. A planning for automatic generation of customized virtual courses. In Proceedings of the ECAI, Stockholm, Sweden, 13–19 July 2018.
6. Ullrich, C.; Melis, E. Pedagogically founded courseware generation based on HTN-planning. *Expert Syst. Appl.* **2017**, *36*, 9319–9332. [CrossRef]
7. Karampiperis, P.; Sampson, D. Adaptive Learning Resources Sequencing in Educational Hypermedia Systems. *J. Educ. Technol. Soc.* **2015**, *8*, 128–147.
8. Rumetshofer, H.; Wöβ, W. XML-based adaptation framework for psychological-driven e-learning systems. *J. Educ. Technol. Soc.* **2003**, *6*, 18–29.
9. Li, G.Q.; Xia, J.L.; Mei, S.Y. Learning Resource Recommendation Method based on Fuzzy Logic. *J. Eng. Sci. Technol. Rev.* **2018**, *51*, 787–814.
10. Chang, Y.C.; Kao, W.Y.; Chu, C.P. A learnings tyle classification mechanism for e-learning. *Comput. Educ.* **2019**, *53*, 273–285. [CrossRef]

Disclaimer/Publisher's Note: The statements, opinions and data contained in all publications are solely those of the individual author(s) and contributor(s) and not of MDPI and/or the editor(s). MDPI and/or the editor(s) disclaim responsibility for any injury to people or property resulting from any ideas, methods, instructions or products referred to in the content.

engineering proceedings

Proceeding Paper

Discussion on Risk Factors of Stadium: A Case Study of Taipei Dome Complex †

Chi-Jan Huang [1] and Ting-Yi Chiang [2,*]

1 General Education Center, Ming Chuan University, Taipei 111013, Taiwan; cjhuang@mail.mcu.edu.tw
2 Department of Computer Science and Engineering, Tatung University, Taipei 104327, Taiwan
* Correspondence: tychiang824@gmail.com

† Presented at the 3rd IEEE International Conference on Electronic Communications, Internet of Things and Big Data Conference 2023, Taichung, Taiwan, 14–16 April 2023.

Abstract: The Taipei Dome Complex is located in Taipei, Taiwan. It is a composite park with a multi-functional dome and incorporates a shopping mall, restaurants, cinemas, and an office building. In 2017, Taipei, Taiwan hosted the Universiade, and thousands of people protested. Smoke bombs were thrown, and the police were attacked. Due to its location in the "Circum-Pacific seismic belt" and the fact it is affected by the subtropical monsoon climate, Taiwan suffers frequent natural disasters. We aim to define the security risk factors for the Taipei Dome Complex by assessing flood simulation from the National Science and Technology Center for Disaster Reduction, earthquake simulation of the Taiwan Earthquake Loss Estimation System (TELES), and a case study of a terrorist attack. We propose 24 risk factors from five major perspectives: equipment safety, traffic accidents, human resources, security management, and disaster events.

Keywords: risk management; disaster prevention; TELES; Taipei Dome Complex

1. Introduction

Because of the spread of terrorism, urban facilities have become the target of terrorist attacks, including transport stations, airports, landmarks of the cities, theaters, critical infrastructures in stadiums, arenas, and domes that hold major sporting events and concerts. On 22 May 2017, at 22:33 British Summer Time, there were explosions at the Manchester Arena where singer Ariana Grande Butera was holding a concert. The nail bombs caused 23 deaths (including that of the attacker) and 119 injuries [1]. It was a terrorist attack by a lone male suicide bomber [2]. After investigation, it was determined that the explosive charges were detonated using a remote-control device and were more likely carried into the venue by being packed in a backpack than in a vest [3]. On 21 August 2017, Taipei, Taiwan hosted the Summer Universiade. Thousands of people protested, a man attacked the police, and several people threw smoke bombs at the venue of the opening ceremony. The accidents prevented athletes from entering the arena on time [4].

The Taipei Dome Complex, the research object of this study, is located in Xinyi District, Taipei City, Taiwan. It is a composite park. The main body is a multi-functional dome, and the other also has constructions such as a shopping mall, restaurant, cinema, and office buildings (Figure 1). The crowd capability of the dome is 59,833 [5], which is approximately double that of the Taipei Arena (currently the largest arena in Taiwan). If an incident like the Manchester Arena terrorist attack occurs in the Taipei Dome Complex, the consequences will be disastrous. Therefore, this study is being carried out to determine the risk factors for disasters and safety events for the Taipei Dome Complex and provides security recommendations for large events.

Citation: Huang, C.-J.; Chiang, T.-Y. Discussion on Risk Factors of Stadium: A Case Study of Taipei Dome Complex. *Eng. Proc.* **2023**, *38*, 22. https://doi.org/10.3390/engproc2023038022

Academic Editors: Teen-Hang Meen, Hsin-Hung Lin and Cheng-Fu Yang

Published: 20 June 2023

Copyright: © 2023 by the authors. Licensee MDPI, Basel, Switzerland. This article is an open access article distributed under the terms and conditions of the Creative Commons Attribution (CC BY) license (https://creativecommons.org/licenses/by/4.0/).

Figure 1. Distribution of venues of Taipei Dome Complex.

2. Literature Review

2.1. Potential Disaster Analysis

The TAIPEI Dome Complex is located in the Xinyi District, Taipei City, Taiwan. We used the disaster-causing analysis template proposed by the Federal Emergency Management Agency (FEMA) to analyze potential disasters in the complex, focusing on fire, flood, earthquake, and terrorist attacks, as shown in Table 1. In terms of earthquakes, Taiwan is in the seismically active area of the "Circum-Pacific seismic belt" and the Shanjiao fault is near the Taipei Dome Complex [6] (Figure 2), meaning that the probability of their occurrence is "possible". According to the simulation results of the Taiwan Earthquake Loss Estimation System (TELES) [7] (Table 2), the number of collapsed houses in the Xinyi District would be 341 (accounting for 2.77% of the entire Taipei City) and the number of resuscitation injuries and deaths would be 66 (accounting for 2.89% of the entire Taipei City), and so the potential impact of an earthquake is "limited". In the Taipei Dome Complex, compared with other buildings, the seismic performance of the baseball dome is the best. Due to the subtropical monsoon climate, the occurrence probability of flooding is "very likely". The occurrence is predicted through torrential rain warnings or typhoon tracking, and disaster losses can be reduced through mitigation and remediation works, and so the potential impact is "negligible". According to the flood simulation of the National Science and Technology Center for Disaster Reduction (NCDR) [6], if the rainfall is more than 130 mm/h, roads in the park will be flooded for 1–3 days (blue patterns showing 0.0–0.3 m flooding and green patterns showing 0.3–1.0 m flooding), as shown in Figure 3. The dome has the best fire performance. If there is a fire in the cinema, shopping mall, or office building, disaster relief assistance will be needed. Based on the political situation between Taiwan and China and the video "No Respite", which named Taiwan as a member of the global anti-ISIS and was released by the Islamic State of Iraq and al-Sham (ISIS) terrorist organization on 25 November 2015, the occurrence probability of a terrorist attack in Taiwan is "perhaps". If the terrorist attack happens, in addition to requiring large amounts of emergency medical resources, it could also create international tensions.

Table 1. Disaster-causing natural disasters on the Taipei Dome Complex.

Item \ Disaster	Earthquake	Flood	Fire	Terrorist Attack
Probability	Possible	Very likely	Possible	Perhaps
Periodicity	No	Yes	No	No
Potential Impact	Limited	Negligible	Serious	Catastrophic

Table 1. *Cont.*

Item \ Disaster	Earthquake	Flood	Fire	Terrorist Attack
Rate of Outbreak	shortest time or too late to have early warning	12–24 h early warning	shortest time or too late to have early warning	shortest time or too late to have early warning
Possibility of Cascading	Yes, might have post-earthquake fire	If the rainfall is more than 130 mm/h some roads in the park will be flooded for 1–3 days	Yes, property damage and casualties	Yes, property damage, casualties, and international disputes

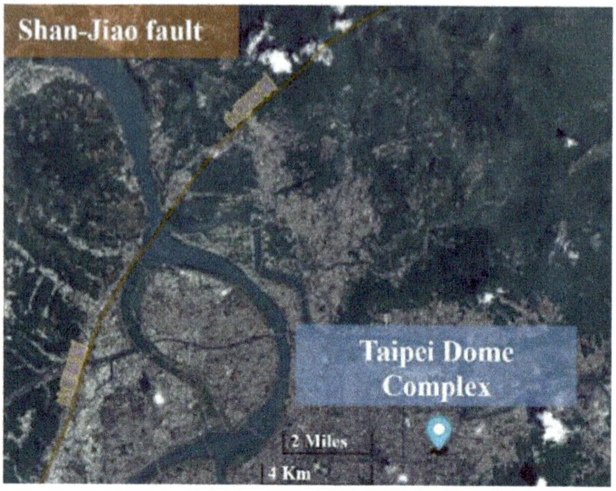

Figure 2. Adjacent fault distribution.

Table 2. Earthquake simulation of Taipei City.

Event	Shan-Jiao Fault
Geographical Coordinates	121.589 25.139
Richter Magnitude Scale	6.9
Depth (Km)	8
PGA (cm/s^2)	411
Simulation system	TELES
Number of collapsed Houses (Xinyi District/Taipei City)	341/12,301
Number of Resuscitation Injuries and Deaths (Xinyi District/Taipei City)	66/2282

2.2. Potential Disaster Analysis

The case of explosions is explored in Manchester Arena, 2017. For disaster relief, human resource deployment for emergency response work is important. The police, emergency services, and ambulance staff entered the building soon after the blast [8]. Conversely, Greater Manchester Fire and Rescue Service crews attended the arena more than two hours and 15 min after the first 999 calls [9], and only three North West Ambulance Service (NWAS) paramedics went into the foyer and triaged and assessed patients [10]. The emergency services showed poor communication across departments and not shared communication channel was established between emergency services [11].

Figure 3. Taipei city flood simulation 130 mm/h.

Medical evacuation measures were inadequate. Police officers decided which patients to take out first, but NWAS lost control of the decision on when to bring patients out. Thus, casualties did not arrive at the hospital [10]. After the rescue operation, expert pointed out that the wounds taken out of the bomb scene were unsafe and unnecessary. More specialist paramedics from the response team could have been sent directly to the bomb scene [10].

Developing and implementing a security plan is important. The Pennsylvania-based company that manages the Manchester Arena described that backpacks were not allowed, and drinks were taken away from people. People had to go through strict security to enter the arena [10]. However, the audience said that there were no security measures as people entered the arena—no wanding, no scanning, and insufficient purse and bag checks [12]. If the metal detector had worked, the explosion might not have happened.

For crime scene handling, crowd evacuation guidance is necessary. A wide area around the venue was quickly cordoned off [13] and a bomb disposal team arrived on the scene [1]. Manchester Victoria station near the arena was evacuated and trains were canceled [2]. Taxis swarmed the area to help take people away [13]. Near the end of disaster relief, the emergency placement of victims was necessary. Shops and takeaways around the arena appeared to be handing out drinks [13]. Hotels opened their doors to concertgoers trapped inside the police cordon, providing them with drinks and phone chargers to enable them to contact family members. Residents also offered stranded concertgoers places to stay in their homes [1].

2.3. Related Articles

For advanced stadium security systems, Wan et al. (2021) proposed the AI-CPS model that can predict anomaly behavior in cyber networks [14]. Facing unconventional emergencies, Ni et al. (2020) created a decision-making repository that contained existing emergency response plans (ERPs) that include many emergency response experiences. Considering the semantic relevance and the scenario consistency of ERP segments, applicable sections form new ERPs rapidly [15]. To maximize the benefits of human resources, Granberg et al. (2020) considered training and equipping human resources from other public service sectors to act as first responders to reduce first response times at a low cost [16]. In order to autonomously deploy and provide communication services in disaster sites, Sanchez-Garcia et al. (2019) proposed an algorithm based on the well-known particle swarm optimization algorithm applied in UAV networks to discover the scenario victims faster [17]. In terms of achieving an efficient prehospital emergency response, the collaborative utilization of regional emergency medical services improves the surge capacity in the field and the practices in field triage and transport capacity, and a transfer strategy can be used in mass burn casualty incidents [18].

3. Establishment of Taipei Dome Complex Disaster Risk Factors

We explore risk factors for disasters and safety events through the analysis of potential disasters, relevant paper references, and cases of terrorist attacks in the arena. The key factors affecting the security of the Taipei Dome Complex are summarized into five aspects: equipment safety, traffic accidents, human resources, security management, and disaster event. The security risk factor structure is shown in Figure 4. Equipment safety includes insufficient fire facilities, incomplete barrier-free facilities, poorly maintained machinery, insufficient surveillance facilities, and insufficient cyber security. Traffic accident includes MRT emergencies, bus emergencies, intercity bus emergencies, and tour bus emergencies tha5t affect the crowd in the complex. Human resources includes insufficient firefighting, insufficient police, insufficient security officer, insufficient medical personnel, and insufficient staff member. The security management includes shortcomings in a moving line, shortcomings in evacuation guidance, unclear escape or fire facilities signage, lack of situational exercise, and lack of staff training. The disaster event includes earthquake, fire terrorist attack, and demonstration/protest event.

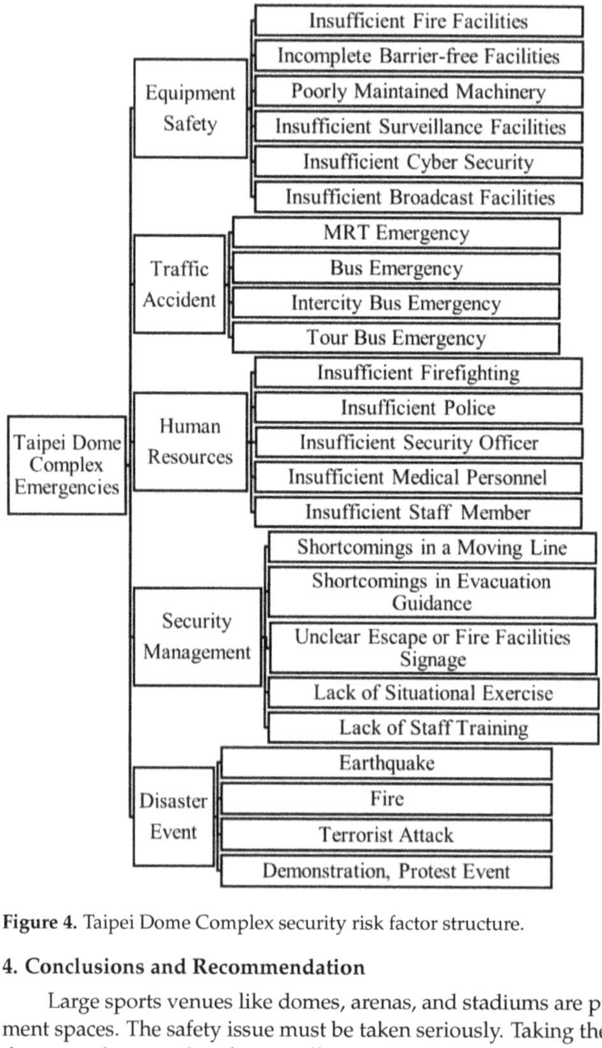

Figure 4. Taipei Dome Complex security risk factor structure.

4. Conclusions and Recommendation

Large sports venues like domes, arenas, and stadiums are public leisure and entertainment spaces. The safety issue must be taken seriously. Taking the Taipei Dome Complex as the research target, key factors affecting the security of the complex are summarized into five aspects: equipment safety, traffic accidents, human resources, security management, and disaster event. In the Taipei Dome Complex, among the baseball dome, shopping mall, restaurant, cinema, and office building, the baseball dome has the best seismic performance and fire performance. To respond to disaster events, emergency response plans must be established, and educational training and situational exercises must be adequate. For terrorist attack prevention, organizers should takes the following measures: tour buses and large transport vehicles must not be allowed to enter the park from two hours before activities start until one hour after activities end to avoid collisions with the crowd. X-ray machines must be installed at every entrance to the dome to detect dangerous goods. From audience entry until the event starts, these should be regularly monitored in the computer room. Disabled toilets, parking lots, and other places are difficult to observe. In terms of preventing suspicious persons from entering the regulated area, it is better to improve identification measures for staff. In Taiwan, no law specifies a hospital specifically responsible for receiving victims of terrorist attacks. The government should develop policies for emergency medical care.

Author Contributions: Conceptualization, C.-J.H.; methodology, C.-J.H.; software, T.-Y.C.; validation, C.-J.H. and T.-Y.C.; formal analysis, T.-Y.C.; investigation, C.-J.H.; resources, C.-J.H.; data curation, T.-Y.C.; writing—original draft preparation, T.-Y.C.; writing—review and editing, C.-J.H.; visualization, T.-Y.C.; supervision, C.-J.H.; project administration, C.-J.H. All authors have read and agreed to the published version of the manuscript.

Funding: This research received no external funding.

Institutional Review Board Statement: Not applicable.

Informed Consent Statement: Not applicable.

Data Availability Statement: Data is contained within the article.

Acknowledgments: The authors do not accept any financial support. Sincerely thanks to all of the researchers for their contributions.

Conflicts of Interest: The authors declare no conflict of interest.

References

1. Ariana Grande Manchester Concert Ends in Explosion, Panic and Death. *The New York Times*, 22 May 2017. Available online: https://www.nytimes.com/2017/05/22/world/europe/ariana-grande-manchester-police.html(accessed on 11 December 2021).
2. Isis Claim Responsibility for the Manchester Attack That Killed 22. *METRO News*, 23 May 2017. Available online: https://metro.co.uk/2017/05/23/isis-claim-responsibility-for-the-manchester-attack-that-killed-22-6655764/(accessed on 11 December 2021).
3. Found at the Scene in Manchester: Shrapnel, a Backpack and a Battery. *The New York Times*, 24 May 2017. Available online: https://www.nytimes.com/interactive/2017/05/24/world/europe/manchester-arena-bomb-materials-photos.html (accessed on 11 December 2021).
4. Zhao, H. Interview with On-Site Personnel: Stop Lying, the Truth about the World Universiade's Anti-Annual Reform Incident Has Gone through. *SHOCUZINE*, 21 August 2017. Available online: https://www.shokuzine.com/2017/08/21/%E7%8F%BE%E5%A0%B4%E4%BA%BA%E5%93%A1%E5%B0%88%E8%A8%AA%EF%BC%9A%E5%81%9C%E6%AD%A2%E8%AC%8A%E8%A8%80%EF%BC%8C%E4%B8%96%E5%A4%A7%E9%81%8B%E5%8F%8D%E5%B9%B4%E6%94%B9%E4%BA%8B%E4%BB%B6%E7%9A%84%E7%9C%9F/(accessed on 28 November 2021).
5. Taipei City Council Bulletin, Vol. 115, Issue 10, Page 1266. Available online: https://perma.cc/Z5VG-UNAH (accessed on 24 January 2022).
6. Disaster Map Website, National Science and Technology Center for Disaster Reduction. Available online: https://dmap.ncdr.nat.gov.tw/1109/map/# (accessed on 10 December 2021).
7. *Taipei City Regional Disaster Prevention and Response Plan*; Taipei City Fire Department: Taipei, Taiwan, 2021.
8. Manchester Arena Attack: Eyewitnesses Describe Blast Horror. *BBC News*, 23 May 2017. Available online: https://www.bbc.com/news/entertainment-arts-40008026(accessed on 10 December 2021).
9. Manchester Arena Inquiry: Senior Fire Officer Ashamed of Bomb Response. *BBC News*, 24 June 2021. Available online: https://www.bbc.co.uk/news/uk-england-manchester-57601517(accessed on 10 December 2021).
10. Manchester Arena Inquiry: Families Call Ambulance Response 'Inadequate'. *BBC News*, 15 December 2021. Available online: https://www.bbc.com/news/uk-england-manchester-59669701(accessed on 10 December 2021).

11. Manchester Arena Inquiry: Greater Manchester Police Admits Mistakes. *BBC News*, 16 December 2021. Available online: https://www.bbc.com/news/uk-england-manchester-59684115(accessed on 10 December 2021).
12. Manchester Arena Claimed Control of Area Where Nail Bomb Exploded. *TMZ News*, 23 May 2017. Available online: https://www.tmz.com/2017/05/23/manchester-arena-bomb-lawsuits/(accessed on 10 December 2021).
13. Manchester Arena Blast: Confusion and Chaos after Explosion at Concert. *BBC News*, 23 May 2017. Available online: https://www.bbc.com/news/uk-england-manchester-40008386(accessed on 10 December 2021).
14. Wan, B.; Xu, C.; Mahapatra, R.P.; Selvaraj, P. Understanding the cyber-physical system in international stadiums for security in the network from cyber-attacks and adversaries using AI. *Wirel. Pers. Commun.* **2022**, *127*, 1207–1224. [CrossRef]
15. Ni, W.; Liu, T.; Chen, S. Rapid generation of emergency response plans for unconventional emergencies. *IEEE Access* **2020**, *8*, 181036–181048. [CrossRef]
16. Granberg, T.A.; Pilemalm, S. Evaluation of new first response initiatives in emergency services. *Socio-Econ. Plan. Sci.* **2020**, *71*, 100829. [CrossRef]
17. Sánchez-García, J.; Reina, D.G.; Toral, S.L. A distributed PSO-based exploration algorithm for a UAV network assisting a disaster scenario. *Future Gener. Comput. Syst.* **2019**, *90*, 129–148. [CrossRef]
18. Lin, C.H.; Lin, C.H.; Tai, C.Y.; Lin, Y.Y.; Shih, F.F. Challenges of burn mass casualty incidents in the prehospital setting: Lessons from the Formosa fun coast park color party. *Prehospital Emerg. Care* **2019**, *23*, 44–48. [CrossRef] [PubMed]

Disclaimer/Publisher's Note: The statements, opinions and data contained in all publications are solely those of the individual author(s) and contributor(s) and not of MDPI and/or the editor(s). MDPI and/or the editor(s) disclaim responsibility for any injury to people or property resulting from any ideas, methods, instructions or products referred to in the content.

engineering proceedings

MDPI

Proceeding Paper

Prison Disaster Factors: A Case Study of Taipei Prison †

Chi-Jan Huang [1], Ting-Yi Chiang [2,*] and Wun-Wu Chen [3]

1 General Education Center, Ming Chuan University, Taipei 111013, Taiwan; cjhuang@mail.mcu.edu.tw
2 Department of Computer Science and Engineering, Tatung University, Taipei 104327, Taiwan
3 Department of Criminal Justice, Ming Chuan University, Taipei 111013, Taiwan; s0956268677@gmail.com
* Correspondence: tychiang824@gmail.com

† Presented at the 3rd IEEE International Conference on Electronic Communications, Internet of Things and Big Data Conference 2023, Taichung, Taiwan, 14–16 April 2023.

Abstract: Prisons have always been considered self-sufficient, and government disaster response plans at all levels rarely mention prisons. Prisons may face emergencies such as earthquakes, floods, fires, prison escapes, or riots. Prisons are located in various disaster potential areas. If prepared, the safety of prison inmates can be secured. If it is not handled properly, society can be threatened. Through a literature review, TELES and SESS earthquake loss estimation system, and other methods, we sorted out three disaster risk factors, personnel risk, equipment risk, and management risk. In the safety part of facilities, old buildings, old prison walls, insufficient monitoring facilities, and insufficient prison space need to be included. The management aspect includes the potential of adjacent disasters. Insufficient regional and disaster prevention materials, medical materials, and connections with surrounding resources need to be solved.

Keywords: prison; TELES; SESS; disaster; risk

Citation: Huang, C.-J.; Chiang, T.-Y.; Chen, W.-W. Prison Disaster Factors: A Case Study of Taipei Prison. *Eng. Proc.* **2023**, *38*, 27. https://doi.org/10.3390/engproc2023038027

Academic Editors: Teen-Hang Meen, Hsin-Hung Lin and Cheng-Fu Yang

Published: 25 June 2023

Copyright: © 2023 by the authors. Licensee MDPI, Basel, Switzerland. This article is an open access article distributed under the terms and conditions of the Creative Commons Attribution (CC BY) license (https://creativecommons.org/licenses/by/4.0/).

1. Introduction

In recent decades, the global prison population has increased significantly. At present, it has exceeded 11 million. More than 60% of the national correction systems in the world are overcrowded. Due to the sharp increase in the number of inmates, they do not receive enough corresponding resources, resulting in overcrowding and an unhealthy environment in correction institutions, which affects the physical and mental health of staff and detainees (Heard, 2019) [1]. At the same time, it also creates a hotbed of infectious diseases. Overcrowding in correctional institutions has always been a problem faced by countries all over the world. Fifty-one correctional institutions affiliated with the correctional department of the Ministry of Justice accommodate more than 54,000 people (the agency of corrections, 2022) [2]. Taking Taipei Prison as an example, the number of inmates is 3955, of which 3401 are approved, 554 are overcharged, and the overcharge rate is 16.3%. There is the Shanchiao fault and Hukou fault nearby the prisons. When an earthquake occurs, it causes a large number of casualties without disaster management.

Prisons have always been regarded as self-sufficient. The disaster response plans of governments at all levels rarely mention prisons. They are often ignored or forgotten in the whole disaster relief process because prisons are not included in the list of disaster work items of the county and municipal governments. If prisons do not have proper disaster response and post-disaster measures, great damage may occur. According to the statistics of the COVID prison project in the United States, 486,585 people were diagnosed with COVID-19 since 14 January 2022. A total of 156,500 prison staff were diagnosed, and 259 died (the COVID prison project, 2022) [3].

Prisons may also face emergencies such as earthquakes, floods, fires, and prison escapes. Prisons are located in various potential areas of disasters. If they are prepared, they can ensure the safety of prison inmates. If they are not, the disaster of prisons threatens society. Adenrele Awotona summarizes the main reasons for ineffective disaster relief in past prison disaster cases in the United States as follows: lack of communication among departments, the staff's inexperience in education, training, and drilling of disaster relief, no supporting contingency plan or unclear plan content and wrong decision-making, improper resettling of sexual offenders, and inappropriate management of prison inmates in disasters [4].

2. Literature Review

The Taiwan Correctional Service was established in 2011 with a long-term shortage of manpower and funds. There is no sufficient manpower in correctional institutions. Therefore, the operation of the institutions is less efficient and even affects the morale of the staff. It has become normal for institutions to have a serious imbalance in the proportion of manpower and workload. The heavy burden and the shortage of manpower make correction work difficult. Therefore, Taiwan's correction institutions are focusing on warehouse management. For maintaining the basic operation of the institutions, the staff is exhausted, which prohibits the personalized treatment of the detainees. During day and night, the supervisor on duty has to face hundreds of inmates alone. The supervisor can only take care of the inmates in special conditions. The overall atmosphere is negative. This attitude undoubtedly affects correction work.

Taipei Prison is located in Guishan District, Taoyuan City. It was established in 1963. In 2017, the Zhishan Building was officially opened as the first high-rise prison in Taiwan. A total of 146 rooms have the capacity to house 1300 people for the Northern Prison (Freedom Times, 2017) [5]. Although there is a new building, each inmate is only allowed to have less than 0.211 m^2. Currently, the number of inmates in Taipei Prison is 3955, of which the approved capacity of inmates is 3401, and the number of overcharged is 554. The overcharge rate is 16.3%. Prisoners between the ages of 40–50 account for 29% (1147), followed by those between the ages of 30–40 (946, 23.9%). A total of 857 are between the ages of 20–30, accounting for 21.7%. A total of 364 are over 60 years old, accounting for 9.2%. The care of the elderly becomes a new problem in future prisons.

We use the FEMA disaster analysis table for fires, landslides, floods, and earthquakes. The analysis result shows that earthquakes are most likely to be large-scale disasters, and fire is low. As shown in Table 1, floods and landslides have obvious cycles, and prevention work can be prepared in advance. The impact time is short, and the impact of serial consequences is not significant. Using the flooding simulation of the National Disaster Prevention and Rescue Technology Center, as shown in Figure 1, the potential flooding area with 300 mm in 12 h is not located near the Taipei Prison. The landslide simulation shows that none of the Taipei prisons are affected (Figure 2).

Table 1. Disaster-causing natural disasters on the Taipei Prison.

	Fire	**Flood**	**Landslide**	**Earthquake**
Probability	Possible	Very likely	Very likely	Possible
Periodicity	No.	Yes, flood control duration (April–October)	Yes, flood control duration (April–October)	Yes, 60 years one cycle
Potential impact	limited	limited	limited	limited
Rate of outbreak	shortest time or too late to have early warning	12–24 h early warning	over 24 h early warning	shortest time or too late to have early warning
Possibility of cascading	None	Yes, landslide, flood	None	Yes, aftershock, hillside collapse, fire

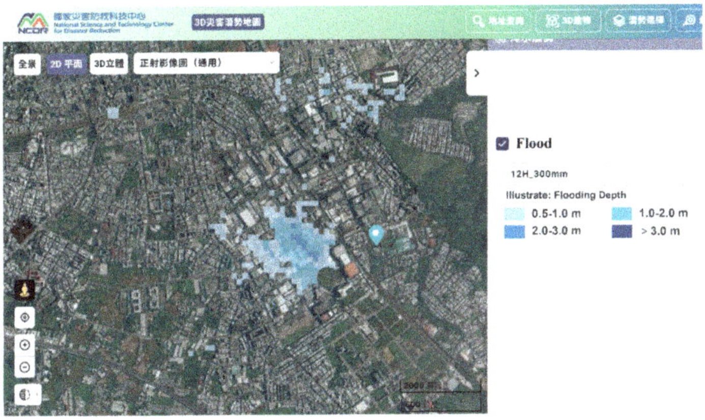

Figure 1. 12 h 300 mm potential inundation map of the Taipei Prison.

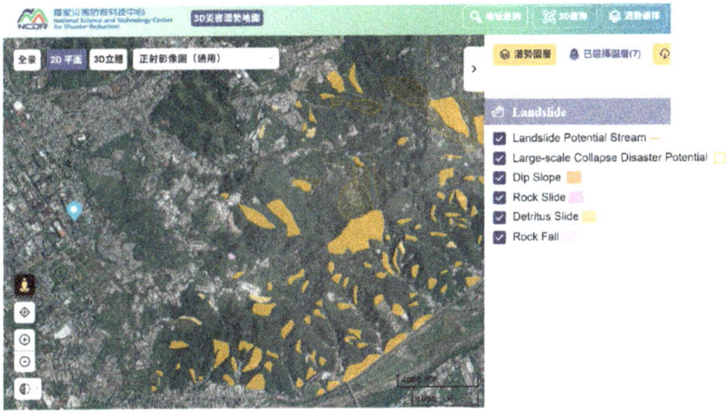

Figure 2. Landslide potential map of the Taipei Prison.

Fires and earthquakes are least likely to occur, but their potential impact is the most serious (Table 1). Although the earthquake can be detected in advance, the onset time is short. Thus, it may be too late to give an early warning. After the earthquake, the cascade consequences are the most serious, including aftershocks, slope collapse, and fire. The location map of the active faults of the National Disaster Prevention and Rescue Science and Technology Center shows that the Hukou fault passes through Taoyuan City without damage. Although the Shanchaio fault does not pass through Taoyuan City (Figure 3), the location is close to where the Taipei Prison is located.

We use the Simple Earthquake Seismic System (SESS), a simple earthquake disaster loss estimation system developed by the Fire Department of the Ministry of the Interior. The system mainly uses the Japanese earthquake experience to compare various elements such as the earthquake occurrence time and related situations. It is easy to complete the disaster simulation in a short time, as the seismic value is the maximum surface acceleration (PGA). When an earthquake occurs, the earthquake information (epicenter and scale) can be obtained from the Bureau of Meteorology. After inputting the point source, the earthquake value (PGA) of each region is obtained. The system estimates the collapse of buildings related to residents' lives and properties, the number of casualties, fires after the earthquake, sheltering and evacuation of people, and other four basic disasters. In terms of calculation, it takes about 1 to 2 min speed to estimate the damage to a county or city. The Taiwan Earthquake Loss Estimation System (hereinafter referred to as TELES) developed by the National Earthquake Center to simulate loss is also used to evaluate the potential danger of earthquakes and disasters and is used for government disaster relief and risk assessment and management. We simulate the fault zones around the selected cases to confirm the damage to the houses around the prison and estimate the casualties around the prison [6].

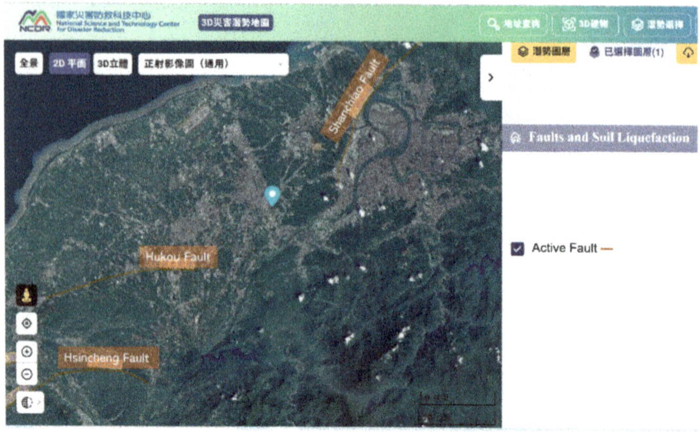

Figure 3. Faults potential map of the Taipei Prison.

According to the simulation results of TELES and SESS, the disaster damage in Taoyuan City is shown in Table 2. The Hukou fault passes through Taoyuan city and belongs to the second type of active fault. It was active 100,000 to 10,000 years ago. Although it belongs to the second type of active fault, the probability of its occurrence is not lower than that of the first type of active fault. (Active in the past 10,000 years) it means that it is still active. Therefore, once the Hukou fault is dislocated, according to the TELES simulation in Table 2, the number of buildings collapsed in Taoyuan City is as high as 14,134, which is the highest among the four faults. In the SESS simulation, the Hukou fault may cause 798 collapses of buildings in Taoyuan City. Although the Shanchiao fault does not pass through Taoyuan City, it is close to Taoyuan near the Guishan District where Taipei Prison is located. According to the TELES simulation in Table 2, the number of buildings that collapse in Taoyuan City reaches 12,495, ranking second among the four faults. In the SESS simulation of the Shanchiao fault, the number of collapsed buildings in Taoyuan City was 1229, which is the highest of the four faults. The Hsinchu fault and Xincheng fault do not pass through Taoyuan City, but it is not near Taoyuan. Therefore, these two faults have little impact on Taoyuan City, and the number of collapsed buildings is zero.

Table 2. Simulation results of earthquake casualties in Taoyuan City event.

	Hukou Fault		Shanchiao Fault		Hsinchu Fault		Hsincheng Fault	
Richter Magnitude Scale	6.6		6.9		6.8		6.8	
Geographical Coordinates	121.127 24.8962		121.415 25.007		120.9588 24.7953		121.019 24.7723	
Depth (Km)	10		10		10		10	
Simulation system	TELES	SESS	TELES	SESS	TELES	SESS	TELES	SESS
Number of Collapsed Houses	14,134	798	2930	73	2930	73	7294	73
Number of Injured	895	633	131	0	131	0	407	0
Number of Deaths	733	2694	107	0	107	0	328	0

3. Establishment of Prison Disaster Risk Factors

Purdum et al. discussed the impact of disasters on inmates and the prison destruction to surrounding communities [7]. He believed that the impact of disasters and the costs of disaster reduction, preparation, response, and recovery from these events are rising. Inmates in prisons are a source of low-cost labor for various tasks before, during, and after the disaster. However, states plan to use inmate labor to meet emergency management, increase urban resilience, and reduce labor costs [8]. Loic et al. discussed the views of New Zealand prisons on disaster risk reduction and emergency management. They explored prison exposure and identified the vulnerability and capacity of detainees and documented the policies and practices of disaster risk reduction and emergency management. It showed that raising awareness at the national level and improving disaster risk reduction policies and emergency management initiatives were not sufficient enough to address the increased risks and the special vulnerability of the rapid growth of inmates. The inherent abilities of these inmates have not been developed. They further highlighted the serious lack of communication in the implementation interface of prison management, disaster reduction, and emergency management in New Zealand [9]. Others proposed the fate of prisons and detainees in disasters in 2012, which attracted very limited attention from scholars and policymakers. However, both prisons and detainees are particularly affected by catastrophic events related to external natural and other disasters. This study focused on three aspects to explore the relationship between imprisonment, marginalization, and vulnerability to natural disasters and other factors: (1) a spatial form of marginalization, through the geography, potential danger, location and secluded nature of the prison, as well as helping to face the danger, (2) a form of social marginalization that deprives inmates who are already poor of further access to economic opportunities, medical resources, and interpersonal relationships, which have proved to be key drivers of people's vulnerability, and (3) a form of political marginalization, including the limited resources provided by the state and its institutions which leads to poor facilities, possible vulnerability and overcrowding, and the lack of visibility of prisons in government policies [10]. Smog is the main cause of toxic substances in Western countries. Meroueh et al. found that there were many tragic fires in prisons. More than 380 inmates died in Honduras in 2012, and 136 inmates died in the Dominican Republic in 2005. Because of the particularity of prisons, fires easily cause a large number of deaths [11].

The purpose of this study is to explore the risk factors of prisons when facing disasters. Through secondary data collection, relevant paper references, and cases, the key factors affecting risk management are summarized. In addition, the earthquake loss estimation system, TELES, and SEES are used to analyze and simulate the number of total and half-collapsed buildings and casualties near a prison due to earthquake disasters. The possible risk factors are summarized and classified. First of all, the personnel structure includes insufficient education and training, an excessive number of detainees, a lack of response system or imperfect system, and insufficient management personnel. The facility safety includes old buildings, old prison exterior walls, insufficient monitoring facilities, and insufficient prison space. The management structure includes adjacent disaster potential areas, and insufficient preparation of disaster prevention materials. The medical materials are not well prepared and the contact with surrounding resources is not enough. The detailed structure is shown in Figure 4.

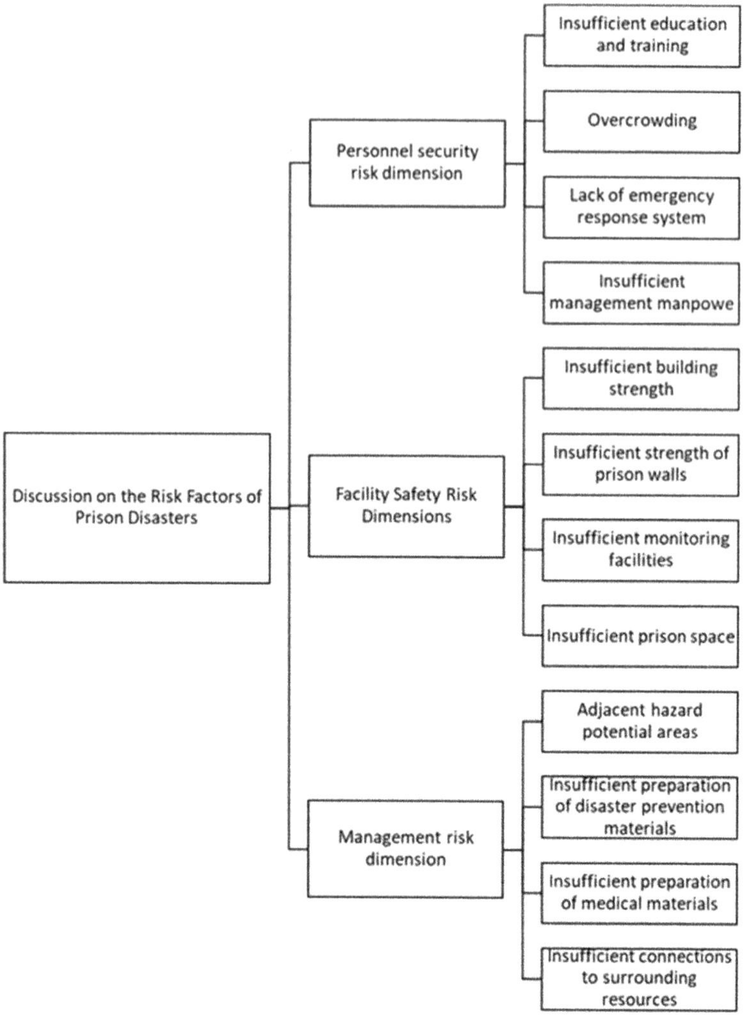

Figure 4. Prison disaster risk factor structure.

4. Conclusions

Prison issues are rarely discussed, especially when prisons face the risk of disasters. We proposed three risks: personnel risk, facility risk, and management risk. The personnel structure includes insufficient education and training, an excessive number of detainees, a lack of response system or imperfect system, and insufficient management personnel. The facility safety includes old buildings, old prison exterior walls, monitoring facilities, and insufficient prison space. The management structure includes the adjacent disaster potential area, the insufficient preparation of disaster prevention materials, the insufficient preparation of medical materials, and the insufficient connection with surrounding resources.

According to the earthquake simulation, both the Hukou fault and the Shanchaio fault may affect Taipei Prison. Therefore, special attention needs to be paid to the planning of earthquake disasters. In a flood, the terrain of Taipei Prison is relatively high, so the impact of flood disasters is limited. There are many types of inmates in the prison. Special consideration needs to be given to the emergency resettlement of inmates of sexual violence, or independent resettlement. Otherwise, it is likely to lead to other criminal acts. Post-earthquake fires or general fires are also the most feared type of prison disaster because prison space is limited and controlled. If it is not handled immediately, the inmates are likely to be hurt and die due to thick smoke. Therefore, it is suggested that fire zoning and smoke exhaust equipment should also be planned for indoor spaces.

In Taiwan, the disaster prevention and rescue law, the basic disaster prevention and rescue plan, and the regional disaster prevention and rescue plan do not include notifying or evacuating prisoners. Therefore, in the future, Taiwan needs to consider prisons to increase the capacity of disaster relief and reduce the impact of disasters.

Author Contributions: Conceptualization, C.-J.H.; methodology, C.-J.H.; software, T.-Y.C. and W.-W.C.; validation, C.-J.H.; formal analysis, T.-Y.C. and W.-W.C.; investigation, C.-J.H., T.-Y.C. and W.-W.C.; resources, C.-J.H.; data curation, T.-Y.C.; writing—original draft preparation, T.-Y.C.; writing—review and editing, C.-J.H.; visualization, T.-Y.C.; supervision, C.-J.H.; project administration, C.-J.H.; funding acquisition, C.-J.H. All authors have read and agreed to the published version of the manuscript.

Funding: This research was funded by Ministry of Science and Technology, Taiwan, MOST 110-2625-M-130-001.

Institutional Review Board Statement: Not applicable.

Informed Consent Statement: Not applicable.

Data Availability Statement: Data is contained within the article.

Acknowledgments: The authors are indebted to Ministry of Science and Technology (MOST 110-2625-M-130-001), Taipei, Taiwan, Republic of China, for financial support.

Conflicts of Interest: The authors declare no conflict of interest.

References

1. HEARD, Catherine. Towards a health-informed approach to penal reform. In *Evidence from Ten Countries*; Institute for Crime & Justice Policy Research: London, UK, 2019.
2. Agency of Corrections, Ministry of Justice, Taiwan. 2022. Available online: https://www.mjac.moj.gov.tw/4786/4923/4927/4931/650115/post (accessed on 10 January 2022).
3. The Covid Prison Project. 2022. Available online: https://covidprisonproject.com/ (accessed on 10 January 2022).
4. Awotona, A. (Ed.) *Planning for Community-Based Disaster Resilience Worldwide: Learning from Case Studies in Six Continents*; Taylor & Francis: Abingdon, UK, 2016.
5. "One Person, One Bed!" Taiwan's First High-Rise Prison Completed, with 1300 More Inmates. *Liberty Times Net NEWS*. October 2017. Available online: https://news.ltn.com.tw/news/society/breakingnews/2211045 (accessed on 21 November 2011).
6. Huang, C.J. Research on Fire Rescue and Emergency Medical Planning Considering Risk after Earthquake. Ph.D. Thesis, Institute of Engineering Science and Technology, National Taipei University of Technology,, Taipei, Taiwan, 2010.
7. Purdum, C.; Henry, F.; Rucker, S.; Williams, D.A.; Thomas, R.; Dixon, B.; Jacobs, F. No justice, no resilience: Prison abolition as disaster mitigation in an era of climate change. *Environ. Justice* **2021**, *14*, 418–425. [CrossRef]
8. Purdum, J.C.; Meyer, M.A. Prisoner labor throughout the life cycle of disasters. *Risk Hazards Crisis Public Policy* **2020**, *11*, 296–319. [CrossRef]
9. Le Dé, L.; Gaillard, J.C. Disaster risk reduction and emergency management in prison: A scoping study from New Zealand. *J. Contingencies Crisis Manag.* **2017**, *25*, 376–381. [CrossRef]
10. Gaillard, J.C.; Navizet, F. Prisons, prisoners and disaster. *Int. J. Disaster Risk Reduct.* **2012**, *1*, 33–43. [CrossRef]
11. Meroueh, F.; Lardeur, J.Y.; Lapostolle, F. Prisons are at highest risk of death by fire and smoke exposure. *Injury* **2017**, *48*, 791. [CrossRef] [PubMed]

Disclaimer/Publisher's Note: The statements, opinions and data contained in all publications are solely those of the individual author(s) and contributor(s) and not of MDPI and/or the editor(s). MDPI and/or the editor(s) disclaim responsibility for any injury to people or property resulting from any ideas, methods, instructions or products referred to in the content.

engineering proceedings

Proceeding Paper

Study of Building Sense of Place in Network World †

Shih-Jen Feng and Si-Ting Li *

Department of Architecture, Chaoyang University of Technology, Taichung 41349, Taiwan; fengshyhren@yahoo.com.tw
* Correspondence: sandy880327@gmail.com; Tel.: +886-972839714

† Presented at the 3rd IEEE International Conference on Electronic Communications, Internet of Things and Big Data Conference 2023, Taichung, Taiwan, 14–16 April 2023.

Abstract: If we look back at the sense of place and deconstruct this concept, it is a combination of "space" and "meaning", and cyberspace may be developed as a "place" that is different from the real world, being unique and charming and more desirable. The purpose of this study is to discuss the unexpected existing space hidden in the Internet under the premise of interlacing life experiences and the Internet. We analyze the relationship between social media-mediated user emotions and cyberspace through a relevant literature review and a questionnaire survey, and summarize the possible factors that influence the relationship between cyberspace and users. The results of the analysis provide suggestions for future designers to consider when creating spaces in the online world and directions for subsequent researchers.

Keywords: sense of place; space; globalization; cyberspace; internet

Citation: Feng, S.-J.; Li, S.-T. Study of Building Sense of Place in Network World. *Eng. Proc.* **2023**, *38*, 60. https://doi.org/10.3390/engproc2023038060

Academic Editors: Teen-Hang Meen, Hsin-Hung Lin and Cheng-Fu Yang

Published: 28 June 2023

Copyright: © 2023 by the authors. Licensee MDPI, Basel, Switzerland. This article is an open access article distributed under the terms and conditions of the Creative Commons Attribution (CC BY) license (https://creativecommons.org/licenses/by/4.0/).

1. Introduction

In today's globalized society, the Internet is a form of a social media with great influence, showing all human beings a different kind of life than that previously imagined. In addition to the world in which we live, by connecting to the Internet through 3C electronic products, such as cell phones, computers, and tablets, we can easily bring our consciousness across time and space to another level of the Internet world, anytime and anywhere, forming a global village in which the whole world lives together. In this regard, Relph suggested that "globalization will transform many places into place-less environments" [1] as a negative effect, while Massey stated that "local specificity will be regenerated continuously" [2] as a positive point.

Given Tuan's thought that "space can become place by giving meaning" [3] and Harrison and Dourish's equation of place, i.e., "place = space + meaning" [4], whether the old theories can also be reflected in the era in which the Internet is inseparable from daily life is an important issue to investigate. We face problems in the age of the Internet, given its inseparability from everyday life. In addition to the emotional deprivation of original space and place, which leads to the proliferation of a sense of placelessness, what lies behind the phenomenon of disappearance may be the creation of a different space. If the specificity of place is not merely repeatedly regenerated in the same place, the possibility of transferring it to cyberspace can be assumed.

A summary report of the broadband usage survey commissioned by the National Communications Commission and conducted by the Taiwan Institute for Economic Research states that "By age, the total number of hours per week that people use social media to browse/read/comment/like/post broadly decreases with age, with 20.9 h being the most for those aged 16–25 and 9.52 h being the least for those aged 66 and younger." [5].

In another report, "Digital Opportunity Survey of Mobile Phone Holders 108", which was also commissioned by the National Communications Commission and conducted by United Marketing Research Corporation (Taipei, Taiwan), states that "The age structure of cell phone users is also stabilizing, with the phenomenon of a significant increase in the elderly population in 107 no longer seen, with 7.0% of cell phone users under the age of 20, 18.3% of cell phone users aged 20–29, and 19.9% of mobile phone users aged 30–39. The cell phone population aged under 20 accounted for 7.0%, 18.3% aged 20–29, 19.9% aged 30–39, 18.7% aged 40–49, 16.5% aged 50–59, and 19.6% aged over 60." [6] It is clear that cell phone users are mainly in the 20–39-year-old age group, and the total number of hours spent on the Internet is highest in of 16–25-year-old the age group, while the second survey report points out that "the cell phone users' mobile Internet access rate is 89.8%, and 31.2% rely solely on cell phones to access the Internet at home" [6]. This evidence shows that cell phones have become the mainstream Internet access tool. To avoid interference from external factors, the target population of this study is confined to college students, who are the largest group in this age group.

2. Literature Review

2.1. Sense of Place

In the book "Sense of Place: Experience, Memory and Imagination of Environmental Space", it is mentioned that the sense of place is composed of five sensory methods: spatial sense, temporal sense, the five senses, awareness, and sensory memory (Table 1). The awareness and sensory memory that human beings have of space forms a certain experience, and that experience has the opportunity to transform space into place, as well as the experience itself into a sense of place [7]. Tuan suggested that experience is a collection of thoughts (mind) and senses [8]. Manzo found that it is the experience-in-place that gives meaning to place [9]. Jorgensen and Stedman argued that the term sense of place contains three components—place identity, place attachment, and place dependence—which correspond to the cognitive, affective, and behavioral domains, respectively [10].

Table 1. Method of forming a sense of place.

Feeling	Sensory Methods	
	Content	Effects
Sense of space	1. Subjective/objective 2. Real/abstract 3. Near/far	Providing stimulation
Sense of time	1. Venue 2. Viewpoint 3. State	Providing stimulation
Five senses	1. Sight 2. Hearing 3. Smell 4. Touch 5. Taste	Receiving stimuli
Awareness, Sensory memory	Memories are formed and feelings are accumulated	Giving meaning

There are many terms related to the sense of place, such as place attachment, place meaning, place dependence, place identity, place estrangement, and place lessness. The reasons for this phenomenon are related to the formation of the sense of place. Since everyone's experience and intentions are different, there are differences in the way they describe their sense of place. Even so, the core of the sense of place cannot be separated from "human emotion about space" and "human-place relationship", and the emotion that transforms space into place, no matter what kind of experience it comes from, can be called sense of place.

Shamai suggested three stages of sense of place, ranging from weak to strong: belonging to a place, attachment to a place, and commitment to a place. Each stage can be further subdivided into two levels, and if a sense of placelessness is added, it becomes a continuous seven-level hierarchy (Table 2) [11].

Table 2. Sense of place level.

Level	Sense of Place Level
	Content
0	Not having any sense of place
1	Knowledge of being located in a place
2	Belonging to a place
3	Attachment to a place
4	Identifying with the place goals
5	Involvement in a place
6	Sacrifice for a place

2.2. Internet

In the late 1950s, during the Cold War, the U.S. Department of Defense established the Advanced Research Project Agency (ARPA) to deal with information technology, and after its formation, ARPA commissioned BBN to assist in the research and development of the system, creating the first network in the western United States in 1969. The network was only connected to computer equipment at four universities, and to prevent other computers from being connected to the network in the event of an emergency, the Network Control Protocol (NCP) was developed; the network was called ARPANET.

In the mid-1980s, it was decreed that all long-distance networks must be TCP/IP compatible, and the network was divided into two separate networks according to their functions. One was ARPANET, which was for research purposes, and the other was MILNET, which was for military purposes. The former network became private and commercialized as time went on, stimulating the creation of later network services, such as WWW, E-MAIL, FTP, and DNS services. At this time, the Internet was viewed as a purely static page, and users were able to access the information provided by the Internet but could not upload information to the web page. Although they could leave messages on the Internet, they could not edit messages or communicate with other users in real time.

Back in 2004, at the O'Reilly Media Web 2.0 conference, the concept of Web 2.0 was not yet established, and only the design and use of websites was regulated, without any restrictions on whom could set up a website. Proponents advocated that people would be able to collaborate and share through the Internet, believing that Web 2.0 was a new way of being online, that this interaction would be generated by universal participation and class equality, and that the concept of decentralization, where "everyone can provide information", was particularly important. Many unprecedented technology products were created with the goal of enabling social networking with real-time online interaction, such as social media, web applications, social networking sites, blogs, and wikis. This era of the Internet, where the viewer is also the publisher, and where passive communication is transformed into active and two-way communication, is called Web 2.0. From 2005 to 2020, social media became the mainstream platform of the Internet era. Taking photos and uploading them to the Internet has become an ordinary daily life, and social media not only provides a platform for everyone to share their lives, but also gives people the best way to directly observe the development of the Internet.

Since the development of the Internet, there are divergent opinions on whether Web 3.0 has arrived. Web 1.0 and Web 2.0 are back-to-basics terms that do not exist as standards for the development of Internet technology, but rather as a term to describe technological changes. Elon Musk, the CEO of Tesla, publicly stated that he thinks Web 3.0 is a marketing gimmick, even saying, "Has anyone seen Web 3.0? I can't find it." Jack Dorsey, one of the founders of Twitter, has also stated that "you don't own Web 3.0, it's owned by Venture Capitals and LPs. You can never escape their control, it's just a labeled entity after all, figure out what you're getting into and what you're getting out of. Web 3.0 is just another Internet marketing gimmick, and the idea of decentralization is also a result of centralized management". Even so, the reason and core purpose of Web 3.0 is to provide users with an unrestricted and freer online experience.

Several scholars mentioned that space cannot be examined independently from society, ignoring time processes. Nowadays, society is becoming more connected to the Internet; thus, the Internet is also crucial in exploring the sense of place. The development, concepts, and relationships of the Internet in its various stages are organized in the following table (Table 3).

Table 3. Comparison of Internet periods.

Item	Internet		
	Web 1.0	**Web 2.0**	**Web 3.0**
Period	1990–2005	2005–2020	2020–now
Ownership	Company	Company–individual	Individual
Symbols	Static web page	Social media, wikipedia, blogs	Cryptocurrency, non-fungible token
Concept	Messaging and data transfer	Real-time interactive communication	Decentralization
Relationship with people	Internet–people	People–internet–people	
Relationship with world	Dependence	Dependence–parallel	

2.3. *Cyberspace*

"Does space exist on the Internet"? This question is common to many scholars and involves more than just asking about the network or space. It includes the boundary between the real and the virtual and is also related to the questioner's experience of cyberspace. It is important to be open minded when confronting this question and to understand that the definition of space will change over time, though it is still important to find common ground in each understanding. The following table lists the views of researchers on space in the Web in each period and analyzes the chronological distribution of Web 1.0, Web 2.0, and Web 3.0, as mentioned in the above section (Table 4).

The metaverse is one of the most talked-about topics in Web 3.0. In the book "Metaverse: Technology giants are competing for investment and unlimited business opportunities are emerging, are you ready?", the author discusses this issue from the point of view of virtual worlds and proposes three world views: the space where human beings use computers is called the "real world", which is the first type; the "ideal world", which simulates the real world in the Internet world and provides Internet services, is the second type; the game space or "fantasy world", which is completely detached from reality and full of imagination, is the third type; and the space that combines the above three types of worlds and reconstructs the real space is the metaverse (Figure 1). The definition of the metaverse is as follows. "An infinite world centered on multiple users entering from the real world made up of the real world that has been expanded to be more useful, the virtual world that has turned imagination into reality, and the multidimensional digital space that has been created by connecting to the Internet" [20].

Table 4. A review of literature on cyberspace.

Web Version	Viewpoint	
	Author	Relevant Research Contents
1.0	S-J Tung (1998) [12]	• Virtual space cannot replace the existence of real space. • Education can be used to increase the sense of community in the virtual space.
	C-I Wu (2001) [13]	• We believe that the existence of cyberspace can overthrow the original class system of real life. • The cyberspace has hollowed out the concrete thinking of the real world, and the understanding of the real world will be reduced. • In terms of life experience, the virtual community and the real community in the cyber world influence each other.
2.0	S-H Weng (2009) [14]	• The Internet is both a space and a platform for various services, with the characteristics and drawbacks of "virtual". • The behavioral model of the Internet is embodied in the physical space, providing convenience and interactivity, but also satisfying practical and spiritual needs.
	C-L Tu; B-K Liang (2009) [15]	• Cyberspace has the socio-spatial characteristic of "no geographic entity". • Social organizations in cyberspace have hierarchical relationships. • It can be linked to local spaces, breaking geographical limitations and allowing people from different regions to come together.
	C-S Sun (2010) [16]	• Space can be "deciphered" or "decoded". In addition to material practices, space also suggests an ideational process, and the symbols are part of the spatial language, the actual relationship between the subject and its environment. • Space embodies the text of the story, referring to the directionality of the relationship with everyday life.
	Jeffrey Li (2010) [17]	• The structure of cyberspace creates new powers and changes the meaning of those who are empowered. • Things in real space can be transformed into digital information through transmission and uploading. • Actions in cyberspace can also become actions that have meaning in real space. • Referring to the concept of "network square", the network is no longer just information, but, gradually, the network will become the world.

Table 4. *Cont.*

Web Version	Viewpoint	
	Author	**Relevant Research Contents**
2.0 – 3.0	W-C Hong (2022) [18]	• Human society will shape a virtual space/metaverse that is "parallel" to the physical world. • The American Foundation for Accelerated Research (ASF) expanded the definition of metaverse to include fully virtual worlds, mirror worlds where external information is integrated into virtual space, life logging where information about activities in physical space is linked to virtual worlds, and life logging where information about activities in physical space is linked to virtual worlds. The metaverse is also used in industrial applications, such as the "sealed universe" and the "augmented reality", which can display information about the physical environment in the virtual space while moving in the physical space.
	S-Y Peng (2022) [19]	• Imaginations range from the physical web, to the full-fledged Internet, and to real–digital virtual worlds. • The online game Roblox creates virtual worlds where users can freely determine their characters, create stories or products with multiple creative tools, and have a separate currency system. • Singers Travis Scott and Ariana Grande have held virtual concerts in the game. • The ultimate goal of the immersive experience is to create a virtual world that looks like reality. • Meta announced the main plan for the metaverse application "Horizon Worlds", including Horizon Workrooms, which is a virtual meeting space; Horizon Home, where friends and family can interact; and Horizon Venues, which will be used for events.

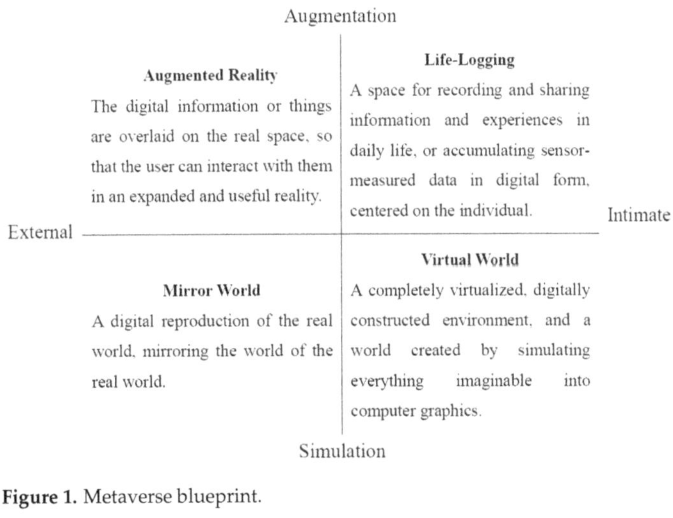

Figure 1. Metaverse blueprint.

It is easy to see that the definition of web space is influenced by the degree of user interaction. In the beginning, in Web 1.0, the Internet was used only as a service, and web space could not replace the existence of real space; in Web 2.0, the behavior of web space may become a behavior with the meaning of real space; and the ambition of Web 3.0 is to create a virtual space/metaverse that is "parallel" to the physical world. The ultimate goal of an immersive experience is to create a virtual world-like reality. With the progress of the times and the participation of users, the existence of space on the Internet has become an indisputable fact.

3. Research Method

Based on the literature review, the attributes of the sense of place, and the applicability of the cyberspace survey, a questionnaire survey was conducted as the research tool. The participants in the questionnaire survey were university students. In total, 12 social media platforms were selected, including Dcard, Facebook, Instagram, Line, Messenger, Pinterest, TikTok, Twitter, WeChat, YouTube, Zenly, and Little Red Book. The questions were used to understand users' current use of social media and spatial cognition measurements. The correlation was analyzed with the chi-square test to explore whether the human emotion of space, as described in the sense of place, can also be applied to online space (Figures 2 and 3).

The questionnaire was designed using a 7-point Likert scale with scores of 0, 1, 2, 3, 4, 5, and 6 from low to high based on Shamai's classification of the sense of place into low, medium, and high levels. One level was allocated for no sense of place (level 0) (Table 5).

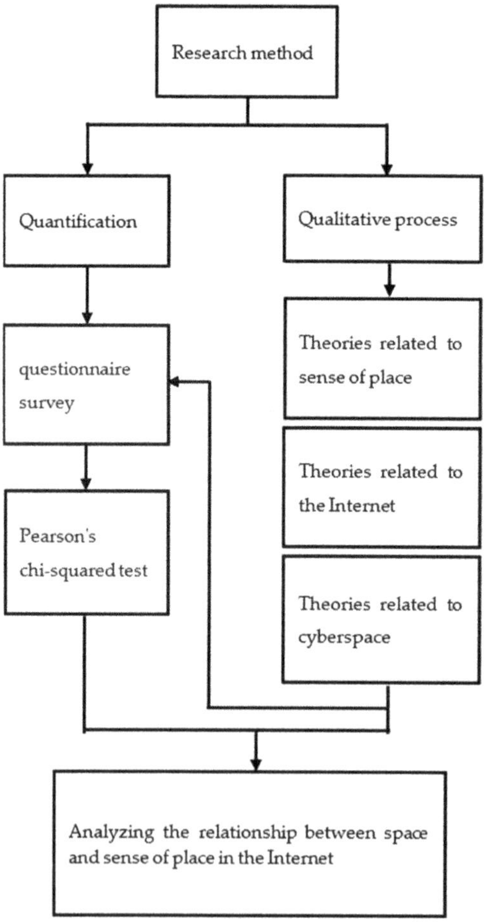

Figure 2. Research flow.

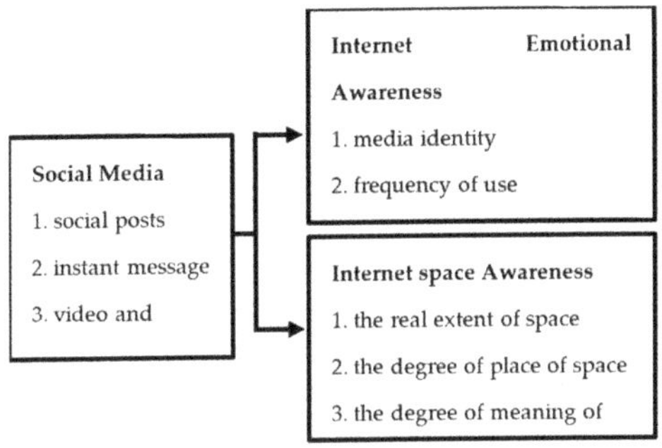

Figure 3. Research framework.

Table 5. Theory of correlation between questionnaire nouns and sense of place and Likert scale correspondence.

Terminology	Questionnaire Design	
	Corresponding Sense of Place Theory	Likert Scale
Media identity	Place identity	(low) 0 1 2 3 4 5 6 (high)
Media dependence	Place dependence	(low) 0 1 2 3 4 5 6 (high)
Frequency of use	Time	(low) 0 1 2 3 4 5 6 (high)
Sense of event reality	Experience	(real) 0 1 2 3 4 5 6 (virtual)
The real extent of space	Space	(real) 0 1 2 3 4 5 6 (virtual)
The degree of meaning of space	Place	(space) 0 1 2 3 4 5 6 (place)
Importance	Meaning	(low) 0 1 2 3 4 5 6 (high)

We selected the social media platforms that repeatedly appeared on the list for this survey of the online space based on five reports on the popularity of social media published every two months from June 2021 to February 2022. The following table shows the cumulative number of times each social media platform appeared in the rankings and the final selection of the 12 study samples (Table 6).

Table 6. Social media ranking list.

Ranking	Social Media	Number of Appearances
1	Dcard	4
2	Facebook Meta	4
3	Instagram	3
4	Line	3
5	Messenger	3
6	TikTok	3
7	Twitter	3
8	YouTube	3
9	Zenly	3
10	Pinterest	2
11	WeChat	2
12	Xiaohongshu	2

4. Research Results

The results of Pearson's chi-square test on the relevance of Social Media in Internet Emotional Awareness and Web space awareness among 418 university students are as follows.

Table 7 shows significant correlations between social media and other factors, except for "sense of event reality (0.590)" and "the degree of meaning of space (0.666)". Tables 8–11 show that the social media post type has intermediate "media identity (56.40)" and "frequency of use (36.30)" and high "media dependence (−72.80)". The instant message type has intermediate "frequency of use (−43.70)" and high "media identity (−44.30)". The video and audio information type had low "frequency of use (−37.80)" and high "media identity (33.50)" and "media dependence (57.70)". Table 11 shows that spatial perceptions are still mainly in favor of real space.

The survey asked about the relevance of social media to the users themselves, as well as what people think it means to turn "space" into "place" on the Internet and how important doing so is to them. Except for "privacy", "dependency", "immersion", and " Amount of information ", all measures were viewed as significant (Table 12), i.e., Interaction (41.20)", "anonymity (−72.80)", "identification (34.20)", "sense of life (30.20)", and "special experience (−49.80)".

Table 7. Results of chi-square test.

Social Media	Media Identity	Frequency of Use	Media Dependence	Sense of Event Reality	Real Extent of Space	Degree of Meaning of Space
Pearson's cardinality	45.240	35.380	55.327	2.812	18.917	2.383
Degrees of freedom	6	6	6	6	6	6
Asymptotic significance (two-tailed test)	0.000	0.000	0.000	0.590	0.001	0.666

Table 8. Results of goodness-of-fit results table (media identity).

Media Identity	Social Posts			Instant Message			Video and Audio Information		
	fo	fe	fo-fe	fo	fe	fo-fe	fo	fe	fo-fe
Low	357	356.7	0.30	311	285.3	25.70	188	214	−26.00
Intermediate	671	614.6	56.40	468	491.7	−23.70	336	368.8	−32.80
High	815	804.2	10.80	599	643.3	−44.30	516	482.5	33.50
Significance		0.000			0.003			0.008	

Table 9. Result of goodness-of-fit results table (frequency of use).

Frequency of Use	Social Posts			Instant Message			Video and Audio Information		
	fo	fe	fo-fe	fo	fe	fo-fe	fo	fe	fo-fe
Low	402	367.9	34.10	298	294.3	3.70	183	220.8	−37.80
Intermediate	523	486.7	36.30	361	389.3	−28.30	284	292	−8.00
High	795	795.8	−0.80	616	636.7	−20.70	499	477.5	21.50
Significance		0.001			0.035			0.020	

Table 10. Result of goodness-of-fit results table (media dependence).

Media Dependence	Social Posts			Instant Message			Video and Audio Information		
	fo	fe	fo-fe	fo	fe	fo-fe	fo	fe	fo-fe
Low	419	371.7	47.30	293	297.3	−4.30	180	223	−43.00
Intermediate	599	527.1	71.90	378	421.7	−43.70	288	316.3	−28.30
High	586	658.8	−72.80	542	527	15.00	453	395.3	57.70
Significance		0.000			0.056			0.000	

Table 11. Result of goodness-of-fit results table (the real extent of space).

Real Extent of Space	Social Posts			Instant Message			Video and Audio Information		
	fo	fe	fo-fe	fo	fe	fo-fe	fo	fe	fo-fe
Close to the real space	511	528.3	−17.30	451	396.4	54.60	268	305.4	−37.40
In between	590	581.1	8.90	401	436	−35.00	362	335.9	26.10
Close to virtual space	641	632.6	8.40	455	474.7	−19.70	377	365.7	11.30
Significance		0.666			0.004			0.031	

Table 12. Results of goodness-of-fit results table (importance).

Importance	Interactivity			Anonymity			Privacy		
	fo	fe	fo-fe	fo	fe	fo-fe	fo	fe	fo-fe
Low	41	57.6	−16.60	87	57.6	29.40	57	57.6	−0.60
Intermediate	129	143.7	−14.70	171	143.7	27.30	140	143.7	−3.70
High	244	202.8	41.20	130	202.8	−72.80	204	202.8	1.20
Significance		0.000			0.000			0.861	
Importance	**Dependency**			**Identity**			**Immersion**		
	fo	fe	fo-fe	fo	fe	fo-fe	fo	fe	fo-fe
Low	61	57.6	3.40	46	57.6	−11.60	62	57.6	4.40
Intermediate	138	143.7	−5.70	128	143.7	−15.70	132	143.7	−11.70
High	209	202.8	6.20	237	202.8	34.20	213	202.8	10.20
Significance		0.624			0.004			0.485	
Importance	**Sense of Life**			**Amount of Information**			**Special Experiences**		
	fo	fe	fo-fe	fo	fe	fo-fe	fo	fe	fo-fe
Low	43	57.6	−14.60	51	57.6	−6.60	70	57.6	12.40
Intermediate	135	143.7	−8.70	151	143.7	7.30	169	143.7	25.30
High	233	202.8	30.20	202	202.8	−0.80	153	202.8	−49.80
Significance		0.007			0.770			0.000	

For the question "Is there anything about "spaces and places on the web" that you think is important that was not mentioned in the questionnaire", there were interesting answers, such as "Satisfaction with use (you want to use it only if it works well)", "Strong rules to limit malicious speech, greater mobilization and rapid dissemination of information", "I actually think the Internet has a certain degree of "influence", "I think immersion and life sense should be included in the definition of place," and "information of a pedagogical nature teaches users the right concepts". These responses showed that users were concerned about the products provided on the Internet, whether in terms of influence or the accuracy of the content, or whether it is smooth to use. Thus, creating a free, but not overly free, space is a potentially important message.

5. Conclusions

The cyberspace of "instant message" and "video and audio information" is "close to the real space". Social media of the instant messaging type is a tool that people use to communicate with each other in their daily lives. Therefore, people are in favor of real space. The bulk of video and audio information is available, which reflects the same situation as an instant message. It is possible that the "social posts" do not fully reflect which type of space is preferred, as people post a variety of content from places of work and study about leisure, video games, animation, and other hobbies. The social media platforms of social posts are not as limited; thus, there are unlimited possibilities in the exploration of "cyberspace". In view of this fact, it is suggested that when developing or establishing various spaces in the online world, we should challenge people to communicate with each other in a virtual world, which is different from the appearance of the real world, to enhance people's view of cyberspace as a virtual space, rather than as an auxiliary tool of the real world. As for the answer to the question of whether the space in the web has a sense of place, cyberspace needs to achieve further steady development. Subsequent research is necessary for the investigation of other types of cyberspace, as this can ensure that more spaces can be identified.

Author Contributions: Writing-original draft preparation, S.-J.F.; Writing-review and editing, S.-T.L.; All authors have read and agreed to the published version of the manuscript.

Funding: This research received no external funding.

Institutional Review Board Statement: Center for Taiwan Academic Research Ethics Education Certificate of Completion No: S110027662.

Informed Consent Statement: Informed consent was obtained from all subjects involved in the study.

Data Availability Statement: The data presented are openly available at: https://docs.google.com/spreadsheets/d/1Qm_bPqyquwcsCR3v71YNLxYHSjBhQoDOkWxn-2X2VTs/edit?usp=drive_link.

Conflicts of Interest: The authors declare no conflict of interest.

References

1. Relph, E.C. *Place and Placelessness*; Pion Limited: London, UK, 1976.
2. Massey, D. *Power-Geometry and a Progressive Sense of Place. Mapping the Futures*; Taylor & Francis: New York, NY, USA, 1994.
3. Tuan, Y.F. *Space and Place: The Perspective of Experience. Bibliovault OAI Repository*; The University of Chicago Press: Chicago, IL, USA, 1978.
4. Harrison, S.; Dourish, P. Re-Place-Ing Space: The Roles of Place and Space in Collaborative Systems. In Proceedings of the ACM Conference on Computer Supported Cooperative Work, Boston, MA, USA, 16–20 November 1996.
5. National Development Council. *Communications Market Development Overview and Trends Survey Analysis Commissioned Research Case, Broadband Usage Survey Summary Report*; National Development Council: Taipei, Taiwan, 2018.
6. National Development Council. *2018 Mobile Phone Users' Digital Opportunity Survey (AE080006)*; Academia Sinica: Taipei, Taiwan, 2019.
7. Tsai, W.C. *Sense of Place: The Experience, Memory and Imagination of Environment Space*; Liwen Publishing Group: Kaohsiung, Taiwan, 2009.
8. Tuan, Y.F. Topophilia: A Study of Environmental Perception, Attitudes, and Values. *Serbiula* **1975**, *65*, 1–18.
9. Manzo, L. For Better or Worse: Exploring Multiple Dimensions of Place Meaning. *J. Environ. Psychol.* **2005**, *25*, 67–86. [CrossRef]
10. Jorgensen, B.; Stedman, R. Sense of Place as an attitude: Lakeshore owners attitudes toward their properties. *J. Environ. Psychol.* **2001**, *21*, 233–248. [CrossRef]
11. Shamai, S. Sense of Place: An Empirical Measurement. *Geoforum* **1991**, *22*, 347–358. [CrossRef]
12. Tung, S.J. The Role and Impact of Virtual Community on Sustainable Rural Development; Review of Agricultural Extension. *Science* **1998**, *14*, 15–34.
13. Wu, C.I. The Social Image of Internet Users vs. Non-Users. *J. Libr. Inform. Sci.* **2001**, *27*, 16–36.

14. Weng, S.H. Web2.0 Interactive Space. Master's Thesis, National Cheng Kung University, Tainan, Taiwan, 2009.
15. Tu, C.L.; Liang, B.K. Connections Between the Cyberspace and the Space of Places—A Case Study of Ford's F.P.R. Club. *Environ. Worlds* **2009**, *20*, 93–118.
16. Sun, C.S. Publishing, Place and Cultural Imagination: Media Spaces and Cultural Landscape of "The Lord of the Rings" Thought and Words. *J. Hum. Soc. Sci.* **2010**, *48*, 103–130.
17. Li, J. The Conflict Between Copyright and Free Flow of Information and the Solutions: From Real World to Cyberspace. Master's Thesis, National Taiwan University, Taipei, Taiwan, 2010.
18. Hong, W.C. Development of Japanese "Metaverse" Industry and Policy Discussion. *Econ. Outlook Bimon.* **2022**, *200*, 69–74.
19. Peng, S.Y. Imagine the development of "metaverse". *Econ. Res. Mon.* **2022**, *45*, 26–32.
20. Choi, H.W. *Meta-Universe: Are You Ready for the Rise of Infinite Opportunities as Technology Giants Compete for Investment?* Global Publishing Group: Taipei, Taiwan, 2021.

Disclaimer/Publisher's Note: The statements, opinions and data contained in all publications are solely those of the individual author(s) and contributor(s) and not of MDPI and/or the editor(s). MDPI and/or the editor(s) disclaim responsibility for any injury to people or property resulting from any ideas, methods, instructions or products referred to in the content.

engineering proceedings

MDPI

Proceeding Paper

Real-Time Safety Warning System for Lifting Operations in Construction Sites †

Hsien-Kuan Chang [1], Li-Ming Lu [2], Hsien-Chou Liao [2], Wen-Der Yu [1] and Zi-Yi Lim [3,*]

1 Department of Construction Engineering, Chaoyang University of Technology, Taichung City 41349, Taiwan; hkchang@cyut.edu.tw (H.-K.C.); wenderyu@cyut.edu.tw (W.-D.Y.)
2 Department of Computer Science and Information Engineering, Chaoyang University of Technology, Taichung City 41349, Taiwan; s10830625@cyut.edu.tw (L.-M.L.); hcliao@cyut.edu.tw (H.-C.L.)
3 Department of Information and Communication Engineering, Chaoyang University of Technology, Taichung City 41349, Taiwan
* Correspondence: zylim@cyut.edu.tw

† Presented at the 3rd IEEE International Conference on Electronic Communications, Internet of Things and Big Data Conference 2023, Taichung, Taiwan, 14–16 April 2023.

Abstract: A construction site is an open and dynamic space. Construction accidents have been the top-ranked occupational accidents among all industries around the world. Due to the limited quality and quantity of occupational safety supervisors on construction sites, it is difficult to control or prevent risks in real-time. Therefore, a real-time safety warning system based on a deep learning technique (DL) is developed for lifting operations in building construction, called a portable lightweight lifting safety control station (PLSCS). Two modes can be switched manually by the supervisor. If the mode is switched to lifting control mode, PLSCS helps to ensure that nobody is in the hazardous area during lifting operations. The advantages and features of this system are as follows: (1) it warns of the potential safety hazards automatically during the operations; (2) it reduces the workload of occupational safety supervisors; (3) the system is self-powered and easy to carry and deploy. The system was tested and verified in the actual construction site. The results show that the system is useful for improving the safety of lifting operations.

Keywords: machine learning; pattern recognition; construction safety; lifting operation; virtual fence

Citation: Chang, H.-K.; Lu, L.-M.; Liao, H.-C.; Yu, W.-D.; Lim, Z.-Y. Real-Time Safety Warning System for Lifting Operations in Construction Sites. *Eng. Proc.* **2023**, *38*, 3. https://doi.org/10.3390/engproc2023038003

Academic Editors: Teen-Hang Meen, Hsin-Hung Lin and Cheng-Fu Yang

Published: 19 June 2023

Copyright: © 2023 by the authors. Licensee MDPI, Basel, Switzerland. This article is an open access article distributed under the terms and conditions of the Creative Commons Attribution (CC BY) license (https://creativecommons.org/licenses/by/4.0/).

1. Introduction

According to the data released by the Ministry of Labor in Taiwan [1], the "construction industry" accounted for the highest proportion of deaths from major occupational disasters in the workplace in 2019. The number of deaths was a total of 168 people, which was an increase of 35% compared with the previous year, and accounted for 53.2% of the 316 people who died that year. The primary reason for the high accident rate is that construction sites are highly open and dangerous. Due to the large-scale and high-rise building projects in the past decades, the utilization of heavy cranes is becoming popular. Moreover, uncertain factors, e.g., a tight schedule, temporary work station, and the moving of lifting tools, have caused the high risk of falling objects and injuries to construction workers. More attention needs to be paid to the safety control of construction lifting operations.

Due to the rapid development of artificial intelligence (AI) technology (such as deep neural network (DNN) and convolutional neural network (CNN)) [2], the traditional on-site occupational safety management problems that are difficult to improve in construction sites have gained an opportunity for improvement.

In this research, a "Real-time Safety Warning System for Lifting Operations in Construction Sites" based on YOLO [3] was developed. YOLO is a CNN-based image detection and recognition technique that is widely used in real-time applications. The YOLO technique is adopted to localize personnel, and then the detected personnel is checked to determine whether they are in the lifting control zone. Then, a warning is triggered to prevent the potential safety hazards.

The remaining sections of this paper are organized as follows. Section 2 collects and analyzes the relevant literature. Section 3 presents our proposed method with the framework. Then, the implementation of the system is given in the following sections. Finally, the last section concludes the findings.

2. Related Works

The definition of "hazard" is a potential factor that causes harm or damage to human health, and "hazard recognition" is the process of identifying the existence of hazards and defining the characteristics of hazards [4]. Generally, the techniques of hazard recognition and risk assessment include methods such as checklists [5]. There is a lack of hazard recognition methods and techniques for on-site implementation [6]. Recently, many scholars have applied advanced information technology to hazard recognition to enhance the effectiveness of automatic safety management [7–9].

Since 2012, deep learning technology (DL) has adopted the graphics processing unit (GPU) meaning that computer recognition capabilities have been improved rapidly. Industry and academia have also paid attention to the application of DL on laborer, machine, and material tracking and the management of construction sites [10,11].

We have reviewed the applications of AI and image recognition technology in the safety management of construction projects. The related literature included workers' safety equipment recognition [4], worker unsafe behavior recognition [10], fall prevention from heights on construction sites [11], and construction site fall risk monitoring [12].

Due to the development of AI, the technology now is applied to real-time image recognition. The technologies involved include deep convolutional neural network (DCNN) [13], region-based convolutional neural network (R-CNN) [14], fast region-based convolutional neural network (Fast R-CNN) [15], faster region-based convolutional neural network (Faster R-CNN) [16], and YOLO image recognition technology (You Only Look Once: Unified, Real-Time Object Detection) [3,17].

Among these technologies, YOLO achieves a higher recognition speed than other technologies to identify object categories quickly. However, its disadvantage is its bad performance in the localization of objects [5]. Due to the requirement of real-time identification in this research, the latest version of YOLO was used for the recognition of workers.

Regarding the relevant research on crane safety management and control, the important studies are as follows. Price et al. [18,19] and Fang et al. [20] proposed to establish a 3D working model of the crane lifting environment through a variety of sensors, simulate the crane movement posture, and provide the operator with lifting assistance in real-time through a graphical interface. The research reduced the operator's limited visibility during the lifting operation, causing collision hazards. Peng et al. [21] proposed the automatic monitoring and early warning of external power failures. Liu et al. [22] proposed an operator fatigue warning system, in which the authors focused on the automatic monitoring of worker safety using artificial intelligence technology for lifting operations.

According to the literature we mentioned before, the safety management of construction sites has evolved from traditional technology to automatic recognition with AI. Our research uses image recognition AI techniques and network communication to construct a safety control system for workers in regard to lifting operations on construction sites as part of the automation of construction management.

3. Proposed Method

3.1. Application Scenario

During the construction phase of a building project (including concrete pouring, formwork, and rebar assembling), it is necessary to employ cranes for material lifting operations frequently. In the lifting operation, if the object falls from the crane, it is easy for it to cause accidents. Therefore, our research will focus on the safety control of lifting operations.

In the traditional practice of occupational safety management, the operation supervisor checks visually whether the construction worker has left the lifting area or the worker is approaching or breaking into the control zone, according to the construction regulations. In recent years, closed-circuit television (CCTV) has been widely installed on construction sites to facilitate safety management. However, due to uncertain factors such as short lifting operations, temporary placement, location changes, and so on, CCTV cannot be relocated efficiently. Moreover, CCTV usually does not have dedicated monitoring personnel when it is online, which makes it impossible to monitor effectively.

To solve the above problem, we proposed a portable lightweight lifting safety control station (PLSCS) to improve the safety problem encountered in traditional lifting operations. The application scenario of the proposed method is shown in Figure 1. The framework of the PLSCS is shown in Figure 2.

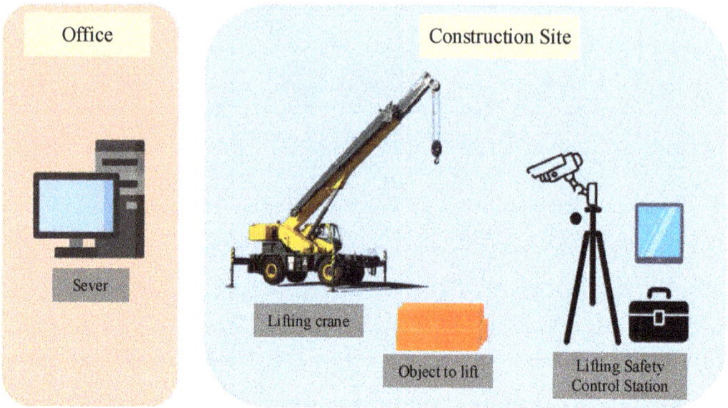

Figure 1. Application scenario.

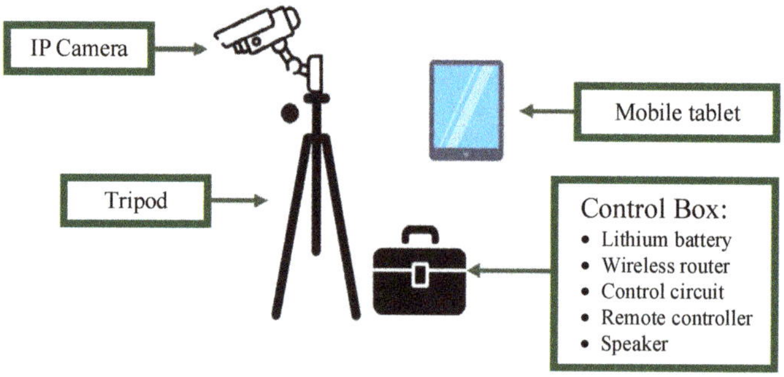

Figure 2. Parts of lightweight lifting safety control station.

3.2. Framework and Flow

In this system, we use YOLO as a technique to recognize the worker, which is applicable in complex construction environments to achieve effective early warning in safety control.

The proposed PLSCS can be deployed quickly and deployed on a large scale. After the deployment of PLSCS is completed, the supervisor only needs to set up the safety control zone (digital fence), according to the environment of the construction site, in order to solve the problem of difficult supervision during lifting operations. The system achieves full-time and real-time safety zone control to ensure the safety of the workers in lifting operations.

The proposed system's software architecture is shown in Figure 3. The flowchart is shown in Figure 4, and the description is shown as follows.

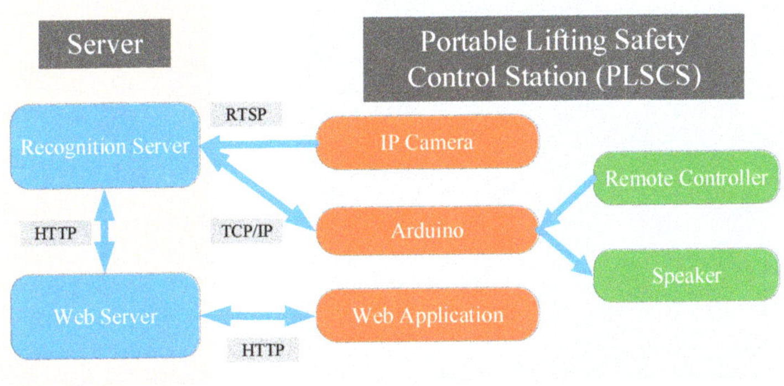

Figure 3. System architecture.

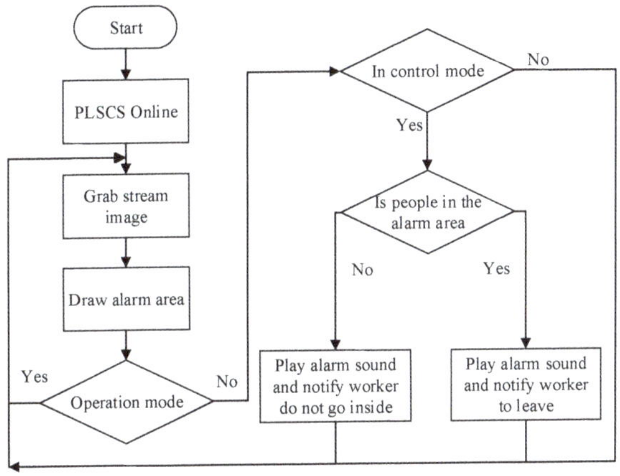

Figure 4. System flowchart.

Firstly, when the PLSCS is powered on, it automatically notifies the server via the TCP/IP that the PLSCS is online. The microcontroller sends the ID of the PLSCS and the online message to the identification server. Once the PLSCS is online, the server obtains the real-time image from the IP camera. The microcontroller and the IP camera must connect to the same 4G router and use the same network to connect to the outside world. Therefore, the recognition server can obtain the IP address of the microcontroller and connect to the IP camera via a specified port to obtain real-time images.

Afterward, a safety control zone needs to be defined. The on-site operation supervisor uses the real-time image provided by the tablet computer and the server to define the safety control zone. The supervisor draws the area through the web application as a safety control zone for lifting operations.

Then, two modes can be set in the system, namely the worker operation mode and lifting control mode. In the worker operation mode, a reasonable phenomenon happens in order for the worker to enter the safety control zone before the lifting operation. In this mode, the image recognition function is turned off. The relevant personnel can watch real-time operation images through the Internet. In the lifting control mode, workers are not allowed to enter the safety control zone. The image recognition function is turned on and the automatic safety warning function is started. The alarm sound "Do not enter the lifting area!" plays through the speaker.

When the recognition system finds that a person has intruded into the safety control zone, an alarm is activated, and the information is provided to the safety personnel through the speaker. The on-site supervisor can choose to reset the status to the worker operation mode or the lifting control mode after the alarm is triggered.

4. System Implementation

The main server's hardware specifications include a CPU with i5-9400 2.90 GHz, memory of 8 GB, and a GPU with Nvidia Geforce GTX 1650. The recognition server and the web server of the web application in this system are both set up in the same server to reduce the cost.

The system was tested and verified at the actual construction site. The on-site implementation of the proposed system is shown in Figure 5. The set of PLSCS hardware equipment included the following.

(a)

(b)

Figure 5. PLSCS in the construction sites: (**a**) two monitoring cameras; (**b**) control box.

1. An IP camera to obtain real-time images of the lifting worksite.
2. A tripod mounted with the IP camera, which can adjust the shooting angle, height, and worksite position.
3. A 4G router placed in a PLSCS control box, providing a webcam, tablet, and microcontroller internet services.
4. A microcontroller responsible for receiving the signal sent by the wireless remote control, sending and identifying server messages, and providing an MP3 (warning audio) playback function.
5. A speaker playing an alarm audio to inform workers of the current control status of the on-site situation.
6. An AC power bank is placed in the control box of the PLSCS to supply power to IP cameras, 4G routers, and microcontrollers.

The system application user interface is shown in Figure 6, which includes the following.

1. An image display area that shows real-time images from the IP cameras, safety control zone, and human recognition results.
2. A warning status area that changes and has four phases: an initialize phase, operation phase, monitoring phase, and hazard warning phase.
3. A safety control zone setting that includes the area type, size adjustment, and area movement.
4. An event log that records hazardous events for the management supervisor to review.

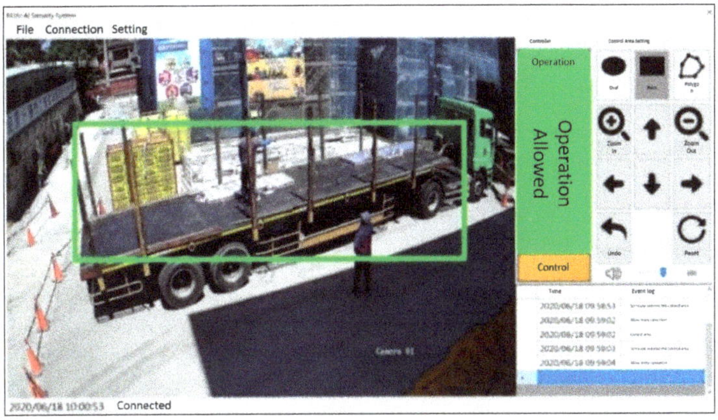

Figure 6. User interface of the proposed system.

The proposed system applies YOLO v3 for human recognition and adopts the latest training model provided by YOLO [17]. The model can detect 80 types of objects, but we only chose the "Person" types from the output from YOLO. Since the lifting operation site is a large area, the camera needs to be elevated and viewed from a slightly downward aerial view to fully capture the operation area. However, the height and viewing angle increase will affect the recognition accuracy, since the human features appear less clear in the image. Therefore, it is recommended to set up the camera with a height less than 4 m and a view angle less than 60° to achieve a balance between the recognition results and the shooting range. In addition, when the confidence threshold was set to 0.6, we found that the triangle cone or canvas was misjudged as a human in the preliminary tests. Finally, the confidence threshold was set to 0.7 to achieve a balance.

Regarding the setting of the safety control zone, the virtual digital fence indicates the safety control zone for lifting operations, which means that when the lifting crane is operating, people are prohibited from entering under or passing by the lifting crane and objects. The safety control zone settings consist of drawing, moving, and zooming operations. The user must select the drawing shape (includes rectangle, ellipse, and irregular shape) before drawing the safety control zone. Next, the user needs to click on the real-time image of the web application to set the coordinates of the shape. The user can move or zoom to fine-tune the drawn area after the shape is set. A simple example of the safety control zone setting and human recognition results is shown in Figure 7; the human is marked with a red color when he/she is in the safety control zone. Otherwise, they are marked with a green color if they are outside the safety control zone.

Figure 7. Schematic diagram of human safety recognition.

5. Conclusions

We present the results of collaborative research between the construction industry and the university to propose a portable lightweight lifting safety control station (PLSCS). The proposed PLSCS integrates servers, tablets, cameras, and edge-computing equipment, and develops a real-time safety warning system for lifting operation in construction sites. The system implemented YOLO v3.0 for human detection, and we set the unsafety alert area with the concept of a virtual digital fence to detect whether any worker is breaking into the safety control zone during lifting operations. The system is triggered by a hazardous situation, i.e., when humans enter the safety control zone, and the system plays a warning alarm to warn the on-site laborers. Moreover, the system also sends a message through the communication application to inform the relevant management supervisor. Therefore, the field operation and the occupational safety management supervisors can monitor the multiple construction sites remotely to reduce the personnel requirements on-site, and also to decrease the occurrence of accidents. As a result, the proposed system has achieved the goal of ensuring the safety of workers in the construction project environment.

Author Contributions: Conceptualization, H.-C.L. and H.-K.C.; methodology, software, validation, L.-M.L. and Z.-Y.L.; investigation, H.-K.C.; writing—original draft preparation, L.-M.L. and H.-K.C.; writing—review and editing, Z.-Y.L.; supervision, H.-C.L.; project administration, W.-D.Y. All authors have read and agreed to the published version of the manuscript.

Funding: This research was partially funded by the National Science and Technology Council under No. MOST 111-2221-E-324-011-MY.

Institutional Review Board Statement: Not applicable.

Informed Consent Statement: Not applicable.

Data Availability Statement: Data are unavailable due to privacy or ethical restrictions.

Acknowledgments: We acknowledge the actual testing site provided by Reiju Construction Co., Ltd. During the testing, the company's Occupational Safety and Health Center and Taichung Green Museumbrary Public Works Office assisted us so that this research could be completed successfully.

Conflicts of Interest: The authors declare no conflict of interest.

References

1. Occupational Safety and Health Administration of Ministry of Labor (Taiwan). 2019 Labor Annual Report. Available online: https://www.osha.gov.tw/1106/1164/1165/1168/29804/ (accessed on 5 November 2021).
2. Russell, S.; Norvig, P. *Artificial Intelligence: A Modern Approach*; Prentice Hall: New York, NY, USA, 2010.
3. Redmon, J.; Divvala, S.; Girshick, R.; Farhadi, A. You Only Look Once: Unified, Real-Time Object Detection. In Proceedings of the 2016 IEEE Conference on Computer Vision and Pattern Recognition, Las Vegas, NV, USA, 27–30 June 2016.
4. Yu, W.-D.; Liao, H.-C.; Hsiao, W.-T.; Chang, H.-K.; Wu, T.-Y.; Lin, C.-C. Application of The Hybrid Machine Learning Techniques for Real-Time Identification of Worker's Personal Safety Protection Equipment. *J. Technol.* **2020**, *35*, 155–165.
5. Ye, Y.-G. Application of Event Tree Analysis in Occupational Safety Risk Assessment. Master's Thesis, Graduate Institute of Environmental Engineering, National Central University, Taoyuan City, Taiwan, 2009.
6. Chang, R.-N. Training of Potential Hazard Identification on Construction Site Using Virtual Reality. Master's Thesis, Department of Civil Engineering, National Yang Ming Chiao Tung University, Taipei, Taiwan, 2011.
7. Brilakis, I.; Park, M.-W.; Jog, G. Automated Vision Tracking of Project Related Entities. *Adv. Eng. Inform.* **2011**, *25*, 713–724. [CrossRef]
8. Park, M.-W.; Brilakis, I. Construction Worker Detection in Video Frames for Initializing Vision Trackers. *Autom. Constr.* **2012**, *28*, 15–25. [CrossRef]
9. Gong, J.; Caldas, C.H. Data Processing for Real-Time Construction Site Spatial Modeling. *Autom. Constr.* **2008**, *17*, 526–535. [CrossRef]
10. Ding, L.; Fang, W.; Luo, H.; Love, P.E.D.; Zhong, B.; Ouyang, X. A Deep Hybrid Learning Model to Detect Unsafe Behavior: Integrating Convolution Neural Networks and Long Short-Term Memory. *Autom. Constr.* **2018**, *86*, 118–124. [CrossRef]
11. Fang, W.; Ding, L.; Luo, H.; Love, P.E.D. Falls from Heights: A Computer Vision-Based Approach for Safety Harness Detection. *Autom. Constr.* **2018**, *91*, 53–61. [CrossRef]

12. Liao, H.-C.; Yu, W.-D.; Hsiao, W.-T.; Chang, H.-K.; Tsai, C.-K.; Lin, C.-C. Application of Image Semantic Segmentation Using the Deep Learning Technique in Monitoring the Fall Risk of Construction Workers in A Building Elevator Shaft. *J. Technol.* **2021**, *36*, 1–12.
13. Szegedy, C.; Toshev, A.; Erhan, D. Deep Neural Networks for Object Detection. In Proceedings of the 26th Advances in Neural Information Processing Systems (NIPS 2013), Lake Tahoe, Nevada, 5–10 December 2013.
14. Girshick, R.; Donahue, J.; Darrell, T.; Malik, J. Region-Based Convolutional Networks for Accurate Object Detection and Segmentation. *IEEE Trans. Pattern Anal. Mach. Intell.* **2016**, *38*, 142–158. [CrossRef] [PubMed]
15. Girshick, R. Fast R-CNN. In Proceedings of the IEEE International Conference on Computer Vision (ICCV), Santiago, Chile, 7–13 December 2015.
16. Ren, S.; He, K.; Girshick, R.; Sun, J. Faster R-CNN: Towards Real-Time Object Detection with Region Proposal Networks. *IEEE Trans. Pattern Anal. Mach. Intell.* **2017**, *39*, 1137–1149. [CrossRef] [PubMed]
17. Redmon, J.; Farhadi, A. YOLOv3: An Incremental Improvement. *arXiv* **2018**, arXiv:1804.02767.
18. Price, L.C.; Chen, J.; Park, J.; Cho, Y.K. Multisensor-Driven Real-Time Crane Monitoring System for Blind Lift Operations: Lessons Learned from a Case Study. *Autom. Constr.* **2021**, *124*, 103552. [CrossRef]
19. Price, L.C.; Chen, J.; Cho, Y.K. Dynamic Crane Workspace Update for Collision Avoidance During Blind Lift Operations. In Proceedings of the 18th International Conference on Computing in Civil and Building Engineering, São Paulo, Brazil, 18–20 August 2020; pp. 959–970.
20. Fang, Y.; Cho, Y.K.; Chen, J. A Framework for Real-Time Pro-active Safety Assistance for Mobile Crane Lifting Operations. *Autom. Constr.* **2016**, *72*, 367–379. [CrossRef]
21. Weifu, P.; Shu, D.; Shaolei, C.; Qing, Z.; Na, T. Research on The Application of Target Detection Based on Deep Learning Technology in Power Grid Operation Inspection. *E3S Web Conf.* **2021**, *236*, 01035. [CrossRef]
22. Liu, P.; Chi, H.-L.; Li, X.; Li, D. Development of a Fatigue Detection and Early Warning System for Crane Operators: A Preliminary Study. In Proceedings of the Construction Research Congress, Tempe, Arizona, 8–10 March 2020.

Disclaimer/Publisher's Note: The statements, opinions and data contained in all publications are solely those of the individual author(s) and contributor(s) and not of MDPI and/or the editor(s). MDPI and/or the editor(s) disclaim responsibility for any injury to people or property resulting from any ideas, methods, instructions or products referred to in the content.

engineering proceedings

Proceeding Paper

Exploratory Study of Preservice Early-Childhood Teachers' Experiences of Electronic Toys and Play Preference †

Poh-Hwa Liang

Department of Early Childhood Development and Education, Chaoyang University of Technology, Taichung 413310, Taiwan; phliang@cyut.edu.tw

† Presented at the 3rd IEEE International Conference on Electronic Communications, Internet of Things and Big Data Conference 2023, Taichung, Taiwan, 14–16 April 2023.

Abstract: Digital play has gradually become popular in young children's families, and electronic toys have become an important type of toy for young children. This study concerns preservice early-childhood teachers' preferences about play and how their experiences with electronic toys in childhood influence their preferences about play. A total of 108 preservice early-childhood teachers participated in this study, and data were collected through questionnaires with them. The results revealed that these preservice teachers rated different play activities as follows: constructive play, expressive play, pretend play, digital play, and gameplay. In addition, preservice teachers who favored traditional toys in childhood preferred constructive play more than those who favored electronic toys. The results of the study provide useful suggestions for the design of electronic toys and preparation programs offered for preservice early-childhood teachers in the future.

Keywords: digital play; electronic toys; play preference; preservice early-childhood teacher

Citation: Liang, P.-H. Exploratory Study of Preservice Early-Childhood Teachers' Experiences of Electronic Toys and Play Preference. *Eng. Proc.* **2023**, *38*, 17. https://doi.org/10.3390/engproc2023038017

Academic Editors: Teen-Hang Meen, Hsin-Hung Lin and Cheng-Fu Yang

Published: 20 June 2023

Copyright: © 2023 by the author. Licensee MDPI, Basel, Switzerland. This article is an open access article distributed under the terms and conditions of the Creative Commons Attribution (CC BY) license (https://creativecommons.org/licenses/by/4.0/).

1. Introduction

Play is important for young children in their everyday lives. Carefully scaffolded, child-directed play can help children in math learning, symbol representation, literacy, decision making, and self-regulation [1]. Although several researchers argue that digital play is different from traditional play, research articles published between 2010 and 2020 indicated that digital play is "real play" [2]. Electronic toys have become an important type of toy for parents to select for children.

In this study, electronic toys are defined as toys or technology products composed of electronic and IC parts with batteries, computer chips, or other technological functions. They include electric cars, toys that make sounds and lights, Pleo robotic dinosaurs, Embassy robots, digital cameras, computer software, language-learning machines, and others. Teachers play a crucial role in young children's play activities, including creating places for play and facilitating them in creating playful and joyful experiences to support educational goals [3]. Teachers also provide suggestions for parents regarding children's play and toy selection. Therefore, preservice early-childhood teachers' preferences about playing were investigated in this study to understand how their experiences with electronic toys in childhood influenced their play preferences. The research questions are as follows. (1) What are preservice early-childhood teachers' preferences about play? (2) Do preservice early-childhood teachers' toy experiences in childhood influence their preferences about play?

2. Teachers' Role and Play Preferences

Teachers play a crucial role in young children's play [3]. The roles that teachers play while interacting with preschool children during free play include onlooker, player, and leader. When teachers play the "onlooker", they usually watch children's play and query about play. However, when they are directly involved in the play, they become the "player" and communicate with the children. Sometimes, teachers become the "leader" to initiate play and demonstrate how to play. Furthermore, the teachers' role is affected by their perceptions of how to stimulate children's learning, and they assume multiple roles in facilitating children's play [4].

In addition to the types of teacher–child interaction during play, the enhanced effects of teacher–child interaction have been reported in studies. Reference [5] examined the relationship between teachers' interaction with children and their play and indicated the enhancing effects of teacher–child interaction. Therefore, after the teacher demonstrated activities, guided children's participation, and interpreted activities, children were able to self-regulate and participate in similar activities by themselves.

As for teachers' play preferences, 515 Australian early-childhood educators were surveyed, and 52% of the respondents indicated that they did not agree that young children benefited from digital technology in terms of learning [6]. Early-childhood teachers' concerns about young children's technology use might include issues with children's development and maturation and the influence of technology on children's social skills, relations, language, as well as brain development [7].

A similar result was found in [8]. They investigated parents' play preferences across the USA, Turkey, China, and South Korea, and found that digital play was the least preferred play for parents. In addition, when parents reported longer technology use for their children, they preferred digital play more. It seemed that most parents preferred a balance of all play activities, and their negative perceptions about the digital play were related to issues of addiction and antisocialization [9].

For several preservice early-childhood teachers, play is more social and physical activity than cognitive activity. Moreover, preservice early-childhood teachers' perceptions about play seem to be influenced by early-childhood teacher education programs [10].

3. Play Types and Digital Play for Young Children

For young children, play can be integrated into the early-childhood curriculum as free play, guided play, and games [3]. Reference [11] modified cognitive play in [12] and classified four categories of children's cognitive play as functional play, constructive play, dramatic play, and games with rules.

With the prevalence of technology, digital play is becoming an important aspect of children's lives. Digital play is now recognized as a form of play [13,14]. Digital play is "related to video and computer games, internet sites and search engines, electronic toys, mobile technologies, smartphones, iPads, and tablets" [8] (p. 133). Reference [9] investigated 21 medium- or high-income Portuguese families with at least one child under 8 years old and found that children used diverse digital media at home, including smartphones, laptops, and tablets. Moreover, more than half of the children owned personal devices. Children mostly used tablets and mainly for playing games and watching videos on YouTube. Technologies can be viewed as cultural tools for 4- and 5-year-old children's play. Young children's epistemic play, such as exploratory behaviors, is often related to the functions of technology. On the other hand, ludic play is associated with the application of technologies for children's symbolic play. Furthermore, young children's digital play begins with exploring the functions of the given technologies through epistemic activities, and then by creating play content through ludic activities [15].

Reference [16] observed preschool children's play with technologies and proposed a digital play framework that consists of three major types: mastery play, imaginative play, and deep play. In mastery play, young children create things in the virtual environment. Young children can also pretend to care for digital pets in their imaginative play. As for deep play, when young children encounter fearful experiences in the digital world, they have more control over the outcomes than they do in the physical environment. Although digital play seems to be similar to traditional play, the educational goals of digital play (e.g., smarter, stronger, and kinder) and the effects of digital games on young children remain unclear [17].

4. Method

In this study, data were collected through questionnaires to investigate preservice early-childhood teachers' preferences about play and how their experiences with electronic toys in childhood influence their play preferences. The participants and instruments of the study are stated in the following.

4.1. Participants

To investigate preservice early-childhood teachers' preferences about play and their toy experiences in childhood, 122 questionnaires were sent to preservice early-childhood teachers at a university in Taiwan and 116 questionnaires were returned, with a return rate of about 95%. A total of 108 questionnaires were analyzed. Three participants were male, and the others were females. Participants' ages ranged from 19 to 24 years old with a mean of 19.30.

4.2. Instruments

The contents of the questionnaire included questions regarding preservice teachers' play preferences and their experiences with electronic toys in childhood. Preservice early-childhood teachers' play preferences were assessed using an adaptation of Johnson's Play Attitude Questionnaire (PA). The reliability of the scale was assessed with a sample of preschool teachers and mothers of preschool children. The test–retest reliability of the instrument was 0.71 [18].

The play preferences questionnaire used for this study consisted of 10 items, 2 items each for the five types of play. These 10 items were for paired comparisons requiring respondents to choose one of the two play activities that they thought was more important for the child to spend time on. The items are listed as follows.

(1) Constructive play: making something from different materials, and building something with table blocks.
(2) Pretend play: make-believe with small figures, and pretending to be a favorite character.
(3) Gameplay: playing a board game, and playing connect-the-dots game.
(4) Digital play: using computers to create things (e.g., funny animals), and imagining with little people on a computer screen (e.g., making them move, jump, or run).
(5) Expressive play: singing children's songs, and dancing and movement expression.

5. Results

The results of the study are discussed based on two aspects: (1) preservice early-childhood teachers' play preferences and (2) preservice early-childhood teachers' toy experiences in childhood. In addition, how preservice early-childhood teachers' experiences of electronic toys in childhood influence their preferences about playing is also discussed.

5.1. Preservice Early-Childhood Teachers' Play Preferences

One of the objectives of the study was to elicit preservice early-childhood teachers' preferences for different play activities for young children. The mean scores on preservice teachers' preferences for constructive play, expressive play, pretend play, digital play, and gameplay are presented in Table 1.

The mean scores on play activities varied from a high of 5.63 for constructive play to a low of 2.12 for gameplay. They rated different play activities in the order of constructive play, expressive play, pretend play, digital play, and gameplay. Preservice teachers endorsed constructive play and regarded it as a proper play activity for young children. They did not prefer gameplay or digital play.

Table 1. Preservice early-childhood teachers' play preferences.

Play Preferences	*M*	*SD*	*Range*
Constructive play	5.63	0.15	1–8
Expressive play	5.02	0.18	1–8
Pretend play	4.40	0.16	0–8
Digital play	2.83	0.20	0–8
Game play	2.12	0.16	0–6

Note. *N* = 108 ("*N*" means the total number of the sample).

5.2. Preservice Early-Childhood Teachers' Toy Experiences in Childhood

As for preservice teachers' toy experiences in childhood, about 21.3% did not have electronic toy experiences in childhood. In addition, 22.6% favored electronic toys, and 86.0% owned more traditional toys in childhood (Table 2).

Table 2. Preservice early-childhood teachers' toy experiences.

Play Experiences		***n* (%)**
Electronic Toy Use in Childhood	Yes	23 (21.3%)
	No	85 (78.7%)
Favorite Toy in Childhood	Traditional toy	82 (77.4%)
	Electronic toy	24 (22.6%)
Number of Toys in Childhood	More traditional toys	92 (86.0%)
	More electronic toys or an equivalent amount	15 (14.0%)

Preservice teachers who favored electronic toys and those who favored traditional toys are compared in terms of play preferences in Table 3. Both groups rated constructive play the highest and gameplay the lowest. Moreover, preservice teachers who favored traditional toys in childhood rated constructive play higher than those who favored electronic toys at t (104) = 2.31 and $p < 0.05$.

Table 3. Differences in play preferences of preservice teachers with different toy experiences in childhood.

Play Preferences	**Favorite Toy in Childhood**	*n*	*M*	*SD*	*t*
Constructive play	Traditional toy	82	5.84	1.51	2.31 *
	Electronic toy	24	5.04	1.43	
Expressive play	Traditional toy	82	5.06	1.84	0.53
	Electronic toy	24	4.83	1.90	
Pretend play	Traditional toy	82	4.28	1.64	−1.11
	Electronic toy	24	4.71	1.71	
Digital play	Traditional toy	82	2.76	2.03	1.00
	Electronic toy	24	3.25	2.42	
Game play	Traditional toy	82	2.06	1.67	−0.27
	Electronic toy	24	2.17	1.81	

Note. * $p < 0.05$.

The data indicated that both groups (preservice teachers who favored electronic toys and traditional toys in childhood) preferred constructive play, but they did not prefer digital play or gameplay, relatively. In addition, preservice teachers who favored traditional toys in childhood preferred constructive play more than preservice teachers who favored electronic toys.

6. Conclusions

The results of the study reveal that preservice teachers endorsed constructive play and regarded it as a proper play activity for young children. They did not prefer gameplay or digital play. The results are consistent with [6,8]. Parents across four countries reported that digital play was the least preferred play [8]. In addition, more than half of the early-childhood educators indicated that they did not agree that young children benefit from digital technology in terms of learning [6]. Parents also preferred a balance of all play activities, and their negative perceptions about digital play were related to children's addiction and antisocialization [9].

The issues concerning young children's technology use might include children's development and maturation and the influence of technology on children's social skills, relations, language, and brain development [7]. Therefore, the reason why preservice early-childhood teachers in this study did not prefer digital play might be due to their concerns about the educational goals of digital play [17].

The results of the study also indicated that preservice teachers who favored traditional toys in childhood preferred constructive play more than preservice teachers who favored electronic toys. Similar results were found in [8]. When parents reported longer technology use for their children, they preferred digital play. However, in this study, regardless of the fact that preservice early-childhood teachers favored electronic toys or traditional toys in childhood, they tended to prefer constructive play. They did not prefer digital play or gameplay. The reasons for their play preferences remain unclear, which needs further investigation.

Funding: This research received no external funding.

Institutional Review Board Statement: Not applicable.

Informed Consent Statement: Not applicable.

Data Availability Statement: The data presented in this study are available on request from the corresponding author.

Conflicts of Interest: The authors declare no conflict of interest.

References

1. Anderson, G.T.; Spainhower, A.R.; Sharp, A.C. Where do the bears go? The value of child-directed play. In *Spotlight on Young Children: Exploring Play*; Bohart, H., Charner, K., Koralek, D., Eds.; NAEYC: Washington, DC, USA, 2015; pp. 37–45.
2. Undheim, M. Children and teachers engaging together with digital technology in early childhood education and care institutions: A literature review. *Eur. Early Child. Educ. Res. J.* **2022**, *30*, 472–489. [CrossRef]
3. Zosh, J.M.; Gaudreau, C.G.; Golinkoff, R.M.; Hirsh-Pasek, K. The power of playful learning in the early childhood setting. *Young Child.* **2022**, *77*, 6–13.
4. Roskos, K.; Neuman, S.B. Descriptive observations of adult's facilitation of literacy in young children's play. *Early Child. Res. Q.* **1993**, *8*, 77–97. [CrossRef]
5. Zukow, P.G. The relationship between interaction with the caregiver and the emergence of play activities during the one-word period. *Br. J. Dev. Psychol.* **1986**, *4*, 223–234. [CrossRef]
6. Zabatiero, J.; Straker, L.; Mantilla, A.; Edwards, S.; Danby, S. Young Children and Digital Technology: Australian Early Childhood Education and Care Sector Adults' Perspectives. *Aust. J. Early Child.* **2018**, *43*, 14–22. [CrossRef]
7. Hatzigianni, M.; Kalaitzidis, I. Early childhood educators' attitudes and beliefs around the use of touchscreen technologies by children under three years of age. *Brit. J. Educ. Technol.* **2018**, *49*, 883–895. [CrossRef]
8. Erdogan, N.I.; Johnson, J.E.; Dong, P.I.; Qiu, Z. Do parents prefer digital play? examination of parental preferences and beliefs in four nations. *Early Child. Educ. J.* **2019**, *47*, 131–142. [CrossRef]
9. Brito, R.; Dias, P.; Oliveira, G. Young children, digital media and smart toys: How perceptions shape adoption and domestication. *Brit. J. Educ. Technol.* **2018**, *49*, 807–820. [CrossRef]
10. Zhulamanova, I.; Raisor, J. Early childhood preservice teachers' perceptions on children's play. *Int. Online J. Prim. Educ.* **2020**, *9*, 128–143.
11. Smilansky, S. *The Effects of Sociodramatic Play on Disadvantaged Preschool Children*; John Wiley: New York, NY, USA, 1968.
12. Piaget, J. *Play, Dreams and Imitation in Childhood*; W. W. Norton: New York, NY, USA, 1962.
13. Disney, L.; Geng, G. Investigating young children's social interactions during digital play. *Early Child. Educ. J.* **2022**, *50*, 1449–1459. [CrossRef] [PubMed]
14. Edwards, S. Digital play and technical code: What new knowledge formations are possible? *Learn. Media Technol.* **2021**, *46*, 306–319. [CrossRef]

15. Bird, J.; Edwards, S. Children learning to use technologies through play: A digital play framework. *Brit. J. Educ. Technol.* **2015**, *46*, 1149–1160. [CrossRef]
16. Marsh, J.; Plowman, L.; Yamada-Rice, D.; Bishop, J.; Scott, F. Digital play: A new classification. *Early Years* **2016**, *36*, 242–253. [CrossRef]
17. Flynn, R.M.; Richert, R.A.; Wartella, E. Play in a digital world: How interactive digital games shape the lives of children. *Am. J. Play* **2019**, *12*, 54–73.
18. Johnson, J.E. Attitudes toward play and beliefs about development. In *Cultural Dimensions of Play, Games, and Sport*; Mergen, B., Ed.; Human Kinetics: Champaign, IL, USA, 1986; pp. 89–101.

Disclaimer/Publisher's Note: The statements, opinions and data contained in all publications are solely those of the individual author(s) and contributor(s) and not of MDPI and/or the editor(s). MDPI and/or the editor(s) disclaim responsibility for any injury to people or property resulting from any ideas, methods, instructions or products referred to in the content.

engineering proceedings

MDPI

Proceeding Paper

Hybrid Courses Based on Basic Sciences and Green Technologies in Engineering Programs †

Huber Nieto-Chaupis

Faculty of Engineering, Universidad Autónoma del Perú, Panamericana Sur Km. 16.3, Villa el Salvador 15842, Peru; hubernietochaupis@gmail.com

† Presented at the 3rd IEEE International Conference on Electronic Communications, Internet of Things and Big Data Conference 2023, Taichung, Taiwan, 14–16 April 2023.

Abstract: An ineluctable topic has emerged and needs to be treated in engineering programs for green energies. Particularly in the programs of electrical and systems engineering, students can acquire the capabilities to envision the arrival of green standards to generate clean energy sources. In this article, the introduction of new forms of green energy is systematically presented through a basic experiment in electricity in which a potato is used to investigate the current and electric power. The experiences were conducted at the 2022-I at a private university in Lima City. The results demonstrated that the topic of biofuel in the framework of green energy can be studied in undergraduate courses. Indeed, prospective students of engineering can be oriented toward research in engineering for the implementation of new electrical systems that provide electricity with low budgets.

Keywords: engineering education; physics; devices and systems

Citation: Nieto-Chaupis, H. Hybrid Courses Based on Basic Sciences and Green Technologies in Engineering Programs. *Eng. Proc.* **2023**, *38*, 49. https://doi.org/10.3390/engproc2023038049

Academic Editors: Teen-Hang Meen, Hsin-Hung Lin and Cheng-Fu Yang

Published: 27 June 2023

Copyright: © 2023 by the author. Licensee MDPI, Basel, Switzerland. This article is an open access article distributed under the terms and conditions of the Creative Commons Attribution (CC BY) license (https://creativecommons.org/licenses/by/4.0/).

1. Introduction

An appropriate time to show the promising green technologies to freshmen [1–3] would be the first semester. In particular, all programs in engineering have to be aggregated in the curricula. Important topics would be the following.

- Searching for green batteries
- Green energy for massive applications
- Software to manage green energies
- Use of ongoing technologies in the green scenario

Thus, the potential ways to teach these topics and the abilities of freshman students to think about them in the future are necessary. With the massive usage of smartphones, tablets, and laptops, the main question about the topic is what green technologies are used to replace conventional batteries [4,5].

The topics about green technologies can be relevant to various factors such as motivation, perspective, compromise in chosen programs, and research aspects. Thus, attention has been paid to the basic courses of physics [6,7], in which a simple scheme is presented for students to gain knowledge to create new circuits based on biological waste. From the well-known experiment to understand electric fields and Ohm's law, the implementation of biological compounds as a replacement for electrical circuits is presented.

The potential usage of waste as a source of energy is examined as a course of physics-II in the system engineering department iofn a private university located in the south pole of Lima City.

The course teaches the following topics.

- The theory of equipotential lines
- The experience of lines of electric fields
- The usage of wasted potatoes

- The demonstration of Ohm law
- Presentation of scenarios in which the waste replaces electrical batteries.

In the experience, students are required to have a solid background in math, particularly calculus, as Gauss law and vectorial analysis are demanded for a solid understanding of the experiment. This encompasses the idea that, in conjunction with theory, math and experiments must be learned together [8–10]. This is the foundation on which students embark to learn new green technology. Thus, the purpose of this study is to present the design of the course in which the topics of physics are taught and to provide capabilities to face the new technologies through basic sessions of experiments.

The paper is structured as follows. In the second section, the content of the session based on theory and experiment is described. In the third section, Ohm's law and a bio-circuit are explained using the peels of potatoes as a source of electric energy. Finally, the conclusion is drawn.

2. Session of Physics

In this section, the design of a session of physics (level 2) in the program of system engineering is presented. According to the content of the course, the topic corresponds to a concrete experience in the verification of equipotential lines. The theory is established on the concepts of vectors and Coulomb's law [11–13].

Students need to understand the sequence of equations.

$$\frac{\boldsymbol{F}}{q} = \boldsymbol{E} = \int d\boldsymbol{E} = \int \frac{dQ(\boldsymbol{r}-\boldsymbol{r}')}{|\boldsymbol{r}-\boldsymbol{r}'|^3} \tag{1}$$

where 4 π ε_0 = 1. With the definition of Coulomb's law, the electric fields can be derived. Through a mathematical procedure based on vectorial calculus, (1) is derived.

$$\boldsymbol{E} = -\nabla\left[\frac{Q}{|\boldsymbol{r}-\boldsymbol{r}'|}\right] \tag{2}$$

Students can recognize the term in brackets which constitutes the electric potential Φ, and has a direct relation with the Coulomb force, which is given by:

$$\boldsymbol{F} = q\boldsymbol{E} = -\nabla\left[\frac{qQ}{|\boldsymbol{r}-\boldsymbol{r}'|}\right]. \tag{3}$$

Direct communication with students specifies the necessity to implement the gradient operator "∇" in (2). It is expected that students pass up to the second course of calculus in which the gradient operator is linked to a geometrical view such as the tangent. Students are asked to simulate the system of two electrodes with opposite signs. Here, the use of an external package in the session combines the usage of math at the basic level [14–16]. In this way, the session employs Wolfram, which allows students to simulate two electrodes emitting lines of electric fields.

Figure 1 shows the 3D representation of three spatial variables. Students use the Gaussian functions and their product to the second-order polynomial. Two well-defined Gaussian functions are separated at a distance between the electrodes. Although it is not a direct mathematical manifestation of two electrodes, the use of mathematical packages allows the simulation of the spatial position of real electrodes.

Once the simulation of electrodes is completed, the next step is the use of Wolfram to reconstruct the lines of the electric field as well as the equipotential lines. The use of directives in physics with Wolfram is to determine the possible topology of lines. Thus, the command is conducted by students (Figure 2).

According to Figures 1 and 2, students can complement the assimilated theory to define the experimental arrangement through in-situ verification. Figure 3 shows the basic arrangement. A bucket, two electrodes, a voltmeter, and a power source to provide 10 V are required in the experiment to understand the field and equipotential lines. It is important to note that the instructor can modify the geometry of electrodes, such as circular, rectangle, and others. Thus, the emission of field lines shows the dependency of the geometry of the source.

```
Plot3D[5 + (10.75 * (x * x - y * y)) Exp[-0.5 (x * x + y * y)], {x, -3, 3},
 {y, -3, 3}, ColorFunction → Function[{x, y, z}, Hue[z]] ,
 AxesStyle → Directive[Black, 35]]
```

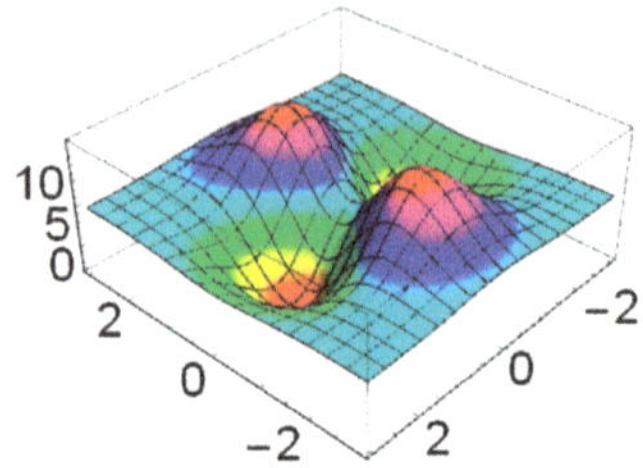

Figure 1. Simulation of two electrodes as a 3d surface of two Gaussian functions by using the package, Wolfram [17].

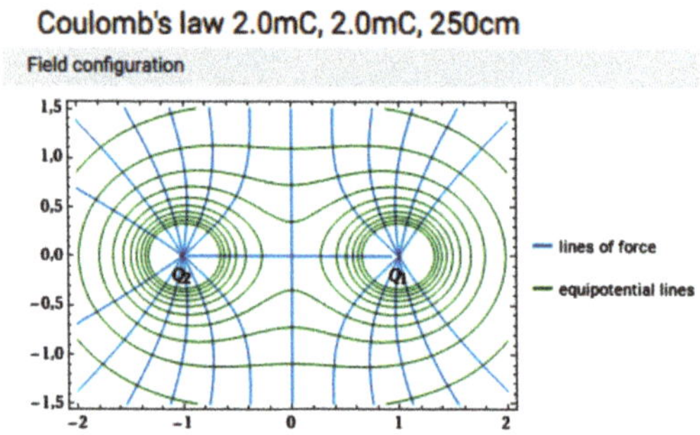

Figure 2. Simulation of Two Electrodes in a 2D Arrangement by Using Installed Commands in Wolfram. This is Conjunction with Figure 1.

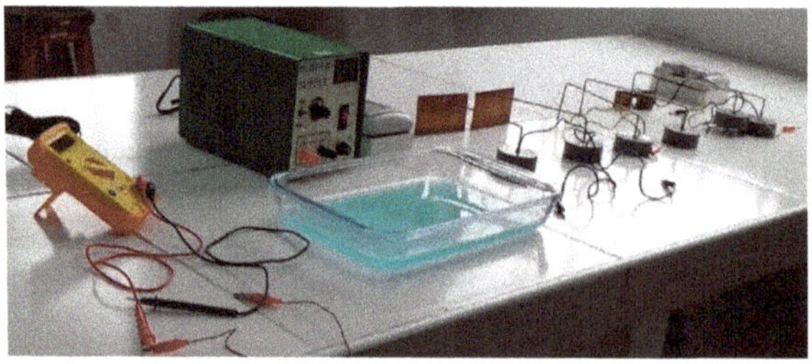

Figure 3. Experimental arrangement of laboratory corresponding to the observation of equipotential lines.

While using Wolfram for a theoretical simulation of involved physics, students witness the physics of Coulomb's law and the gradient operator, which constitutes an educational objective in the program.

3. Usage of Potato Peels

In South America, the consumption of potatoes is general, so there are lots of peels. Thus, the idea is to reuse the peel as biological materials for producing electricity [18–20].

Therefore, the peel in the experiment of equipotential lines is used to measure the main electrical parameters and test the potential capability in a real circuit. From (3), the electric field is defined as

$$\mathbf{dE} = \frac{dQ}{|R|^2}. \quad (4)$$

From $\rho = dQ/dV$,

$$\mathbf{dE} = \frac{\rho dV}{|R|^2}. \quad (5)$$

By keeping a spherical symmetry,

$$\frac{1}{4\pi}\mathbf{dE} = \frac{\rho(R,\alpha)dRR^2}{|R|^2}. \quad (6)$$

Then, the electric field can be written as:

$$\mathbf{E} = \frac{1}{4\pi}\int \mathbf{dE} = \int \frac{\rho(R,\alpha)R^2dR}{|R|^2}. \quad (7)$$

$\rho(R,\alpha)$ is a volumetric charge density depending on the radius, and α is a parameter characterizing the density of electric charge.

The measurement of electricity before and after adding the peel to water is performed to see if the peel is forming an electric field and is changing the electrical configuration.

Figure 4 shows that in presence of a charged compound, the lines of the field are distorted, this suggests the deformation of equipotential lines [21]. Thus, such a phenomenon triggers the deformation of equipotential lines with the peel of potatoes. Figure 5 shows the experiment with 150 g of the peel. The peel is introduced in 0.3 L of water. With a power source of 10 V, the equipotential lines are dramatically changed yielding 5.54 and 4.96 V. The electric field is affected as

$$\mathbf{E} = \lambda\ell + \int \frac{\rho(R,\alpha)R^2dR}{|R|^2} \quad (8)$$

where "λ" is the linear charge density and l is the length of the metallic bar in the bucket. A mathematical form for $\rho(R,\alpha)$ is a nonlinear function. Electric charges can be found in the proteins of potatoes. This is the reason why the deformation of field lines is presented in Figure 4. The arrow indicates the place of the peel of the potato emitting asymmetric field lines without following any pattern. From the above, students understand that the peel of the potato exhibits electrical properties.

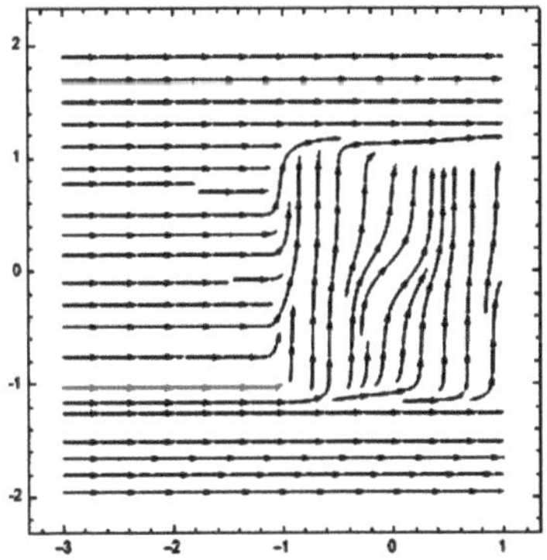

Figure 4. Use of command stream points in Wolfram to simulate the incorporation of peel of potato into bucket. The arrow indicates the position of peel deforming the lines of field.

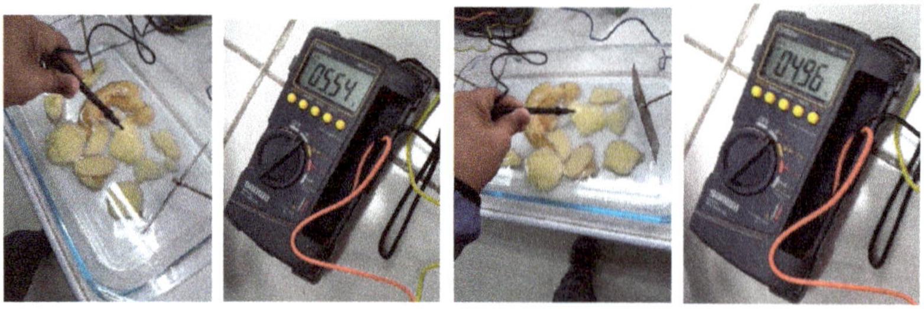

Figure 5. Inclusion of Peel Potato with Water in A Bucket. The Electrodes Are a Pointer. The Power Source is 10 V.

4. Ohm Law and Proposal for Green Energy Circuit

Once the peel of the potato is recognized by students as an inherent bio-electrical component, in the next session, the introduction of circuit laws such as Ohm's law is explained. Students are encouraged to pay attention to the theoretical and experimental points of Ohm's law. For this, the difference in electrical potential is discussed (see Figure 5), and resistance is introduced. The tool turns out to be useful for observing Ohm's law [22]. Figure 6 shows two resistors, R_2 and R_3, in parallel and in series to R_1. Theoretically, Ohm's law is expressed as

$$V = \frac{R_2 R_3}{R_2 + R_3} I, \tag{9}$$

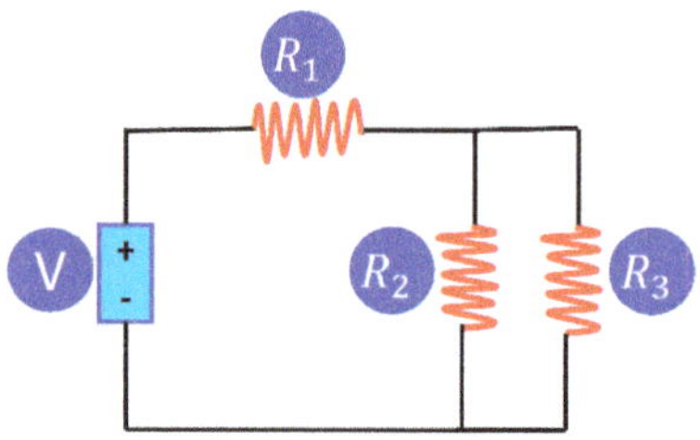

Figure 6. Basic circuit with 2 resistors in parallel being in series with one near to power source. The peel of the potato is expected to be R_3.

While factorizing "R_2",

$$V = \frac{R_2 R_3}{R_2\left(1 + \frac{R_3}{R_2}\right)} I = \frac{R_3}{\left(1 + \frac{R_3}{R_2}\right)} I. \tag{10}$$

The denominator has the below form if $\left(\frac{R_3}{R_2}\right)^q \approx 0$ for $q \geq 2$ then

$$Exp\left(\frac{R_3}{R_2}\right) \approx 1 + \left(\frac{R_3}{R_2}\right) + \frac{1}{2!}\left(\frac{R_3}{R_2}\right)^2 + \frac{1}{3!}\left(\frac{R_3}{R_2}\right)^3 + \cdots \tag{11}$$

Therefore, (10) is written as

$$V = Exp\left(-\frac{R_3}{R_2}\right) I R_3. \tag{12}$$

Equation (12) behaves like a Weibull distribution from the fact that these probability distribution functions can be written as $f(x) = x^{k+1} e^{-x^k}$. Thus, the replacement of "R_3" by peel potato is interesting for students who verify that biological compounds can serve as biodevices. The peel of the potato can also be an "extra" power source in water, as shown in Figure 6.

In Figure 7 the equivalent electrical circuits done with PASCO is displayed. When "R_3" becomes a power source, an advanced theory of circuits is applied. The instructor suggests demonstrating the theorem of Thevenin and experimenting. When "R_3" is a power source, the Thevenin voltage and current are determined, then "R_2" is removed. Thus, when "R_3" = V_B the biological power source is applied with Kirchhoff's law as follows:

$$I = \frac{R_2 + R_3}{R_2 R_3}(V_B - V). \tag{13}$$

Figure 7. Experimental Design with PASCO of Figure 6. The Resistance R3 had Two Values: 33 Ohm and 100 Ohm.

The student can check out the linearity between the Thevenin's current and the difference in voltage "$V_B - V$". Any difference between the current from (13) and the derived equation from (11) is written as:

$$I = \frac{V}{R_3} Exp\left(\frac{R_3}{R_2}\right), \tag{14}$$

Mass and humidity are relevant parameters. When "R_3" is replaced by the peel of the potato, a current is obtained depending on the potato-like resistor exhibiting a two-zone behavior when I(R_3) is plotted. Thus, the student can try other forms with the waste of fruits, spoiled food, or another biological compound for searching new green technologies. Finally, in Table 1, the outcomes of the whole experience are listed. It is noteworthy that variables have shown to keep a direct relationship to student satisfaction, being this very import to avoid desertion. Thus, communications among students were optimal, task and workshop were done in home. However, it is not clear if students after lab have had the chance to improve the gained knowledge. In addition, the lab is clearly a healthy space to demonstrate sociability among students. Future work, shall be analyzed the issue of impact of physics laboratory onto the formation of engineers, as well as the period after the finishing the grade of bachelor of engineering.

Table 1. Outcomes of Experience.

Variable	Impact	In Campus	Outside Campus
Communication	Middle	Yes	Yes
Workshop	Middle	Yes	No
Review of literature	High	Yes	Probable
Sociability	High	High	High

5. Conclusions

In this study, experiences on the use of the peel of the potato as a potential candidate for a green device in electrical circuits are obtained with the theoretical foundation about the role of the biological resistor. This can be introduced in the second course of physics to motivate students to understand the construction of electrical circuits with the prospective use of biomaterials [23–25].

Funding: The Universidad Autónoma del Perú has funded totally the present research.

Institutional Review Board Statement: Not applicable.

Informed Consent Statement: Not applicable.

Data Availability Statement: Not applicable.

Conflicts of Interest: The author declares no conflict of interest.

References

1. Schoenung, A.J.M. Teaching green topics to chemical engineers: Innovation in the process design curriculum. In Proceedings of the 2009 IEEE International Symposium on Sustainable Systems and Technology, Tempe, AZ, USA, 18–20 May 2009; pp. 1–2.
2. Duerden, S.; Doak, B.; Garland, J.; Green, M.; Roedel, R.; Williams, P.; McCarter, J. Scaling up Arizona State University's first-year integrated program in engineering: Problems and solutions. Proceedings of Frontiers in Education 1997 27th Annual Conference. Teaching and Learning in an Era of Change, Pittsburgh, PA, USA, 5–8 November 1997; Volume 3, pp. 1442–1449.
3. Hesketh, R.P.; Hollar, K.A.; Dericks, B.J. Incorporating green engineering in the chemical engineering curriculum. In Proceedings of the SMC'03 Conference Proceedings. 2003 IEEE International Conference on Systems, Man and Cybernetics. Conference Theme—System Security and Assurance (Cat. No.03CH37483), Washington, DC, USA, 8 October 2003; Volume 2, pp. 1745–1752.
4. Manas, K.; Gupta, J.; Singh, B. Green Energy Based Multipurpose EVs Charging Station with Flexible DC Charging Functionality for Residential/Commercial Building Applications. In Proceedings of the 2022 IEEE Global Conference on Computing, Power and Communication Technologies (GlobConPT), New Delhi, India, 23–25 September 2022; pp. 1–6.
5. Arsalis, A.; Papanastasiou, P.; Georghiou, G.E. Integration of a Green Hydrogen Subsystem in a Photovoltaic-Battery Nanogrid System. In Proceedings of the 2022 2nd International Conference on Energy Transition in the Mediterranean Area (SyNERGY MED), Thessaloniki, Greece, 17–19 October 2022; pp. 1–5.
6. Kattayat, S.; Josey, S. Implementation of digital platform in teaching physics concepts applied to engineering education and student's performance: An exploratory study. In Proceedings of the 2018 Advances in Science and Engineering Technology International Conferences (ASET), Dubai, Sharjah, Abu Dhabi, United Arab Emirates, 6 February–5 April 2018; pp. 1–5.
7. Srinivasa, A.; Bassichis, W.H. Work in Progress: The Role of Physics in an Engineering Education. In Proceedings of the Frontiers in Education: 36th Annual Conference, San Diego, CA, USA, 27–31 October 2006; pp. 27–28.
8. Tan, S.; Xu, C.; Huo, J.; Wang, X. The remote education assistant system for the virtual physics experiment teaching system. In Proceedings of the 2009 4th International Conference on Computer Science & Education, Nanning, China, 25–28 July 2009; pp. 1451–1454.
9. SenthilKumar, R. Work in Progress: Use of Interactive Simulations in the Active Learning Model in Physics Education for Engineering Students at a College in Oman. In Proceedings of the 2019 IEEE Global Engineering Education Conference (EDUCON), Dubai, United Arab Emirates, 8–11 April 2019; pp. 1359–1362.
10. Larkin-Hein, T. The write touch in physics and engineering education. In Proceedings of the 31st Annual Frontiers in Education Conference. Impact on Engineering and Science Education. Conference Proceedings (Cat. No.01CH37193), Reno, NV, USA, 10–13 October 2001; Volume 3, pp. S3F–S31.
11. Serway, R.A. *Physics for Scientists & Engineers with Modern Physics*; Saunders College Pub.: Philadelphia, PA, USA, 2005.
12. Physics, U. *Vol-1 Sanny, Loyola Marymount University Samuel Ling*; Truman State University: Kirksville, MO, USA, 2016.
13. Griffiths, D.J. (Ed.) *Introduction to Electrodynamics*; Prentice Hall: Hoboken, NJ, USA, 1999.
14. Klug, M.; Hausberger, P. Motivaton of students for further education in simulation by an applied example in a related other course in engineering education—A case study. In Proceedings of the 2009 Winter Simulation Conference (WSC), Austin, TX, USA, 13–16 December 2009; pp. 248–255.
15. Todorova, S. Using Simulation as an Experiential Learning Tool for Enhancing Students Learning in Environmental Engineering. In Proceedings of the 2019 IEEE Global Engineering Education Conference (EDUCON), Dubai, United Arab Emirates, 8–11 April 2019; pp. 1384–1385.
16. Van Til, R.; Banachowski, M.; Wagner, C.; Sengupta, S.; Hillberg, P. Using a discrete event simulation program in an engineering probability and statistics course. In Proceedings of the 2009 39th IEEE Frontiers in Education Conference, San Antonio, TX, USA, 18–21 October 2009; pp. 1–5.
17. Available online: https://www.wolframalpha.com/input?i=Coulomb%27s+law+2.0mC%2C+2.0mC%2C+250cm&lang=es (accessed on 1 April 2023).
18. Núñez-Delgado, A.; Dominguez, J.R.; Zhou, Y.; Race, M. New trends on green energy and environmental technologies, with special focus on biomass valorization, water and waste recycling: Editorial of the special issue. *J. Environ. Manag.* **2022**, *316*, 115209. [CrossRef] [PubMed]
19. Wang, Z.; Chen, H.; Xia, R.; Han, F.; Ji, Y.; Cai, W. Energy, exergy and economy (3E) investigation of a SOFC-GT-ORC waste heat recovery system for green power ships. *Therm. Sci. Eng. Prog.* **2022**, *32*, 101342. [CrossRef]

20. Dafiqurrohman, H.; Safitri, K.A.; Setyawan, M.I.B.; Surjosatyo, A.; Aziz, M. Gasification of rice wastes toward green and sustainable energy production: A review. *J. Clean. Prod.* **2022**, *366*, 132926. [CrossRef]
21. Elmoursi, A.A.; Castle, G.S.P. Mapping of field lines and equipotential contours in electric field problems using the charge-simulation technique. *J. Electrost.* **1987**, *19*, 221–234. [CrossRef]
22. Available online: https://www.pasco.com/resources/lab-experiments.
23. Nieto-Chaupis, H. Experiences on the quality of education assessment in a network engineering program in Lima city. In Proceedings of the 2017 IEEE URUCON, Montevideo, Uruguay, 23–25 October 2017; pp. 1–4.
24. Rubiano, S.M.M.; López-Cruz, O.; Soto, E.G. Teaching computer programming: Practices, difficulties and opportunities. In Proceedings of the 2015 IEEE Frontiers in Education Conference (FIE), El Paso, TX, USA, 21–24 October 2015; pp. 1–9. [CrossRef]
25. Palacios, C.A.; Reyes-Suárez, J.A.; Bearzotti, L.A.; Leiva, V.; Marchant, C. Knowledge Discovery for Higher Education Student Retention Based on Data Mining: Machine Learning Algorithms and Case Study in Chile. *Entropy* **2021**, *23*, 485. [CrossRef] [PubMed]

Disclaimer/Publisher's Note: The statements, opinions and data contained in all publications are solely those of the individual author(s) and contributor(s) and not of MDPI and/or the editor(s). MDPI and/or the editor(s) disclaim responsibility for any injury to people or property resulting from any ideas, methods, instructions or products referred to in the content.

engineering proceedings

MDPI

Proceeding Paper

Teaching Experience of Microchip Education during COVID-19 †

Yu-Cherng Hung

Department of Electronic Engineering, National Chin-Yi University of Technology, Taichung 411030, Taiwan, China; ychung@ncut.edu.tw; Tel.: +886-4-23924505-7374

† Presented at the 3rd IEEE International Conference on Electronic Communications, Internet of Things and Big Data Conference 2023, Taichung, Taiwan, 14–16 April 2023.

Abstract: A teaching experience on microchip (chip) education during COVID-19 is shared in this article. The program outcomes obtained and executed in the Teaching Practice Research Program supported by the Ministry of Education (MOE) of Taiwan are described as the teaching experience. In this program, two courses were offered in a semester: (1) Analog IC Design and (2) Full Custom IC Layout. The COVID-19 pandemic interrupted the teaching process, and determining how to maintain the normal teaching process was a significant challenge. The experience in teaching chips during COVID-19 was recorded in the teaching practice research program and shared for the future development of teaching resources.

Keywords: chip design; chip layout; IC education; IC learning; COVID-19 impaction

1. Introduction

The semiconductor industry in Taiwan was valued at TWD 4 trillion in 2022, and the value was over 15% of the gross domestic product (GDP) of Taiwan. Thus, the semiconductor industry is critical to Taiwan's national economy. As machine learning, artificial intelligence (AI), 5G, self-driving cars, virtual reality, cloud computing, and information security continuously develop, the semiconductor demand is growing rapidly. Therefore, human resource development for integrated chip (IC) technology and personnel training for IC design are important tasks. Since 2019 when the first case of COVID-19 was found in China [1,2], its impact has influenced daily life as well as the economy and the teaching-and-learning process in education. Recently, daily life has gradually returned to normal but the impact on education still exists, which is worth reviewing.

The project, Teaching Practice Research Program, was initiated by the Ministry of Education (MOE) of Taiwan in 2020 [3] aiming to improve the teaching quality in university education [4]. In this one-year project, we promoted the basic knowledge and skills of IC design and layout. Two courses including Analog IC Design and Full Custom IC Layout were offered each semester. In Analog IC Design (Course 1), the circuit operation and the principle of the metal–oxide–semiconductor field effect transistor (MOS FET) were introduced and taught. The related sub-blocks of the MOS transistor and circuit such as the current source, current mirror, voltage reference, and single-stage amplifier were described, too. In Full Custom IC Layout (Course 2), lectures on the concept of semiconductor manufacturing and the IC layout technique were given. Students were required to become familiar with related software tools such as IC electronic design automation (EDA). Course 1 was mainly related to theory, while Course 2 was practical training for IC layout skills.

The article is organized into five sections. Section 1 introduces the motivation of the research program. Section 2 describes the background of the semiconductor industrial chain and chip-realization technology. Section 3 describes the lecture of two courses during COVID-19 in detail. The feedback of students is compared with the same courses after COVID-19. Section 4 shows the class activities to improve the student learning achievements. Finally, Section 5 concludes this study.

Citation: Hung, Y.-C. Teaching Experience of Microchip Education during COVID-19. *Eng. Proc.* **2023**, *38*, 75. https://doi.org/10.3390/engproc2023038075

Academic Editors: Teen-Hang Meen, Hsin-Hung Lin and Cheng-Fu Yang

Published: 10 July 2023

Copyright: © 2023 by the author. Licensee MDPI, Basel, Switzerland. This article is an open access article distributed under the terms and conditions of the Creative Commons Attribution (CC BY) license (https://creativecommons.org/licenses/by/4.0/).

2. Background on Microchip Education

Resistors and inductors in nano-Henry, capacitors in pico-Faraday, and bi-polar junction transistors (BJTs) or metal–oxide–semiconductor field-effect transistors (MOSFETs) can be shrunk in dimension and integrated into a small thin silicon or other material, which is known as an integrated circuit or chip. There are various chips in different packaged forms within modern electronic products. For example, the iPhone 13 contains a 64-bit ARM-based system-on-a-chip (SoC) named Apple A15 Bionic designed by Apple Inc. The A15 is fabricated by Taiwan Semiconductor Manufacturing Company (TSMC) by using 5 nm technology in which 15 billion transistors are included in a die area of one square centimeter. The chip design and manufacturing process are more innovative than conventional sub-micro-fabricated technologies.

The chip fabrication is implemented by a succession of photo-lithography procedures. According to circuit functionality and performance, the circuit schematic is designed by IC designers upon market demand. The corresponding layout diagram is precisely drawn with a computer by IC layout engineers. The layout diagram comprises various colorful forms of rectangles depending on the circuit schematic, in which each color represents one thin-film material or doped action. After this step, the layout diagrams are transferred and fabricated in a set of photomasks (or reticles). The generated photomasks are used in photolithography processes to create the specific structure of related thin-film material. The photo-lithography procedures are repeated until the whole chip structure is constructed. Finally, the finished chip wafer is packaged and tested. The aims of the chip's packaging include protection of the bare die, heat removal, and electrical and signal connection, whereas that of the IC test includes general performance verification. Figure 1 shows the whole process of chip production from the circuit design to a standalone device element. A functional chip is manufactured through a complicated process. However, in university education, priority is given to determining how to increase the students' learning interests and promote IC layout design knowledge and skills. As the IC industry is so important, many researchers are studying to improve the teaching and education of chip-related knowledge [5–11].

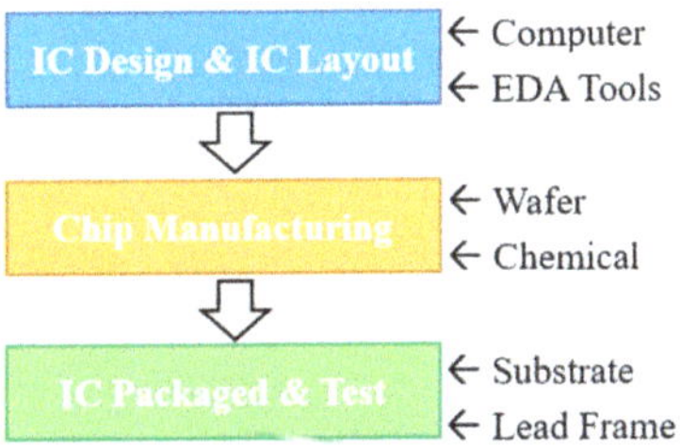

Figure 1. Chip realization process.

3. Teaching Process and Experiences in Chip Education

Lectures for two courses were given from August 2020 to July 2021. Lectures for Course 1 and Course 2 were given in one semester sequentially. In universities in Taiwan, each semester has eighteen weeks, and a course has one to three hours each week. In this research program, the two courses had three lessons each week with fifty-four hours of lectures in one semester. The teaching process was impacted by COVID-19.

3.1. Course 1: Analog IC Design

Course 1 was an elective course for junior or senior students, and 42 students enrolled in the course. The major content of the course included the basic concept of semiconductor technology, the semiconductor manufacturing process, the principle of metal–oxide–semiconductor (MOS) field-effect transistor (FET) operation, a small-signal equivalent circuit, a sub-block analog switch, a bias circuit, a current source, a current mirror, and the design of a single-stage amplifier. The course content was integrated with two courses to give students the whole outline of analog IC design: Microelectronic Circuits and IC technology. Due to the circuit complexity, we helped students understand the basic operating theorem of the MOS transistor. When students had the correct concept of MOSFET operation, that knowledge could become useful and available. Based on the correct knowledge, students can understand the MOS circuit design and analysis. Therefore, in the teaching strategy, it was planned to give a two-hour instruction in the beginning and a one-hour practice later. All class assignments and additional materials were included in the learning portfolio. Each student in this course was assigned a personal portfolio. After students completed their assignments, the portfolio was submitted for examination and scoring. Praise and encouragement with candies or beverages to students with high achievements encouraged them to be motivated to learn. Encouraging rather than criticizing seems to be a better educational idea.

In the class, teachers gave instructions in the first two lessons, and then students practiced class assignments in the third lesson. In the meantime, students were encouraged to discuss the assignments with classmates to find the proper solutions from information on the internet. The learning outcomes were evaluated from the individual portfolio. Finally, teachers evaluated and encouraged students for their performance and achievement. Three learning models with instruction, practice, and praise were used in the learning cycle, as shown in Figure 2. In the model, students were not passively receiving information or unidirectionally learning but actively surveyed related information. Compared to traditional teaching, a positive learning recycle was created for higher learning efficiency. Even during COVID-19, teaching activities for Course 1 were not impacted significantly, due to the preventive measure.

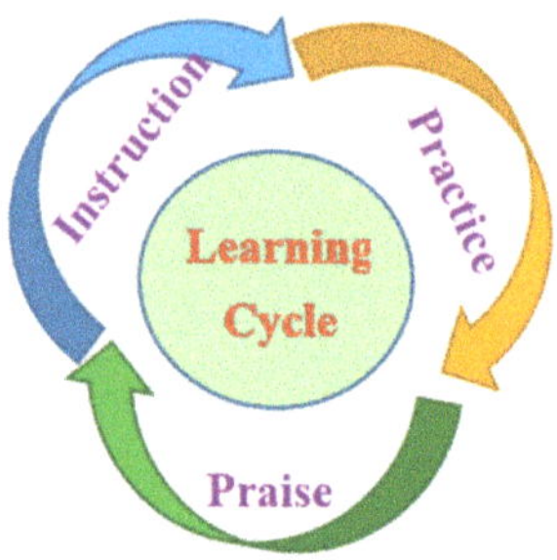

Figure 2. Three learning situational models in our teaching strategy.

3.2. *Course 2: Full Custom IC Layout*

In Course 2, it was also planned to have three lessons every week with 40 students. The course focused on practicing IC layout skills. Electronic design automation (EDA) software authorized and approved by Taiwan Semiconductor Research Institute (TSRI) was used in the laboratory as this course demanded students to manipulate the EDA software. The content of the course included the basic concept of the semiconductor manufacturing process, the various methodologies of chip realization, UNIX command, a cross-sectional view of the CMOS transistor, a circuit schematic, an IC layout drawn in the EDA software environment, and IC verifications using design-rule-check (DRC) and layout-vs.-schematic (LVS). In the course, many teaching aids were provided such as Microsoft PowerPoint animations to show the three-dimensional structure of the MOS transistor and its overall structure. A micrographic image of the real chip was observed by using optical microscopes. For the understanding of the CMOS chip process, CMOS logic gate, stick diagram of the layout, and the design of adders, assignments were given in lectures including building a physical layout diagram for the MOS transistor. Teaching aids such as color papers and glue were used to construct the MOS transistor. The auxiliary teaching activities increased students' interest. Since the course aimed to train students using the IC layout EDA tool, every student was required to use computers. The whole practice was instructed to teach the background and conception and to demonstrate how to build the chip layout step by step by teaching assistants (TAs).

From May 2021, the outbreak of COVID-19 made all schools practice online education in Taiwan. There was still a quarter of the semester until the end of the course. The remaining subjects were important as students needed to finalize their projects and practice computer skills. Thus, how to continue the course was a significant challenge for skill training. The following principles were followed for the course to be continued in distance learning, too.

1. In distance learning, we used Google Meet, Microsoft Teams, or Cisco WebEx to connect and share information with all students at home.
2. Although students could not come to school, the practice of the IC layout skill was continued by using a virtual private network (VPN). The student at home used a personal computer to connect to the layout server through the school network gateway to accomplish the final layout project. Even though the bandwidth was limited, the course continued as normal as before.
3. Lectures on layout skills were provided as video files with the help of teaching assistants. Students repeatedly watched the video during the course.
4. An information platform for the course was offered for exchanging information and announcing the submission status of the final project. We used LINE and Google Drive to construct an information platform. LINE is a freeware app for instant communications, operating on smartphones and personal computers. Google Drive was developed by Google to allow users to store files in the cloud via internet access. The two services are popular in Taiwan. Therefore, the enrolled students and teachers exchanged information and communicated conveniently on the platform.
5. Using the above methods, course-related information on the platform became open and transparent, and students were encouraged and supervised in the course. Each student understood the status of all classmates' final project progress. At the end of the course, 38 enrolled students remained. A total of 36 students completed the final project and fulfilled the requirements with the achievement rate for the IC layout project of 95%. Teachers and students put a lot of effort into learning the course.

A questionnaire was created for the evaluation including various aspects such as teaching strategy and attitude, lecturing skill, learning assistance, learning outcomes, course attention, and self-evaluation. Twenty-one items were contained in this questionnaire. A five-point Likert scale was used for this questionnaire with scores for strongly disagree, disagree, neither agree nor disagree, agree, and strongly agree. The feedback data for teaching were extracted from the school's teaching historical database. Table 1 shows the results of the students' evaluation for the courses, with scores of 4.68 and 4.87.

Table 1. Feedback results of student questionnaires.

Year_Semester	Course 1	Course 2
2014_2	4.79	4.79
2015_2		4.37
2016_1	4.14	
2016_2		4.42
2017_2	4.12	
2018_1	4.39	
2019_1	4.49	
This Study * 2020_1	4.68	
This Study * 2021_2		4.87

* The period of the project execution is from the first semester year of 2020 to the second semester year of 2021.

"Slow and steady wins the race", "Learn to walk before you run", and "Practice makes perfect" were also applied to the teaching of the courses. Since the state-of-the-art IC technologies have been scaled down to the order of nanometers, to maintain high-sensitivity performance, the chip design and layout have become complicated. Therefore, the learning process must be exciting to improve the learning efficiency. Students need to have a firm and clear grasp on basic concepts of the IC processes, design, and layout technology. Step-by-step learning and deep thinking in circuit theory and transistor operation are essential to enhance basic abilities. In the practice experiments of distance learning, video software and apps for instant communications between teachers and students were useful to exchange information.

4. Related Activities

In the research program, many exercises for IC circuit analysis were organized for students. Every student had an individual learning portfolio. According to the progress, every student practiced related skills after the teacher's instruction. During the practice, students were encouraged to find the proper solutions by themselves through discussion with classmates. Figure 3 shows the learning material and small rewards for students. Figure 4 shows a snapshot of the students' practice during Course 1. To relieve study fatigue, the practice was usually scheduled on the third hour after giving a two-hour lecture. Figure 5 shows students using personal computers to practice IC layout skills in Course 2. In Course 2, TAs helped students manipulate the layout software and answered questions.

Figure 3. Learning material and rewards for students.

Figure 4. Students' practice in Course 1.

Figure 5. Students used computers to practice IC layout skills in Course 2.

Figure 6 shows the TA's demonstration of using the IC layout software. To verify the learning outcomes, a midterm practice test was carried out. All students were grouped and sequentially tested for building the circuit schematics. Figure 7 shows the midterm test in Course 2. In the program, teaching aids were self-made for the demonstration of the CMOS transistor structure and the corresponding layout diagram. Figures 8 and 9 show the structure of the MOS transistor in a three-dimensional view and visual aids of the IC layout.

Figure 6. Teaching assistant (TA) instructs and guides the students for IC layout.

Figure 7. The midterm practice test of Course 2 (Full Custom IC Layout).

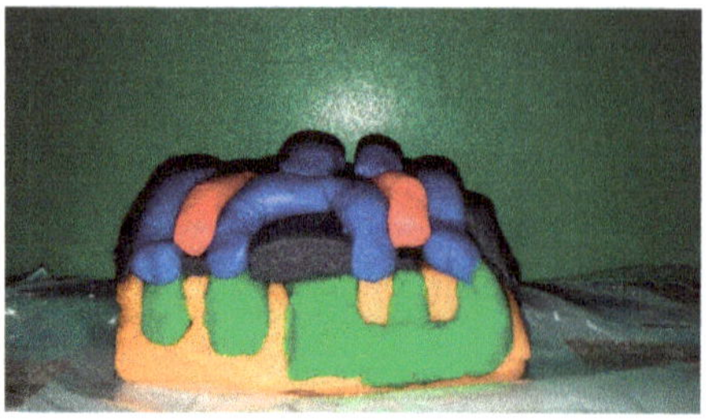

Figure 8. The teaching aid for structure of MOS transistor in three-dimensional view.

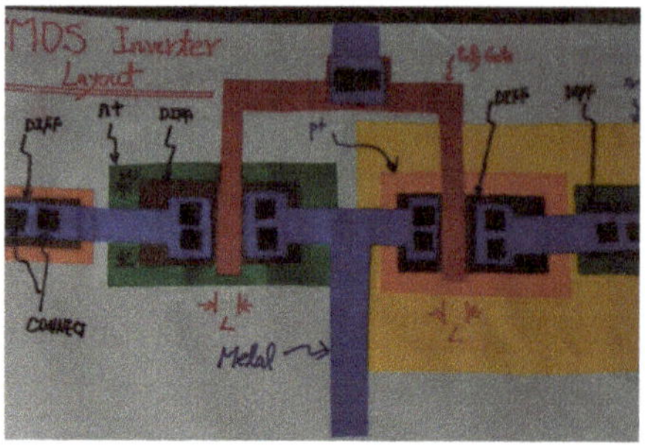

Figure 9. The demonstration aid for IC layout.

Figure 10 shows the samples of students' assignments. In Course 2, silicon wafers, packaged chips, and bare dies are shown. Figure 11 shows the teaching aids and their demonstration. Figures 12 and 13 show students' layout diagrams under the requirements of design-rule-check (DRC) and layout-versus-schematic (LVS). The two IC layout diagrams for the eight-bit adder were sampled from students' final projects. Due to the COVID-19 pandemic, the lecture was difficult to offer. Part of the layout was created in on-site classes, and the remaining part was completed in distance learning.

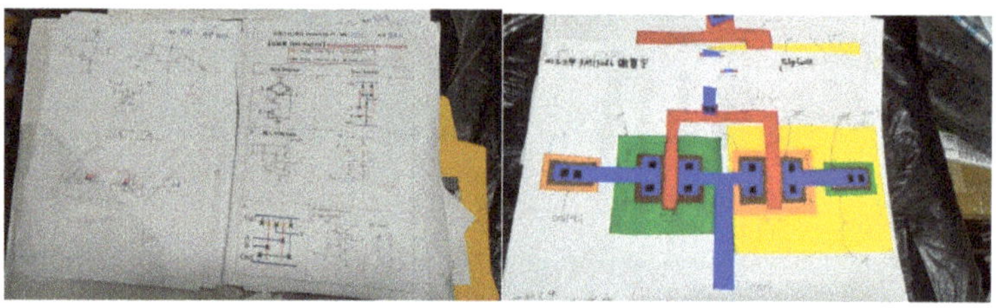

Figure 10. Students' assignments during Course 2.

Figure 11. Wafer and chip physical demonstration by using an optical microscope.

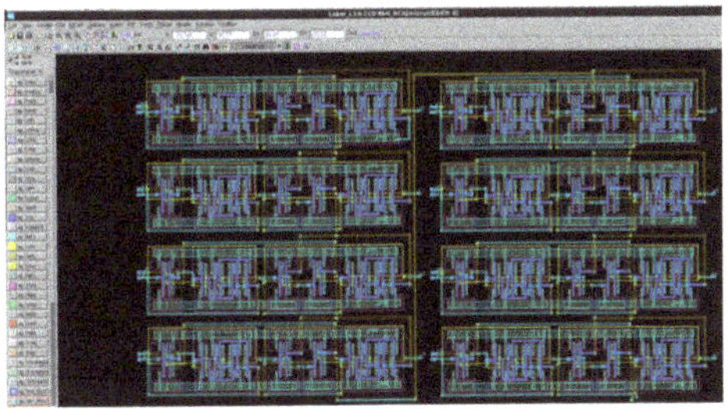

Figure 12. Student's report for the final project.

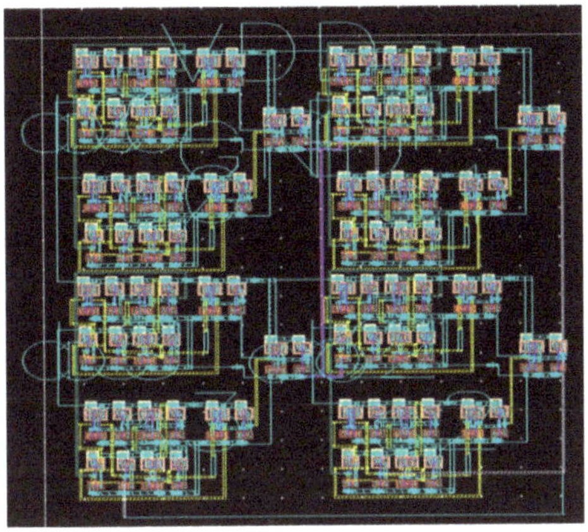

Figure 13. Student's report for the final project.

How to resolve the layout problem efficiently online was a challenge for effective learning. During distance learning, students were required to make a self-test via the internet and self-evaluate the learning outcomes. The quizzes were conducted by using Google Forms. Figure 14 shows the samples of these quizzes created in Google Forms. To improve students' practical ability in the IC layout, we invited IC engineers to give lectures. Figure 15 shows the lecture given by the IC engineers. The overall goal of the research program was achieved with rich and heuristic class activities, and students achieved satisfactory learning results in IC technology, circuit analysis, and IC layout skill.

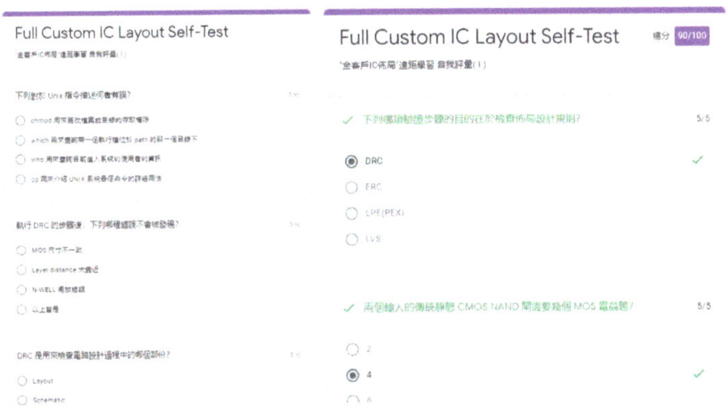

Figure 14. Student's self-test on the internet to self-evaluate the learning outcomes.

Figure 15. An industry IC engineer was invited to give lectures.

5. Conclusions

Due to the outbreak of COVID-19 in the late stage of Course 2, a proper teaching method needed to be found for distance learning. With internet technology, TAs, and the joint efforts of students, the course continued without interruption. The achieved rate of the IC layout project reached 95%. The student's evaluation results showed that the teaching achievement of the courses was the best compared to the historical data. A new and interesting teaching process in the courses during the period of COVID-19 was achieved in cooperation with the school, students, and teachers owing to internet technology.

Funding: This research was funded by the Teaching Practice Research Program, 2020, Ministry of Education (MOE), Taiwan, R.O.C., grant number No. PEE1090359.

Informed Consent Statement: Informed consent was obtained from all subjects involved in the study.

Data Availability Statement: Data Availability Statements are available in the National Chin-Yi University of Technology, Taichung, Taiwan, R.O.C. at https://evaluation.ncut.edu.tw/ (accessed on 1 September 2021).

Acknowledgments: This work was supported by the Teaching Practice Research Program (No. PEE1090359), Ministry of Education (MOE), Taiwan, R.O.C. We would like to express our sincere thanks to the National Chin-Yi University of Technology, Taiwan for full support and all students who participated in this project. Thanks to teaching assistant Wei-Jen Lin for the video recording and help with the IC layout in the classroom.

Conflicts of Interest: The authors declare no conflict of interest.

References

1. Wikipedia. Available online: https://en.wikipedia.org/wiki/COVID-19 (accessed on 14 May 2023).
2. COVID-19 Dashboard. Available online: https://covid-19.nchc.org.tw/index.php (accessed on 14 May 2023).
3. Hung, Y.-C. *By Teaching Guide for Chip Learning by Modular and Self-Built for Technical and Vocational Education System*; Project Report for MOE Teaching Practice Research Program; Ministry of Education: Taiwan, China, 2021.
4. Ministry of Education. Available online: https://tpr.moe.edu.tw/index (accessed on 14 May 2023).
5. Bertrand, Y.; Azais, F.; Flottes, M.-L.; Lorival, R. A successful distance-learning experience for IC test education. In Proceedings of the IEEE International Conference on Microelectronic Systems Education, Arlington, MA, USA, 19–21 July 1999.
6. Prevedello, P.; Aita, A.L. Virtual collaborative environment for support of teamwork-based education of analog IC design. In Proceedings of the 2018 IEEE International Symposium on Circuits and Systems (ISCAS), Florence, Italy, 27–30 May 2018; pp. 1–4.
7. Zezin, D. Modern open source IC design tools for electronics engineer education. In Proceedings of the VI International Conference on Information Technologies in Engineering Education, Moscow, Russia, 12–15 April 2022.
8. Yasuda, T.; Tanaka, T.; Fujimura, N.; Yamada, A.; Hattori, T. Suggestion of an IC design education method for understanding intuitively and feelingly. In Proceedings of the International Conference on Biometrics and Kansei Engineering, Tokyo, Japan, 5–7 July 2013.
9. Aikio, J.P.; Rahkonen, T. CAD tools utilized in microelectronic IC design education in the University of Oulu. In Proceedings of the 10th European Workshop on Microelectronics Education, Tallinn, Estonia, 14–16 May 2014.

10. Campi, F.; Ancill, J. Introducing IC reliability elements in digital circuits and systems design education. In Proceedings of the IEEE International Symposium on Circuits and Systems, Montreal, QC, Canada, 22–25 May 2016.
11. Grout, I. Digital IC test development engineering teaching and learning in higher education. In Proceedings of the 30th Annual Conference of the European Association for Education in Electrical and Information Engineering, Prague, Czech Republic, 1–3 September 2021.

Disclaimer/Publisher's Note: The statements, opinions and data contained in all publications are solely those of the individual author(s) and contributor(s) and not of MDPI and/or the editor(s). MDPI and/or the editor(s) disclaim responsibility for any injury to people or property resulting from any ideas, methods, instructions or products referred to in the content.

engineering proceedings

Proceeding Paper

A Swimming Goggles Optical Design by Fresnel Lenses †

Feng-Ming Yeh [1,*], Liang-Ying Huang [2], Chao-Kai Chang [3], Ya-Hui Hsieh [4], Hsuan-Fu Wang [5], Rong-Seng Chang [2] and Der-Chin Chen [1]

1 Department of Optometry, Yuanpei University of Medical Technology, Hsinchu 300044, Taiwan; kanatasan.tw@yahoo.com.tw
2 Department of Optics and Photonics, National Central University, Taoyuan 320317, Taiwan; 108286001@cc.ncu.edu.tw (L.-Y.H.); rschang2000@hotmail.com (R.-S.C.)
3 Department of Optometry, Da-Yah University, Changhua 515006, Taiwan; chaokai@mail.dyu.edu.tw
4 Department of Healthcare Information and Management, Ming Chuan University, Taoyuan 333001, Taiwan; hsieh_yahui@hotmail.com
5 Department of Aeronautical Engineering, National Formosa University, Yunlin 632301, Taiwan; hfwang@nfu.edu.tw
* Correspondence: optfmy@yahoo.com.tw; Tel.: +886-36102322
† Presented at the 3rd IEEE International Conference on Electronic Communications, Internet of Things and Big Data Conference 2023, Taichung, Taiwan, 14–16 April 2023.

Abstract: Currently, many swimming goggle lenses use optical plates to maintain zero refractive power in air and water. However, people's widespread use of 3C products has increased myopia significantly, so lenses have a demand for refractive power. Lenses will have different refractive power problems in water and air media. Therefore, we solved the refractive power change in air and water by using a plane Fresnel lens with a diopter to replace plano-concave lenses. In this study, a first-order design was created and then the microstructure of the Fresnel lens was optimized using optical software. The Fresnel lens simulation results showed that the error was within 5%, which was compared with the data using the lensmaker's equation calculation. For swimming goggles, this error value is tolerable for human vision.

Keywords: Fresnel lens; refractive power; refractive index

Citation: Yeh, F.-M.; Huang, L.-Y.; Chang, C.-K.; Hsieh, Y.-H.; Wang, H.-F.; Chang, R.-S.; Chen, D.-C. A Swimming Goggles Optical Design by Fresnel Lenses. *Eng. Proc.* **2023**, *38*, 90. https://doi.org/10.3390/engproc2023038090

Academic Editors: Teen-Hang Meen, Hsin-Hung Lin and Cheng-Fu Yang

Published: 3 August 2023

Copyright: © 2023 by the authors. Licensee MDPI, Basel, Switzerland. This article is an open access article distributed under the terms and conditions of the Creative Commons Attribution (CC BY) license (https://creativecommons.org/licenses/by/4.0/).

1. Introduction

In summer, many people go to the beach or swimming pools to cool off, and most swimmers wear swimming goggles to improve their visibility and protect their eyes from harmful chemicals. Thus, the choice of swimming goggles is crucial [1,2]. Swimming goggles separate the eyes from the water and restore air contact in the swimming goggles, which helps people to see objects in the water better than they would without goggles. The most common swimming goggles have a spherical surface, because the surface of the lens is slightly convex for the refractive power in the water. These lenses have a similar effect to a concave lens. Without myopia, it is similar to wearing myopia glasses. When swimming, eyes come to have a high degree of astigmatism in a normal field of vision due to repeated entry into the water over a short period; this causes eye damage and fatigue, with dizziness, vomiting, and other symptoms in the body. A proper pair of swimming goggles must maintain consistent vision in and out of the water without causing dizziness and with the functions of being waterproof, anti-fog, low resistance, causing no pressure on the eyeballs, a wide field of vision, and a high light transmittance. Swimming goggles with vision correction must have the same function as general land myopia glasses in water [3,4]. They have a refractive error correction function, accurately fusing left and right eye images and maintaining the same pupillary distance. The lens must be light, so that the field of vision is not distorted. Swimming goggles adopt an afocal lens, which solves the diopter's problems of spherical and curved swimming goggles. People with myopia and astigmatism must carefully choose goggle lenses with a suitable refractive power of the lens.

In this study, we designed the refractive power of a swimming goggles lens with Fresnel lenses, which are widely used in various optical applications and whose most significant advantages are their lightweight and flat optical surfaces. Basic design rules were applied in the first-order design and feasibility studies. Fresnel lenses work on the principle that the refractive index of the lens is contained only on the surface of the lens, removing as much optical material as possible while maintaining the curvature of the surface. Practical methods for compressing the refractive power of the lens surface to a flat surface require concentric grooves with a small prism pitch, oblique component, and draft component. The general method is to specify a tilt angle relative to the lens plane and a draft angle close to the normal. By using the Plano-Fresnel lens with a diopter to replace the plano-concave lens of swimming goggles, the different refractive power in the air and water due to different media can be overcome. By using the optimization design of the Fresnel lens microstructure, lenses with vision correction have the same refractive power in two various media.

2. Principle of Fresnel Lens

A Fresnel lens, also known as a threaded lens, is designed with a large aperture and characterized by a short focal length, less material consumption, and a smaller weight and volume than ordinary lenses [5,6]. Figure 1 shows that (A) is a convex lens and (B) is a Fresnel lens with the same optical characteristics. When a Fresnel lens is used in afocal optical components, attention must be paid to minimizing the micro-grooves' impact on vision. In the design, this is first realized by selecting the facet spacing to be less than or equal to the resolution of the human eye, which makes the human eye almost unable to see the existence of micro prisms. A healthy human eye has a visual acuity of around one arc-minute (or 1/60°). This can be used to compute the maximum pitch size that will be visible for a given distance from the eye.

$$w \leq z \times 2.91 \times 10^{-4} \quad (1)$$

where w is the maximum pitch size that can be visually resolved and z is the effective distance of the lens to the eye. To ensure that the pitch of the Fresnel lens does not produce a beat frequency limited by the minimum resolvable pitch of any other micro-prismatic components of the system, the rule to select the pitch is defined as:

$$d_1 = (m + 0.35) \times d_2 \quad (2)$$

where d_1 is the prism pitch of the Fresnel lens, m is an integer, where the larger it is the better, and d_2 is the pitch of the other micro-prismatic component in the lens. Geometrically, the smaller the prism size, the closer the small flat prism slope facets come to approximating the idealized aspherical surface. However, diffractive effects operate contrarily.

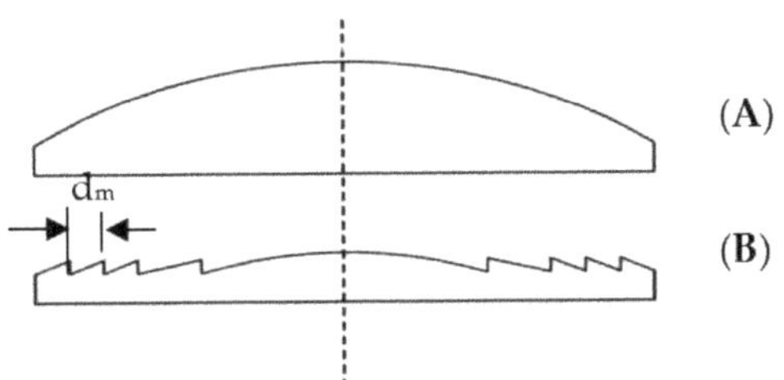

Figure 1. From a convex lens (**A**), to a Fresnel lens (**B**).

The object is beamed onto the image plane after passing through the Fresnel lens and is classified according to the relative position of the output beam and the groove on its optical surface. The Fresnel lenses are organized into a converging lens, collimator, and diverging lens. For example, the Fresnel lens with a positive focal length is used to converge light.

Thus, it is used as a converging lens and collimating lens. The Fresnel lens with a negative focal length is used to diverge light as a diverging lens. Many researchers have proposed algorithms to simulate the parameters of the focusing Fresnel micro-prism structure and successfully and easily fabricated this element using diamond-cutting technology. The Zemax optical design software was used in this study to design a radial mode. The microstructure on the optical plane has many concentric grooves. It radiates outward from the center and the cross-sectional shape of the lens is a miniature right-angle prism. The value of the shape parameters, + Depth and -Frequency, are essential for the groove. If this parameter is positive, it corresponds to the depth of each groove in shot units. If negative, it corresponds to the frequency of the groove. Equations (3)–(5) optimize the pitch size of the grooves (w) and the lens diameter (Φ).

$$\Phi = (m \times h)/[1 + (m \times s_1)/(l - s_1)] \quad (3)$$

$$w = 1.5 \times (\lambda \times f)^{1/2} \quad (4)$$

$$w \leq p \times \tan[(1/60)^\circ] \quad (5)$$

where λ, f, s_1, m, l, and p are the wavelength of the light source, the focal length of the lens, the distance between the lens and the object, the magnification of the lens, the distance between the object and the observer, and the distance between the observer and the lens [4–6].

3. Design and Experiment Results

The design process included the subject requirement, first-order design, optimization of the plano-concave lens, transformation into a Fresnel lens, optimization of the microstructure parameter, image evaluation, and tolerance analysis. As there are many types of refractive errors, we considered myopia only, so the subject requirement was the diopter of the myopia. The first-order design discussed the relationship between the object, image, and lens's focal length. The thin lens formula decided the image position and size. If the distances from the object to the lens and from the lens to the image are S_1 and S_2, respectively, and the focal length of the thin lens is f in air, then the relationship between the three parameters is as follows.

$$(1/S_1) + (1/S_2) = (1/f) \quad (6)$$

This can also be put into another form.

$$X_1 \times X_2 = f^2 \quad (7)$$

where $X_1 = S_1 - f$ and $X_2 = S_2 - f$.

With myopia, a user requirement includes pupil distance, the refractive error (diopter) of the lens, the plano-concave lens, and the face size for making the goggles. The parameter of the first-order design includes the object distance, object high, image high, material of the lens, and other constraints. Optimizing the plano-concave lens makes the optical lens have no diopter difference in different object space mediums, water, and air. Other conditions, such as the back focal length, remain fixed, and the image quality and resolution are the same. In addition to obtaining the best spherical radius of the curvature, it is also possible to use aspheric surfaces. The purpose of transforming into a Fresnel lens is to use a flat optical plate to reduce the diopter difference between the lens in water and air, so the spherical surface of the plano-concave lens after optimization is converted into a Fresnel surface. In the above two steps, the focal length is calculated from the known refractive error and the radius of the curvature of the plano-concave lens can be further calculated using Equation (8).

$$S = (n_2 - n_1)/R \tag{8}$$

where R is the curvature radius of the spherical surface, S is the refractive error, and n_2 and n_1 are the refractive indices on both sides of the spherical surface.

When the unit of R is m, the unit of S is diopter (D). The curvature radius of the spherical surface of the plano-concave lens is 1.0 m and the refractive index of the lens material is 1.5. If it is used in air, S is 2.0 D. The Fresnel lens adopts a discontinuous surface profile composed of a hundred concentric rings, and tiny serrated prisms are arranged on these small rings to retain the rings' ability to focus the light toward the center or divergence. The parameters for controlling the tiny saw tooth (in μm) are the pitch angle, depth, and pitch. By optimizing the microstructure parameter of the Fresnel surface, the optical power of the Fresnel surface and plano-concave lens become the same, still with an error. The Fresnel surface replaces the spherical surface, which is only the initial condition unchanged. However, the error value needs to be within the acceptable range of vision to find the best microstructure parameters of the Fresnel surface here and meet the requirements and image quality. In the optimization method, the microstructure's frequency and pitch angle are changed to optimize the minimum light spot. After the above steps are completed, the image evaluation is performed for various aberrations and spot size values. The aberration has a spherical aberration, coma, astigmatism, distortion, field curvature, and chromatic aberration.

After the image evaluation is qualified, the tolerance analysis is performed, and the entire design is completed. Then, the pilot run is performed. Tolerance is the limitation of the processing accuracy in the lens production and manufacturing, so that after the finished product is completed, it meets the expected specifications at the design time. To meet the tolerance requirements, manufacturers must know what manufacturing and testing methods are used when producing these lenses to make the production process model consistent with reality. These engineering problems have corresponding theoretical models for analyses. For statistical tolerance, the variation in a set of inputs is taken to calculate the expected variation in the outputs of interest. In mechanical engineering, product designs are composed of features whose tolerance values are related to various aspects of these features. In this study, we conducted a preliminary design of a Fresnel lens for swimming goggles and did not elaborate on the image evaluation and tolerance analysis.

The design method was as follows. The non-sequential method of infinite grooves was used to simulate the Fresnel lens of a −2.0 D flat and concave lens. For the relevant conditions, the radius, thickness, material, and imaging spot diameter of the radial Fresnel lens were set as 60 mm, 1 mm, BK7, and 117 μm, respectively. The distance between the lens and the image plane was 500 mm. The refractive index of BK7 was 1.517 and the V number was 64.169. Table 1 lists the nine different refractive diopter lenses commonly used in stores and the related microstructure parameters used in the simulation process. These parameter values were optimized with the minimum light spot as the target. Figure 2 shows the Flow chart of Fresnel lens for swimming goggles design. Figure 3 shows the cross-sectional structure of the Plano-Fresnel lens for the swimming goggles using the layout of zemax optical design program. The blue lines are marginal rays of 0° field of view in object space, and the light ray is parallel to the optical axis. A Fresnel lens with a negative focal length diverges light for people with myopia.

Table 1. Optical parameters of nine different diopter lenses.

Plano-Concave Lens			Plano-Fresnel Lens			
Diopter (D)	f (mm)	R (mm)	Freq.	Pitch Angle (°)	Spot Size (μm)	BFL (mm)
−2	−500.00	258.50	−2	0.5	117	−500.69
−3	−333.33	172.33	−1	0.5	98	−332.48
−4	−250.00	129.25	−1	0.5	117	−250.8
−5	−200.00	103.40	−1	0.5	245	−201.5
−6	−166.67	86.17	−1	0.5	466	−166.3
−7	−142.86	73.86	−1	0.5	733	−141.3
−8	−125.00	64.63	−1	0.5	941	−127.3
−9	−111.11	57.44	−1	0.5	1306	−110.6
−10	−100.00	51.70	−1	0.5	1528	−103.1

Note: The thickness of plano-concave lens and plano-Fresnel lens are 2 mm.

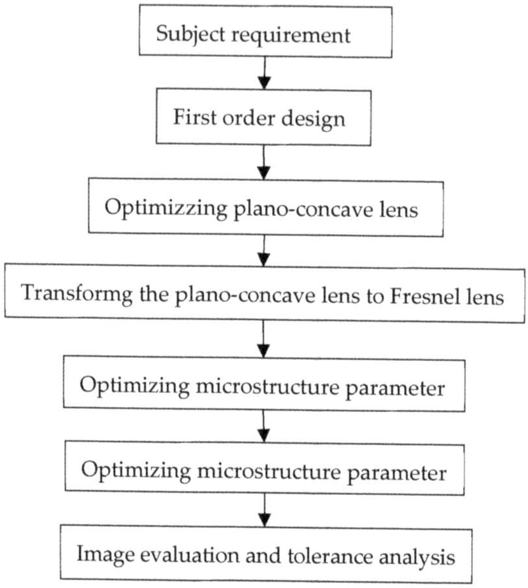

Figure 2. Flow chart of Fresnel lens for swimming goggles design.

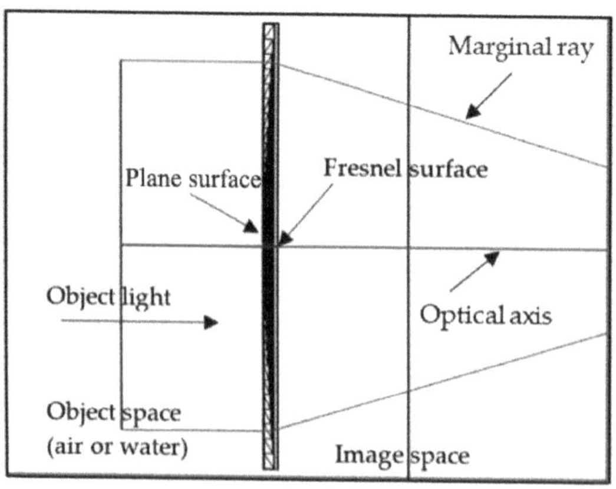

Figure 3. The cross-sectional structure of the Plano-Fresnel lens of swimming goggles [7].

4. Conclusions

The purpose of using a Plano-Fresnel lens for vision correction instead of a plano-concave lens is to solve the refractive power change in swimming goggles in the air and water. For people with high myopia, the thickness of the goggle lenses' edge is too thick and too heavy, and the production yield decreases. The product was designed with an innovative method that was used in this study and avoided the above problems. The thickness of the lens' edge was the same as that of the center. In this study, the lens' diopter of the swimming goggles was calculated by satisfying the object space of water and air at the same time. Afterward, the spherical optical surface of the plano-concave lens was calculated and converted into a Fresnel surface. Finally, the microstructure of the Fresnel surface was optimized using the optical design program. During this process, the lens's diopter of the swimming goggles in two different media was highly close to the same value. The simulation results were compared to those calculated with the lens maker's formula and the error of the two data was within 5%, which is acceptable for swimming goggle lenses.

Author Contributions: Conceptualization, F.-M.Y. and L.-Y.H.; methodology, C.-K.C. and Y.-H.H.; software, H.-F.W.; validation, F.-M.Y. and R.-S.C.; investigation, C.-K.C.; data curation, F.-M.Y.; writing—review and editing, L.-Y.H. and D.-C.C. All authors have read and agreed to the published version of the manuscript.

Funding: This research project was supported by the National Science and Technology Council, under Grant No. 110-2622-8-008-006-TS1.

Institutional Review Board Statement: Not applicable.

Informed Consent Statement: Not applicable.

Data Availability Statement: Not applicable.

Conflicts of Interest: The authors declare no conflict of interest.

References

1. Kumar, A.; Bisht, D.S. A User-Centred Design Approach to Investigate the Design Parameters for Prescription Swimming Goggles. In *Research into Design for a Connected World*; Springer: Singapore, 2019; pp. 475–485.
2. Decorato, F. Optically Corrected Swimming Goggles. U.S. Patent 3994345A, 16 March 1976.
3. Miller, O.E.; McLeod, J.H.; Sherwood, W.T. Thin sheet plastic Fresnel lenses of high aperture. *J. Opt. Soc. Am.* **1951**, *41*, 807–815. [CrossRef]
4. Languy, F.; Fleury, K.; Lenaerts, C.; Loicq, J.; Regaert, D.; Thibert, T.; Habraken, S. Flat Fresnel doublets made of PMMA and PC: Combining low cost production and very high concentration ratio for CPV. *Opt. Express* **2011**, *19*, A280–A294. [CrossRef] [PubMed]
5. Arthur, D.; Frank, K. Optical Design using Fresnel Lenses [Tutorial]. *Opt. Photon.* **2007**, *2*, 52–55.
6. Xie, W.T.; Dai, Y.; Wang, R.Z.; Sumathy, K. Concentrated solar energy applications using Fresnel lenses: A review. *Renew. Sustain. Energy Rev.* **2011**, *15*, 2588–2606. [CrossRef]
7. ZEMAX®. *Optical Design Program. User's Manual*; ZEMAX: Canonsburg, PA, USA, 8 July 2011.

Disclaimer/Publisher's Note: The statements, opinions and data contained in all publications are solely those of the individual author(s) and contributor(s) and not of MDPI and/or the editor(s). MDPI and/or the editor(s) disclaim responsibility for any injury to people or property resulting from any ideas, methods, instructions or products referred to in the content.

engineering proceedings

MDPI

Proceeding Paper

Exploring Research on Elevating Male's Multiple Sexual Climaxes †

Chun-Ming Shih [1], Yung-Jui Huang [1], Jiin-Chyuan Mark Lai [2,*] and Ming-Yuan Hsieh [1,3,*]

1 Graduate School of Human Sexuality, Shu-Te University, Kaohsiung 82445, Taiwan; cmshih0901@stu.edu.tw (C.-M.S.); bbhouse@stu.edu.tw (Y.-J.H.)
2 Department of Applied Foreign Languages, TransWorld University, Douliu 640302, Taiwan
3 Department of International Business, National Taichung University of Education, Taichung 40306, Taiwan
* Correspondence: marklai07@gmail.com (J.-C.M.L.); s22871101@stu.edu.tw (M.-Y.H.)

† Presented at the 3rd IEEE International Conference on Electronic Communications, Internet of Things and Big Data Conference 2023, Taichung, Taiwan, 14–16 April 2023.

Abstract: No matter how old people are, the ability to feel the sexual climax does not change. Whether a male can enjoy sexual pleasure in multiple sexual climaxes without pressures of the physical erection and ejaculations is a topic to search for, though. A young man can enjoy the thrill of riding on the sexual waves, while an old man is also supposed to also enjoy it in the sexual pool without any pressure of erection and ejaculations. The frequency of ejaculation gradually decreases with age, but satisfaction brought by sexual climax does not decrease. The physiological response to sexual climax may vary with age. As a result, the sexual climaxes of the male and the female in the sexual response cycle are worth exploring to evaluate the quality of sexual life. With this, interfering factors are also necessary to determine including the physical erection and ejaculation, the sexual pleasure, motivation, emotion, and satisfaction of the sexual climax for males. Then, the best solution (decision) for the development of the closed couple relationship can be found.

Keywords: sexual intercourse; multiple male orgasm (MMO); erection; ejaculation

Citation: Shih, C.-M.; Huang, Y.-J.; Lai, J.-C.M.; Hsieh, M.-Y. Exploring Research on Elevating Male's Multiple Sexual Climaxes. *Eng. Proc.* **2023**, *38*, 31. https://doi.org/10.3390/engproc2023038031

Academic Editors: Teen-Hang Meen, Hsin-Hung Lin and Cheng-Fu Yang

Published: 21 June 2023

Copyright: © 2023 by the authors. Licensee MDPI, Basel, Switzerland. This article is an open access article distributed under the terms and conditions of the Creative Commons Attribution (CC BY) license (https://creativecommons.org/licenses/by/4.0/).

1. Introduction

Sexual climax is a manifested sexual pleasure from the sudden release or exposure of the nervous system response and rhythmic muscle contractions in the pelvic region due to accumulated sexual stimulus in the sexual response cycle. In detail, the sexual climax of both men and women is created and controlled by the autonomic nervous system with involuntary actions and physical responses including muscle rigidities in multiple areas of the body, feelings of euphoria and satisfaction, and frequent vocalizations and incapability in autonomy of body movements [1]. Most males express silence in reaching sexual climax because they are taught to control and hide their true exciting and emotion [2]. However, several males display a series of strong physiological reactions such as loud shouting or violent body twitching reaching the sexual climax [3]. According to the professional research of [4–6], both males and females have four phases in the sexual response cycle including the initial excitement, growing plateau, highest sexual (orgasmic), and last resolution phases as displayed in Figure 1 [7–9].

Previous research described the rhythmic contraction in the sexual climax and confirmed that males and females feel the rhythmic contraction with an interval of 0.8 s, and the speed and intensity of the post-contraction gradually slow down after the sexual climax as the neurohormones, oxytocin and prolactin, are released in the brain to create a sense of relaxation which leads to entering the sexual refractory period [10].

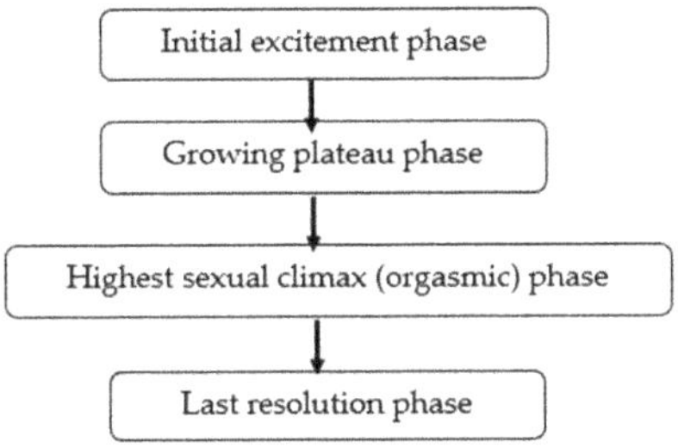

Figure 1. Human sexual response cycle.

2. Literature Review

The mechanism of the sexual climax is triggered by endorphins (brain endorphins). In the sexual refractory period, the majority of males feel exhausted and an inability for erection, while the majority of females feel indescribable pleasures and happiness with uncontrollable physiological muscle spasms. The empirical research confirmed that the majority of males reaching the sexual climax with ejaculation enter a refractory period immediately after that. In most cases, they are not able to ejaculate in the refractory period. The difference between the male and female is that the female enters the refractory period after sexual climax, so they can have multiple orgasms. However, the male enters a refractory period immediately after reaching the sexual climax, which results in the inability to ejaculate. When a male has an orgasm, he does not necessarily ejaculate. For example, when a man is drunk, stressed, or too tired or just wants to have quick intercourse with his partner, he can pretend that he reaches an orgasm with ejaculation. Men also fake orgasms as women do. However, as the "orgasm" occurs in the brain, men have a "deep" feeling which is the source of real happiness. When the male masturbates, or his emotion and body are closely connected with a partner, even without physical contact, orgasms can be achieved. The closer the desire to reach an orgasm is, the easier the sexual pleasure is gained.

Most males consider that sexual climax and ejaculation appear and are achieved simultaneously in their experiences. In the traditional concept and physical activities to reach orgasm, males take a series of stimuli and excited actions on their penis, especially on the glans, to reach ejaculation and sexual climax. Apparently, in terms of the male's ejaculation, the sense is the "reflex physical action". Several males ejaculate even without a sense of sexual climax while executing sexual stimuli in various sexual activities such as masturbation. This is just a physical behavior caused by uncontrollable physiological muscle spasms.

Many males have the misunderstanding that ejaculation is orgasm, and orgasm is ejaculation. However, in fact, ejaculation and sexual climax are different. For many males, the pleasure feelings of sexual climax accompany ejaculation and, then, the sexual refractory period. They are semi-forced to feel exhausted and unable to keep erect during the release of the neurohormones oxytocin and prolactin. A minority of males still maintain the strong feeling of sexual climax after ejaculation and can ejaculate without any sense of the sexual climax. Furthermore, another group of males can experience short-term or weak orgasms during sexual intercourse in addition to ejaculation. Several males deliberately stop sexual actions for a while to consciously bear the physical pleasures of not reaching sexual climax or ejaculate to obtain more intensive pleasures for the next climax.

Nevertheless, as a physiological concept, the majority of males decrease their frequency of ejaculation due to the increase in age and the decrease in hormones. The secretion of semen in middle and old ages in the sexual climax does not disappear, and old males have the feeling of sexual climax without ejaculation. However, most males tend to have unnecessary pressures on erection and ejaculation during the sexual climax. Even the amount of semen is regarded as a test of attractiveness, which inadvertently affects males to ejaculate as long as they have sex. Under this pressure, it is difficult to enjoy sex and to remain erect [11–15]. This situation wrecks the sexual interest and relationship of a couple as shown in Figure 2.

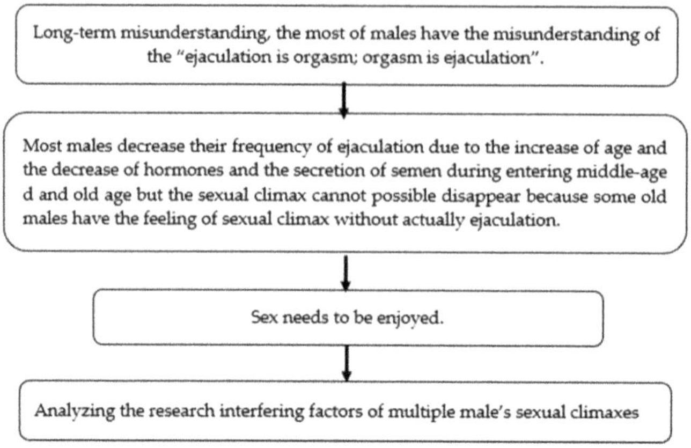

Figure 2. Human sexual response cycle.

3. Research Method

We researched the sexual climax of the males and females in the sexual response cycle and evaluated the quality of the males' sexual climax. The research content is described in Figure 3 [16–24].

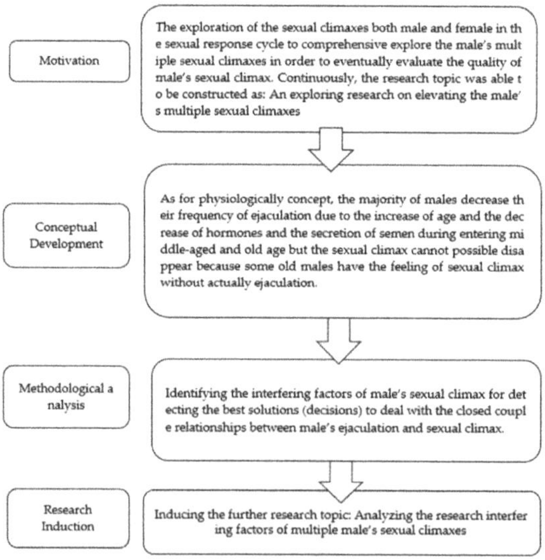

Figure 3. Research method.

4. Discussion

The exploration of the sexual climax of the male and female in the sexual response is explored to evaluate the quality of sexual life. Future research is needed to investigate the interfering factors: the physical erection and ejaculations, sexual pleasure, sexual motivation, sexual emotion of the sexual climax, and sexual satisfaction of the sexual climax for multiple sexual climaxes as shown in Figure 4 [25–29] to find the best solution (decision) for the close couple relationship for achieving sexual climax.

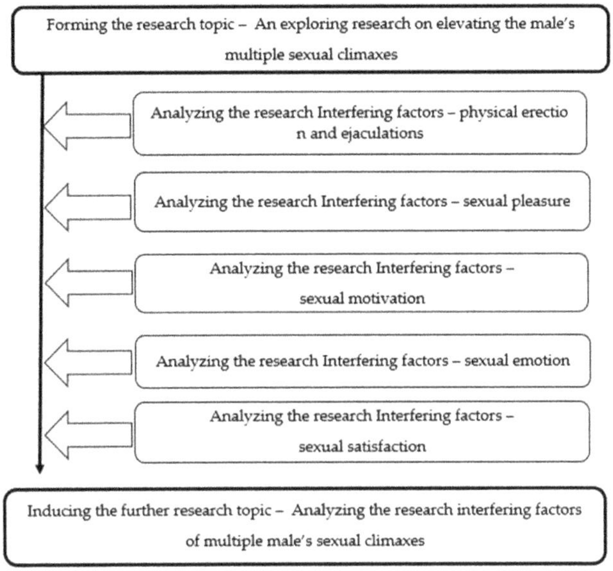

Figure 4. Human sexual response cycle.

5. Conclusions and Recommendation

Regardless of age, people feel sexual climax. It is important to find how a male can enjoy sexual pleasures in multiple sexual climaxes without the pressures of erection and ejaculation. Not only young men but old men also need to enjoy an active sexual life without any pressure. The ejaculation gradually decreases with age, but the satisfaction does not. Therefore, knowing how to have an active sexual life is important in life.

Author Contributions: Conceptualization, C.-M.S., Y.-J.H. and M.-Y.H.; methodology, M.-Y.H.; validation, J.-C.M.L.; formal analysis, M.-Y.H.; investigation, Y.-J.H.; resources, J.-C.M.L.; writing—original draft preparation, M.-Y.H.; writing—review and editing, M.-Y.H.; visualization, C.-M.S. and Y.-J.H.; supervision, M.-Y.H.; project administration, C.-M.S. and M.-Y.H. All authors have read and agreed to the published version of the manuscript.

Funding: This research was supplied by the research supporting research projects (MOST 110-2420-H-002-003-MY3-Y11209) of the Ministry of Science and Technology and (NTCU111103) from the National Taichung University of Education.

Institutional Review Board Statement: This research did not execute the questionnaires and interview and hence, this research did not require ethical approval.

Informed Consent Statement: Not applicable because this research did not involve any humans.

Data Availability Statement: No new data were created.

Conflicts of Interest: There is no conflict of interest in this research.

References

1. Levin, R.J. Critically revising aspects of the human sexual response cycle of Masters and Johnson: Correcting errors and suggesting modifications. *Sex. Relatish. Ther.* **2008**, *23*, 393. [CrossRef]
2. Chivers, M.L.; Bailey, J.M. A sex difference in features that elicit genital response. *Biol. Psychol.* **2005**, *70*, 115. [CrossRef] [PubMed]
3. Chivers, M.L. A brief review and discussion of sex differences in the specificity of sexual arousal. *Sex. Relatish. Ther.* **2005**, *20*, 377. [CrossRef]
4. Masters, W.H.; Johnson, V.E. *Human Sexual Response*; Bantam Books: New York, NY, USA, 1966.
5. Brody, S.; Weiss, P. Simultaneous penile–vaginal intercourse orgasm is associated with satisfaction (sexual, life, partnership, and mental health). *J. Sex. Med.* **2011**, *8*, 734. [CrossRef]

6. Chadwick, S.B.; van Anders, S.M. Do women's orgasms function as a masculinity achievement for men? *J. Sex. Res.* **2017**, *54*, 1141–1152. [CrossRef]
7. Fahs, B.; Plante, R. On 'Good Sex' and other dangerous ideas: Women narrate their joyous and happy sexual encounters. *J. Gen. Stud.* **2017**, *26*, 33. [CrossRef]
8. Freihart, B.K.; Sears, M.A.; Meston, C.M. Relational and interpersonal predictors of sexual satisfaction. *Curr. Sex. Health Rep.* **2020**, *12*, 136. [CrossRef]
9. Hamilton, L.; Armstrong, E.A. Gendered sexuality in young adulthood: Double binds and flawed options. *Gen. Soc.* **2009**, *23*, 589. [CrossRef]
10. Masters, W.H.; Johnson, V.E.; Kolodny, R.C. *Masters and Johnson on Sex and Human Loving*; Little, Brown and Company: Boston, MA, USA, 1988.
11. Masters, W.H.; Johnson, V.E.; Kolodny, R.C. *Heterosexuality*; Harper Collins: New York, NY, USA, 1994.
12. Huang, C.C.; Chan, Y.-K.; Hsieh, M.Y. The Determinants of ESG for Community LOHASism Sustainable Development Strategy. *Sustainability* **2022**, *14*, 11429. [CrossRef]
13. Chan, Y.-K.; Hsieh, M.Y. An Empirical Study on Higher Education C-ESG Sustainable Development Strategy in Lower-Birth-Rate Era. *Sustainability* **2022**, *14*, 12629. [CrossRef]
14. Hsieh, M.Y. The Sustainable Development and Strategic Approaches for Contemporary Higher Education. *Sustainability* **2022**, *14*, 12925. [CrossRef]
15. Huang, C.-C.; Chan, Y.-K.; Hsieh, M.-Y. Preliminary Research on the Sustainable Determinants of Taiwanese Ecotourism with the International Standards. *Int. J. Environ. Res. Public Health* **2022**, *19*, 14489. [CrossRef] [PubMed]
16. Chan, Y.-K.; Hsieh, M.-Y.; Usak, M. A Concrete Study on Social-Media Connection of Global Literacy Abilities in MOOCs under the Dual Impacts of Lower Birth-Rate and COVID-19. *Sustainability* **2021**, *13*, 2203. [CrossRef]
17. Usak, M.; Hsieh, M.-Y.; Chan, Y.-K. A Concertizing Research on Making Higher Education Sustainability Count. *Sustainability* **2021**, *12*, 1806–1821.
18. Duran, M.; Usak, M.; Hsieh, M.-Y.; Uygun, H. A New Perspective on Pedagogical Content Knowledge: Intellectual and Emotional Characteristics of Science Teachers. *Rev. Cercet. Si Interv. Soc.* **2021**, *72*, 9–32. [CrossRef]
19. McCracken, H. The End of the Zero-Sum Game. 27 January 2011. Available online: http://technologizer.com/2011/01/27/the-end-of-the-zero-sum-game/ (accessed on 12 November 2022).
20. Greenwald, A.; Littman, M.L. Introduction to the special issue on learning and computational game theory. *Mach Learn* **2007**, *67*, 3. [CrossRef]
21. Wu, T.-L.; Hsieh, M.-Y.; Min, K.-W.; Yu, M.-T.; Ho, C.-T. Use of Sensor Technologies in Online Courses in Post-COVID-19 Era. *Sens. Mater.* **2021**, *33*, 2045–2062. [CrossRef]
22. Hsieh, Y.-M. Online learning era: Exploring the most decisive determinants of MOOCs in Taiwanese higher education. *Eurasia J. Math. Sci. Technol. Educ.* **2016**, *12*, 1163. [CrossRef]
23. Hsieh, Y.-M.; Usak, M. High Education Radical Transformation Era: How Teachers' Competency can Enhance the Students' Employability. *Rev. Cercet. Si Interv. Soc.* **2020**, *68*, 95. [CrossRef]
24. Hsieh, Y.-M. Employing MCDM methodology to verify correlationship between social media and service quality in the dynamic m-commerce era. *J. Internet Technol.* **2018**, *19*, 225.
25. Epstein, M.; Calzo, J.P.; Smiler, A.P.; Ward, L.M. Anything from making out to having sex: Men's negotiations of hooking up and friends with benefits scripts. *J. Sex Res.* **2009**, *46*, 414. [CrossRef] [PubMed]
26. Bay-Cheng, L.Y. The agency line: A neoliberal metric for appraising young women's sexuality. *Sex Roles* **2015**, *73*, 279. [CrossRef]
27. Braun, V.; Gavey, N.; McPhillips, K. The fair deal'? Unpacking accounts of reciprocity in heterosex. *Sexualities* **2003**, *6*, 237. [CrossRef]
28. Kleinplatz, P.J.; Charest, M.; Rosen, L.A.; Ménard, A.D. Optimal couple sexuality: A review of the (Limited) literature. *Curr. Sex. Health Rep.* **2022**, *14*, 63. [CrossRef]
29. Kleinplatz, P.J.; Ménard, A.D.; Paquet, M.P.; Paradis, N.; Campbell, M.; Zuccarino, D.; Mehak, L. The components of optimal sexuality: A portrait of "great sex. *Can. J. Hum. Sex.* **2009**, *18*, 1.

Disclaimer/Publisher's Note: The statements, opinions and data contained in all publications are solely those of the individual author(s) and contributor(s) and not of MDPI and/or the editor(s). MDPI and/or the editor(s) disclaim responsibility for any injury to people or property resulting from any ideas, methods, instructions or products referred to in the content.

engineering proceedings

MDPI

Proceeding Paper

Parametric Study on Performance of Straight Type of Internally Finned Tube †

Chun-Lang Yeh *, Dong-Long Wu, Qi-Ying Cai, Yu-Xi Chung, Chang-Xin Liu and Zhi-Qian Wu

Department of Aeronautical Engineering, National Formosa University, Yunlin 632, Taiwan; 10978104@nfu.edu.tw (D.-L.W.); 11078104@nfu.edu.tw (Q.-Y.C.); 11178121@nfu.edu.tw (Y.-X.C.); 40831246@nfu.edu.tw (C.-X.L.); 40831239@nfu.edu.tw (Z.-Q.W.)

* Correspondence: clyeh@nfu.edu.tw; Tel.: +886-5-6315527

† Presented at the 3rd IEEE International Conference on Electronic Communications, Internet of Things and Big Data Conference 2023, Taichung, Taiwan, 14–16 April 2023.

Abstract: The performance of a straight type of internally finned tube (SIFT) is studied using computational fluid dynamics (CFD). It is found that a longer fin yields a larger pressure drop that is nearly proportional to the fin length. When the fin length is larger than 30 d, the pressure drop is greater than that of a bare tube. The temperature uniformity decreases with the fin length of the SIFT. Furthermore, a smaller fin angle yields a larger pressure drop, while a larger fin angle yields better temperature uniformity. A larger contraction yields a larger pressure drop, too.

Keywords: SIFT; temperature uniformity; pressure drop; fin length; fin angle; fin amplitude

Citation: Yeh, C.-L.; Wu, D.-L.; Cai, Q.-Y.; Chung, Y.-X.; Liu, C.-X.; Wu, Z.-Q. Parametric Study on Performance of Straight Type of Internally Finned Tube. *Eng. Proc.* **2023**, *38*, 37. https://doi.org/10.3390/engproc2023038037

Academic Editors: Teen-Hang Meen, Hsin-Hung Lin and Cheng-Fu Yang

Published: 26 June 2023

Copyright: © 2023 by the authors. Licensee MDPI, Basel, Switzerland. This article is an open access article distributed under the terms and conditions of the Creative Commons Attribution (CC BY) license (https://creativecommons.org/licenses/by/4.0/).

1. Introduction

Mixing in a pipe system is important in industrial applications. There are two major mixing-enhanced methods: active and passive methods. The former uses an external energy source, while the latter includes no moving parts. The passive mixing-enhanced devices increase an internal surface area and enhance the flow mixing. The solid fin and the straight type of internally finned tube (SIFT) have an increased surface area [1–5]. With enhanced flow mixing, radial mixing occurs in the tube which consequently yields a more homogeneous process. The radial mixing is found in the helical type of internally finned tube (HIFT), the mixing element radiant tube (MERT) by Kubota, the intensified heat transfer (IHT) by Lummus and Sinopec, and the SIFT technology. In Ref. [6], the HIFT was investigated numerically. In this study, the performance of the SIFT concerning heat transfer enhancement and pressure drop gain is investigated using computational fluid dynamics (CFD) to obtain better thermal uniformity and minimize pressure drop.

2. Numerical Methods

In this research, we use ANSYS FLUENT V.17 [7] to analyze the flow development in the SIFT. The SIMPLE algorithm is used for the solution algorithm [8]. In turbulence modeling, we adopt the transition shear stress transport (SST) model. Considering the accuracy and stability, we used the discrete ordinate radiation model [9] for radiation simulation.

3. Results and Discussion

In this study, the parameters include fin angle (2α), fin amplitude (in terms of radiuses R_a and R_i), and fin length (L) to obtain better thermal uniformity and minimize pressure drop. Figure 1 illustrates the investigated SIFT. The tube length is 132 d (d is a pipe diameter of 50.7 mm). Fourteen different helix lengths ranging from L = 0 (i.e., bare tube) to L = 132 d are investigated to study the influence of helix length. Three different fin angles (2α), including $2\alpha = 30°$, 45°, and 60° and four different R_a, including 0.99, 0.9, 0.8, and 0.7 R_t, as well as four different R_i, including 0.9, 0.8, 0.7 and 0.6 R_i, where $R_t = d/2$ are used for the investigation. The boundary conditions are outlined as follows.

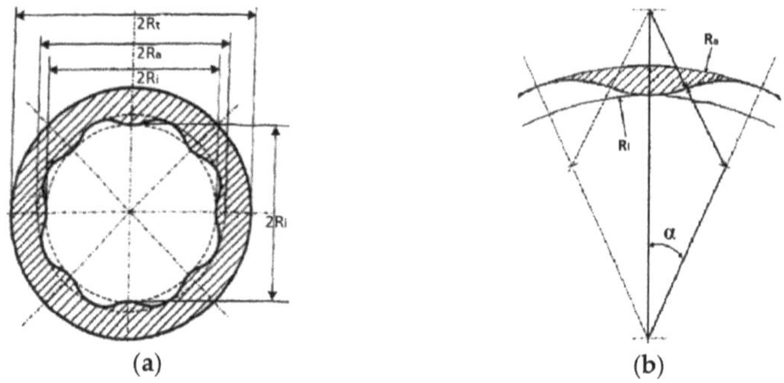

Figure 1. Illustration of the SIFT investigated: (**a**) cross-sectional view; (**b**) zoom-in view of a fin.

The tube inlet (Figure 2) includes four parts. In Region 1, V = 56.2 m/s and T = 273.15 K, in Region 2, V = 61.3 m/s and T = 298.15 K, in Region 3, V = 66.5 m/s and T = 323.15 K, and in Region 4, V = 71.6 m/s and T = 348.15 K. The inlet turbulence kinetic energy (k) is assumed to be 10% of $V^2/2$. The turbulence dissipation rate is modeled by Equation (1).

$$\varepsilon = C_\mu^{3/4}\frac{k^{3/2}}{l}. \quad (1)$$

In Equation (1), the constant $C_\mu = 0.09$, parameter $l = 0.07L$, and the hydraulic diameter L is equal to the pipe diameter d. At the wall boundaries, the transition SST model automatically takes the wall effects into account. Furthermore, adiabatic walls are assumed. At the tube exit, the gauge pressure is zero. The outflow diffusion flux for the other flow variables is zero, and the conservation of mass is satisfied.

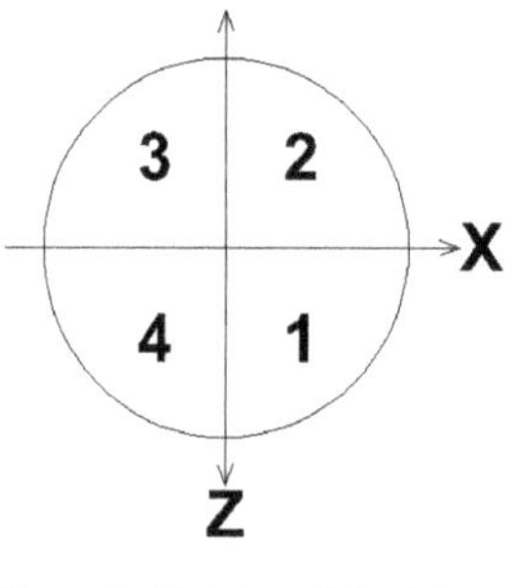

Figure 2. Division of the inlet.

The effect of fin length on the pressure variation along the tube is shown in Figure 3. The pressure drop increases with the fin length nearly in proportion. The pressure drop is closely related to the tube's inner surface friction, which is connected to the fin length. When the fin length is larger than 30 d, the pressure drop is greater than that of a bare tube.

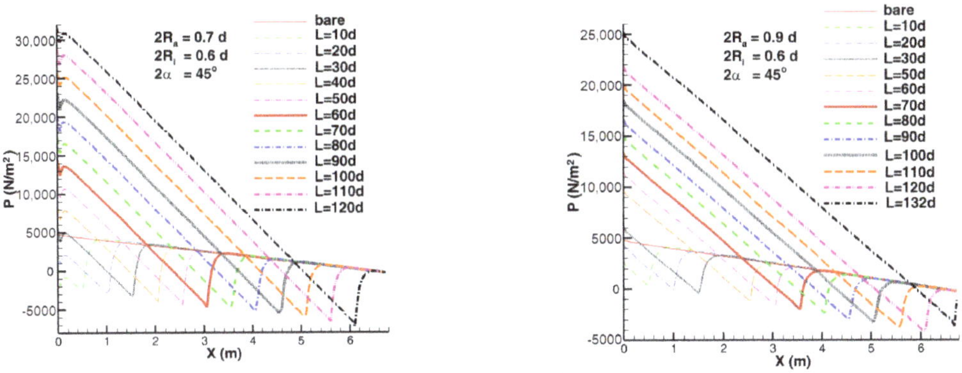

(**a**) (**b**)

Figure 3. Variation of the cross-sectional average pressure: (**a**) 2Ra = 0.7d, 2Ri = 0.6d, 2α = 45°; (**b**) 2Ra = 0.9d, 2Ri = 0.6d, 2α = 45°.

The effect of fin length on the cross-sectional average temperature distribution along the tube is shown in Figure 4. The temperature of the SIFT is higher than that of a bare tube. The temperature of the SIFT rises abruptly at the fin inlet due to the contraction of the tube and descends abruptly at the fin outlet due to the expansion of the tube. In addition, the temperature becomes higher for a longer fin because the tube wall frictional heating and the flow acceleration in the fin region are larger.

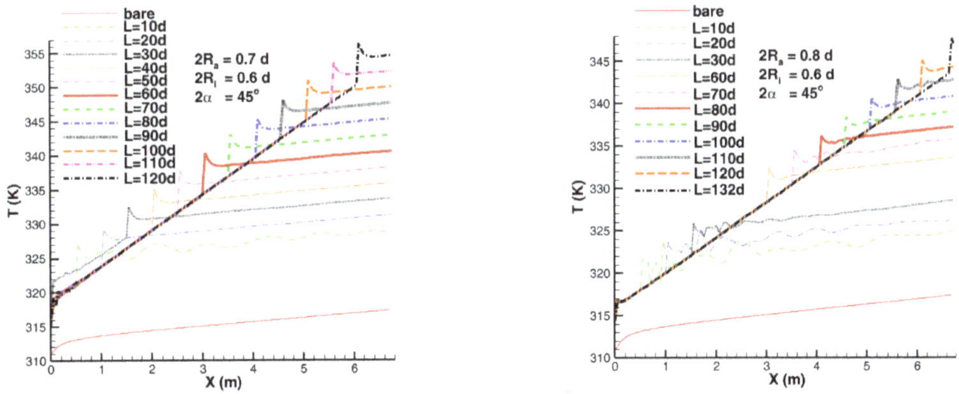

(a) (b)

Figure 4. Variation of the cross-sectional average temperature: (**a**) 2Ra = 0.7d, 2Ri = 0.6d, 2α = 45°; (**b**) 2Ra = 0.8d, 2Ri = 0.6d, 2α = 45°.

The effect of fin length on the area-weighted temperature uniformity index defined as Equation (2) [7] is shown in Figure 5.

$$\gamma_a = 1 - \frac{\sum_{i=1}^{n}\left[\left(\left|T_i - \overline{T}_a\right|\right)\right]A_i}{2\left|\overline{T}_a\right|\sum_{i=1}^{n}A_i}. \quad (2)$$

In Equation (2), i is the facet index and n is the number of facets of a surface. $\overline{T}_a$ is the surface average temperature.

$$\overline{T}_a = \frac{\sum_{i=1}^{n}T_iA_i}{\sum_{i=1}^{n}A_i}. \quad (3)$$

A value of one indicates optimal uniformity. Figure 5 shows that SIFT improves temperature uniformity as compared to a bare tube. The change in the temperature uniformity index is similar to fin lengths. In the fin, the temperature uniformity index for different fin lengths nearly coincides, while after leaving the fin region, the temperature uniformity increases to another coincident value for different fin lengths. Therefore, there is no need to use an SIFT with a long fin. A fin length of 10 d is enough to obtain satisfactory temperature uniformity.

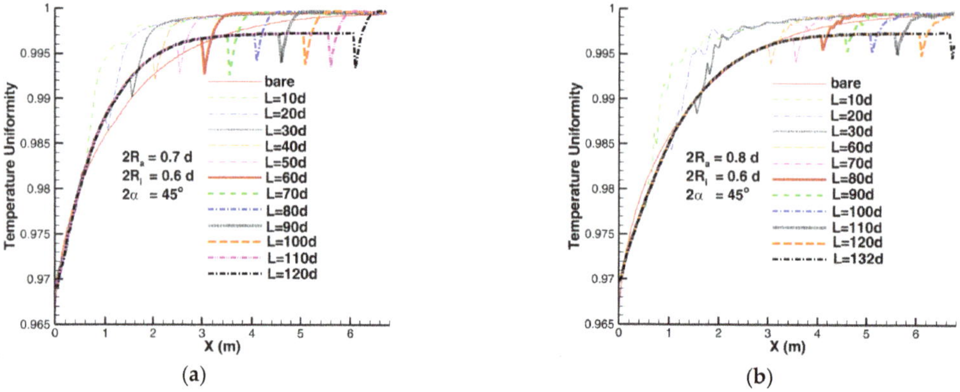

Figure 5. Temperature uniformity index along the tube: (**a**) 2Ra = 0.7d, 2Ri = 0.6d, 2α = 45°; (**b**) 2Ra = 0.8d, 2Ri = 0.6d, 2α = 45°.

The effect of the fin angle on the pressure variation along the tube is shown in Figure 6. A smaller fin angle yields a larger pressure drop. The pressure drop is closely related to the tube's inner surface friction. Because the wall area is connected to the fin, a smaller fin angle corresponds to more fins, which in turn have a larger inner surface area and larger pressure drop.

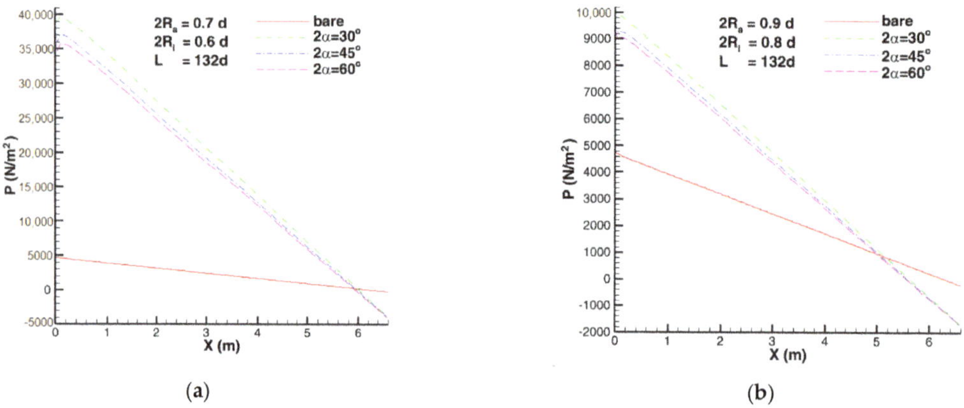

Figure 6. Variation of the cross-sectional average pressure: (**a**) 2Ra = 0.7d, 2Ri = 0.6d, L = 132d; (**b**) 2Ra = 0.9d, 2Ri = 0.8d, L = 132d.

The effect of the fin angle on the cross-sectional average temperature along the tube is shown in Figure 7. A larger fin angle yields a lower temperature. This is because a larger fin angle corresponds to fewer fins, which in turn have a smaller inner surface area and lower frictional heating effect.

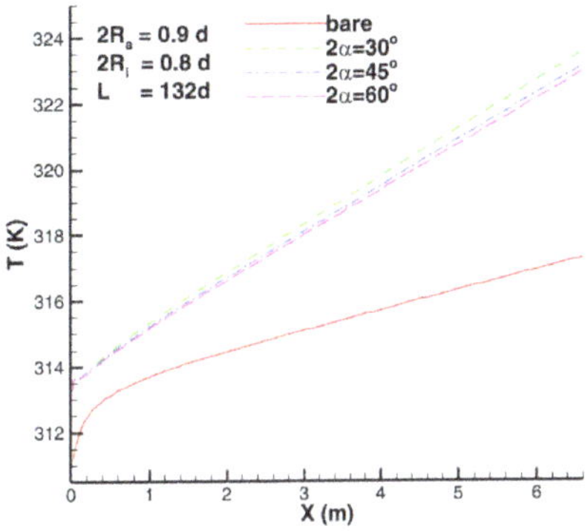

(a)

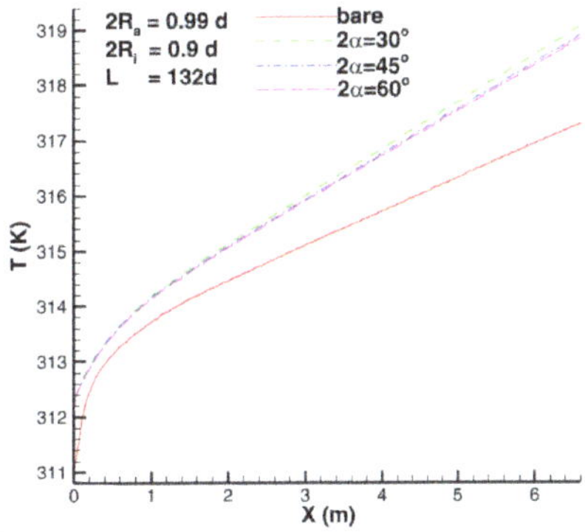

(b)

Figure 7. Variation of the cross-sectional average temperature: (**a**) 2Ra = 0.9d, 2Ri = 0.8d, L = 132d; (**b**) 2Ra = 0.99d, 2Ri = 0.9d, L = 132d.

The effect of the fin angle on the temperature uniformity index along the tube is shown in Figure 8. Compared to a bare tube, SIFT improves its temperature uniformity. However, the influence of the fin angle on the temperature uniformity is not significant, although a careful observation reveals that a larger fin angle yields better uniformity. This is because more fins (smaller fin angles) result in faster flow acceleration in the fin region and alleviates the secondary flow (Dean Vortex) at the cross-section, which leads to worse mixing. The smaller the fin angle is, the worse the temperature uniformity will be.

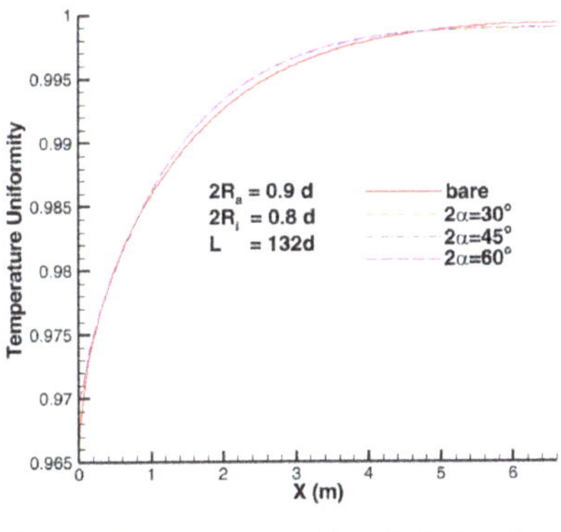

Figure 8. Temperature uniformity index along the tube.

The effect of fin amplitude on the pressure variation along the tube is shown in Figure 9. It can be seen that a larger contraction, i.e., a smaller R_a for a given R_i or a smaller R_i for a given R_a, yields a larger pressure drop. This is because a larger contraction results in a higher flow velocity, and therefore, the pressure drop is higher.

The effect of fin amplitude on the cross-sectional average temperature distribution is shown in Figure 10. The fin amplitude has an obvious influence on the temperature variation. A larger contraction, i.e., a smaller R_i for a given R_a, yields a higher temperature. This is because a larger contraction results in a higher flow velocity. Therefore, the temperature increases due to the higher wall friction.

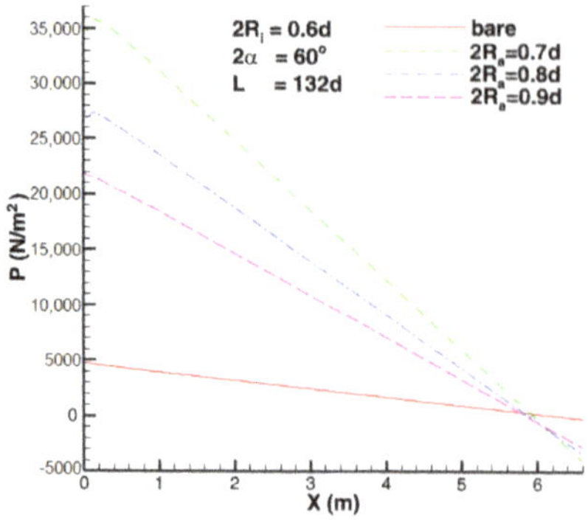

Figure 9. Variation of the cross-sectional average pressure.

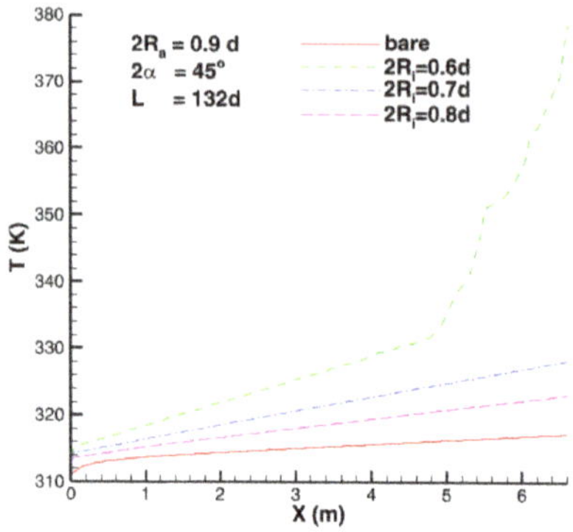

Figure 10. Variation of the cross-sectional average temperature.

The effect of the fin amplitude on the temperature uniformity index is shown in Figure 11. SIFT improves the temperature uniformity index for a smaller fin amplitude, i.e., a smaller 2(R_a-R_i), as compared to a bare tube. This is because a larger fin amplitude may alleviate the secondary flow (Dean Vortex) at the cross-section because of a higher flow acceleration, which leads to worse mixing. The larger the fin amplitude is, the worse the temperature uniformity will be.

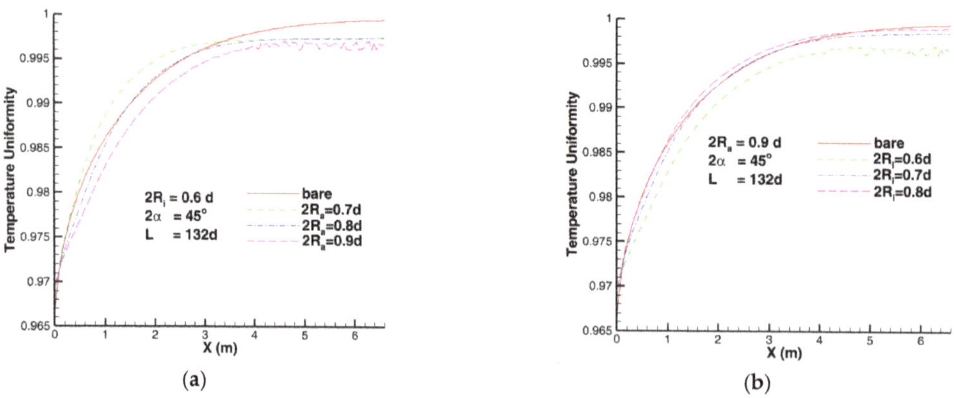

Figure 11. Temperature uniformity index along the tube: (**a**) 2Ri = 0.6d, 2α = 45°, L = 132d; (**b**) 2Ra = 0.9d, 2α = 45°, L = 132d.

4. Conclusions

The performance of the SIFT concerning the heat transfer enhancement and the pressure drop was investigated using CFD to obtain optimal temperature uniformity and reduce pressure drop. A longer fin yields a larger pressure drop, which is nearly proportional to the fin length. When the fin length is larger than 30 d, the pressure drop becomes greater than that of a bare tube. The temperature of the SIFT is higher than that of a bare tube. The temperature increases with the fin length, rises abruptly at the fin inlet, and descends abruptly at the fin outlet. The SIFT improves the temperature uniformity compared to a bare tube. However, there is no need to use an SIFT with a long fin. A fin length of 10 d is enough to obtain satisfactory temperature uniformity. A smaller fin angle yields a larger pressure drop. The temperature decreases with the fin angle. A larger fin angle yields better temperature uniformity. A larger contraction yields a larger pressure drop and a higher temperature rise. Finally, the temperature uniformity is improved by the SIFT for a smaller fin amplitude.

Author Contributions: Conceptualization, C.-L.Y.; methodology, C.-L.Y.; validation, C.-L.Y.; formal analysis, C.-L.Y.; data curation, C.-L.Y., D.-L.W., Q.-Y.C., Y.-X.C., C.-X.L. and Z.-Q.W.; writing—original draft preparation, C.-L.Y.; writing—review and editing, C.-L.Y., D.-L.W., Q.-Y.C., Y.-X.C., C.-X.L. and Z.-Q.W.; visualization, C.-L.Y., D.-L.W., Q.-Y.C., Y.-X.C., C.-X.L. and Z.-Q.W. All authors have read and agreed to the published version of the manuscript.

Funding: This research received no external funding.

Institutional Review Board Statement: Not applicable.

Informed Consent Statement: Not applicable.

Data Availability Statement: The data presented in this study are available on request from the corresponding author.

Conflicts of Interest: The authors declare no conflict of interest.

References

1. Albano, J.; Rhoe, A.; Sundaram, K.; Summer, C. Pyrolysis Heater. European Patent 0,305,799, B1, 23 October 1991.
2. Albano, J.; Sundaram, K.; Maddock, M. Application of extended surface in pyrolysis coils. *Eng. Prog.* **1988**, *8*, 160–168.
3. Brown, D. Internally finned radiant coils: A valuable tool for improving ethylene plant economics. In Proceedings of the 6th EMEA Petrochemicals Technology Conference, London, UK, 23–26 June 2004.
4. Schietekat, C.M.; Cauwenberge, D.V.; Geem, K.V.; Marin, G.B. Computational fluid dynamics based design of finned steam cracking reactors. In Proceedings of the AIChE Spring Meeting: Ethylene Producers Conference–Fundamentals of Technology Session, San Antonio, TX, USA, 2 May 2013.
5. Wölpert, P.; Ganser, B.; Jakobi, D.; Kirchheiner, R. Finned Tube for the Thermal Cracking of Hydrocarbons, and Process for Producing a Finned Tube. U.S. Patent 7,963,318 B2, 21 June 2011.
6. Yeh, C.L.; Wu, D.L.; Cai, Q.Y.; Yang, S.Y.; Lin, X.J.; Liu, G.H. Parametric study on the performance of a helical type of internally finned tube. In Proceedings of the 2022 IEEE 4th Eurasia Conference on IOT, Communication and Engineering, Yunlin, Taiwan, 28–30 October 2022.
7. Fluent Inc. *ANSYS FLUENT 17 User's Guide*; Fluent Inc.: New York, NY, USA, 2017.
8. Patankar, S.V. *Numerical Heat Transfer and Fluid Flows*; McGraw-Hill: New York, NY, USA, 1980.
9. Siegel, R.; Howell, J.R. *Thermal Radiation Heat Transfer*; Hemisphere Publishing Corporation: Washington, DC, USA, 1992.

Disclaimer/Publisher's Note: The statements, opinions and data contained in all publications are solely those of the individual author(s) and contributor(s) and not of MDPI and/or the editor(s). MDPI and/or the editor(s) disclaim responsibility for any injury to people or property resulting from any ideas, methods, instructions or products referred to in the content.

engineering proceedings

MDPI

Proceeding Paper

Design and Analysis of Compliant Rack and Pinion Using Compliant Contact Rolling Joint †

Lakshmipathy Panchanathan and Shyh-Chour Huang *

Department of Mechanical Engineering, National Kaohsiung University of Science and Technology, Kaohsiung 807618, Taiwan; lakshmipathy351@gmail.com
* Correspondence: shuang@nkust.edu.tw

† Presented at the 3rd IEEE International Conference on Electronic Communications, Internet of Things and Big Data Conference 2023, Taichung, Taiwan, 14–16 April 2023.

Abstract: This paper presents the design of a compliant rack and pinion mechanism with the support of a compliant rolling contact joint (CRCJ). The flexible strips are mounted to connect the rack and pinion to accelerate the mechanism. Two types of rack and pinion mechanism are proposed: rack and pinion mechanism and double rack and pinion actuator. The 3D model is created with finite element analysis (FEM) for structural analysis to observe the maximum strength and the deformation of the flexible strip. A 3D-printed model is proposed by using a fused deposition method (FDM) by printing layer by layer for experimental and demonstration purposes. In this study, we propose a set of examples to convert a traditional rigid body mechanism to complete a complaint mechanism to accumulate the actual function and also to increase the high possibilities of mounting.

Keywords: compliant mechanism; rack and pinion mechanism; compliant contact rolling joint; rack and pinion actuator

Citation: Panchanathan, L.; Huang, S.-C. Design and Analysis of Compliant Rack and Pinion Using Compliant Contact Rolling Joint. *Eng. Proc.* **2023**, *38*, 36. https://doi.org/10.3390/engproc2023038036

Academic Editors: Teen-Hang Meen, Hsin-Hung Lin and Cheng-Fu Yang

Published: 26 June 2023

Copyright: © 2023 by the authors. Licensee MDPI, Basel, Switzerland. This article is an open access article distributed under the terms and conditions of the Creative Commons Attribution (CC BY) license (https://creativecommons.org/licenses/by/4.0/).

1. Introduction

The rack and pinion mechanism is the traditional mechanism to convert rotational motion into linear motion. By bearing the pinion, we transform line motion to rotational motion [1,2]. Rack and pinion is a well-known steering mechanism. A pinion is attached to the end of the steering shaft in the rack and pinion type of steering mechanism. Because the pinion and the rack are mated, the movement of the handle causes the pinion to rotate, which in turn causes the rack to move sideways. Another application of rack and pinion is actuators, which contain dual racks and single pinions. In pipeline transport, valve control is accomplished with pneumatic rack and pinion actuators [3].

The compliant mechanism holds the potential for flexibility. The mobility of the compliant mechanism at least gains from the deflection of flexible members rather than from movable joints only [4]. This design is inspired by the compliant rolling contact joint (CRCJ) mechanism [5,6]. The conventional CRCJ consists of four similar semi-cylindrical cams connected in a row at alternative positions, as shown in Figure 1. The flexure straps deform until both cams are aligned and connected [7,8]. To achieve zero angular stiffness, the compliant rolling contact joint has maintained the strain energy at a constant level while in action [9,10].

We introduce two compliant rack and pinion mechanisms: a rack and pinion steering mechanism and a rack and pinion actuator. With the knowledge of compliant rolling, the design of the mechanism is proposed. The vivid functions of steering and actuator mechanisms are presented in this study. The following design is created with FEM for structural analysis.

Figure 1. Core [4].

2. Design and Fabrication

2.1. Design and Modeling

The model was designed by using Solidworks. Figure 2 represents the components of the pinion, rack, flexible strip, and casing. The pinion was designed with four pairs of grooves, shafts, and pivotal holes. The extrusion of the rib around the pinion (1 mm) was placed in the center of the pinion, as shown in Figure 2a. The rack contained four grooves, and the slot was designed in the middle of the rack, as shown in Figure 2b. In Figure 2c, the flexible strip was designed according to the circumference of the pinion. The geometry of the flexible strip was designed to place in-between pinion and rack grooves. The casing was modeled to mount all the parts inside. To prevent slippage during running, 2 pathways were presented to mount the rack and for the sliding motion, as shown in Figure 2d. The dimensions of all the components are shown in Table 1.

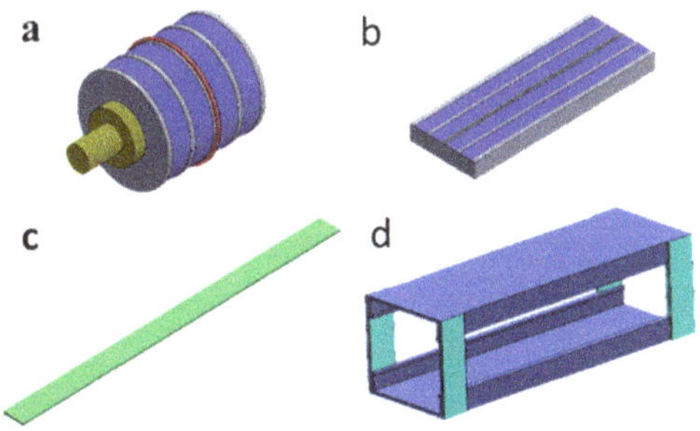

Figure 2. Model of components: (**a**) pinion; (**b**) rack; (**c**) flexible strip; (**d**) casing.

Table 1. Dimensions of the components.

Objects	Length	Thickness	Diameter
Pinion	43 mm	-	40 mm
Rack	125 mm	10 mm	-
Flexible strip	125 mm	0.6 mm	-
Casing	160 mm	64 mm	-

The flexible strip was modeled according to the circumference of the pinion. The radius of the pinion R was 20 mm. The perimeter of the circle is denoted as,

$$C = 2\pi R \text{ or } \pi D \quad (1)$$

The total length of the flexible strip was 125 mm, and the length of the flexible strip and the circumference of the pinion were relative to each other when the pinion diameter was increased as the total length of the flexible strip also increased.

The CRCJ mechanism is constrained by the contact surface to have a constant curvature. This constant curvature creates well-defined moments in the flexures. The relationship between the curvature of a beam and the internal moments is defined by the Bernoulli–Euler equation.

$$M = EI\,\frac{d\theta}{ds} \tag{2}$$

where E is the modulus of elasticity, I is the moment of inertia, and $d\theta/ds$ is the curvature.

The Bernoulli–Euler equation is valid for large deflections. The beam with a constant radius of curvature creates a moment that is also constant along the beam length. This means the segment has the same behavior as a beam with a moment applied to the free end.

$$M = \frac{EI}{R'} \tag{3}$$

Thus, the maximum stress in the CRCJ depends on the modulus of elasticity of the chosen material, the thickness of the flexure, and the effective radius of curvature [4,11].

2.2. *Manufacturing*

The parts were manufactured with polylactic acid (PLA) material. The FDM method was used for the 3D models, which were printed layer-by-layer using thermoplastic filaments to create the 3D-printed model. After the development of the model, the model is exported into Ultimaker Cura 5.1, Taipei, Taiwan. the slicing process was done by Cura 5.1. The infill rate was 100%, the printing temperature of the PLA material was 220 °C, and the bed temperature was 200 °C. Then, the file is transferred to JGmaker 3d printing machine, Taipei, Taiwan for fused deposition method (FDM) layer-by-layer printing. The material used for simulation was polylactic acid (PLA) with a density of 1.28 g/cm^3, Young's modulus was 2.7 GPa, and tensile strength was 62.9 MPa.

3. Results

3.1. *Pinion*

Normally, the rolling contact joint mechanism is considered to be a semi-cylindrical type. The movement we obtained was less than 180°. The rotation occurred at half of the circumference. The compliant pinion played a vital role in this design to obtain the maximum linear motion. We achieved the maximum circumference of the pinion rotation with the help of compliant flexible strips.

The compliant pinion contained four grooves (0.4 mm), and in the center, an extruded part was placed to prevent pinion slippage, as shown in Figure 3. To actuate the pinion, two types of input and output methods were designed. For shafts and holes, we obtained the linear motion by actuating the rack and the placement of the bearing into the hole to achieve the linear rotation motion.

Figure 3. Pinion.

3.2. Compliant Rack

The compliant rack had the same grooves in the pinion; the groove depth of the compliant rack was 0.2 mm. The center groove was presented to mount the pinion in that precise groove to avoid slippage, as shown in Figure 4. The rolling compliant contact joint had two types of cylinder parts placed opposite to each other, but the grooves were placed on either side of the cylinder part. We increased the efficiency of the design; the grooves in the rack were placed to minimize the strain on the flexible strip. The pinion rotated in two directions: clockwise and counterclockwise. While in the clockwise direction, the rack moved forward; in the counterclockwise direction, a backward linear motion was observed.

Figure 4. Rack.

3.3. Flexible Strip

The flexible strips play a vital role in this design. The rack and pinion had four grooves placed on each other. The rack had 0.2 mm grooves, and the pinion had 0.4 mm grooves. The total thickness of the flexible strip was 0.6 mm, as shown in Figure 5. Four strips were connected to the grooves to operate the rack and pinion. The two strips were connected in the clockwise grooves to make the clockwise and forward motion of the compliant rack and pinion. The other two strips were connected in the center or counterclockwise grooves of the compliant rack and pinion to make the backward linear motion. By altering the positions of the flexible strips, we created two types of mechanism, a steering rack and pinion mechanism and a rack and pinion actuator.

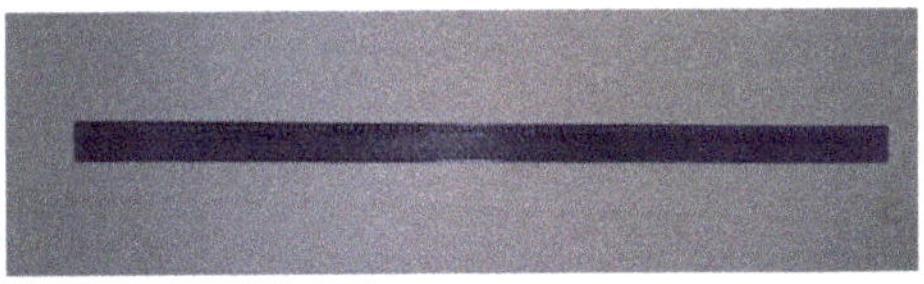

Figure 5. Flexible strip.

The pair of the rack was actuated by pinion with the help of compressed air to obtain the clockwise and the anti-clockwise motion. The placement of both clockwise and counterclockwise strips in the bottom of the rack achieved the steering rack and pinion mechanism. Placing the clockwise strips at the bottom of the rack and the counterclockwise strip at the top of the rack with the support of the casing achieved the rack and pinion actuator.

3.4. Casing

The pinion, rack, and flexible strips were mounted into the casing. The casing covered all the parts and prevented slippages during the working condition, as shown in Figure 6. To obtain the changes in the steering and actuator rack and pinion mechanism, the rack and the flexible strip placement were different in the mounting positions. The rack was positioned inside the casing ribs so that the rack moved independently. To obtain the steering rack and pinion mechanism, both racks were connected, and the flexible strips were mounted on the bottom of the rack.

Figure 6. Casing.

The actuation of the pinion moved the bottom rack and the supporting rack to linear motion. For the rack and pinion actuator, the top rack and the bottom rack moved independently to acquire the parallel forward and backward motion. The clockwise flexible strips were connected to the bottom of the rack, and the counterclockwise strips were connected to the top of the rack to achieve the mechanism by the support of the casing.

3.5. Rack and Pinion Steering Mechanism

The assembled view of the compliant rack and pinion mechanism is presented in Figures 7 and 8. Figure 7 represents the isometric view of the rack and pinion mechanism. The casing contains a slot to hold the top and base rack for sliding motive. The pinion is placed in between both racks. The flexible strips play a vital role in connecting the rack and pinion for the motions. Four grooves are made in the rack and pinion; a pair of grooves represent clockwise grooves placed on both ends, and a further pair denotes counterclockwise grooves located center. The base rack and the pinion are connected by two flexible strips in clockwise grooves, and the other pair of flexible strips connect the base rack and pinion in counter-clockwise grooves. For the first type of rack and pinion mechanism, both racks are connected by a supporting plate to obtain the parallel line motion. Actuating the base rack leads a way to act together. The counterclockwise rotation of the pinion assists the base rack in moving forward with the help of flexible strips connected in the counterclockwise grooves. The clockwise rotation motion of the pinion tends to move the rack backward in linear motion by the guidance of a flexible strip connected in the clockwise grooves. By bearing the pinion, giving the liner input to the racks allows the output of rotational motion.

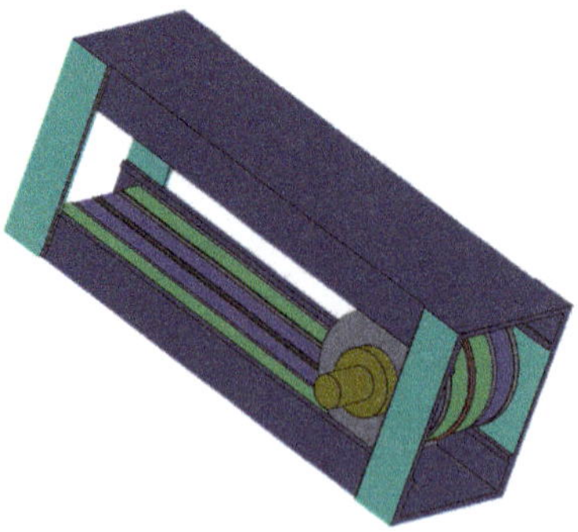

Figure 7. Isometric view of rack and pinion mechanism.

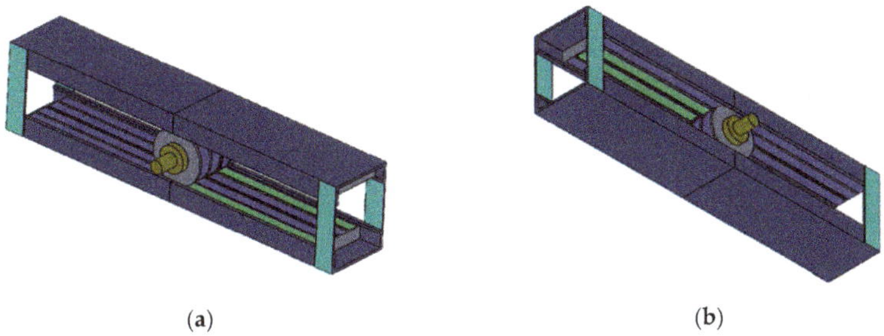

Figure 8. Isometric view of the double rack and pinion actuator: (**a**) base rack; (**b**) top rack.

3.6. Rack and Pinion Actuator

Figure 8 represents the design of the rack and pinion actuator and the placement of the flexible strip and the rack in the actuator mechanism. The base rack holds the position at the bottom of the casing as well as the top rack in the top slot of the casing. The pair of flexible strips connect the pinion and the base rack at the clockwise grooves, as denoted in Figure 9. The second pair of flexible strips connect the top rack and the pinion in counterclockwise grooves, as represented in Figure 9. The clockwise motion of the pinion makes the base rack move backward, and the top rack moves forward. The counterclockwise motion leads a base rack to move forward and the top rack to slide backward. By pivoting the pinion and the input of the rack, linear motion turns into rotational motion.

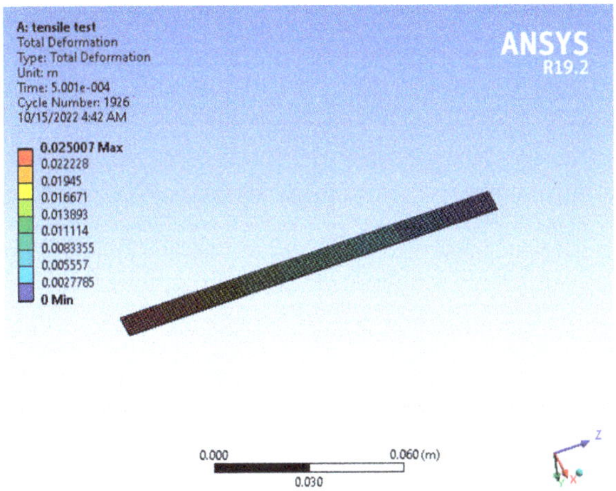

Figure 9. Total deformation.

3.7. Finite Element Analysis

The finite element method (FEM) was conducted using ANSYS. The maximum torque of the pinion and the load increment of the rack increases the tension of the flexible cable [12]. The FEM was used to figure out the maximum stress and deformation of the flexible strips. Figure 9 presents the computation of total deformation and the maximum deformation of the flexible strip of 0.025 mm. The plot of maximum deformation is shown in Figure 10. The material deformation occurs in the fixed position of the rack. When converting the linear motion into rotational motion, the load is applied to the rack. The flexible strip connects the rack and pinion. The rack is subjected to the maximum load, and the stress also increases in the flexible strip. However, the compressive force acting on the flexible strip reduces the tension occurrence. The stress is induced by the flexible cable. To observe the stress activity of the flexible strip, von-mises stress was exerted. The maximum stress of the flexible strip is 4.4593 × 108 Pa, and the value of minimum stress obtained at 2.8991 × 108 Pa is shown in Figures 11 and 12.

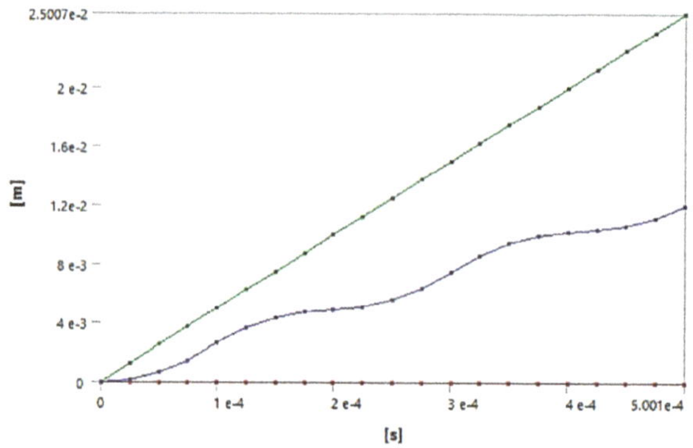

Figure 10. FEA Total deformation.

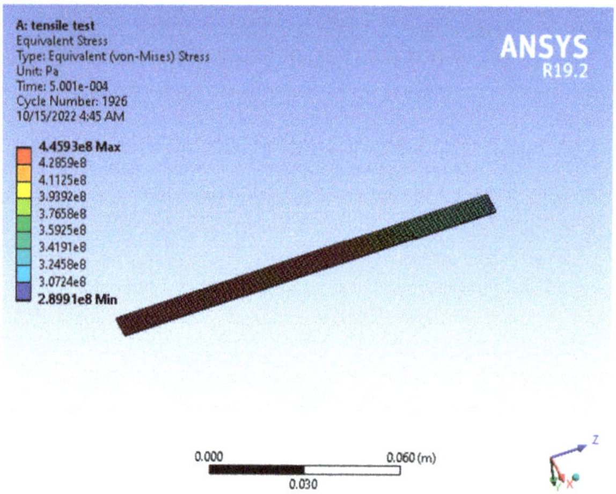

Figure 11. Von-mises stress.

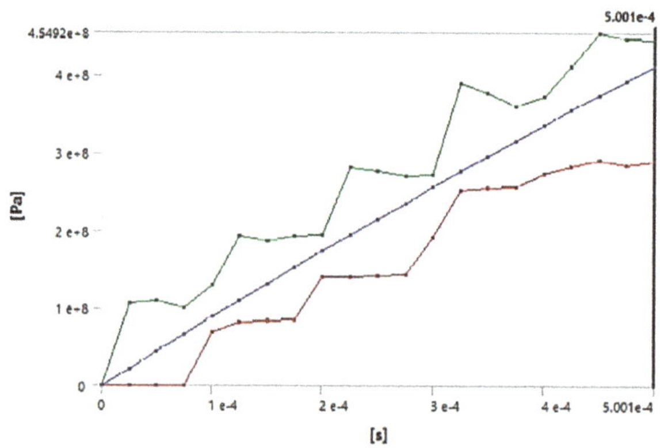

Figure 12. FEA Equivalent (von-mises).

4. Discussion

For the replacement of the traditional rack and pinion method, a novel method was developed to achieve the respective motions. The compliant rolling contact joint (CRCJ) is the major inspiration for this model. The CRCJ joint acts as a gear. Instead of the gear teeth, the flexible strips are connected as two rigid links to keep no-slip contact. The flexure avoids being stretched or becoming loose. Further, it achieves zero angular stiffness to maintain a constant level of strain energy in rotation. However, the conditions are impossible for the CRCJ joint to achieve the 360° full rotation, as the internal stress does not increase. To obtain the total length of the rack movement, the pinion approximately rotates less than (>360°), acquiring the internal stress and maintaining the constant strain energy. The position of the flexible strip in between the rack and pinion grooves reduces the flexible strip tension.

5. Conclusions

We presented the two types of compliant rack and pinion methods. The movement of linear motion depends on the circumference of the pinion of 40 mm in diameter. The total length of linear motion is 125 mm. This method is inspired by CRCJ, and the rotation motion of the pinion is >360°. The maximum deformation is 0.025 mm, and the maximum stress of flexible cable is 4.4593 × 108 Pa. The design successfully achieves linear motion as well as rotational motion by using the flexible component. The speed of the motions increases with the traditional rack and pinion mechanism. In the future, the movement of the rack needs to increase using the threading technique. For example, the total length of linear motion depends on the circumference of the pinion. By introducing the threading technique, the pinion achieves twice the linear motion.

Author Contributions: Conceptualization, S.-C.H.; methodology, S.-C.H.; software, L.P.; validation, L.P. and S.-C.H.; formal analysis, L.P.; writing—original draft preparation, L.P.; writing—review and editing, S.-C.H. All authors have read and agreed to the published version of the manuscript.

Funding: This research received no external funding.

Institutional Review Board Statement: Not applicable.

Informed Consent Statement: Not applicable.

Data Availability Statement: Not applicable.

Acknowledgments: The authors acknowledge and thank the Ministry of Science and Technology of the Republic of China for their partial financial support of this study under Contract Number MOST 111-2221-E-992-068.

Conflicts of Interest: The authors declare no conflict of interest.

References

1. Torvi, S.; Ingale, V.; E, R.K. Design and Analysis of Rack and Pinion Mechanism in Automobile Applications Using Structural Steel and PLA. *Int. J. Prod. Eng.* **2018**, *4*, 13–22. [CrossRef]
2. Akash, K.; Sangavi, S.; Venkatesan, M. Double acting syringe pump using a rack and pinion mechanism—Simulink model. In Proceedings of the 2016 IEEE International Conference on Computational Intelligence and Computing Research (ICCIC), Tamil Nadu, India, 15–17 December 2016; pp. 1–4.
3. Mustalahti, P.; Mattila, J. Nonlinear model-based controller design for a hydraulic rack and pinion gear actuator. In *Fluid Power Systems Technology*; American Society of Mechanical Engineers: New York, NY, USA, 2018; Volume 51968, p. V001T01A020.
4. Cannon, J.R.; Craig, P.L.; Larry, L.H. Compliant rolling-contact element mechanisms. In Proceedings of the International Design Engineering Technical Conferences and Computers and Information in Engineering Conference, Long Beach, CA, USA, 24–28 September 2005; Volume 47446, pp. 3–13.
5. Zhang, D.; Sun, Y.; Lueth, T.C. Design of a novel tendon-driven manipulator structure based on monolithic compliant rolling-contact joint for minimally invasive surgery. *Int. J. Comput. Assist. Radiol. Surg.* **2021**, *16*, 1615–1625. [CrossRef] [PubMed]
6. Hu, Y.; Zhang, L.; Li, W.; Yang, G.-Z. Design and Fabrication of a 3-D Printed Metallic Flexible Joint for Snake-Like Surgical Robot. *IEEE Robot. Autom. Lett.* **2019**, *4*, 1557–1563. [CrossRef]

7. Shaw, L.A.; Chizari, S.; Dotson, M.; Song, Y.; Hopkins, J.B. Compliant rolling-contact architected materials for shape reconfigurability. *Nat. Commun.* **2018**, *9*, 4594. [CrossRef] [PubMed]
8. Kim, S.-H.; Hyunki, I.; Song, J.-R.; Cho, K.-J. Force characteristics of rolling contact joint for compact structure. In Proceedings of the 2016 6th IEEE International Conference on Biomedical Robotics and Biomechatronics (BioRob), Singapore, 26–29 June 2016; pp. 1207–1212.
9. Nelson, T.G.; Lang, R.J.; Magleby, S.P.; Howell, L.L. Curved-folding-inspired deployable compliant rolling-contact element (D-CORE). *Mech. Mach. Theory* **2016**, *96*, 225–238. [CrossRef]
10. Jeanneau, A.; Herder, J.; Laliberté, T.; Gosselin, C. A compliant rolling contact joint and its application in a 3-dof planar parallel mechanism with kinematic analysis. In Proceedings of the International Design Engineering Technical Conferences and Computers and Information in Engineering Conference, Salt Lake City, UT, USA, 28 September–2 October 2004; Volume 46954, pp. 689–698.
11. Halverson, P.A.; Howell, L.L.; Jensen, B.D.; Magleby, S.P. Concepts for achieving multi-stability in compliant rolling-contact elements. In Proceedings of the International Design Engineering Technical Conferences and Computers and Information in Engineering Conference, Las Vegas, NV, USA, 4–7 September 2007; Volume 48094, pp. 67–73.
12. Halverson, P.A.; Howell, L.L.; Magleby, S.P. Tension-based multi-stable compliant rolling-contact elements. *Mech. Mach. Theory* **2010**, *45*, 147–156. [CrossRef]

Disclaimer/Publisher's Note: The statements, opinions and data contained in all publications are solely those of the individual author(s) and contributor(s) and not of MDPI and/or the editor(s). MDPI and/or the editor(s) disclaim responsibility for any injury to people or property resulting from any ideas, methods, instructions or products referred to in the content.

engineering proceedings

Proceeding Paper

A Parametric Study on the Performance of a Helical Type of Internally Finned Tube †

Chun-Lang Yeh *, Dong-Long Wu, Qi-Ying Cai, Shi-Yi Yang, Xuan-Jun Lin and Geng-Hong Liu

Department of Aeronautical Engineering, National Formosa University, Yunlin 632, Taiwan; 10978104@nfu.edu.tw (D.-L.W.); 11078104@nfu.edu.tw (Q.-Y.C.); 40831126@nfu.edu.tw (S.-Y.Y.); 40831151@nfu.edu.tw (X.-J.L.); 40831227@nfu.edu.tw (G.-H.L.)

* Correspondence: clyeh@nfu.edu.tw; Tel.: +886-5-6315527

† Presented at the 3rd IEEE International Conference on Electronic Communications, Internet of Things and Big Data Conference 2023, Taichung, Taiwan, 14–16 April 2023.

Abstract: We study the performance of a helical type of internally finned tube (HIFT), using computational fluid dynamics (CFD) to obtain optimal temperature uniformity and reduce the pressure drop. In comparison with a bare tube, the HIFT improved the temperature uniformity with a helix length of 10 d, which was enough to obtain satisfactory temperature uniformity. A smaller helix interval angle and a smaller helix pitch yielded a larger pressure drop. The temperature of the HIFT decreased with an increase in the helix pitch.

Keywords: HIFT; temperature uniformity; pressure drop; helix length; helix interval angle; helix pitch

Citation: Yeh, C.-L.; Wu, D.-L.; Cai, Q.-Y.; Yang, S.-Y.; Lin, X.-J.; Liu, G.-H. A Parametric Study on the Performance of a Helical Type of Internally Finned Tube. *Eng. Proc.* **2023**, *38*, 40. https://doi.org/10.3390/engproc2023038040

Academic Editors: Teen-Hang Meen, Hsin-Hung Lin and Cheng-Fu Yang

Published: 26 June 2023

Copyright: © 2023 by the authors. Licensee MDPI, Basel, Switzerland. This article is an open access article distributed under the terms and conditions of the Creative Commons Attribution (CC BY) license (https://creativecommons.org/licenses/by/4.0/).

1. Introduction

Fluid mixing achieves process homogeneity and affects flow system performance significantly. Improvement in mixing allows mixture homogenization and enhances heat and mass transfer for chemical reactions. Poor thermal uniformity may cause equipment fatigue and damage and influence the piping geometry and operating conditions. Thermal and fluid flow prediction is therefore important for equipment safety. In this study, we apply computational fluid dynamics (CFD) to study the performance of a helical type of internally finned tube (HIFT) and its heat transfer and pressure drop. As a result, the optimal temperature uniformity is determined to reduce the pressure drop.

2. Methods

In this research, we used ANSYS FLUENT V.17 [1] to analyze the flow development in an HIFT. The SIMPLE algorithm was used as a solution algorithm [2]. In turbulence modeling, we adopted the transition SST model. Considering the accuracy and stability, we used the discrete ordinate radiation model [3] for radiation simulation.

3. Results and Discussion

The parameters investigated included the helix interval angle ($\Delta\alpha$), helix pitch (P), and helix length (L) to obtain better thermal uniformity and minimize the pressure drop. Figure 1 illustrates the investigated HIFT. The tube length was 132 d, where d is a pipe diameter of 50.7 mm. Fourteen helix lengths ranging from L = 0 (i.e., a bare tube) to L = 132 d were tested to understand the influence of the helix length. In addition, seven helix interval angles ($\Delta\alpha$) ranging from 36° to 180° and three helix pitches (2d, 3d, and 4d) were inspected as well. The boundary conditions were as follows.

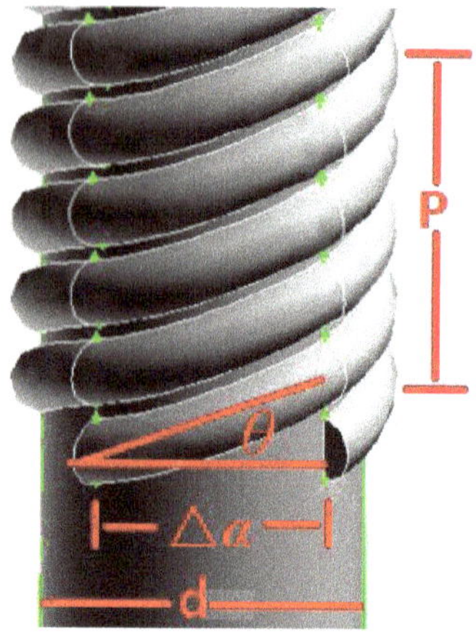

Figure 1. Illustration of the HIFT investigated.

The tube inlet (Figure 2) included four parts. In Region 1, V = 56.2 m/s and T = 273.15 K. In Region 2, V = 61.3 m/s and T = 298.15 K. In Region 3, V = 66.5 m/s and T = 323.15 K. Finally, in Region 4, V = 71.6 m/s and T = 348.15 K. The inlet turbulence kinetic energy (k) was assumed to be 10% of $V^2/2$. The turbulence dissipation rate is modeled by Equation (1).

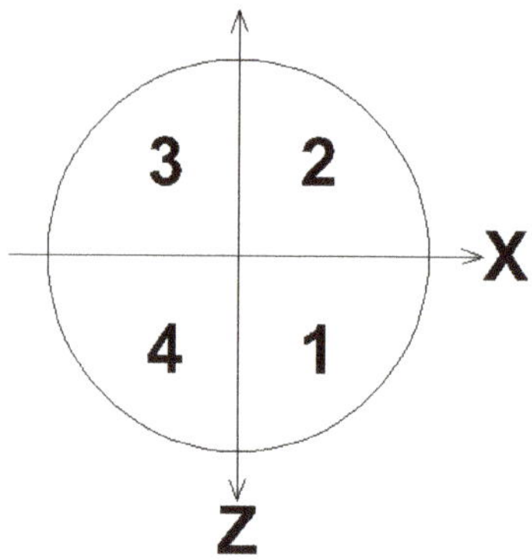

Figure 2. Division of the inlet.

$$\varepsilon = C_{\mu}^{3/4}\frac{k^{3/2}}{l}. \quad (1)$$

In Equation (1), the constant $C_{\mu} = 0.09$, parameter $l = 0.07L$, and the hydraulic diameter L is equal to the pipe diameter d. At the wall boundaries, the transition SST model automatically takes the wall effects into account. In this study, adiabatic walls were assumed. At the tube's exit, the gauge pressure was zero. For the other flow variables, the outflow diffusion flux was zero, and the conservation of mass was satisfied. The effect of the helix length on the pressure variation along the tube is shown in Figure 3. A longer helix yielded a larger pressure drop which was nearly proportional to the helix length. The pressure drop was closely related to the tube's inner surface friction, which was connected with the helix length. The pressure drop was greater than that of the bare tube when the helix length was larger than 10 d.

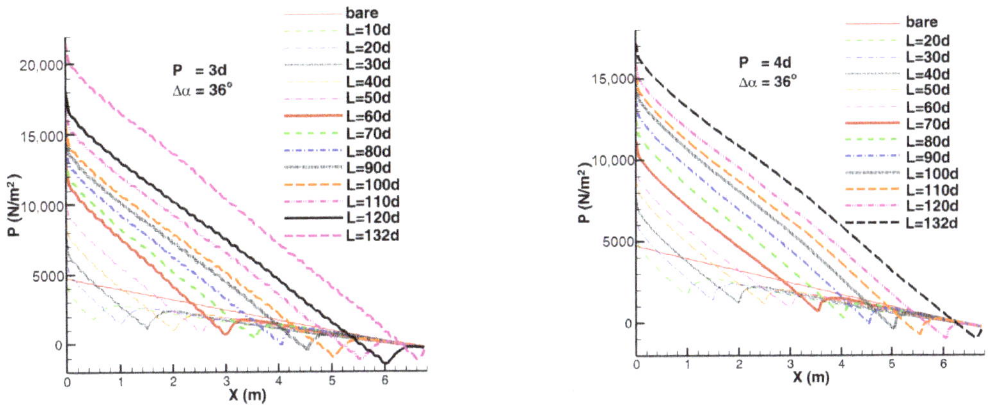

(**a**) (**b**)

Figure 3. Variation in the cross-sectional average pressure. (**a**) P = 3d and Δα = 36°. (**b**) P = 4d and Δα = 36°.

The effect of the helix length on the cross-sectional average temperature is shown in Figure 4. The temperature of the HIFT was higher than that of the bare tube. In addition, the temperature of the HIFT increased with the helix length because of the greater frictional heating of the tube wall.

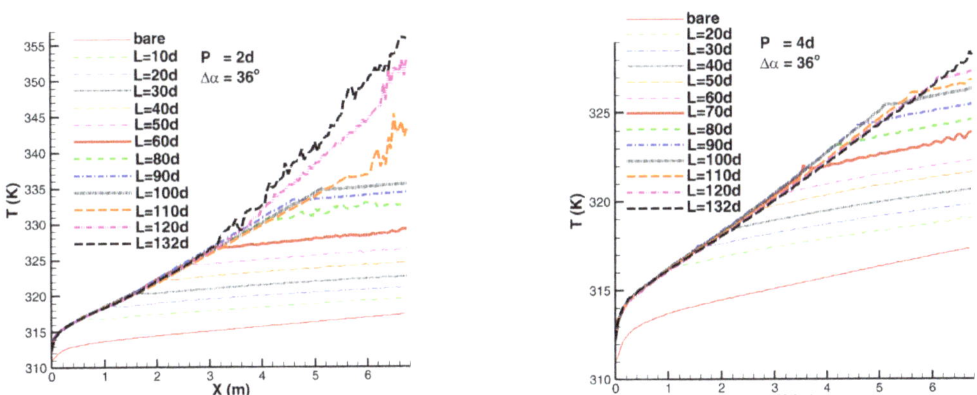

(**a**) (**b**)

Figure 4. Variation in the cross-sectional average temperature. (**a**) P = 3d and Δα = 36°. (**b**) P = 4d and Δα = 36°.

The effect of the helix length on the area-weighted temperature uniformity index is defined as follows (Figure 5) [1]:

$$\gamma_a = 1 - \frac{\sum_{i=1}^{n} \left[\left(\left| T_i - \bar{T}_a \right| \right) \right] A_i}{2 \left| \bar{T}_a \right| \sum_{i=1}^{n} A_i} \tag{2}$$

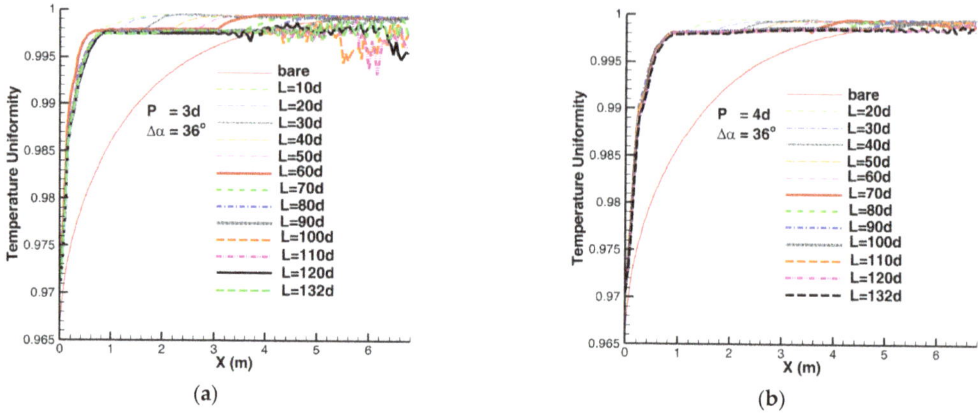

Figure 5. Temperature uniformity index along the tube. (**a**) P = 3d and Δα = 36°. (**b**) P = 4d and Δα = 36°.

In Equation (2), i is the facet index, and n is the number of facets of a surface. $\bar{T}_a$ is the surface average temperature:

$$\bar{T}_a = \frac{\sum_{i=1}^{n} T_i A_i}{\sum_{i=1}^{n} A_i}. \tag{3}$$

A value of one indicates optimal uniformity. Figure 5 shows that the HIFT could improve the temperature uniformity index when compared with a bare tube. The evolution of the temperature uniformity index depended on the helix length. In the helix, the temperature uniformity is nearly constant at a specific value, while after leaving the helix, the temperature uniformity increases to another specific value. Therefore, using the HIFT with a long helix is unnecessary. A helix length of 10 d is enough to obtain satisfactory temperature uniformity.

The effect of the helix interval angle on the pressure variation is shown in Figure 6. It was observed that a smaller helix interval angle yielded a larger pressure drop which was nearly inversely proportional to the helix interval angle. Because the pressure drop was closely related to the tube's inner surface friction, a smaller helix interval angle led to more helixes and hence a greater inner surface area and more friction in the tube.

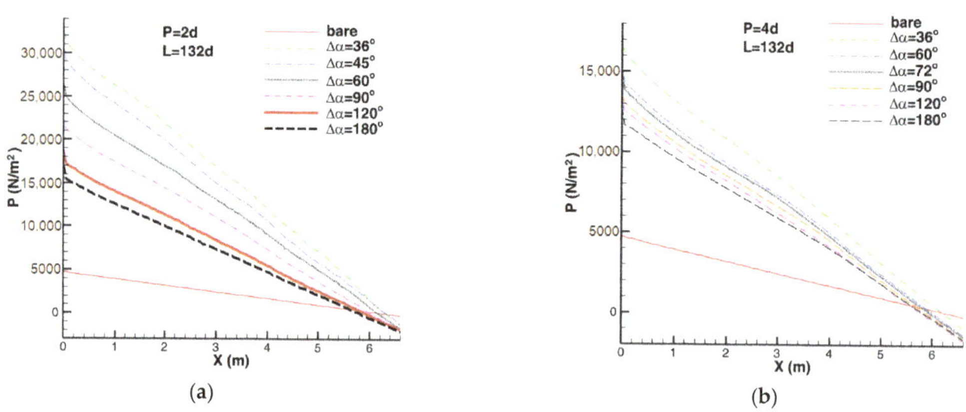

Figure 6. Variation in the cross-sectional average pressure. (**a**) P = 2d and L = 132d. (**b**) P = 4d and L = 132d.

The effect of the helix interval angle on the cross-sectional average temperature distribution is shown in Figure 7. A larger helix interval angle yielded a lower temperature. This is because a larger helix interval angle corresponds to fewer fins, which in turn have a smaller inner surface area and lower frictional heating effect.

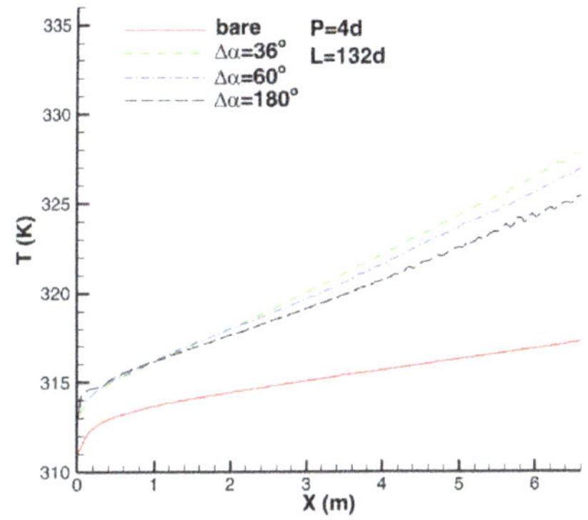

Figure 7. Variation in the cross-sectional average temperature.

The effect of the helix interval angle on the temperature uniformity index is shown in Figure 8. The larger the helix interval angle was, the worse the temperature uniformity would be. This is because a larger helix interval angle leads to fewer helixes and worse mixing.

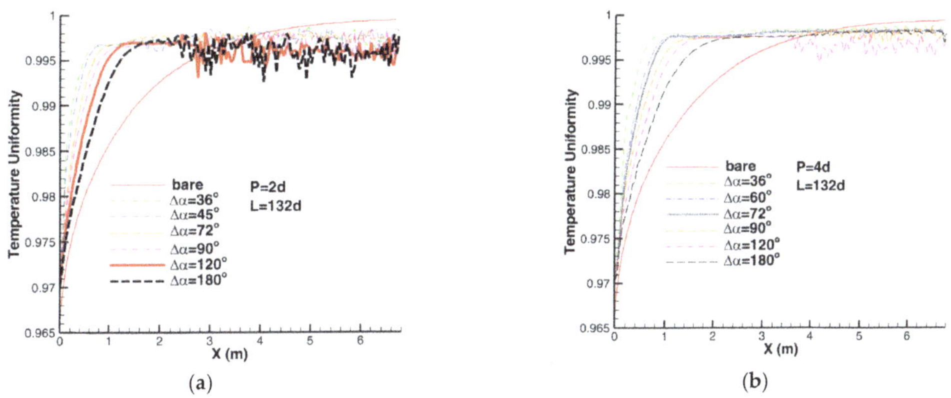

Figure 8. Temperature uniformity index along the tube. (**a**) P = 2d and L = 132 d. (**b**) P = 4d and L = 132 d.

The effect of the helix pitch on the pressure variation is shown in Figure 9. A smaller helix pitch yielded a larger pressure drop which was nearly inversely proportional to the helix pitch. Because the pressure drop was closely related to the tube's inner surface friction, a smaller helix pitch led to more helixes and hence a larger inner surface area and more friction in the tube.

The effect of the helix pitch on the cross-sectional average temperature distribution is shown in Figure 10. A larger helix pitch yielded a lower temperature for the HIFT. This is because a larger helix pitch results in fewer helixes and hence a smaller inner surface area and lower frictional heating effect on the tube.

The effect of the helix pitch on the temperature uniformity index is shown in Figure 11. The influence of the helix pitch on temperature uniformity was not significant. However, careful observation revealed that a smaller helix pitch yielded better temperature uniformity.

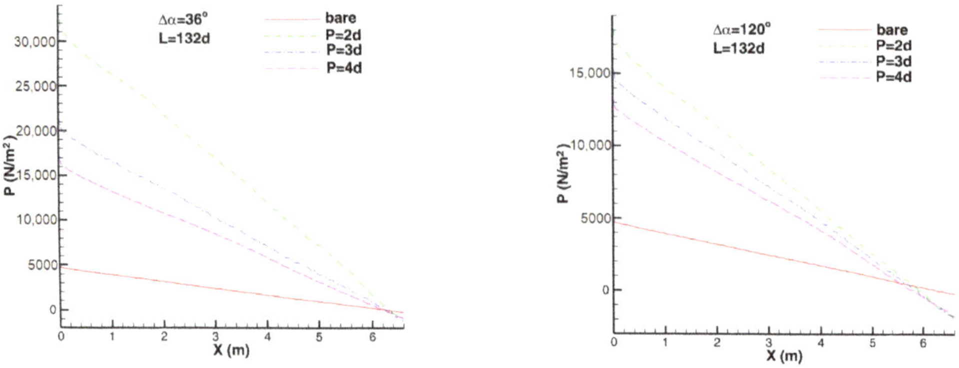

(a) (b)

Figure 9. Variation in the cross-sectional average pressure. (**a**) $\Delta\alpha = 36^\circ$ and L = 132 d. (**b**) $\Delta\alpha = 120^\circ$ and L = 132 d.

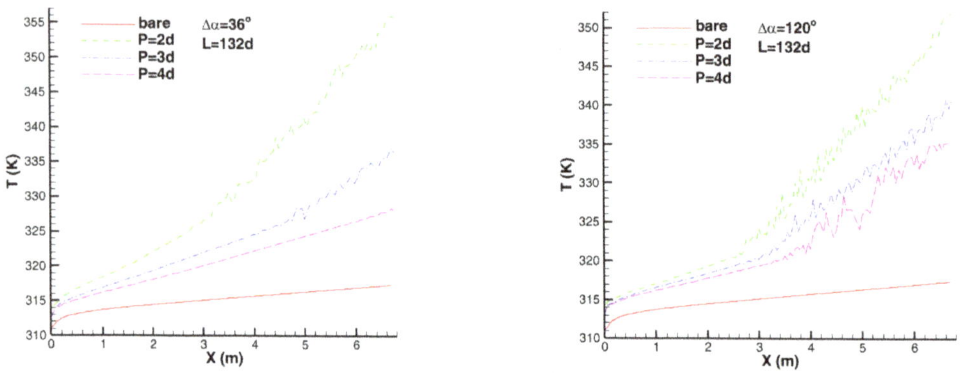

(a) (b)

Figure 10. Variation in the cross-sectional average temperature. (**a**) $\Delta\alpha = 36^\circ$ and L = 132 d. (**b**) $\Delta\alpha = 120^\circ$ and L = 132 d.

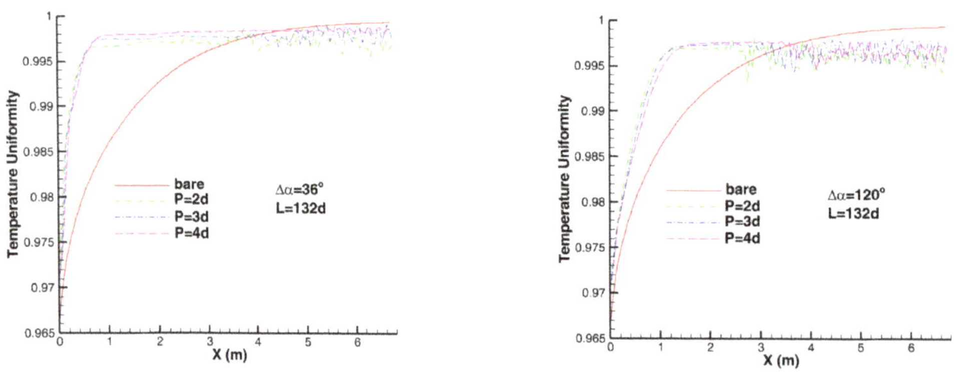

(a) (b)

Figure 11. Temperature uniformity index along the tube. (**a**) $\Delta\alpha = 36^\circ$ and L = 132 d. (**b**) $\Delta\alpha = 120^\circ$ and L = 132 d.

4. Conclusions

The performance of the HIFT was investigated by using CFD to obtain optimal temperature uniformity and reduce the pressure drop. We found that a longer helix yielded a larger pressure drop which was nearly proportional to the helix length. When the helix length was larger than 10 d, the pressure drop was greater than that of the bare tube. The temperature of the HIFT was higher than that of the bare tube. The temperature of the HIFT increased with the helix length. In comparison with the bare tube, the HIFT improved the temperature uniformity with the helix length. However, using the HIFT with a long helix was unnecessary. A helix length of 10 d was enough to obtain satisfactory temperature uniformity. A smaller helix interval angle yielded a larger pressure drop which was nearly inversely proportional to the helix interval angle. The temperature of the HIFT decreased with the helix interval angle. In addition, the larger the helix interval angle was, the worse the temperature uniformity would be. A smaller helix pitch yielded a larger pressure drop. The temperature of the HIFT decreased with the helix pitch, but the influence of the helix pitch on the temperature uniformity was not significant. However, careful observation revealed that a smaller helix pitch yielded better temperature uniformity.

Author Contributions: Conceptualization, C.-L.Y.; methodology, C.-L.Y.; validation, C.-L.Y.; formal analysis, C.-L.Y.; data curation, C.-L.Y., D.-L.W., Q.-Y.C., S.-Y.Y., X.-J.L. and G.-H.L.; writing—original draft preparation, C.-L.Y.; writing—review and editing, C.-L.Y., D.-L.W., Q.-Y.C., S.-Y.Y., X.-J.L. and G.-H.L.; visualization, C.-L.Y., D.-L.W., Q.-Y.C., S.-Y.Y., X.-J.L. and G.-H.L. All authors have read and agreed to the published version of the manuscript.

Funding: This research received no external funding.

Institutional Review Board Statement: Not applicable.

Informed Consent Statement: Not applicable.

Data Availability Statement: The data presented in this study are available on request from the corresponding author.

Conflicts of Interest: The authors declare no conflict of interest.

References

1. Fluent Inc. *ANSYS FLUENT 17 User's Guide*; Fluent Inc.: New York, NY, USA, 2017.
2. Patankar, S.V. *Numerical Heat Transfer and Fluid Flows*; McGraw-Hill: New York, NY, USA, 1980.
3. Siegel, R.; Howell, J.R. *Thermal Radiation Heat Transfer*; Hemisphere Publishing Corporation: Washington, DC, USA, 1992.

Disclaimer/Publisher's Note: The statements, opinions and data contained in all publications are solely those of the individual author(s) and contributor(s) and not of MDPI and/or the editor(s). MDPI and/or the editor(s) disclaim responsibility for any injury to people or property resulting from any ideas, methods, instructions or products referred to in the content.

engineering proceedings

Proceeding Paper

Surface Rejuvenation Model for Turbulent Thermophoresis Velocity †

Ming-Chi Chiou and Hung-Yih Tsai *

Department of Vehicle Engineering, National Formosa University, Yunlin 632, Taiwan; achi@gs.nfu.edu.tw
* Correspondence: hungyih@nfu.edu.tw

† Presented at the 3rd IEEE International Conference on Electronic Communications, Internet of Things and Big Data Conference 2023, Taichung, Taiwan, 14–16 April 2023.

Abstract: A mathematical presentation of the surface rejuvenation model is used to relate the mean velocity and temperature distribution to the mean vortex dwell time and approach distance. Coupled with a proper estimate of these modeling parameters, it provides quantitative forecasts of the mean thermophoresis velocity. As expected, the particles obtain inertia along the wall from the eddies in the turbulent core and are trapped in the viscous sublayer where coherent temperature gradients exist in the wall region. Playing an important role in particle transfer, this results in particle transfer along the wall. Particle thermophoresis with small molecule diffusion may be the only mechanism that enhances the particle transfer process in the viscous wall region.

Keywords: surface rejuvenation model; thermophoresis; residence time; approach distance

1. Introduction

The introduction should briefly place the study in a broad context and define the purpose of the work and its significance. A theoretical model of the thermal drift mechanism depends on a description of the thermophoretic velocity of a particle with a temperature gradient. This was first observed for particles that are large compared to the mean free path in the gaseous environment [1]. There was an important parameter affecting thermophoresis, which is defined by the Knudsen number K_n and the mean free path λ of gas molecules and the particle radius r_p.

$$K_n = \frac{\lambda}{r_p} \tag{1}$$

Particle sizes in the mean free paths fall into the continuum, transition, or free molecular regimes, respectively, regardless of the size. Hidy and Brock [2] classified these three regimes according to the values of K_n and suggested that in the slip region, $K_n << 1.0$, in the transition region, $0.25 \leq K_n \leq 10.0$, and in the free molecular environment, $K_n > 10.0$. The mean free path of liquids is orders of magnitude smaller than the mean free path of gases between 0.04 and 0.1 µm under ambient conditions [3].

For large aerosol particles ($K_n << 1.0$), Epstein [1] expressed the thermophoretic force, F_T, in an appropriate form for spherical particles in gas at rest as

$$F_T = -9\pi r_p \frac{\mu^2}{\rho}\left(\frac{K_g}{2K_g + K_p}\right)\frac{\nabla T}{T} \tag{2}$$

where K_p is the particle's thermal conductivity; K_g and ∇T are the thermal conductivity and temperature gradient of gas, respectively. Under the steady-state condition, the thermophoretic force balances the drag force, i.e., $F_T = 6\pi r_p \mu V_t$. Therefore,

$$V_t = -\frac{3}{2}\nu\left(\frac{K_g}{2K_g + K_p}\right)\frac{\nabla T}{T} \tag{3}$$

Citation: Chiou, M.-C.; Tsai, H.-Y. Surface Rejuvenation Model for Turbulent Thermophoresis Velocity. *Eng. Proc.* **2023**, *38*, 42. https://doi.org/10.3390/engproc2023038042

Academic Editors: Teen-Hang Meen, Hsin-Hung Lin and Cheng-Fu Yang

Published: 26 June 2023

Copyright: © 2023 by the authors. Licensee MDPI, Basel, Switzerland. This article is an open access article distributed under the terms and conditions of the Creative Commons Attribution (CC BY) license (https://creativecommons.org/licenses/by/4.0/).

In Epstein's calculations, the thermal force for particles of high thermal conductivity is significantly underestimated.

Brock [4] indicated that this arises due to a lack of boundary conditions appropriate for the slip flow and the convective terms. For this reason, the modification of thermophoretic force was considered for the case of inclusion of Knudsen number corrections of the temperature increase at the solid-fluid interface and the isothermal slip. In such a way, the corresponding thermophoretic velocity can be appropriately estimated by

$$V_t = -\frac{3}{2}\nu \frac{(\frac{K_g}{K_p} + C_t K_n)}{(1 + 3C_m K_n)(1 + 2\frac{K_g}{K_p} + C_t K_n)} \frac{\nabla T}{T} \quad (4)$$

Brock [4] chose reasonable ranges for the thermal jump coefficient C_t = 1.875~2.48 and for the hydrodynamic slip coefficient C_m = 1.00~1.27. Equation (4) is induced to Equation (3) when $K_n = 0$. Although an agreement was found in the experimental measurements obtained by Jacobsen and Brock [5], the proposed relationship for V_t failed, while for particles with high thermal conductivity, the influence of the values selected for the slip coefficient was significant. However, the discrepancy between experimental and theoretical results is much less than that obtained by using Epstein's model. Derjaguin and Yalamov [6] argued that in the deviation of Epstein's equation and in Brock's work, there are two questions. The first is that the heat flux in the gas volume is ignored, and the second is that boundary conditions are used to assume that, for most gases before hitting the interface, the distribution of the velocity of gas molecules is the same. They modified the boundary conditions to account for the temperature jump and gave a modified expression:

$$V_t = -\frac{1}{2}\nu \frac{8K_g + K_p + 2C_t K_p K_n}{2K_g + K_p + 2C_t K_p K_n} \frac{\nabla T}{T} \quad (5)$$

Derjaguin et al. [7] experimentally confirmed this modification within the limits of experimental error as compared to both the Epstein and Brock models.

The particle radius of small aerosol particles ($K_n >> 1.0$) is small for the mean free path of the gas. Thus, in the free molecular environment, the particles do not affect the distribution of gas velocity and the force on the moving particle is directly obtained by calculating the momentum transport per unit of time. Waldmann [8] applied this idea for calculating the diffusional velocity of the particle, i.e., the motion of particles due to existing concentration gradients. Since the thermal conductivity is not significant for smaller particles, the thermal force can result from the net impulse in the direction of the temperature gradient imparted to the particles (Equation (6)).

$$F_T = -\frac{32}{15} r_p^2 \frac{K}{\overline{V}} \nabla T \quad (6)$$

where $\overline{V}$ is the average thermal velocity of gas molecules and $K = \frac{15K_g \mu}{4m}$ is the translational part of thermal conductivity with the molecular mass m of gas. Jacobsen and Brock [5] compared the experiment with the thermophoretic force equation of Waldmann [8] and indicated that the equation of F_T was in error by approximately 5% at $\frac{r_p}{\lambda} = 10$ and 10% at $\frac{r_p}{\lambda} = 5$. The corresponding thermophoretic velocity was given by Waldmann and Schmitl [9] as follows.

$$V_t = -\frac{1}{5(1 + \frac{\pi\alpha}{8})} \frac{K_g}{P} \nabla T \quad (7)$$

where P is gas pressure and α is gas thermal diffusivity. It has been demonstrated that this consideration is valid for Knudsen numbers greater than 10 [10].

Talbot et al. [11] modified the Brock model with an improved thermal slip coefficient, which agrees well with experimental measurements for $K_n \leq 0.1$. They have also compared Zernik's concept with fitting the available experimental data accurately over Knudsen numbers K_n, $0 \leq \frac{\lambda}{r_p} \leq \infty$. In consequence, they proposed a general equation for the thermophoretic force as

$$F_T = -12\pi r_p \mu \nu \frac{(\frac{K_g}{K_p} + C_t K_n)}{(1+3C_m K_n)(1+2\frac{K_g}{K_p} + 2C_t K_n)} \frac{\nabla T}{T} \quad (8)$$

Correspondingly, an improved expression for the thermophoretic velocity can be written as

$$V_t = -2C_s \nu \frac{(\frac{K_g}{K_p} + C_t K_n)}{(1+3C_m K_n)(1+2\frac{K_g}{K_p} + 2C_t K_n)} \frac{\nabla T}{T} \quad (9)$$

in which the velocity jump coefficient $C_m = 1.146$, the temperature increase coefficient is $C_t = 2.18$, and the thermal creep coefficient is $C_s = 1.147$ [12].

2. Surface Rejuvenation Model

The unsteady viscous sublayer surface regeneration model is based on the assumption that fluid eddies occurring at different times reach the core of the turbulence in different distances from the interface. In unsteady momentum and heat transfer, it is assumed that heat transfer is controlled within the wall region such that a single vortex resides near the wall. Neglecting convection and pressure gradient effects, the unsteady transport process of a fluid with constant properties can be written as a single regeneration cycle as

$$\frac{\partial \varphi}{\partial t} = \Re \frac{\partial^2 \varphi}{\partial y^2} \quad (10)$$

where the transport parameter φ represents the axial mean velocity or temperature and t is the instantaneous contact time of eddy. $\Re$ stands for turbulence inertia, $\Re = \nu + \varepsilon_t$, and for the heat flux, $\Re = \alpha + \alpha_t$. Since both processes of momentum and heat are governed by the same equation, the eddy thermal conductivity α_t and the eddy viscosity ε_t due to the diffusive action of turbulence are equal. The initial condition and boundary condition are given as

$$\begin{gathered} \varphi = \gamma(y)[1 - U(y-H)] + \varphi_\infty[U(y-H)]att = 0 \\ \varphi = \varphi_w aty = 0 \\ \varphi = \varphi_\infty asy \to \infty \end{gathered} \quad (11)$$

where $U(y)$ is the unit step function, φ_∞ represents the bulk stream velocity u_∞ and temperature T_∞, and φ_w is a specified temperature T_w of the wall where $u_w = 0$. H is the approach distance of the instantaneous eddy from the wall or the thickness of the viscous sublayer, and $\gamma(y)$ is the initial profile at the first instant of rejuvenation for $y < H$. The quantities t, H, and $\gamma(y)$ follow the statistical distribution with relevant distribution density function $p_t(t)$, $p_H(H)$, and $p_\gamma(\gamma)$. Therefore, by definition, the mean profile can be written as

$$\overline{\varphi}(y) = \int_0^\infty \int_0^\infty \int_0^\infty p_t(t) p_\gamma(\gamma) p_H(H) d\tau d\gamma dH \quad (12)$$

The differential equation in initial-boundary conditions is converted to a domain prior with each term multiplied by the distribution density function. Danckwerts [13] proposed that the contact time is exponentially distributed as

$$p_t(t) = \frac{1}{\overline{\tau}} e^{-\frac{t}{\overline{\tau}}} dt \quad (13)$$

where $\overline{\tau}$ express the mean residence time between two continuous eddies. Because the predictions for the mean transport properties obtained by the Hanratty [14] model are inherently independent of time distribution, the analogical forms of $p_\gamma(\gamma) = e^{-\gamma/\overline{\gamma}}/\overline{\gamma}\, d\gamma$

and $p_H(H) = e^{-\overline{H}/H}/\overline{H}\,dH$ are used in this study. The transformed mean profile becomes

$$\overline{\varphi} - \overline{\gamma}(y)e^{-\frac{y}{H}} - \varphi_\infty\left(1 - e^{-\frac{y}{H}}\right) = \Re\overline{\tau}\frac{\partial^2\overline{\varphi}}{\partial y^2} \tag{14}$$

where $\overline{\varphi}$ is the mean profile for $p_\gamma(\gamma)$ and $p_H(H)$.

If the contact time t is replaced by that of the residence time τ, $\overline{\gamma}(y)$ can be calculated by

$$\overline{\gamma}(y) = \int_0^\infty\int_0^\infty\int_0^\infty p_\tau(\tau)p_\gamma(\gamma)p_H(H)d\tau d\gamma dH \tag{15}$$

The contact time t in the Danckwert random distribution and residence time τ is related by $\overline{\gamma}(y) = \overline{\varphi}(y)$. This implies that the mean distribution of the residence time at the first moment is the same as the mean distribution of the residence time average. Equation (14) can then be written as

$$(\overline{\varphi} - \varphi_\infty)(1 - e^{-\frac{y}{H}}) = \Re\overline{\tau}\frac{\partial^2\overline{\varphi}}{\partial y^2} \tag{16}$$

and the corresponding boundary conditions are

$$\begin{aligned} \overline{\varphi} &= \varphi_w aty = 0 \\ \overline{\varphi} &= \varphi_\infty aty \to \infty \end{aligned} \tag{17}$$

Accurate analytical solutions of the turbulent convection integrated surface rejuvenation model are obtained as

$$\frac{\overline{\varphi} - \varphi_\infty}{\varphi_w - \varphi_\infty} = \frac{J_{2\beta}(2\beta e^{-\frac{y}{2H}})}{J_{2\beta}(2\beta)} \tag{18}$$

where $\beta = \frac{\overline{H}}{\sqrt{\Re\overline{\tau}}}$.

Talbot's equation is adopted to calculate the mean thermophoretic velocity, $\overline{V}_t$, of the particles, which is expressed as

$$\overline{V}_t = -k\frac{\nu}{\overline{\overline{T}}}\frac{\partial\overline{T}}{\partial y} \tag{19}$$

where the thermophoretic coefficient k is a ratio of the Knudsen number and other properties of particles. Then,

$$k = \frac{2C_sC_c(\frac{K_p}{K_g} + C_tK_n)}{(1 + 3C_mK_n)(1 + 2\frac{K_p}{K_g} + 2C_tK_n)} \tag{20}$$

where C_c is the Cunningham correction factor; k is the thermophoretic coefficient, typically varying from 0.2 to 1.2. The temperature gradient is calculated according to Equation (18), which is the same as the result shown in Ref. [15] for the mean temperature distribution in turbulent convective heat transfer, that is,

$$\frac{\overline{T}(y) - T_\infty}{T_w - T_\infty} = \frac{J_{2\vartheta}(2\vartheta e^{-\frac{y}{2H}})}{J_{2\vartheta}(2\vartheta)} \tag{21}$$

where $\vartheta = \frac{\overline{H}}{\sqrt{\alpha\overline{\tau}}}$. Thus, the distribution of the mean thermophoretic velocity on the wall is obtained by

$$\overline{V}_t = \frac{k\nu}{\sqrt{\alpha\overline{\tau}}}\left(\frac{e^{-(\frac{1}{2H})y}J_{2\vartheta-1}(2\vartheta e^{-(\frac{1}{2H})y}) - J_{2\vartheta}(2\vartheta e^{-(\frac{1}{2H})y})}{\frac{T_\infty}{T_w - T_\infty}J_{2\vartheta}(2\vartheta) + J_{2\vartheta}(2\vartheta e^{-(\frac{1}{2H})y})}\right) \tag{22}$$

In the wall area, the expression of the average velocity distribution is taken from Equation (19) as follows.

$$\overline{u}(y) = u_\infty\left(1 - \frac{J_{2v}(2ve^{-(\frac{1}{2H})y})}{J_{2v}(2v)}\right) \tag{23}$$

where $v = \frac{\overline{H}}{\sqrt{\nu\overline{\tau}}}$. $\overline{\tau}$ is assumed to represent the mean residence time throughout sublayer development. According to the local mean friction velocity, the mean residence time is expressed as u_* from the definition of the wall shear stress $\overline{\sigma}_0 = \rho u_*^2 = \mu\frac{\partial\overline{u}}{\partial y}\Big|_{y=0}$. As a result, the mean residence time is calculated in respect to the commonly used dimensionless quantities by

$$\overline{\tau}^+ = \sqrt{\frac{2}{f}}\left(\frac{J_{2v^+-1}(2v^+) - J_{2v^+}(2v^+)}{J_{2v^+}(2v^+)}\right) \tag{24}$$

where $\overline{\tau}^+ = u_*\sqrt{\frac{\overline{\tau}}{\nu}}$, $v^+ = \frac{\overline{H}^+}{\overline{\tau}^+}$, and $\overline{H}^+ = \frac{\overline{H}u_*}{\nu}$. The friction factor f is required and is accurately calculated as $f = 0.07725\left[log(\frac{Re}{7})\right]^{-2}$ for the flow by turbulence in smooth tubes [16]. Following this, the dimensionless distribution of the mean thermophoretic velocity on the wall is obtained by

$$\overline{V}_t^+ = \frac{\overline{V}_t}{u_*} = \frac{kPr^{\frac{1}{2}}}{\overline{\tau}^+}\left(\frac{e^{-(\frac{1}{2\overline{H}^+})y^+}J_{2\vartheta^+-1}(2\vartheta^+e^{-(\frac{1}{2\overline{H}^+})y^+}) - J_{2\vartheta^+}(2\vartheta^+e^{-(\frac{1}{2\overline{H}^+})y^+})}{\frac{T_\infty}{T_w - T_\infty}J_{2\vartheta^+}(2\vartheta^+) + J_{2\vartheta^+}(2\vartheta^+e^{-(\frac{1}{2\overline{H}^+})y^+})}\right). \tag{25}$$

where $\vartheta^+ = Pr^{\frac{1}{2}}(\frac{\overline{H}^+}{\overline{\tau}^+})$ and $y^+ = \frac{yu_*}{\nu}$.

The Bessel function $J_n(x)$ is defined as

$$J_n(x) = \frac{1}{\pi}\int_0^\pi cos(xsin\theta - n\theta)d\theta \tag{26}$$

The function is based on the Sookne's code [17]. The order of the Bessel function must be a positive integer with a backward recursive procedure and strict error control. The relationship between the two passed parameters is defined as

$$\vartheta^+ = Pr^{\frac{1}{2}}v^+ \tag{27}$$

The numerical predictions of thermophoretic velocity distribution are restricted to the positive integers $Pr^{\frac{1}{2}}$ because the orders of the Bessel function v and ϑ of Equations (24) and (25) must be positive integers and n.

3. Results and Discussions

The radial resolute of the fluctuation velocity near the wall region is used to build a viscous sublayer randomly. The analysis result of momentum transfer shows a relationship between the mean residence time $\overline{\tau}^+$ of the mean wall shear stress and the mean velocity profile. A sequence of $\overline{\tau}^+$ and distance $\overline{H}^+$ is depicted in Figures 1 and 2. Relative to the Reynolds number and dimensionless pipe diameter, the sequence was calculated by

$$d^+ = Re\sqrt{\frac{f}{2}} \tag{28}$$

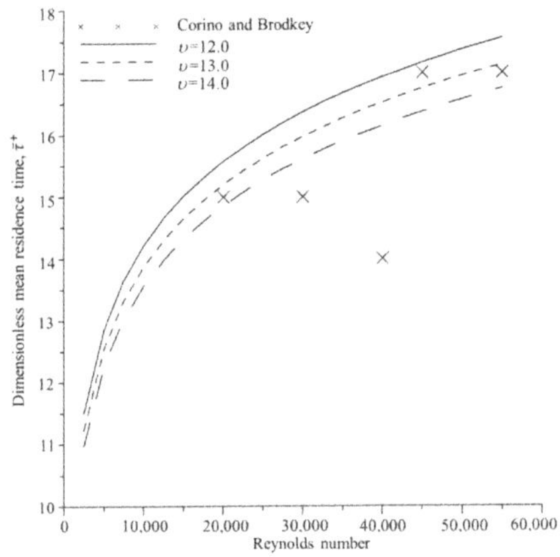

Figure 1. Dimensionless mean residence time with Reynolds number.

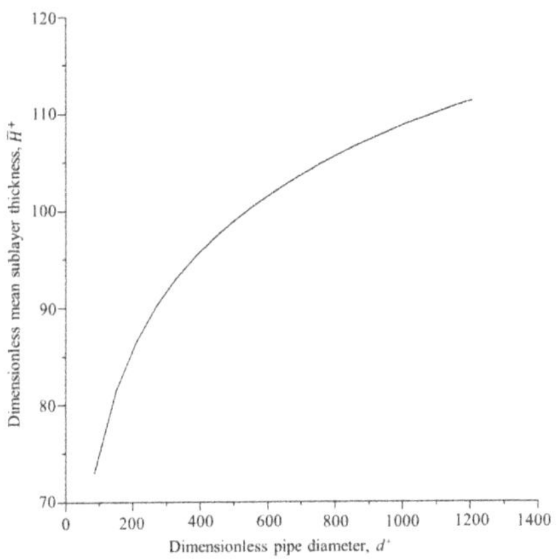

Figure 2. Variation of dimensionless mean sublayer thickness with pipe diameter and Reynolds number.

The eddy appears so frequently at lower Reynolds numbers, and the sublayer development is maintained almost at the wall region. Reversely, the sublayer is progressively established from the wall and seems to be evicted out of the wall region. The variable is a dimensionless mean residence time of turbulence, $\overline{\tau}^+$. The result of the sublayer model by Meek and Baer [18] shows that for Reynolds numbers greater than 10^4, the mean residence time is constant at 18.0. The visual observations of Corino and Brodkey [19] indicated that the dimensionless cycle period of the viscous sublayer is $14 \leq \overline{\tau}^+ \leq 17$ in the smooth tube flow of $2 \times 10^4 < Re < 5.5 \times 10^4$. The surface rejuvenation model calculation shows that the transfer parameter v is around 13.0; its coinciding with the visual observations (Figure 1) seems satisfactory. Therefore, the prediction of thermophoretic velocity distribution relative to Prandtl number and temperature gradient is calculated with $v = 13.0$ and $\overline{\tau}^+ = 17$ for $Re = 5.0 \times 10^4$.

With fixed values of Reynolds numbers and v, a plot of the calculated thermophoretic velocity relative to Prandtl number is shown in Figure 3 for $T_w - T_\infty = \pm 50$ °C. The thermophoretic velocity becomes greatest near the surface by the steep temperature fluctuation since the sublayer oscillations generate wall temperature fluctuations. The velocity is closer to the surface with increasing Prandtl number and diminishes rapidly in the center of the pipe. The amplitude distribution is broad, especially at low Prandtl numbers, and is sharper at higher Prandtl numbers. A discrepancy in slope near the edge of the sublayer indicates an enhancing coupling of the thermal and turbulent mechanisms or appears due to the semi-infinite boundary condition.

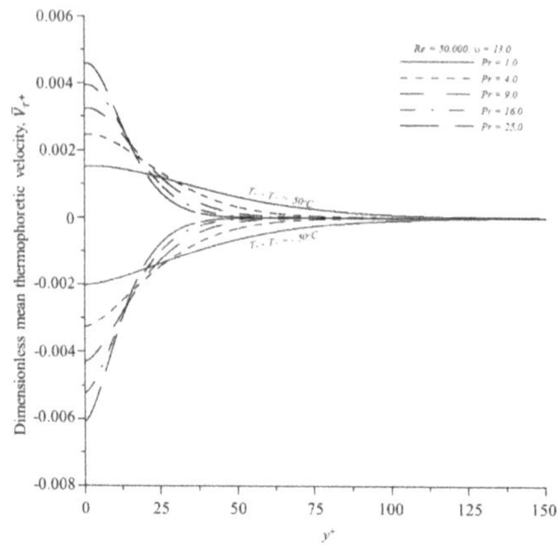

Figure 3. Dimensionless mean thermophoretic velocity distribution with various Prandtl numbers.

If particles are suspended in turbulent flow, turbulence governs the particle transport in the core, which becomes steadily weak when approaching the wall. When the eddy impaction induced by turbulence in the outer boundary layer is acting alone, the particles tend to gather in the viscous sublayer. Coherent temperature gradients are observed along the wall, playing a critical role in particle transport. Figure 4 demonstrates the intensity of the thermophoretic mechanism and indicates a pronounced increase in the thermophoretic velocity with increasing temperature difference, $T_w - T_\infty$. The difference widens at the surface where the thermal gradient is greatest. As expected, thermophoresis of particles with the diffusion of small molecules may be the only mechanism that increases particle transfer near the viscid wall. If thermal fluctuations are higher in the area near the wall, the actual transfer rate is expected to be higher. Changing the temperature gradient in the cooling section significantly affects thermophoretic deposition. When heated, reverse thermophoresis moves particles away from the wall, while the transport by turbulence makes particles move towards the wall. As these two effects interact, particles gather in certain areas close to the wall.

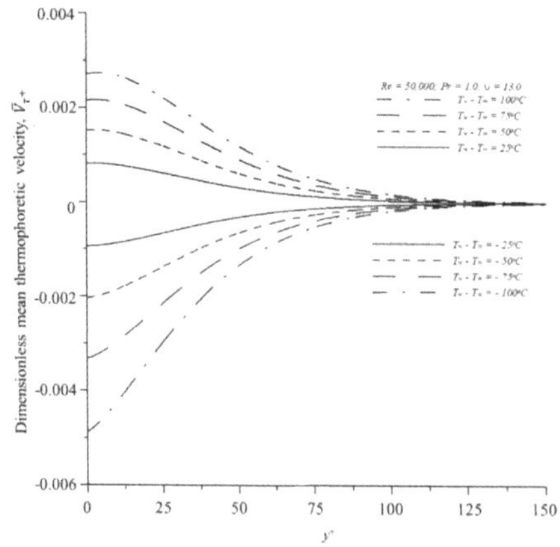

Figure 4. Dimensionless mean thermophoretic velocity distribution with various temperature difference.

Author Contributions: Conceptualization, M.-C.C. and H.-Y.T.; methodology, M.-C.C.; software, M.-C.C.; validation, M.-C.C.; data curation, M.-C.C.; writing—original draft preparation, H.-Y.T.; writing—review and editing, M.-C.C. and H.-Y.T.; supervision, M.-C.C.; project administration, M.-C.C.; All authors have read and agreed to the published version of the manuscript.

Funding: This research received no external funding.

Institutional Review Board Statement: Not applicable.

Informed Consent Statement: Not applicable.

Data Availability Statement: Not applicable.

Conflicts of Interest: The authors declare no conflict of interest.

References

1. Epstein, P.S. Zur theorie des radiometers. *Z. Phys.* **1929**, *54*, 537–563. [CrossRef]
2. Hidy, G.M.; Brock, J.M. *The Dynamics of Aerocolloidal Systems*; Pergman Press: Oxford, UK, 1970.
3. Whitmore, P.J.; Meisen, A. Estimation of thermo- and diffusionphoretic particle deposition. *Can. J. Chem. Eng.* **1977**, *55*, 279–285. [CrossRef]
4. Brock, J.R. On the theory of thermal forces acting on aerosol particles. *J. Colloid Sci.* **1962**, *17*, 768–780. [CrossRef]
5. Jacobsen, S.; Brock, J.R. The thermal force on spherical sodium chloride aerosol. *J. Colloid Sci.* **1965**, *20*, 544–554. [CrossRef]
6. Derjaguin, B.V.; Yalamov, Y. Theory of thermophoresis of large aerosol particles. *J. Colloid Sci.* **1965**, *20*, 555–570. [CrossRef]
7. Derjaguin, B.V.; Storozhilova, A.I.; Rabinovich, Y.I. Experimental verification of the theory of thermophoresis of aerosol particles. *J. Colloid Interface Sci.* **1966**, *21*, 35–58. [CrossRef]
8. Walderman, L. Uber die kraft eines inhomogenen gases auf kliene suspendierte. *Z. Nat. A* **1959**, *14*, 588–599.
9. Walderman, L.; Schmitl, K.H. Thermophoresis and diffusiophoresis of aerosol. In *Aerosol Science*; Davies, C.N., Ed.; Academic Press: London, UK, 1966.
10. Tong, T.N.; Bird, G.A. The thermal force in the low density limit. *J. Colloid Interface Sci.* **1971**, *35*, 403–408. [CrossRef]
11. Talbot, L.; Cheng, R.K.; Schefer, R.W.; Willis, D.R. Thermophoresis of Particles in a Heated Boundary Layer. *J. Fluid Mech.* **1980**, *101*, 737–758. [CrossRef]
12. Batcheler, G.K.; Shen, C. Thermophoretic deposition of particles in gas flowing over cold surfaces. *J. Colloid Interface Sci.* **1985**, *107*, 21–37. [CrossRef]
13. Danckwerts, P.V. Significance of liquid-film coefficients in gas absorption. *Ind. Eng. Chem.* **1951**, *43*, 1460–1467. [CrossRef]
14. Hanratty, T.J. Turbulent exchange of heat and momentum with a boundary. *AIChE J.* **1956**, *2*, 359–362. [CrossRef]
15. Thomas, L.C.; Gingo, P.J.; Chung, B.T.F. The surface rejuvenation model for turbulent convective transport-an exact solution. *Chem. Eng. Sci.* **1975**, *30*, 1239–1242. [CrossRef]
16. Eck, B. *Technische Stromungslehre*; Springer: New York, NY, USA, 1973.
17. Sookne, D.J. Bessel functions of real argument and integer order. *J. Res. Natl. Bur. Stand. J. Res. B* **1973**, *77*, 125–132. [CrossRef]
18. Meek, R.I.; Baer, A.D. The periodic viscous sublayer in turbulent flow. *AIChE J.* **1970**, *16*, 841–848. [CrossRef]
19. Corino, E.R.; Brodkey, R.S. A visusal investigation of the wall region on turbulent flow. *J. Fluid Mech.* **1969**, *37*, 1–30. [CrossRef]

Disclaimer/Publisher's Note: The statements, opinions and data contained in all publications are solely those of the individual author(s) and contributor(s) and not of MDPI and/or the editor(s). MDPI and/or the editor(s) disclaim responsibility for any injury to people or property resulting from any ideas, methods, instructions or products referred to in the content.

engineering proceedings

Proceeding Paper

Preparation of High-Performance Cementitious Materials from Industrial Solid Wastes †

Peng Yang [1], Yueqiang Qi [1] and Renshan Chen [2,*]

[1] Shandong Huanghe Shuncheng Water Conservancy and Hydro-Power Engineering Co., Jinan 250000, China; 17861417182@163.com (P.Y.); hna626@163.com (Y.Q.)
[2] School of Transportation and Logistics Engineering, Shandong Jiaotong University, Jinan 250357, China
* Correspondence: 401006@sdjtu.edu.cn
† Presented at the 3rd IEEE International Conference on Electronic Communications, Internet of Things and Big Data Conference 2023, Taichung, Taiwan, 14–16 April 2023.

Abstract: Based on coal gangue, sintered red mud, and fly ash, which are primary industrial waste products, a high-performance cementitious material was developed that considered its composition, physical and chemical properties, and microstructure. The optimal ratio of the materials with an appropriate exciter content was determined through the production and testing of samples composed of various ratios. It was found that the compressive strength reached its maximum when the ratio of coal gangue: red mud: fly ash was 40:40:20 and the exciter content was 8%. The synergistic effect of the waste-based cementitious material and the exciter content was confirmed, which lays a foundation for the development of diverse waste-based industrial materials. The use of such materials helps reduce abandoned industrial waste and encourages the recycling of materials for various purposes.

Keywords: coal gangue; red mud; fly ash; solid-waste-based cementitious materials

Citation: Yang, P.; Qi, Y.; Chen, R. Preparation of High-Performance Cementitious Materials from Industrial Solid Wastes. *Eng. Proc.* **2023**, *38*, 50. https://doi.org/10.3390/engproc2023038050

Academic Editors: Teen-Hang Meen, Hsin-Hung Lin and Cheng-Fu Yang

Published: 27 June 2023

Copyright: © 2023 by the authors. Licensee MDPI, Basel, Switzerland. This article is an open access article distributed under the terms and conditions of the Creative Commons Attribution (CC BY) license (https://creativecommons.org/licenses/by/4.0/).

1. Introduction

The shortage of resources such as cement and gravel and the environmental effect of using them has been a concern for road construction for a long time. Therefore, the research on the curing agent based on recycled solid wastes has received a lot of attention. Solid waste used in construction is mainly collected from wasted rocks from mining, beneficiation, and tailings, as well as from waste slag from fuel production or smelting [1,2]. Liu [3] et al. prepared a highway pavement subgrade material using red mud waste from Bayer process waste products, coal gangue, and fly ash. The material was found to have appropriate mechanical and environmental properties that met the national standards. Liu [4] et al. studied the effects of fly ash addition on the early strength and volumetric stability of a mixture of desulfurization gypsum, lime, and the exciter of red-mud-filled materials. They found that the desulfurization gypsum promoted the generation of calcium alumina, the lime enhanced the effect of the fly ash, and the exciter accelerated the process of the hydration reaction of red mud and fly ash. The synergistic effect of the three materials improved the strength of the red-mud-filled body.

Wei [5] pointed out that coal gangue and red mud could be used in road construction, and Wang et al. [6] prepared alkali-activated cementitious materials with fly ash, titanium gypsum, and carbide slag; they found the optimal ratio of the mixture to have the best mechanical property as a cementitious material. Wang et al. [7] selected coal gangue and fly ash as base materials to prepare geopolymer foaming materials. The compressive strength, density, porosity, and thermal conductivity of the foaming materials prepared under different blending ratios of coal gangue and fly ash were studied to find an appropriate blending ratio to satisfy the expected engineering effects. Ye et al. [8] tested aluminosilicate solid waste as cementitious materials and found that the performance of the materials was enhanced significantly when red mud and coal gangue were mixed in a ratio of 39 and

26 wt.%, respectively. Gao [9] developed multi-source solid-waste-based marine grouting materials using Bayer red mud, calcium carbide slag, and silica fume as raw materials and proved that the compressive requirements of the material met the industrial standard. In an indoor test, Wang et al. [10] found that the solid-waste-based soil curing agent composed of ISW and Portland cement improved mechanical properties and durability compared to solidified clay, expansive soil, solidified silty soil, and fine sandy soil. The mixture's unconfined compressive strength after 28 days reached above 3.3 MPa. By comparing the soil curing agent made from a mixture of industrial wastes, including fly ash, tailings, slag, and slaked lime and PF32.5, Wang [11] proved that the industrial-waste-based soil curing agent was effective for curing clay soils of a low liquid limit. Wei [12] optimized the ratio and performance of grouting material that was prepared with alkali-excited cement-based material for mines and road construction. They found that, as the ratio of slag powder/fly ash increased, the precipitation rate gradually decreased, and the compressive strength gradually increased and stabilized.

Referring to the previous study results of various cementitious materials using industrial wastes, we developed a high-performance waste-based cementitious material using coal gangue, sintered red mud, and fly ash. Coal gangue is one of the most-produced industrial solid wastes in China due to the enormous use of coal, and red mud is a well-known material for its great potential as a new alkali-inspired cementitious material [13]. Cement production requires a huge amount of various raw materials. Thus, if coal gangue and red mud are used in cement production, abandoned coal gangue and red mud can be reused to save production costs and increase product reliability, as well as to support environmental protection [14]. In this study, the physical and chemical properties, as well as the microstructure of a waste-based cementitious material, were investigated to understand its mineral synergy and excitation characteristics. The optimal ratio of each raw material and an exciter was determined by experimenting with the performance. The result provides a reference for recycling and using industrial wastes for various purposes to develop diverse, environment-friendly construction materials.

2. Materials and Methods

2.1. Analysis Method

The major component of raw materials was analyzed using X-ray fluorescence analysis (XRF) and X-ray powder diffraction (XRD) analysis methods. In XRD analysis, the powder method was adopted with a Rigaku Ultima type IV ray diffractometer. The conventional wide angle was 10–80°, and the conventional test rate was 2°/min. The size of the particles of the materials did not exceed 75 μm, and 0.5 g of each material was used for the analysis. The fluidity was measured with the conventional spiral fluidity test method for cement slurry. The penetration resistance test (ASTM C803) was carried out for compressive strength with $4 \times 4 \times 4$ cubes made from cementitious materials with various mixing ratios (explained in Section 3.2).

2.2. Coal Gangue

In this study, burned coal gangue was selected as the main component of the high-performance composite gel material and was collected from the Jinan Huaiyin western heat source plant. The analyzed content of metals and minerals are shown in Tables 1 and 2. Rare metals such as gallium, vanadium, titanium, and cobalt were contained in the coal gangue. The content of titanium was the highest (110.9670 mg/kg on average). The average contents of gallium, vanadium, and cobalt were 44.0505, 4.7352, and 4.8384 mg/kg, respectively. In the coal gangue, the content of titanium was about 2.5 times higher than that of gallium and about 23 times higher than those of vanadium and cobalt, while the content of vanadium and cobalt was similar. It is important to know their contents, as they may cause environmental pollution and radioactivity, which requires continuous environmental monitoring to use them.

Table 1. Metal contents of coal gangue used for producing high-performance waste-based cementitious material.

Sample Number	Gallium (mg/kg)	Vanadium (mg/kg)	Titanium (mg/kg)	Cobalt (mg/kg)
1	45.1848	5.2487	109.7672	5.0205
2	42.8322	3.9336	117.7885	4.5892
3	44.1345	5.0234	105.3452	4.9056
Average	44.0505	4.7352	110.9670	4.8384

Table 2. Mineral contents of coal gangue used for producing high-performance composite gel material.

Analysis Number	Mineral (Chemical Formula)	Content (wt%)			Average Content (wt%)
		Sample 1	Sample 2	Sample 3	
1	Silicate (SiO_2)	27.9	25.6	28.0	27.2
2	Muscovite ($(KF)_2(Al_2O_3)_3(SiO_2)_6$)	7.0	6.2	8.0	7.1
3	Titanium oxide (TiO_2)	0.4	1.0	0.9	0.8
4	Albite ($NaAlSi_3O_8$)	7.3	5.7	8.6	7.2
5	Kaolinite ($2SiO_2 \cdot Al_2O_3 \cdot 2H_2O$)	28.2	27.5	26.4	27.4
6	Iron oxide (Fe_2O_3)	5.6	3.5	4.9	4.7
7	Chlorite ($(Mg,Fe)_5Al(Si_3Al)O_{10}(OH)_8$)	7.2	13.5	6.4	9.0
8	Montmorillonite ($(Na,Ca)_3(Al,Mg)_2Si_4O_{10}(OH)_2 \cdot n(H_2O)$)	16.5	17.0	16.8	16.8
	Total (weight %)	100	100	100	100

In the coal gangue, kaolinite was the most abundant (27.4% on average), followed by silicate (27.2%), while iron oxide was the least abundant (4.78%). Kaolinite, silicate, montmorillonite, chlorite, and muscovite were major minerals in the coal gangue. SiO_2 and Al_2O_3 comprised 55% of the coal gangue. TiO_2 is a harmful substance that causes environmental pollution and radioactivity, and it plays a role as a main control factor for pollution. The montmorillonite content was 16.8% on average. It reacts chemically with rainwater, which expands gangue particles, as it changes Al_2O_3 into $Al(OH)_3$. The reaction also attenuates the mechanical strength of coal gangue in general. Thus, attention needs to be paid to the leakage of coal gangue in use.

2.3. *Red Mud*

The red mud was obtained from three sources, including the second stockpile of Chinalco Shandong Branch (Shandong red mud) and Shandong Weiqiao Aluminum Company (Weiqiao red mud) after the sintering process and the red mud stockpile of Shandong Aluminum (Zibo red mud) after the Bayer process. The samples were stored and dried for over half a year. The average moisture content of the red mud was between 40.7 and 44.5%. The average liquid limit, plastic limit, and plasticity index values were 45.2–50.9%, 31.4–44.0%, and 6.9–14.6%, respectively. There were significant differences in those properties for different red mud samples, as shown in Table 3 below.

Table 3. Several properties of red mud collected from three sources.

Samples	Average Moisture Content (%)	Average Liquid Limit (%)	Average Plastic Limit (%)	Average Plasticity Index
Shandong red mud	41.4	50.9	44.0	6.9
Weiqiao red mud	40.7	48.0	33.4	14.6
Zibo red mud	44.5	45.2	31.4	13.8

The typical particle composition of red mud is shown in Figure 1. In general, the red mud particles smaller than 75 μm (in diameter) comprised more than 80% of the red mud. The particles smaller than 10 μm amounted to about 10% of the content, while the sizes of the rest ranged between 20 and 50 μm. The average unevenness coefficient was 29.0, and the curvature coefficient was 0.31. A large unevenness coefficient results in a wide particle size distribution, a better gradation, and a small curvature coefficient. In this case, particles were easy to compact, as the red mud lacked large particles. The curvature coefficient of the samples in this study was small, thus indicating that the red mud lacked large particles.

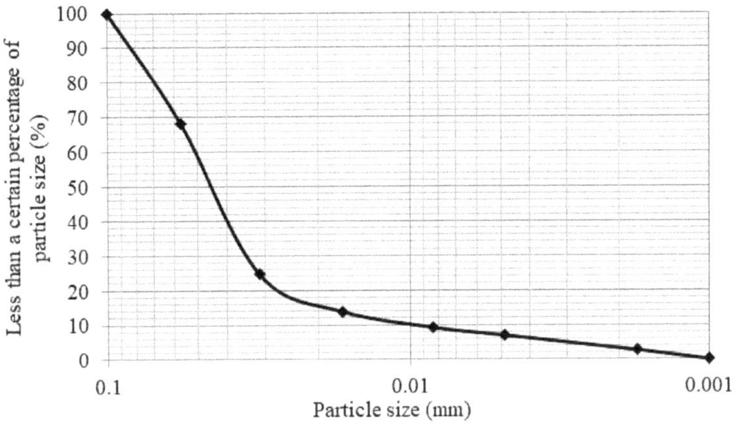

Figure 1. Conventional particle composition of red mud.

The red mud samples were analyzed with XRF to identify their compositions after being dried at 105 °C. The chemical compositions of Shandong red mud and Weiqiao red mud after the sintering process were similar in the content of chemical elements (carbon, sodium, magnesium, phosphorus, sulfur, chloride, potassium, titanium, vanadium, and manganese). The contents of silicon, calcium, aluminum, and iron were different in the two red mud samples. Shandong red mud contained higher contents of iron and aluminum than Weiqiao red mud. In particular, the iron content of Shandong red mud was 33.3% higher than that of Weiqiao red mud. Zibo red mud produced after the Bayer process showed higher contents of Al_2O_3, Fe_2O_3, SiO_2, and other oxides than those of Shandong red mud and Weiqiao red mud but a lower content of CaO. This difference in the chemical composition led to the difference in the color of Zibo red mud, which was dark red. Sintered red mud had higher CaO and SiO_2 contents and lower Fe_2O_3 content than the Bayer-processed red mud, which made it mostly white or gray in color. The differences in the chemical compositions of the red muds with different processes are shown in Figure 2. The analysis result showed a significant difference in the contents of SiO_2, Al_2O_3, CaO, and Fe_2O_3. Considering the environmental effect and cementing performance, the red mud with the sintering process calcined at 1200 °C was used in this study.

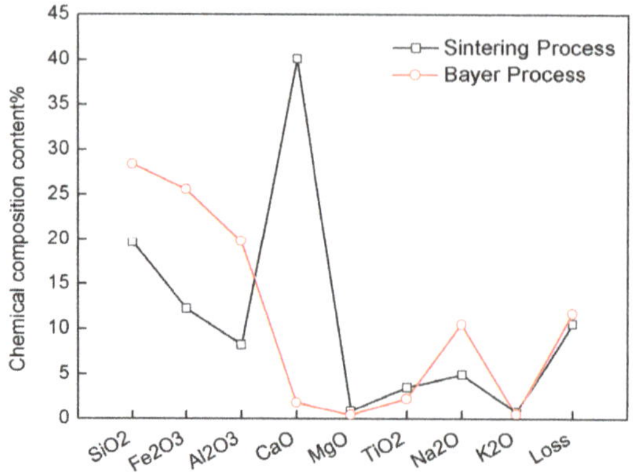

Figure 2. Comparison of chemical compositions of red mud samples produced with the sintering process (the black line) and the Bayer process (the red line).

2.4. Fly Ash

The fly ash was obtained from Jinan Thermal Power Company. XRF analysis of fly ash collected from the landfill site was performed to determine its chemical composition as shown in Table 4. The main components of the fly ash were SiO_2 (50.76%) and Al_2O_3 (20.12%). In general, more SiO_2 increases the activity of the fly ash regarding the mixture of cementitious materials.

Table 4. Chemical composition of fly ash collected from the landfill site in this study.

Composition	SiO_2	CaO	SO_3	Fe_2O_3	Al_2O_3
Content (wt.%)	50.76	16.8	6	6.22	20.12

The SEM image in Figure 3 shows that the fly ash was composed of a variety of particles, of which spherical particles accounted for more than 50% of the total particles, and irregular-shaped particles accounted for about 35% of the total particles. It is known that irregular-shaped particles have higher chemical reactivity in high-temperature calcination. However, all particles store high chemical energy after high-temperature calcination, so fly ash maintains a high activity. As the total content of SiO_2, Al_2O_3, CaO, and Fe_2O_3 reached 80%, and the particle size was small enough at the micrometer level, the fly ash turned out to have a high possibility of synergistic polymerization with the C-S-H gel structure.

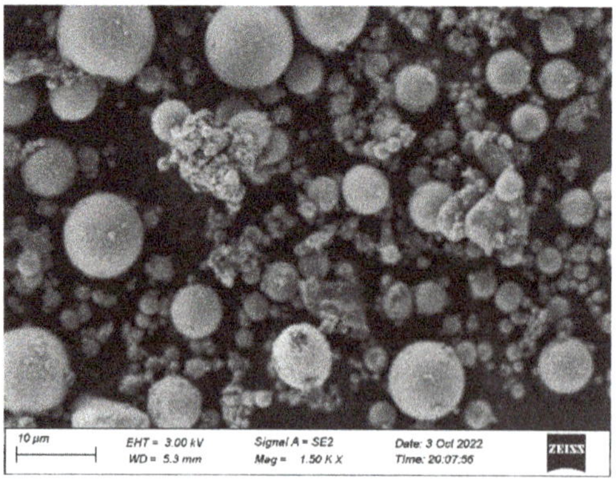

Figure 3. SEM image of fly ash collected from landfill site.

3. Preparation and Characterization of Waste-Based Cementitious Material

3.1. Preparation of Exciter

To prepare the high-performance waste-based cementitious material, we used sodium silicate, triterpene saponin, polymerized aluminum sulfate, magnesium fluorosilicate, sodium hydroxide, hydroxypropyl methylcellulose, carboxymethyl starch ether, sodium tripolyphosphate, and sulfonated melamine formaldehyde resin at the pure reagent level. A total of 35 g sodium silicate was dissolved in distilled water, and sodium hydroxide was added when the temperature was raised to 70–80 °C. The modulus of the sodium silicate was controlled at about 1.5, and the temperature was cooled to 60–65 °C after stirring for 1–2 h. Then, 6.4 g of polymerized aluminum sulfate, 35 g of triterpenoid saponins, 3 g of magnesium fluorosilicate, and 0.3 g of sulfonated melamine formaldehyde resin were added to the solution. After stirring for 1–2 h and leaving it to cool down to room temperature, 0.8 g hydroxypropyl methylcellulose, 0.4 g carboxymethyl starch ether, and 2 g sodium tripolyphosphate were added to the solution and dissolved completely. The solution was dried to obtain the alkaline composite exciter.

3.2. Preparation of Material

The alkaline composite exciter was added to the mixture of coal gangue, red mud, and fly ash in various ratio quantities (Table 5). First, the exciter was dissolved in distilled water according to the designated content shown in Table 5. After the exciter was completely dissolved, the solution was cooled to a room temperature of 20 °C. Then, coal gangue, red mud, and fly ash were added to the solution while the solution was stirred continuously so that the materials could be fully mixed until the color became uniform and no longer changed. The prepared solution was poured into the net slurry mixer container to obtain 4 × 4 × 4 cm cubes. The cubes (samples) were cured for 1 day to be hardened. The samples were hydrated in anhydrous ethanol before being used for the experiment. A total of 10 different samples were produced according to the different mixing rates of coal gangue, red mud, fly ash, and different exciter contents, which were determined through preliminary experiments.

Table 5. Coal gangue/red mud/fly ash ratio.

Sample Number	Ratio in Weight (Coal Gangue:Red Mud:Fly Ash)	Water-to-Solid Ratio	Exciter Content (%)
A0	0:0:100	0.4	6
A1	5:5:90	0.4	6
A2	10:10:80	0.37	8
A3	10:20:70	0.35	10
A4	20:20:60	0.4	6
A5	20:30:50	0.37	8
A6	30:30:40	0.35	10
A7	30:40:30	0.4	6
A8	40:40:20	0.37	8
A9	40:50:10	0.35	10
A10	50:50:0	0.4	6

3.3. Performance Assessment and Proportional Optimization

With reference to the preliminary experiments of this study, the mechanical property was tested, and microscopic characterization was conducted within 28 days after curing. Experiments were carried out to optimize the coal gangue/red mud/fly ash ratio and exciter content and to obtain the waste-based cementitious material with optimal performance using eleven different samples with different mixing ratios and exciter contents (Table 5).

Figures 4 and 5 show the degrees of fluidity and the compressive strengths of the samples, respectively. With the increase in amounts of coal gangue and red mud, the degree of fluidity decreased, but the compressive strength increased. The compressive strength increased until the ratio of coal gangue and red mud reached 40% in the composite (A0–A8). A higher ratio of coal gangue and red mud decreased the compressive strength. Sample A8 showed the highest compressive strength, with a ratio of 40:40:20 (coal gangue: red mud: fly ash) and an exciter content of 8%. Thus, the ratio and exciter content allowed for the best synergistic performance among the samples tested.

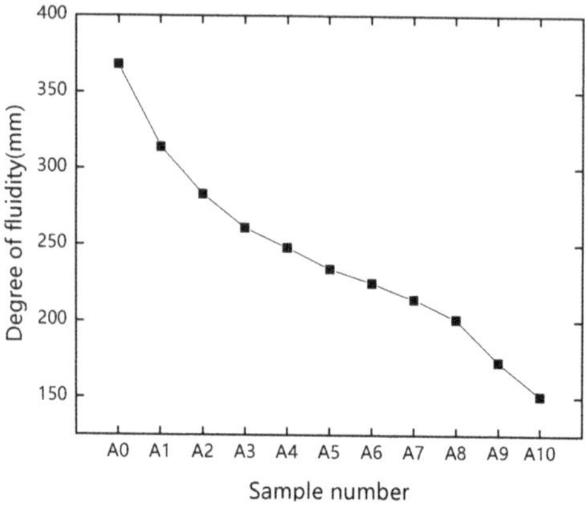

Figure 4. Effect of ratios of coal gangue, red mud, and fly ash and exciter content on degrees of fluidity of cementitious materials.

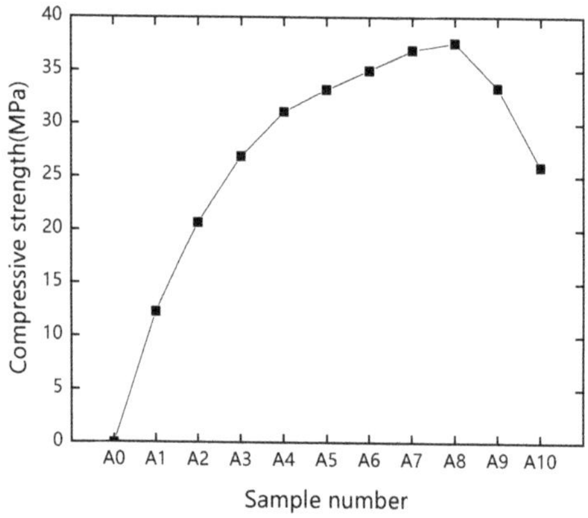

Figure 5. Effect of ratios of coal gangue, red mud, and fly ash and exciter content on compressive strengths of cementitious materials.

When coal gangue, red mud, and fly ash were used as the sole cementitious material, their hydration was too low to have enough cementation reaction. When the three materials were combined, the compressive strength increased considerably. When the contents of coal gangue and red mud were between 0–20% (A0–A4), the compressive strength increased greatly, which corresponded to a decrease in the degree of fluidity. With the contents of 30–40% of coal gangue and red mud, the compressive strength growth still increased but tended to be stabilized at a plateau. The compressive strength reached a maximum value of 37.6 MPa with the contents of coal gangue and red mud at 40%.

The influence of the exciter on the compressive strength was also observed. When the summed content of coal gangue and red mud exceeded 50% (A5–A8), the compressive strength increased, regardless of the exciter content. When the summed content was higher than 90%, the compressive strength decreased, even with a higher content of the exciter. However, the exciter content did not affect the degree of fluidity significantly, as the degree kept decreasing with the higher content of the exciter. The cross-section image shows white coarse dots with an uneven distribution in the samples with higher contents of coal gangue and red mud. This indicates a fast-setting phenomenon and incomplete reaction, which leads to reduced internal stress and mechanical properties.

In summary, the optimal ratio of coal gangue, red mud, and fly ash was found to be 40:40:20 with an exciter content of 8%, wherein the compressive strength reached a maximum value of 37.6 MPa, and the degree of fluidity was below 200 mm. This indicates that the three materials had a synergistic effect, which is appropriate as a cementitious material.

4. Conclusions

With the increase in the use of natural resources such as coal and aluminum for various industrial activities, the amount of waste from using them also increases considerably. The construction of roads also has been increasing due to the development of industry and society, which demands an enormous amount of construction materials. To reduce the amount of abandoned industrial wastes and newly exploited materials, various waste-based materials have been developed for industrial use, but many of them do not have the appropriate mechanical property to be used. Therefore, we proposed a new cementitious material made from coal gangue from mining, red mud from aluminum production, and fly ash from incineration processes. With an alkaline exciter, the optimal ratio of the materials was determined through the experiment of testing the compression strength and degree of fluidity. The chemical compositions and microstructure of the new material with various mixture ratios and exciter content were analyzed and observed with XRF, XRD, SEM, and other methods. The result showed that the material of the mixture ratio of 40:40:20 (coal gangue: red mud: fly ash) with an exciter content of 8% had the optimal mechanical property, wherein it showed a compressive strength of 37.6 MPa and a degree of fluidity that was less than 200 mm. The newly developed cementitious material can be used as a reference for the study of the curing mechanism and performance analysis of the solid-waste-based binder, as well as the development of waste-based industrial materials.

Author Contributions: Methodology, P.Y.; test, Y.Q. and R.C.; data curation, R.C.; writing-original draft preparation, R.C.; writing-review and editing, P.Y. All authors have read and agreed to the published version of the manuscript.

Funding: This research received no external funding.

Institutional Review Board Statement: Not applicable.

Informed Consent Statement: Not applicable.

Data Availability Statement: Not applicable.

Conflicts of Interest: The authors declare no conflict of interest.

References

1. Sun, J.; Geng, C.; Zhang, Z.; Wang, X.; Xu, L. Comprehensive utilization of industrial solid waste resources. *Mater. Bull.* **2012**, *26*, 105–109.
2. Wang, X.; Li, Q.; Chen, S.; Yue, G. Application research and prospect of industrial solid waste in the field of new building materials. *Silic. Bull.* **2019**, *38*, 3456–3464.
3. Liu, X.; Tang, B.; Yin, H.; Emile, M. Durability and environmental performance of red mud-highway pavement base materials. *J. Eng. Sci.* **2018**, *40*, 438–445.
4. Liu, J.; Zhou, Z.; Wu, A.; Wang, Y. Preparation and hydration mechanism of low concentration Bayer red mud filling materials. *J. Eng. Sci.* **2020**, *42*, 1457–1464.
5. Wei, J. Application of solid waste materials in road base. *Synth. Mater. Aging Appl.* **2022**, *51*, 162–164.
6. Wang, Y.; Yuan, H.; Chen, S. Preparation of alkali-initiated gelling materials from solid waste. *Chem. Miner. Process.* **2022**, 1–11. [CrossRef]
7. Wang, R.; Wang, C.; Han, T. Experimental study on the preparation of geopolymer foam from coal gangue-fly ash. *Build. Mater. Technol. Appl.* **2022**, 11–14. [CrossRef]
8. Ye, Z.; Wang, Q. Preparation and properties of aluminosilicate solid waste gelling materials. *Gold Sci. Technol.* **2020**, *28*, 658–668.

9. Gao, Y. *Research on the Development and Performance of Multi-Source Solid Waste-Based Offshore Grouting Materials;* Shandong University: Jinan, China, 2021.
10. Wang, X.; Yang, B.; Zhao, Y.; Wang, Y. Road performance of solid waste-based gelling agent stabilized soil. *J. Wuhan Univ. Technol. (Transp. Sci. Eng. Ed.)* **2021**, 1–9. [CrossRef]
11. Wang, Q. Construction technology of industrial waste soil curing agent for soil improvement. *Henan Sci. Technol.* **2022**, *41*, 109–112.
12. Wei, W. *Optimization of Alkali-Inspired Cement-Based Grouting Material Ratio and Performance Study;* North China University of Water Resources and Hydroelectricity: Zhengzhou, China, 2021.
13. Liu, Q.; Wang, Q.; Wu, P.; Wang, J.; Lv, J. Research progress on the application of red mud in cementitious materials. *J. Shandong Univ. Sci. Technol. (Nat. Sci. Ed.)* **2022**, *41*, 66–74.
14. Zhang, N.; Liu, X.; Sun, H. XPS analysis of the hydration process of red mud-gangue-based cementitious materials. *Met. Min.* **2014**, *453*, 171–176.

Disclaimer/Publisher's Note: The statements, opinions and data contained in all publications are solely those of the individual author(s) and contributor(s) and not of MDPI and/or the editor(s). MDPI and/or the editor(s) disclaim responsibility for any injury to people or property resulting from any ideas, methods, instructions or products referred to in the content.

engineering proceedings

MDPI

Proceeding Paper

Structure of Parallel Mechanism Combined with Waterbomb-Base-Inspired Origami †

Lulu Al Marjan and Shyh-Chour Huang *

Department of Mechanical Engineering, National Kaohsiung University of Science and Technology, Kaohsiung 807618, Taiwan; f109142190@nkust.edu.tw
* Correspondence: shuang@nkust.edu.tw

† Presented at the 3rd IEEE International Conference on Electronic Communications, Internet of Things and Big Data Conference 2023, Taichung, Taiwan, 14–16 April 2023.

Abstract: Structure from the geometry and analysis of the three-spherical kinematic chain-base parallel mechanism have been studied. The parallel mechanism evolved from an origami fold as chain legs with three spherical kinematic chains becoming rigid bodies. The parallel mechanism with a three 6R kinematic chain as three chain legs is complicated. The reconfiguration of the parallel mechanism with full tilt–circle movement, kinematic, and workspace are investigated, too. This parallel mechanism can be applied in specific applications with certain treatments.

Keywords: origami inspires; parallel mechanism; spherical mechanism

Citation: Al Marjan, L.; Huang, S.-C. Structure of Parallel Mechanism Combined with Waterbomb-Base-Inspired Origami. *Eng. Proc.* **2023**, *38*, 51. https://doi.org/10.3390/engproc2023038051

Academic Editors: Teen-Hang Meen, Hsin-Hung Lin and Cheng-Fu Yang

Published: 26 June 2023

Copyright: © 2023 by the authors. Licensee MDPI, Basel, Switzerland. This article is an open access article distributed under the terms and conditions of the Creative Commons Attribution (CC BY) license (https://creativecommons.org/licenses/by/4.0/).

1. Introduction

The parallel mechanism is the system that converts the motions of several bodies into constrained motions of other bodies [1]. The parallel mechanism has a closed-loop as a type of mechanism [1], and that is made of an end-effector (mobile platform) [2] and a fixed platform, linked together by independent kinematic chains [1]. Furthermore, structure, workspace considerations, singularities, and link interference need to be considered in the design [2]. The parallel mechanism has much potential in several fields, including industrial, space, medical science, and miscellaneous applications [2].

Origami is Japanese cultural art, which is the art of folding paper. In general, origami starts with flat paper (2D object), then folds become a 3D object with various shapes and forms without stretching, cutting, or gluing [3]. Moreover, origami has become an inspiration for engineers in various fields. Especially waterbomb-base origami has become an inspiration in engineering for several applications. Fonseca and Savi [3] presented an investigation of the origami waterbomb-base pattern from its unit cell and explored the different formulations for origami structure. Liang, Gao, Huang, and Li [4] presented the design of a pneumatic rigid–flexible coupling origami gripper from a waterbomb-base origami pattern. Salerno, Zhang, Menciassi, and Dai [5] proposed the concept of a miniaturized surgical tool grasper as a 3-DOF parallel module inspired by the waterbomb-base origami.

It is common to see the creases of origami mechanisms as the compliant mechanism joint with flexible material. However, the origami mechanism becomes rigid if the creases are replaced with non-flexible material. One of the rigid kinematic origami models describes the creases with the revolute joint mechanism [6].

2. Geometry of Parallel Mechanism

A parallel mechanism consists of a base, a platform, and legs to sustain the platform. The leg structure is obtained from waterbomb-base origami, which involves the spherical kinematic chain with a close loop structure. The parallel mechanism consists of three legs and one actuator in the middle of the mechanism.

2.1. Waterbomb-Base Origami for Spherical Mechanisms

As shown in Figure 1a,b, origami starts with flat paper and then folds to become creases. The creases consist of two types. The first type has a convex shape called Mountain creases, denoted by M, and the second type has a concave shape called Valley creases, represented by V [7,8].

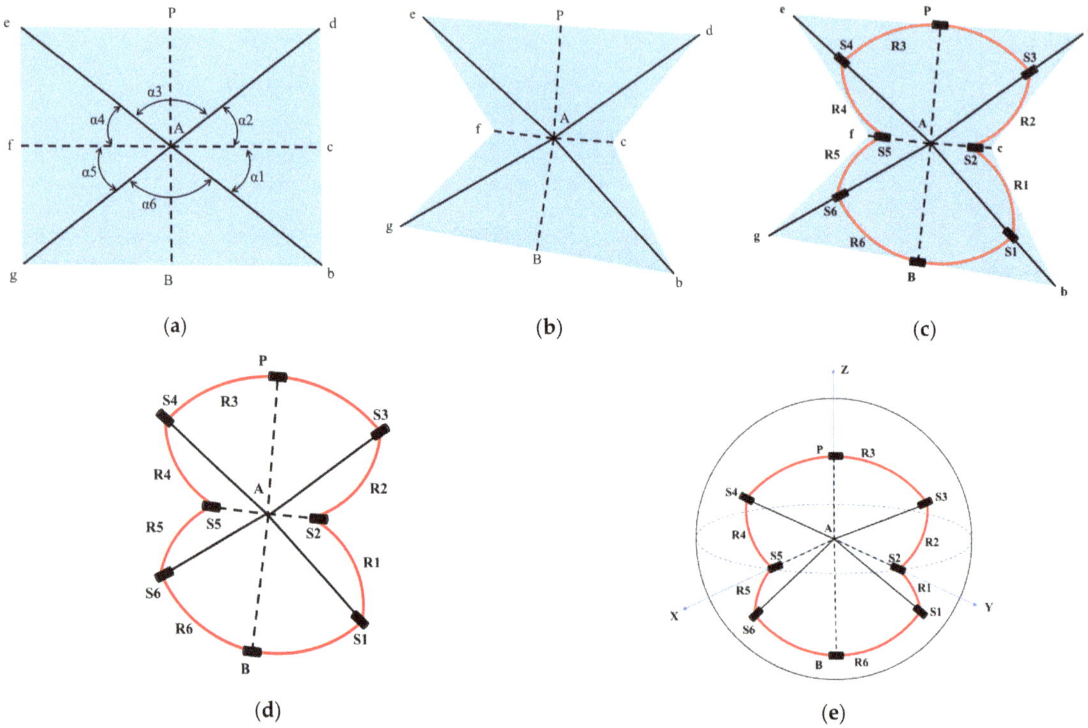

Figure 1. Waterbomb-base origami involves a spherical mechanism. (**a**) origami starts with flat paper; (**b**) origami starts fold become a crease; (**c**) the creases and the planes on the waterbomb-base structure; (**d**) the figure mapping the spherical mechanism from the origami evolved; (**e**) the spherical mechanism.

Waterbomb-base origami consists of six planes and six creases. Of the six creases, Ac and Af are the valley-crease types, and Ab, Ad, Ae, and Ag are the mountain-crease types. As shown in Figure 1c, the creases and the planes on the waterbomb-base structure are represented for the revolute pairs and the links, respectively [7]. Thus, the waterbomb-base origami form contains six revolute pairs and six links. Barreto et al. [8] made an analogy between the spherical mechanism and the origami vertex as a concept of mechanism design.

Maekawa's-Justin's Theorem [9] is as follows. Let M be the number of mountain creases and V be the number of valley creases adjacent to a vertex in a flat origami crease pattern. Then, M − V = ±2. Moreover, Kawasaki–Justin's Theorem [9] lets v be a vertex of degree $2n$ is an origami crease pattern and $\alpha_1 \ldots, \alpha_{2n}$ be the consecutive angles between the creases. Then, the creases adjacent to v (locally) fold flat only if

$$\alpha_1 - \alpha_2 + \alpha_3 - \ldots - \alpha_{2n} = 0 \tag{1}$$

That is, the two theorems correspond with waterbomb-base origami form. As shown in Figure 1d,e, the figure mapping and the spherical mechanism from the origami evolve. In the waterbomb base origami, Point A is the intersection of the whole of the connected creases, while in the spherical mechanism, point A is the virtual point intersection of the axis from the whole revolute pair in the spherical kinematic chain. The revolute pair and links are denoted with S and R, respectively. The link R_i $(i = 1, 2, \ldots, 6)$ is deformed by revolute joints S_{i-1} and S_i. Due to this close loop structure, link R_6 forms by revolute joint S_6 and S_1. R_3 and R_6 have a revolute joint for connecting mobile and base platforms. Link R_i $(i = 1, 2, \ldots, 6)$ has the angle α_i $(i = 1, 2, \ldots, 6)$. The relationship between the angles in waterbomb-base origami is $\alpha_1 = \alpha_2 = \alpha_4 = \alpha_5$; meanwhile, $\alpha_3 = \alpha_6$.

The initial shape of waterbomb-base origami can be a square ($bd = de$) or a rectangular ($bd < de$), depending on the intended use, application, and other parameters. Determining it requires further discussion.

2.2. Parallel Mechanism

The parallel structure mechanism consists of a mobile platform, a fixed platform, three legs, and a truss in the middle of the mechanism. Meanwhile, the legs are represented by the spherical kinematic chain involved in waterbomb-base origami, and then, the structure has three spherical kinematic chains for the legs [7]. The truss consists of a prismatic connector hinge and two offset universal hinges to connect to the mobile platform and base platform. The truss component has supported the movement of the mobile platform concerning the base platform.

Figure 2 shows the revolute pair S_{ij} and the link R_{ij}, where i $(i = 1, 2, 3)$ is the number of legs and j $(i = 1, 2, \ldots, 6)$ is the number of a revolute pair. P and B are the revolute joints paired with the mobile and base platforms.

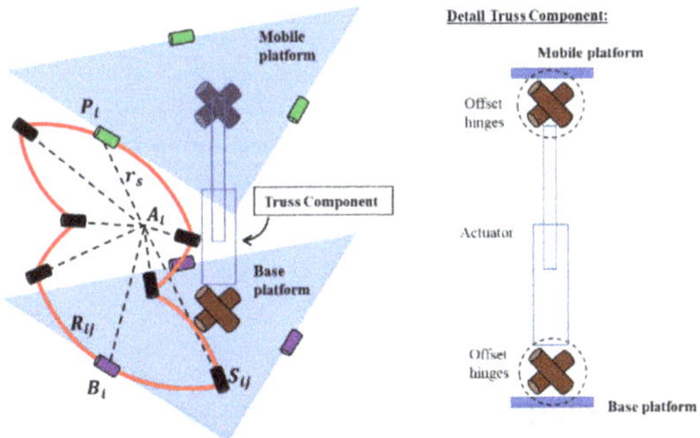

Figure 2. Structure of parallel mechanism.

3. Kinematic of Structure Mechanism

In this section, the displacement analysis of the parallel mechanism is presented. This parallel mechanism establishes global reference and local reference. Point Ob–XYZ is established as the global reference frame at the base platform. The X-axis and the Y-axis are perpendicular and parallel to the axis of the revolute hinge with the center point B_1, respectively. At the same time, the Z-axis is normal to the base following the right-hand rule [7]. In addition, the mobile platform has a point Op-XYZ as the midpoint, and the legs and the truss component have the local reference frame. Table 1 explains the procedure of the DH Convention for knowing the forward kinematics [10].

The table presents the DH parameter determined by four transformation parameters [10] from the link and joint parameters [11]. Link parameters are the length of the link (a_i) and the angle of a twist of a link (α_i); meanwhile, joint parameters as the offset of link (d_i) and joint angle (θ_i) represent the relative positions of the following links [11].

Table 1. Table of Procedure based on DH Convention.

Step 1	Determining and labeling of the joint axis $Z_0, \ldots, Z_{n-1}$.
Step 2	Determine the base frame. Set the origin on Z_0-axis. X_0-axis and Y_0-axis are chosen conveniently, according to the right-hand rule frame. For $i = 1, \ldots, n-1$
Step 3	The origin is O_i, where the common is normal to Z_i and intersects between Z_{i-1} and Z_i. If Z_i intersects with Z_{i-1}, then the O_i is located at that intersection, or if Z_i and Z_{i-1} are parallel, the origin O_i is located in any convenient position along Z_i.
Step 4	Determine the X_i from the origin O_i along the common normal between Z_{i-1} and Z_i, or if Z_{i-1} and Z_i intersects, the direction normal to the $Z_{i-1} - Z_i$ plane.
Step 5	Determine Y_i to complete the frame according to the right-hand rule.
Step 6	Determine the end-effector frame $O_n - X_n Y_n Z_n$. Define the origin O_n along Z_n, preferably at the center of the end-effector, and set X_n and Y_n following the right-hand rule.
Step 7	Create a table of DH parameters.
Step 8	Substituting the DH parameters into the equation to obtain the form of the homogeneous transformation matrices.
Step 9	The product result from step 8 is then given the position and orientation of the end-effector with respect base coordinate frame.

In Ref. [10],

- a_i = distance from the intersection of the x_i, and z_{i-1}-axis to point O_i are measured along x_i;
- α_i = angle from z_{i-1} to z_i is measured about x_i.
- d_i = distance from O_{i-1} to the intersection of the x_i and z_{i-1}-axis are measured along z_{i-1}. If joint i is prismatic, d_i is variable.
- θ_i = angle from x_{i-1} to x_i, and is measured about z_{i-1}. If joint i is revolute, θ_i i is variable.

3.1. Leg

In Ref. [3], the waterbomb chain is related to a close chain, and the last revolute joint pairs with the first revolute joint. Therefore, waterbomb chain does not have an end effector. In this regard, evaluating the close loop equation is necessary, rather than considering the end-effector [3].

3.2. Truss Parallel Mechanism

The truss component section consists of a prismatic hinge connector, and two offset universal hinges connect to the mobile platform and the base platform. Figure 2 shows the detail of the truss component. Point O_6 and point O_1 are the center of the hinge axe connected to the mobile platform and the base platform, respectively. Point O_2 and point O_5 are the centers of the two universal hinge axes. At the same time, point O_5 and point O_4 are directly connected to the prismatic hinge [12]. l represents the prismatic hinge length, the distance between the center point O_3 and point O_4. Each hinge axe variable has the offset distance between the center of the local coordinate hinge axe. The distance between point Ob and point Op represents the height of the mechanism (h).

Table 1 shows the procedure of the DH convention. As shown in Figure 3, the axis of each joint is already available, which corresponds from step 1 to step 7. Table 2 shows the Denavit–Hartenberg (DH) parameters [13] of the truss component for each link.

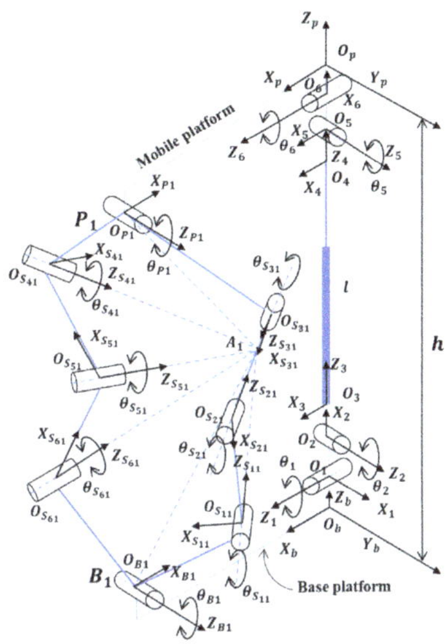

Figure 3. Kinematic model with global coordinate frame and local coordinate frame.

Table 2. Denavit–Hartenberg (DH) parameter of the truss component.

Link i	a_i	α_i	d_i	θ_i
1	0	90	d_1	90 + θ_1
2	a_2	90	0	90 + θ_2
3	0	90	d_3	90
4	0	90	d_4	θ_4
5	a_5	90	0	90 + θ_5
6	0	90	0	90

According to step 8 in Table 1, substituting Equation (3) obtains the form of the homogeneous transformation matrices.

$$T_i^{i-1} = T_{Rz}(\theta).T_z(d).T_{Rx}(\alpha).T_x(a) \tag{2}$$

$$T_i^{i-1} = \begin{bmatrix} c\theta & -s\theta.c\alpha & s\theta.s\alpha & a.c\theta \\ s\theta & c\theta.c\alpha & -c\theta.s\alpha & a.s\theta \\ 0 & s\alpha & c\alpha & d \\ 0 & 0 & 0 & 1 \end{bmatrix} \tag{3}$$

$$T_0^6 = \prod_{i=1}^{6} T_i^{i-1} \tag{4}$$

$$T_0^6 = T_0^1.T_1^2.T_2^3.T_3^4.T_4^5.T_5^6 \tag{5}$$

Equation (5) gives the position and orientation of the mobile platform with respect base platform, as shown in Figure 4.

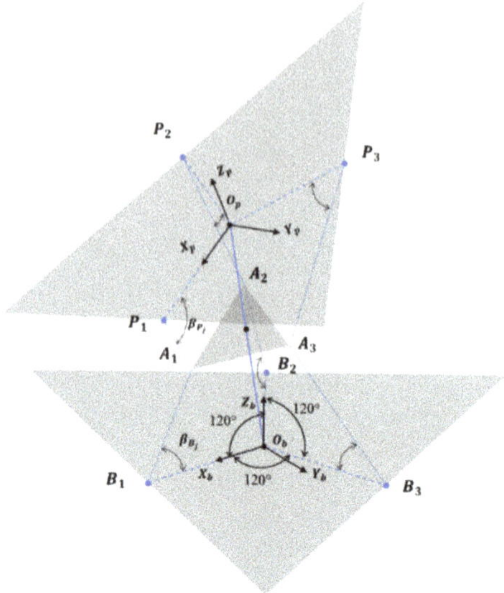

Figure 4. Schematic of the parallel mechanism.

4. Motion Characteristics of a Structure Mechanism

In this structure, the truss component is active, and the legs are passive. Therefore, the active component moves as the structure drives, while the passive component limits the driven movement.

4.1. Rotation around X-axis

Figure 5 shows the rotation around the X-axis of the global coordinate frame in the base platform. The rotation happens due to the rotation of the Z_1-axis. The rotation of the Z_6-axis determines the orientation of the end effector. The revolute hinge rotates clockwise or counterclockwise. Activating the prismatic hinge adjusts the radius of the structure's rotation. The minimum and maximum radii are the prismatic hinge's minimum and maximum strokes, respectively.

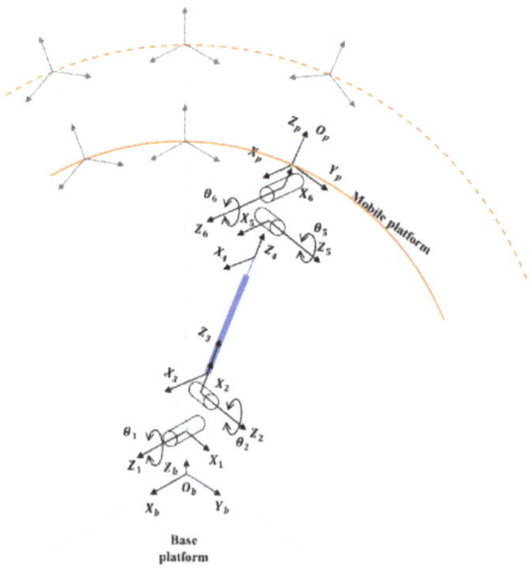

Figure 5. Rotation around the X-axis.

4.2. Rotation around Y-axis

Figure 6 shows the rotation around the Y-axis of the global coordinate frame in the base platform. The rotation happens due to the rotation of the Z_2-axis. The rotation of the Z_5-axis determines the orientation of the end effector. The revolute hinge rotates clockwise or counterclockwise. Activating the prismatic hinge adjusts the radius of the structure's rotation. The minimum and maximum radii are the prismatic hinge's minimum and maximum strokes, respectively.

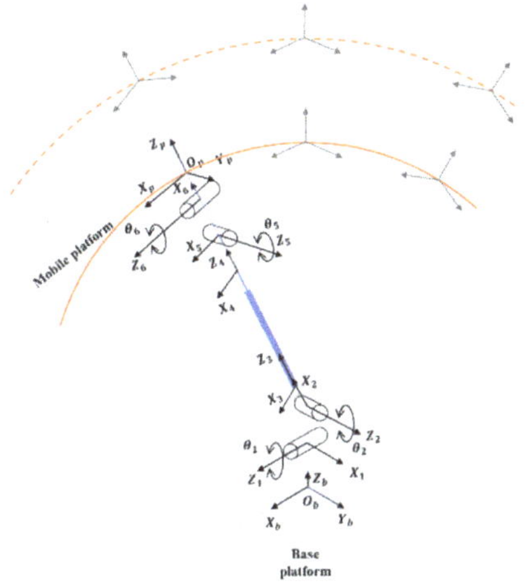

Figure 6. Rotation about the Y-axis.

4.3. Translation along X-axis

Figure 7 shows the translation along the X-axis of the global coordinate frame in the base platform. The translation movement activates the combination of two rotations and a prismatic hinge. Meanwhile, the rotations occur on the Z_2-axis and the Z_5-axis. The translational moving away depends on the stroke of a prismatic hinge. Furthermore, the angle degree of the Z_2-axis is the same as with the Z_5-axis in the opposite direction. Thus, the rotation on the revolute hinge can rotate clockwise or counter-clockwise.

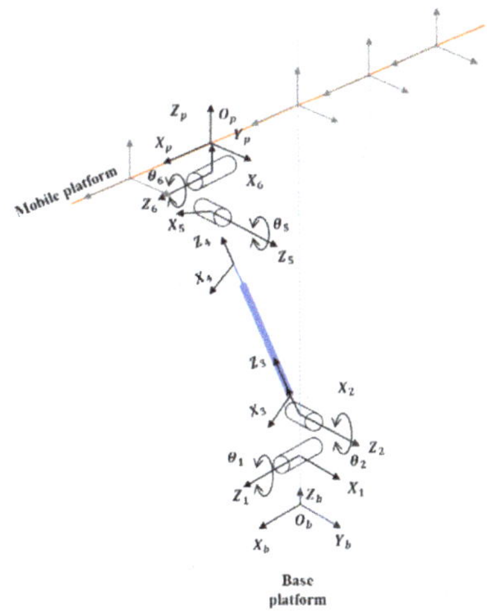

Figure 7. Translation along the X-axis.

4.4. Translation along Y-axis

Figure 8 shows the translation along the Y-axis of the global coordinate frame in the base platform. The translation movement activates the combination of two rotations and a prismatic hinge. Meanwhile, the rotation happens on the Z_1-axis and the Z_6-axis. The translational moving away depends on the stroke of a prismatic hinge. Moreover, the angle degree of the Z_1-axis is the same as with the Z_6-axis in the opposite direction. Thus, the revolute hinge rotates clockwise or counterclockwise.

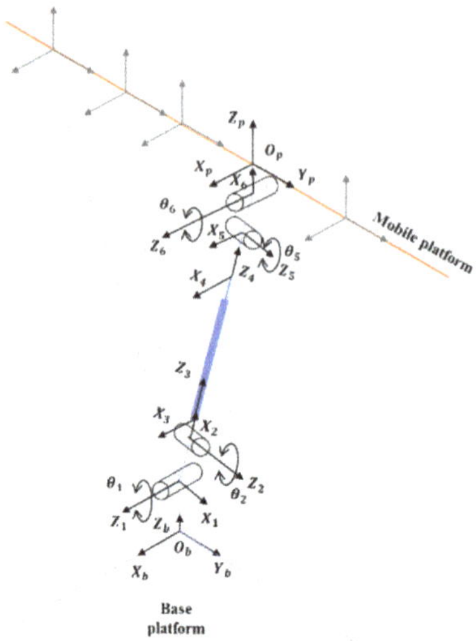

Figure 8. Translation along the Y-axis.

4.5. Translation along Z-axis

Figure 9 shows the translation along the Z-axis of the global coordinate frame in the base platform. The translation movement only activates a prismatic hinge. The high of the structure's parallel mechanism depends on the stroke of the prismatic joint.

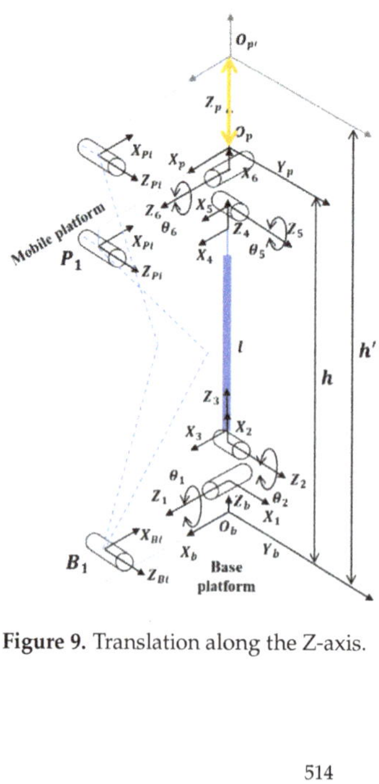

Figure 9. Translation along the Z-axis.

4.6. Rotation with Tilt Movement

As shown in Figure 10, there is a combination of the structure movement parallel mechanism. The mechanism shows a tilt with moving around. All motors can be active and need control, depending on the requirements.

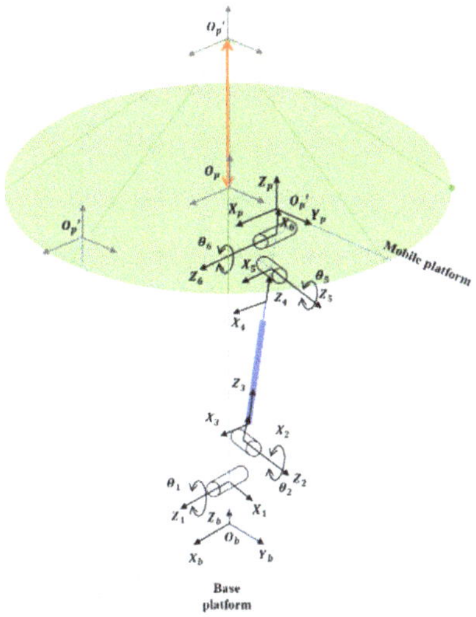

Figure 10. Rotation with tilt movement.

5. Conclusions

The parallel structure mechanism consists of a mobile platform, a fixed platform, three legs, and a truss in the middle of the mechanism. Point Ob–XYZ is established as the global reference frame at the base platform. Movement of the structure to the Ob–XYZ is observed in the base platform. There are six motion characteristics of the structure mechanism.

Author Contributions: Conceptualization, S.-C.H.; methodology, L.A.M. and S.-C.H.; software, L.A.M.; validation, S.-C.H.; formal analysis, L.A.M.; investigation, S.-C.H.; writing—original draft preparation, L.A.M.; writing—review and editing, S.-C.H.; project administration, S.-C.H. All authors have read and agreed to the published version of the manuscript.

Funding: This research received no external funding.

Institutional Review Board Statement: Not applicable.

Informed Consent Statement: Not applicable.

Data Availability Statement: Not applicable.

Acknowledgments: The authors acknowledge and thank the Ministry of Science and Technology of the Republic of China for their partial financial support of this study under Contract Number MOST 111-2221-E-992-068.

Conflicts of Interest: The authors declare no conflict of interest.

References

1. Liu, X.; Wang, J. Classification of Parallel Mechanisms. In *Parallel Kinematics*; Springer: Berlin/Heidelberg, Germany, 2014; pp. 3–29.
2. Patel, Y.D.; George, P.M. Parallel Manipulators Applications—A Survey. *Mod. Mech. Eng.* **2012**, *2*, 57–64. [CrossRef]
3. Fonseca, L.M.; Savi, M.A. On the symmetries of the origami waterbomb pattern: Kinematics and mechanical investigations. *Meccanica* **2021**, *56*, 2575–2598. [CrossRef]
4. Liang, D.; Gao, Y.; Huang, H.; Li, B. Design of a Rigid-Flexible Coupling Origami Gripper. In Proceedings of the 2021 IEEE International Conference on Robotics and Biomimetics (ROBIO), Sanya, China, 27–31 December 2021; pp. 1283–1287. [CrossRef]

5. Salerno, M.; Zhang, K.; Menciassi, A.; Dai, J.S. A Novel 4-DOF Origami Grasper with an SMA-Actuation System for Minimally Invasive Surgery. *IEEE Trans. Robot.* **2016**, *32*, 484–498. [CrossRef]
6. Bowen, L.A. A Study of Action Origami as Systems of Spherical Mechanisms. Ph.D. Thesis, Brigham Young University, Provo, UT, USA, 2013.
7. Zhang, K.; Fang, Y.; Fang, H.; Dai, J. Geometry and Constraint Analysis of the Three-Spherical Kinematic Chain Based Parallel Mechanism. *J. Mech. Robot.* **2010**, *2*, 031014. [CrossRef]
8. Barreto, R.L.P.; Morlin, F.V.; Souza, M.B.D.; Carboni, A.P.; Martins, D. Multiloop origami inspired spherical mechanisms. *Mech. Mach. Theory* **2021**, *155*, 104063. [CrossRef]
9. Hull, T. Counting Mountain-Valley Assignments for Flat Folds. *Ars Comb.* **2003**, *67*, 175–188.
10. Spong, M.W.; Hutchinson, S.; Vidyasagar, M. *Robot Modeling and Control*; John Wiley & Sons: Hoboken, NJ, USA, 2006.
11. Pratihar, D.K. *Fundamentals of Robotics: Manipulators, Wheeled and Legged Robots*; Alpha Science International Ltd.: Oxford, UK, 2017.
12. Zhang, Y.; Han, H.; Zhang, H.; Xu, Z.; Xiong, Y.; Han, K.; Li, Y. Acceleration analysis of 6-RR-RP-RR parallel manipulator with offset hinges by means of a hybrid method. *Mech. Mach. Theory* **2022**, *169*, 104661. [CrossRef]
13. Denavit, J.; Hartenberg, R.S. A Kinematic Notation for Lower-Pair Mechanisms Based on Matrices. *ASME J. Appl. Mech.* **1955**, *22*, 215–221. [CrossRef]

Disclaimer/Publisher's Note: The statements, opinions and data contained in all publications are solely those of the individual author(s) and contributor(s) and not of MDPI and/or the editor(s). MDPI and/or the editor(s) disclaim responsibility for any injury to people or property resulting from any ideas, methods, instructions or products referred to in the content.

engineering proceedings

MDPI

Proceeding Paper

Investigation of Relationship between Kansei Design Elements of Taiwanese Wooden Furniture and Consumer Demand by Fuzzy Theory †

An-Sheng Lee

Department of Wood-Based Materials and Design, National Chiayi University, Chiayi City 600355, Taiwan; sheng1887@mail.ncyu.edu.tw

† Presented at the 3rd IEEE International Conference on Electronic Communications, Internet of Things and Big Data Conference 2023, Taichung, Taiwan, 14–16 April 2023.

Abstract: This study aimed to investigate the affective Kansei/affective design elements of Taiwanese wooden furniture and the consumer demand and perception for the design using Kansei engineering methodology and fuzzy theory. An online survey was conducted to carry out triangular fuzzy linguistic analysis from fuzzy theory and understand the consumers' demands and perceptions of the design elements of wooden seat shapes. The research results showed that consumers agreed with "sturdy" the most, indicating the element needs to comply with the demand and purchase intention of consumers. In terms of the Kansei design elements of a rectangular seat shape, consumers' agreement on "elegant" was medium, while consumers agreed with "novel" the least. The research results provide the Taiwanese wooden furniture industry and designers with a reference for decision-making, design, and manufacturing.

Keywords: wooden furniture; Kansei engineering; fuzzy theory; consumer demand

Citation: Lee, A.-S. Investigation of Relationship between Kansei Design Elements of Taiwanese Wooden Furniture and Consumer Demand by Fuzzy Theory. *Eng. Proc.* **2023**, *38*, 64. https://doi.org/10.3390/engproc2023038064

Academic Editors: Teen-Hang Meen, Hsin-Hung Lin and Cheng-Fu Yang

Published: 29 June 2023

Copyright: © 2023 by the author. Licensee MDPI, Basel, Switzerland. This article is an open access article distributed under the terms and conditions of the Creative Commons Attribution (CC BY) license (https://creativecommons.org/licenses/by/4.0/).

1. Introduction

Facing the ever-changing demands in the consumer market, the furniture industry is urgently seeking how to create and enhance the added value of furniture. Researchers have addressed the current situation of the wooden furniture industry in Taiwan in three major aspects: product materials, production technology, and business philosophy [1]. For product materials, solid wood is used as the main material in the industry. As a retro style obtains popularity in home decoration, the sales of easy-to-process wood furniture continue to grow steadily and account for approximately 50% of the total domestic sales. Wood furniture thus becomes the mainstream of domestic furniture sales [2], and materials for the furniture play an important role in design practice [3].

Human beings attach great importance to feelings. 97% that people's senses are stimulated visually to observe the simulated image [4]. In addition, the appearance of a furniture product attracts consumers and influences their repurchase behavior significantly. It is argued that the functions, aesthetic elements, and design of furniture are important, but, meanwhile, consumers' emotions and demands must also be valued, too, as an important factor [5]. Kansei/affective design explores consumers' minds to abstract perceptual concepts and adds value to the furniture. The development of products with Kansei/affective design is a mandatory development direction in the consumer market with specific Kansei/affective design elements of wooden furniture. Therefore, the application of Kansei/affective design for the wooden furniture design is an issue worth exploring.

At present, Kansei engineering is extensively applied to the Kansei/affective design of products. The Kansei engineering method is used to analyze people's emotional changes and their perception of things, help conduct quantitative analyses of consumer demand,

and decide the design methods of different design elements based on the customer-centered design concept [6].

This study aims to provide the furniture industry with criteria for the design elements of wooden furniture by applying fuzzy theory to understand consumers' affective perception of wooden furniture. Consumers' perceptions of the Kansei/affective design of Taiwanese wooden furniture are analyzed with the correlation between consumer demand and the Kansei/affective design elements. The design elements reflecting consumers' affective perception can be considered as Kansei/affective design criteria for Taiwanese wooden furniture. Such research result can be used to enhance the business performance of Taiwanese wooden furniture and provides the furniture industry and designers with a reference for decision-making in manufacturing and designing the furniture.

2. Literature Review

2.1. Wooden Furniture

Compared with other materials such as plastic and steel, wood is a material more appropriate for human life. Wood has excellent insulation and is humidity-proof. It has natural textures and colors which are favored by the public. In addition, it is easy to process. Therefore, wood is used considerably in furniture and home decoration. Research on the American market showed that consumers value environmental protection by purchasing eco-friendly products made of wood [7]. However, wood is imported by many countries. Thus, the wooden furniture industry is vulnerable to the changes in economy, politics, and tariffs. Therefore, the furniture industry is sensitive to consumer demand to have a win-win situation. The furniture industry in Taiwan seeks the concept of innovative design for life. Representative furniture brands in Taiwan emphasize the design concept to value consumers' emotions and needs [5]. Consequently, in developing and designing furniture, functions, and technologies must be valued for aesthetic styles and creativities to compete in the global market. This is an important management direction of the furniture industry in Taiwan [8].

2.2. Consumer Demand

Demand drives consumer behavior. Maslow, a psychologist, introduced the concept of a hierarchy of needs, in which needs were classified into five levels of physiological needs, safety needs, love, belonging needs (social needs), esteem, and self-actualization. Nevertheless, demand presents a state of psychological deprivation, and can be divided into needs and wants. Therefore, demands generate consumption to meet the demand of consumers. Consumption is influenced by psychological and substantial factors [9]. Consumers tend to buy items to satisfy their "needs", for example, rice, blanket, and other necessities for their life. These products are purchased according to physiological needs and safety needs. The choice of a product to purchase depends on the demand for the product according to the "consumption value" [10]. Hence, the motivation of consumers' consumption changes by the functions of a product and other factors such as the beauty of a product and emotions for products.

2.3. Kansei Engineering

Kansei is the psychological feeling and image generated when a person touches things through the senses. As a psychologically abstract, Kansei in Japanese represents the expression of emotions such as "perception", "feeling", and "impression" [11]. In 1970, Mitsuo Nagamachi, a Japanese scholar, predicted that people would want to satisfy their emotional needs in the age of materialization and civilization. "Kansei" is interpreted as the feeling or image that people have for things, that is, psychological expectation from things [12].

At present, consumers' affective demand is increasing. There has been much research related to Kansei engineering applied to wood which explored the affective imagery of metal, rubber, marble, and wood [3]. The visual features of wood texture by Kansei Engineering were researched [13], and a Kansei engineering evaluation system for furniture design was constructed [6]. The relationship between floor materials and living space was analyzed [14], and the emotional experience of recliners was investigated based on Kansei/affective design [15]. The previous research results reveal that the applications and methods of Kansei engineering can be used in design evaluation and as a product design method.

2.4. Fuzzy Theory

Many attributes in real life cannot be precisely explained by traditional concepts due to the existence of fuzziness. The fuzzy theory is thus proposed to deal with fuzzy phenomena. Zadeh first proposed fuzzy sets that were eventually developed into the fuzzy theory to explain the uncertainty and fuzziness of phenomena [16]. A comprehensive fuzzy evaluation is appropriate for what is affected by fuzzy factors and fuzzy transformation and is featured by multiple criteria and ambiguity [17]. The fuzzy theory is used in this research to analyze the Kansei/affective design elements of Taiwanese wooden furniture and the affective perception of consumer demand.

3. Research Methodology

The relationship between a decision-making model of the Taiwanese wooden furniture industry and consumers' affective demand was analyzed with Kansei engineering and fuzzy system in this research. Focusing on the design and manufacture of wooden furniture and product shapes in Taiwan, wooden chairs were selected as research objects to explore the Kansei/affective design elements of furniture that meet "consumer demand". Previous literature was reviewed to compile adjectives describing affection, including unique, masculine, futuristic, unconventional, luxurious, fun, modern, crazy, honorable, playful, trendy, distinctive, professional, cute, urban, bold, formal, joyful, splendid, precise, novel, mature, serious, fashionable [18], attractive, fashionable, technical, pleasant, reliable, valuable, convenient, unique, innovative, practical [19], artistic, sturdy, eco-friendly, civilized, rare, elegant, relaxed, harmless, contemporary, high-grade [20], comfortable, stable, soft, relaxed, happy, tall, big, long, perfect, simple, solid, pleasant, hard, pretty, cute, happy, amazing, quiet, cheap, strong, beautiful, and surprising [15]. Experts in furniture design were invited to select the adjectives, and they added the following adjectives: masculine, honorable, playful, novel, attractive, fashionable, artistic, sturdy, elegant, and relaxed.

A complete chair consists of four fundamental elements: backrest, seat, arm, and leg [21]. Among them, seat shapes were chosen for a questionnaire survey and analysis. Appropriate adjectives were selected from the collected adjectives for the seat shape of wooden chairs including round, rectangular, and square (Figure 1). The images of wooden chairs were combined with adjectives. Adopting a convenience sampling method, an online survey was carried out to collect data on consumer demand and perception of the seat shapes of wooden chairs. Maslow classified needs into five levels: physiological needs, safety needs, love and belonging needs, esteem, and self-actualization [22]. low-level physiological needs were first analyzed to identify the connection between the connotation and product design as "the product itself having basic technology or functions, which meet the basic physiological needs of users and satisfy them" [23].

(**a**) (**b**) (**c**)

Figure 1. Seat shapes of wooden chairs in this research. (**a**) Round seat (**b**) Square seat (**c**) Rectangular seat (The chairs were made by the wood factory of National Chiayi University, and the photos were taken in this research).

In the questionnaire survey, a five-point Likert scale was used. The linguistic fuzziness of the questionnaire items was analyzed by fuzzification mechanism, fuzzy rule base, fuzzy inference engine, and defuzzification [24]. The fuzzy linguistic values were calculated with fuzzy equations to obtain the perception of the Kansei/affective design of wooden chairs by fuzzy evaluation. The values were analyzed to understand the demand of the participants. Triangular fuzzy numbers and the $\check{J}$ defuzzification equation are described as follows.

$$\text{Assume } \check{J} = (a_1, a_2, a_3), C_{\check{J}} = (a_1 + 2a_2 + a_3)/4x5 \tag{1}$$

An equation is derived from the fuzzy descriptive statistical equation of the fuzzy questionnaire.

$$C_{\check{J}} = (a_1 + 2a_2 + a_3)/4x5 \tag{2}$$

Equation (2) is transformed into the fuzzy linguistic mean which is

$$\sum_1^n (a_1 + 2a_2 + a_3)/4N_X 5(\text{N : number of participants}) \tag{3}$$

Based on the fuzzy equation, the absolute value of utility of each evaluation of the Kansei/affective design element of the shapes of the wooden chair was calculated, and a triangular fuzzy number chart was created for the future analysis of the Kansei/affective design elements.

4. Data Analysis

A convenience sampling method was adopted to design a questionnaire. The online questionnaire was distributed through Google Forms in June and July 2022. 102 questionnaires were retrieved for analysis (Table 1).

Table 1. Statistical results of scores of seat shapes of wooden chairs.

	N	Minimum	Maximum	Sum	Mean		Standard Deviation
6-4 Consumers' affective perception and opinion of a round seat shape. [Novel]	102	1.0	4.0	233.0	2.284	0.0716	0.7227
6-2 Consumers' affective perception and opinion of a round seat shape. [Honorable]	102	1.0	5.0	260.0	2.549	0.0820	0.8281
6-6 Consumers' affective perception and opinion of a round seat shape. [Fashionable]	102	1.0	5.0	261.0	2.559	0.0819	0.8274
6-3 Consumers' affective perception and opinion of a round seat shape. [Playful]	102	1.0	5.0	296.0	2.902	0.0898	0.9066
8-9 Consumers' affective perception and opinion of a rectangular seat shape. [Elegant]	102	1.0	5.0	299.0	2.931	0.0851	0.8590
7-6 Consumers' affective perception and opinion of a square seat shape. [Fashionable]	102	1.0	5.0	301.0	2.951	0.0949	0.9583
6-10 Consumers' affective perception and opinion of a round seat shape. [Relaxed]	102	1.0	5.0	391.0	3.833	0.0925	0.9343
8-8 Consumers' affective perception and opinion of a rectangular seat shape. [Sturdy]	102	2.0	5.0	407.0	3.990	0.0830	0.8384
6-8 Consumers' affective perception and opinion of a round seat shape. [Sturdy]	102	2.0	5.0	432.0	4.235	0.0699	0.7061

The mean score of "6-8 Consumers' affective perception and opinion of a round seat shape is 'sturdy'" was 4.235, which was the highest. The mean score of "8-8 Consumers' affective perception and opinion of a rectangular seat shape is 'sturdy'" was 3.990, which was the second highest. The mean score of "6-10 Consumers' affective perception and opinion of a round seat shape is 'relaxed'" was 3.833, which was the third highest. The mean score of "6-3 Consumers' affective perception and opinion of a round seat shape is 'playful'" was 2.902, which was the median of the whole score. The mean score of "8-9 Consumers' affective perception and opinion of a rectangular seat shape is 'elegant'" was 2.931, which was the median, too. The mean score of "7-6 Consumers' affective perception and opinion of a square seat shape is 'fashionable'" was 2.951, which was the median of the next ranking.

The mean score of "6-4 Consumers' affective perception and opinion of a round seat shape is 'novel." was 2.284, which was the lowest. The mean score of "6-2 Consumers' affective perception and opinion of a round seat shape is 'honorable'" was 2.549, which was the second lowest. The mean score of "6-6 Consumers' affective perception and opinion of a round seat shape is 'fashionable'" was 2.559, which was the highest.

Descriptive statistics were transformed by the fuzzy theory to obtain the triangular fuzzy linguistic values of consumers' affective perception. An equation is derived from the fuzzy descriptive statistical equation of the fuzzy questionnaire.

$$C_J = (a_1 + 2a_2 + a_3)/4x5 \tag{4}$$

Equation (4) is transformed into the fuzzy linguistic mean, that is,

$$\sum_{1}^{n}(a_1 + 2a_2 + a_3)/4N_X5 \tag{5}$$

When u(X) = 1, the maximum value is 3.20 whereas the minimum value is 1.00. The obtained triangular fuzzy linguistic values and membership function graph of the fuzzy theory are shown in Figure 2. The triangular fuzzy linguistic shows that the triangle is the largest for the mean score (4.235) of "6-8 Consumers' affective perception and opinion of a round seat shape is 'sturdy'", which is the highest, and the weight of the triangular fuzzy linguistic value is also high. The mean score of "8-8 Consumers' affective perception and opinion of a rectangular seat shape is 'sturdy'" is 3.990, which is the second highest, and the weight of the triangular fuzzy linguistic value is also the second highest. The mean score of "6-10 Consumers' affective perception and opinion of a round seat shape is 'relaxed'" is 3.990, which is the third highest, and the weight of the triangular fuzzy linguistic value is also the third highest. Consumers agreed with "sturdy" for the Kansei/affective design elements of a round seat shape. Therefore, it is necessary to design a "sturdy" chair in the future, and it is suggested that a round shape be used to meet the demand of consumers' affective perception.

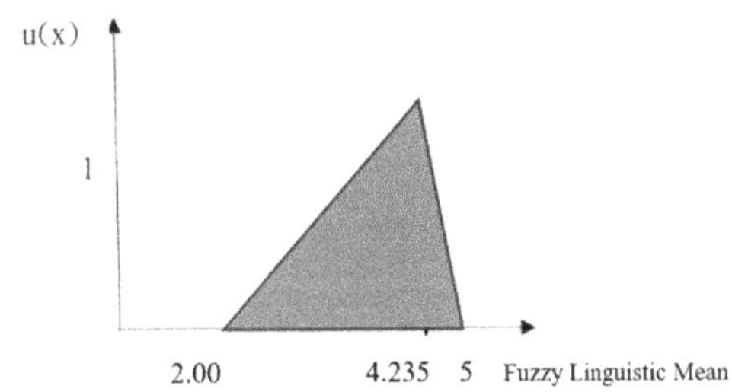

Figure 2. Triangular fuzzy linguistic value (maximum) and membership function graph of "Consumers' affective perception and opinion of a round wooden chair seat shape is 'sturdy'.".

Figure 3 shows that consumers' agreement on "elegant" is medium in terms of the Kansei/affective design elements of a rectangular seat shape. Hence, it is necessary to design an "elegant" chair in the future, and it is suggested that a rectangular seat shape be used as the Kansei/affective design element.

Figure 3. Triangular fuzzy linguistic value (median) and membership function graph of "Consumers' affective perception and opinion of a rectangular wooden chair seat shape is 'elegant'".

Consumers agreed with "novel" the least in terms of the Kansei/affective design elements of a round seat shape. Thus, designing a "novel" chair seat needs to be considered carefully in the future. A round seat shape may not be used as the Kansei/affective design element (Figure 4).

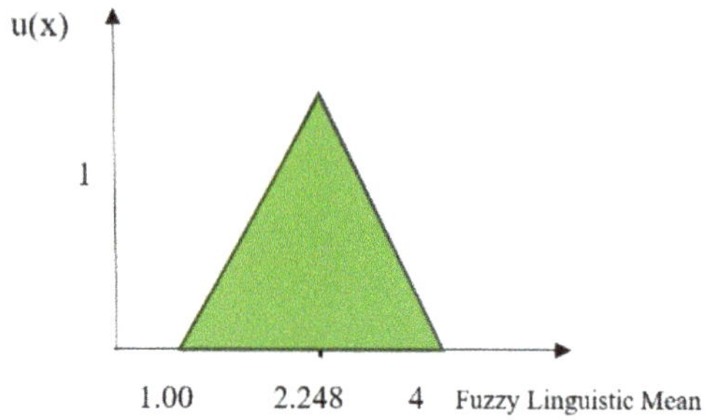

Figure 4. Triangular fuzzy linguistic value (minimum) and membership function graph of "Consumers' affective perception and opinion of a round wooden chair seat shape is 'novel'".

5. Conclusions and Suggestions

The Kansei/affective design elements of wooden chairs with different shapes were analyzed to meet consumers' demands by fuzzy theory. The research result is summarized as follows. Consumers agreed with "sturdy" the most in terms of the Kansei/affective design elements of a round wooden chair seat shape. Therefore, it is necessary to design a "sturdy" chair in the future with a round seat shape to meet the demand and purchase intention of consumers' affective perception. "Elegant" also needs to be considered to be used as the Kansei/affective design element. However, a "novel" design would not be used for wooden chairs as was agreed on the least.

Funding: This research received no external funding.

Institutional Review Board Statement: Not applicable.

Informed Consent Statement: Not applicable.

Data Availability Statement: Not applicable.

Acknowledgments: This research is supported by the National Science Council (NSTC 111-2313-B-415-).

Conflicts of Interest: The author declares no conflict of interest.

References

1. Wang, L.C.; Guan, S.S.; Chen, T.L. A Study on Image and Status Survey of Taiwan Wooden Furniture Industry. *J. Des. Res.* **2011**, *4*, 16–32.
2. Current Situation of Taiwan Furniture Industry. Available online: http://www.interwood.tw/industryNewsContent.php?ID=838andln=twzh (accessed on 12 May 2023).
3. Hsiao, K.A. The Affective Images between Shapes and Materials. *Kaohsiung Norm. Univ. J. Sci. Technol.* **2018**, *45*, 21–42.
4. McKellar, P. *The Investigation of Mental Images*; Penguin Science Survey: Harmondsworth, UK, 2006.
5. Wang, L.C. A Study on Design Image Analysis of Taiwan Furniture Cultural and Creative Brands. *J. Des. Res.* **2018**, *11*, 37–55.
6. Chen, M.; Lyu, J.H.; Li, S.G.; Wu, X. Construction and implementation of a panel furniture design evaluation system at the design stage. *Adv. Mech. Eng.* **2017**, *9*, 1–8. [CrossRef]
7. Roper Green Gauge Report Summary Brochure. Available online: http://www.scjohnson.com/Libraries/Download_Documents/SCJ_and_GfK_Roper_Green_Gauge.sflb.ashx (accessed on 12 May 2023).
8. Fong, W.C.; Chen, T.L. The Influences of the Bauhaus style on Furniture Design in Taiwan: Using the Furniture Brand "TW.U.C.M." as an Example. *Int. J. Des.* **2019**, *24*, 17–40.
9. Liu, S.N.; Chu, W.C. A Study on Customers' Recognition and Demand for Aromatheraphy. *J. Beauty Technol.* **2009**, *6*, 43–61.
10. Sheth, J.N.; Newman, B.I.; Gross, B.L. Consumption values and market choice: Theory and application. *J. Mark. Res.* **1991**, *29*, 487–489.
11. Chen, K.S. Kansei Engineering. *Qual. Mon.* **2009**, *45*, 21–22.
12. Chen, K.S.; Guan, S.S.; Deng, Y.S.; Chang, Y.M. Kansei Engineering—A Method of Rationalizing Sensibility. *Ind. Des.* **2001**, *29*, 2–16.
13. Yanfeng, M.; Huiyuan, G. Research on Visual Characteristics about Wood Mountain Peak Texture Based on Kansei Engineering. *Furnit. Inter. Des.* **2013**, 58–60. (In Chinese)
14. Akiyama, A.; Araki, Y.; Hosoya, S. Research into the Relationship between Floor Materials and Living Spaces as Revealed by KANSEI Evaluations. *Int. J. Affect. Eng.* **2017**, *16*, 113–120. [CrossRef]
15. Kim, W.; Ko, T.; Rhiu, I.; Yun, M.H. Mining affective experience for a Kansei/affective design study on a recliner. *Appl. Ergon.* **2019**, *74*, 145–153. [CrossRef] [PubMed]
16. Zadeh, L.A. Fuzzy sets. *Inf. Control* **1965**, *8*, 338–353. [CrossRef]
17. Chen, W.L.; Chen, T.H. Applying Fuzzy Theory for Design Evaluation of the Gentleman Fashion Design—A Case Study on POLO Shirts. *Taiwan Text. Res. J.* **2016**, *20*, 64–73.
18. Chen, W.L.; Lin, J.J.; Shen, M.C.; Chang, S.M. A Study on the Relationship between Product Form and Image Perception for Personal Digital Assistant. *J. Prof. Mech. Eng.* **2008**, *1*, 38–45.
19. Guo, F.; Qu, Q.X.; Nagamachi, M.; Duffy, V.G. A proposal of the event-related potential method to effectively identify Kansei words for assessing product design features in Kansei engineering research. *Int. J. Ind. Ergon.* **2020**, *76*, 102940. [CrossRef]
20. Wan, Q.; Li, X.; Zhang, Y.; Song, S.; Ke, Q. Visual perception of different wood surfaces: An event-related potentials study. *Ann. For. Sci.* **2021**, *78*, 25. [CrossRef]
21. Mortenson, M.E. *Geometric Modeling*; John Wiley and Sons: New York, NY, USA, 1985; pp. 450–451.
22. Maslow, A.H. A theory of human motivation. *Psychol. Rev.* **1943**, *50*, 370. [CrossRef]
23. Chen, T.Y.; Wu, C.H.; Chen, Y.M. A Maslow's Hierarchy of Needs-based Networking Reviews Analysis Mechanism for Product Positioning. *J. e-Bus.* **2015**, *17*, 393–422.
24. Klir, G.J.; Yuan, B. *Fuzzy Sets and Fuzzy Logic: Theory and Application*; Prentice Hall: Hoboken, NJ, USA, 1995.

Disclaimer/Publisher's Note: The statements, opinions and data contained in all publications are solely those of the individual author(s) and contributor(s) and not of MDPI and/or the editor(s). MDPI and/or the editor(s) disclaim responsibility for any injury to people or property resulting from any ideas, methods, instructions or products referred to in the content.

engineering proceedings

MDPI

Proceeding Paper

Numerical Simulations on Heat Transfer Enhancement of Nanofluids in Microchannel Using Vortex Generator †

Yong-Bin Lee and Chuan-Chieh Liao *

Department of Mechanical Engineering, Chung Yuan Christian University, Taoyuan 32023, Taiwan
* Correspondence: chuanchiehliao@cycu.edu.tw

† Presented at the 3rd IEEE International Conference on Electronic Communications, Internet of Things and Big Data Conference 2023, Taichung, Taiwan, 14–16 April 2023.

Abstract: Vortex-induced vibration (VIV) is the periodic motion of a bluff body caused by fluid flow and is widely discussed in the engineering field. With the advancement of science and technology, miniaturization and integration have become the mainstream trends in biomedical chips and electronic systems, resulting in higher heat dissipation requirements per unit area. Therefore, the improvement of the heat dissipation effect of movable structures in the flow channel has been widely discussed. Among them, adding VIV motion in the microchannel generates a vortex structure, which improves heat transfer efficiency. Different from the direct displacement method of active vibration, the passive displacement of VIV is a multi-physics problem. It needs to integrate the flow field and the spring-mass system of the object for fluid–solid coupling, which greatly increases the difficulty of analysis. In this study, the Immersed-boundary method (IBM) combined with the equation of motion is used to numerically study a vortex generator that is elastically installed in a microfluidic channel and is then used to enhance the convective heat transfer of nanofluids in the channel. Unlike the common body-fitted mesh, IBM greatly reduces the computational resources required for mesh regeneration when simulating the problem of object movement in fluid–structure interaction. In addition, Buongiorno's two-phase mixing model is used to simulate the convective heat transfer of nanofluids in microchannels by considering the Brownian motion and thermophoretic diffusion of nanoparticles in the carrier liquid. By changing the important parameters such as nanofluid concentration, Reynolds number, mass ratio, and Ur, the influence of the response characteristics of vortex-induced vibration on the heat flow field in the microfluidic channel is discussed, and the key factors for enhancing heat transfer are found out.

Keywords: vortex-induced vibration; nanofluid; fluid–structure interaction; Immersed-boundary method; heat-transfer enhancement

Citation: Lee, Y.-B.; Liao, C.-C. Numerical Simulations on Heat Transfer Enhancement of Nanofluids in Microchannel Using Vortex Generator. *Eng. Proc.* **2023**, *38*, 68. https://doi.org/10.3390/engproc2023038068

Academic Editors: Teen-Hang Meen, Hsin-Hung Lin and Cheng-Fu Yang

Published: 29 June 2023

Copyright: © 2023 by the authors. Licensee MDPI, Basel, Switzerland. This article is an open access article distributed under the terms and conditions of the Creative Commons Attribution (CC BY) license (https://creativecommons.org/licenses/by/4.0/).

1. Introduction

With the advancement of science and technology, miniaturization and integration have become the mainstream trends in biomedical chips and electronic systems. As biochips are designed and manufactured with the principles of microelectromechanical systems combined with biochemistry and bioinformatics, the advantages of biochips are similar to computer chips. Both complete a large amount of calculation and analysis in a short time, but the disadvantages are similar. That is, with the miniaturization of the motor system, the heat dissipation demand per unit area becomes higher [1]. Therefore, with the improvement of the micro heat exchange, heat dissipation efficiency has been discussed recently. There are many ways to improve heat dissipation efficiency, such as adding fans to increase the flow of the working fluid, adding cooling fins to increase the contact area, and replacing the radiator or working fluid with materials with high thermal conductivity. However, adding devices such as fans or fins increases the volume and weight of the product, which is contrary to the recent trend of shrinking electronic components. Micro-channels solve the above problems. A micro-channel heat sink (MHS) is one of the heat exchange technologies proposed by Tuckerm and Pease [2] to solve the heat dissipation problem of very large-scale integration (VLSI). This technology is now widely used in MEMS.

In addition to the micro-channel radiator that can improve heat transfer efficiency, adding a fixed or movable object in the micro-channel can increase the turbulent flow to mix the fluid on the boundary layer and promote the destruction of the thermal boundary layer (TBL). Yeom et al. [3] used piezoelectric stacked actuators to vibrate objects within the flow channel to enhance convective heat transfer on the flow channel wall. At the same time, the destruction of this TBL is similar to the effect of actively oscillating cylinders on heat transfer enhancement in the flow channel oscillation [4], which has a significant effect on the heat transfer enhancement of the wall. Celik [4] also mentioned that when the cylinder oscillates in the cross-flow direction at a frequency close to the natural vortex shedding frequency, the vortex shedding is synchronized with the cylinder motion. This synchronization is called Lock-in. Kumar et al. [5] also studied the effect of oscillating cylinders on enhanced heat transfer in the flow channel. However, the movement method of the movable cylinder studied by Kumar is passive. The effect of the periodic oscillation of the object on the heat dissipation of the wall is observed by changing the reduced velocity. The effect of the periodic oscillation of the object on the heat dissipation of the wall is observed by changing the reduced velocity. Compared with the active type, the passive type does not require other actuators or equipment, but simple passive elements such as springs and damping can make the cylinder perform the periodic motion. The periodic motion of this type of bluff body caused by the flow of fluid is called vortex-induced vibration. Different from the direct displacement method of active vibration, the passive displacement of VIV is a multi-physics problem. It is necessary to integrate the flow field and the spring-mass system of the object for fluid–solid coupling and greatly increases the difficulty of analysis.

The vortex shedding of a stationary cylinder is greatly affected by the Reynolds number. Lienhard [6] described the state of vortex shedding at various Reynolds numbers. When Re < 5, the fluid detaches from the surface of the cylinder and flows along the contour of the cylinder. When $5 \leq \mathrm{Re} < 45$, the fluid starts to fall off from the back of the cylinder, forming a pair of symmetrical vortices. When $45 \leq \mathrm{Re} < 150$, the vortices start to fall off periodically from both sides. All are still laminar flow; until $150 \leq \mathrm{Re} < 300$, the fluid is still laminar at the cylinder boundary. However, the shed wake becomes turbulent, and $\mathrm{Re} \geq 300$ is completely turbulent. Zhao [7] studied the flow of elastically mounted cylinders at different Reynolds numbers through three-dimensional simulations. When Re = 250, the wake of the vibrating cylinder is two-dimensional, and when Re = 300, it is three-dimensional. We assume that the two are laminar flows between plates. From the above-mentioned Reynolds number setting and laminar turbulence change, Re is set as 100.

The above-mentioned active or passive cylindrical vibration heat dissipation uses pure fluid. However, with the advancement of science and technology, electronic equipment has increased heating power, reduced volume, and reduced working space with fluids. The thermal conductivity of pure fluids can no longer meet the heat dissipation needs of today's electronic components. Thus, we add tiny solid particles into the fluid to form nanofluids by combining the good thermal conductivity of solids with the convection properties of fluids, thereby improving the overall heat transfer efficiency. With Buongiorno's two-phase model [8], the convective heat transfer of the nanofluid in a microchannel is simulated. By changing, we can observe the effect of amplitude on heat transfer.

2. Materials and Method

The schematic diagram of the system considered in this study is shown in Figure 1. The model is a two-dimensional double-plate flow, which contains a passive vortex generator. D is the diameter of the vortex generator at 7.5×10^{-5} m, the inlet condition is a fully developed flow, the temperature is 293 K, and the flow velocity distribution is in Ref. [5]. The lower wall has a section of heat generator whose temperature is 373 K, and the rest of the walls are adiabatic. In contrast, the diameter of the vortex generator is 7.5×10^{-5} m, the nanofluid is mainly composed of basic fluid (water) and nanoparticles (Al_2O_3), and the selected force is 25 nm with the basic properties proposed by Lao [9].

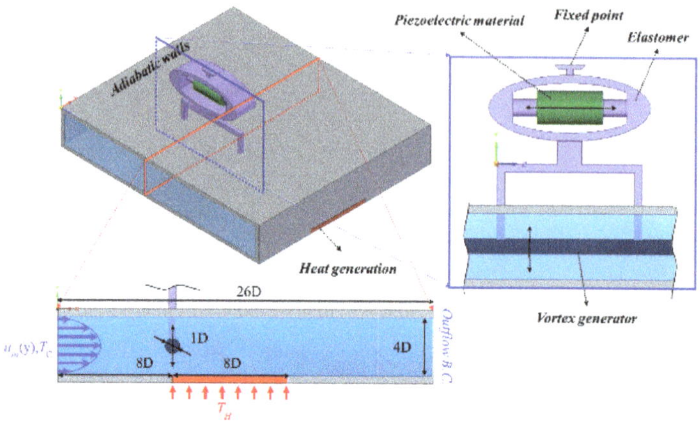

Figure 1. Schemes follow the same formatting.

In this study, a non-uniform grid is used to simulate the heat transfer characteristics of nanofluid in the flow channel, and the finite-volume method discretized in Cartesian coordinates is used to solve the flow field and thermal field. In order to deal with the motion of the vortex generator that is elastically installed in the microfluidic channel, we adopt the Immersed-boundary method (IBM) [10] combined with the equation of motion to deal with the interaction between the fluid and the solid. IBM computes the velocities of adjacent grid points outside the solid region through linear difference. Different from the common body-fitted mesh, the computational resources required for mesh regeneration can be greatly reduced when simulating the problem of object movement in fluid–structure interaction. The flow boundary condition is applied at the structural boundary to study the effect of the vortex generator on the heat transfer and flow of the nanofluid in the flow channel. Thermophoresis and Brownian motion of nanofluid are considered in the Boungiorno two-phase model. The fluid is assumed to be a Newtonian, viscous incompressible fluid, and the flow field is laminar, unsteady, and two-dimensional.

According to the above conditions, the governing equation includes the continuity equation, the momentum equation, the energy equation, and nanoparticle transport (volume fraction of nanoparticles) equations [9]

$$\frac{\partial \rho_{nf} u}{\partial x} + \frac{\partial \rho_{nf} v}{\partial y} = 0 \tag{1}$$

$$\rho_{nf}\left(u\frac{\partial u}{\partial x} + v\frac{\partial u}{\partial y}\right) = -\frac{\partial p}{\partial x} + \mu_{nf}\left(\frac{\partial^2 u}{\partial x^2} + \frac{\partial^2 u}{\partial y^2}\right) + f_{Mx} \tag{2}$$

$$\rho_{nf}\left(u\frac{\partial v}{\partial x} + v\frac{\partial v}{\partial y}\right) = -\frac{\partial p}{\partial y} + \mu_{nf}\left(\frac{\partial^2 v}{\partial x^2} + \frac{\partial^2 v}{\partial y^2}\right) + g(\rho\beta)_{nf}(T - T_c) + f_{My} \tag{3}$$

$$u\frac{\partial T}{\partial x} + v\frac{\partial T}{\partial y} = \alpha_{nf}\left(\frac{\partial^2 T}{\partial x^2} + \frac{\partial^2 T}{\partial y^2}\right) + f_E \tag{4}$$

$$u\frac{\partial \varphi}{\partial x} + v\frac{\partial \varphi}{\partial y} = \left[\frac{\partial}{\partial x}\left(D_B\frac{\partial \varphi}{\partial x}\right) + \frac{\partial}{\partial y}\left(D_B\frac{\partial \varphi}{\partial y}\right)\right] + \left[\frac{\partial}{\partial x}\left(D_T\frac{\partial T}{\partial x}\right) + \frac{\partial}{\partial y}\left(D_T\frac{\partial T}{\partial y}\right)\right] + f_C \tag{5}$$

where x, y are the Cartesian coordinates in x and y directions, u, v is the velocity components in x and y directions, p is the pressure, g is the gravitational acceleration, α_{nf} is the thermal diffusivity of the nanofluid, T is the temperature, φ is the nanoparticle volume fraction, D_B is the Brownian coefficient, D_T is the thermophoresis coefficient, f_{Mx}, f_{My} is the discrete-time momentum forcing in x and y directions, f_E is the discrete-time energy forcing, f_C is the discrete-time volume fraction forcing. $\rho_{nf}, \beta_{nf}, (Cp)_{nf}$, the effective density, the thermal expansion coefficient, and the thermal diffusivity of the nanofluid are obtained as follows.

$$\rho_{nf} = (1-\varphi)\rho_f + \varphi\rho_p \tag{6}$$

$$(\rho\beta)_{nf} = (1-\varphi)(\rho\beta)_f + \varphi(\rho\beta)_p \tag{7}$$

$$(\rho c_p)_{nf} = (1-\varphi)(\rho c_p)_f + \varphi(\rho c_p)_p \tag{8}$$

where μ_{nf}, k_{nf} are the effective viscosity and the thermal conductivity of the nanofluid, respectively, and are predicted by the experiments-based empirical correlations [11]. The effective viscosity and thermal conductivity of the carrier liquid are obtained from the equations studied by Meis et al. [12].

$$\frac{\mu_{nf}}{\mu_f} = \frac{1}{1-34.87\left(d_p/d_f\right)^{-0.3}\varphi^{1.03}} \tag{9}$$

$$\mu_f = 1 - 1.1292T + 0.4904T^2 \tag{10}$$

$$\frac{k_{nf}}{k_f} = 1 + 4.4\mathrm{Re}^{0.4}\mathrm{Pr}^{0.66}\left(\frac{T}{T_{fr}}\right)^{10}\left(\frac{k_p}{k_f}\right)^{0.03}\varphi^{0.66} \tag{11}$$

$$k_f = 1 + 0.1572T - 0.0470T^2 \tag{12}$$

As presented in Equation (5), the slip mechanisms of Brownian motion and thermophoresis are included to mimic the interaction between the nanoparticles and the based fluid, where the Brownian diffusion coefficient (D_B) and thermophoresis coefficient (D_T) are defined in Ref. [8].

$$D_B = \frac{k_b T}{3\pi\mu_f d_p} \tag{13}$$

$$D_T = 0.26\frac{k_f}{2k_f + k_p}\frac{\mu_f}{\rho_f T}\varphi \tag{14}$$

The undamped spring-mass system is constrained to one translational degree of freedom motion (in the vertical direction). The dynamics of this system are governed by the following equation [5]:

$$\ddot{Y} + \frac{4\pi^2}{U_r^2}Y = \frac{C_L}{2M} \tag{15}$$

where C_L and M is the lift coefficient and mass ratio.

3. Result and Discussion

3.1. Code Validation and Mesh Independence Studies

In order to determine the correctness of the current simulation, the current simulation results are compared with the simulation of a stationary cylinder [13] and the simulation [5] of an elastically mounted cylinder [5].

Figure 2A shows the simulation data of Soti et al. [13] compared with the current simulation results. The results of this study are in good agreement with Soti's. Figure 2B shows the variation in maximum oscillation amplitude ($A_{\max}$) for an elastically mounted cylinder with Ur as for Kumar et al. [5]. Figure 2B also presents that the Lock-in range in this study ($A_{\max} > 0.1$) is similar to Kumar, as both are in $3 \leq Ur \leq 5$. ($A_{\max}$ is dimensionless by the diameter of the cylinder).

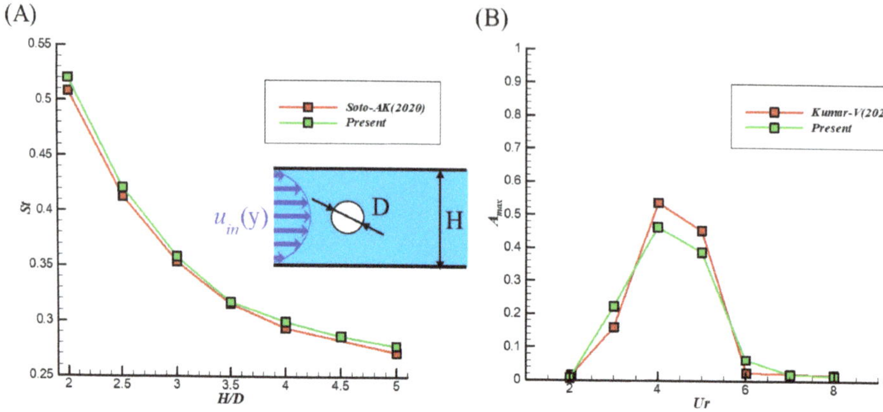

Figure 2. (**A**) Compared with the data in Soti [13], the Strouhal number (St) with the stationary circular cylinder of diameter (D) kept inside a channel of height (H) at Re = 100. (**B**) compared with the data in Kumar [5], $A_{\max}$ with Ur at Re = 100.

A mesh independence study is then performed by using the Kumar domain [5]. The computational domain uses a combination of a uniform mesh of size Δx = Δy (covering the area around the cylinder) and a non-uniform stretched mesh (covering the rest of the domain). Mesh independence studies were performed by changing the mesh size ΔX = Δx/D to 1/5 (3116), 1/10 (9590), 1/20 (31,122), 1/30 (63,012), 1/40 (105,564), and 1/60 (221,028) in a uniform mesh area. Figure 3 shows the maximum amplitude $A_{\max}$ and frequency ratio (f_v/f_n) for cylinders $Ur = 4$ of six mesh sizes. A mesh size of ΔX = 1/30 was chosen for the simulation because refinement beyond 1/30 did not change the maximum oscillation amplitude and frequency ratio by more than 1%. (f_v and f_n is vortex shedding frequency of vibrating cylinder and natural frequency of cylinder).

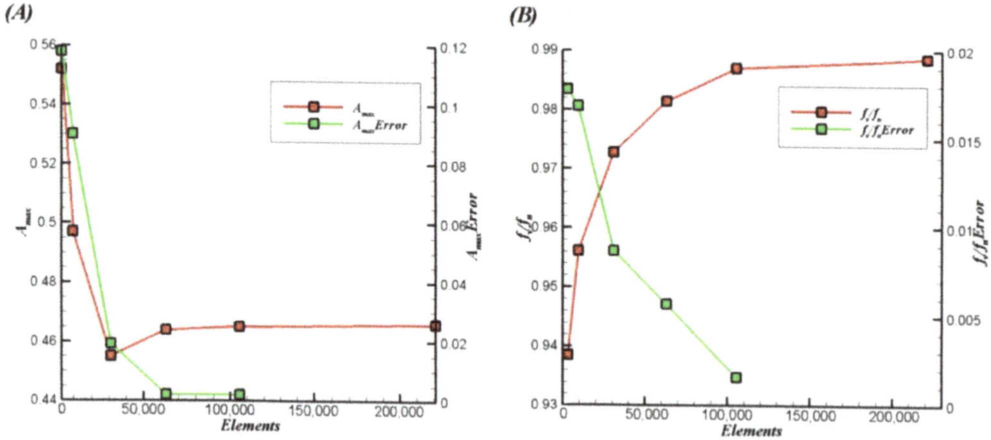

Figure 3. Mesh independence study for Ur = 4. Effect of elements on (**A**) maximum amplitude $A_{\max}$ and (**B**) frequency ratio f_v/f_n.

3.2. Result

Figure 4 shows the vorticity and isotherm diagrams in $t/t_p = 1.00$, $t/t_p = 0.25$, $t/t_p = 0.50$, $t/t_p = 0.75$, by observing the movement of the cylinder in the flow channel under the conditions of $Ur = 3$, $\varphi = 0\%$. In Figure 4A,B, at $t/t_p = 1.00$, the thermal plume structure and the vortex structure are located at equal positions, and it is inferred that the TBL is destroyed due to the interaction with the convective vortex. The same goes for the remaining t/t_p because the eddies periodically pass through the TBL, forcing the fluid near the TBL to exchange towards the middle, destroying the TBL and ultimately improving heat transfer (t,t_p is time, time period of the oscillation cycle.).

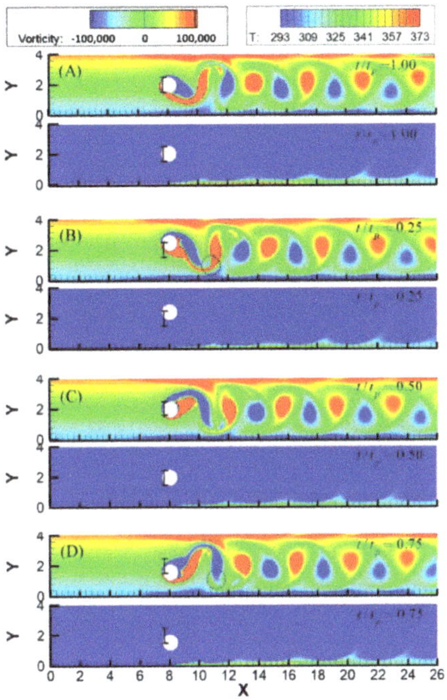

Figure 4. (**A**) mean position (moving upward), (**B**) upper extreme, (**C**) mean position (moving downward), and (**D**) lower extreme positions of the cylinder vorticity plot and isotherm plot. The legend "vorticity" is used for the upper plot, and the legend "T" is used for the lower plot.

When discussing the effect of heat transfer, we pay attention to the oscillation amplitude of the cylinder. When the cylinder oscillates, it drags the position of the rear vortex so that the two ends of the newly formed vortex are closer to the TBL, and the fluid convection at this position eventually increases the heat transfer, as shown by the dotted line in Figure 4B,D.

Figure 5A shows the relationship between Ur and $A_{\max}$ for different φs, while Figure 5B shows the relationship between Ur and $Nu/Nu_{stationary,\varphi=0\%}$ for different φs. Figure 5A also presents that the increase in the concentration reduces the Lock-in range. However, Figure 5B shows that concentration increase in $Nu/Nu_{stationary,\varphi=0\%}$ and amplitude has a positive correlation with $Nu/Nu_{stationary,\varphi=0\%}$. The decrease in the Lock-in range means that under the same Ur condition, the increase in φ makes the cylinder change from large $A_{\max}$ to small $A_{\max}$, as indicated by $Ur = 4.5$. Small amplitude is unfavorable for heat transfer, as indicated by $Ur = 4.5$ in Figure 5B.

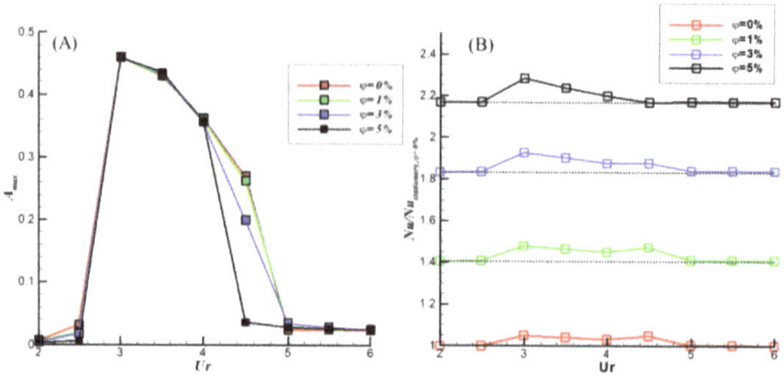

Figure 5. (**A**) the $A_{\max}$ curve of different Ur and φ. (**B**) the $Nu/Nu_{stationary}$ curve of different φs.

4. Conclusions

By using nanoparticle transport equations and considering the Brownian motion and thermophoresis of the particles, an immersed boundary method is used to simulate the elastically mounted cylinder moving inside the nanofluid-filled microchannel. The good thermal conductivity of solids is combined with the convective properties of fluids to improve overall heat transfer efficiency. In combination with the vortex-induced vibration of the cylinder, the heat transfer efficiency is improved again. The main results in this study with bullet points are presented as follows.

(1) In the Lock-in range, the heat transfer efficiency can be improved. For example, when $Ur = 3$ and $\varphi = 0\%$, the heat transfer efficiency is increased by about 5.08% compared to the fixed cylinder case.
(2) Increasing the concentration reduces the Lock-in range.
(3) Increasing the concentration results in heat transfer enhancement. When $Ur = 3$, φ increases from 0 to 5%, and the heat transfer efficiency increases by 117.3%.

Author Contributions: Y.-B.L.: Validation, Investigation, Data curation, Writing—original draft preparation, Visualization; C.-C.L.: Conceptualization, Methodology, Formal analysis, Investigation, Resources, Writing—original draft preparation, Writing—review and editing, Supervision, Project administration, Funding acquisition. All authors have read and agreed to the published version of the manuscript.

Funding: This research was funded by the Ministry of Science and Technology (NSTC 112-2221-E-033-026, MOST 111-2221-E-033-031).

Institutional Review Board Statement: Not applicable.

Informed Consent Statement: Not applicable.

Data Availability Statement: Not applicable.

Acknowledgments: This research was funded by the Ministry of Science and Technology, Taiwan, ROC via Grant No. MOST 111–2221-E-033–031.

Conflicts of Interest: The authors declare no conflict of interest.

References

1. Ju, W.J.; Fu, L.M.; Yang, R.J.; Lee, C.L. Distillation and detection of SO_2 using a microfluidic chip. *Lab. Chip.* **2012**, *12*, 622–626. [CrossRef] [PubMed]
2. Tuckerman, D.B.; Pease, R.F.W. High-performance heat sinking for VLSI. *Electron Device Lett.* **1981**, *2*, 126–129. [CrossRef]
3. Yeom, T.; Simon, T.W.; Yu, Y.; North, M.T.; Cui, T. Convective heat transfer enhancement on a channel wall with a high frequency, oscillating agitator. In Proceedings of the ASME International Mechanical Engineering Congress and Exposition, Denver, CO, USA, 11–17 November 2011; pp. 875–884.
4. Celik, B.; Raisee, M.; Beskok, A. Heat transfer enhancement in a slot channel via a transversely oscillating adiabatic circular cylinder. *Int. J. Heat Mass Transf.* **2010**, *53*, 626–634. [CrossRef]

5. Kumar, V.; Garg, H.; Sharma, G.; Bhardwaj, R. Harnessing flow-induced vibration of a D-section cylinder for convective heat transfer augmentation in laminar channel flow. *Phys. Fluids* **2020**, *32*, 19. [CrossRef]
6. Lienhard, J.H. *Synopsis of Lift, Drag, and Vortex Frequency Data for Rigid Circular Cylinders*; Technical Extension Service; Washington State University: Pullman, WA, USA, 1966; Volume 300.
7. Zhao, M.; Cheng, L.; An, H.; Lu, L. Three-dimensional numerical simulation of vortex-induced vibration of an elastically mounted rigid circular cylinder in steady current. *J. Fluids Struct.* **2014**, *50*, 292–311. [CrossRef]
8. Buongiorno, J. Convective transport in nanofluids. *J. Heat Transf.-Trans. ASME* **2006**, *128*, 240–250. [CrossRef]
9. Liao, C.C. Heat transfer transitions of natural convection flows in a differentially heated square enclosure filled with nanofluids. *Int. J. Heat Mass Transf.* **2017**, *115*, 625–634. [CrossRef]
10. Liao, C.-C.; Chang, Y.-W.; Lin, C.-A.; McDonough, J. Simulating flows with moving rigid boundary using immersed-boundary method. *Comput. Fluids* **2010**, *39*, 152–167. [CrossRef]
11. Corcione, M. Empirical correlating equations for predicting the effective thermal conductivity and dynamic viscosity of nanofluids. *Energy Conv. Manag.* **2011**, *52*, 789–793. [CrossRef]
12. Meis, M.; Varas, F.; Velazquez, A.; Vega, J.M. Heat transfer enhancement in micro-channels caused by vortex promoters. *Int. J. Heat Mass Transf.* **2010**, *53*, 29–40. [CrossRef]
13. Soti, A.K.; De, A. Vortex-induced vibrations of a confined circular cylinder for efficient flow power extraction. *Phys. Fluids* **2020**, *32*, 16. [CrossRef]

Disclaimer/Publisher's Note: The statements, opinions and data contained in all publications are solely those of the individual author(s) and contributor(s) and not of MDPI and/or the editor(s). MDPI and/or the editor(s) disclaim responsibility for any injury to people or property resulting from any ideas, methods, instructions or products referred to in the content.

engineering proceedings

MDPI

Proceeding Paper

Numerical Simulation Analysis of Penetration Performance of Armor-Piercing Fin-Stabilized Discarding Sabot to Steel Plate †

Yu-Shi Wang * and Cheung-Hwa Hsu *

Department of Mold and Die Engineering, National Kaohsiung University of Science and Technology, Kaohsiung 807618, Taiwan
* Correspondence: navyhata1228@gmail.com (Y.-S.W.); chhsu@nkust.edu.tw (C.-H.H.)

† Presented at the 3rd IEEE International Conference on Electronic Communications, Internet of Things and Big Data Conference 2023, Taichung, Taiwan, 14–16 April 2023.

Abstract: An armor-piercing fin-stabilized discarding sabot is used with a spare shell for 30 mm chain artillery. Due to the limited firepower of the bomb, the limited domestic shooting range, and the high public awareness of environmental protection, it is not easy to test the intrusion force. Therefore, an armor-piercing fin-stabilized discarding sabot is simulated and analyzed to optimize the design of tungsten composition to save research and development costs and improve ammunition performance. The key work of this research includes three parts: pre-work, armor-piercing fin-stabilized discarding sabot drawing, and numerical simulation analysis. Establishing a basic reference data system for ammunition specifications is currently being developed through finite element simulation analysis including the establishment of an armor-piercing fin-stabilized discarding sabot model and performing optimization design evaluation analysis.

Keywords: armor-piercing fin-stabilized discarding sabot; intrusion force; finite element simulation analysis

Citation: Wang, Y.-S.; Hsu, C.-H. Numerical Simulation Analysis of Penetration Performance of Armor-Piercing Fin-Stabilized Discarding Sabot to Steel Plate. *Eng. Proc.* **2023**, *38*, 76. https://doi.org/10.3390/engproc2023038076

Academic Editors: Teen-Hang Meen, Hsin-Hung Lin and Cheng-Fu Yang

Published: 10 July 2023

Copyright: © 2023 by the authors. Licensee MDPI, Basel, Switzerland. This article is an open access article distributed under the terms and conditions of the Creative Commons Attribution (CC BY) license (https://creativecommons.org/licenses/by/4.0/).

1. Introduction

At present, the production accuracy of 30 mm wing stable shelling armor-piercing tracers is stable [1–3]. In order to make the tracer suitable for use in various countries, it is necessary to improve the bullet's appearance to improve the speed and accuracy of the bullet [4,5]. The cost of mold production is relatively high, considering factors such as saving the defense budget and saving costs. The focus of this research is to establish a 30 mm wing-stabilized shelling armor-piercing tracer and different steel plate thickness models through finite element simulation analysis. Finite element analysis is used to set the density and impact speed of each tungsten rod, and the optimized mathematical calculation module is used. This research conducts analysis and uses materials and processes to construct physical bullets for comparative analysis, and through the big data database, to verify the reliability of the mathematical parameter calculation module of this research, and to implement the database construction and function integration.

2. Numerical Analysis

2.1. 30 mm Armor-Piercing Fin-Stabilized Discarding Sabot Geometry

Using the ANSYS Workbench finite element analysis software graphics design tool, and according to the size of the 30 mm wing-stabilized shell-piercing tracer projectile, the 2D and 3D drawing files of this type of projectile were constructed, as shown in Figure 1.

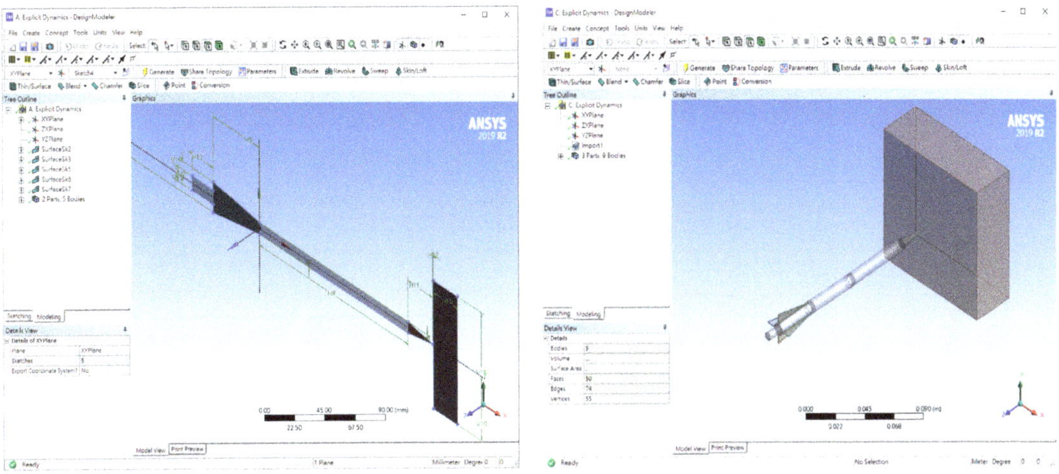

(**a**) (**b**)

Figure 1. Build model: (**a**) 2D model and (**b**) 1D model.

2.2. Reference Armor Firing Rate

According to the US military technical manual MIL-STD-367 and MIL-STD-662F specifications as shown in Table 1, the shooting muzzle velocity is 1430 m/s, the simulated vertical target position is 1000 m away from the muzzle, and the impact velocity is 1306 m/s.

Table 1. Distance and impact speed comparison table.

Range (Meter)	Remain Velocity (m/s)
0	1430
100	1417.6
200	1405.2
300	1392.9
400	1380.5
500	1368.1
600	1355.7
700	1343.3
800	1330.9
900	1318.5
1000	1306.1

2.3. Two-Dimensional and Three-Dimensional Models for Simulation Analysis

According to the 30 mm wing stable shelling armor-piercing tracer material to select tungsten alloy material from the built-in database of ANSYS in Table 2, it is confirmed that the material properties are consistent with this analysis, and the 2D and 3D image files are imported to use the ANSYS Explicit Dynamics module for mesh analysis. The number of nodes is 21,198 and the number of grids is 20,422. The analysis and calculation module function confirm that the grid has converged and set the boundary condition parameters for the armor-piercing tracer and steel plate.

Table 2. Tungsten rod material parameters.

	A	B	C
1	Property	Value	Unit
2	Density	17,790	kg mˆ − 3
3	Young's Modulus	3.9×10^{11}	Pa
4	Poisson's Ratio	0.28	
5	Bulk Modulus	2.9545×10^{11}	Pa
6	Shear Modulus	1.5234×10^{11}	Pa
7	Yield Strength	84,000	psi
8	Tangent Modulus	130	psi
9	Specific Heat, C_p	184	J kgˆ − 1ˆ − 1

3. Results and Discussion

Through ANSYS finite element analysis, the armor-piercing tracer and the steel plate are set to be 1000 m away from the muzzle, and the thickness of the steel plate is set to 25, 35, 40, and 50 mm, and various mechanical analyses are performed through the post-processing module function. The strain data obtained from tracer bullets and steel plates of various thicknesses are shown in Figure 2.

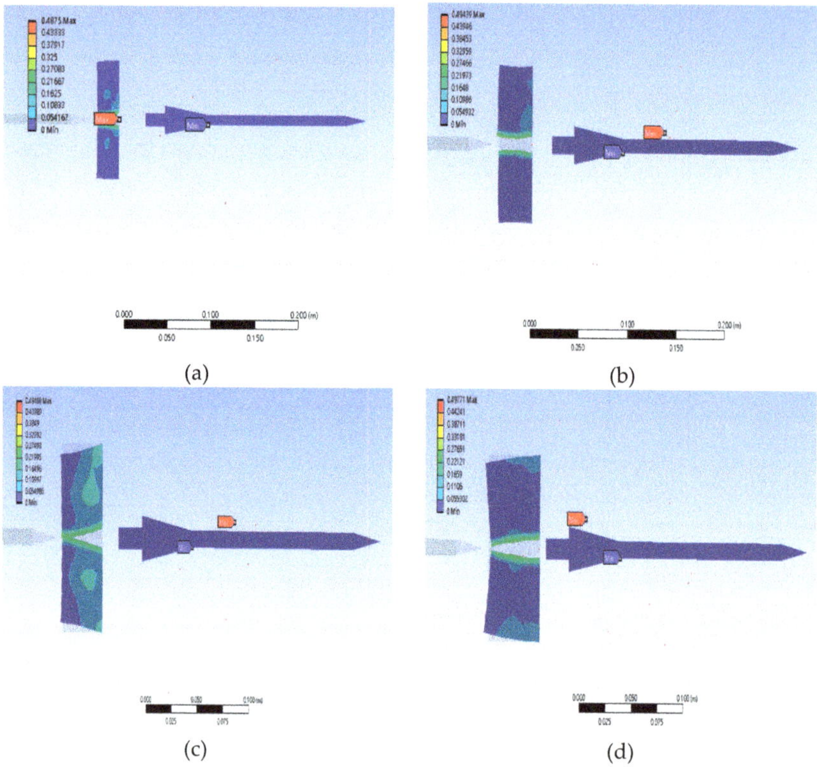

Figure 2. Steel plate thickness and strain analysis. (**a**) Thickness: 25 mm; (**b**) thickness: 35 mm; (**c**) thickness: 40 mm; (**d**) thickness: 50 mm.

If the density of the armor-piercing projectile does not change, by changing the boundary condition setting of the impact speed from 1000 to 1300 m/s, the armor-piercing projectile is affected by the speed, and the penetration depth continues to increase. This causes certain damage to various types of tank armor. The effect is shown in Table 3.

Table 3. Comparison table of relationship between impact speed and penetrable depth.

Rate of Fire (m/s)	Distance (m)	Density (g/cm^3)	Penetration Depth (mm)
1000	1000	17.79	52
1100	1000	17.79	54
1200	1000	17.79	55
1300	1000	17.79	56

The penetration depth of the steel plate is analyzed and observed after changing the density of the armor-piercing tracer, and the density is adjusted from 17.5 to 17.79 g/cm^3, 18 and 18.5 g/cm^3 through ANSYS finite element analysis. When the density of the tungsten rod increases, the penetration depth increases, and the results are shown in Table 4.

Table 4. Penetration depth with tungsten rod density.

Rate of Fire (m/s)	Distance (m)	Density (g/cm^3)	Penetration Depth (mm)
1306	1000	17.5	52
1306	1000	17.79	57
1306	1000	18	60
1306	1000	18.5	65

Taking a 25 mm-thick steel plate as an example, the influence of the incident angle is discussed. The inclination angle is defined as the angle α, the impact velocity of the armor-piercing projectile is set to 1306 m/s, and the inclination angle of the steel plate is set to 75, 60 and 45 degrees. As shown in Figure 3, when the angle of the steel plate is 45°, the armor-piercing projectile cannot penetrate completely through the steel plate; this angle is the extreme design angle.

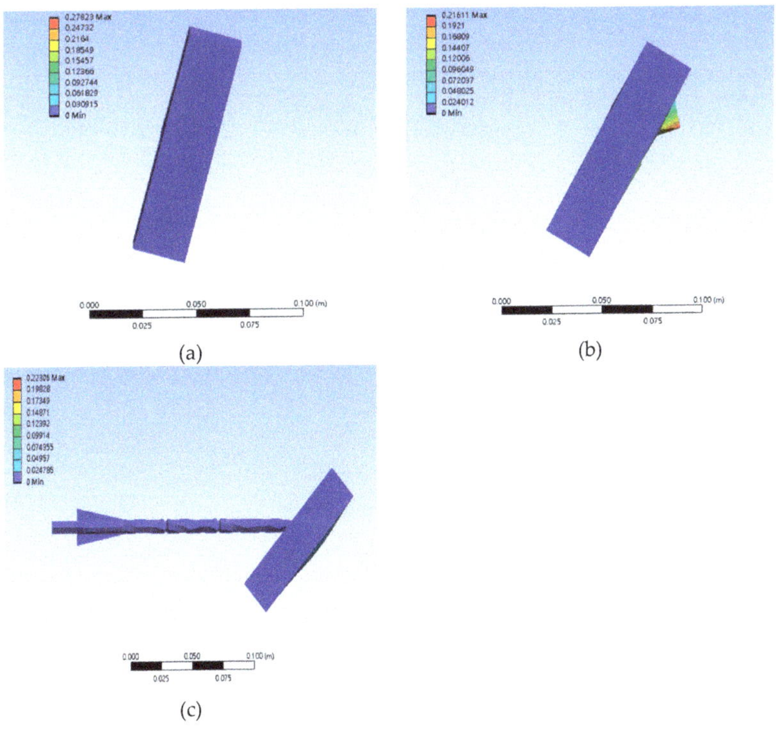

Figure 3. Effect of inclination angle. (**a**) Inclination: 75°; (**b**) inclination: 60°; (**c**) inclination: 45°.

4. Conclusions

In this study, an armor-piercing tracer with a 30 mm stabilized wing is developed. The software ANSYS Workbench is used for analysis. The density of the tungsten rod, the impact velocity, and the angle position of the steel plate designed for the armor-piercing tracer are analyzed by the Explicit Dynamics module. The test has led to the following conclusions.

1. Through numerical simulation analysis, we calculated the penetration depth of the steel plate at different impact speeds. The greater the impact speed, the greater the depth of the steel plate that can be penetrated.
2. The analysis of changing the density of the tungsten rod shows that the density of the tungsten rod increases, and the increased thickness of the armor-piercing tracer that can penetrate the steel plate increases.
3. By changing the design and installation angle of the steel plate under the stable firing muzzle velocity, the armor-piercing tracer cannot penetrate the steel plate and only causes a certain degree of damage around the surface of the steel plate when it is placed at an inclination of 45°. The steel plate can be penetrated only by firing at the same range of shooting points. The analysis and research data can be used for evaluation and reference by the developers of the subsequent design of armored vehicles.

Author Contributions: Conceptualization, Y.-S.W. and C.-H.H.; methodology, Y.-S.W.; software, Y.-S.W.; validation, Y.-S.W. and C.-H.H.; formal analysis, Y.-S.W.; investigation, Y.-S.W.; resources, C.-H.H.; data curation, Y.-S.W.; writing—original draft preparation, Y.-S.W.; writing—review and editing, C.-H.H.; visualization, Y.-S.W.; supervision, C.-H.H.; project administration, C.-H.H. and Y.-S.W. All authors have read and agreed to the published version of the manuscript.

Funding: This research received no external funding.

Institutional Review Board Statement: Not applicable.

Informed Consent Statement: Not applicable.

Data Availability Statement: Not applicable.

Conflicts of Interest: The authors declare no conflict of interest.

References

1. MIL-STD-367 Armor Test Data Reporting. 1986. Available online: http://everyspec.com/MIL-STD/MIL-STD-0300-0499/MIL-STD-367_29115/ (accessed on 17 May 2019).
2. MIL-STD-662F Department of Defense Test Method Standard v50 Ballistic Test for Armor. 1997. Available online: http://everyspec.com/MIL-STD/MIL-STD-0500-0699/MIL-STD-662F_6718/ (accessed on 21 May 2019).
3. MIL-A-12560H (MR) w/INT. Amendment 4 Armor Plate, Steel, Wrought, Homogeneous (for Use in Combat-Vehicles and for Ammunition Testing). 2007. Available online: http://everyspec.com/MIL-SPECS/MIL-SPECS-MIL-A/MIL-A-12560H_AMENDMENT-3_10355/ (accessed on 21 May 2019).
4. Kuo, H.-T. Simulations for the Dynamics of Sabot Discard and the Projectile Trajectory. Master's Thesis, Chung Yuan Christian University, Taoyuan, Taiwan, 2004.
5. Fu, S.-M. *Study of Aerodynamic Characteristics of a Tail Fin-Stablized Projectile*; Chung Cheng Institute of Technology, National Defense University Department of Mechanical and Aerospace Engineering: Taoyuan, Taiwan, 2013.

Disclaimer/Publisher's Note: The statements, opinions and data contained in all publications are solely those of the individual author(s) and contributor(s) and not of MDPI and/or the editor(s). MDPI and/or the editor(s) disclaim responsibility for any injury to people or property resulting from any ideas, methods, instructions or products referred to in the content.

engineering proceedings

Proceeding Paper

Optimum Design of Polymer Composite Reactor Pressure Vessels †

Bai-Chau Chen and Jui-Hung Cheng *

Department of Mold and Die Engineering, National Kaohsiung University of Science and Technology, Kaohsiung 807618, Taiwan; f111147130@nkust.edu.tw
* Correspondence: rick.cheng@nkust.edu.tw

† Presented at the 3rd IEEE International Conference on Electronic Communications, Internet of Things and Big Data Conference 2023, Taichung, Taiwan, 14–16 April 2023.

Abstract: Based on DBR, a pressure vessel and standard vessel are designed using a public view and diagrams. They are formulated to a standard specification using the ASME pressure vessel code rules and the European standard EN13445. The pressure is determined under influence of a closed end. There are limitations to the structure of the container. Using DBA, several analyses were performed based on FEM to provide parameters to complete the DOE, and combined with DFA and DFM to verify that the analysis was conducive to giving users a good analysis plan. The core premise of this study is to simplify operation, to provide an exemplary user interface, observe various parameters according to requirements, meet pressure requirements during the reaction of polymer composite materials, and ensure that the temperature, strength, and manufacturing cost are within safety factors, to provide a good user experience.

Keywords: DBR; DBA; FEM; DFA; DFM; UI/UX

Citation: Chen, B.-C.; Cheng, J.-H. Optimum Design of Polymer Composite Reactor Pressure Vessels. *Eng. Proc.* **2023**, *38*, 39. https://doi.org/10.3390/engproc2023038039

Academic Editors: Teen-Hang Meen, Hsin-Hung Lin and Cheng-Fu Yang

Published: 27 June 2023

Copyright: © 2023 by the authors. Licensee MDPI, Basel, Switzerland. This article is an open access article distributed under the terms and conditions of the Creative Commons Attribution (CC BY) license (https://creativecommons.org/licenses/by/4.0/).

1. Introduction

A polymer composite reaction pressure vessel is a reaction vessel for heating, pressure detection, control, automatic stirring, etc., and automatic operation. They are manufactured to different parameters according to additional product requirements [1]. Nowadays, Industry 4.0 and IoT have become the trend. In addition to accurately controlling incoming and outgoing raw materials through intelligent monitoring, human errors can be reduced to improve the stability of production [2].

1.1. Design Requirements

In addition to conforming to the Design for Assembly (DFA), being easy to assemble and disassemble, and having a structural strength that allows operation at different temperatures and pressures, a polymer composite reaction pressure vessel must be designed to accurately control the dosage of reagents and have a simple easy-to-understand operation panel [3,4]. The goal is to optimize the user's operation through UI/UX (User Interface/User Experience) so that the designer can understand all aspects of the user's requirements, and improve the cross-domain communication efficiency most appropriately [5].

1.2. FEM-Design

The Finite Element Method (FEM) is a widely used method to numerically solve differential equations in engineering and to produce mathematical models. Problems include traditional areas such as structural analysis, heat transfer, fluid flow, mass transfer, and electromagnetic potential. The simple equations that model finite elements are combined into a broader system of equations to model the entire problem [6]. FEM then approximates the solution by minimizing the relevant error function by variational methods, and using Ansys analysis, desired results, and given parameters.

1.3. Introduction to DBA

Design by Analysis (DBA) is intended to complement and replace traditional DBR: this restricted route uses steels and steel castings with sufficient toughness for calculating temperatures below the creep range. The term 'action', which replaces the old term 'load', means a parameter is applied to the structure, such as force (including pressure), temperature, changes, and applied displacements, resulting in pressure or tension [7]. A simple, straightforward composition with environmental pressure is possible, while the flexibility expected of modern code adjusts the safety margin according to the difference of the change in movement, the possibility of a combination of actions, the consequences of failure, differences in structural behavior, consequences of different failure modes, and uncertainty in analysis, which all contribute to the concept of a safety factor. This concept uses different actions, different local safety factors, the different roles of the structure, other failure modes, and the corresponding resistance [8]. The safety factor takes into account actions as well as the resistance of the structure.

2. Quality Improvement Project

2.1. DBA & DBR Approach

The study investigated a double-layer vertical cylindrical pressure vessel (shell) with skirt supports and annular pressure heads, using DBA and DBR, and using ANSYS to classify the study into elastic and inelastic, to analyze structural integrity. Table 1 shows the design specifications. According to the specification, two extremely different steels were selected for investigation: high-strength pressure vessel steel P500-QT and low-alloy steel P355. The proposed work flow is shown in Figure 1. The stress resistance of the design, according to the ASME code, is represented by Equation (1). The allowable design stress resistance is in accordance with the EN13445 European standard. The thickness of the outer cylindrical shell is given by Equation (3). The axial stress resistance (σxy) is obtained using Equation (4). According to the specification, the meaning of a symbol is derived as follows: Rp(0.2/t) = Yield value, $R(m/20)$ = UTS, t = Corrosion Allowance, PR = Pressure of Cylindrical Shell Inner Radius, PD = Pressure of Inner/Shell Diameter, SE = Young's Coefficient of Maximum Allowable Design Application:

$$f = \min\left(\frac{R_{p(\frac{0.2}{t})}}{1.5}; \frac{R_{\frac{m}{20}}}{2.14}\right), \tag{1}$$

$$f = \min\left(\frac{R_{p(\frac{0.2}{t})}}{1.5}; \frac{R_{\frac{m}{20}}}{2.4}\right), \tag{2}$$

$$T = \frac{PR}{SE - 0.6P}, \tag{3}$$

and

$$\sigma xp = \frac{PD}{4t}. \tag{4}$$

Table 1. Cylindrical shell design specification.

Material Properties	P500-QT	P355
Young's modulus, E(GPa)	210	200
Poison Ratio (γ)	0.29	0.28
UTS, σ_u ($R_{M/20}$), Mpa	640	600
Yield value, σ_y ($R_{P0.2/t}$), Mpa	580	380
Density, ρ(Kg/m^3)	7872	7850

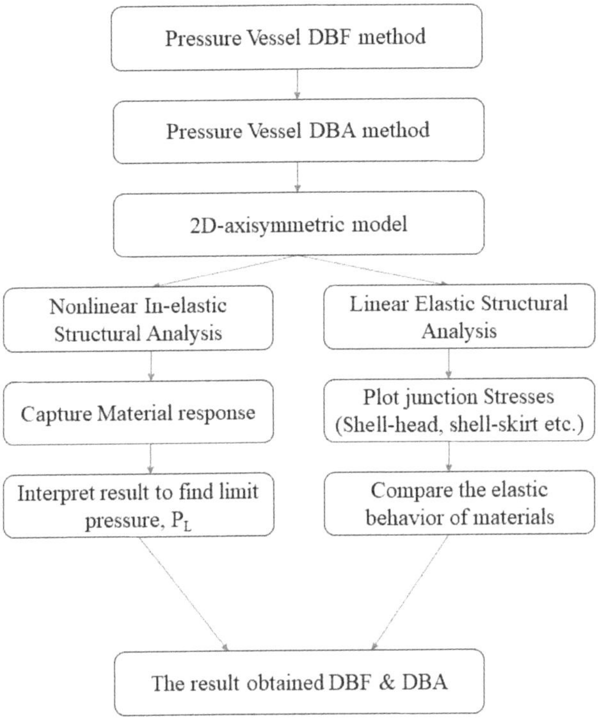

Figure 1. Pressure vessel design by DBA route.

2.2. QFD Result

Customer requirements for QFD include ease of disassembly, durability, and temperature and pressure control [9]. DFM design-optimized processing methods are used to manufacture ideal products and reduce assembly parts through DFA, which is convenient for users during the operation process [10]. In this way, the user can quickly and efficiently take out the experimental product ready for the next experiment. On the other hand, the product is designed following the pressure vessel specification, which makes the product structure more robust, as such it instils confidence in the experimental process, and the equipment is monitored at the same time. It also provides more intuitive and accurate data to users for an accurate and efficient user experience [11].

The DBA/DBR is verified and analyzed to obtain results. Because it is impossible to understand the usage habits and environments of all users, it is necessary to thoroughly discuss how to meet user needs in the design stage to improve the overall product. The cross-comparison of 10 vessels with results from Columns #7 and 6 is in line with customer needs, so it can improve the reliability of this analysis for market users.

3. Conclusions

According to the design method of QFD, it is understood how to meet customer's expectations; DBA/DBF are used to produce an equation for research vessel design that meets design specifications and to obtain corresponding design parameters, and the products can have detailed data to enable comparison. The pressure vessel is optimized through FEM. Feedback for real-time monitoring of temperature and pressure provides a variety of I/O interfaces, adjusts the stirring speed as needed, and avoids many problems encountered by traditional reaction vessels. The visual window is used to observe the reaction process of chemical materials, accurately control the input and output of materials, avoid parameter errors that lead to poorly finished products, automatically stir the reaction, and make the reaction process more complete. Finally, the IoT function gives users an entire real-time operating experience, which reduces human errors, increases the controllability of the finished product, and improves the fluency of the overall operation. Finally, an integral anti-corrosion material is made to avoid damage to the pressure vessel caused by different chemical materials through high-temperature and high-pressure reactions.

Author Contributions: Conceptualization, B.-C.C. and J.-H.C.; methodology, B.-C.C. and J.-H.C.; software, B.-C.C.; validation, B.-C.C. and J.-H.C.; formal analysis, B.-C.C.; investigation, B.-C.C.; resources, J.-H.C.; data curation, B.-C.C.; writing—original draft preparation, B.-C.C.; writing—review and editing, J.-H.C.; visualization, B.-C.C.; supervision, B.-C.C.; project administration, J.-H.C.; funding acquisition, J.-H.C. All authors have read and agreed to the published version of the manuscript.

Funding: This research received no external funding.

Institutional Review Board Statement: Not applicable.

Informed Consent Statement: Not applicable.

Data Availability Statement: Not applicable.

Conflicts of Interest: The authors declare no conflict of interest.

References

1. Pous-Romero, H.; Lonardelli, I.; Cogswell, D.; Bhadeshia, H.K.D.H. Austenite grain growth in a nuclear pressure vessel steel. *Mater. Sci. Eng. A* **2013**, *567*, 72–79. [CrossRef]
2. Moss, D.R. *Pressure Vessel Design Manual*; Elsevier: Amsterdam, The Netherlands, 2004.
3. Gill, S.S. (Ed.) *The Stress Analysis of Pressure Vessels and Pressure Vessel Components: International Series of Monographs in Mechanical Engineering*; Elsevier: Amsterdam, The Netherlands, 2004; Volume 3.
4. Pickering, E.J.; Bhadeshia, H.K.D.H. Macrosegregation and microstructural evolution in a pressure-vessel steel. *Metall. Mater. Trans. A* **2014**, *45*, 2983–2997. [CrossRef]
5. Liu, P.F.; Zheng, J.Y.; Ma, L.; Miao, C.J.; Wu, L.L. Calculations of plastic collapse load of pressure vessel using FEA. *J. Zhejiang Univ.-Sci. A* **2008**, *9*, 900–906. [CrossRef]
6. Baylac, G.; Kiesewetter, N.; Zeman, J.; Handtschoewercker, A.; McFarlane, R.; Delle Site, C.; Tonti, A.; Holdsworth, S. Creep amendments in the European standard EN 13445. In Proceedings of the ASME Pressure Vessels and Piping Conference, San Antonio, TX, USA, 22–26 July 2007; pp. 541–546.
7. Karthikeyan, K.M.B.; Balasubramanian, T.; Bruce, A.R.; Premkumar, P. Pressure Vessel Design by Design by Analysis Route. In *IOP Conference Series: Materials Science and Engineering*; IOP Publishing: Bristol, UK, 2020; Volume 923, p. 012020.
8. Ramakrishnan, M.; Muthupandi, V. Application of submerged arc welding technology with cold wire addition for drum shell long seam butt welds of pressure vessel components. *Int. J. Adv. Manuf. Technol.* **2013**, *65*, 945–956. [CrossRef]
9. Cristiano, J.J.; Liker, J.K.; White, C.C., III. Key factors in the successful application of quality function deployment (QFD). *IEEE Trans. Eng. Manag.* **2001**, *48*, 81–95. [CrossRef]
10. Gündoğdu, F.K.; Kahraman, C. A novel spherical fuzzy QFD method and its application to the linear delta robot technology development. *Eng. Appl. Artif. Intell.* **2020**, *87*, 103348. [CrossRef]
11. Deveci, M.; Öner, S.C.; Canıtez, F.; Öner, M. Evaluation of service quality in public bus transportation using interval-valued intuitionistic fuzzy QFD methodology. *Res. Transp. Bus. Manag.* **2019**, *33*, 100387. [CrossRef]

Disclaimer/Publisher's Note: The statements, opinions and data contained in all publications are solely those of the individual author(s) and contributor(s) and not of MDPI and/or the editor(s). MDPI and/or the editor(s) disclaim responsibility for any injury to people or property resulting from any ideas, methods, instructions or products referred to in the content.

engineering proceedings

Proceeding Paper

Research on Road Performance of Solid-Waste-Based-Gelling-Agent-Stabilized Sub-Base †

Tao Guo and Yueqiang Qi *

Shandong Huanghe Shuncheng Water Conservancy and Hydro-Power Engineering Co., Jinan 250000, China; wxc13792680392@163.com
* Correspondence: hna626@163.com

† Presented at the 3rd IEEE International Conference on Electronic Communications, Internet of Things and Big Data Conference 2023, Taichung, Taiwan, 14–16 April 2023.

Abstract: To determine the road performance and optimum dosing of solid-waste-based gelling agents in road subgrade, we conducted unconfined compressive strength testing for clay, sandy soil, weathered sand, and crushed stone mixed with different proportions of solid-waste gelling agents and compared the results with the road performance of cement-stabilized crushed stone materials. The results show that the optimum admixture of cementitious materials for clay and clay is 8–10%. The optimum admixture of cementitious materials for weathered sand is 5–7%. The optimum admixture of cementitious materials for stabilized gravel is 5.5%. The late strength growth of stabilized gravel with solid-waste-based cementitious agents is significantly better than that of cement-stabilized gravel.

Keywords: solid-waste-based cementitious materials; optimum admixture; road performance

Citation: Guo, T.; Qi, Y. Research on Road Performance of Solid-Waste-Based-Gelling-Agent-Stabilized Sub-Base. *Eng. Proc.* **2023**, *38*, 55. https://doi.org/10.3390/engproc2023038055

Academic Editors: Teen-Hang Meen, Hsin-Hung Lin and Cheng-Fu Yang

Published: 28 June 2023

Copyright: © 2023 by the authors. Licensee MDPI, Basel, Switzerland. This article is an open access article distributed under the terms and conditions of the Creative Commons Attribution (CC BY) license (https://creativecommons.org/licenses/by/4.0/).

1. Introduction

Deelwal et al. [1] found that the maximum dry density of red mud was less than 1.75 g/cm^3, which is the minimum requirement for use in road substrates for India's national highways, state highways, and major regional arterials and other heavily trafficked roads [2]. The liquid limit and plasticity index were much higher than the minimum requirements of the Indian standard. Sahoo et al. [3] and Deelwal et al. [1] obtained CBR (California bearing ratio) values for red mud and submerged CBR values of 4.2 and 4%, respectively, under submerged conditions, which are 20–30% lower than those required by Indian standards for pavement subgrade road construction. Although higher than the minimum requirement of 2.5% for Irish standards and 3% for major road design in Queensland, Australia, they are 10% lower than those required by statewide urban road design specifications [4]. Li et al. conducted a detailed experimental study on the distribution and mechanical performance indexes of coal gangue in cold regions such as the three northeastern provinces and found problems and testing indexes when using coal gangue as road base fill material in construction in cold regions [5–7]. Liu [8–10] studied the change law of test values of the compaction test and unconfined compressive strength of saline soils stabilized by a soil conglomerate in different salt content rates and different maintenance age conditions. Yue [11] adopted monomer ethylene glycol, monovinyl polyethylene glycol ether, acrylic acid, and hydroxypropyl acrylate in a polymerization reaction at room temperature to produce solid-waste-based cementitious materials, polycarboxylic acid water reducing agent.

2. Raw Materials

2.1. Earth and Rock Materials and Solid-Waste-Based Cementitious Agents

For inorganic bonded solid-waste-based cementitious stabilization (sub-base), the type of soil, the nature of the solid-waste-based cementitious agent, and the construction process affect the performance of the stabilized sub-base. Four types of typical soil and rock materials, namely clay, sand, weathered sand, and gravel, were used in Shandong, and the chemical composition of the selected materials was analyzed. The corresponding solid-waste-based gelling agents (in different ratios) were configured for testing.

The gelling material is solid-waste-based gelling from Beijing Donghao Technology Co. (Beijing, China).The main ingredients are sintered red mud micronized powder, fly ash, calcined gangue micronized powder, and a compound exciter consisting of sodium silicate, triterpene saponin, polymeric aluminium sulphate, magnesium fluorosilicate, and sodium hydroxide. The characteristics of each type of raw material are shown in Table 1, Table 2 amd Table 3.

Table 1. Performance indicators for solid-waste-based gelling agents.

Indicators / Type	Fineness (%)	Density (kg/m^2)		Specific Surface Area (m^2/kg)
Solid-waste-based gelling	4.6	2870		681
Water consumption at standard consistency (%)		Coagulation time (min)		Sand flow rate (mm)
30.4		Initial condensation	198	182
		Final condensation	269	
Flexural strength at different ages (MPa)		Compressive strength at different ages (MPa)		
3 d	4.3	3 d		21.1
28 d	10.6	28 d		57.7

Table 2. Engineering properties of clay, sandy soil, and weathered sand raw materials.

	Maximum Dry Density (g/cm^3)	Optimum Moisture Content (%)	Plastic Limit (%)	Liquid Limit (%)	Plasticity Index
Clay	1.98	11.7	14.8	28.2	13.4
Sandy soil	1.88	9.9	12.0	21.8	9.8
Weathered sand	2.11	6.9	-	-	

Table 3. Composition of crushed stone particle gradation.

Aperture (mm)	10	5	2	1	0.5	0.25	0.075
Pass rate (%)	100	93.6	62.7	44.9	22.6	11.3	2.5

2.2. *Cement-Stabilized Aggregates*

To compare the difference in performance effects between solid-waste-based-gelling-agent-stabilized aggregates and cement-stabilized aggregates, aggregates with the same gradation were selected for testing. The proportions of each grade of aggregate and the synthetic gradation are shown in Tables 4 and 5, respectively. The cement used is P.C-32.5, a type commonly used in substrates. The feedstock properties are shown in Table 6.

Table 4. Stabilized gravel mix ratio.

Aggregate Sizes (mm)	Proportion (%)
20–30	17
10–20	38
5–10	18
0–5	27

Table 5. Synthetic gradation.

Sieve Hole Size (mm)	31.5	26.5	19	9.5	4.75	2.36	0.6	0.075
Pass rate (%)	100	-	74.2	45.3	27.8	19.2	8.4	3.8

Table 6. Physical properties of cement.

Indicators / Surface	Fineness (%)	Density (kg/m^2)	Specific Surface Area (m^2/kg)	Water Consumption at Standard Consistency (%)	
P.C-32. 5	1.5	3000	387.6	29.0	
Coagulation time (min)		Flexural strength at different ages(MPa)		Compressive strength at different ages (MPa)	
Initial condensation	260	3 d	3.39	3 d	11.52
Final condensation	310	28 d	8.05	28 d	49.32

3. Tests and Results

According to the Test Procedure for Inorganic Binding Material Stabilization for Highway Engineering (JTGE51-2009), four typical materials—clay, sand, weathered sand, and gravel—were mixed with different proportions of the solid-waste-based cementing agent. The standard specimens were prepared according to the maximum dry density and optimum moisture content obtained from compaction tests. They were tested for unconfined compressive strength with 7 days of standard conditions to determine the optimum amount of solid-waste-based cementing agent.

3.1. Compaction Tests

Table 7 presents the results of compaction tests on different types of soils stabilized by different doses of solid-waste-based gelling agents and cement.

Table 7. Results of compaction tests.

Curing Materials	Types of Soil	Curing Material Dosing (%)	Optimum Moisture Content (%)	Maximum Dry Density (g/cm^3)
Solid-waste-based gelling agents	Clay	4	11.9	1.99
		6	12.1	2.01
		8	11.8	2.04
		10	12.4	2.05
		12	12.0	2.08
	Sandy soil	4	9.8	1.91
		6	10.0	1.92
		8	9.7	1.95
		10	9.9	1.95
		12	10.1	1.97
	Weathered sand	4	6.9	2.12
		7	6.8	2.13
		10	6.9	2.15
	Gravel	6	4.9	2.35
Cement	Gravel	5	5.0	2.33

The optimum water content does not change much with the increase in the cementitious material, and the maximum dry density increases slightly with the increase in the cementitious material admixture. This is because the density of the cementitious material is greater than that of the clay, sandy soil, and weathered sand. Comparing the results of optimum water content and maximum dry density of cement-stabilized gravel and cement-stabilized gravel, it is found that there is little difference between the results of the two at 6% admixture.

3.2. Lateral Limitless Compressive Strength Test

For the asphalt pavement sub-base, the unconfined compressive strength index is one of the most important indicators of road performance, indicating the strength value that the specimen can withstand when placed under unconfined lateral conditions. The unconfined compressive strength of clay, sand, weathered sand, and macadam stabilized by the solid-waste base binder for 7 days and the unconfined compressive strength of cement-stabilized macadam base for 7 days were tested. The results are shown in Table 8 and analyzed concerning the amount of solid-waste-based cementitious binder incorporated (Figure 1).

Table 8. Results of 7 d unconfined compressive strength tests for different materials with different amounts of cementitious materials.

Curing Materials	Types of Soil	Curing Material Dosing (%)	7 d Unconfined Compressive Strength Average Value (MPa)	Representative Value for 7 d Unconfined Compressive Strength (MPa)
Solid-waste-based gelling agents	Clay	4	1.58	1.21
		6	2.49	2.15
		8	3.38	3.01
		10	4.72	4.38
		12	5.59	5.22
	Sandy soil	4	1.73	1.36
		6	2.77	2.40
		8	3.51	3.14
		10	4.95	4.58
		12	6.16	5.79
	Weathered sand	4	2.95	2.58
		7	4.97	4.59
		10	6.21	5.74
	Gravel	6	5.72	5.26
Cement	Gravel	5	9.86	9.03

For the same raw material, the unconfined compressive strength tended to increase significantly with the increase in the amount and the strength values of the aggregates stabilized with the solid-waste-based gelling agent. This indicates that the amount of solid-waste-based gelling agent plays a decisive role in the unconfined compressive strength.

For the same raw material, the unconfined compressive strength tended to increase significantly with the increase in the amount of solid-waste-based gelling agent. This indicates that the amount of solid-waste-based gelling agent plays a decisive role in the unconfined compressive strength.

2.5–3.0 MPa of the grassroots level for 7 d standard maintenance is appropriate to determine the best mixture of materials as a sub-base in solid-waste-based cementitious materials (Table 9).

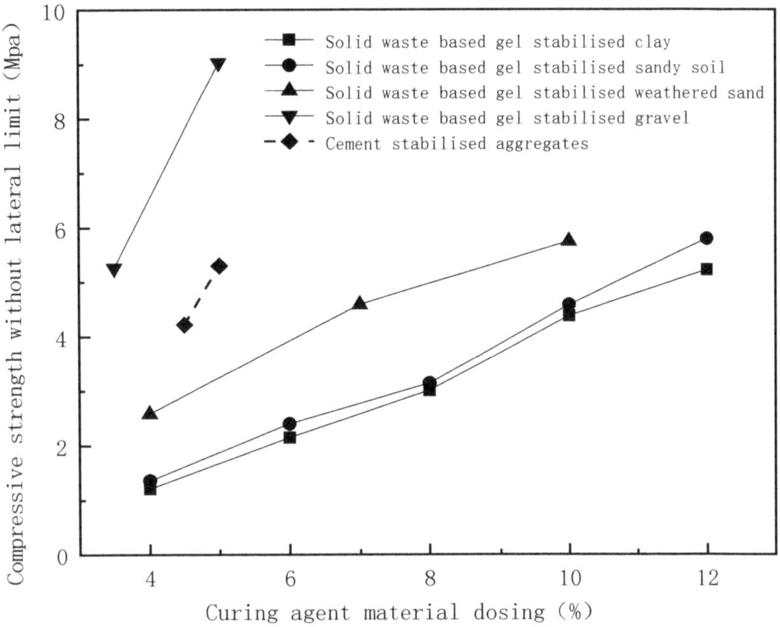

Figure 1. Correlation between the unconfined compressive strength of four types of road materials stabilized by curing agents and the amount of admixture.

Table 9. Optimal dosing of solid-waste-based gelling agents in several typical mixed materials.

Types of Mixed Materials		Optimum Dosing of Gel Material (%)
Solid-waste-based-gelling-agent-stabilized clay	Grassroots	10
	Substrate	8
Solid-waste-based-gelling-agent-stabilized sandy soil	Grassroots	10
	Substrate	8
Solid-waste-based gelling agent to stabilize weathered sand	Grassroots	7
	Substrate	5
Solid-waste-based-gelling-agent-stabilized gravel	Grassroots	5.5
Cement-stabilized aggregates	Grassroots	5

The variation of the unconfined compressive strength with the curing age was investigated for different material specimens with the optimum solid-waste-based gelling agent admixture. The results are shown in Figure 2.

The unconfined compressive strength of the four types of stabilized materials increased significantly with the increase in the curing age. The strength of the sandy soil materials stabilized with the solid-waste-based gelling agent increased relatively slowly, while that of weathered sand and clay increased in a similar trend. For the stabilized aggregates, the late strength growth (14–28 days) of the solid-waste-based-gelling-agent-stabilized aggregates was significantly better than that of the cement-stabilized aggregates.

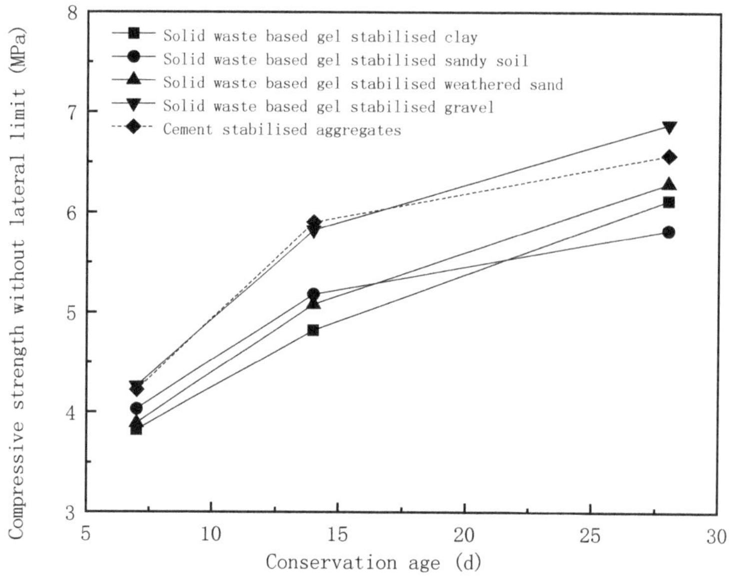

Figure 2. Variation of unconfined compressive strength with the age of curing for various types of stabilized materials.

4. Conclusions

The optimum water content is not related to the amount of cementitious material admixture, and the maximum dry density increases slightly with the increase in cementitious material admixture. At the admixture of 6%, the optimum water content and maximum dry density of crushed stone and cement-stabilized crushed stone with the solid-waste-based cementitious stabilized are almost the same. The optimum admixture of cementitious material for clay and clay is 8 to 10%. The optimum admixture of cementitious material for weathered sand is 5 to 7%. The optimum admixture of cementitious material for stabilized gravel is 5.5%. The unconfined compressive strength of each type of stabilized material increases linearly with the increase in solid-waste-based cementing admixture. Under the condition of a certain admixture of cementing material, the unconfined compressive strength of each type of stabilized material increases with the increase in the curing age.

Author Contributions: Methodology, T.G.; test, Y.Q.; data curation, Y.Q.; writing-original draft preparation, Y.Q.; writing-review and editing, T.G.; All authors have read and agreed to the published version of the manuscript.

Funding: This research received no external funding.

Institutional Review Board Statement: Not applicable.

Informed Consent Statement: Not applicable.

Data Availability Statement: Not applicable.

Conflicts of Interest: The authors declare no conflict of interest.

References

1. Wang, K.; Wang, H.; Li, S. Hydrochloric acid leaching of red mud for aluminium recovery. *Nonferrous Met.* **2012**, *7*, 16–18. (In Chinese)
2. Chen, D. Recovery of titanium dioxide from red mud. *Ind. Environ.* **1994**, *3*, 42–45. (In Chinese)
3. Agatzini-Leonardou, S.; Oustadakis, P.; Tsakiridis, P.; Markopoulos, C. Titanium leaching from red mud by diluted sulfuric acid at atmospheric pressure. *J. Hazard. Mater.* **2008**, *157*, 579–586. (In Chinese) [CrossRef] [PubMed]

4. Zhang, J. Preliminary study on the recovery of titanium dioxide from red mud. *China Resour. Compr. Util.* **2003**, *1*, 28–30. (In Chinese)
5. Niu, W. Technical points of highway coal gangue filling roadbed. *Heilongjiang Sci. Technol. Inf.* **2013**, *14*, 241. (In Chinese)
6. Jan, X. Application of coal gangue backfill for highway roadbed. *TranspoWorld* **2013**, *291*, 187–188. (In Chinese)
7. Cui, H.; Zhu, C.; Yang, S.; Hu, S. Study on the strength characteristics of new improved soil containing red mud curing agent for cold zone road base. *J. Hebei Inst. Archit. Civ. Eng.* **2021**, *39*, 18–26. (In Chinese)
8. Xiao, Z.; Dai, A.; Zhang, Q.; Sui, N. Research on the construction technology of cold regenerated base layer of asphalt pavement. *J. Highw. Transp. Res. Dev. (Appl. Technol. Ed.)* **2020**, *16*, 91–92+96. (In Chinese)
9. Liu, J.; Wei, B.; Yang, S. Experimental study on the unconfined compressive strength of saline soils modified with soil condensate. *J. Hebei Inst. Archit. Civ. Eng.* **2021**, *39*, 29–33. (In Chinese)
10. Liu, J.; Zhang, G.; Zhang, N. Experimental study on salt swelling characteristics of saline soils improved by curing agents under freeze-thaw cycles. *J. Hebei Inst. Archit. Civ. Eng.* **2020**, *38*, 60–63. (In Chinese)
11. Yue, G.; Xie, R.; Guo, W.; Peng, L.; Tang, Y.; Yu, P.; Mo, L. Preparation and properties of polycarboxylate superplasticizer for solid waste-based cementitious materials. *New Build. Mater.* **2022**, *49*, 120–123+128. (In Chinese)

Disclaimer/Publisher's Note: The statements, opinions and data contained in all publications are solely those of the individual author(s) and contributor(s) and not of MDPI and/or the editor(s). MDPI and/or the editor(s) disclaim responsibility for any injury to people or property resulting from any ideas, methods, instructions or products referred to in the content.

engineering proceedings

MDPI

Proceeding Paper

Degassing of Medical Powder Plastics in Fused Deposition 3D Printing †

Tzu-Han Chao and Chuan-Chieh Liao *

Department of Mechanical Engineering, Chung Yuan Christian University, Taoyuan 32023, Taiwan
* Correspondence: chuanchiehliao@cycu.edu.tw

† Presented at the 3rd IEEE International Conference on Electronic Communications, Internet of Things and Big Data Conference 2023, Taichung, Taiwan, 14–16 April 2023.

Abstract: Compared with traditional manufacturing methods, 3D printing is designed according to the needs of patients at a lower cost. Therefore, it has recently been developed vigorously in the application of medical equipment. The printing of biomedical materials with powder plastic (PEG-PCL) in the form of fused deposition modeling is one of the most recent key development projects. In this method, the plastic was heated and melted at the nozzle of the bucket and then printed. However, due to the existence of voids between the particles, the powder was melted, and the gas was coated in the melt. Due to the high viscosity of the melt, the air bubbles could not escape freely depending on the density difference, which led to a discontinuous plastic output during the printing process and, in turn, affected the appearance and material strength of the printed product. In this study, the Volume of Fluid method was used to simulate the rising process of bubbles in the molten liquid. By studying the internal flow of the liquid, the influence of the viscosity of different fluids on the rising of bubbles was discussed. In addition, the reliability of the simulation results was obtained through experimental verification. In the future, the rotating conical agitator should be used to generate forced convection inside the liquid to accelerate a rise in bubbles, and the influence of the conical agitator on the rise of bubbles at various speeds needs to be further studied. Finally, a suitable speed range was found for the most effective degassing effect.

Keywords: biomedical material; 3D printing; fused deposition modeling (FDM); degassing

Citation: Chao, T.-H.; Liao, C.-C. Degassing of Medical Powder Plastics in Fused Deposition 3D Printing. *Eng. Proc.* **2023**, *38*, 69. https://doi.org/10.3390/engproc2023038069

Academic Editors: Teen-Hang Meen, Hsin-Hung Lin and Cheng-Fu Yang

Published: 30 June 2023

Copyright: © 2023 by the authors. Licensee MDPI, Basel, Switzerland. This article is an open access article distributed under the terms and conditions of the Creative Commons Attribution (CC BY) license (https://creativecommons.org/licenses/by/4.0/).

1. Introduction

The development of 3D printing technology was made to cope with the advantages of complex structures and rapid manufacturing so that fields of application could continue to increase. PEG-PCL is a biocompatible plastic [1]. It is manufactured by 3D printing and applied to medical equipment to design and print what is needed for patient treatment. Because it needs to be implanted into human tissue, the material requires good biocompatibility. This material is used to build a biocomposite scaffold for bone cell tissue to treat cartilage tissue with poor self-repairing ability. Due to the characteristics of the material, it cannot be pressed into a thin wire shape that is convenient for feeding; therefore, it needs to be heated and melted before printing.

There are several types of 3D printing, and the appropriate printing method can be selected according to different needs. Jihua et al. [2] listed several common types, such as fused deposition modeling, powder bed inkjet printing, stereolithography, and selective laser sintering. Fused deposition modeling sends the polymer plastic into the nozzle, heats it into a semi-liquid state, extrudes it to the top of the platform to form layer by layer, and then cools and solidifies. Powder bed inkjet printing has a liquid binder on the powder through a nozzle to form a pattern. The printing methods of stereolithography and selective laser sintering both polymerize or melt powder materials in the form of a laser. This report cites the existing fused deposition modeling printers for biomedical materials. Plastic

PEG-PCL is available in powder form. During the preheating process, the air is surrounded by particles during the melting process, causing bubbles to form inside the barrel. Due to the high viscosity of the material, the density difference does not help air bubbles escape from the molten liquid, as shown in Figure 1. These air bubbles affect the printing process and the quality of the molded product. Therefore, a suitable degassing operation needs to be selected to improve the print quality.

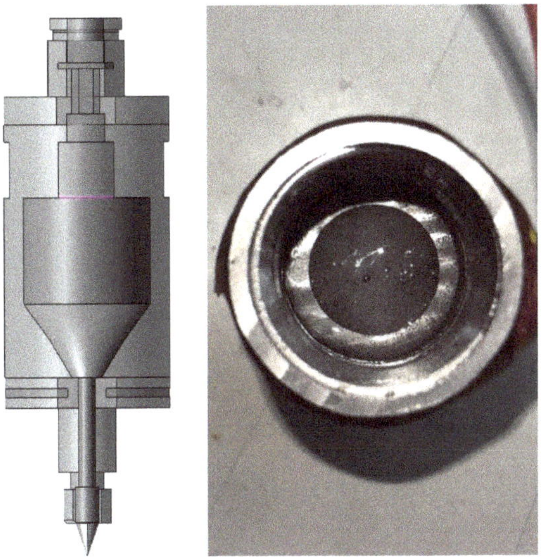

Figure 1. The 3D printer structure and bubble problems.

There are many examples of current 3D printing practices that are used in medical equipment, but there is little research on the problem of bubbles in the melting of medical plastics in the barrel despite the fact that these bubbles can easily affect the quality of the finished product. There are many degassing problems of high-viscosity fluids in the industrial field, and these degassing methods include a chemical clarifying agent, ultrasonic degassing, low-frequency vibrations, purging gas, blade stirring, and rotary stirring [3–6].

However, these bubbles come from the gaps between plastic particles, which are encapsulated by a high viscous force during the melting process. They do not need to rely on the cavitation pressure difference generated by the purge gas or ultrasonic vibration to release the dissolved gas molecules. Powder plastics are a form of medical equipment; therefore, it is not advisable to use excessively high temperatures to increase the circulating convection or add clarifying agents to it.

In terms of numerical models, Xu et al. [7] used axisymmetric simulations to simulate the bubbles breaking through the surface of the molten metal, which coincided with the experimental results. Moreover, Xu et al. [8] used VOF to simulate the bubbles in water and verified its feasibility from the average velocity and experiments.

Maniruzzaman et al. [9] used axisymmetric geometry to simulate the flow patterns, bubble distribution, inclusion trajectories, and turbulent flow structures in a ladle. Kuncewicz et al. [10] used a 3D/2D mixing model to simulate the operation of the blade in laminar flow. Based on dividing the entire mixer space into two regions, the velocity distribution was determined based on the model, which was in good agreement with the experiment. Therefore, it is feasible to use a 2D model to simulate the rotating flow field.

Researchers have carried out degassing research on air bubbles in 3D printing material barrels. However, there are many ways to apply degassing in the industry. We used a cone stirrer to form forced convection and accelerate the rise in air bubbles while reducing air bubbles during printing. In order to reduce this computational complexity, a two-dimensional axisymmetric model was used to establish a two-phase mathematical model and simulate the movement behavior of bubbles in the degassing unit.

2. Geometric Description

The powdered plastic was placed inside the barrel, preheated, and melted through the heater on the wall surface. Then, the heat source was transferred inward from the wall surface, which coated the air that had not been diffused in the particle. In the axisymmetric simulation, a two-dimensional geometric model is shown in Figure 2. The height of the bucket was 0.04 m, the symmetrical width was 0.01 m, and the depth of the conical agitator was 0.01 m. The rotation radius was 0.0025, 0.005, and 0.0075 m, respectively.

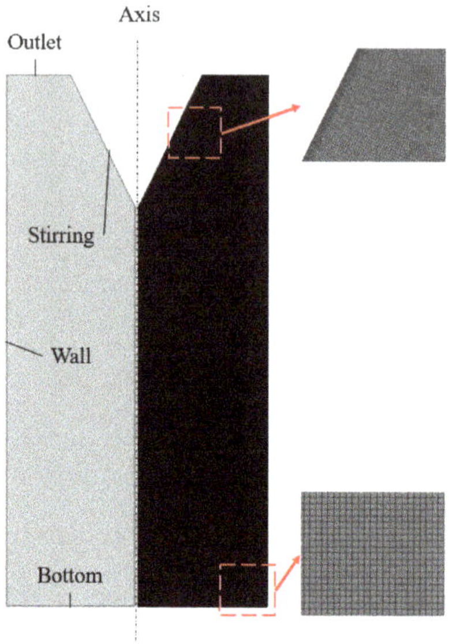

Figure 2. Simulation geometry and mesh.

3. Governing Equation

In this model, the flow behavior of the bubbles in the melt was predicted by numerically solving the governing equations. Both the gas phase and the liquid phase obeyed the continuity equation; the momentum equation is called the Navier–Stokes equation. The continuity equation is shown as follows:

$$\nabla\cdot\left(\rho\vec{u}\right)=0 \quad (1)$$

where the non-rotating momentum equations are:

$$\begin{aligned}&\frac{\partial}{\partial t}(\rho u)+\frac{1}{r}\frac{\partial}{\partial x}(r\rho u^2)+\frac{1}{r}\frac{\partial}{\partial r}(r\rho uv)\\&=-\frac{\partial p}{\partial x}+\frac{1}{r}\frac{\partial}{\partial x}[\mathrm{r}\mu(2\frac{\partial u}{\partial x})]+\frac{1}{r}\frac{\partial}{\partial r}[\mathrm{r}\mu(\frac{\partial u}{\partial r}+\frac{\partial v}{\partial x})]+\rho g_x+F_X\end{aligned} \quad (2)$$

$$\begin{aligned}&\frac{\partial}{\partial t}(\rho v)+\frac{1}{r}\frac{\partial}{\partial x}(r\rho uv)+\frac{1}{r}\frac{\partial}{\partial r}(r\rho v^2)\\&=-\frac{\partial p}{\partial x}+\frac{1}{r}\frac{\partial}{\partial r}[\mathrm{r}\mu(2\frac{\partial v}{\partial r})]+\frac{1}{r}\frac{\partial}{\partial x}[\mathrm{r}\mu(\frac{\partial u}{\partial r}+\frac{\partial v}{\partial x})]-2\mu\frac{v}{r^2}+\rho\frac{\omega^2}{r}+\rho g_r+F_r\end{aligned} \quad (3)$$

where u is the axial velocity, v is the radial velocity, x is the axial coordinate, r is the radial coordinate, p is the pressure in the fluid domain, and ω is the swirl velocity.

F represents the surface tension per unit volume between the air and water. The surface tension source term in the momentum equation could be defined as:

$$F = \sigma \frac{\rho \kappa \nabla \alpha}{0.5(\rho_l + \rho_g)} \tag{4}$$

where σ is surface tension; $\kappa = \nabla \times \hat{n}$, $\hat{n} = \frac{n}{|n|}$; and $n = \nabla \alpha_q$.

The VOF model introduced each mesh into the volume fraction of each phase. The volume fractions of all phases were then added to the units in the control volume. Combined with the laminar flow, this method had the advantages of easy implementation and high accuracy. If α corresponds to the liquid volume fraction, the phases and interface could be determined by the following definitions.

$$\begin{cases} \alpha = 0 & \text{, stands for gas phase} \\ 0 < \alpha < 1 & \text{, stands for interface} \\ \alpha = 1 & \text{, stands for liquid phase} \end{cases}$$

The transport equation of each volume fraction α_1 and α_2 in an incompressible two-fluid system is given by:

$$\frac{\partial \alpha_i}{\partial t} + \nabla \cdot (\vec{u}\ \alpha_i) = 0,\ i = 1,\ 2 \tag{5}$$

with u_i being the velocity of component i.

Based on the volume fraction, the density and viscosity in the mixing zone were calculated by:

$$\rho = \rho_l \alpha + \rho_g (1 - \alpha) \tag{6}$$

$$\mu = \mu_l \alpha + \mu_g (1 - \alpha) \tag{7}$$

4. Numerical Validation

In order to calculate this partial differential equation, a two-dimensional axisymmetric model was established. The equations were discretized by the finite volume method. The coupling term of pressure and velocity adopted the PISO algorithm in a transient simulation. In discrete governing equations, the pressure term used the PRESTO! (PREssure STaggering Option) scheme. The momentum term adopted QUICK (Quadratic Upwind Interpolation for convection Kinetics). The volume fraction term was Geo-Reconstruct.

Before the numerical simulation, we verified the correctness of selecting the VOF model for calculation. In Figure 3, Li et al. [11] studied the rising deformation of bubbles in glycerol. The bubbles in glycerol showed a change of 61.23% from the sphere to the ellipse gradually. The fluid properties in this study refer to the rising process of bubbles under 100% glycerol, in which the viscosity of 100% glycerol was 1.407 kg/ms, which is a thousand times higher than that of water.

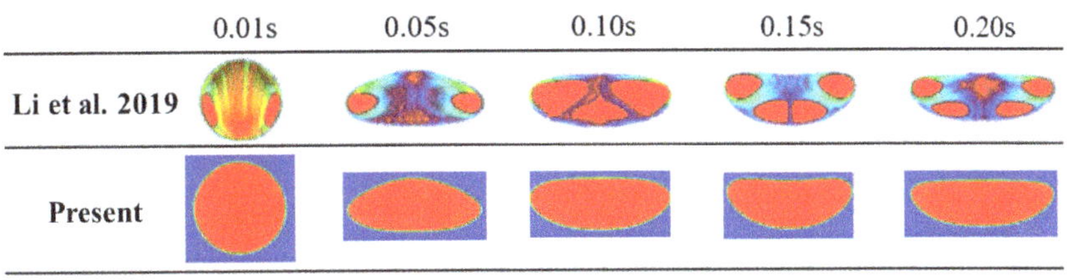

Figure 3. Deformation process of a single central circular bubble [11].

5. Result and Discussion

High viscous forces slow the fluid flow, which, in turn, causes the bubbles to rise slowly. The most direct way is to make the fluid produce forced convection to guide the bubble to accelerate upward. At first, the bubbles were not significantly affected because they were at a certain distance from the cone agitator and maintained a steady rise. As shown in Figure 4, the position map of the bubbles with and without rotation speeds are compared at different times. When the air bubbles entered the influence range of the agitator, the distance between the rotating cone and the air bubbles at no rotation speed gradually increased.

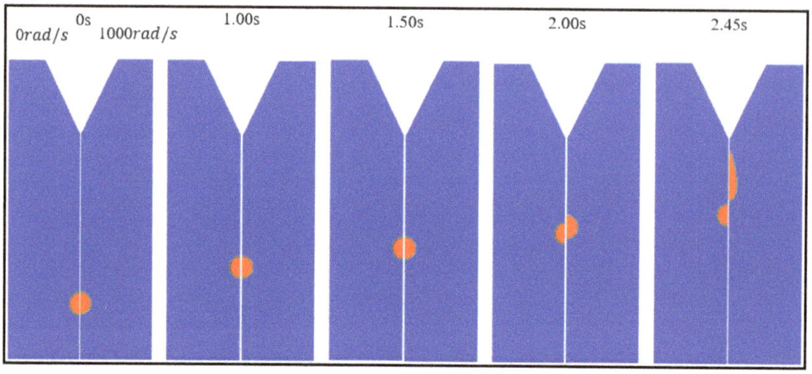

Figure 4. Comparison of bubbles at different time points at 1000 rad/s and no rotation speed.

In the viscosity fluid, the cone agitator drove the flow of the liquid. Due to the centrifugal force generated by the rotation, forced convection was formed in the fluid with the center upward and the upper edge of the bubble closer to the cone. The faster flow and viscosity of the original fluid made the bubbles gradually take on the shape of water droplets.

5.1. Rotation Speed Effect

In the case of different rotational speeds, the effect of convection on the air bubbles increased. As shown in Figure 5, the horizontal axis and the vertical axis, represented the position of the bubbles, which rose at different times. When the stirring radius of the bubbles was 5 mm, the positions of the bubbles were at different speeds. As the rotation speed increased, this effect became obvious, but the height of the bubbles rose only in the later stage. This was because the influence range of stirring was still limited by the viscosity. Therefore, when the bubbles rose to the influence range obvious effect, the distance that the bubbles moved in 2 s and 2000 rad/s increased by 7.3% compared with no rotation speed under the action of increasing the rotation speed. With the increase in the rotation effect, it increased to 13% at 2.2 s. The increase in the rotational speed had the effect of increasing the bubbles.

5.2. Rotation Radius Effect

At a fixed rotational speed, different rotation radius changes took place. As shown in Figure 6, under the rotation speed of 1000 rad/s and the influence of a different radius, the position of bubbles at each time varied. At the minimum rotation radius of 0.0025 m (2.5 mm), the convection effect caused by the rotation radius was small, and the curve observed on the ascent height map was closer to that without rotation. In the observation at 2 s, when the rotation radius of the bubble increased, the rotation radius of the bubble increased to 0.0075 m (7.5 mm), and the distance that the bubble moved was 13.4% higher than that of the bubble without rotation. Figure 7 shows that the bubbles rose under the influence of a different stirring radius at 2 s.

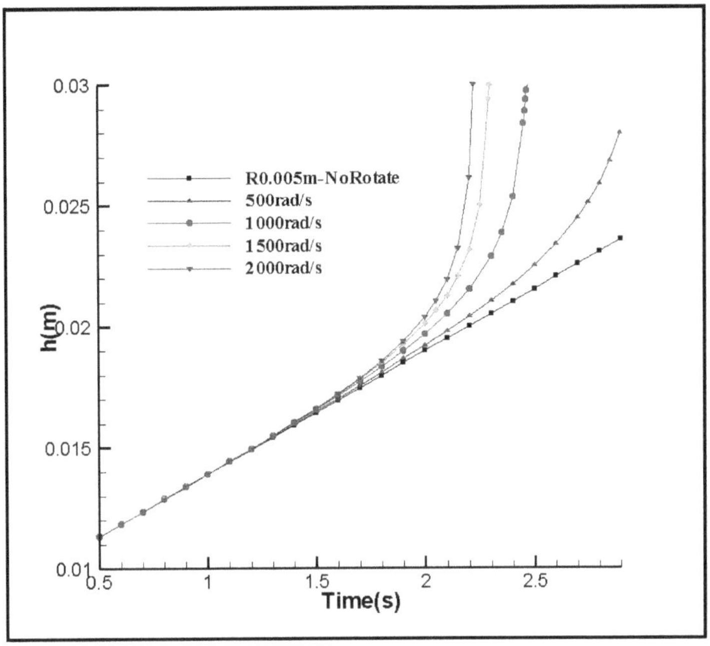

Figure 5. Height of bubbles rising at different rotational speeds.

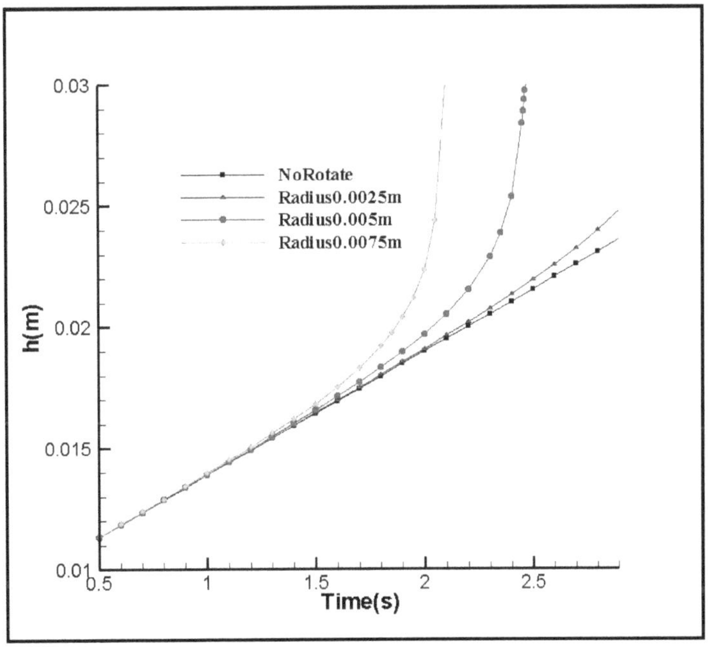

Figure 6. Effect of different radius at speed of 1000 rad/s.

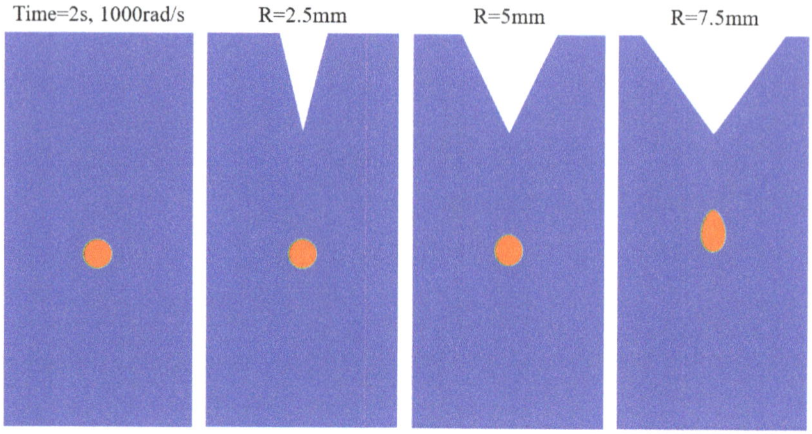

Figure 7. Effect of a different stirring radius at 2 s.

6. Conclusions

VOF can be used to simulate the flow of bubbles in the viscous fluid, and the effect of bubbles rising was observed through the influence of the agitator. The influence of the cone agitator on the bubbles was studied, including the speed of the agitator (500−2000 rad/s) and the size (5−7.5 mm). At 2s, the rotation speed increased under the fixed rotation radius (5 mm). When the bubble's rising speed increased, the moving distance increased by 13%. At a fixed rotation speed (1000 rad/s), the moving distance of the bubble with a large rotation radius increased by 13.4%, which showed the effect of the cone agitator on the bubble rise gain. Subsequently, this cone-stirring model was applied to a higher-viscosity fluid. In addition to observing the effect on the bubbles, the rotation radius and depth of the cone were changed at the same time, and the volume of the cone was controlled at the minimum proportion to obtain the maximum effect.

Author Contributions: T.-H.C.: Validation, Investigation, Data curation, Writing—original draft preparation, Visualization; C.-C.L.: Conceptualization, Methodology, Formal analysis, Investigation, Resources, Writing—original draft preparation, Writing—review and editing, Supervision, Project administration, Funding acquisition. All authors have read and agreed to the published version of the manuscript.

Funding: This research was funded by the Ministry of Science and Technology (MOST 110-2221-E-033-016, MOST 111-2221-E-033-031).

Institutional Review Board Statement: Not applicable.

Informed Consent Statement: Not applicable.

Data Availability Statement: Not applicable.

Acknowledgments: This research was funded by the Ministry of Science and Technology, Taiwan, ROC via Grant Nos. MOST 110-2221-E-033-016 and MOST 111–2221-E-033–031.

Conflicts of Interest: The authors declare no conflict of interest.

References

1. Hsieh, Y.H.; Shen, B.Y.; Wang, Y.H.; Lin, B.; Lee, H.M.; Hsieh, M.F. Healing of osteochondral defects implanted with biomimetic scaffolds of poly (ε-caprolactone)/hydroxyapatite and glycidyl-methacrylate-modified hyaluronic acid in a minipig. *Int. J. Mol. Sci.* **2018**, *19*, 1125. [CrossRef] [PubMed]
2. Wang, X.; Jiang, M.; Zhou, Z.W.; Gou, J.H.; Hui, D. 3D printing of polymer matrix composites: A review and prospective. *Compos. Part B Eng.* **2017**, *110*, 442–458. [CrossRef]
3. Zhan, X.; He, Y.; Shen, B.; Sun, Z.; Shi, T.; Li, X. Removal of gas bubbles from highly viscous non-Newtonian fluids using controlled vibration. *Chem. Eng. Sci.* **2018**, *185*, 76–83. [CrossRef]

4. Liu, X.; Zhang, Z.; Hu, W.; Le, Q.; Bao, L.; Cui, J.; Jiang, J. Study on hydrogen removal of AZ91 alloys using ultrasonic argon degassing process. *Ultrason. Sonochem.* **2015**, *26*, 73–80. [CrossRef] [PubMed]
5. Gómez, E.R.; Zenit, R.; Rivera, C.G.; Trápaga, G.; Ramírez-Argáez, M.A. Mathematical Modeling of Fluid Flow in a Water Physical Model of an Aluminum Degassing Ladle Equipped with an Impeller-Injector. *Met. Mater. Trans. B* **2012**, *44*, 423–435. [CrossRef]
6. Ma, Q.; Li, C.; Zhang, G.; Fang, H. A novel method for accelerating bubble rising in highly viscous molten quartz. *Int. J. Heat Mass Transf.* **2019**, *138*, 1359–1367. [CrossRef]
7. Xu, Y.; Ersson, M.; Jönsson, P. Numerical Simulation of Single Argon Bubble Rising in Molten Metal Under a Laminar Flow. *Steel Res. Int.* **2015**, *86*, 1289–1297. [CrossRef]
8. Xu, Y.; Ersson, M.; Jönsson, P.G. A Numerical Study about the Influence of a Bubble Wake Flow on the Removal of Inclusions. *ISIJ Int.* **2016**, *56*, 1982–1988. [CrossRef]
9. Maniruzzaman, M.; Makhlouf, M. Mathematical modeling and computer simulation of the rotating impeller particle flotation process: Part I. Fluid flow. *Met. Mater. Trans. B* **2002**, *33*, 297–303. [CrossRef]
10. Kuncewicz, C.; Rieger, F.; Pietrzykowski, M.; Stelmach, J. 3D/2D hybrid model for ribbon impellers operating in laminar regime. *Chem. Eng. Process. Process. Intensif.* **2013**, *73*, 50–58. [CrossRef]
11. Li, X.; Zhang, P.; Li, J.; Wang, W.; Chen, G. Analysis of deformation and internal flow patterns for rising single bubbles in different liquids. *Chin. J. Chem. Eng.* **2019**, *27*, 745–758. [CrossRef]

Disclaimer/Publisher's Note: The statements, opinions and data contained in all publications are solely those of the individual author(s) and contributor(s) and not of MDPI and/or the editor(s). MDPI and/or the editor(s) disclaim responsibility for any injury to people or property resulting from any ideas, methods, instructions or products referred to in the content.

engineering proceedings

MDPI

Proceeding Paper

The Fabrication of Porous ZnO Nanorods through Two-Step Aqueous Synthesis, and Their Properties †

Chih-Feng Yen *, Hung-Chang Hsu, Chung-Hung Lin, Yu-De Lin and Shih-Fong Chao *

Department of Microelectronics Engineering, National Kaohsiung University of Science and Technology, No. 142, Haizhuan Rd., Nanzi Dist., Kaohsiung City 811213, Taiwan; f110187113@nkust.edu.tw (H.-C.H.); f110187118@nkust.edu.tw (C.-H.L.); f110187105@nkust.edu.tw (Y.-D.L.)

* Correspondence: cfyen@nkust.edu.tw (C.-F.Y.); sfchao@nkust.edu.tw (S.-F.C.); Tel.: +886-7-3617141 (ext. 23374) (C.-F.Y.); +886-7-3617141 (ext. 23365) (S.-F.C.)

† Presented at the 3rd IEEE International Conference on Electronic Communications, Internet of Things and Big Data Conference 2023, Taichung, Taiwan, 14–16 April 2023.

Abstract: In this study, ZnO nanorods with a porous structure were successfully prepared using the chemical bath deposition (CBD) method. According to the surface and cross-sectional images photographed by scanning electron microscopy, we found that the diameters of samples with concentrations of 10, 30, 50, and 70 mM were 90, 141, 214, and 259 nm, respectively, with 6 h of deposition. The height of the nanorods was maintained at approximately 1.3 μm. The EDS material analysis showed that the ratio of zinc atoms and oxygen atoms in samples with different concentrations changed with the concentration of the growth solution. The column diameter of the zinc oxide nanorods prepared by the chemical bath deposition method was closely related to the concentration of the growth solution. The higher the concentration, the larger the column diameter. The height of the nanorods was directly proportional to the deposition time and was not influenced by solution concentration.

Keywords: CBD; ZnO nanorods; porous; zinc oxide

Citation: Yen, C.-F.; Hsu, H.-C.; Lin, C.-H.; Lin, Y.-D.; Chao, S.-F. The Fabrication of Porous ZnO Nanorods through Two-Step Aqueous Synthesis, and Their Properties. *Eng. Proc.* **2023**, *38*, 79. https://doi.org/10.3390/engproc2023038079

Academic Editors: Teen-Hang Meen, Hsin-Hung Lin and Cheng-Fu Yang

Published: 10 July 2023

Copyright: © 2023 by the authors. Licensee MDPI, Basel, Switzerland. This article is an open access article distributed under the terms and conditions of the Creative Commons Attribution (CC BY) license (https://creativecommons.org/licenses/by/4.0/).

1. Introduction

With the development of technology and the evolution of the semiconductor industry, the requirements for the integration of integrated circuits are higher, and the evolution of materials has moved from the micron level to the nanometer level. However, zinc oxide (ZnO) has attracted much attention and is one of the most widely used metal materials. Zinc oxide has three types of crystallite structures, rock salt structure, sphalerite, and hexagonal wurtzite [1]. The most stable structure of zinc oxide is the hexagonal wurtzite structure; this has a forbidden band width of 3.37 eV, and a large exciton binding energy of 60 MeV [2,3]. In the semiconductor industry, due to its high oxidizing power, stability, and low toxicity, zinc oxide is widely used. At present, the common preparation methods of zinc oxide include metal–organic chemical vapor deposition (MOCVD) [4], molecular beam epitaxy (MBE) [5], pulsed excimer vapor deposition (PVD) [6], vapor-phase epitaxy (VPE) [7], the physical vapor transport method (VPT) [8], and the gas–liquid–solid-phase method (VLS) [9].

Zinc oxide is a material that has been widely used to make inexpensive, non-toxic, and high-performance photocatalysts, to degrade a wide variety of organic chemicals and organic dyes [10]. It can be used in ceramics, gas sensing, light sensing, photocatalyst, and other applications. Nanorods (NRs) are one-dimensional structures of nanomaterials, and zinc oxide nanorods are more outstanding in the above-mentioned aspects.

The seed layers of zinc oxide were successfully prepared using the sol–gel method, and the ZnO nanorods were built by the non-vacuum chemical bath deposition method to prepare uniformly distributed and porous ZnO nanorods. These pores can expand the specific surface area, being very useful for photocatalytic reactions and gas-sensing detection [11].

2. Experimental Section

P-Si(100) was chosen as the substrate in this study. First, the silicon substrate was cleaned with DI water, acetone, and ethanol by ultrasonic shock for 10 min in sequence, and then shock-washed with DI water again to remove the surface of the substrate residual ethanol. Finally, blow drying with nitrogen was performed to remove moisture. The experiment consisted of two parts: (a) the seed layer of zinc oxide was prepared on a silicon substrate by the sol–gel method and spin coater to form a seed layer; (b) ZnO NRs were deposited on the seed layer by non-vacuum chemical bath deposition.

Subsequently, 0.01 mol of $Zn(CH_3COO)_2{\cdot}2H_2O$ and appropriate amounts of $CH_3OCH_2CH_2OH$ were fully stirred at 60 °C for 20 min; then, $NH_2CH_2CH_2OH$ was slowly added as a stabilizer for the reaction, and after stirring at the same temperature for 2 h, the solution was aged at 25 °C for 24 h. The preparation steps of the seed layer growth solution are shown in Figure 1. A seed layer was prepared using a spin coater; spin-coating was performed at 3000 rpm for 30 s; and then soft-baking was performed. The process was repeated 3 times to ensure that the seed layer was successfully attached to the silicon substrate, and finally placed in a high-temperature quartz furnace tube for annealing at 500 °C for 2 h.

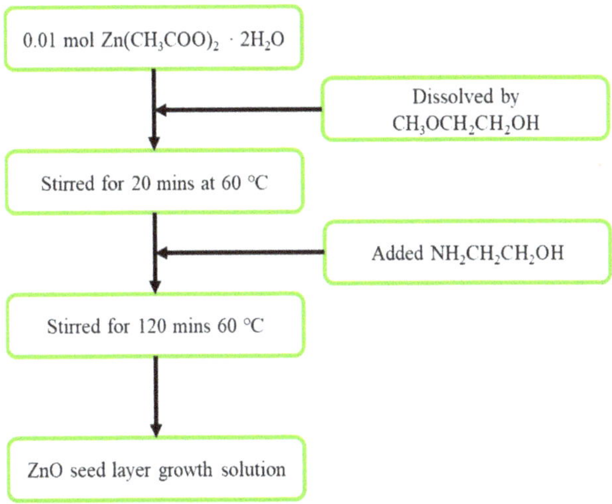

Figure 1. Seed layer growth solution preparation.

The zinc oxide nanorod growth solution used in the liquid deposition method was prepared by mixing $Zn(NO_3)_2 \cdot 6H_2O$ and $C_6H_{12}N_4$ (HMT). First, an appropriate amount of $Zn(NO_3)_2 \cdot 6H_2O$ was stirred with deionized water for 20 min. HMT was stirred with deionized water at the same time. Then, the mixed growth solution was placed in a constant-temperature water bath for 20 min at 90 °C to preheat; then, the silicon substrate with the seed layer was completely deposited in the growth solution for 6 h. After deposition, the samples were taken out and cleaned with deionized water, and the chemical reaction was terminated. After drying with N_2, the deposited zinc oxide nanorods were annealed at 500 °C for 2 h in a high-temperature quartz furnace tube with air. Finally, the annealed samples were subjected to surface analysis and material composition detection. The preparation steps of the NR layer growth solution are shown in Figure 2.

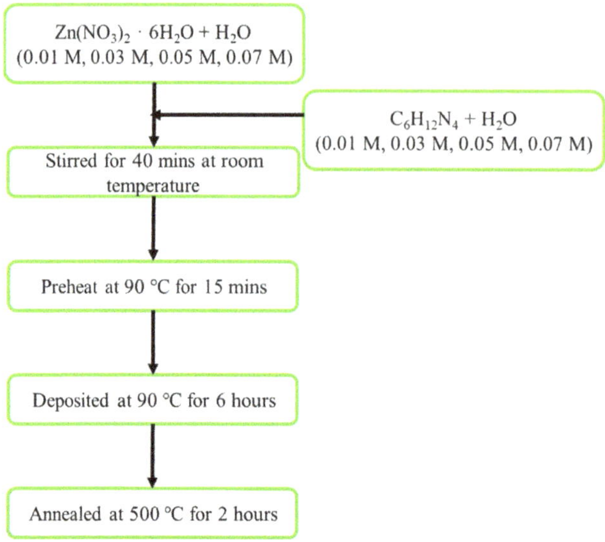

Figure 2. Nanorod growth solution preparation.

When the nanorods were initially deposited, HMT began to decompose into ammonia and produced hydroxide ions or OH^-; the detailed reaction is shown in Formula (1) and (2) [12]:

$$(CH)_6N_4 + 6H_2O \rightarrow 6HCHO + NH_3 \quad (1)$$

$$H_2O + NH_3 \rightleftharpoons OH^- + NH^{4+} \quad (2)$$

The reactive formation of the ZnO core is due to the reaction of Zn^{2+} cations with NH_3 and OH^- anions; the detailed reaction is depicted in Formula (3) [13]:

$$Zn^{2+} + 2OH^- \rightarrow Zn(OH)_2 \text{ or } Zn(NH_3)_4{}^{2+} + 2OH^- \rightarrow Zn(OH)_2(NH_3)_4 \quad (3)$$

Under the influence of specific temperatures and OH ions, the crystal core degrades into ZnO core nanorods. Over time, the ZnO core grows and finally forms nanorods, as detailed by the reaction in Formula (4) [13]:

$$Zn(OH)_2 \rightarrow ZnO + H_2O \text{ or } Zn(OH)_2(NH_3)_4 \rightarrow ZnO + H_2O + 4NH_3 \quad (4)$$

3. Results and Discussion

In this study, the porous nanorods of zinc oxide were successfully deposited on the silicon substrate using the chemical bath deposition method. The material composition was confirmed by EDS. Whether the concentration of the growth solution was related to the change in the surface appearance of the NRs was confirmed by cross-sectional SEM images.

According to the EDS composition analysis results, it can be determined that the nanorods in those samples were composed of Zn atoms and O atoms. The proportion of O atoms in the ZnO nanorod sample accounted for 44.77%, and the proportion of Zn atoms accounted for 55.23%, as shown in Figure 3a. The proportion of O atoms in the ZnO nanorod sample accounted for 47.15%, and the proportion of Zn atoms accounted for 52.85%, as shown in Figure 3b. According to the above results, the ratio of oxygen atoms to zinc atoms changed with the concentration of the growth solution.

Figure 4a–h show the SEM images of ZnO NRs under different concentrations of growth solution. The reason why the liquid-phase deposition method can grow ZnO nanorods is due to the higher surface energy of zinc oxide along the [0001] direction, which will have a higher growth rate. Figure 4a–d show high-magnification images (magnification

20 K) of samples with concentrations of 10, 30, 50, and 70 mM, respectively; Figure 4e–h are the top views of samples with corresponding concentrations. The image in Figure 4 shows that all grown ZnO nanorods were also in the shape of hexagonal columns, and most of the ZnO nanorods grew vertically upward. The column diameters of the nanorods of zinc oxide were 90, 141, 214, and 259 nm, respectively. This shows that the column diameter of the grown nanorods of zinc oxide increased when the concentration of the growth solution increased. The higher the concentration of the growth solution, the faster the growth rate in the [0001] direction. Many small pores can be observed from the surface of the sample, which is likely to be caused by the high-temperature gasification of acetate ions and intermediate reactants in the sample during the annealing process. The specific surface area of the meter column is beneficial for photocatalysis.

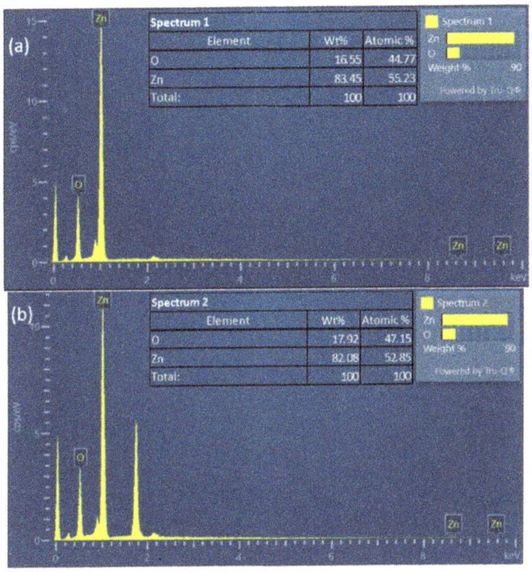

Figure 3. EDAX spectrum of ZnO NRs: (**a**) 10 mM and (**b**) 50 mM.

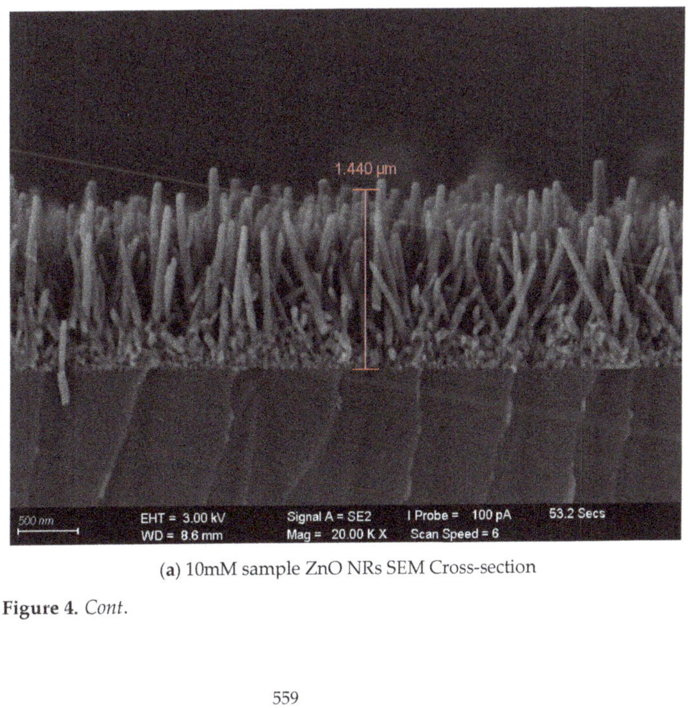

(**a**) 10mM sample ZnO NRs SEM Cross-section

Figure 4. *Cont.*

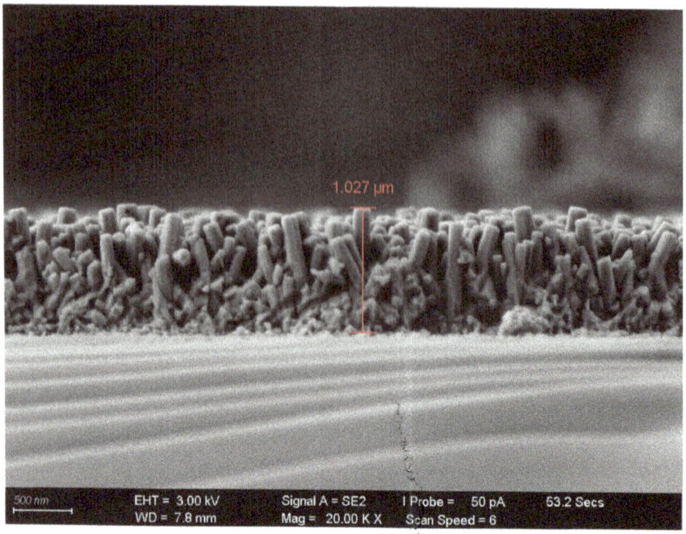

(**b**) 30mM sample ZnO NRs SEM Cross-section

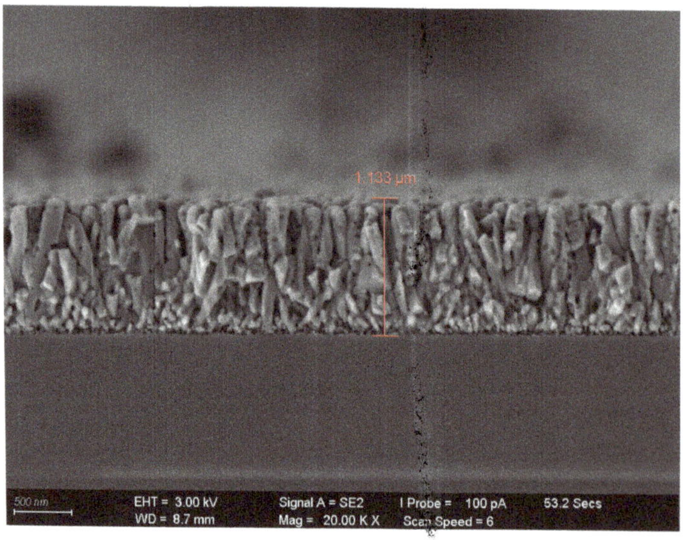

(**c**) 50mM sample ZnO NRs SEM Cross-section

Figure 4. *Cont.*

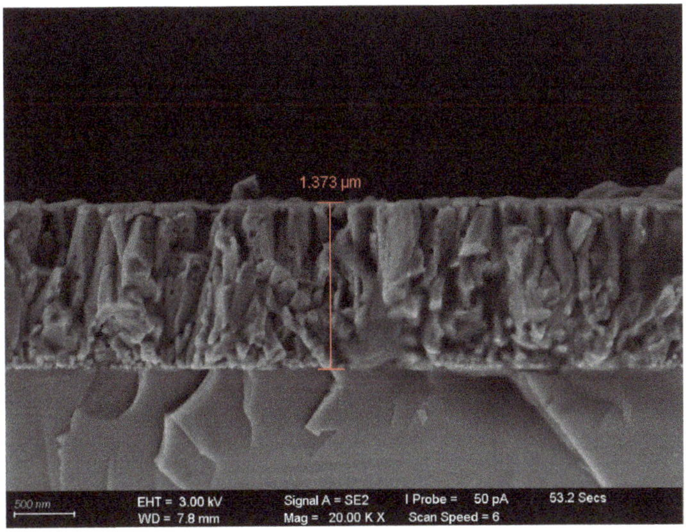

(**d**) 70mM sample ZnO NRs SEM Cross-section

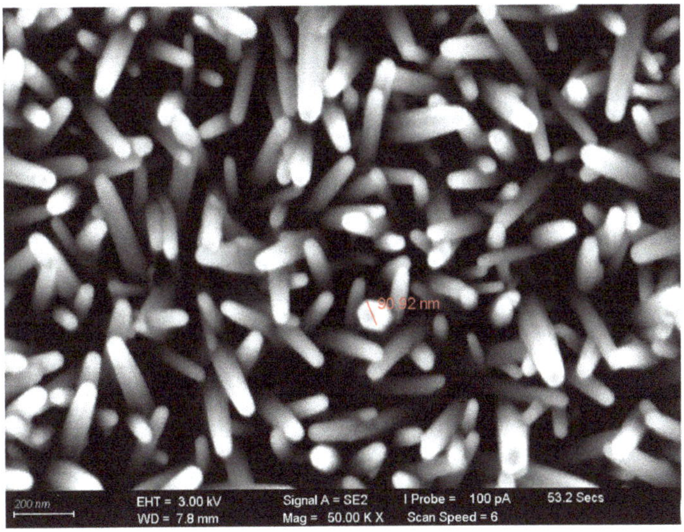

(**e**) 10mM sample ZnO NRs SEM Top View

Figure 4. *Cont.*

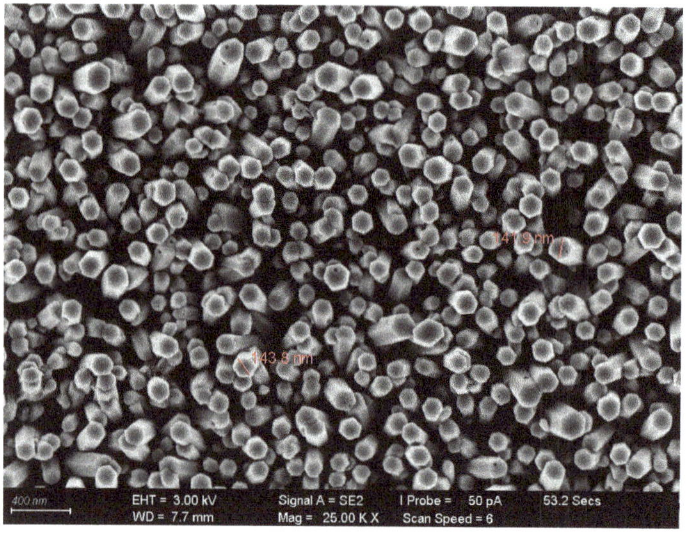

(**f**) 30mM sample ZnO NRs SEM Top View

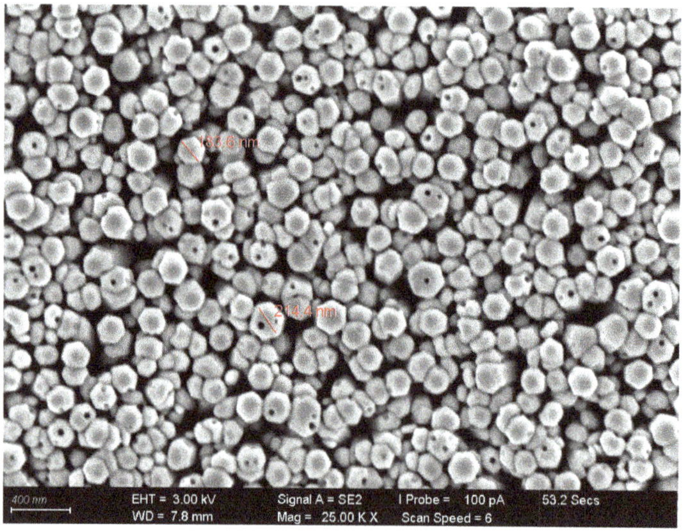

(**g**) 50mM sample ZnO NRs SEM Top View

Figure 4. *Cont.*

(**h**) 70mM sample ZnO NRs SEM Top View

Figure 4. FE-SEM images of ZnO NRs with different concentrations of growth solution: (**a**,**e**) 10 mM; (**b**) and (**f**) 30 mM; (**c**) and (**g**) 50 mM; (**d**) and (**h**) 70 mM.

4. Conclusions

In this study, the chemical bath method and subsequent annealing process were used to successfully prepare porous zinc oxide nanorod arrays on the P-Si(100) substrate. According to the SEM images, it was found that the samples of 10 mM, 30 mM, 50 mM, and 70 mM were deposited over 6 hours; the column diameters were 90 nm, 141 nm, 214 nm, and 259 nm, respectively. The height of the nanorods was maintained at about 1.3 μm. This indicates that the growth solution of zinc oxide nanorods prepared by the chemical bath deposition method will affect the diameter of ZnO NRs, and the height of ZnO NRs is related to the deposition time rather than the concentration of the growth solution. Additionally, according to the EDS material analysis, it was found that nanorods grown from growth solutions with different concentrations were composed of different ratios of zinc atoms and oxygen atoms. The pores created by annealing greatly increased the specific surface area of the nanorods, making them advantageous for future applications in photocatalysis.

Author Contributions: Conceptualization, C.-F.Y. and H.-C.H.; methodology, C.-F.Y.; software, H.-C.H.; validation, C.-F.Y., S.-F.C. and H.-C.H.; formal analysis, H.-C.H.; investigation, H.-C.H.; resources, H.-C.H., C.-H.L. and Y.-D.L.; data curation, H.-C.H.; writing—original draft preparation, H.-C.H.; writing—review and editing, C.-F.Y.; visualization, C.-F.Y.; supervision, C.-F.Y. and S.-F.C.; project administration, C.-F.Y. and S.-F.C.; funding acquisition, C.-F.Y. and S.-F.C. All authors have read and agreed to the published version of the manuscript.

Funding: This research received no external funding.

Institutional Review Board Statement: Not applicable.

Informed Consent Statement: Not applicable.

Data Availability Statement: Not applicable.

Conflicts of Interest: The authors declare no conflict of interest.

References

1. Chinnasamy, M.; Balasubramanian, K. Enhanced UV photodetection behavior of Cr doped wurtzite ZnO crystalline nanorods. *Opt. Mater.* **2020**, *110*, 110492. [CrossRef]
2. Liu, B.; Zeng, H.C. Hydrothermal Synthesis of ZnO Nanorods in the Diameter Regime of 50 nm. *J. Am. Chem. Soc.* **2003**, *125*, 4430–4431. [CrossRef] [PubMed]
3. Jiang, Z.-Y.; Xu, T.; Xie, Z.-X.; Lin, Z.-W.; Zhou, X.; Xu, X.; Huang, R.-B.; Zheng, L.-S. Molten Salt Route toward the Growth of ZnO Nanowires in Unusual Growth Directions. *J. Phys. Chem. B* **2005**, *109*, 23269–23273. [CrossRef] [PubMed]
4. Wu, B.; Zhang, Y.; Shi, Z.; Li, X.; Cui, X.; Zhuang, S.; Zhang, B.; Du, G. Different defect levels configurations between double layers of nanorods and film in ZnO grown on c-Al2O3 by MOCVD. *J. Lumin.* **2014**, *154*, 587–592. [CrossRef]
5. Tien, L.; Norton, D.; Pearton, S.; Wang, H.; Ren, F. Nucleation control for ZnO nanorods grown by catalyst-driven mo-lecular beam epitaxy. *Appl. Surf. Sci.* **2007**, *253*, 4620–4625. [CrossRef]
6. Kek, R.; Ong, G.; Yap, S.; Lim, L.; Koh, S.; Nee, C.; Tou, T.; Yap, S. Growth of Al-doped ZnO nanostructures in low pres-sure background gas by pulsed laser deposition. *Mater. Sci. Semicond. Process.* **2022**, *145*, 106636. [CrossRef]
7. Maejima, K.; Ueda, M.; Fujita, S.; Fujita, S. Growth of ZnO Nanorods on A-Plane (1120) Sapphire by Metal-Organic Vapor Phase Epitaxy. *Jpn. J. Appl. Phys.* **2003**, *42*, 2600–2604. [CrossRef]
8. Gray, C.; Trefflich, L.; Röder, R.; Ronning, C.; Henry, M.O.; McGlynn, E. Growth of 18 O isotopically enriched ZnO nanorods by two novel VPT methods. *J. Cryst. Growth* **2017**, *460*, 85–93. [CrossRef]
9. Suh, D.-I.; Byeon, C.C.; Lee, C.-L. Synthesis and optical characterization of vertically grown ZnO nanowires in high crystallinity through vapor–liquid–solid growth mechanism. *Appl. Surf. Sci.* **2010**, *257*, 1454–1456. [CrossRef]
10. Irani, M.; Mohammadi, T.; Mohebbi, S. Photocatalytic Degradation of Methylene Blue with ZnO Nanoparticles; a Joint Experimental and Theoretical Study. *J. Mex. Chem. Soc.* **2016**, *60*, 218–225.
11. Duan, X.; Li, C.; Fu, L.; Wu, Y.; Chen, G.; Xu, K.; Shao, L.; Yang, P.; Yu, Z.; Ding, P.; et al. Structural and optical properties of porous ZnO nanorods synthesized by a simple two-step method. *Superlattices Microstruct.* **2019**, *128*, 30–36. [CrossRef]
12. Kumar, V.; Gupta, R.; Bansal, A. Hydrothermal Growth of ZnO Nanorods for Use in Dye-Sensitized Solar Cells. *ACS Appl. Nano Mater.* **2021**, *4*, 6212–6222. [CrossRef]
13. Ichikawa, T.; Shiratori, S. Fabrication and Evaluation of ZnO Nanorods by Liquid-Phase Deposition. *Inorg. Chem.* **2010**, *50*, 999–1004. [CrossRef] [PubMed]

Disclaimer/Publisher's Note: The statements, opinions and data contained in all publications are solely those of the individual author(s) and contributor(s) and not of MDPI and/or the editor(s). MDPI and/or the editor(s) disclaim responsibility for any injury to people or property resulting from any ideas, methods, instructions or products referred to in the content.

engineering proceedings

MDPI

Proceeding Paper

Big Data Analysis on the Characteristics of Drought Disaster Changes in Hebei Province in the Past 30 Years †

Zhihui Tian * and Shuo Wang

Department of Landscape Design & Forestry, Beijing University of Agricultural, Beijing 102206, China; shawnwangbj@163.com
* Correspondence: tzhui@126.com

† Presented at the 3rd IEEE International Conference on Electronic Communications, Internet of Things and Big Data Conference 2023, Taichung, Taiwan, 14–16 April 2023.

Abstract: Based on the monthly precipitation data of 139 meteorological stations from 1981 to 2015, the trend line method was used to analyze the variation of the drought area and sown area from 1981 to 2015 and the characteristics of drought in Hebei Province. The results showed the following. (1) the frequency of drought was the highest in winter and the lowest in summer, which was 43.2 and 26.2%, respectively. (2) The ratio of drought stations was highest in winter and lowest in summer. The frequency of global drought was 37 and 17% in summer and winter, respectively. Except for the increasing trend of drought in summer, all other seasons showed a decreasing trend which was significant in autumn ($\alpha = 0.05$). (3) The drought intensity of the four seasons was all less than 1. The drought intensity in spring and summer showed an upward trend, while that in autumn and winter showed a downward trend. The drought intensity in autumn showed a significant downward trend ($\alpha = 0.05$). (4) Drought disaster rate was positively correlated with summer drought intensity and drought ratio.

Keywords: drought; percentage of precipitation anomaly; the ratio of drought stations; drought intensity; Hebei Province

Citation: Tian, Z.; Wang, S. Big Data Analysis on the Characteristics of Drought Disaster Changes in Hebei Province in the Past 30 Years. *Eng. Proc.* **2023**, *38*, 52. https://doi.org/10.3390/engproc2023038052

Academic Editors: Teen-Hang Meen, Hsin-Hung Lin and Cheng-Fu Yang

Published: 27 June 2023

Copyright: © 2023 by the authors. Licensee MDPI, Basel, Switzerland. This article is an open access article distributed under the terms and conditions of the Creative Commons Attribution (CC BY) license (https://creativecommons.org/licenses/by/4.0/).

1. Introduction

Drought is the main natural disaster in China with a wide range of impacts on the social economy, production, and the life of people [1]. According to statistics, from 1990 to 2016, the average annual grain loss caused by drought in China was as high as 25.2 billion kg [2]. With the change in global climate, the occurrence and development of drought disasters showed new characteristics [3]. Therefore, drought disasters had become an attractive research issue for scientists all over the world. Among them, the grasp of the spatio-temporal variation of drought disasters was particularly important for disaster prevention and reduction of losses caused by disasters. Thus, scholars have carried out extensive research in this regard and achieved important results [2,4–11]. In the process of research on drought, the following types of drought indicators were usually used. (1) Single index, mainly including precipitation anomaly percentage [12], standardized precipitation index [4], relative humidity index [5], soil relative humidity drought index [13], and the Palmer Drought Index [8]. These indexes require appropriate data that is easy to obtain and calculate. (2) Comprehensive indexes such as the Z index [9] and crop water deficit index [10] require data that is difficult to obtain as their calculation is complicated. However, they effectively reflect the actual drought and flood situation. (3) Disaster indicators include crop disaster, disaster, and harvest area, casualties, direct economic losses, and various derivative indicators reflect disaster-causing factors, disaster-causing environment, disaster effect, and disaster resistance.

In any case, it is difficult to obtain small-scale spatial and temporal data. Many scholars [10,11] have used disaster indicators to study drought and made important achievements. The reduction of precipitation was the most fundamental reason for the occurrence of drought. Among these indicators, precipitation anomalies directly reflect the drought. The precipitation data was easy to obtain as it showed temporal continuity and spatial distribution. Therefore, with the precipitation anomaly as a drought indicator, we systematically analyze the characteristics of drought in Hebei Province in the previous 30 years to provide a theoretical basis for drought prevention and mitigation.

2. Overview of Study Area

Hebei Province is one of the 13 major grain-producing areas according to the '13th Five-Year Plan' and 'Science and Technology Innovation for High Grain Yield and Efficiency'. The website of the National Bureau of Statistics (http://data.stats.gov.cn/ (accessed on 31 December 2018)) shows that the planting area of maize and wheat in Hebei Province accounts for 9 and 10% of the national area from 2000 to 2018.

Drought accounts for about 62% of meteorological disasters in Hebei Province and is the main disaster for agriculture. The average annual drought in Hebei Province from 1981 to 2015 affected an area of 1,760,000 km^2, which was 18.4% of the whole crop planting area in China. The interannual variation of the drought was shown in Figure 1. The rate in Hebei Province in the late 1980s and 1990s was relatively high. In 1999 and 1997, the drought disaster rate was 34.1 and 32.7%. After the m-k trend test, the downward trend of drought reached the significant indigenous level of 0.01. To further explore the spatial and temporal distribution of drought in Hebei Province, we take the precipitation anomaly as an indicator to discuss the frequency of drought, drought ratio, and drought intensity.

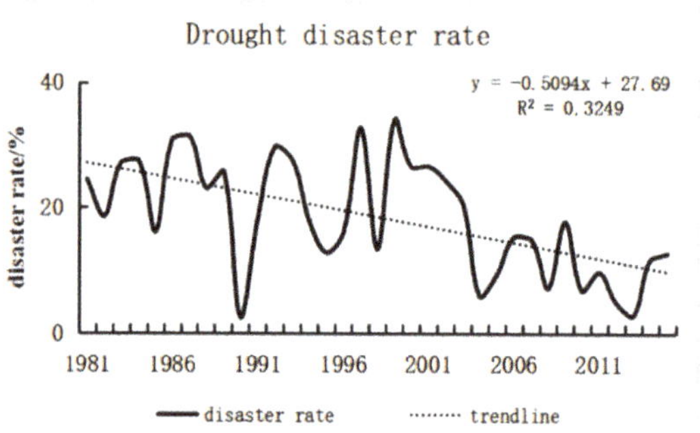

Figure 1. Interannual variation of drought disaster rate in Hebei Province.

3. Data Sources and Research Methods

3.1. Data Source

The monthly precipitation of 139 meteorological stations in Hebei Province from 1981 to 2015 was obtained from the 13th Five-Year National Key Research and Development Plan (*2017YFD0300400*). The data was based on the area of drought and crop production in Hebei Province during the same period. The area and crop production were collected from the China Civil Administration Statistical Yearbook. To eliminate the influence of the interannual variation of crop planting area on the disaster area, we used the drought index such as the crop disaster rate in Hebei Province. Crop disaster rate refers to the ratio of the affected area and drought and sowing area in the same period. The crop sowing area is obtained from the 60-year statistics of agriculture in New China (1981–1998) of the Ministry of Agriculture and Rural Affairs of the People's Republic of China and the website of the National Bureau of Statistics (http://data.stats.gov.cn (accessed on 31 December 2018)) (1999–2015). The seasons are divided as follows: spring from March to May, summer from June to August, Autumn from September to November, and Winter from December to February.

3.2. Research Method

Precipitation anomaly (*PA*) refers to the ratio of precipitation anomaly in a certain period and the average annual precipitation. In this study, the precipitation anomaly was used as an indicator to evaluate the drought of Hebei Province in four seasons each year. The evaluation criteria were collected from the national standards of the People's Republic of China (*GB/T20481-2017*), and the seasonal scale standard was adopted from it, too as follows. When $PA > -25\%$, there was no drought. When $-50\% < PA \leq -25\%$, light drought. When $-70\% < PA \leq -50\%$, moderate drought. When $-80\% < PA \leq -70\%$, severe drought. When $PA \leq -80\%$, extreme drought.

The frequency of drought in each scale (light, moderate, severe, and extreme drought) in Hebei Province from 1981 to 2015 was expressed by the interannual occurrence of drought. The influence range and drought intensity were expressed by the ratio of drought and drought intensity. The calculation methods of drought frequency, the ratio of drought, and drought intensity followed Ref. [14].

The analysis of the temporal change of drought is based on the trend line method in which the climate propensity rate [14] is used to express the changing trend and rate of drought disaster indicators by establishing a univariate linear regression equation between disaster factors and time series. The rank-based nonparametric Mann-Kendall statistical test method was used to test the significance of the changing trend of each time series. The calculation method of Ref. [14] was used in this study.

4. Results and Analysis

4.1. Spatial-Temporal Distribution Characteristics of Drought Occurrence Frequency in Hebei Province

4.1.1. Frequency of Spring Drought in Hebei Province

The average frequency of spring drought in Hebei Province was 37.2%, mainly mild and moderate drought. Table 1 shows the areas of high-frequency drought on different scales in Hebei Province. The average frequency of light drought was 21.7%. Shijiazhuang, Hengshui, and Baoding were the centers of high-frequency drought. The maximum value appeared at Pingshan, which was 40%. The average frequency of moderate drought was 11.0%. The areas with high frequency were mainly concentrated in Tangshan and Qinhuangdao. In the center of Qian' an, the frequency was 25.7%. The average frequency of severe drought was 2.9%. The areas of high frequency were concentrated in Cangzhou, Xingtai, and northeast Baoding, with a frequency of 14.3%. The frequency of spring drought was 1.7%, and the high frequency occurred in Handan and north-central Langfang.

Table 1. High-frequency regions of spring drought in Hebei Province.

Drought Level	High Frequency Spring Drought Occurrence Area/City (Station)	Drought Frequency/%
Mild drought	Baoding City (Baoding, Xushui, Mancheng, Li County), Hengshui City (Shenzhou, Anping, Wuqiang, Raoyang), Shijiazhuang City (Pingshan, Jinzhou, ZhaoCounty), Chengde City (Kuancheng, Chengde County), Cangzhou City (Wuqiao), Zhangjiakou City (Huai'an).	31.4~40
Moderate drought	Qinhuangdao City (Lulong, Qinglong, Changli, Qinhuangdao), Tangshan City (Qian'an, Luannan, Fengnan, Luanxian, Tanghai); Xingtai City (Longyao); Baoding City (Gaobeidian).	17.1~25.7
Severe drought	Baoding City (Yi County, Zhuozhou), Cangzhou City (Cangzhou), Xingtai City (Wei County, Xinhe, Lincheng), Cangzhou City (Cangzhou), Hengshui City (Jizhou), Langfang City (Dachang).	8.6~4.3
Extreme drought	Handan City (Yongnian, Wu'an, Jize, Handan, Ci County, Linzhang); Langfang City (Xianghe, Langfang, Wen'an, Yongqing); Xingtai City (Shahe, Guangzong, Ren County, Julu, Nanhe, Pingxiang); Tangshan City (Yutian).	5.7~8.6

4.1.2. Frequency of Summer Drought in Hebei Province

The average frequency of drought in summer in Hebei Province was 26.0%, and mild drought was dominant in summer. Table 2 shows the high-frequency occurrence areas of drought in summer. The average frequency of light drought was 20.3%, and the high frequency appeared in the central region of Hebei Province, including southern Tangshan, Langfang, Baoding, Shijiazhuang, and western Xingtai western Handan. Gaocheng and Yongnian showed 34.3% of the frequency. The frequency of middle drought was 5.0%, and the area of the high frequency region was mainly distributed in the southern part of Hebei Province, with the highest value in Nanpi (17.1%). The frequency of severe drought and extreme drought was low in summer. The area was mainly distributed in the south of Hebei Province, and the area of the high frequency of extreme drought was mainly distributed in the west and Langfang of Handan City and the eastern part of Cangzhou City.

Table 2. High frequency regions of summer drought in Hebei Province.

Drought Level	High Frequency Summer Drought Occurrence Area/City (Station)	Drought Frequency/%
Mild drought	Baoding City (Gaoyang, Baoding, Gaobeidian, Xiongxian, Zhuozhou); Cangzhou City (Suning, Dingzhou); Handan City (Yongnian, Jize, Qiu County); Hengshui City (Raoyang); Qinhuangdao City (Qinglong), Shijiazhuang City (Gaocheng, Yuanshi, Wuji, Luancheng), Tangshan City (Luannan, Tanghai), Xingtai City (Shahe).	28.6~34.3
Moderate drought	Baoding City (Mancheng), Cangzhou City (Nanpi, Botou, Xianxian County), Dingzhou City (Dingzhou), Handan City (Cheng'an, Linzhang), Langfang City (Gu'an, Sanhe), Shijiazhuang City (Wuji, Xingtang), Tangshan City (Yutian), Xingtai City (Neiqiu, Nangong).	11.4~17.1
Severe drought	Baoding City (Shunping, Li County), Cangzhou City (Qing County, Cangzhou, Yanshan), Handan City (Daming, Fengfeng, Quzhou), Hengshui City (Wuqiang, Wuyi, Shenzhou, Raoyang, Jizhou), Shijiazhuang City (Zanhuang, Shenze, Xinle), Xingtai City (Ren County, Longyao, Ningjin, Pingxiang, Xinhe, Qinghe).	2.9
Extreme drought	Cangzhou City (Suning, Hejian City); Hengshui City (Shenzhou, Anping, Hengshui City); Shijiazhuang City (Gaoyi, Jinzhou, Gaocheng); Xinji City (Xinji); Xingtai City (Lincheng, Baixiang, Julu).	2.9

4.1.3. Frequency of Autumn Drought in Hebei Province

The average frequency of autumn drought in Hebei Province was 383%, and the drought was mainly mild and moderate. The frequency of autumn drought is shown in Table 3. The average frequency of mild drought was 24.1%, and the area was distributed in western Handan and Xingtai. The frequency of Jinzhou was the highest, which was 42.9%. The average frequency of moderate drought was 9.5%, which was mainly concentrated in the southern part of Hebei Province, especially northeast Handan and eastern Xingtai. The highest frequency was 20.0% in Handan. The high-incidence areas of severe drought were mainly concentrated in the southeast of Hebei Province. especially Cangzhou and Handan. The highest frequency appeared in Guangping and Gu' an, showing 11.4%. The area of extreme drought was mainly concentrated in the southern part of Hebei Province, especially in Handan.

Table 3. High frequency regions of autumn drought in Hebei Province.

Drought Level	High Frequency Summer Drought Occurrence Area/City (Station)	Drought Frequency/%
Mild drought	Cangzhou City (Haixing, Renqiu), Chengde City (Weichang), Handan City (Linzhang, Ci County, Fengfeng, Feixiang, Handan, Xiangxian County, Yongnian, Jize, Qiu County), Hengshui City (Anping, Wuqiang), Shijiazhuang City (Jinzhou, Gaocheng, Zhao County), Xinji City (Xinji), Xingtai City (Shahe, Lincheng, Nangong, Ningjin, Neiqiu, Nanhe, Linxi), Zhangjiakou City (Chongli, Shangyi).	31.4~42.9
Moderate drought	Handan City (Handan, Yongnian, Quzhou, Wu'an, Daming); Hengshui City (Jizhou); Xingtai City (Wei County, Julu, Qinghe, Xinhe); Zhangjiakou City (Huai'an, Zhuolu, Zhangbei).	17.1~20.0
Severe drought	Cangzhou City (Botou, Wuqiao, Dongguang, Xianxian County), Handan City (Guangping, Guantao, Feixiang), Hengshui City (Jizhou), Langfang City (Gu'an, Yongqing), Shijiazhuang City (Shenze), Xingtai City (Lincheng).	8.6~11.4
Extreme drought	Baoding City (Tang County), Cangzhou City (Qing County), Handan City (Guantao, Daming, Wei County, Jize, Qiu County, Handan, Quzhou County), Langfang City (Dacheng), Shijiazhuang City (Luancheng, Yuanshi), Xingtai City (Linxi, Pingxiang).	5.7

4.1.4. Frequency of Winter Drought in Hebei Province

The average frequency of winter drought in Hebei Province was 43.2%, and the frequency of severe drought and extreme drought was higher. The high frequency of winter drought was shown in Table 4. The average frequency of mild drought was 14.7%, mainly concentrated in the northern part of Chengde, Pingquan and Xinglong station had the highest frequency of 31.4%. The average frequency of moderate drought was 10.9%. Tangshan and Qinhuangdao had the highest frequency, and the highest frequency was 25.7% in Qian'an. The average frequency of severe drought was 5.1%, mainly concentrated in the southern area of Tangshan and Qinhuangdao, the highest value was 14.3%, appeared in Luanxian and Luannan station. The frequency of extreme drought was much higher than that in spring, summer and autumn, with an average of 12.6%, which was mainly concentrated in the southern part of Hebei Province, and the highest value was 22.9% in the Anping, Hejian, Cangzhou. and Zhaoxian stations.

Table 4. High frequency regions of winter drought in Hebei Province.

Drought Level	High Frequency Summer Drought Occurrence Area/City (Station)	Drought Frequency/%
Mild drought	Baoding City (Anguo, Yi County), Cangzhou City (Yanshan, Huanghua County), Chengde City (Xinglong, Pingquan, Chengde County), Handan City (Feixiang, Qiu County), Langfang City (Xianghe), Zhangjiakou City (Chongli), Shijiazhuang City (Luancheng, Yuanshi, Zanhuang, Jingxing), Tangshan City (Tanghai), Xingtai City (Baixiang).	22.9~31.4
Moderate drought	Langfang City (Dacheng); Qinhuangdao City (Lulong, Funing, Changli City); Tangshan City (Qian'an, Tangshan, Yutian); Zhangjiakou City (Zhangjiakou, Zhuolu).	20.0~25.7
Severe drought	Cangzhou City (Nanpi City), Hengshui City (Zaoqiang), Qinhuangdao City (Funing, Changli City), Tangshan City (Luan County, Luannan, Fengnan, Tanghai), Zhangjiakou City (Zhangbei).	11.4~14.3
Extreme drought	Baoding City (Quyang, Xiong County), Cangzhou City (Cangzhou, Hejian County, XianXian County), Hengshui City (Anping, Hengshui, Wuyi, Shenzhou, Raoyang, Fucheng), Shijiazhuang City (Zhao County, Zhengding, Luancheng, Jinzhou, Gaoyi), Xinji City (Xinji), Xingtai City (Ningjin, Baixiang, Longyao, Neiqiu, Nanhe).	20.0~22.9

4.2. *Trend Analysis of the Ratio of Drought Stations in Hebei Province*

The ratio of drought stations could indicate the range of drought influence. The average ratio of drought stations in spring, summer, autumn. and winter in Hebei Province was 37%, 26%, 38%, and 43%, respectively. It could be found that the range of drought influence in winter was the largest, and the range of drought influence in spring and autumn was similar, reaching the regional drought degree. The variation trend of drought impact range in the four seasons was shown in Figure 2. In addition to the increase in the ratio of drought stations in summer, all other seasons showed a decreasing trend. After the m-k trend test, it could be seen that the decrease trend of the ratio of drought stations in autumn was significant ($\alpha = 0.05$), and the other seasonal change of trends were not significant.

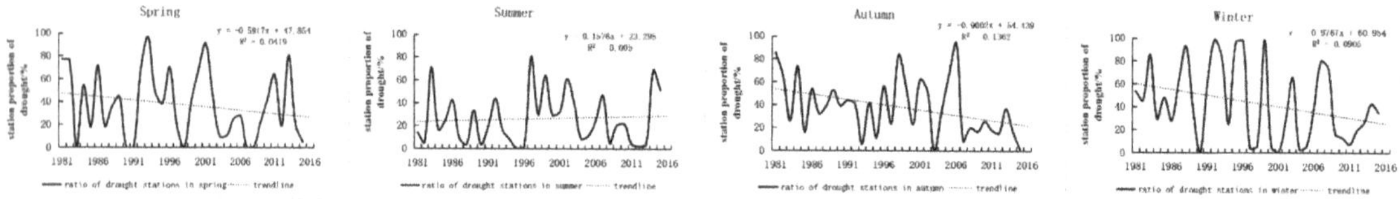

Figure 2. Interannual variation of the ratio of drought stations in hebei province over four seasons.

4.3. *Trend Analysis of Drought Intensity in Hebei Province*

The drought intensity index indicates the change in drought intensity in Hebei Province. The average drought intensity in spring, summer, autumn, and winter in Hebei Province was 0.38, 0.36, 0.42, and 0.52. The average drought intensity in winter was the highest. The drought intensity at the regional level in the four seasons was lower than 1, and the regional drought intensity was mild (Figure 3). Figure 3 shows that the spring and summer drought intensity in Hebei Province was on the rise, while autumn and winter drought intensity was on the decline. The m-k trend test result showed the decreasing trend of drought intensity in autumn at the significant level of 0.05. The trend in other seasons was not significant.

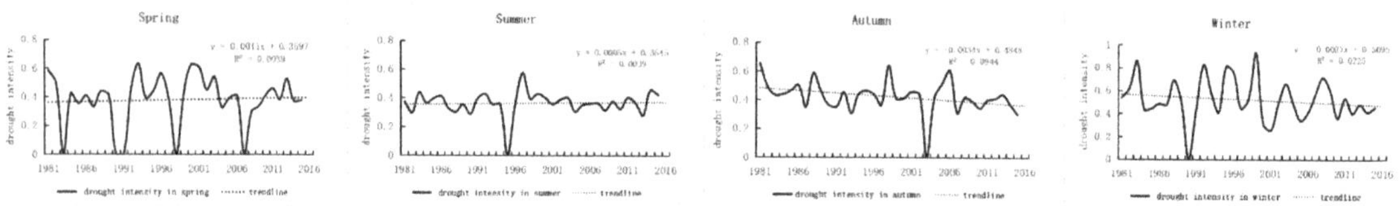

Figure 3. Interannual variation of drought intensity in four seasons in Hebei Province.

4.4. *Correlation Analysis of Drought Disaster Rate with Drought Intensity and the Ratio of Drought Stations*

The disaster indicator reflects the intensity of disaster-causing factors, and exposure, and is related to the vulnerability of disaster-affected bodies and human resilience [13]. Therefore, we used the disaster index of drought to conduct a correlation analysis with the drought intensity and the ratio of drought in season. The results are shown in Table 5. The drought disaster rate in Hebei Province was significantly positively correlated with the drought intensity and the ratio of drought in summer, but not significantly correlated in other seasons. Therefore, the summer drought intensity and the expansion of drought were the main reasons for the increase in the drought disaster rate. The ratio of drought and drought intensity in summer showed an increasing trend. Although the increasing trend did not pass the m-k trend test, the increasing trend was obvious. This helps for drought prevention and control in the future.

Table 5. Correlation analysis of drought disaster rate with drought intensity and the ratio of drought stations.

Project		Drought Intensity				the Ratio of Drought Stations			
		Spring	Summer	Autumn	Winter	Spring	Summer	Autumn	Winter
disaster rate of drought	Pearson relevance	0.247	0.398 *	0.026	0.325	0.326	0.496 **	0.188	0.32
	Significance (bilateral)	0.153	0.018	0.883	0.057	0.056	0.002	0.279	0.061

Note: **. Significant correlation at the 0.01 level (bilateral); *. Significant correlation at the 0.05 level (bilateral). Discussion.

We analyzed the variation characteristics of drought disaster rate, the ratio of drought stations, and drought intensity in Hebei Province. The results showed that the drought disasters in autumn in Hebei Province significantly weakened, and the trend of drought disasters in other seasons did not reach the level of anomaly by the m-k trend test. However, the ratio of drought stations and drought intensity increased in summer. Summer drought increased by an average rate of 1.576% for the previous 10 years, and drought intensity increased at an average rate of 0.005. The drought disaster rate was significantly positively correlated with the ratio of drought and drought intensity in summer. Although the frequency of drought in summer was the lowest and was dominated by mild drought, the frequency of drought was still 26%. Moderate, severe, and extreme droughts also occur, and the frequency of moderate and severe droughts was 5.7%. The areas of high frequency of summer drought are located in Shijiazhuang, Xingtai, Cangzhou, Hengshui, Baoding, and Handan. According to the Economic Yearbook of Hebei Province, the planting area of grain crops in these five cities accounted for 74.2% of the province in 2017. Therefore, combined with the above factors, autumn and winter drought showed a decreasing trend. The characteristics and trends of summer drought were still issues, and they need special attention for drought prevention and disaster reduction.

5. Conclusions

The frequency of drought in Hebei Province was the highest in winter (43.2%), followed by spring and autumn, and the lowest in summer (26.2%). The drought characteristics were as follows: mild and moderate droughts were dominant in spring and autumn, mild droughts were dominant in summer, and severe droughts and extreme droughts occurred most frequently in winter. The spatial distribution of drought in four seasons showed different characteristics. The frequency of extreme drought in Handan and Xingtai in southern Hebei Province was higher. The ratio of drought in Hebei Province winter was highest, the influence was biggest with a ratio of 37%. In spring and autumn, the frequency of regional drought was 31 and 34%. The ratio of drought in summer was the smallest, but the frequency of regional drought was still 17%. Except for the increase of the ratio of drought in summer, it showed a decreasing trend in other seasons. The decreasing trend in autumn reached a significant level of 0.05. The drought intensity values of Hebei Province in the four seasons were all less than 1, with the largest in winter, followed by spring and autumn, and the smallest in summer, with mild drought at the regional level. The drought intensity showed an upward trend in spring and summer, a downward trend in autumn and winter. The downward trend in autumn reached a significant level of 0.05. The drought disaster rate in Hebei Province was positively correlated with summer drought intensity and the ratio of drought stations.

Author Contributions: Conceptualization, Z.T.; methodology, Z.T.; software, S.W.; validation, S.W.; formal analysis, Z.T.; investigation, S.W.; resources, Z.T.; data curation, S.W.; writing—original draft preparation, S.W.; writing—review and editing, Z.T.; visualization, Z.T.; supervision, Z.T.; project administration, Z.T.; funding acquisition, Z.T. All authors have read and agreed to the published version of the manuscript.

Funding: Funded by the National Key R & D Program "Major grain-producing areas the main meteorological disaster process and the key technology of disaster reduction and production maintenance regulation" (2017YFD0300400).

Conflicts of Interest: The authors declare no conflict of interest.

References

1. United Nations Secretariat of the International Strategy for Disaster Reduction. *Disaster Risk Reduction Frame-Work and Practices: Contributing to the Hyogo Framework for Action*; UNISDR: Geneva, Switzerland, 2009.
2. Qu, Y.; Lv, J.; Su, Z. Review and prospect of research on drought resistance and disaster reduction. *J. Hydraul. Eng.* **2018**, *49*, 115–125. (In Chinese)
3. Huang, J.P.; Yu, H.; Dai, A.G.; Wei, Y.; Kang, L. Drylands face potential threat under 2 °C global warming target. *Nat. Clim. Chang.* **2017**, *7*, 417–422. [CrossRef]
4. Han, H.; Liu, Z.; Liu, C.; Chen, Q.; Xie, G. Drought characteristics analysis and short-term prediction in Ji'an area from 1960 to 2018. *J. Irrig. Drain.* **2019**, *38*, 85–92. (In Chinese)
5. Wang, M.; Wang, X.; Huang, W.; Zhang, Y.; MA, J. Spatial-temporal distribution characteristics of seasonal drought in southwest China based on relative humidity index. *Trans. Chin. Soc. Agric. Eng.* **2012**, *28*, 85–92. (In Chinese)
6. Zhao, S.; Gong, Z.; Liu, X. Correlation analysis between vegetation cover and drought conditions in North China from 2001 to 2013. *Acta Geogr. Sin.* **2015**, *70*, 717–729. (In Chinese)
7. Huang, D.; Peng, T.; Wang, K.; Chen, G. Analysis of results of using Z-index method to judge seasonal drought in southern China. *Chin. J. Agrometeorol.* **2003**, *24*, 12–15. (In Chinese)
8. He, B.; Liu, Z.; Yang, X.; Sun, S. Spatio-temporal distribution characteristics of agro-meteorological disasters of major crops in China under the background of climate change (II.): Drought of major grain crops in Northwest China. *Chin. J. Agrometeorol.* **2017**, *38*, 31–41. (In Chinese)
9. Liu, Y.; Wang, S.; Wang, J.; Wang, Z. Risk assessment of drought disasters in Southwest China under the background of climate warming. *J. Nat. Resour.* **2018**, 33, 325–336. (In Chinese)
10. Tian, Z.; Li, X. Analysis of drought change trend in China's main grain-producing areas from 1949 to 2016. *J. China Agric. Univ.* **2019**, *24*, 159–167. (In Chinese)
11. Wang, M.; Yan, J. 1960–2012Characteristics of precipitation changes and heavy flooding in the southeast coastal areas of China in 2008Harmful trendsjudgement. *Earth Environ.* **2015**, *43*, 667–674. (In Chinese)
12. Shen, L.; He, J.; Zhou, X.; Chen, L.; Zhu, C. Study on the characteristics of summer precipitation and water vapor transport in China in the past 50 years. *Acta Meteorol. Sin.* **2011**, *68*, 918–931. (In Chinese)
13. Wang, S.P.; Zhang, C.J.; Song, L.C.; Li, Y.H.; Feng, J.Y.; Wang, J.S. Relationship between soil redative humidity and the multiscale meteorological drought indexes. *J. Glaciol. Geocryol.* **2013**, *35*, 865–873.
14. Wei, F. *Modern Climate Statistical Diagnosis and Prediction Technology*; Meteorological Press: Beijing, China, 2007; pp. 1–20. (In Chinese)

Disclaimer/Publisher's Note: The statements, opinions and data contained in all publications are solely those of the individual author(s) and contributor(s) and not of MDPI and/or the editor(s). MDPI and/or the editor(s) disclaim responsibility for any injury to people or property resulting from any ideas, methods, instructions or products referred to in the content.

engineering proceedings

MDPI

Proceeding Paper

Environmental Aspects of Vernacular Settlement of Xinpu: A Case Study for Social Responsibility †

Shaofu Li

Department of Architecture and Urban Planning, Chung Hua University, Hsinchu 30012, Taiwan; shaofu@g.chu.edu.tw

† Presented at the 3rd IEEE International Conference on Electronic Communications, Internet of Things and Big Data Conference 2023, Taichung, Taiwan, 14–16 April 2023.

Abstract: The architectural heritage of a remote village not only forms an important part of its local history but is also a sustainable source of knowledge. The purpose of this study is to introduce the environmental factors in the design of a traditional ancestral hall located in Xinpu Township, Hsinchu County, Taiwan. The content presented covers all scales of architecture, starting from large-scale urban design to small-scale building components. First, it presents the urban form of Xinpu. The environmental investigation is based on the nine main ancestral halls, corresponding to the local climatic conditions, the sun and shade conditions of the courtyard, and the orientation of the remaining traditional houses. Then, the architecture of traditional dwellings is introduced. Environmental studies are based on a computer analysis using Ecotect. Field studies include measurements of winter and summer air temperatures in a house, while computer analysis is conducted for sunlight and shading conditions for nine ancestral buildings. Combining Ecotect Analysis, Desktop Radiance, and WinAir4, daylight levels and airflow in the houses are simulated. The results are used to explain the university's practice of social responsibility in rural areas.

Keywords: traditional ancestral halls; daylight conditions; wind field; Xinpu; university social responsibility

Citation: Li, S. Environmental Aspects of Vernacular Settlement of Xinpu: A Case Study for Social Responsibility. *Eng. Proc.* **2023**, *38*, 45. https://doi.org/10.3390/engproc2023038045

Academic Editors: Teen-Hang Meen, Hsin-Hung Lin and Cheng-Fu Yang

Published: 27 June 2023

Copyright: © 2023 by the author. Licensee MDPI, Basel, Switzerland. This article is an open access article distributed under the terms and conditions of the Creative Commons Attribution (CC BY) license (https://creativecommons.org/licenses/by/4.0/).

1. Introduction

For a long time, the application of computer graphics in learning architecture has been for technical training. With no clear need or purpose, students only learn how to use the technology [1]. Often, there is not a strong desire for students to complete a project when they seldom use the knowledge of computer graphics. In computer-aided design (CAD) courses, the traditional learning method needs to be changed as innovation and creativity are required nowadays, and it is required to assess the learning outcome of students from a practical perspective. Fortunately, with the advent of the university social responsibility (USR) program in recent years, there seems to be a chance.

This study aims to present an experimental process of teaching and learning in a CAD course that provides a sense of purpose to students who are involved in activities of the USR program. The town of Xinpu, with its local industrial history, is a target area for the student's architectural design. Students first learn its environment from the scale of the town and then study the reasons for its orientation from the perspective of buildings [2]. Taking the traditional ancestral hall buildings as a model, an analysis of the actual urban environment is carried out to consider wind and heat factors for an appropriate design.

2. General Background

In the USR program promoted by the Ministry of Education (MOE), the Xinpu area is classified as a rural area that needs investment in research and support for development. In the early nineteenth century, Xinpu had become a distribution center for the three major economic crops of cane sugar, tea, and camphor. At that time, the commercial prosperity of Xinpu was almost as high as that of Hsinchu City. Moreover, 99.1% of its land was owned privately in Xinpu Township, which caused problems in the promotion of public construction. Thus, there are not many public facilities in the area still. The terrain of Xinpu is hilly and sloped, accounting for 76.17% of the total area. However, 23.83% of the flat area includes the urban planning area, agricultural area, and industrial area under a restriction for the development of Xinpu. The geographical situation is slightly on the fringe of economic development.

2.1. Topography

Xinpu Township is located on the south side of the Dianzi Lake platform with many mountains and few plains. Geographically, it is run through the river Fengshan from the east and the river Xiaoli, which converges from the northeast. The main settlements in the town are found along the river steps of the valley. The land develops to the eastern and western sides and gathers into streets in the narrow alluvial plain (Figure 1).

Figure 1. Topography of the vernacular settlement of Xinpu. Source: Google Map, https://www.google.com.tw/maps/@24.8272863,121.0759617,1459m/data=!3m1!1e3?hl=zh-TW&entry=ttu (accessed on 2 October 2022).

2.2. Climate Analysis

The climate of Xinpu has an annual rainfall of about 2322 mm. From February to September, the rainy season with more rainfall occurs, while the dry season lasts from October to January with less rainfall. The dry and wet seasons are not obvious. Although the wind continues to blow, the air is not dry, due to the high humidity. From January to April and from October to December, clouds cover the sky almost every day. The total annual evaporation is 1041 mm, which is less than the annual rainfall. Only in October is the evaporation greater than the rainfall, so there is no shortage of irrigation water.

Hsinchu's monsoon index is the highest among the counties and cities in the west, so it is called "Wind City". This weather is closely related to its geographical location because Hsinchu is only 150 km away from the narrowest point of Pingtan, Fujian Province, across the Taiwan Strait. The wind speed increases at the narrowest part of the gorge, especially during the northeast monsoon season. Especially during September, the speed of the wind reaches 20 m/s. Xinpu has a subtropical oceanic climate with an annual average temperature of 21.3 °C. Higher temperatures are observed in July and August with average temperatures of 28.1 °C and 27.6 °C. The temperature is the lowest in February with an average of about 13 °C. Figure 2 shows the monthly temperatures and daylight hours.

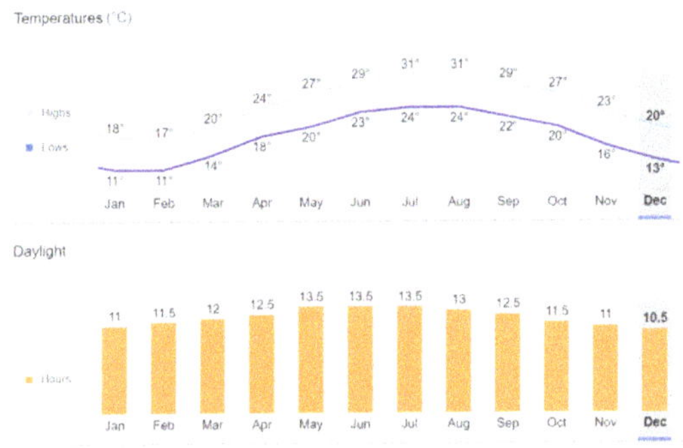

Figure 2. Monthly values of temperature and daylight hours. Sourse: Weather and Climate, https://weather-and-climate.com/average-monthly-Rainfall-Temperature-Sunshine,hsinchu-city-hsinchu-county-tw,Taiwan (accessed on 5 May 2022).

3. Environment

Xinpu was a prosperous town for the Hakka people in the Qing Dynasty. According to the topography and the size of the population at that time, a spatial pattern of Three Streets, Six Lanes, and Nine Ancestral Halls (Figure 3) was formed [3]. Along the old streets of Xinpu today, it looks like a clan museum without walls, and traces of Hakka's efforts can be seen everywhere.

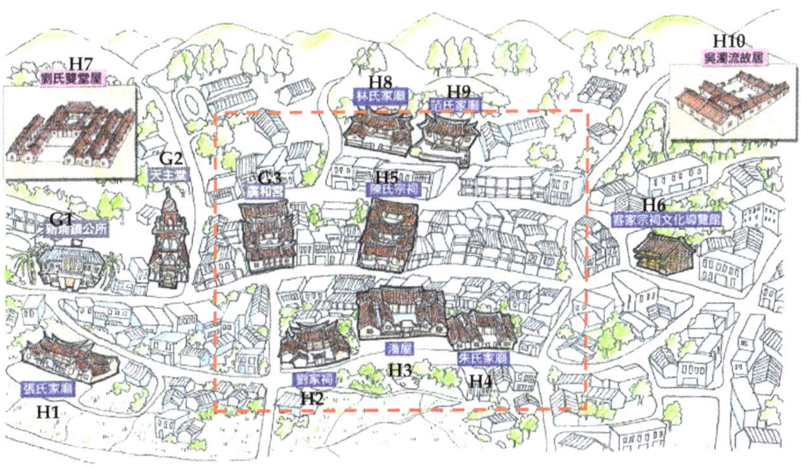

Figure 3. Three Streets, Six Lanes, and Nine Ancestral Halls in Xinpu. Source: Xinpu Ancestral Museum Promotion Association, http://210.63.200.70/exhib.asp. (accessed on 2 October 2022) (G1: Xinpu Town Office (新埔鎮公所); G2: Cathedral (天主堂); G3: Guanghe Temple (廣和宮); H1: Chang Family Hall (張氏家廟); H2: Liu Family Hall (劉家祠); H3: Peng Family Hall (潘屋); H4: Chu Family Hall (朱氏家廟); H5: Chen Family Hall (陳氏宗祠); H6: Hakka Culture Guide Hall (客家宗祠文化導覽館); H7: Liu Family Twin Hall (劉氏雙堂屋); H8: Lin Family Hall (林氏家廟); H9: Fan Family Hall (范氏家廟); H10: Wu's former residence (吳濁流故居); The red dotted line marking the location of the core testing area (shown in Figure 4)).

Figure 4. The location of the core building group in Figure 3. (Source: Google Map, 2022).

3.1. *Urban Area*

The three parallel east–west streets are the core of the life of the people of Xinpu. There are more than six small alleys intermittently, connecting the three main streets into a network in the north–south direction. There are the ancestral hall buildings of the major families, which are scattered on these three streets with the buildings facing south. Affected by the northern mountain's shape and the direction of the stream, the three parallel streets are located close together. The streets in the middle naturally become the core commercial streets. The main street on the south side accommodates the people's livelihood and consumption places in the town. The largest fruit and vegetable market and most restaurants are located on the south side. The north side is dominated by stores selling daily necessities, government agencies, and cultural and educational palaces and temples. On the other hand, the low-rise buildings at the foothills on the north side are dotted with small wholesale stores, all of which are mixed residential and commercial buildings.

3.2. *Orientation and Shading Conditions of Buildings*

In such a long and narrow road network, almost all buildings face the street with a south or north orientation. Except for the nine ancestral halls and the cultural and educational institutions, most of them are built in the form of shophouses. Of course, what we see today are modern, long, multi-story buildings, not traditional shophouses with several internal courtyards. Figure 4 shows the range of the red dotted line marking the location of the core building group. All family ancestral halls and temples must face south. In addition to conforming to the topography and the direction of traffic flow, it also fully complies with the requirements of traditional Chinese Fengshui.

There are two types of traditional courtyard houses in Chinese villages: closed and open. Building materials are either brick or wood, depending on local conditions. Moreover, most of the houses in Taiwan face south. In the middle of the north side, there is the main room, and on the left and right, the living room and the granary are located. The east wing is used as a kitchen and dining room, and the west wing is used as a bedroom. There are walls built to connect the east and west wing rooms, and the doors are opened at the opposite side of the wall. The main room is close to a courtyard, and rooms without walls are used as open courtyards. The open space in the middle is mainly used as a drying area, and flowers, plants, and fruit trees are planted around it. With the development of the local economy, more rural brick houses and urban buildings are under construction.

Using Ecotect software, the shading conditions of the environment of Xinpu settlements were analyzed, and a later rendering was carried out. The three-dimensional model was drawn and tested for the summer solstice (22 June) (Figure 5a,b), the winter solstice (22 December) (Figure 5c,d), and the spring/autumn equinox at representative hours of the day.

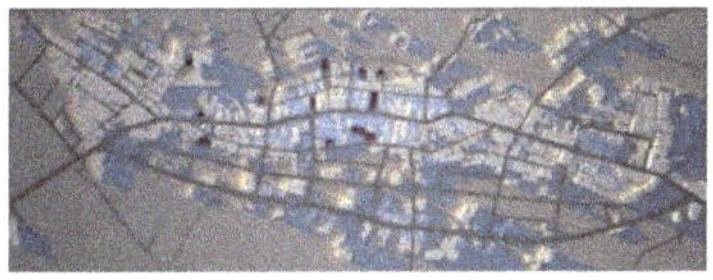

(**a**) Shadows at 5:00, Summer Solstice (**b**) Shadows at 19:00, Summer Solstice

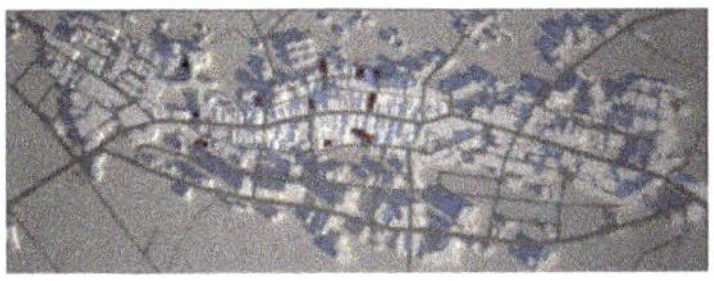

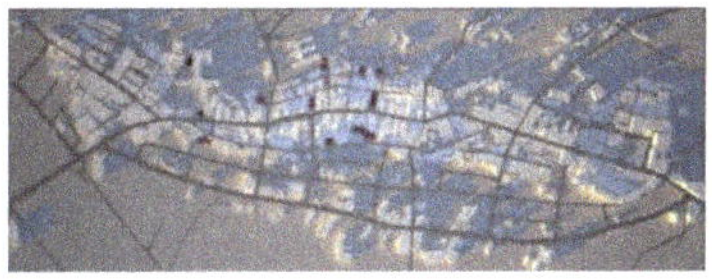

(**c**) Shadows at 7:00, Winter Solstice (**d**) Shadows at 19:00, Winter Solstice

Figure 5. Ecotect rendering for the Old Street of Xinpu Township in the daytime. (Source: Author 2022).

Lower sun elevations in the winter shade significant portions of the atrium, while in summer, shadows are minimized. Since the vast majority of settlements are on the south–southwest side of the valley, the main direction of the housing is north to southwest. Additionally, many houses face south or north. However, the ancestral hall buildings must be oriented towards the favorable south.

4. Environmental Aspects of Daylight

On the whole, the sunshine duration in Xinpu is sufficient and even intense. The ancestral hall buildings are relatively concentrated. Figure 6 shows the distribution of the air temperature under fluid analysis. Although they are all low-rise buildings, the air temperature is relatively high due to the large and dense shophouse buildings [4]. Therefore, the combination of warmer temperatures and long hours of sunshine makes Xinpu a warmer area.

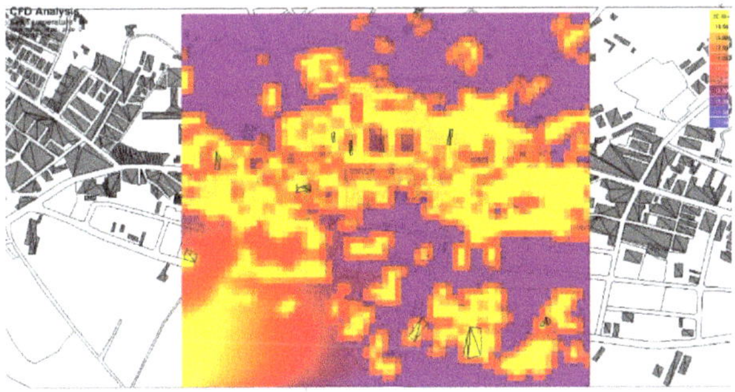

Figure 6. Cell temperature by CFD analysis in the area of the ancestral hall buildings. (Source: Author 2022).

The Pan Family House on the southernmost side of Old Street is shown in Figure 7. In summer, there are up to 8 h of strong sunshine from 9 am to 4 pm. The average sunlight intensity is as high as 25,000 lumens. However, it is soothing in the winter, and the average intensity is about 11,000 lumens (Figure 8). The plants on Old Street are distributed on the hillside on the north side, and the rest are scattered on the southeast side. The larger and older plants are distributed around the ancestral hall buildings.

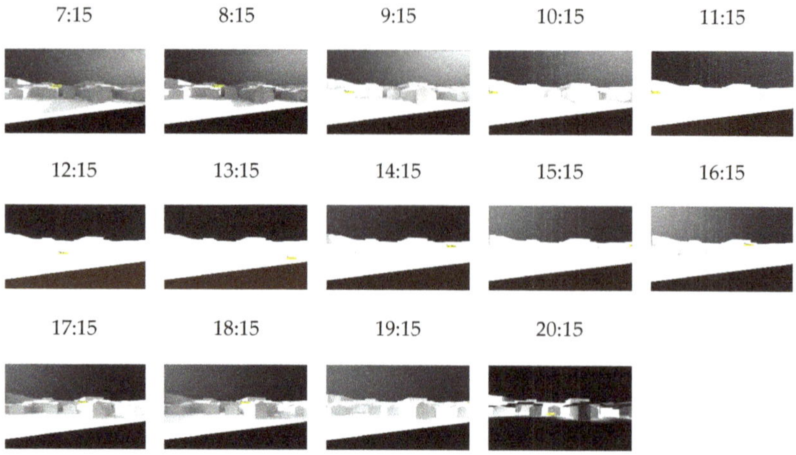

Figure 7. An ancestral hall (the Pan Family House) building facing south, in the sunshine of the summer solstice. (Source: Author 2022).

Figure 8. Facing south, the Pan Family House building in the sunshine of the winter solstice at 1 pm. (Source: Author 2022).

5. Environmental Aspects of Wind Flow

The wind flow in the town is interesting. The east–west slope on the north side blocks the famous 'Nine-Falling Wind.' However, the gentle slope on the south side and the open stream on the east side allow the northeast monsoon in September to blow on the old streets of Xinpu. Figures 9 and 10 show the period when the Nine-Falling Wind prevails with the change in building volume. The flow and distribution of the air change drastically and in an instant. The change in wind direction is observed from the vector diagram (Figure 11). The simulation and analysis of the wind field are carried out with Cardiff University's WinAir4, and the visual performance is controlled with Ecotect Analysis. The disadvantage of WinAir4 is the limitation of the number of grids, and its three-dimensional proportions cannot be completely consistent. Therefore, the estimated results can only be used as a preliminary reference. However, it has the advantage of ease of use and free use for research purposes.

Figure 9. The airflow rate of Xinpu Old Street under the action of the Nine-Falling Wind. (Source: Author 2022).

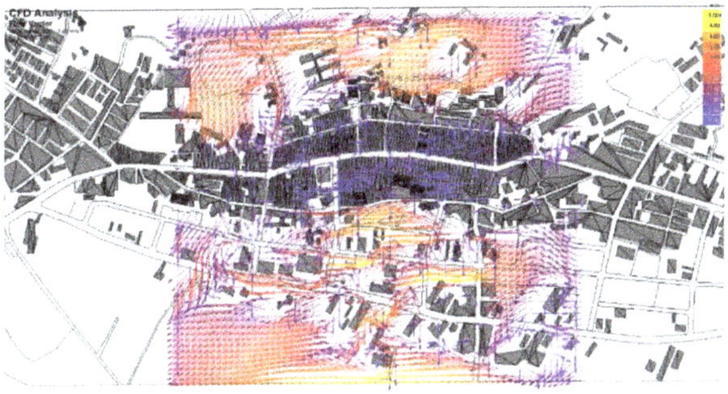

Figure 10. Diagram of the Nine-Falling Wind in flow vector format. (Source: Author 2022).

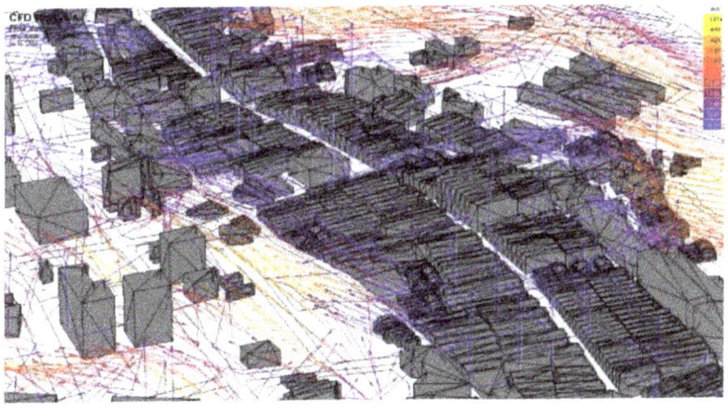

Figure 11. Flow vector diagram of the Nine-Falling Wind on the side of north–south. (Source: Author 2022).

6. Conclusions

This research project in Xinpu Township was carried out for three years. In the first year, a small space renovation plan was performed for a small corner of the town. In the second year, we conducted a preliminary analysis of pedestrian flow to obtain complete basic data for subsequent research. Xinpu has a rich cultural atmosphere. The ancestral hall buildings are scattered in the Three Streets area with a reputation as an open museum owing to their good preservation. The traditional shophouse architecture has almost been replaced by modern concrete buildings. At first glance, the whole old street area can seem like a modern town. Without this batch of traditional buildings, the significance of the research may be lost.

The USR Project is provided for student learning. Therefore, we designed a learning road map (Figure 12) based on a physical environmental analysis [5]. Software including AutoCAD 2D drafting, SketchUp 3D modeling, Revit modeling, Ecotect Analysis, Desktop Radiance, and the CFD of WinAir4 was used.

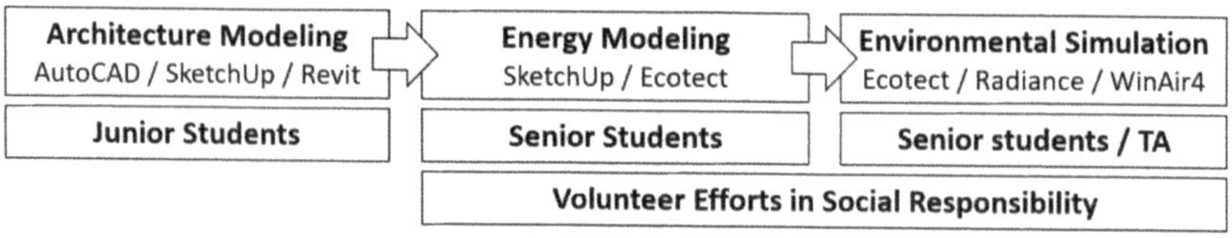

Figure 12. Diagram of the teaching/learning road map of the environmental aspect of the vernacular settlements. (Source: Author 2022).

In terms of the construction of basic spatial models, students can handle it freely. The analysis of the light environment involves a lot of theories. Because the models provide tangible objects, students can master the essentials of manipulation. The application of ventilation is the most difficult for students to learn. The theory is not easy to understand, and the calculation process is tedious. Even if the analytics can be calculated, they may not have the corresponding ability to explain it. In the USR Project, students learn the process of local development and problem-solving based on local-based courses and activities [6]. To guide students to learn independently and promote a cooperative mode of field practice learning and sharing among students, the integration and reproduction of knowledge need to be encouraged, which allows the development of new teaching models and research.

Funding: This experiential study is partly supported by the Ministry of Education of Taiwan, Republic of China, through grants of the University Social Responsibility Program (Phase II, 112-ZP0-TEC-001-03-B, 2020–2022) for Chung Hua University, Taiwan.

Institutional Review Board Statement: Not applicable.

Informed Consent Statement: Not applicable.

Data Availability Statement: Not applicable.

Acknowledgments: These courses once again thank the students who participated in the action of learning-by-doing, especially the students who were trying to solve some sort of problem in computer vision. The author of this paper is currently services for the USR Office and the first batch of AI seed teachers in the school.

Conflicts of Interest: The author declares no conflict of interest.

References

1. Ruxandra, V.; Barna, C.; Epure, M.; Baicu, C. Developing university social responsibility: A model for the challenges of the new civil society. *Procedia-Soc. Behav. Sci.* **2010**, *2*, 4177–4182.
2. Tom, W.; Kimmins, S. *Green Building Handbook: A Guide to Building Products and Their Impact on the Environment*; Routledge: New York, NY, USA, 2003; Volume 5.
3. Peng, Y. Kinship networks and entrepreneurs in China's transitional economy. *Am. J. Soc.* **2004**, *109*, 1045–1074. [CrossRef]
4. Ibarra, D.I.; Reinhart, C.F. Daylight factor simulations—How close do simulation beginners get. *Build. Simul.* **2009**, *196*, 203.
5. Randall, T. *Environmental Design: An Introduction for Architects and Engineers*, 3rd ed.; Taylor & Francis: Oxfordshire, UK, 2006.
6. Adam, L.; Swaen, V. Corporate social responsibility. *Int. J. Manag. Rev.* **2010**, *12*, 1–7.

Disclaimer/Publisher's Note: The statements, opinions and data contained in all publications are solely those of the individual author(s) and contributor(s) and not of MDPI and/or the editor(s). MDPI and/or the editor(s) disclaim responsibility for any injury to people or property resulting from any ideas, methods, instructions or products referred to in the content.

engineering proceedings

MDPI

Proceeding Paper

Preliminary Research on Carbon-Neutral Sustainable Development Strategies in Community Development Education †

Jiin-Chyuan Mark Lai [1], Chiung-Ling Wang [2] and Ming-Yuan Hsieh [3,*]

1 Department of Applied Foreign Languages, Trans World University, Douliu 64063, Taiwan; marklai07@gmail.com
2 Department of Industrial Education, National Taiwan Normal University, Taipei City 106010, Taiwan; lin8332317@mail.edu.tw
3 Department of International Business, National Taichung University of Education, Taichung City 40306, Taiwan
* Correspondence: cpawisely@mail.ntcu.edu.tw; Tel.: +886-975-118-922

† Presented at the 3rd IEEE International Conference on Electronic Communications, Internet of Things and Big Data Conference 2023, Taichung, Taiwan, 14–16 April 2023.

Abstract: Owing to the severe development and progression of climate change and the greenhouse effect, carbon neutrality has attracted much research interest in environment-related social research. Specifically, in Article 12 of the Regulations on Community Development Work (RCDW) for more than 25,000 Taiwanese community development associations, carbon neutrality is the main purpose of environmental education, environmental protection and infrastructures, and social responsibilities. Carbon neutrality has to be instituted into efficient governance and operations to provide effective and diversified services for residents' lives. Thus, sustainable strategies (environmental, social, and governance, ESG) need to induce effective and efficient carbon-neutral sustainable development in order to improve environmental conditions for the community's residents without long-term government compensation and resources. Environmentally sustainable strategies focus on environmental protection and education for the whole community. Social strategies need to be associated with social responsibility and public services, while governance needs to be concentrated on sustainable operation and management. In this study, effective and efficient carbon-neutral sustainable development strategies were proposed for the community sustainability of community development associations.

Keywords: carbon neutral sustainable development strategy; community development education; carbon-neutral sustainable development strategy; ESG (environmental, social and governance)

Citation: Lai, J.-C.M.; Wang, C.-L.; Hsieh, M.-Y. Preliminary Research on Carbon-Neutral Sustainable Development Strategies in Community Development Education. *Eng. Proc.* **2023**, *38*, 53. https://doi.org/10.3390/engproc2023038053

Academic Editors: Teen-Hang Meen, Hsin-Hung Lin and Cheng-Fu Yang

Published: 27 June 2023

Copyright: © 2023 by the authors. Licensee MDPI, Basel, Switzerland. This article is an open access article distributed under the terms and conditions of the Creative Commons Attribution (CC BY) license (https://creativecommons.org/licenses/by/4.0/).

1. Introduction

With rapid economic development and social evolution, people pursue financial and economic success as well as mental and spiritual accomplishment for self-satisfaction in harmonious co-existence between humans and nature. As an essential foundational goal of Taiwanese society, the advancement of lifestyles of health and sustainability (LOHAS) is regarded as important. LOHAS originated from the LOHAS doctrine (LOHASism) for sustainable happiness, health, and self-sufficiency. Health, happiness, protection, environmental protection, and sustainability are core principles in environmental issues. Organic foods, natural cotton and linen, secondhand houseware, bicycle riding, regular sports and fitness, mind-soothing music, a friendly personality, and a sustainable lifestyle have been the focused of human lives following these principles. LOHAS is important to support environmental protection in consumption. The three essential principles of "do good, feel good, and look good" are also emphasized [1–4].

Taiwan has a land area of 36,000 km^2 and is located near the southeastern coast of China, bordering the Taiwan Strait and facing the Pacific Ocean. Taiwan is near Okinawa and the Philippines with an annual average temperature of 22–25 °C. There are 268 mountains on Taiwan island, among which the highest one is Yushan Mountain, at 3952 m. The population is approximately 23 million, and the official languages are Mandarin, Taiwanese, and Hakka. Buddhism and Taoism are the dominant religions. Taiwan has rich natural resources, and so eco-tourism including Taiwanese aboriginal tribe culture trips, local agricultural products tours, and local art tours is very popular. Therefore, LOHAS is involved in the diverse communities and associations in Taiwan.

There have been systematic surveys on regional development and urban–rural disparities due to the different levels of economic and industrial development in Taiwan [5–12]. Such disparities are believed to create a gap between rich and poor communities. The concept of utilitarianism has been introduced for the purposes of attain the maximum happiness among the most people. Most communities propose that economic outcomes from various industries need to be shared through government grants, subsidies, and incentives. Therefore, most urban and rural community development associations commit themselves to local industrial and economic development through local product development and the tourism industry. Some have claimed that the profits of local development need to be invested into minority groups, including local enterprises. However, most communities focus only on tourism. Hence, issues have been raised regarding pollution, contamination, transportation problems, and noise [6–12].

Therefore, LOHAS must be implemented for the development of communities to advance life quality from the perspective of environmental, social, and governance (ESG) perspectives. ESG policies support environmental protection in communities under mutualism and strengthen social responsibility and government performance. These policies also pay attention to environmental issues, including climate change, scarcity of natural resources, pollution prevention and control, and waste disposal. Social benefits such as human rights and interests, product duty, public information security, and life risks are also emphasized in ESG [13]. In governance, the issues of administrative performance, managerial effectiveness, and efficiency, developmental sustainability, and administrative executive leadership are considered in ESG. ESG policies are also required to diminish the disparities between urban and rural communities.

2. Literature Review

LOHAS reflects the rise of sustainable development, focusing on healthy, self-sufficient, environmentally protected, organic, natural, cultural, and creative lives. In modernization and globalization, people can experience emptiness, estrangement, and resource shortages. LOHAS has corresponded to social development since being popularized in Europe and America in the 2000s. The concrete statement of LOHAS was proposed in the 2008 Young Chinese Fashion Culture Forum in Ningbo City, China, which was hosted by the Party Central Committee, All-China Youth Federation, Ningbo City Hall, and Ningbo Municipal Party Committee. The statement embodied twelve notions for self-improvement, including a happy life, free creation, a strong body, sustainable food, simple consumption, contented peace, good treatment, a closeness with nature close, environment protection, a public spirit, and proactive sharing. In the definition, the concise concept of LOHAS is defined as a "happy, pleasurable, and sustainable lifestyle". Approximately one-fourth of Americans have embraced the LOHAS lifestyle, while one-third of Europeans have adopted the LOHAS lifestyle. LOHAS is not only a lifestyle, but also a cultural habit and creation. The natural, healthy, harmonious, and environmentally protected lifestyle in LOHAS is related to dieting, clothing, accommodation, transportation, education, and recreation [14,15].

Unfortunately, rapid economic development has caused a series of environmental problems such as pollution in the air, water, and land, with climate anomalies, food contamination, deracination, global warming, and deforestation. As a result, people have realized the importance of environmental protection and reduced their consumption to advocate for eco-consumerism including green products, carbon reduction, green delivery, and so on. Considerably more consumers have begun to implement the LOHAS concept in their lives. They seek green, carbon-reduced, recyclable, and sustainable packaging for products and recycle wastes actively. Therefore, the essential concept and philosophies of LOHAS have been applied to consumer psychology, art theory, material science, manufacturing technology, fabrication technology, product logistics, and customer services. LOHAS helps to create impressive economic outputs, too. To diminish the disparities between urban and rural communities in Taiwan through LOHAS, the unilateral development of the economy and the industry with LOHAS ideas and activities is required for a better quality of life.

3. Conclusions and Future Direction

To mitigate the urban–rural disparities in Taiwan, attention must be paid to regional economic and industrial development by developing local products and boosting the local economy. However, Taiwanese communities should also consider environmental issues seriously to achieve cultural, educational, and social benefits in order to enhance the quality of life of community residents. Under Article 12 of the Regulations on Community Development Work, there are over 25,000 Taiwanese community development associations. In consideration of their concepts, purposes, and functions, environmental protection and education, social responsibility and services, and efficient governance and operations are required for them to provide more effective and diversified public services. Hence, it is necessary to focus on the advanced role of the associations from the perspective of ESG to maintain fundamental operations and offer their services even without long-term government compensation. For sustainable development and operations, considerations of the "environment" need to focus on the preservation and development of local natural resources and culture, "social" policies need to involve social responsibility and public services, "governance" needs to foster efficiency and sustainability. Creative research is required to deeply and comprehensively help the community develop and strengthen the sustainability of community development under ESG.

Author Contributions: Conceptualization, J.-C.M.L. and M.-Y.H.; methodology, M.-Y.H.; validation, J.-C.M.L.; formal analysis, M.-Y.H.; investigation, J.-C.M.L.; resources, J.-C.M.L. and C.-L.W.; writing—original draft preparation, M.-Y.H. and C.-L.W.; writing—review and editing, M.-Y.H.; visualization, J.-C.M.L.; supervision, M.-Y.H.; project administration, M.-Y.H.; funding acquisition, and C.-L.W. All authors have read and agreed to the published version of the manuscript.

Funding: This research was supplied by the research supporting research projects (MOST 110-2420-H-002-003-MY3-Y11017) of Ministry of Science and Technology and (NTCU109105) from the National Taichung University of Education.

Institutional Review Board Statement: This research did not execute the questionnaires and interview and hence, this research did not require ethical approval.

Informed Consent Statement: Not applicable because this research did not involve any humans.

Data Availability Statement: No new data were created.

Conflicts of Interest: There is no conflict of interest in this research.

References

1. Raihan, A. Nexus between economic growth, natural resources rents, trade globalization, financial development, and carbon emissions toward environmental sustainability in Uruguay. *Electeonic J. Educ. Soc. Econ. Technol.* **2023**, *4*, 55. [CrossRef]
2. Raihan, A.; Muhtasim, D.A.; Farhana, S.; Hasan, M.A.U.; Paul, A.; Faruk, O. Toward environmental sustainability: Nexus between tourism, economic growth, energy use and carbon emissions in Singapore. *Glob. Sustain. Res.* **2022**, *1*, 53. [CrossRef]
3. Raihan, A.; Muhtasim, D.A.; Farhana, S.; Hasan, M.A.U.; Paul, A.; Faruk, O. Dynamic linkages between environmental factors and carbon emissions in Thailand. *Environ. Process.* **2023**, *10*, 5. [CrossRef]
4. Raihan, A.; Tuspekova, A. Toward a sustainable environment: Nexus between economic growth, renewable energy use, forested area, and carbon emissions in Malaysia. *Conserv. Recycl. Adv.* **2022**, *15*, 200096. [CrossRef]
5. Wang, Q.J.; Feng, G.F.; Wang, H.J.; Chang, C.P. The influence of political ideology on greenhouse gas emissions. *Glob. Environ. Chang.* **2022**, *74*, 102496. [CrossRef]

6. Hsieh, Y.M. The most sustainable niche principles of social-media education in a higher education contracting era. *Sustainability* **2020**, *12*, 339. [CrossRef]
7. Huang, Y.-M.; Hsieh, Y.M.; Usak, M. A Multi-Criteria Study of Decision-Making Proficiency in Student's Employability for Multidisciplinary Curriculums. *Mathematics* **2020**, *8*, 897. [CrossRef]
8. Chan, Y.-K.; Hsieh, M.-Y.; Usak, M. A Concrete Study on Social-Media Connection of Global Literacy Abilities in MOOCs under the Dual Impacts of Lower Birth-Rate and COVID-19. *Sustainability* **2021**, *13*, 2203. [CrossRef]
9. Usak, M.; Hsieh, M.-Y.; Chan, Y.-K. A Concertizing Research on Making Higher Education Sustainability Count. *Sustainability* **2021**, *13*, 2724–2738. [CrossRef]
10. Duran, M.; Usak, M.; Hsieh, M.-Y.; Uygun, H. A New Perspective on Pedagogical Content Knowledge: Intellectual and Emotional Characteristics of Science Teachers. *Rev. Cercet. Interv. Soc.* **2021**, *72*, 9. [CrossRef]
11. Wu, T.-L.; Hsieh, M.-Y.; Min, K.-W.; Yu, M.-T.; Ho, C.-T. Use of Sensor Technologies in Online Courses in Post-COVID-19 Era. *Sens. Mater.* **2021**, *33*, 2045–2062. [CrossRef]
12. Huang, C.C.; Chan, Y.-K.; Hsieh, M.Y. The Determinants of ESG for Community LOHASism Sustainable Development Strategy. *Sustainability* **2022**, *14*, 11429. [CrossRef]
13. Chan, Y.-K.; Hsieh, M.Y. An Empirical Study on Higher Education C-ESG Sustainable Development Strategy in Lower-Birth-Rate Era. *Sustainability* **2022**, *14*, 12629. [CrossRef]
14. Hsieh, M.Y. The Sustainable Development and Strategic Approaches for Contemporary Higher Education. *Sustainability* **2022**, *14*, 12925. [CrossRef]
15. Huang, C.-C.; Chan, Y.-K.; Hsieh, M.-Y. Preliminary Research on the Sustainable Determinants of Taiwanese Ecotourism with the International Standards. *Int. J. Environ. Res. Public Health* **2022**, *19*, 14489. [CrossRef] [PubMed]

Disclaimer/Publisher's Note: The statements, opinions and data contained in all publications are solely those of the individual author(s) and contributor(s) and not of MDPI and/or the editor(s). MDPI and/or the editor(s) disclaim responsibility for any injury to people or property resulting from any ideas, methods, instructions or products referred to in the content.

engineering proceedings

MDPI

Proceeding Paper

Phytoremediation Efficiency of Weathered Petroleum-Contaminated Soils by *Vetiveria zizanioides* and *Cymbopogon nardus itle* †

Pei-Cheng Cheng [1,2], Yuan-Chung Lin [1,2], Min-Siou Lin [3], Sun-Long Lin [4], Yin-Hsiu Hsiao [4], Chin-Yuan Huang [5], Pei-Chun Tu [6,*] and Shu-Fen Cheng [5,*]

1 Institute of Environmental Engineering, National Sun Yat-Sen University, Kaohsiung 804, Taiwan; eric62795@gmail.com (P.-C.C.); yclin@faculty.nsysu.edu.tw (Y.-C.L.)
2 Center for Emerging Contaminants Research, National Sun Yat-Sen University, Kaohsiung 804, Taiwan
3 Institute of Biochemical Technology, Chaoyang University of Technology, Taichung 413, Taiwan; minsiou.lin@gmail.com
4 CPC Exploration & Development Research Institute, Miaoli City 360, Taiwan; 155381@cpc.com.tw (S.-L.L.); 965090@cpc.com.tw (Y.-H.H.)
5 Department of Environmental Engineering and Management, Chaoyang University of Technology, Taichung 413, Taiwan; sanada.r88686@msa.hinet.net
6 Department of Horticulture and Landscape Architecture, National Taiwan University, Taipei 106, Taiwan
* Correspondence: topage2@gmail.com (P.-C.T.); shufen@cyut.edu.tw (S.-F.C.)
† Presented at the 3rd IEEE International Conference on Electronic Communications, Internet of Things and Big Data Conference 2023, Taichung, Taiwan, 14–16 April 2023.

Abstract: Weathered petroleum-contaminated soil was treated with *Vetiveria zizanioides* (Vetiver) and *Cymbopogon nardus* (Lemongrasss) to investigate the efficiency of phytoremediation. The initial total petroleum hydrocarbon (TPH) concentration of soil was 3000−8000 mg/kg, and after 6 months, the TPH concentrations were degraded by 50−75% under the action of soil native microbial. Planting vetiver and lemongrass stabilized soil pH and electrical conductivity, and it accelerated the decomposition of TPH in soil. Planting vetiver showed a better effect. After 6 months of planting, the TPH decomposition efficiency reached about 90%, and most of the easily decomposed TPH has been decomposed. The results of rhizosphere soil microbiota analysis also showed that planting vetiver increased the abundance of soil microbiota.

Keywords: TPH; phytoremediation; vetiver; lemongrass

Citation: Cheng, P.-C.; Lin, Y.-C.; Lin, M.-S.; Lin, S.-L.; Hsiao, Y.-H.; Huang, C.-Y.; Tu, P.-C.; Cheng, S.-F. Phytoremediation Efficiency of Weathered Petroleum-Contaminated Soils by *Vetiveria zizanioides* and *Cymbopogon nardus itle*. *Eng. Proc.* **2023**, *38*, 63. https://doi.org/10.3390/engproc2023038063

Academic Editors: Teen-Hang Meen, Hsin-Hung Lin and Cheng-Fu Yang

Published: 29 June 2023

Copyright: © 2023 by the authors. Licensee MDPI, Basel, Switzerland. This article is an open access article distributed under the terms and conditions of the Creative Commons Attribution (CC BY) license (https://creativecommons.org/licenses/by/4.0/).

1. Introduction

Petroleum is the main fuel for human life. Thus, petroleum contamination of soil is often caused by accidental spills of petroleum from manufacturing, storage, and transportation [1–3]. The main components of petroleum are alkane hydrocarbons, which are therefore referred to as total petroleum hydrocarbon (TPH). A lot of the components in petroleum have adverse effects on human health. Therefore, petroleum is an important pollutant for soil in many countries. Remediation technologies for TPH-contaminated soil at present include biological remediation, soil vapor extraction, thermal desorption, and chemical oxidation. Compared with physical and chemical remediation technologies, the bioremediation method using microorganisms is more economical with low energy consumption and is environmentally friendly [3–5]. However, for TPH-contaminated soil with high carbon numbers and weathering, the decomposition efficiency of bioremediation was often not high.

Phytoremediation is a remediation technology that uses plants to remove pollutants [6,7]. It is used to remediate various pollutants, including heavy metals, inorganic salts, and organic pollutants. The mechanism of phytoremediation for the removal of organic pollutants includes phytoextraction, phytodegradation, phytovolatilization, and rhizosphere bioremediation [8–12]. Phytoextraction is to uptake and accumulate pollutants in plants through the absorption of water and nutrients by plants in order to remove pollutants from the soil. The degradation of pollutants through the metabolism of plants is called phytodegradation or phytotransformation. The evaporation of pollutants into the atmosphere through the evapotranspiration of plant leaves is called phytovolatilization [8–10]. In addition, plant roots release exudates, such as low-molecular-weight organic acids, to the soil to stimulate the degradation of organic chemicals, the growth of new species, and/or increase soluble substrate concentrations for all microorganisms, which is known as phytostimulation or plant-assisted bioremediation [11,12].

In addition to the advantages of less energy consumption, low cost, and less impact on soil properties, phytoremediation greens landscapes, and it is the most accepted remediation technology by the public. Corresponding to the international goal of promoting net zero by 2050, phytoremediation reduces the energy required for remediation and achieves the effect of carbon reduction. Thus, it needs to be used more. Therefore, with vetiver and lemongrass that have economic values and strong growth ability, we carried out the phytoremediation experiment of petroleum-contaminated soil to study the efficiency of remediation of the two plants on weathered petroleum-contaminated soil. Vetiver is an evergreen perennial herb and is propagated by divisions. It grows in an environment with an annual rainfall of 300−6,000 mm and survives for 8 months in a water-soaked environment and for 5 months in the arid desert edge. It can survive in harsh soil environments such as acidic, alkaline, saline land, sandy land, gravel land, and mining spoil, and in a temperature of −15 to 55 °C. Vetiver has strong ecological adaptability and resistance to adversity and has the advantages of fast growth, a well-developed root system, and easy cultivation. Lemongrass is a perennial herb and is propagated by seeds or divisions. Lemongrass prefers a sunny and warm climate with strong drought tolerance. It does not restrict a type of soil with a well-developed root system. The leaves contain lemon fragrance, which can be distilled to extract essential oils and which have economic value.

2. Methods

The soil in this study was collected from the site of the Taichung oil supply center of China Petroleum Corporation in Taiwan in the coordinates of 120.5432 E and 24.2993 N The soil was classified into three soils according to the TPH concentrations. The average concentration of the low TPH concentration was about 3000 mg/kg, the middle concentration was about 4500 mg/kg, and the high concentration was about 8000 mg/kg. The experiment was carried out with potted plants, each of which was filled with about 30 kg of soil. Three experimental soils were tested: control, vetiver, and lemongrass groups. Each experimental group had three soils. The control group did not have any plants, and the same method was applied with vetiver and lemongrass planted, with the same growth conditions of water and sunshine. The rhizosphere soil was collected every two months for the analysis of pH, electrical conductivity, and TPH concentration. For the measurement of soil pH, 20 g of air-dried soil was taken and added 20 mL of reagent water. After mixing and stirring, the solution was left to take the supernatant liquid to measure pH with a pH electrode. Soil conductivity was measured as the conductivity of the filtrate with a conductivity meter after taking 10 g of air-dried soil, to which was added 50 mL of deionized water and shaken at 140 rpm for 1 h and filtered with Whatman No.5 filter paper. For the analysis of soil TPH concentration, 2 g of soil was taken and 10 mL of n-hexane was added, which was extracted by ultrasonic wave, concentrated under reduced pressure, and then quantified by GC-FID. Before the test and 5 months after the test, the soil microbiological analysis was carried out by the Next Generation Sequencing method.

3. Results and Discussion

3.1. Soil Properties

The particle size analysis result of the soils with the three concentrations of TPH showed that the proportion of sand particles accounted for about 98.0% and silt for about 2.0%. Thus, the soil was classified to be sandy soil. The average concentration of TPH in the low-concentration soil was 3029 mg/kg, pH was 6.97, electrical conductivity was 171.2 μS/cm, and organic matter content was 7.43%. The average concentration of TPH in the medium-concentration soil was 4617 mg/kg, pH was 7.39, electrical conductivity was 169.8 μS/cm, and organic matter content was 7.90%. The average concentration of TPH in high-concentration soil was 7865 mg/kg, pH was 7.60, electrical conductivity was 167.7 μS/cm, and organic matter content was 8.67%.

The soil pH after 6 months of the test was shown in Figure 1. After 6 months of the test, the pH of the soil increased. However, the increase in pH in the soil planted with vetiver and lemongrass was lower than that in the control group. Figure 1b shows the measured results of soil electrical conductivity after 6 months of the test. The electrical conductivity of all soils increased significantly after 6 months of testing. The increase in the soil with vetiver was the smallest, followed by the soil with lemongrass, and the soil without plant showed the largest increase in electrical conductivity. Such results show that planting plants slowed the changes in soil pH and electrical conductivity.

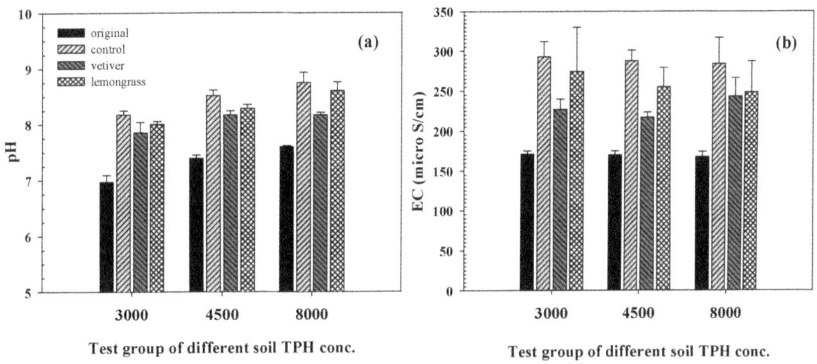

Figure 1. Properties of soil samples. (**a**) pH and (**b**) electrical conductivity.

3.2. TPH Concentration

Figure 2 shows the change in the TPH concentration in the control soil. The initial average concentration of TPH in low-concentration soil was 3029 mg/kg. After 2 months, the average concentration was 1455 mg/kg, and the degradation rate of TPH was about 52%. After 6 months, the average concentration of TPH was about 904 mg/kg, and the degradation rate of TPH was about 70%. During the period of 6−15 months, the residual TPH concentration in the soil did not change much. At 15 months, the average concentration of TPH was about 827 mg/kg, and the average removal rate was about 73%. The initial average concentration of TPH in the medium-concentration soil was 4617 mg/kg. After 2 months, the average concentration was 2533 mg/kg, and the TPH degradation rate was about 45%. After 6 months, the average TPH concentration was about 2000 mg/kg, and the TPH degradation rate was about 56%. At 15 months, the average concentration of TPH was about 1109 mg/kg, and the degradation rate of TPH was about 76%. The average initial TPH concentration of high-concentration soil was 7865 mg/kg, and after 2 months, the average soil TPH concentration was 5125 mg/kg, and the TPH degradation rate was about 35%. At 6 months, the average TPH concentration was about 3893 mg/kg, and the TPH degradation rate was about 50%/. After 15 months, the average TPH concentration was about 1285 mg/kg, and the TPH degradation rate was about 83%.

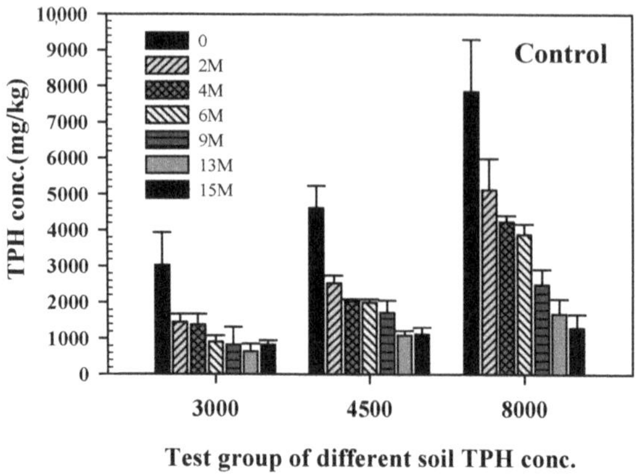

Figure 2. The change of soil TPH concentration in the control soil.

Figure 3 showed the change in the TPH concentration in the soil with vetiver. After 2 months in low-concentration soil, the average concentration of TPH in the soil was 897 mg/kg, and the degradation rate of TPH was about 70%. After 6 months, the average concentration of TPH was about 278 mg/kg, and the degradation rate of TPH was about 90%. During 6−15 months, the soil TPH concentration did not change much, indicating that most of the easily decomposed TPH had been decomposed at 6 months. In medium-concentration soil, after 2 months, the average concentration of TPH in the soil was 1431 mg/kg, and the TPH degradation rate was about 70%. After 6 months, the average concentration of TPH was about 578 mg/kg, and the TPH degradation rate was 88%. After 15 months, the average TPH concentration was about 528 mg/kg, and the TPH degradation rate increased to 89%. For high-concentration soil, after 2 months, the average concentration of TPH in the soil was 4478 mg/kg, and the degradation rate of TPH was about 43%. After 6 months, the average concentration of TPH was about 1616 mg/kg, and the degradation rate of TPH was about 80%. After 15 months, the average concentration of TPH was about 927 mg/kg, and the degradation rate of TPH was close to 90%.

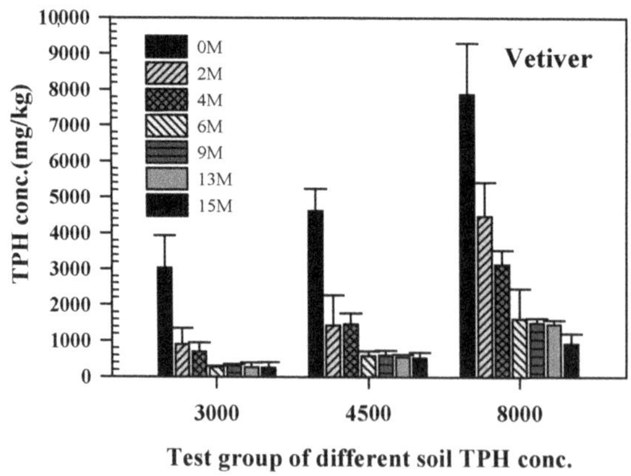

Figure 3. The change of soil TPH concentration with vetiver.

Figure 4 showed the variation of TPH concentration in the soil with lemongrass. For the low-concentration soil, after 2 months, the average concentration of TPH in the soil was 669 mg/kg, and the degradation rate of TPH was about 78%. After 6 months, the average concentration of TPH was about 263 mg/kg, and the degradation rate of TPH was about

91%. From 6 to 15 months, the residual TPH concentration in the soil did not change much. In the medium-concentration soil, after 2 months, the average concentration of TPH in the soil was 1015 mg/kg, and the degradation rate of TPH was about 78%. After 6 months, the average concentration of TPH was about 587 mg/kg, and the degradation rate of TPH was about 87%, and little change in TPH concentration was observed up to 15 months. In the high-concentration soil, after 2 months, the average concentration of TPH in the soil was 3669 mg/kg, and the TPH degradation rate was about 53%. After 6 months, the average concentration of TPH was about 2245 mg/kg, and the TPH degradation rate was about 71%. After 15 months, the TPH average concentration was about 1126 mg/kg, and the degradation rate of TPH was about 86%.

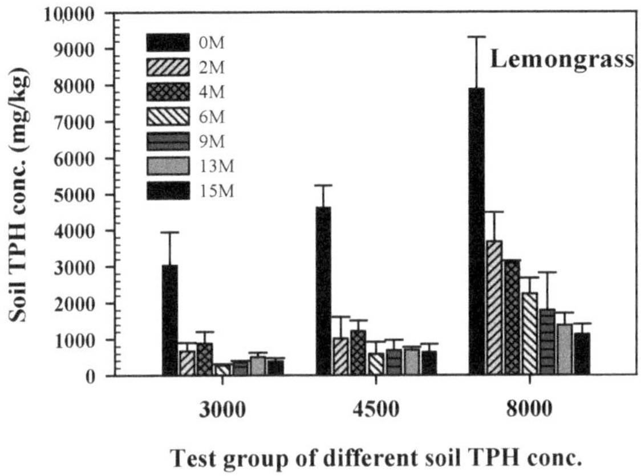

Figure 4. The change of soil TPH concentration in the soil with lemongrass.

In the control soil without planting plants, the soil TPH slowly degraded. During the 15-month test period, the soil removal rate of the low-concentration soil was about 73%, that of the medium-concentration soil was about 76%, and that of the high-concentration soil was about 83%. The TPH degradation efficiency of the high-, medium-, and low-concentration soil with vetiver reached about 90% in 6 months, and the TPH degradation rate remained almost similar during 6 to 15 months. It showed that TPH was easily degraded by up to 90%. With lemongrass, the degradation rate of TPH was close to 90% in 6 months for the medium-and low-concentration soils. The degradation rate during 6 to 15 months was similar to that with vetiver without a significant increase. However, the degradation efficiency of TPH in the high-concentration soil was relatively slow, showing a degradation rate of about 71% in 6 months and 86% in 15 months. For the high-concentration soil, the TPH degradation efficiency with lemongrass was less than that with vetiver.

3.3. Soil Bacteria

Table 1 showed the distribution of bacteriophage in the soils before the test and after 5 months of the test. For the low-concentration soil (3000 mg/kg), before the test, the bacteria in the soil included 715 genera and 1092 species. After 5 months, 604 genera and 772 species remained. There were 865 genera and 1123 species in the soil with vetiver while 612 genera and 768 species in that with lemongrass. After 5 months of experimentation, except for the soil with vetiver, the bacteriophage in the soils with lemongrass and without plant decreased. For the medium-concentration soil (4500 mg/kg), before the test, the soil bacteria included 735 genera and 1148 species. After 5 months, the control soil had 640 genera and 832 species, 823 genera, and 1086 species with vetiver, and 706 genera and 909 species with lemongrass. For the high-concentration soil (8000 mg/kg), before the test, the soil bacteria included 695 genera and 1078 species. After 5 months of testing, the control soil had 627 genera and 827 species, 721 genera and 947 species in the soil with vetiver, and 670 genera and 877 species in the soil with lemongrass. The three TPH concentration soils showed that the bacteriophage with vetiver was the most abundant.

Table 1. Changes in soil bacterial phase abundance before and after 5 months of the experiment.

Test Group	TPH 3000 mg/kg		TPH 4500 mg/kg		TPH 8000 mg/kg	
	Genus	Species	Genus	Species	Genus	Species
Before test	715	1092	735	1148	695	1078
Control	604	772	640	832	627	827
Vetiver	865	1123	823	1086	721	947
Lemongrass	612	768	706	909	670	877

The abundance of bacteria in the control soil decreased after 5 months of the test. It is speculated that the longer the test time, the fewer available nutrients in the soil, resulting in a decrease in the abundance of bacteria. The bacterial abundance of the soil with lemongrass also decreased after 5 months. However, the soils with plants had more abundant bacteria than the control soil. The soil bacterial abundance with lemongrass did not increase, which may be related to the poor growth of the lemongrass. After 5 months of the experiment, bacteria abundance in the soil with vetiver increased compared with other soils, which showed that vetiver helped increase the richness of the microbiota in the soil.

4. Conclusions and Suggestion

The results of this study showed that the remediation of the petroleum-contaminated soil by the native microorganisms was slow, and the degradation rate of TPH was about 50−70% in 6 months. Planting vetiver and lemongrass stabilized soil pH and electrical conductivity and accelerated the degradation of TPH in soil with vetiver having a better effect. In this study, phytoremediation by vetiver was more effective with a TPH degradation rate of about 90% in 6 months. The bacteriophage in the rhizosphere of the soils also showed that planting vetiver increased the abundance of soil microbiota. Phytoremediation with vetiver was helpful to remediate petroleum-contaminated soil. Other than the remediation effect, vetiver has economic value as it is used to extract aromatic oil and used to produce handicraft materials, papers, and fuels. Thus, it is recommended to use the vetiver for the phytoremediation of petroleum-contaminated soils with multiple benefits.

Author Contributions: Conceptualization, P.-C.C. and S.-F.C.; methodology, P.-C.C. and P.-C.T.; investigation, M.-S.L., S.-L.L., Y.-H.H. and C.-Y.H.; writing—original draft preparation, P.-C.C.; writing—review and editing, Y.-C.L. and S.-F.C.; project administration, S.-F.C. All authors have read and agreed to the published version of the manuscript.

Funding: This research was funded by the Fund of Soil and Groundwater Contamination Remediation, Environmental Protection Administration of Taiwan.

Institutional Review Board Statement: Not applicable.

Informed Consent Statement: Not applicable.

Data Availability Statement: Not applicable.

Conflicts of Interest: The authors declare no conflict of interest.

References

1. Lv, Y.; Bao, Y.J.; Zhu, L. A comprehensive review of recent and perspective technologies and challenges for the remediation of oil-contaminated sites. *Energy Rep.* **2022**, *8*, 7976–7988. [CrossRef]
2. Ambaye, T.G.; Chebbi, A.; Formicola, F.; Prasad, S.; Gomez, F.H.; Franzetti, A.; Vaccari, M. Remediation of soil polluted with petroleum hydrocarbons, and their reuse for agriculture: Recent progress, challenges, and perspectives. *Chemosphere* **2022**, *293*, 133572. [CrossRef] [PubMed]

3. Sui, X.; Li, Y.; Ji, H. Remediation of petroleum-contaminated soils with microbial and microbial combined methods: Advances, mechanisms, and challenges. *Sustainability* **2021**, *13*, 9267. [CrossRef]
4. Zahed, M.A.; Salehi, S.; Madadi, R.; Hejabi, F. Biochar as a sustainable product for remediation of petroleum contaminated soil. *Curr. Res. Green Sustain. Chem.* **2021**, *4*, 100055. [CrossRef]
5. Kumari, B.; Singh, G.; Sinam, G.; Singh, D.P. Microbial remediation of crude oil-contaminated sites. In *Environmental Concerns and Sustainable Development*; Shukla, V., Kumar, N., Eds.; Springer: Singapore, 2020; pp. 333–351. [CrossRef]
6. Curiel-Alegre, S.; Velasco-Arroyo, B.; Rumbo, C.; Khan, A.H.A.; Tamayo-Ramos, J.A.; Rad, C.; Gallego, J.L.R.; Barros, R. Evaluation of biostimulation, bioaugmentation, and organic amendments application on the bioremediation of recalcitrant hydrocarbons of soil. *Chemosphere* **2022**, *307*, 135638. [CrossRef] [PubMed]
7. Wang, R.; Wu, B.; Zheng, J.; Chen, H.; Rao, P. Biodegradation of total petroleum hydrocarbons in soil: Isolation and characterization of bacterial strains from oil contaminated soil. *Appl. Sci.* **2020**, *10*, 4173. [CrossRef]
8. Liu, Y.; Li, C.; Huang, L.; He, Y.; Zhao, T.; Han, B.; Jia, X. Combination of a crude oil-degrading bacterial consortium under the guidance of strain tolerance and a pilot-scale degradation test. *Chin. J. Chem. Eng.* **2017**, *25*, 1838–1846. [CrossRef]
9. Mara, K.; Decorosi, F.; Viti, C.; Giovannetti, L.; Papaleo, M.C.; Maida, I.; Perrin, E.; Fondi, M.; Vaneechoutte, M.; Nemec, A. Molecular and phenotypic characterization of *Acinetobacter* strains able to degrade diesel fuel. *Res. Microbiol.* **2012**, *163*, 161–172. [CrossRef] [PubMed]
10. Rong, L.; Zheng, X.; Oba, B.T.; Shen, C.; Wang, X.; Wang, H.; Luo, Q.; Sun, L. Activating soil microbial community using bacillus and rhamnolipid to remediate TPH contaminated soil. *Chemosphere* **2021**, *275*, 130062. [PubMed]
11. Basim, Y.; Mohebali, G.; Jorfi, S.; Nabizadeh, R.; Moghadam, M.A.; Ghadiri, A.; Fard, N.J.H. Bacterial strains diversity in contaminated soils and their potential for bioremediation of total petroleum hydrocarbons in southwest of Iran. *J. Environ. Health Sci. Eng.* **2022**, *20*, 601–608. [CrossRef] [PubMed]
12. Fondi, M.; Maida, I.; Perrin, E.; Orlandin, V.; Torre, L.; Bosi, E.; Negroni, A.; Zanaroli, G.; Fava, F.; Decorosi, F. Genomic and phenotypic characterization of the species Acinetobacter venetianus. *Sci. Rep.* **2016**, *6*, 21985. [PubMed]

Disclaimer/Publisher's Note: The statements, opinions and data contained in all publications are solely those of the individual author(s) and contributor(s) and not of MDPI and/or the editor(s). MDPI and/or the editor(s) disclaim responsibility for any injury to people or property resulting from any ideas, methods, instructions or products referred to in the content.

engineering proceedings

MDPI

Proceeding Paper

Assessment of Inverse Distance Weighting and Local Polynomial Interpolation for Annual Rainfall: A Case Study in Peninsular Malaysia †

Ren Jie Chin [1,*], Sai Hin Lai [2], Wing Son Loh [3], Lloyd Ling [1,*] and Eugene Zhen Xiang Soo [1]

1 Department of Civil Engineering, Lee Kong Chian Faculty of Engineering and Science, Universiti Tunku Abdul Rahman, Kajang 43000, Malaysia; szx.eugene@gmail.com
2 Department of Civil Engineering, Faculty of Engineering, Universiti Malaya, Kuala Lumpur 50603, Malaysia; laish@um.edu.my
3 Department of Mathematical and Actuarial Sciences, Lee Kong Chian Faculty of Engineering and Science, Universiti Tunku Abdul Rahman, Kajang 43000, Malaysia; lohws@utar.edu.my
* Correspondence: chinrj@utar.edu.my (R.J.C.); linglloyd@utar.edu.my (L.L.)
† Presented at the 3rd IEEE International Conference on Electronic Communications, Internet of Things and Big Data Conference 2023, Taiwan, China, 14–16 April 2023.

Abstract: Rainfall data are crucial in hydrology models. In this study, the assessment of two spatial interpolation approaches of Inverse Distance Weighting (IDW) and Local Polynomial Interpolation (LPI) for rainfall in Peninsular Malaysia was conducted. The daily precipitation for 515 rainfall stations across Peninsular Malaysia during 2011–2020 was used as the reference data. The performance of IDW and LPI was evaluated by the computation of the coefficient of determination (R^2), the mean absolute error (MAE), and the root mean square error (RMSE). The results show that LPI methods surpass IDW methods on the annual scale rainfall interpolations in Peninsular Malaysia by exhibiting a better statistical evaluation.

Keywords: interpolation; Inverse Distance Weighting; Local Polynomial Interpolation; rainfall

Citation: Chin, R.J.; Lai, S.H.; Loh, W.S.; Ling, L.; Soo, E.Z.X. Assessment of Inverse Distance Weighting and Local Polynomial Interpolation for Annual Rainfall: A Case Study in Peninsular Malaysia. *Eng. Proc.* **2023**, *38*, 61. https://doi.org/10.3390/engproc2023038061

Academic Editors: Teen-Hang Meen, Hsin-Hung Lin and Cheng-Fu Yang

Published: 29 June 2023

Copyright: © 2023 by the authors. Licensee MDPI, Basel, Switzerland. This article is an open access article distributed under the terms and conditions of the Creative Commons Attribution (CC BY) license (https://creativecommons.org/licenses/by/4.0/).

1. Introduction

Rainfall data are crucial for hydrological modeling when anticipating extreme precipitation events such as droughts and floods and assessing the quantity and quality of the surface and groundwater. However, in most situations, the precipitation measurement station network is poor, and the data supplied are insufficient to define the highly variable precipitation and its geographical distribution. This is particularly true in underdeveloped nations such as Algeria, where the complexity of rainfall distribution is compounded by measuring challenges. As a result, the methods for estimating rainfall in regions where rainfall has not been recorded must be established based on data from nearby meteorological stations [1–3].

One of the ways to forecast rainfall is by using spatial interpolation techniques. In environmental management, geographic continuous data (or spatial continuous surfaces) are important for planning, risk assessment, and decision-making. They are, however, not always widely available and can be difficult and expensive to obtain, especially in hilly or deep-water places. During field surveys, environmental data are frequently collected from point sources. Environmental managers, on the other hand, typically require precise geographic continuous data throughout an area to make effective and confident choices, whereas scientists require accurate spatial and continuous data across a region to make justified conclusions [4–6].

The spatial continuous data of environmental variables have become more significant as geographic information systems (GIS) and modeling approaches have become more powerful for the conservation of natural resources and biological conservation. As a result, attribute values at unsampled places must be inferred, requiring spatial interpolation from data sets for spatial continuous data. Once the variational surface has various degrees of resolution, the cell density, or inclination other than what is required, is also necessary [7]. Furthermore, when a continuous region is presented by a different information type than what is required, and the confirming data do not completely cover the region of interest, spatial interpolation is required [8]. Therefore, this study aimed to investigate accurate and efficient spatial interpolation methods to evaluate rainfall data.

2. Methodology

The historical daily precipitation data (2011–2020) of 550 rain gauges were obtained from the Department of Irrigation and Drainage Malaysia (DID). Since rain gauges only show the point sampling of a storm's areal spread, before using the data, it is necessary to undergo the process of a quality check, which is important to ensure that rainfall data are consistent. Stations with no missing data were acceptable in this study, while stations with missing data were further categorized into categories of less than 10% and more than 10%. According to Burhanuddin et al. [9], only data with a low quantity of missing data (less than 10%) could be considered good quality data, and thus, stations with more than 10% of missing data were directly eliminated from the study. Apart from this, according to Chow et al. [10], for station X with less than 10% missing data, the arithmetic procedure could be adopted to estimate the missing observation of station X.

$$\bar{p} = \frac{\sum_{i=1}^{n} p}{n} \tag{1}$$

$\bar{p}$ = combined mean for the rainfall station
n = number of the rainfall station
i = individual rainfall station

Hence, in this study, missing data were filled with precipitation values from the nearest stations, which tended to have similar characteristics using the arithmetic procedure. Next, the daily precipitation was aggregated into yearly, monthly, and daily scales for better comparison. After the process of data acquisition, a total of 515 stations were applied to the research.

The methods of spatial interpolation were created for specific data types or variables. Li and Heap [11] analyzed the essential elements of the most often utilized approaches. The precipitation value at a location with no recorded data could be determined using known precipitation readings at nearby weather stations through spatial interpolation. Spatial interpolation is a technique for generating surface data from a set of sample points, which can then be used for analysis and modeling. In this study, ArcMap 10.8 was used to create maps, compile geographic data, and analyze mapped information. Two spatial interpolation methods were used in this study, which included Inverse Distance Weighted (IDW) and Local Polynomial Interpolation (LPI).

All interpolation methods were based on the assumption that points closer together could have more correlations and similarities than those further apart. The rate of these correlations and similarities between neighbors was proportional to the distance between them in the IDW approach, which could be defined as the distance reverse function of each location from the surrounding points. It is vital to remember that the specification of the nearby radius and the accompanying power to the distance reverse function were considered significant difficulties in this approach. A state with a sufficient number of sample sites (at least 14) and an appropriate degree of dispersion in local scale levels was essential to apply this strategy. The value of the power parameter was the most important element impacting the accuracy of the inverse distance interpolator. IDW made use of Equation (2).

$$Z_o = \frac{\sum_{i=1}^{s} Z_i \frac{1}{d_i^K}}{\sum_{i=1}^{s} \frac{1}{d_i^K}} \quad (2)$$

Z_o = Predicted value at the unsampled site
Z_i = Observed value
D_i = The distance between the prediction and measured locations
s = The number of measured sampling points within the neighborhood
K = Power parameter defining the rate of reduction in weights as the distance increase

IDW must be an accurate interpolator to avoid division by zero at the sample points when $d_{i0} = 0$. Furthermore, the interpolated surface's maximum and lowest values can occur only at the data points. Although IDW is a quick interpolation method, it is prone to outliers and data clustering. Furthermore, this technique does not provide an implicit evaluation of the forecast's accuracy.

LPI is used to fit each polynomial within a particular overlapping neighborhood. The search neighborhood can be chosen using the search neighborhood conversation. The form, the maximum and lowest number of points, and the sector organization are all selectable. Surfaces that capture a short-range variation can be produced through LPI. As an alternative, a slider can be used to select the neighborhood's width and a power parameter which, based on the neighborhood's sample points' distance, lessens their weights. As a result, LPI creates surfaces that account for more local variation.

In this research, statistical analysis was used to assess the competency of a model on unknown data. The selected statistical parameters in this study were the coefficient of determination (R^2), the mean absolute error (MAE), and the root mean square error (RMSE).

3. Results and Discussion

This section compares the spatial interpolation results using IDW and LPI with the ground-based rainfall data on an annual scale. Both IDW and LPI interpolation results are shown in their graphical form (Figure 1). Additionally, the performance of IDW and LPI were compared using statistical analyses covering R^2, (MAE), and RMSE. These results are presented in Table 1. The R^2 values after the interpolation procedures ranged between 0.35 and 0.69. For the IDW method, the lowest R^2 value was recorded at 0.35 in the year 2015, and the highest R^2 value was 0.65 in the year 2011. Meanwhile, for the LPI method, the lowest R^2 value was 0.39 in the year 2015, and the highest was 0.69 in the year 2011. For MAE, the value ranged between 0.79 and 1.16 for the IDW method, where the lowest value was recorded in the year 2016, and the highest value was shown in the year 2011. Meanwhile, for the LPI method, the lowest MAE value was 0.75 in the year 2016, and the highest was 1.12 in the year 2011. In terms of RMSE, the lowest value for the IDW method was 1.09, which is slightly higher than that of the LPI method, with a value of 1.07. The lowest RMSE value for both methods was recorded in the year 2016. The highest value of RMSE for the IDW method was 1.66 in the year 2014, while for the LPI method, the highest value was 1.60, recorded in the year 2017. By comparing the result of R^2, MAE, and RMSE calculations on an annual scale for the IDW and LPI methods, it was concluded that the LPI method outperformed the IDW method. This was because the MAE and RMSE values of LPI were always lower than that of IDW. At the same time, a higher R^2 value could also be found in LPI compared to IDW.

Table 1. Statistical analyses of IDW and LPI compared to ground-based rainfall data.

Year	IDW			LPI		
	RMSE	MAE	R^2	RMSE	MAE	R^2
2011	1.64	1.16	0.65	1.55	1.12	0.69
2012	1.51	1.08	0.49	1.44	1.04	0.53
2013	1.57	1.11	0.58	1.50	1.06	0.61
2014	1.66	1.13	0.54	1.58	1.09	0.57
2015	1.31	0.91	0.35	1.25	0.88	0.39
2016	1.09	0.79	0.59	1.07	0.75	0.61
2017	1.64	1.13	0.64	1.60	1.09	0.65
2018	1.32	0.96	0.48	1.26	0.93	0.52
2019	1.40	1.06	0.47	1.33	0.99	0.52
2020	1.33	1.00	0.51	1.32	0.98	0.52

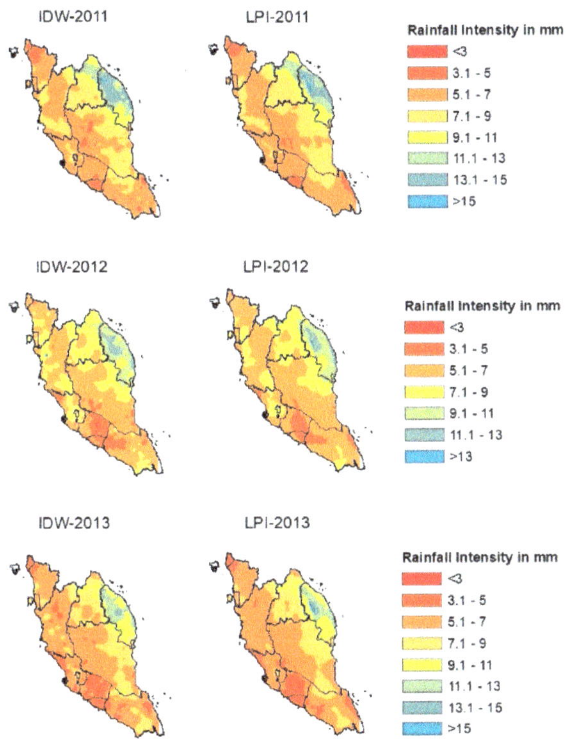

(a)

Figure 1. *Cont.*

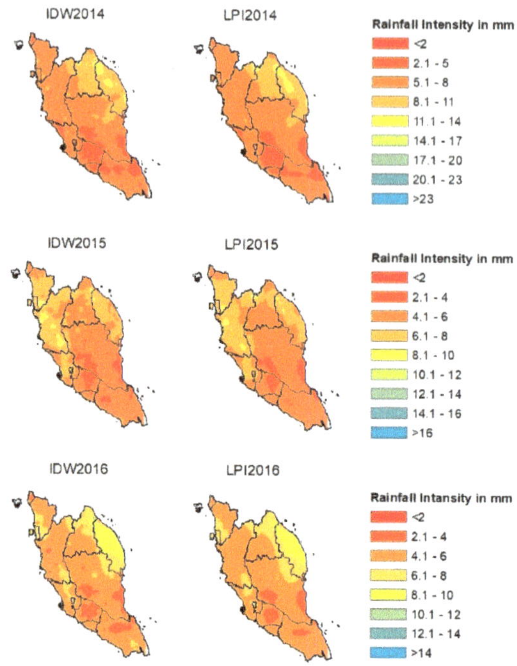

(b)

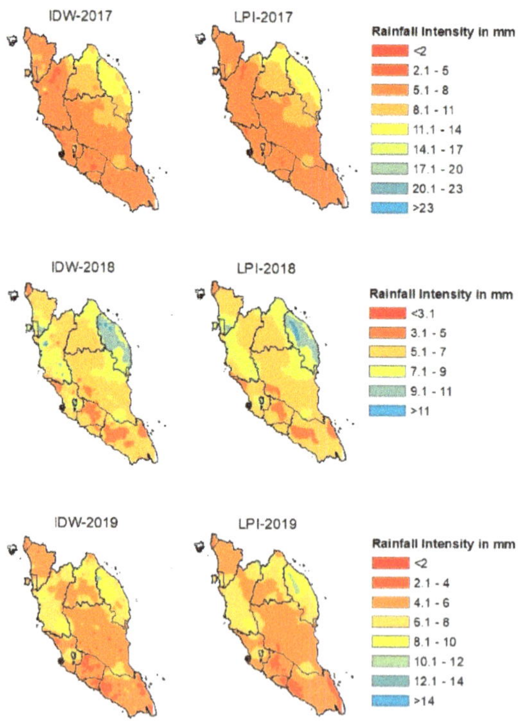

(c)

Figure 1. *Cont.*

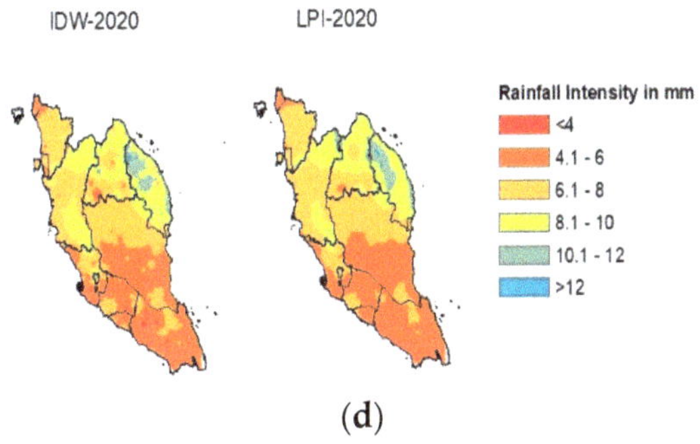

Figure 1. (**a**) Comparison of Annual Rainfall Intensity in mm using IDW and LPI methods with the ArcGIS Map during 2011–2013. (**b**) Comparison of Annual Rainfall Intensity in mm using IDW and LPI methods with the ArcGIS Map during 2014–2016. (**c**) Comparison of Annual Rainfall Intensity in mm using IDW and LPI methods with the ArcGIS Map during 2017–2019. (**d**) Comparison of Annual Rainfall Intensity in mm using IDW and LPI methods with the ArcGIS Map in 2020.

4. Conclusions

The maps of rain data were generated for each year from the year 2011 to 2020. The statistical analyses, including R^2, MAE, and RMSE, were implemented to test the performance of IDW and LPI as interpolation methods. In conclusion, the statistical analyses showed that the LPI method exhibited better performance than the IDW method as it had lower MAE and RMSE values but high R^2. In future research, the study duration can be expanded to 10 or 15 years to obtain more reliable data and reduce performance errors. In addition, a variety of widely used methodologies, such as triple collocation analysis, which does not require real values, could be proposed for more accurate study.

Author Contributions: Conceptualization, R.J.C. and S.H.L.; methodology, R.J.C. and S.H.L.; formal analysis, W.S.L. and E.Z.X.S.; writing—original draft preparation, W.S.L. and E.Z.X.S.; writing—review and editing, R.J.C. and L.L.; funding acquisition, R.J.C. and L.L. All authors have read and agreed to the published version of the manuscript.

Funding: This research was funded by Ministry of Higher Education (MoHE) Malaysia through the Fundamental Research Grant Scheme project (FRGS/1/2021/WAB07/UTAR/02/1), Kurita Asia Research Grant (8128/0002) provided by Kurita Water and Environment Foundation and Universiti Tunku Abdul Rahman Research Fund (IPSR/RMC/UTARRF/2020-C2/C04; IPSR/RMC/UTARRF/2022-C2/C04).

Institutional Review Board Statement: Not applicable.

Informed Consent Statement: Not applicable.

Data Availability Statement: The data presented in this study are available on request from the corresponding author.

Acknowledgments: The authors would like to thank Lee Kong Chian, Faculty of Engineering and Science, Universiti Tunku Abdul Rahman, and Faculty of Engineering, Universiti Malaya, for all the technical support provided for this study.

Conflicts of Interest: The authors declare no conflict of interest.

References

1. Ceron, W.L.; Andreolo, R.V.; Kayano, M.T.; Canchal, T.; Carvajal-Escobar, Y.; Souza, R.A.F. Comparison of spatial interpolation methods for annual and seasonal rainfall in two hotspots of biodiversity in South America. *An. Acad. Bras.Ciên.* **2021**, *93*, 1–22. [CrossRef] [PubMed]
2. De Silva, R.P.; Dayawansa, N.D.K.; Ratnasiri, M.D. A comparison of methods used in estimating missing rainfall data. *J. Agric. Sci.* **2007**, 3, 101–108. [CrossRef]

3. Keblouti, M.; Ouerdachi, L.; Boutaghane, H. Spatial Interpolation of Annual Precipitation in Annaba-Algeria—Comparison and Evaluation of Methods. *Energy Procedia* **2012**, *18*, 468–475. [CrossRef]
4. Ahrens, B. Distance in spatial interpolation of daily rain gauge data. *Hydrol. Earth Syst. Sci.* **2006**, *10*, 197–208. [CrossRef]
5. Jamaludin, S.; Sayang Mohd, D.; Wan, W.Z.; Abdul Aziz, J. Trends in peninsular Malaysia rainfall data during the southwest monsoon and northeast monsoon seasons: 1975–2004. *Sains Malays.* **2021**, *39*, 533–542.
6. Tanjung, M.; Syahreza, S.; Rusdi, M. Comparison of interpolation methods based on Geographic Information System (GIS) in the spatial distribution of seawater intrusion. *J. Nat.* **2020**, *20*, 24–30. [CrossRef]
7. Liang, J.; Tan, M.L.; Hawcroft, M.; Catto, J.L.; Hodges, K.L.; Haywood, J.M. Monsoonal precipitation over Peninsular Malaysia in the CMIP6 HighResMIP experiments: The role of model resolution. *Clim. Dyn.* **2021**, *1432*, 1–23. [CrossRef]
8. Burrough, P.A.; McDonnell, R.A.; Lloyd, C.D. *Principles of Geographical Information Systems*, 3rd ed.; Oxford University Press: Oxford, UK, 2015.
9. Burhanuddin, S.N.Z.A.; Deni, S.M.; Ramli, N.M. Imputation of Missing Rainfall Data Using Revised Normal Ratio Method. *Adv. Sci. Lett.* **2017**, *23*, 10981–10985. [CrossRef]
10. Chow, C.W.; Cooper, J.C.; Waller, W.S. Participative Budgeting: Effects of a Truth-Inducing Pay Scheme and Information Asymmetry on Slack and Performance. *Account. Rev.* **1988**, *63*, 111–122.
11. Li, J.; Heap, A.D. Spatial interpolation methods applied in the environmental sciences: A review. *Environ. Model. Softw.* **2014**, *53*, 173–189. [CrossRef]

Disclaimer/Publisher's Note: The statements, opinions and data contained in all publications are solely those of the individual author(s) and contributor(s) and not of MDPI and/or the editor(s). MDPI and/or the editor(s) disclaim responsibility for any injury to people or property resulting from any ideas, methods, instructions or products referred to in the content.

engineering proceedings

MDPI

Proceeding Paper

Applied Research on Optimal Scale and M&A Efficiency in the Natural Gas Energy Industry †

Ting-Kun Liu

Department of Finance Chaoyang, University of Technology, Taichung 41349, Taiwan; tkliu@cyut.edu.tw

† Presented at the 3rd IEEE International Conference on Electronic Communications, Internet of Things and Big Data Conference 2023, Taichung, Taiwan, 14–16 April 2023.

Abstract: Currently, the number of manufacturers in the natural gas energy industry in several regions is stagnant. However, to improve operating efficiency, it is necessary to review the current optimum number of public gas fuel utilities. Therefore, using the manufacturers of Taiwan's natural gas energy industry and the number of natural gas energy users as a network variable from 1995 to 1999 as panel data, the consideration of network effects of the translog cost function model is incorporated in this study in order to understand the difference from the previous research.

Keywords: natural gas energy; industry application; translog cost function; M&A; operating efficiency; optimal scale

1. Introduction

Compared with coal and oil, natural gas is a low-carbon and clean energy source, causing less air pollution. With the growing awareness of the importance of environmental protection, the world's energy policies are gradually changing to increase the proportion of low-polluting energy sources. At present, natural gas energy in Taiwan only accounts for 5% of total domestic energy, so there is still considerable room for flexibility in promoting the expansion of natural gas use. The natural gas energy utilization rate in the northern region is higher than that in the southern region. This is because population concentration and urbanization in the northern region are higher than those in the southern region, as pipeline erection cost is lower, gas supply price is lower, and public acceptance is also higher in the northern region. All natural gas use areas are located in the western part of Taiwan. Due to geography, there is currently no public gas manufacturer in the eastern part.

Since the industry requires a considerable amount of capital to be invested, it must endure more losses at the beginning of its operation. According to the survey data of this research, most natural gas energy manufacturers in Taiwan are currently making profits but not to a reasonable level. Given that manufacturers are unable to introduce new technologies, this limits the effective improvement of the quality of production and service. In addition, given that the number of natural gas energy companies in several regions is stagnating, it is necessary to review the current optimal number of public gas fuel companies to improve operational efficiency, and the discussion of the optimal scale is particularly critical.

There is no relevant empirical research on the optimal operating scale of this industry in Taiwan. There are few foreign studies on this industry [1–3]. Only Ref. [2] has tested the economic scale of this industry. It was found that, besides technological changes, the factor with the most significant contribution to the productivity increase in the U.S. natural gas energy industry is the economy of scale. Network industries such as aviation, transportation, hydropower, electricity, and telecommunications, were also found to affect the economies of scale of the industry [4–7]. Therefore, in this study, the panel data of the manufacturers of Taiwan's natural gas energy industry from 1995 to 1999 were constructed.

Citation: Liu, T.-K. Applied Research on Optimal Scale and M&A Efficiency in the Natural Gas Energy Industry. *Eng. Proc.* **2023**, *38*, 66. https://doi.org/10.3390/engproc2023038066

Academic Editors: Teen-Hang Meen, Hsin-Hung Lin and Cheng-Fu Yang

Published: 29 June 2023

Copyright: © 2023 by the author. Licensee MDPI, Basel, Switzerland. This article is an open access article distributed under the terms and conditions of the Creative Commons Attribution (CC BY) license (https://creativecommons.org/licenses/by/4.0/).

Using the translog cost function model, the economies of scale, density, and scope of industries were investigated.

2. Research Methodology

In the literature, MES is mainly discussed in terms of engineering and cost analysis [8–10]. In Ref. [11], the MES was estimated and compared with the cost analysis method and the engineering method, and it was found that the MES estimated by the engineering method was overestimated. In natural gas energy, product quality is uniform, which increases the applicability of the cost analysis method. Therefore, in the natural gas energy industry [1–3], the translog cost function is used as a model for empirical analysis. Therefore, the research framework of the cost analysis method was referred to from the literature, and the translog cost function was constructed as an empirical model for exploring the natural gas energy industry in this study.

2.1. Definition of Variables

The data of 24 manufacturers from 1995 to 1999 for 5 years were obtained without unsuitable samples. Therefore, the data of 21 natural gas energy manufacturers from 1995 to 1999 for 5 years were used to construct the manufacturer-specific panel data for this empirical study.

Short-term variable cost (VC) was obtained by adding the cost of intermediate inputs to the cost of labor. The intermediate input price (P_1) was calculated by the annual cost of raw gas purchased by natural gas energy manufacturers and then deducted from the amount of raw gas purchased. The labor price (P_2) was calculated using the average price method, and the annual personnel cost of each natural gas energy manufacturer was divided by the total number of employees. Gas output (Q_1) was measured based on the annual gas sale volume of each natural gas energy manufacturer, installation output (Q_2) is the annual installation sales volume of natural gas energy manufacturers, and real capital (K) is the fixed assets of natural gas energy manufacturers. In terms of network variables in the model, the number of the manufacturer's natural gas energy users (N) in the current year was represented. The above basic statistics were arranged as shown in Table 1. Time trends (T) were dummy variables (dummy variable T = 1,2, ... ,5 for 1995−1999). The other variables were deflated by the relevant price index of 1995 as the base period.

Table 1. Results of short-term translog cost function estimation (1995–1999).

Explanatory	Parameter	Explanatory	Parameter
Constant	−6.983 (−1.92) [c]	$\ln P_1 \ln N$	−0.094 (−4.35) [a]
$\ln P_1$	0.411 (6.14) [a]	$\ln P_1 \cdot T$	0.018 (4.52) [a]
$\ln P_2$	0.590 (8.81) [a]	$\ln P_2 \ln Q_1$	−0.640 (−2.01) [b]
$\ln Q_1$	3.961 (1.87) [c]	$\ln P_2 \ln Q_2$	0.064 (0.70)
$\ln Q_2$	−0.466 (−0.77)	$\ln P_2 \ln K$	0.266 (2.76) [a]
$\ln K$	−1.156 (−1.81) [c]	$\ln P_2 \ln N$	0.241 (0.70)
$\ln N$	0.777 (0.31)	$\ln P_2 \cdot T$	0.026 (0.66)
T	0.056 (0.25)	$\ln Q_1 \ln Q_2$	0.511 (4.11) [a]
$\ln P_1 \ln P_1$	−0.266 (−15.60) [a]	$\ln Q_1 \ln K$	−0.303 (−1.71) [c]
$\ln P_2 \ln P_2$	−0.253 (−12.75) [a]	$\ln Q_1 \ln N$	−0.079 (−0.20)
$\ln Q_1 \ln Q_1$	0.171 (0.36)	$\ln Q_1 \cdot T$	0.170 (2.69) [a]
$\ln Q_2 \ln Q_2$	0.033 (0.43)	$\ln Q_2 \ln K$	−0.087 (−1.78) [c]
$\ln K \ln K$	0.190 (2.89) [a]	$\ln Q_2 \ln N$	−0.408 (−3.37) [a]
$\ln N \ln N$	0.112 (0.30)	$\ln Q_2 \cdot T$	0.030 (1.73) [c]
$T \cdot T$	−0.084 (−5.05) [a]	$\ln K \ln N$	0.071 (0.32) [a]
$\ln P_1 \ln P_2$	−0.262 (−16.30) [a]	$\ln K \cdot T$	−0.025 (−1.70) [c]
$\ln P_1 \ln Q_1$	0.165 (7.36) [a]	$\ln N \cdot T$	−0.147 (−2.37) [b]
$\ln P_1 \ln Q_2$	0.012 (1.75) [c]		

Table 1. *Cont.*

Explanatory	Parameter	Explanatory	Parameter
R2			
Cost function 0.958			
Intermediate input share function 0.858			

Note: 1. Values in parentheses are T values. 2. [a], [b], and [c] represent significance at the 1%, 5%, and 10% statistical test levels, respectively.

2.2. Empirical Model

In empirical research, it is not suitable for industries with a large number of fixed factors to use the long-term cost function to directly analyze since the production technology is not in the optimal equilibrium state. It is appropriate to use the short-term cost function that considers fixed factors to measure [12,13]. According to the setting of the translog cost function by Ref. [14], the second-order approximation equation of the cost function is expressed as follows.

$$\begin{aligned} \ln VC &= \beta_0 + \sum_{i=1}^{2} \beta_{P_i} \ln P_i + \sum_{i=1}^{2} \beta_{Q_i} \ln Q_i + \beta_K \ln K + \beta_N \ln N + \beta_T T \\ &+ \tfrac{1}{2} \sum_{i=1}^{2} \sum_{j=1}^{2} \beta_{P_i P_j} (\ln P_i \ln P_j) + \tfrac{1}{2} \sum_{i=1}^{2} \sum_{j=1}^{2} \beta_{Q_i Q_j} (\ln Q_i \ln Q_j) \\ &+ \tfrac{1}{2} \beta_{KK} (\ln K)^2 + \tfrac{1}{2} \beta_{NN} (\ln N)^2 + \tfrac{1}{2} \beta_{TT} (T)^2 \\ &+ \sum_{i=1}^{2} \sum_{j=1}^{2} \beta_{P_i Q_j} \ln P_i \ln Q_j + \sum_{i=1}^{2} \beta_{P_i K} \ln P_i \ln K \\ &+ \sum_{i=1}^{2} \beta_{P_i N} \ln P_i \ln N + \sum_{i=1}^{2} \beta_{P_i T} \ln P_i T + \sum_{i=1}^{2} \beta_{Q_i K} \ln Q_m \ln K \\ &+ \sum_{i=1}^{2} \beta_{Q_i N} \ln Q_i \ln N + \sum_{i=1}^{2} \beta_{Q_i T} \ln Q_i T + \beta_{KN} \ln K \ln N \\ &+ \beta_{KT} T \ln K + \beta_{NT} T \ln N + \varepsilon \end{aligned} \quad (1)$$

Adopting the concept of the augmented scale of economies (ASCE) [15], the augmented and estimated annual short-run cost function was used for total regional consolidated firms after merging ASCE to expand the number of users at a given price (the average price of consolidated manufacturers in each region).

$$\sum_i VC_i = \sum_i f_i(\overline{P}_1, \overline{P}_2, Q_{1i}, Q_{2i}, K_i, N_i, T),\ i = 1, \ldots, n \quad (2)$$

3. Results

The parameter estimation of the short-term translog cost function is shown in Table 2. The results show that the R2 of the short-term translog cost function and the intermediate input share function are 0.96 and 0.86, respectively, and the explanatory power is more than 85%. The results indicate that the regression model in this study had an ideal adaptation state. Therefore, in this study, the estimated parameters were used to measure the short-term economies of scale, density economies, category economies, and the Allen partial substitution elasticity of the natural gas energy industry. The estimated results are shown in Table 2.

The estimated value of short-term economies of scale in Taiwan's natural gas energy industry during 1995–1999 is greater than 1, indicating that Taiwan's natural gas energy industry does have the characteristics of economies of scale when considering the network effect. When faced with a given market factor price and real capital volume, increasing output and network density at the same time can reduce costs. The estimated value of density economics is 3.230, which is greater than 1. This means that, under the given input price, real capital volume, the number of users, and the output of Taiwan's natural gas energy industry increase, and the effect of cost reduction is extremely significant. In addition, the category economy was 15.247, indicating that Taiwan's natural gas energy industry produces two outputs of gas and equipment at the same time, which showed the advantages of the scope economy or diversification economy compared with the mode of only producing a single output.

Table 2. Short-term economies of scale, density, scope and Allen partial substitution elasticity.

	Estimates	Standard Deviation
Economies of scale (*SE*).	1.127	0.944
Economies of density (*DE*).	3.230	2.585
Category economy (*SCOPE*).	15.247	9.668
Intermediate input-labor (σ_{12}).	−0.423	0.367
Intermediate input-Intermediate input (σ_{11}).	−1.503	1.128
Labor-labor (σ_{22}).	−7.178	5.243

Note: Values are the average of sample estimates.

In addition, Allen's estimation of the elasticity of substitution showed complementarity between intermediate inputs and labor. That is, when the price of intermediate inputs rises, natural gas energy manufacturers cannot use labor to replace intermediate inputs to save costs. Thus, when the prices of intermediate inputs rise, they cannot slow down the rise in variable costs by increasing labor input. Therefore, with soaring raw material prices, manufacturers need to consider increasing the network effect or expanding the scale to reduce the pressure and risk of costs. The optimal production scale of the natural gas energy industry is discussed in this article later.

The comparison of the augmented scale economy coefficient form before and after M&A is shown in Table 3. The ASCE values of all regions after M&A, except for Northern Region III, were significantly higher than those of the manufacturers before the merger, indicating that the manufacturers in all regions after the merger had the benefits of expanded economies of scale. Among them, the northern region II had the most significant benefit after M&A. While the estimated value of ASCE after the merger in the southern region is less than 1, it is also close to the level of expanding fixed economies of scale. Therefore, the results of the ex-ante M&A imply that when the manufacturer faces scale expansion and market expansion after M&A, it does not harm the cost side, but increases the output and improves the user network through M&A to achieve the goal of reducing costs and pursuing optimal scale.

Table 3. Comparison of augmented economies of scale coefficients before and after M&A.

	Augmented Economies of Scale	
Regional Markets	**Before M&A**	**After M&A**
Northern Region I.	0.698 (0.1435)	1.877 (1.1732)
Northern Region II	0.4291 (0.399)	1.811 (0.032)
Northern Region III.	1.218 (0.677)	1.242 (1.427)
Central region	0.445 (0.282)	1.027 (0.693)
Southern region	0.448 (0.475)	0.955 (0.734)

Note: Estimates are the mean of the sample, and values in parentheses are standard deviations.

4. Conclusions

The results of this study show that the industry has the characteristics of scale economy, density economy, and category economy effect in the mode of simultaneously selling gas and equipment. Allen's estimates of partial elasticity of substitution show that there is complementarity between intermediate inputs and labor. This implies that when the price of intermediate inputs rises, natural gas energy manufacturers cannot use labor to replace intermediate inputs to save costs. According to the minimum effective scale output and the actual demand for natural gas energy, it is estimated that the optimal number of natural gas energy manufacturers is three in the north, one in the middle, and one in the south. Finally, the expanded economies of scale coefficients were compared before and after the merger, showing that the merged regional firms had the benefit of expanding economies of scale. Through the merger, they increased output and improved user networks, reducing costs and pursuing optimal scale.

Due to the limitation of the research sample investigated, only 5 years of panel data were used empirically. Whether structurally competitive utilities need to be merged, MES may not be the only consideration. It is necessary to continue to collect and construct more complete sample data to provide more detailed economic and policy implications. In this study, the annual data were used as the observation value, and the gas output varied due to different seasons and time points. The change in the cost structure at these time points could not be represented in the model of this paper. In addition, Ref. [16] proposed the AGEM (additive general error model) model, pointing out that, compared with the estimation of the translog cost function and the share equation, the AGEM model uses the demand function instead of the cost share function for estimation. The more informative the model, the more the error term can be explained by the input quantity. Empirical research results also showed that the AGEM model outperformed the translog cost function and cost share estimation models [17]. There authors of [18] also used the AGEM model [16] for empirical analysis in their research on technological changes, learning effects, and structures in the production of machinery and equipment in the United States. The above-mentioned parts were not included in the model, though they are important research directions. Thus, they will be further considered and adjusted in the future.

Funding: This research received no external funding.

Institutional Review Board Statement: Not applicable.

Informed Consent Statement: Not applicable.

Data Availability Statement: Taiwan natural gas energy manufacturers survey in 1995.

Conflicts of Interest: The author declares no conflict of interest.

References

1. Sing, M. Are combination gas and electric utilities multiproduct natural monopolies. *Rev. Econ. Stat.* **1987**, *69*, 392–398. [CrossRef]
2. Aivazian, V.A.; Jeffrey, L.C.; Chan, M.W.L.; Mountain, D.C. Economies of scale versus technological change in the natural gas energy transmission industry. *Rev. Econ. Stat.* **1987**, *69*, 556–561. [CrossRef]
3. Chermak, J.M.; Patrick, R.H. A well-based cost function and the economics of Exhaustible resources: The case of natural gas energy. *J. Environ. Manag.* **1995**, *28*, 174–189. [CrossRef]
4. Nemoto, T.; Nakanishi, Y.; Madono, S. Scale economies and over-capitalization in Japanese electric utilities. *Int. Econ. Rev.* **1993**, *34*, 431–440. [CrossRef]
5. Salvanes, K.G.; Tjotta, S. Productivity differences in multiple output industries: An empirical application to electricity distribution. *J. Product. Anal.* **1994**, *5*, 23–43. [CrossRef]
6. Bhattacharyya, A.; Harris, T.R.; Narayanan, R.; Raffiee, K. Specification and estimation of the effect of ownership on the economic efficiency of the water utilities. *Reg. Sci. Urban Econ.* **1995**, *25*, 759–784. [CrossRef]
7. Zheng, X.L.; Niu, H.W.; Wang, X.O. Research on long- and short-term scale economy and density economy of Tai-power corporation. *Humanit. Soc. Sci.* **1997**, *9*, 125–159.
8. Chenery, H.B. Engineering production functions. *J. Econ.* **1949**, *63*, 507–531. [CrossRef]
9. Heathfield, D.F.; Wibe, S. *An Introduction to Cost and Production Functions, London Macmillan Education*; Houghton Mifflin Press: London, UK, 1987.
10. Scherer, F.M.; Ross, D. *Industrial Market Structure and Economic Performance*; Macmillan Education Press: London, UK, 1990.
11. Fuss, M.A.; Gupta, V.K. A cost function approach to the estimation of minimum efficient scale, returns to scale, and suboptimal capacity: With an application to Canadian manufacturing. *Eur. Econ. Rev.* **1981**, *15*, 123–135. [CrossRef]

12. Nelson, R.A. Returns to scale from variable and total cost functions. *Econ. Lett.* **1985**, *18*, 217–276. [CrossRef]
13. Mocan, H.N. The child care industry: Cost functions, efficiency, and quality. *Natl. Bur. Econ. Res. Work. Pap.* **1995**, *xviii*, 5293.
14. Christensen, L.R.; Jorgenson, D.W.; Lau, L.J. Transcendental logarithmic production frontiers. *Rev. Econ. Stat.* **1973**, *55*, 28–45. [CrossRef]
15. Benston, G.J.; Hanweck, G.A.; Humphrey, D.B. Scale economies in banking. *J. Money Credit. Bank.* **1982**, *14*, 435–456. [CrossRef]
16. McElroy, M.B. Additive general error models for production, cost, and derived demand or share systems. *J. Political Econ.* **1987**, *95*, 737–757. [CrossRef]
17. Norsworthy, J.R.; Jang, S.L. Quantity-based measurement of technological change in economic models. In *Empirical Measurement and Analysis of Productivity and Technological Change: Application in High-Technology and Service Industries*; Emerald Publishing Limited: Bingley, UK, 1992; Chapter 3; pp. 59–82.
18. Norsworthy, J.R.; Tsai, D.H. Technological change, learning-by-doing, and the structure of production in the U.S. machine tool industry. In *Macroeconomic Policy as Implicit Industrial Policy: Its Industry and Enterprise Effects*; Kluwer Academic Publisher: Boston, MA, USA; Dordrecht, The Netherlands; London, UK, 1998; Chapter 6; pp. 96–121.

Disclaimer/Publisher's Note: The statements, opinions and data contained in all publications are solely those of the individual author(s) and contributor(s) and not of MDPI and/or the editor(s). MDPI and/or the editor(s) disclaim responsibility for any injury to people or property resulting from any ideas, methods, instructions or products referred to in the content.

2. Study Area and Method

2.1. Study Area

The Taichung Irrigation Area is located in central Taiwan. The area stretches across 3 river basins of the Daan River, Dajia River, and Wu River, from north to south. The irrigation area of the Daan River, Dajia River, and Wu River is 10,575 ha, 12,222 ha, and 2894 ha, respectively [9,10]. According to the data from the soil and groundwater pollution remediation fund management board [4], there are 738 contaminated land sites (84.8 ha) under control, whereby most of which are in the Dali area (578 sites) and regarded as seriously contaminated areas (Figure 1 and Table 1). Through the efforts of the related bureau, none of the 738 contaminated land sites in Taichung have been under control yet. In this research, 2 potential pollution irrigation areas (Yuanli and Rinan) in the Daan River basin, 5 areas (Babao, Xitun, Nantun, Shalu, and Tunzijiao) in the Dajia River basin, and 3 areas (Dali, Wangtian, and Dadu) in the Wu River basin were chosen for investigation (Figure 2).

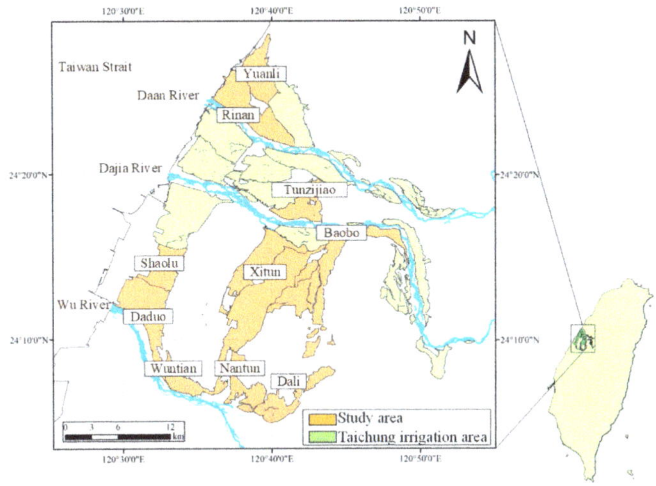

Figure 2. Study area.

2.2. Properties of Ion Exchange Resin Capsules

Chang tested the adsorption capacity of the ion exchange resin capsules using the bottle-cup test (JAR-TEST) [6]. The results showed that it took 20 min to reach the resin's adsorption rate of above 35% for Pb, Ni, Ca, Mn, and Zn, and 30 min to reach the resin's adsorption rate of above 40% for Cu, Ni, and Ti. In an open environment, the ion exchange resin adsorbs heavy metals. The resin capsule is a yellow-brown granule (Figure 3) similar to an agent for hard water softening or the demineralization of water. It is made of gel polystyrene with a polymer structure, a functional group characterized by sulfonic acid, and sodium-form (Na^+) ion exchange. The cation exchange capacity is 2.0 eq/L at a water content of about 44–48%. A single particle size is about 300–1200 μm, with a particle density of about 1.29 g/cm^3 and an enduring temperature of up to 120 °C [7].

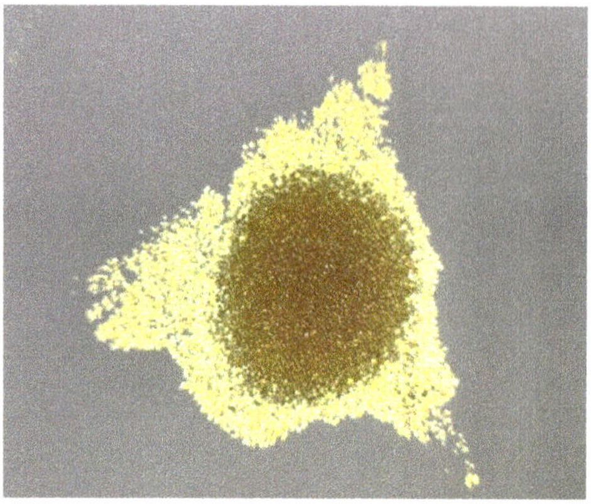

Figure 3. The resin capsules.

Before the resin capsule was placed, a site survey was needed to confirm the exact location of a suspicious pollution source. To avoid loss by flushing or human destroying, it needed to be placed in a hidden place such as under waterweed shade or on the canal side and be fixed with steel nails or iron wires. It is suggested that it be placed for 3–7 days for insufficient absorption to occur. Biofilm may be formed on its surface and decrease the absorption if placed for longer than 7 days [6]. After heavy metal absorption, the resin capsule was retaken and cleaned to remove the surface biofilm and other material on the surface. Then, X-ray fluorescence (XRF) was used to detect the content of different heavy metals adsorbed by the ion exchange resin [7,11].

First, the Ca concentration of the resin capsule was determined. If the Ca concentration was between 40,000 and 50,000 mg/kg, the absorption time needed to be shortened. If the Ca concentration was less than 3000 mg/kg, data were considered to be invalid as this represents a low water flow rate. Equation (1) was used to calculate the ratio of different metals' concentrations and Sr concentration (Sr is the environmental background element and was used to represent the flux of the background element). Then, whether the concentrations exceeded the standard was determined using Equation (2) [7].

$$X = \frac{C_x}{C_{Sr}} \quad (1)$$

$$X \geq \overline{X} + 2\sigma \quad (2)$$

where X is the ratio of a heavy metal concentration to Sr concentration, C_x is the concentration of a heavy metal detected via XRF, C_{Sr} is the concentration of Sr detected via XRF, $\overline{X}$ is the average of the ratio of a heavy metal concentration to Sr concentration, and σ is the standard deviation. When the ratio of a metal concentration to that of Sr was more than the average ratio plus 2 standard deviations, the sample was regarded as being polluted with the metal. The resin capsule was suitable for finding pollution sources. However, the resin capsules had inconveniences, such as easy loss and recovery failure. The placement, arrangement, and concealment of the capsule need to be improved. The resin capsules need to use the relative Sr ratio of the upstream and downstream to judge whether there is pollution. When the result was not obvious, it was difficult to decide on water quality monitoring, and traditional water quality sampling and laboratory testing are required.

3. Results

In 2018, the resin capsule was used to be implemented to determine the high or possible pollution potential of canals. The recovery ratio was about 85.4%. The investigation was continued in 2019. A total of 140 resin capsules were planted, of which 8 were lost, and

132 were successfully recovered with a recovery ratio of 94.3%. In 2020, 620 capsules were planted with 19 capsules lost and 601 capsules recovered, and the recovery ratio was 96.9%. In 2021, 217 capsules were planted. Briefly, 24 were lost and 193 were recovered with a recovery ratio of 88.9%. In total, 1025 resin capsules were planted with a recovery ratio of 94.3%. The calculation result of the ratios of heavy metals to the concentration of Sr is shown in Table 2. The concentration of Ca and Sr in the resin capsules is displayed in Figure 4. The correlation between Sr and Ca was 0.835. The Sr/Ca ratio in Taoyuan was 0.017, which was slightly lower than in the Taichung area.

Table 2. The criteria of the resin capsule in different heavy metals.

Ratio of Metal to Sr	Average	σ	Criteria
Cu/Sr	0.007	0.053	0.114
Pb/Sr	0.013	0.189	0.392
Zn/Sr	0.095	0.295	0.684
Cr/Sr	0.026	0.452	0.929
Ni/Sr	0.011	0.069	0.149

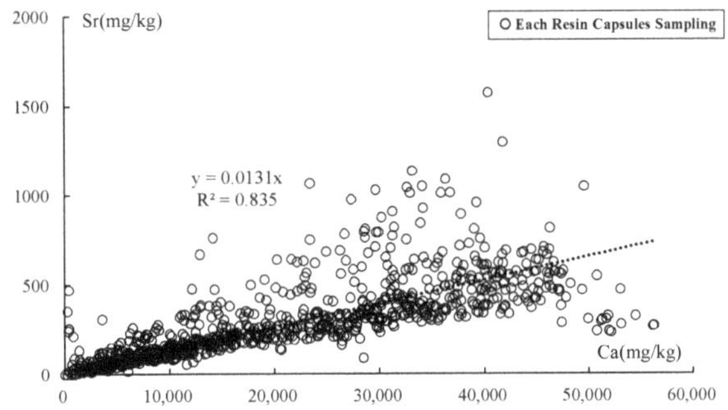

Figure 4. The relationship between Ca and Sr in resin capsules.

In 2018, a survey was conducted in the polluted potential irrigation area in Taichung. The results showed that Pb and Zn were high near the rainwater outlet in the drainage system of the industrial park in the Yuanli irrigation area. The Dali area was slightly polluted by heavy metals such as Zn. The results in 2019 showed that the Dali irrigation area was slightly polluted by Zn and Cr. In the Niuchou branch canal of the Babao irrigation area and the Shenpi branch canal of the Xitun irrigation area, Cu and Ni polluted the downstream of the specific suspected factory. In 2020, it was found that the industrial drainage system of the Rinan and Yuanli irrigation areas was polluted by wastewater containing a high concentration of Cu and Zn. The Tunzijiao irrigation area was investigated too, and the result showed Cu and Zn pollution in the downstream area of the steel plant neighboring a rice paddy. An investigation in the Niuchou branch canal in the Babao irrigation area and the Shenpi branch canal in the Xitun irrigation area showed high concentrations of Cu and Zn in the wastewater from factories near the irrigation areas. The Amili canal in the Dali area showed Zn pollution. The survey results in 2021 showed Cu pollution in the Niuchou branch canal of the Babao irrigation area and Zn pollution in the Dali area. Other irrigation areas such as Nantun, Shalu, Wangtian, and Dadu did not show any pollution.

In terms of pollution source, when the industrial zone is located nearby, discharged wastewater may contain heavy metals of high concentrations. If the downstream agricultural area is irrigated, heavy metal pollution may be observed. The pollution in the Tunzijiao area originated from nearby steel factories, but the degree of pollution was relatively low. The Xitun and Babao areas were affected by metal surface treatment factories. The Dali area was heavily polluted since the upstream water contained the regional drainage and the wastewater of factories. In the Dali irrigation area, the wastewater from factories seems to contaminate water and soil.

4. Conclusions

The resin capsule was used to investigate water pollution in the Taichung area. The research was conducted for 4 years from 2018 to 2021. The results showed that the Babao, Xitun, and Dali areas were seriously polluted by heavy metals. The pollutants in the Babao area were Cu and Ni, while Zn and Cr pollution was found in the Dali area. Other areas such as the Yuanli and Rinan areas were also affected by water pollution. The Shalu, Nantun, Wangtian, and Dadu areas showed a lower degree of pollution. The resin capsule is convenient for the investigation of pollution sources and can be a useful tool to replace time- and resource-consuming operations of traditional water quality monitoring and conduct long-term water quality monitoring.

Author Contributions: Conceptualization and methodology, F.-W.C.; software, formal analysis, investigation, and writing, Y.-W.T. All authors have read and agreed to the published version of the manuscript.

Funding: This research received no external funding.

Institutional Review Board Statement: Not applicable.

Informed Consent Statement: Not applicable.

Data Availability Statement: Data is unavailable due to privacy restrictions.

Conflicts of Interest: The authors declare no conflict of interest.

References

1. Chinese National Federation of industries and China credit information service Ltd. *50 Years of Industrial Development in Taiwan*; Industrial Development Bureau, MOEA: Taipei, Taiwan, 2000.
2. AERC—Agricultural engineering research center. *The Water Quality Monitoring of High Contaminated Canals, Advanced Investigation of Nantun Irrigation Area and Improvement Planning of Water Quality in Datuliao Canal, Taichung Irrigation Area in 2020*; AERC-20-RR-14; Taichung Management Office, Irrigation Agency, Council of Agriculture, Executive Yuan: Taipei, Taiwan, 2000.
3. Chien, H.C.; Hsiung, P.H.; Huang, S.H.; Wang, J.; Tsai, H.C.; Chen, Y.H.; Wang, T.H. *2020 Annual Report Soil and Groundwater Pollution Remediation*; Soil, and Groundwater Pollution Remediation Fund Management Board; Executive yuan: Taipei, Taiwan, 2001.
4. Soil and Groundwater Pollution Remediation Fund Management Board. Available online: https://sgw.epa.gov.tw/ContaminatedSitesMap/Default.aspx (accessed on 20 February 2022).
5. Lin, S.C. *Measuring Device with Ion-Exchange Resin and Measuring Method Thereof*; National Pingtung University of Science Technology: Pingtung, Taiwan, 2016.
6. Chang, T.K. *Design of Time-Lapse Capsules for Heavy Metal Monitoring by Using Ion Exchange Resin*; Soil and Groundwater Pollution Remediation Fund Environmental Protection Agency Executive Yuan SUBSIDY research and Model Field Test Project Plan; Executive Yuan: Taipei, Taiwan, 2015.
7. Shih, P.K.; Chiang, L.C.; Lin, S.C.; Chang, T.K.; Hsu, W.C. Application of Time-Lapse Ion Exchange Resin Sachets (TIERS) for Detecting Illegal Effluent Discharge in Mixed Industrial and Agricultural Areas, Taiwan. *Sustainability* **2019**, *11*, 3129. [CrossRef]
8. Agriharvest. *Wastewater Discharge Has Nowhere to Hide! This Magical "Capsule" Can Get the Source through Time and Space*; Agriharvest: Taipei, Taiwan, 2000.
9. AERC—Agricultural engineering research center. *The Investigation of Water Pollution in Irrigation Suspended Period in Taichung Management Office Area in 2021*; AERC-21-RR-11; Taichung Management Office, Irrigation Agency, Council of Agriculture, Executive Yuan: Taipei, Taiwan, 2021.
10. AERC—Agricultural engineering research center. *The Investigation and Assessment of Irrigation Water Quality Monitoring and Potential Pollution Characteristics in Taichung Management Office Area in 2022*; AERC-22-RR-07; Taichung Management Office, Irrigation Agency, Council of Agriculture, Executive Yuan: Taichung, Taiwan, 2022.
11. Environmental protection agency executive yuan. *National Control, and Investigation Plan for Farmland with High Heavy Metal Pollution Potential (Phase 5)*; EPA-105-GA02-03-A262; Executive Yuan: Taipei, Taiwan, 2000.

Disclaimer/Publisher's Note: The statements, opinions and data contained in all publications are solely those of the individual author(s) and contributor(s) and not of MDPI and/or the editor(s). MDPI and/or the editor(s) disclaim responsibility for any injury to people or property resulting from any ideas, methods, instructions or products referred to in the content.

engineering proceedings

MDPI

Proceeding Paper

Outdoor Thermal Comfort Study on Urban Areas with Various Densities in Taipei †

Ying-Ming Su and Janmichael Antoni *

Department of Architecture and Urban Design, National Taipei University of Technology, Taipei 106, Taiwan; ymsu@ntut.edu.tw
* Correspondence: janmichael_antoni@yahoo.com
† Presented at the 3rd IEEE International Conference on Electronic Communications, Internet of Things and Big Data Conference 2023, Taichung, Taiwan, 14–16 April 2023.

Abstract: With relatively high temperatures and evenly distributed precipitation throughout the year, outdoor thermal comfort is important in Taiwan. Thus, we discuss thermal comfort in different density areas with A: a low density (average buildings height/Street width ratio: 1.3, SVF: 0.28), B: a mid density (H/W: 2.75, SVF: 0.14), and C: a high density (H/W: 6.88, SVF: 0.11). Data were acquired through field meteorological measurements from the three sites on typical summer days from 15 to 17 July 2022. Around the Wanhua district of Taipei City, thermal comfort was obtained through the calculation of thermal comfort indices: physiological equivalent temperature (PET) and Universal Thermal Climate Index (UTCI). Results showed that different urban density areas had different thermal comforts; streets in A were more exposed to sunlight, causing the mean radiant temperature to be the highest; C had lower solar radiation and stronger wind with the highest comfort level; and B had higher air temperature but lower radiation than A. PET contributed to uncomfortableness in A. Thermal sensation vote results showed that B had the most "uncomfortable" votes because wind speed in B was the lowest. Medium-density urban areas had enough solar exposure on the pedestrian level, but not enough wind to dissipate the heat. PET was more sensitive to mean radiant temperature. These findings provide a guideline and technical reference for urban designers to mitigate the high temperatures of summer in urban areas and improve thermal comfort to create a healthy environment and encourage people to utilize outdoor environments.

Keywords: outdoor thermal comfort; urban density; questionnaire survey; thermal comfort indices; urban outdoor environment; environmental engineering; climate change

Citation: Su, Y.-M.; Antoni, J. Outdoor Thermal Comfort Study on Urban Areas with Various Densities in Taipei. *Eng. Proc.* **2023**, *38*, 77. https://doi.org/10.3390/engproc2023038077

Academic Editors: Teen-Hang Meen, Hsin-Hung Lin and Cheng-Fu Yang

Published: 10 July 2023

Copyright: © 2023 by the authors. Licensee MDPI, Basel, Switzerland. This article is an open access article distributed under the terms and conditions of the Creative Commons Attribution (CC BY) license (https://creativecommons.org/licenses/by/4.0/).

1. Introduction

Urban outdoor spaces accommodate pedestrian passage and provide space for outdoor activities; thus, they are important components of urban life. Encouraging urban residents to visit outdoor spaces and streets benefits the city economically, physically, environmentally, and socially [1–4]. Taiwan entered the era of an aged society with more than 14% of the population being 65 years old and above. A study in Japan [5] concluded that walkable urban spaces improved the lives of senior citizens. Outdoor spaces can also provide health benefits and offer a feeling of social support. However, global warming and the rapid urbanization process of big cities have caused tremendous changes to the climate, especially in urban areas, intensifying the urban heat island (UHI) effect and affecting outdoor thermal comfort. The UHI effect is a phenomenon where heat accumulates in urban areas caused by urban development and activities [6]. In the past decade, the urban temperature has been continuously rising and has caused a serious impact on urban residents. Exposure to extremely high temperatures leads to fatigue and dizziness, and is even life threatening [7,8]. According to the Central Weather Bureau, Taiwan is located in a humid–subtropical region with relatively high temperatures and heavy precipitation with

204 rainy days per year. The summer of 2020 was recorded as the hottest season ever in Taiwan, with the highest temperature of 40.2 °C in Taitung and 39.7 °C in Taipei. According to Taiwan Climate Change Projection Information and the Adaptation Knowledge Platform, by the end of this century, there is a high possibility that summer will be lengthened by up to 210 days per year, with winter shortened by less than 50 days if global warming continues to escalate.

To encourage citizens to utilize urban outdoor spaces, thermal comfort is a necessary factor to be considered. Thermal comfort depends on the satisfaction of the human body in relation to the thermal environment. Global climate change has made people pay more attention to outdoor thermal comfort. Outdoor thermal comfort is affected by meteorological factors (air temperature, humidity, wind, sun, and exposure) and personal factors (gender, age, clothing insulation, and metabolism). Thermal comfort can be assessed through thermal indices and subjective questionnaires. Thermal indices can be calculated through meteorological data acquired from field monitoring while conducting questionnaire surveys around the monitoring area.

Physiological equivalent temperature (PET) and Universal Thermal Climate Index (UTCI) are the most commonly used thermal comfort indices. PET was developed in the 1990s by Hoppe, and UTCI was developed by the International Society for Biometeorology (ISB) [9]. Both PET and UTCI include factors affecting thermal comfort in the calculation: meteorological parameters and personal parameters. UTCI has the advantage of presenting different climates, weather, and locations [10]. Reference [11] indicated that PET was more sensitive in the heat of summer; thus, both PET and UTCI were considered suitable thermal indices to be utilized in this research. Several studies on urban planning showed that different urban densities had different impacts on the wind ventilation and microclimate of the area. Reference [12] concluded that stronger wind had the effect of dispersing and diluting both pollutants and heat. Urban densities have significant impacts on the urban climate, where good ventilation can improve thermal comfort. Studies in the past have shown that urban densities are affected by factors such as sky view factor (SVF) and height-to-width ratio (H/W), which is the ratio of building heights to street width [13]. Lower density urban areas are proven to have more wind [14]. In urban areas with taller buildings, the wind is accelerated around the corners [15]. During the daytime, air temperature and SVF have a positive correlation, where lower SVF and H/W areas can block direct solar exposure, thus resulting in cooler temperatures [16–18]. Studies in Canada showed [14] that SVF between 0.3 and 0.85 caused a 1.5 K temperature difference. However, compared to high-density areas, low-density areas can lose more heat and have lower temperatures in the absence of solar radiation. Mean radiant temperature (Tmrt) is also proven to be lower in narrow streets, where Tmrt increased with SVF value [19]. Research in Tainan [18] studied the thermal environment of a school area and found that PET was lower in low SVF areas. SVF and wind speed directly affect outdoor thermal comfort [20]. In a humid–subtropical climate, lowering radiation and air temperature can result in higher thermal comfort, and accelerating wind speed can improve urban ventilation and thermal comfort. Therefore, overall thermal comfort can be evaluated through PET and UTCI with relevant climate factors in the calculations.

This study aims to explore the relationship between urban density and the influences on pedestrian thermal comfort in Taiwan through field monitoring and questionnaire surveys. Thermal comfort in areas is compared with different urban densities in the Wanhua district of Taipei on hot summer days. The effect of various urban densities on human thermal comfort is investigated to determine urban density with the highest thermal comfort level in summer, and to provide a guideline for urban planners to optimize the thermal comfort of outdoor spaces in Taiwan.

2. Methodology

2.1. Experiment Site

Taiwan has a humid–subtropical climate with high temperatures, especially in the summer season, and a high level of humidity. Urban areas selected for this study are located in Ximen, a part of the northeast side of the Wanhua district, which is located in the west of Taipei City, the northern part of Taiwan (Figure 1). The pedestrian area in Ximen was the first to be built in Taipei, and the largest in Taiwan. Ximen was selected as the location of this study due to the variety of age groups and a large number of pedestrians, making this place ideal for an outdoor human thermal comfort study. Three different locations were selected as experimental sites for meteorological measurement and questionnaire surveys in all three sites (A-C) were located in the same district to minimize the climate difference (Figure 2). The H/W ratio and SVF value were used to determine the urban density of this study. Fisheye images were taken and SVF was calculated using RayMan Pro software [21–25]. Three types of urban densities (low, medium, and high) were compared in this study. Site A was located in a low-density area that has an H/W ratio of 1.3 with an SVF value of 0.28. Site B was in a medium-density area with an H/W of 2.75 and an SVF value of 0.14. Site C had the highest density of all three, with an H/W of 6.88 and an SVF value of 0.11. Taking solar angle into account, all three sites were north–south oriented to minimize differentiation between sites (Table 1).

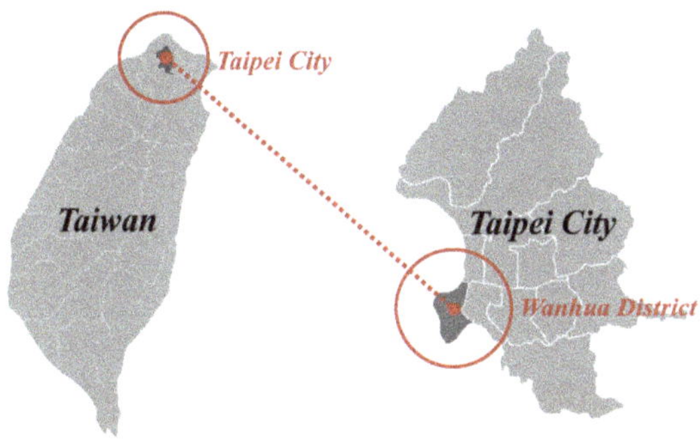

Figure 1. Location of Wanhua District.

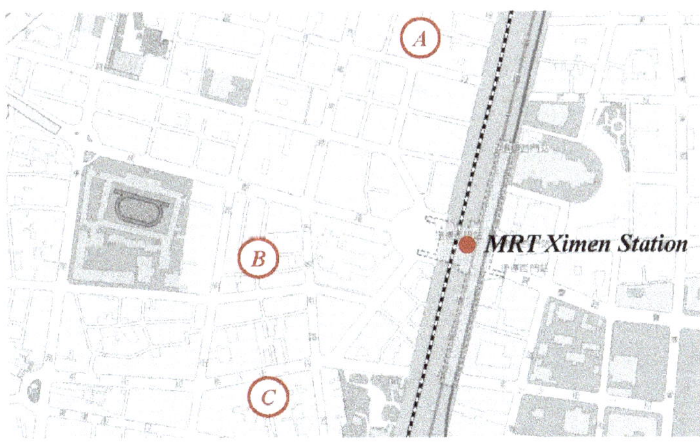

Figure 2. Location of the three experiment sites in Ximen.

Table 1. SVF and H/W Ratio of three different sites.

Site A (Low Density)	**Site B (Medium Density)**	**Site C (High Density)**
	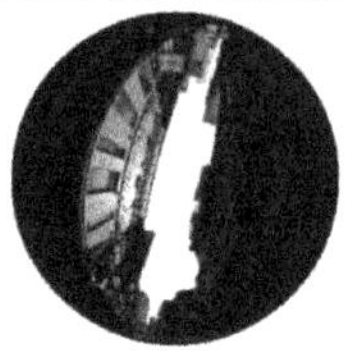	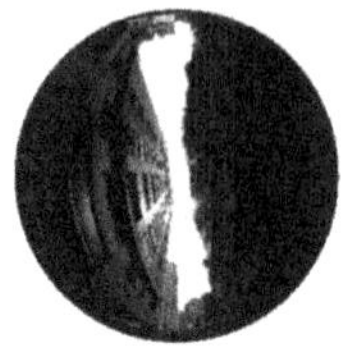
SVF = 0.287 H/W = 1.3	SVF = 0.140 H/W = 2.75	SVF = 0.114 H/W = 6.88

2.2. Meteorological Measurement

Site monitoring and meteorological data measurements of this study were conducted during the summer season from the 15th to the 17th of July 2022. The dates chosen were the 196th to 198th days of the year, which were the typical hottest summer days in the Wanhua district. The majority of the shops in Ximen opened around 10 AM; thus, this study was conducted from 10:00 to 18:00 when the typical daytime activity occurred in Ximen to ensure the number of questionnaire surveys in this study. Meteorological factors such as air temperature (Ta), relative humidity (RH), wind speed (v), and black globe temperature (Tg) were measured using the testo-480 measuring equipment. The equipment and probes were placed at a height of 1.5 m, and thermal data were collected every 1 min.

2.3. Questionnaire Survey

A questionnaire survey was conducted simultaneously with field data measurement, within a radius of 5m of the measuring equipment. Interviewees were randomly selected. Taking thermal adaptation into account, interviewees stayed outdoors (preferably in the same area) for at least 10 min to be eligible to fill out the questionnaire. The interview method was used for the questionnaire, in which interviewees were asked questions in the questionnaire. The questionnaire comprised three sections. The first section included questions for information about the interview, such as date, time, and location. The second section was for basic information on the interviewees, such as gender, age group, upper and lower body clothing, and activities. The third section was for the thermal sensation vote. This section included 6 questions. The first question was to determine the thermal sensation of the interviewees according to their sensation at the moment of the interview with a 5-point scale (−2, hot; −1, warm; 0, neutral; +1, cool; +2, cold). The second through fifth questions were for the preference vote towards the climate parameters, including air temperature, relative humidity, wind speed, and radiation with a 3-point scale (−1, 0, +1). Interviewees were asked about their expectations of the parameters. The last question was about the overall thermal comfort of the interviewee using a 3-point scale (−1, uncomfortable; 0, neutral; +1, comfortable). A total of 1177 valid questionnaires were completed. Personal parameters such as age, clothing, and activities were collected to calculate thermal comfort with PET and UTCI.

2.4. Thermal Comfort Indices

UTCI and PET are two of the most commonly used thermal comfort indices to assess outdoor thermal comfort. Both indices were considered with every parameter influencing thermal comfort: meteorological parameters (air temperature, relative humidity, wind speed, and radiation), and personal parameters (clothing and metabolic rate) [9]. PET was developed by Hoppe using the Munich Personal Energy Balance Model or MEMI, defined as an environment where human skin and body temperature reach the typical indoor environment where Tmrt = Ta, vp = 12 hPa, and v = 0.1 m/s. UTCI was developed by ISB, using equivalent temperature where RH = 50%, Tmrt = Ta, v < 0.15 m/s, and vp < 20 hPa.

With meteorological data obtained through field measurement and personal data obtained through questionnaire surveys, outdoor thermal comfort was calculated through PET and UTCI. Both indices were calculated with RayMan software, an applied meteorology and climatology software developed by Andreas and Rutz [21–25]. The time, date, longitude, and latitude of the experiment sites, meteorological data, and personal data were input into the software to calculate thermal indices. Fisheye images were used to calculate the SVF value for mean radiant temperature calculations. The average metabolic rates and clothing insulation of interviewees were used in this study.

3. Results and Discussions

3.1. Meteorological Measurement

Among the three sites, the maximum mean air temperature of sites A and B were both 39.5 °C (Table 2). Site B had a slightly higher mean air temperature than A. Site B showed the highest mean air temperature. The main reason was that the medium-density area had enough sun exposure and solar radiation to the pedestrian level, but did not have a high enough wind speed to disperse the heat. The mean air temperature and maximum air temperature of site C were the lowest of all three. Site C also had the lowest black globe temperature because C had the highest H/W ratio and lowest SVF value. Site A, with the lowest H/W ratio and highest SVF value, showed the highest black globe temperature. Although site A was more exposed to solar radiation, the openness allowed stronger wind than B to lower the air temperature. Wind speed was the strongest in site C, but relative humidity in site C was the highest. Humidity was defined as the concentration of water vapor in the air. Higher-density areas could cause wind speed to be stronger around tall buildings, but could also cause humid air to take longer to evaporate. In conclusion, site C had the lowest mean air temperature and lowest black globe temperature. There was a positive correlation between air temperature and wind speed, and between urban density and black globe temperature. The higher the H/W ratio, the lower the radiation. The wind speed in site B was the weakest, and the mean air temperature was the highest.

Table 2. Meteorological measurement results.

Site	SVF	Tg (°C)			Ta (°C)			RH (%)			v (m/s)		
		Mean	Max.	Min.	Mean	Max.	Min.	Mean	Max.	Min.	Mean	Max.	Min.
A	0.287	37.9	42.3	34.4	36.5	39.5	33.0	47.1	56.1	40.9	0.9	3.8	0.1
B	0.14	37.2	40.0	33.8	36.7	39.5	34.4	45.7	53.0	39.9	0.7	2.6	0.1
C	0.114	36.1	43.1	32.5	35.7	38.2	33.1	47.7	57.1	41.1	1.5	5.8	0.2

3.2. Questionnaire Result

A total of 1177 valid questionnaires were collected: 457 in site A (208 males, 249 females), 384 in site B (178 males, 206 females), and 336 in site C (155 males, 181 females). Average clothing insulation slightly differed from sites, which ranged from 0.28 to 0.29. Activity level was also typical, which ranged from 2.03 met to 2.11 met. The majority of the interviewees belonged to the 19–30 age group, with the 13–18 age group being the second largest, and the 31–40 age group being the third. Tables 3 and 4 summarize the data of the interviewees. Most of the interviewees voted "hot = −2" on the thermal sensation vote, and most preferred a lower temperature. More than half of the interviewees voted "lower = +1" on the relative humidity vote. Stronger wind speed was preferred on sites A and B, but remained unchanged in site C. Regarding solar radiation, most voted for "lower = +1". Sites A and B had a majority vote of "uncomfortable = −1", while site C had a majority of "neutral = 0". The air temperature preference vote result showed that the thermal environment in Ximen was extreme. The result showed that wind speed had a direct correlation with the thermal sensation vote of the interviewees.

Table 3. Summary of the interview data.

Site	Gender		No. of Subjects	Avg. Clothing (clo)	Avg. Metabolism (met)
A	M	208	457	0.28	2.03
	F	249			
B	M	178	384	0.28	2.11
	F	206			
C	M	155	336	0.29	2.06
	F	181			

Table 4. Age groups of interviewees.

Age Range	<12	13–18	19–30	31–40	41–50	51–60	61–65	65<	Total
No. of Subjects	33	163	539	123	111	100	41	67	1177
Percentage (%)	2.8	13.8	45.8	10.5	9.4	8.5	3.5	5.7	100.0

3.3. *Relationship between Urban Densities and Pedestrian Thermal Comfort*

3.3.1. Thermal Sensation Vote (TSV)

The thermal sensation vote (Figure 3) showed that the majority of pedestrians felt "hot = −2", as the summer thermal environment of Ximen was extremely hot. In site B, the highest percentage of 82% was obtained, and in site C, the lowest percentage of 61% was recorded. This result proved that high-density areas (H/W = 6.88, SVF = 0.11) had significantly better thermal sensation in the summer season. The medium-density areas of site B (H/W = 2.75, SVF = 0.14) had the worst thermal comfort of all three sites. Low-density areas (H/W = 1.3, SVF = 0.28) had 75% of the vote for "hot = −2". Comparing the percentages of thermal sensation votes, different urban densities showed different thermal comfort, with thermal comfort ranking as site C > site A > site B.

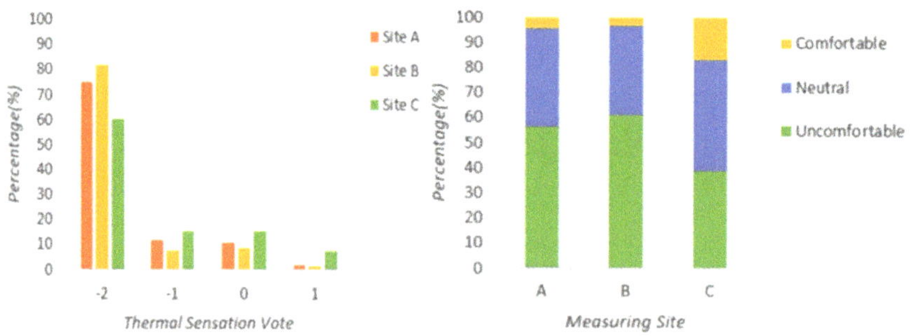

Figure 3. TSV (**left**) and OTC (**right**) on three different sites.

3.3.2. Meteorological Preferences Vote

Table 5 and Figure 4 show a majority of "lower = +1" on all sites, ranked as site C (78.9%) < site B (91.9%) < site A (92.3%). Site A had the highest vote of pedestrians who preferred the temperature to be "lower = +1". While sites A and B had different results with a difference of 0.4%, site C had 78.9% votes for lower temperatures, and 20.8% votes for "unchanged = 0", which were significantly higher than the other two sites. The relationship between Ta preference and TSV was similar, but the ranking was not similar. The v has a direct relation to thermal sensation vote, where the stronger wind was preferred on lower TSV sites. The v preference of "stronger = +1" was ranked as site C (46.1%) < site A (56%) < site B (63.5%). Solar radiation preference of "lower = +1" was ranked as site A (75.1%) > site B (72.4%) > site C (63.1%), which was similar to the measurement results. RH preference between sites was slightly different, as the human body was not sensitive to changes in humidity.

Table 5. Summary of interview responses.

No.	Question	Variable	Option	Statistics and Percentage					
				A		B		C	
1	How do you feel?	TSV	Hot = −2	344	75.3%	315	82.0%	204	60.7%
			Warm = −1	55	12.0%	30	7.8%	53	15.8%
			Neutral = 0	49	10.7%	34	8.9%	53	15.8%
			Cool = +1	9	2.0%	5	1.3%	26	7.7%
			Cold = +2	0	0.0%	0	0.0%	0	0.0%
2	How would you prefer the air temperature to be?	Ta preference	Higher = −1	2	0.4%	0	0.0%	1	0.3%
			Unchanged = 0	33	7.2%	31	8.1%	70	20.8%
			Lower = +1	422	92.3%	353	91.9%	265	78.9%
3	How would you prefer the relative humidity to be?	RH preference	Higher = −1	25	5.5%	27	7.0%	9	2.7%
			Unchanged = 0	199	43.5%	146	38.0%	131	39.0%
			Lower = +1	233	51.0%	211	55.0%	196	58.3%
4	How would you prefer the wind speed to be?	v preference	Weaker = −1	30	6.6%	14	3.7%	12	3.6%
			Unchanged = 0	171	37.4%	126	32.8%	169	50.3%
			Stronger = +1	256	56.0%	244	63.5%	155	46.1%
5	How would you prefer the solar radiation to be?	Sun exposure preference	Higher = −1	11	2.4%	7	1.8%	3	0.9%
			Unchanged = 0	103	22.5%	99	25.8%	121	36.0%
			Lower = +1	343	75.1%	278	72.4%	212	63.1%
6	How is your overall thermal comfort?	Thermal Comfort	Uncomfortable = −1	259	56.7%	236	61.5%	130	38.7%
			Neutral = 0	178	38.9%	135	35.1%	149	44.3%
			Comfortable = +1	20	4.4%	13	3.4%	57	17.0%

In conclusion, both Ta and solar radiation of Ximen in summer were extreme. Site C had the highest percentage of "unchanged = 0" in Ta, solar radiation, and v preferences compared to the other three sites. RH showed no significant correlation to thermal comfort. Lower solar radiation and Ta improved thermal comfort, but in a situation where Ta and radiation were slightly different, wind speed showed a more direct correlation to thermal comfort.

3.3.3. Overall Thermal Comfort (OTC)

OTC was used to evaluate TSV and the meteorological preference votes' correlation to thermal comfort. The majority of votes in sites A and B were "uncomfortable = −1" (A = 56.7%, B = 61.5%). Site C had the highest thermal comfort, where the majority of pedestrians voted "neutral = 0" (44.3%). OTC was ranked as site C > site A > site B. The result correlated with TSV results, which was also consistent with wind speed preferences.

3.4. *Thermal Comfort Indices*

3.4.1. PET and UTCI Assessments

Meteorological data combined with average clothing insulation and metabolic rates on each site were used to calculate PET and UTCI. Thermal indices calculations were conducted in three periods to analyze the meteorological data and its impact on thermal indices (Table 6). Tmrt was the highest on every period at site A, because A had the largest street area and the most sun exposure. In the period "10:00–12:30", both PET and UTCI in site B presented the most uncomfortable site. The ranking of PET and UTCI was site B < site A < site C, with C being the most comfortable site. As can be seen in Figure 5, during "13:00–15:30" and "15:30–18:00", site C remained the most comfortable, and site A was the most uncomfortable site. The difference between PET and UTCI during "15:30–18:00" was small. During "13:00–15:30", site A was significantly more uncomfortable than B due to the direct sun exposure and high Tmrt. PET was more sensitive to heat when the Tmrt level was high, and PET had a narrow range compared to UTCI. On average, PET was higher than expected from UTCI. Sites A and B were "very hot" from 10:00 to 15:30, and "hot" from 15:30 to 18:00. The result was consistent with TSV and OTC, where C was the most comfortable site, being "hot" from 10:00–15:30, and "warm" from 15:30–18:00. The TSV and OTC showed that site B had the most votes of "uncomfortable". This proved that air temperature and mean radiant temperature had a bigger influence on thermal indices, but subjective thermal comfort in this study was affected by wind speed.

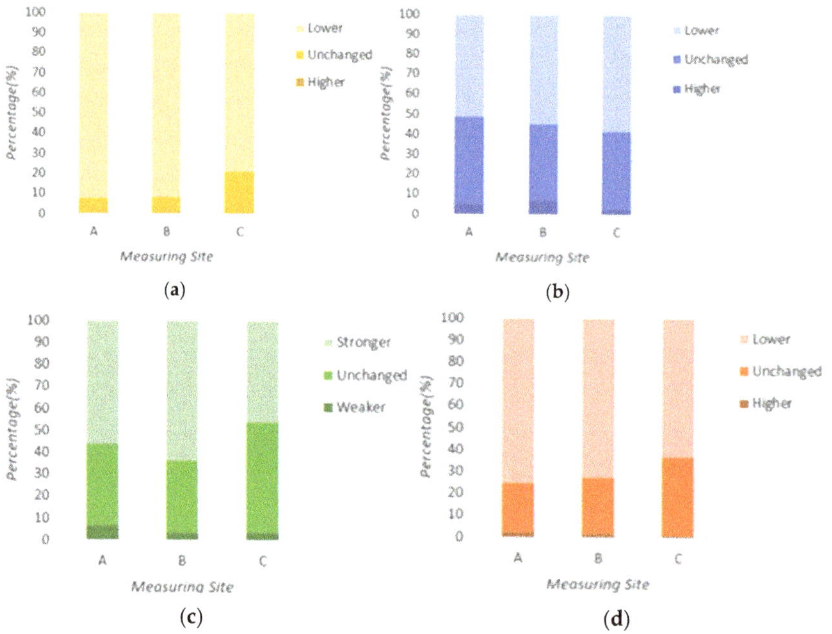

Figure 4. Preference votes on three different sites: (**a**) Ta, (**b**) RH, (**c**) v, (**d**) Solar radiation.

Table 6. Summary of Tmrt, PET, and UTCI.

Site	10:00–12:30			13:00–15:30			15:30–18:00			Average		
	Tmrt	PET	UTCI	Tmrt	PET	UTCI	Tmrt	PET	UTCI	Tmrt	PET	UTCI
A	47.9	42.4	39.9	50.7	45.6	42.4	39.7	38.8	38.5	46.1	42.2	40.3
B	46.8	42.9	40.6	45.2	42.3	40.6	38.8	38.1	38.1	43.6	41.1	39.8
C	44.7	39.7	38.3	44.0	40.3	39.1	38.0	36.9	36.8	42.3	38.9	38.0

In summary, the result showed that PET was more sensitive to thermal and radiation changes than UTCI. In this study, PET was defined to be at three levels; "warm", "hot", and "very hot", while UTCI had two levels of "strong heat stress" and "very strong heat stress". PET and UTCI showed that the thermal environment in the Ximen area in summer was extreme.

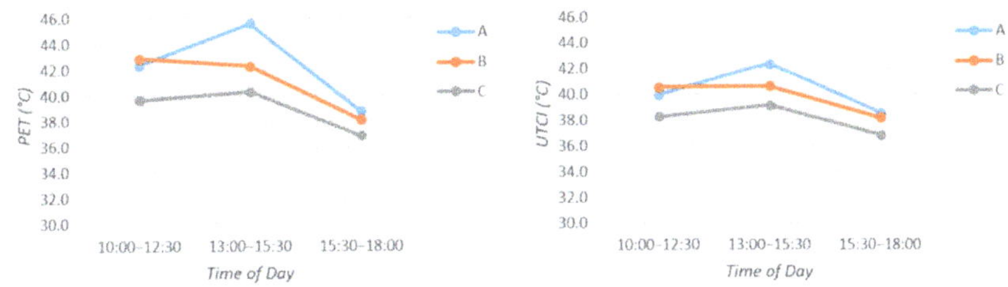

Figure 5. Comparison of PET (**left**) and UTCI (**right**) throughout the day.

3.4.2. Linear Regression Analysis

The relationship between thermal indices and questionnaire results was calculated and analyzed using the linear regression analysis method to identify which thermal index has more impact on subjective thermal sensation. Mean thermal sensation votes on each site were obtained through TSV data from the questionnaire, ranging from −2 (hot) to +2 (cold), which were defined as the dependent variable of the analysis. To achieve higher accuracy in the analysis, data from three days of surveys were calculated separately, with each day consisting of three different times of day on each site, resulting in 27 cases of data. PET and UTCI values of each site at the same time of day were calculated and defined as the independent variable (see Figure 6).

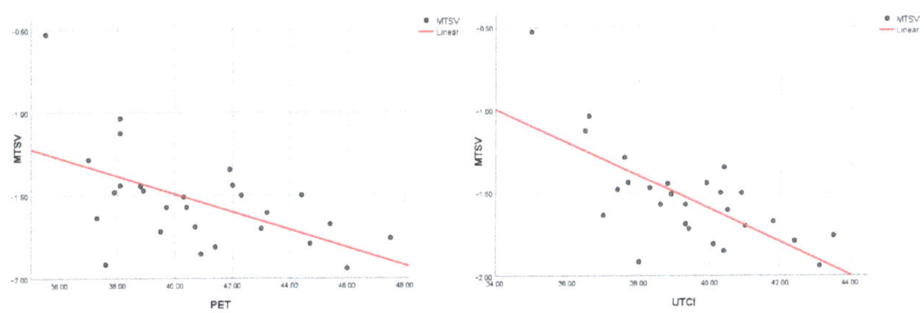

Figure 6. Correlation between TSV and PET (**left**) and UTCI (**right**).

The coefficient of determination (defined as R-squared) is used to determine which of the thermal indices fit the subjective TSV data. Linear regression of PET and MTSV has an R-squared value of 0.307, while that of UTCI and MTSV has an R-squared value of 0.486. Thus, by comparing the relationship between TSV and thermal indices, UTCI has a higher significance of linear regression, which was shown by a higher R-squared value of 0.486, which determines UTCI to be more suitable for outdoor thermal comfort evaluation in this study.

4. Conclusions

Three different urban density areas were investigated to explore the relationship between urban density and outdoor thermal comfort. The H/W ratio and SVF were used to determine different urban densities. Through meteorological measurement and questionnaire surveys, PET and UTCI were calculated, and questionnaire data were analyzed to determine which site has the highest comfort level.

Urban densities affected the pedestrian thermal environment. Different H/W ratios and SVFs had different impacts on pedestrian thermal comfort. In this study, site C (H/W = 6.88, SVF = 0.11) showed the best thermal comfort in summer. The questionnaire survey result showed that the thermal comfort ranking of the three sites was site C > site A > site B, with B being the most uncomfortable site. Thermal comfort ranking was dependent on wind speed, as wind speed had a bigger influence on subjective thermal comfort in this study, while relative humidity had the least influence on thermal comfort.

PET and UTCI showed that the thermal environment in Ximen during the summer season was extreme. Mean radiant temperature had a higher influence on both PET and UTCI than wind speed. PET was more sensitive in extreme heat, and UTCI had a wider range than PET. Both PET and UTCI showed that site C had the highest thermal comfort level. UTCI had a higher significance towards the subjective thermal sensation vote result. Considering the calculation result of PET and UTCI, further study is required to understand the difference between the two.

Although site A had the highest solar radiation and Tmrt, site B had the highest mean air temperature. This was because site A had a wider street width and lower buildings, which caused sun exposure to be the highest, but stronger wind ventilated the pedestrian. Site B had a higher average building height, but was not tall enough that sun could penetrate the street. On the other hand, wind speed in site B was the lowest of all three sites, causing the mean air temperature to be the hottest. Site C showed the highest thermal comfort because tall buildings provided shading and stronger wind around the corners. Urban ventilation was an important factor for outdoor thermal comfort. The findings of this study provide a technical benchmark for urban planners, especially in hot summer areas, such as Taiwan, to mitigate the effects of climate change and global warming that cause extreme outdoor thermal environments for urban residents, and to create a healthier outdoor environment.

Author Contributions: Conceptualization, Y.-M.S. and J.A.; methodology, J.A.; site survey, J.A.; writing—original draft preparation, J.A.; writing—review and editing, Y.-M.S.; supervision, Y.-M.S. All authors have read and agreed to the published version of the manuscript.

Funding: This research was funded by The Ministry of Science and Technology of Taiwan, R.O.C (Project number: 110-2221-E-027-013-MY2) and the Research Center of Energy Conservation for New Generation of Residential, Commercial, and Industrial Sectors.

Institutional Review Board Statement: Not applicable.

Informed Consent Statement: Informed consent was obtained from all subjects involved in the study.

Data Availability Statement: Not applicable.

Conflicts of Interest: The authors declare no conflict of interest.

References

1. Hakim, A.A.; Petrovitch, H.; Burchfiel, C.M.; Ross, G.W.; Rodriguez, B.L.; White, L.R.; Yano, K.; Curb, J.D.; Abbott, R.D. Effects of Walking on Mortality among Nonsmoking Retired Men. *N. Engl. J. Med.* **1998**, *338*, 94–99. [CrossRef] [PubMed]
2. Hass-Klau, C. Impact of pedestrianization and traffic calming on retailing: A review of the evidence from Germany. *Transp. Policy* **1993**, *1*, 21–31. [CrossRef]
3. Jacobs, J. *The Death and Life of Great American Cities*; Penguin: Harmondsworth, UK, 1961.
4. Kaplan, R.B.A.; Whyte, W.H. City: Rediscovering the Center. *J. Arch. Educ.* **1991**, *44*, 241. [CrossRef]
5. Takano, T.; Nakamura, K.; Watanabe, M. Urban residential environments and senior citizens' longevity in megacity areas: The importance of walkable green spaces. *J. Epidemiol. Community Health* **2002**, *56*, 913–918. [CrossRef]
6. Liu, L.; Lin, Y.; Liu, J.; Wang, L.; Wang, D.; Shui, T.; Chen, X.; Wu, Q. Analysis of local-scale urban heat island charac-teristics using an integrated method of mobile measurement and GIS-based spatial interpolation. *Build. Environ.* **2017**, *117*, 191–207. [CrossRef]
7. Xiong, J.; Lian, Z.; Zhou, X. Investigation of Subjectively Assessed Health Symptoms and Human Thermal Perceptions in Transient Thermal Environments. *Procedia Eng.* **2015**, *121*, 212–216. [CrossRef]
8. Wang, Y.; Wang, A.; Zhai, J.; Tao, H.; Jiang, T.; Su, B.; Yang, J.; Wang, G.; Liu, Q.; Gao, C.; et al. Tens of thousands additional deaths annually in cities of China between 1.5 °C and 2.0 °C warming. *Nat. Commun.* **2019**, *10*, 3376–3377. [CrossRef]
9. Zhang, L.; Wei, D.; Hou, Y.; Du, J.; Liu, Z.; Zhang, G.; Shi, L. Outdoor Thermal Comfort of Urban Park—A Case Study. *Sustainability* **2020**, *12*, 1961. [CrossRef]
10. Huang, J.; Zhou, C.; Zhuo, Y.; Xu, L.; Jiang, Y. Outdoor thermal environments and activities in open space: An experiment study in humid subtropical climates. *Build. Environ.* **2016**, *103*, 238–249. [CrossRef]
11. Cheung, P.K.; Jim, C. Comparing the cooling effects of a tree and a concrete shelter using PET and UTCI. *Build. Environ.* **2018**, *130*, 49–61. [CrossRef]
12. Hu, T.; Yoshie, R. Effect of atmospheric stability on air pollutant concentration and its generalization for real and idealized urban block models based on field observation data and wind tunnel experiments. *J. Wind. Eng. Ind. Aerodyn.* **2020**, *207*, 104380. [CrossRef]

13. Lai, D.; Liu, W.; Gan, T.; Liu, K.; Chen, Q. A review of mitigating strategies to improve the thermal environment and thermal comfort in urban outdoor spaces. *Sci. Total Environ.* **2019**, *661*, 337–353. [CrossRef] [PubMed]
14. Wang, Y.; Akbari, H. Effect of Sky View Factor on Outdoor Temperature and Comfort in Montreal. *Environ. Eng. Sci.* **2014**, *31*, 272–287. [CrossRef]
15. Berardi, U.; Wang, Y. The Effect of a Denser City over the Urban Microclimate: The Case of Toronto. *Sustainability* **2016**, *8*, 822. [CrossRef]
16. Johansson, E. Influence of urban geometry on outdoor thermal comfort in a hot dry climate: A study in Fez, Morocco. *Build. Environ.* **2006**, *41*, 1326–1338. [CrossRef]
17. Kakon, A.N.; Mishima, N.; Kojima, S. Simulation of the urban thermal comfort in a high density tropical city: Analysis of the proposed urban construction rules for Dhaka, Bangladesh. In *Building Simulation*; Springer: Berlin, Germany, 2009; Volume 2, pp. 291–305.
18. Shih, W.-M.; Lin, T.-P.; Tan, N.-X.; Liu, M.-H. Long-term perceptions of outdoor thermal environments in an elementary school in a hot-humid climate. *Int. J. Biometeorol.* **2017**, *61*, 1657–1666. [CrossRef]
19. Krüger, E.L.; Minella, F.O.; Rasia, F. Impact of urban geometry on outdoor thermal comfort and air quality from field measurements in Curitiba, Brazil. *Build. Environ.* **2011**, *46*, 621–634. [CrossRef]
20. Mahmoud, A.H.A. Analysis of the microclimatic and human comfort conditions in an urban park in hot and arid regions. *Build. Environ.* **2011**, *46*, 2641–2656. [CrossRef]
21. Matzarakis, A.; Rutz, F.; Mayer, H. Modelling radiation fluxes in simple and complex environments—Application of the RayMan model. *Int. J. Biometeorol.* **2007**, *51*, 323–334. [CrossRef]
22. Matzarakis, A.; Rutz, F.; Mayer, H. Modelling radiation fluxes in simple and complex environments: Basics of the RayMan model. *Int. J. Biometeorol.* **2010**, *54*, 131–139. [CrossRef]
23. Solovchenko, A.; Lukyanov, A.; Nikolenko, A.; Shurygin, B.; Akimov, M.; Gitelson, A. Physiological foundations of spectral imaging-based monitoring of apple fruit ripening. *Acta Hortic.* **2021**, *1314*, 419–428. [CrossRef]
24. Matzarakis, A.; Gangwisch, M.; Fröhlich, D. RayMan and SkyHelios Model. *Urban Microclim. Model. Comf. Energy Stud.* **2021**, *2021*, 339–361. [CrossRef]
25. Fröhlich, D.; Gangwisch, M.; Matzarakis, A. Effect of radiation and wind on thermal comfort in urban environ-ments—Application of the RayMan and SkyHelios model. *Urban Clim.* **2019**, *27*, 1–7. [CrossRef]

Disclaimer/Publisher's Note: The statements, opinions and data contained in all publications are solely those of the individual author(s) and contributor(s) and not of MDPI and/or the editor(s). MDPI and/or the editor(s) disclaim responsibility for any injury to people or property resulting from any ideas, methods, instructions or products referred to in the content.

engineering proceedings

MDPI

Proceeding Paper

Precursory Study on Sustained Development in Food and Agriculture Education in the Post-Legislative Era †

Jiin-Chyuan Mark Lai [1], Chiung-Ling Wang [2] and Ming-Yuan Hsieh [3,*]

1 Department of Applied Foreign Languages, TransWorld University, Douliu City 640302, Taiwan; marklai07@gmail.com
2 Department of Industrial Education, National Taiwan Normal University, Taipei City 244014, Taiwan; lin8332317@mail.edu.tw
3 Department of International Business, National Taichung University of Education, Taichung City 403061, Taiwan
* Correspondence: capwisely@mail.ntcu.edu.tw; Tel.: +886-975-118-922

† Presented at the 3rd IEEE International Conference on Electronic Communications, Internet of Things and Big Data Conference 2023, Taichung, Taiwan, 14–16 April 2023.

Abstract: We analyze the recent media reports related to food and agriculture education (FAE) in Taiwan and present the main aspects of the FAE in different ways such as historical society and ethics, food and agriculture industry and responsibility, food and agriculture culture preservation, and education and health promotion. The reports are classified into seven categories: culture, life, agriculture, campus, society, environment, and industry. The results point out the enhancement of the sustained development of the FAE in the post-legislative era due to the increased consolidated relationships among the sustainable development goals (SDGs) and FAE in the food (production and marketing certification), agriculture (commercializing organic agriculture certification), and travel (friendly conservation tourism). Further research needs to focus on the main evaluated aspects of FAE that appear more frequently in the related research for the promotion of education and health, and historical and ethical justice. FAE seems to be preferred in the metropolitan area while teaching in the cultural field is more important in the non-metropolitan area than in the metropolitan area, which needs further study.

Keywords: food and agriculture education (FAE); developed sustainability; sustainable development goals (SDGs)

Citation: Lai, J.-C.M.; Wang, C.-L.; Hsieh, M.-Y. Precursory Study on Sustained Development in Food and Agriculture Education in the Post-Legislative Era. *Eng. Proc.* **2023**, *38*, 83. https://doi.org/10.3390/engproc2023038083

Academic Editors: Teen-Hang Meen, Hsin-Hung Lin and Cheng-Fu Yang

Published: 20 July 2023

Copyright: © 2023 by the authors. Licensee MDPI, Basel, Switzerland. This article is an open access article distributed under the terms and conditions of the Creative Commons Attribution (CC BY) license (https://creativecommons.org/licenses/by/4.0/).

1. Introduction

The global environment has been affected by human and non-human factors such as climate change, the COVID-19 epidemic, and the war between Russia and Ukraine. Thus, issues related to food and agriculture education (FAE) have been important in human life. The public has paid extra attention to the security of food and agriculture, the self-sufficiency rate of food, and agricultural sustainability through social media reports, videos, and articles. People consider food an important source of nutrition and energy and the production, quality, supply, and distribution of food are critical. Countries emphasize FAE to ensure the stability of production and sales of foods and confirm safe food sources and origins such as organic farming, soil optimization, farming method, and others. The ultimate goal of the sustainable development of agriculture cannot be achieved with food safety problems and polluted agricultural production [1].

FAE has dealt with complexities in the economy, society, politics, culture, and communication as food and diet is important for human beings. Food helps humans avoid sickness and weakness and affects human relationships and even the management of political power. Food and diet influence a person's attitude and define human identity, as what people eat has affected cultural development. Food culture has transformed history and civilization as it is related to botany, physics and chemistry, agriculture, animal science, agronomy, ecology, anthropology, sociology, geopolitics, political economy, trade, technology, cooking, physiology, medicine, and philosophy. Food and diet impact various social, environmental, and economic phenomena as they are combined with culture, tradition, and community sustainability [2].

Therefore, sustainable development goals (SDGs) have been proposed to solve those problems. As a result, FAE-related initiatives have been promoted among governments, industries, academia, and civil societies all over the world. Societal innovation and integration are integrated into the SDGs to propose the "sociSDGs". The consolidation of the SDGs and FAE covers the issues of a diversified economy, environment, and social dimensions, including hunger, malnutrition, desertification, water resource use, loss of biodiversity, overconsumption, obesity, and public health. A new FAE, then, is proposed to achieve the first, second, third, fourth, fifth, eighth, tenth, eleventh, and twelfth goals of the SDGs [3–9]. To connect SDGs and FAE, the United Nations Committee on World Food Security established the International Agri-Food Network in which FAE is combined with the seventeen goals of the SDGs from multiple perspectives of sustainable development [10–13].

Recently, international communities and Taiwan have set up social enterprises to advocate eco-friendly agriculture and organic farming integrating educational, industrial, and social entities for sustainable development in a new research mainstream. Empirically, each government has an obligated responsibility to establish the comprehensive development of the FAE system by constructing a certification system for food (production and marketing certification), agriculture (commercializing organic agriculture certification), and travel (conservation-friendly tourism) [14–16]. Frequent food safety incidents over the past ten years have allowed widespread attention and support for FAE from the public, especially the melamine-tainted milk powder incident that occurred in September 2008 that affected mainland China and Taiwan and raised awareness of food safety in the entire Taiwanese society. Other infamous food safety-related incidents occurred including the poisoned oil mixed with low-cost sunflower and cottonseed oils in 2011, the copper chlorophyll added coloring in 2013, and waste oil blended into edible oils in 2014.

FAE has been adopted and developed in many countries without educational consideration for many years but food and agriculture environment education (FAEE) is still necessary to create sustainability based on the original FAE that has educational functions and social movements. Innovation is required in "unsustainable solutions" to solve the unsustainability of FAE. The most significant reason is that FAE has been only a duty of the government but in fact, is an obligation of the public. Thus, the Taiwanese Food and Drug Administration (FDA) of the Ministry of Health and Welfare has taken practical action to promote the importance of food hygiene and safety through websites, social media, and exhibitions. At the same time, each educational institution has strengthened activities regarding FAE. In addition, the Taiwanese National Health Administration of the Ministry of Health and Welfare has supervised the catering industry and schools to reduce chronic diseases caused by diet through multiple channels including regulations, rules, and policies.

To effectively allay the deep doubts about food safety, governments have increased their concerns and the supervision of the agriculture and food industries. In Taiwan, the Food Safety Office is in charge of food safety based on the "Eight Food Strengthening Measures" and the "Food Safety Five Rings" legislated in 2014 and 2016. Since food and agricultural issues are related to health and the environment and food safety is related to the welfare of people, the Ministry of Health and Welfare, the Environmental Protection Agency (EPA), the Ministry of the Interior, and the Ministry of Education are involved in the "education" of food and agriculture. Since 2007, the Council of Agriculture, which is closely related to the FAE has been planning and promoting the "do-it-yourself" approach and the "four health" system in the framework of FAE. Therefore, the responsibility of manufacturers and farmers has increased with the increase in people's awareness of food safety.

Non-governmental organizations (NGOs) pay more attention to FAE issues for eco-friendly agriculture, the fair trade of food and agricultural products, and a low-carbon diet. The "Four Health Association of the Republic of China" has implemented "Food Education in Primary School" through the "Seed Teacher Camp" for mentor training of FAE from 2013 [17–19]. According to the demand for sustainable development, the Ministry of Education has revised the "School Health Law" to add an article of "health-related courses" in Article 16: "The first health-related courses should include healthy eating education to establish correct eating habits to develop respect for life and nature and to enhance awareness of environmental protection, deepen understanding of the source of food materials, and understand the food culture of the country and region. Schools should encourage students to participate in the school meal preparation process" [20]. In 2015, the article was stipulated in the law, providing a legal foundation for schools at all levels to promote FAE [21]. The government, academia, and civil society are trying to execute FAE and conduct all the related legislative processes, as the "Food and Agriculture Education Law" (FAEL) has passed and taken effect as of 2022 under the request of the society. Thus, FAE promotes a complete legislative foundation that gives accountability, rights, and responsibilities to all relevant competent authorities and the implementation procedures. The FAEL combines educational concepts for sustainability.

This research explores the concepts related to FAE in the world and investigates how sustainable development can be developed in the "educational concepts" of the Taiwanese food and agriculture environment and education (TFAEE). TFAEE is defined as a framework to construct food and agriculture education to help sustainable development. Based on the above review, the elements of food (production and marketing certification), agriculture (commercializing organic agriculture certification), and travel (eco-friendly conservation and tourism) are determined to discuss the sustained development in FAE under the research framework shown in Figure 1 [22].

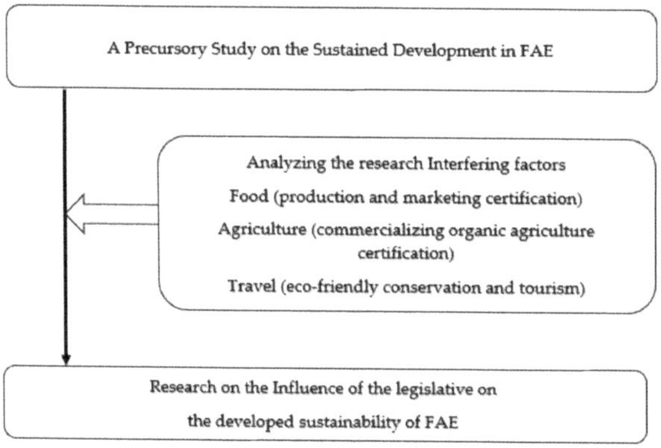

Figure 1. Research framework.

2. Research Concept

The SDGs were set by the United Nations to replace the Millennium Development Goals. SDGs are being implemented from 2016 to 2030 with 17 goals and 169 sub-targets [23,24]. The 17 goals of the SDGs are as follows:

(1) No poverty: eradicating poverty in all its forms everywhere;
(2) Zero hunger: ensuring food security, eradicating hunger, and promoting sustainable agriculture;
(3) Good health and well-being: ensuring and promoting healthy living and well-being at all ages;
(4) Quality education: ensuring access to education without discrimination, fairness, and high-quality education, and promoting lifelong learning;

(5) Gender equality: achieving gender equality and empowering women;
(6) Clean water and sanitation: ensuring access to water, sanitation, and its sustainable management for all;
(7) Affordable and clean energy: ensuring access to affordable, reliable, sustainable, and modern energy for all;
(8) Decent work and economic growth: promoting inclusive and sustainable economic growth with good jobs for everyone;
(9) Industry, innovation, and infrastructure: building resilient infrastructure, promoting inclusive and sustainable industries, and accelerating innovation;
(10) Reduced inequalities: reducing inequality within and between countries;
(11) Sustainable cities and communities: building cities and villages that are inclusive, safe, resilient, and sustainable;
(12) Responsible consumption and production: promoting a green economy and ensuring sustainable consumption and production patterns;
(13) Climate action: completing mitigation and adaptation actions to address climate change and its impacts;
(14) Life below water: executing conservation and sustainable use of marine ecosystems to ensure biodiversity and prevent degradation of the marine environment;
(15) Life on land: implementing conservation and sustainable use of terrestrial ecosystems, ensuring biodiversity, and preventing land degradation;
(16) Peace, justice, and strong institutions: promoting peaceful and pluralistic societies, ensuring judicial equality, and building credible and inclusive systems;
(17) Partnerships for the goals: establishing multiple partnerships and working together to promote a sustainable vision.

3. Conclusions and Future Direction

We review the media reports related to the FAE in Taiwan and present the main aspects of FAE in historical and societal ethics, the food and agriculture industry and responsibility, food and agriculture culture preservation, and education and health promotion. Seven categories are defined as culture, life, agriculture, campus, society, environment, and industry. The enhancement of the sustained development in FAE with legislation is found according to the increasingly consolidated relationship between SDGs and FAE. Further research is needed to focus on the issues that appear more frequently in the current food and agriculture education and research. FAE may be preferred in the metropolitan area while teaching cultural subjects is more acceptable in the non-metropolitan area. Therefore, the cross-analysis of teaching and training is also necessary.

Author Contributions: Conceptualization, J.-C.M.L. and M.-Y.H.; methodology, M.-Y.H.; validation, J.-C.M.L.; formal analysis, M.-Y.H.; investigation, J.-C.M.L.; resources, J.-C.M.L. and C.-L.W.; writing—original draft preparation, M.-Y.H. and C.-L.W.; writing—review and editing, M.-Y.H.; visualization, J.-C.M.L.; supervision, M.-Y.H.; project administration, M.-Y.H.; funding acquisition, C.-L.W. All authors have read and agreed to the published version of the manuscript.

Funding: This research was supported by the research projects (MOST 110-2420-H-002-003-MY3-Y11209) of the Ministry of Science and Technology and (NTCU111103) and the National Taichung University of Education.

Institutional Review Board Statement: This research did not execute the questionnaires and interview and hence, this research did not require ethical approval.

Informed Consent Statement: Not applicable because this research did not involve any humans.

Data Availability Statement: No new data were created.

Conflicts of Interest: There are no conflict of interest in this research.

References

1. FAO. Available online: http://www.fao.org/documents/card/en/c/ca8996en (accessed on 23 March 2023).
2. FAO. Available online: http://www.fao.org/3/y4671e/y4671e06.htm (accessed on 11 January 2023).
3. Ashraf, S.A.; Siddiqui, A.J.; Elkhalifa, A.E.O.; Khan, M.I.; Patel, M.; Alreshidi, M. Innovations in nanoscience for the sustainable development of food and agriculture with implications on health and environment. *Sci. Total Environ.* **2021**, *768*, 144990. [CrossRef] [PubMed]
4. Bryngelsson, D.; Wirsenius, S.; Hedenus, F.; Sonesson, U. How can the EU climate targets be met? A combined analysis of technological and demand-side changes in food and agriculture. *Food Policy* **2016**, *59*, 152. [CrossRef]
5. OECD/FAO. Available online: https://doi.org/10.1787/19991142 (accessed on 23 March 2023).
6. Baig, I.A.; Chandio, A.A.; Ozturk, I.; Kumar, P.; Khan, Z.A.; Salam, M. Assessing the long-and short-run asymmetrical effects of climate change on rice production: Empirical evidence from India. *Environ. Sci. Pollut. Res.* **2022**, *29*, 34209. [CrossRef] [PubMed]
7. Gupta, R.; Mishra, A. Climate change induced impact and uncertainty of rice yield of agro-ecological zones of India. *Agric. Syst.* **2019**, *173*, 1–11. [CrossRef]
8. Hossain, M.; Islam, M.; Sujan, M.; Khan, H.; Tuhin, M.; Bekun, F.V. Towards a clean production by exploring the nexus between agricultural ecosystem and environmental degradation using novel dynamic ARDL simulations approach. *Environ. Sci. Pollut. Res.* **2022**, *29*, 53768–53784. [CrossRef] [PubMed]
9. Praveen, B.; Sharma, P. Climate change and its impacts on Indian agriculture: An econometric analysis. *J. Public Aff.* **2020**, *20*, e1972. [CrossRef]
10. Hsieh, Y.M. Interdisciplinarily Exploring the Most Potential IoT Technology Determinants in the Omnichannel E-Commerce Purchasing Decision-Making Processes. *Appl. Sci.* **2020**, *10*, 603. [CrossRef]
11. Hsieh, Y.M. The most sustainable niche principles of social-media education in a higher education contracting era. *Sustainability* **2020**, *12*, 339. [CrossRef]
12. Huang, Y.-M.; Hsieh, Y.M.; Usak, M. A Multi-Criteria Study of Decision-Making Proficiency in Student's Employability for Multidisciplinary Curriculums. *Mathematics* **2020**, *8*, 897. [CrossRef]
13. Chan, Y.-K.; Hsieh, M.-Y.; Usak, M. A Concrete Study on Social-Media Connection of Global Literacy Abilities in MOOCs under the Dual Impacts of Lower Birth-Rate and COVID-19. *Sustainability* **2021**, *13*, 2203. [CrossRef]
14. Usak, M.; Hsieh, M.-Y.; Chan, Y.-K. A Concertizing Research on Making Higher Education Sustainability Count. *Sustainability* **2021**, *13*, 2724. [CrossRef]
15. Duran, M.; Usak, M.; Hsieh, M.-Y.; Uygun, H. A New Perspective on Pedagogical Content Knowledge: Intellectual and Emotional Characteristics of Science Teachers. *RCIS* **2021**, *72*, 9. [CrossRef]
16. Wu, T.-L.; Hsieh, M.-Y.; Min, K.-W.; Yu, M.-T.; Ho, C.-T. Use of Sensor Technologies in Online Courses in Post-COVID-19 Era. *Sens. Mater.* **2021**, 33, 2045–2062. [CrossRef]
17. Huang, C.C.; Chan, Y.-K.; Hsieh, M.Y. The Determinants of ESG for Community LOHASism Sustainable Development Strategy. *Sustainability* **2022**, *14*, 11429. [CrossRef]
18. Chan, Y.-K.; Hsieh, M.Y. An Empirical Study on Higher Education C-ESG Sustainable Development Strategy in Lower-Birth-Rate Era. *Sustainability* **2022**, *14*, 12629. [CrossRef]
19. Hsieh, Y.-M. Online learning era: Exploring the most decisive determinants of MOOCs in Taiwanese higher education. *Eurasia J. Math. Sci. Technol. Educ.* **2016**, *12*, 1163. [CrossRef]
20. Hsieh, Y.-M. Employing MCDM methodology to verify correlationship between social media and service quality in the dynamic m-commerce era. *J. Internet Technol.* **2018**, *19*, 225.
21. Hsieh, Y.-M.; Usak, M. High Education Radical Transformation Era: How Teachers' Competency can Enhance the Students' Employability. *RCIS* **2020**, *68*, 95. [CrossRef]
22. Huang, Y.-M.; Hsieh, Y.M. An Interdisciplinary Research on Students' Employability in Technology Education to Advance Higher Education Enrollment Sustainability. *Sustainability* **2020**, *12*, 1806. [CrossRef]
23. Hsieh, M.Y. The Sustainable Development and Strategic Approaches for Contemporary Higher Education. *Sustainability* **2022**, *14*, 12925. [CrossRef]
24. Huang, C.-C.; Chan, Y.-K.; Hsieh, M.-Y. Preliminary Research on the Sustainable Determinants of Taiwanese Ecotourism with the International Standards. *Int. J. Environ. Res. Public Health* **2022**, *19*, 14489. [CrossRef] [PubMed]

Disclaimer/Publisher's Note: The statements, opinions and data contained in all publications are solely those of the individual author(s) and contributor(s) and not of MDPI and/or the editor(s). MDPI and/or the editor(s) disclaim responsibility for any injury to people or property resulting from any ideas, methods, instructions or products referred to in the content.

MDPI AG
Grosspeteranlage 5
4052 Basel
Switzerland
Tel.: +41 61 683 77 34

Engineering Proceedings Editorial Office
E-mail: engproc@mdpi.com
www.mdpi.com/journal/engproc

Disclaimer/Publisher's Note: The statements, opinions and data contained in all publications are solely those of the individual author(s) and contributor(s) and not of MDPI and/or the editor(s). MDPI and/or the editor(s) disclaim responsibility for any injury to people or property resulting from any ideas, methods, instructions or products referred to in the content.

www.ingramcontent.com/pod-product-compliance
Lightning Source LLC
Chambersburg PA
CBHW061449220926
51850CB00004B/52
9783725814473